Annual Review of
Genetics

Production Editor: Absolom J. Hagg
Managing Editor: Linley E. Hall
Bibliographic Quality Control: Mary A. Glass
Electronic Content Coordinators: Suzanne K. Moses, Erin H. Lee
Illustration Editor: Christie Newman

Annual Review of
Genetics

Volume 48, 2014

Bonnie L. Bassler, *Editor*
Princeton University

Michael Lichten, *Associate Editor*
National Cancer Institute

Gertrud Schüpbach, *Associate Editor*
Princeton University

www.annualreviews.org • science@annualreviews.org • 650-493-4400

Annual Reviews
4139 El Camino Way • P.O. Box 10139 • Palo Alto, California 94303-0139

Annual Reviews
Palo Alto, California, USA

International Standard Serial Number: 0066-4197
International Standard Book Number: 978-0-8243-1247-3
Library of Congress Control Number: 67029891

All Annual Reviews and publication titles are registered trademarks of Annual Reviews.

⊗ The paper used in this publication meets the minimum requirements of American National Standards for Information Sciences—Permanence of Paper for Printed Library Materials, ANSI Z39.48-1992.

TYPESET BY APTARA
PRINTED AND BOUND BY SHERIDAN BOOKS, INC., CHELSEA, MICHIGAN

Contents

Annual Review of Genetics

Volume 48, 2014

Errata

An online log of corrections to *Annual Review of Genetics* articles may be found at
http://www.annualreviews.org/errata/genet

Related Articles

From the ***Annual Review of Biochemistry***, Volume 83 (2014)

Human RecQ Helicases in DNA Repair, Recombination, and Replication
Deborah L. Croteau, Venkateswarlu Popuri, Patricia L. Opresko, and Vilhelm A. Bohr

From the ***Annual Review of Cell and Developmental Biology***, Volume 30 (2014)

Bacterial Pathogen Manipulation of Host Membrane Trafficking
Seblewongel Asrat, Dennise A. de Jesús, Andrew D. Hempstead, Vinay Ramabhadran, and Ralph R. Isberg

The Rhomboid-Like Superfamily: Molecular Mechanisms and Biological Roles
Matthew Freeman

Determinants and Functions of Mitochondrial Behavior
Katherine Labbé, Andrew Murley, and Jodi Nunnari

Zygotic Genome Activation During the Maternal-to-Zygotic Transition
Miler T. Lee, Ashley R. Bonneau, and Antonio J. Giraldez

From the ***Annual Review of Microbiology***, Volume 68 (2014)

Bacterial Sigma Factors: A Historical, Structural, and Genomic Perspective
Andrey Feklístov, Brian D. Sharon, Seth A. Darst, and Carol A. Gross

L1 Retrotransposons and Somatic Mosaicism in the Brain

Sandra R. Richardson,[1] Santiago Morell,[1] and Geoffrey J. Faulkner[1,2]

[1]Mater Research Institute, The University of Queensland, Translational Research Institute, Woolloongabba QLD 4102, Australia; email: faulknergj@gmail.com

[2]School of Biomedical Sciences, The University of Queensland, Brisbane QLD 4072, Australia

Annu. Rev. Genet. 2014. 48:1–27

First published online as a Review in Advance on July 14, 2014

The *Annual Review of Genetics* is online at genet.annualreviews.org

This article's doi: 10.1146/annurev-genet-120213-092412

Keywords

LINE-1, transposon, retrotransposition, neuron, neurogenesis

Abstract

Long interspersed element 1 (LINE-1 or L1) retrotransposons have generated one-third of the human genome, and their ongoing mobility is a source of inter- and intraindividual genetic diversity. Although retrotransposition in metazoans has long been considered a germline phenomenon, recent experiments using cultured cells, animal models, and human tissues have revealed extensive L1 mobilization in rodent and human neurons, as well as mobile element activity in the *Drosophila* brain. In this review, we evaluate the available evidence for L1 retrotransposition in the brain and discuss mechanisms that may regulate neuronal retrotransposition in vivo. We compare experimental strategies used to map de novo somatic retrotransposition events and present the optimal criteria to identify a somatic L1 insertion. Finally, we discuss the unresolved impact of L1-mediated somatic mosaicism upon normal neurobiology, as well as its potential to drive neurological disease.

INTRODUCTION

Transposable elements are found in virtually all eukaryotic life. Their activity rearranges and adds to the genetic instructions encoded by DNA and, as such, is a key component of genotypic variation and evolutionary selection. That transposons, retrotransposons, and other mobile DNA contribute to phenotype is clear; striking examples of variegation in plants and animals, such as morning glory flower pigmentation (56), agouti mouse coat color (94), and, in arguably the prototypical case, maize kernel mosaicism explained by Barbara McClintock (85), are all manifestations of transposable element activity (reviewed in 120). As established by Kazazian et al. (62) in 1988, retrotransposition continues to occur in humans and, as shown repeatedly then and since, is a notable source of mutagenesis leading to disease (10, 20, 34, 53). Thus, transposable elements are an integral part of genome evolution in humans and other species, with the capacity to fundamentally impact a wide range of biological phenomena.

The autonomous retrotransposon long interspersed element 1 (LINE-1 or L1) is arguably the most impactful—and still active—human transposable element. First, L1 sequences account for approximately 17% of our DNA. Second, the L1-encoded enzymatic machinery mobilizes nonautonomous retroelements, such as *Alu* and SINE-VNTR-*Alu* (SVA), and generates processed pseudogenes (31). L1-mediated retrotransposition has therefore resulted in the accumulation of at least one-third of the human genome (26, 69) and continues to shape its landscape, as witnessed by the estimated 400 million polymorphic retrotransposon insertions in the global human population (reviewed in 8, 34, 110).

L1 Retrotransposition Mechanism

A retrotransposition-competent human L1 is approximately 6 kb in length (28, 115) (**Figure 1a**). The L1 5' end comprises a 5' untranslated region (UTR) that harbors an internal promoter and an antisense promoter of unknown function (122, 125). The L1 internal promoter contains *cis*-acting binding sites for several transcription factors, including YY1 and RUNX3 as well as SOX family transcription factors (5, 9, 68, 89, 129, 136). L1 encodes two open reading frames (28, 115), ORF1 and ORF2, the protein products of which are ORF1p, a ~40-kDa nucleic acid–binding protein (55, 63, 81), and ORF2p, a ~150-kDa protein with demonstrated endonuclease (EN) and reverse transcriptase (RT) activities (29, 30, 36, 82, 128). Both proteins are required for L1 retrotransposition (93). The 3' end of the L1 element consists of a 3' UTR, followed by a poly-A tail thought to facilitate efficient L1 translation and reverse transcription (28, 115). New L1 insertions are typically flanked by variable length target-site duplications (TSDs), which are a structural hallmark of the L1 integration process (48) and distinguish retrotransposition from other types of genomic rearrangement (e.g., translocation).

The process of retrotransposition begins with transcription of a full-length L1 from its internal promoter (**Figure 1b**). The resulting L1 mRNA is exported to the cytoplasm, where it is translated by an unconventional termination-reinitiation mechanism (1, 27). Multiple ORF1p molecules and potentially as few as one ORF2p molecule bind back to their encoding L1 mRNA in a phenomenon known as *cis* preference, giving rise to the L1 ribonucleoprotein particle (RNP), a hypothesized retrotransposition intermediate (31, 55, 66, 67, 80, 133). The L1 RNP then enters the nucleus by a mechanism that is not completely understood but that can take place independently of cell division (65).

In the nucleus, the L1-encoded EN activity creates a single-strand nick in genomic DNA, with a loose preference for 5'-TTTT/A-3' motifs (60). The nick liberates a free 3' hydroxyl residue, which is in turn used as a primer from which the L1 RT initiates reverse transcription of its associated

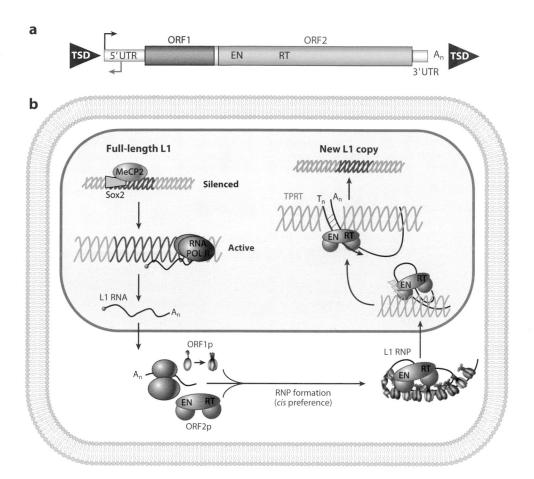

Figure 1

(*a*) The structure of a retrotransposition-competent L1 (long interspersed element 1). The L1 5′ UTR (untranslated region) contains an internal promoter (*black arrow*). L1 harbors two open reading frames: ORF1 (*dark blue rectangle*) and ORF2 (*light blue rectangle*); ORF2 encodes L1 endonuclease (EN) and reverse transcriptase (RT) activities. The L1 sequence terminates in a 3′ UTR and a poly-A tail (A_n). L1s in the genome are frequently flanked by target-site duplications (TSDs; *black triangles*). (*b*) L1 retrotransposition mechanism. L1 expression in neuronal cell types is dynamically regulated by factors such as MeCP2 and Sox2. In a round of retrotransposition, the L1 RNA is transcribed by RNA polymerase II and then translated, giving rise to multiple copies of ORF1p and as few as one copy of ORF2p. Both proteins associate with their encoding RNA (*cis* preference) to assemble the L1 ribonucleoprotein particle (RNP). The L1 RNP enters the nucleus, where the L1 endonuclease activity nicks the genomic DNA at the consensus 5′-TTTT/A-3′ to liberate a 3′ hydroxyl residue from which the L1 reverse transcriptase initiates reverse transcription of its associated L1 mRNA. This process, termed target-site-primed reverse transcription (TPRT), generates a new, frequently 5′ truncated L1 insertion.

mRNA (21, 22, 36). In vitro studies have elucidated a complex set of rules governing target-site selection due to L1 RT preference for genomic poly-T tracts and perfect terminal complementarity for L1 RNA-genomic DNA complexes (91). This process, known as target-site-primed reverse transcription (TPRT), was first elucidated by biochemical studies of the *Bombyx mori* non-LTR retrotransposon R2 (77). Second-strand target DNA cleavage usually takes place some distance downstream of first-strand cleavage, giving rise to the aforementioned TSDs. Second-strand

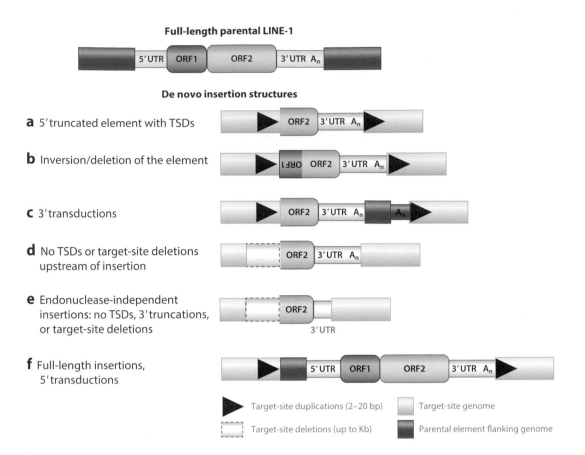

Full-length parental LINE-1

De novo insertion structures

a 5′ truncated element with TSDs

b Inversion/deletion of the element

c 3′ transductions

d No TSDs or target-site deletions upstream of insertion

e Endonuclease-independent insertions: no TSDs, 3′ truncations, or target-site deletions

f Full-length insertions, 5′ transductions

Target-site duplications (2–20 bp)

Target-site deletions (up to Kb)

Target-site genome

Parental element flanking genome

Figure 2

Features of L1 (long interspersed element 1) retrotransposition events. Above: a full-length parental L1. (*a*) A typical L1 insertion is 5′ truncated, ends in a poly-A tail (A_n), and is flanked by target-site duplications (TSDs; *black triangles*). (*b*) L1 insertions frequently contain internal inversions and deletions of the retroelement sequence. (*c*) Bypass of the L1 polyadenylation signal in favor of a strong genomic polyadenylation signal can lead to 3′ transductions of genomic DNA from the donor L1 locus. (*d*) The position of second-strand cleavage can lead to insertions lacking TSDs or small deletions of target-site DNA. (*e*) Endonuclease-independent insertions lack TSDs, are frequently 3′ truncated, and are associated with deletion of target-site DNA. (*f*) Transcription of a parental element from an upstream genomic promoter can lead to full-length insertions with 5′ transductions of the genomic sequence. Abbreviation: UTR, untranslated region.

cleavage can also occur directly opposite the first-strand cleavage site, leading to blunt insertions lacking TSDs, or upstream of the first-strand cleavage site, leading to small deletions of target-site DNA (40, 41, 45) (**Figure 2**). The subsequent steps of insertion formation, including second-strand DNA synthesis and integration, are not completely understood. The majority of new L1 insertions are 5′ truncated (48, 69) and often contain internal rearrangements, such as inversions and deletions of the L1 sequence (40, 41) (**Figure 2**).

Host factors almost certainly play roles in L1 retrotransposition by interacting with the L1 mRNA-encoded proteins. Recent work by Taylor et al. (128) identified 37 proteins that interact with the L1 RNP, including the polymerase-δ-associated sliding clamp PCNA (proliferating cell nuclear antigen) and the nonsense-mediated decay factor UPF1. PCNA interacts with L1 ORF2p via a PIP-box motif and is proposed to influence L1 integration during or immediately after TPRT

(128). In addition, Goodier et al. (44) identified 96 L1 ORF1p-interacting proteins, 19 of which could restrict L1 retrotransposition in a cultured cell assay by greater than 60%.

Previous studies have demonstrated that cellular factors, including members of the APOBEC3 family of cytidine deaminases (13, 14, 18, 64, 97, 103, 113, 123), the nuclease Trex1 (124), the antiretroviral factor SAMHD1 (138), and the Mov10 putative helicase (4, 43, 75), can restrict retrotransposition in cultured cells and may play a role in regulating L1 activity in vivo. RNA-based mechanisms have also been implicated in L1 regulation. Heras et al. (54) demonstrated that the miRNA biogenesis factor Microprocessor/Drosha-DGCR8 specifically binds L1, *Alu*, and SVA retroelement RNAs, and can cleave L1 RNA in vitro. Indeed, experiments in cultured cells revealed that Microprocessor negatively regulates L1 and *Alu* retrotransposition. In addition, Ciaudo et al. (19) demonstrated a role for both Dicer-dependent and Ago2-dependent RNAi in L1 regulation in mouse embryonic stem cells. Along with epigenetic silencing, these mechanisms defend a host genome from the likely deleterious consequences of unrestrained retrotransposition (120).

Under certain conditions, L1 is capable of generating new insertions by a noncanonical, EN-independent (ENi) pathway (23, 95, 96). ENi retrotransposition has been demonstrated in Chinese hamster ovary cells deficient in both p53 and components of the nonhomologous end joining pathway of DNA repair. In these cells, EN-deficient L1s can retrotranspose efficiently, presumably by exploiting pre-existing genomic DNA lesions resulting from delayed or impaired DNA repair. ENi retrotransposition events are often distinguished by a lack of TSDs as well as by L1 3′ truncation and deletion of flanking genomic DNA (96) (**Figure 2**). Notably, the human genome reference sequence incorporates 21 L1 insertions bearing structural features consistent with ENi retrotransposition, indicating that ENi retrotransposition can occur in the human germline (116). Thus, although in wild-type cells the vast majority of L1 retrotransposition occurs via TPRT, exceptional cases of ENi mobilization do occur, especially in cellular environments associated with elevated DNA damage.

Consequences of Retrotransposition for the Host Genome

In addition to duplication and deletion of target-site DNA, L1 retrotransposition is frequently associated with various genomic alterations, including addition of nontemplated nucleotides and, occasionally, large-scale rearrangements, such as chromosomal translocations (40, 41, 90, 126, 127). In 10–20% of cases, L1 retrotransposition is associated with 3′ transduction, in which transcription of the donor element bypasses the canonical L1 polyadenylation signal and uses a genomic signal some distance downstream, causing the non-L1 sequence to be retrotransposed to a new genomic location (47, 92, 109) (**Figure 2**). Furthermore, L1 retrotransposition proximal to and within coding regions can disrupt gene function and regulation. Exonic retrotransposition events can act as insertional mutagens (62); additionally, insertions within introns can induce missplicing or premature polyadenylation due to cryptic splice and polyadenylation signals within the AT-rich L1 sequence (11, 107). Intronic insertions can also impact RNA polymerase processivity through host genes, which has led to the hypothesis that L1 can act as a molecular rheostat to effect subtle changes in gene expression levels (52) and even engage in gene breaking (134). The L1 antisense promoter and other transcription initiation sites in the L1 3′ end can generate transcripts with the potential to impact regulation of adjacent genes (35, 83, 104, 122, 135). A recent study has implicated L1-derived stable nuclear RNA in regulating chromatin state, suggesting an expanded impact of L1 activity on global gene expression (50). L1 insertions are also subject to epigenetic regulation; in some cases, e.g., in PA-1 embryonal carcinoma cells, epigenetic marks may be targeted specifically to nascent L1 insertions during TPRT (38). Epigenetic silencing of L1 insertions may impact the expression of nearby genes if chromatin modifications spread from the L1 sequence into surrounding DNA, as seen for LTR retrotransposons (111). L1 retrotransposition can therefore impact the host genome, epigenome, and transcriptome via numerous routes, any of which may be sufficient to subtly or grossly alter organismal phenotype.

Retrotransposition in vivo: Where and When?

L1 has historically been viewed as a molecular parasite that must be transmitted to subsequent generations to ensure its propagation. As such, the host germline has long been considered as the primary milieu for retrotransposition (reviewed in 73). Recent studies using tissue culture systems, animal models, and human samples have reinforced this view, revealing that L1 retrotransposition

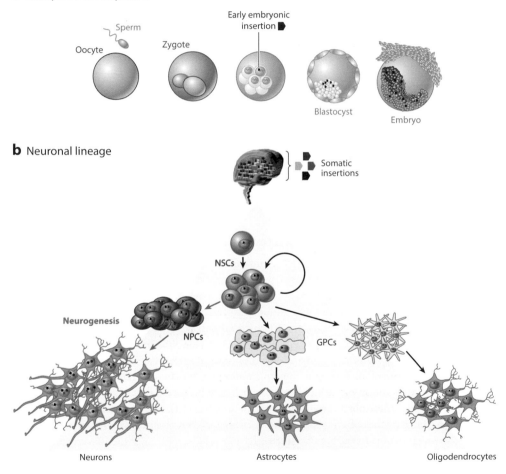

Figure 3

L1 (long interspersed element 1) activity during development. (*a*) Embryogenesis. L1 retrotransposition in the early embryo can generate somatic mosaicism (*black pentagons*) and can contribute to the germ lineage. (*b*) Neuronal lineage. L1 is active in both neural stem cells (NSCs) (*blue*) and neural precursor cells (NPCs) (*blue-green*), thus playing a role of potential importance in genetic mosaicism and perhaps neuronal plasticity in mature neurons (*bright green*). Glial precursor cells (GPCs) support markedly less L1 activity than NPCs. Each colored pentagon represents a different somatic insertion.

can occur in the mammalian germline and in the early embryo prior to germline specification (15, 37, 61, 105, 131, 132) (**Figure 3a**). By contrast, L1 sequences are normally transcriptionally repressed in somatic cells by DNA methylation, with these patterns established in the primordial germline (12). However, in 1992 Miki et al. (88) linked a case of colorectal cancer to a de novo somatic retrotransposition event in the archetypal tumor suppressor gene APC, implicating a likely somatic retrotransposition as an agent of oncogenesis and also demonstrating incomplete epigenetic suppression of retrotransposition in somatic cells. Furthermore, recent studies have uncovered L1-mediated retrotransposition in a variety of tumor types, including lung, liver, colorectal, prostate, and ovarian cancers (57, 71, 118, 121). Thus, deregulated L1 retrotransposition in the soma has the potential to alter cellular phenotype, as evidenced by overt consequences such as tumorigenesis. More subtle, as yet undescribed effects of somatic retrotransposition are also likely to exist.

One of the most intriguing recent findings in L1 biology is that elevated levels of retrotransposition can occur in normal cells in the mammalian brain. In this review, we present the seminal works leading to the discovery of neuronal retrotransposition, and discuss mechanisms by which neuronal retrotransposition is regulated. We then present criteria that should be met in order to classify a potential somatic insertion as a bona fide retrotransposition event, and examine recent reports that have exploited high-throughput sequencing strategies to map and characterize such insertions in neuronal cells. Finally, we highlight important unanswered questions regarding the timing, frequency, impact on human disease, and ultimately the normal physiological role, if any, of neuronal retrotransposition.

L1 RETROTRANSPOSITION IN NEURONS

The first evidence that L1 retrotransposition occurs in the mammalian brain came from studies employing an engineered L1-EGFP (enhanced green fluorescent protein) reporter transgene in cultured cells and animal models. Neural progenitor cells (NPCs), which can be derived in vitro from several brain regions, including the hippocampus and subventricular zone, are multipotent cells capable of giving rise to diverse cell types of the neuronal lineage (87) (**Figure 3b**). In 2005, Muotri et al. (99) reported the unexpected observation that L1 mRNAs were abundant in NPCs derived in vitro from adult hippocampus neural stem cells. L1 expression was further demonstrated to be repressed in neural stem cells by the transcription factor Sox2. Intriguingly, a decrease in Sox2 expression during lineage commitment correlated with derepression of the L1 promoter and increased L1 transcription in NPCs. The L1-EGFP reporter readily mobilized in NPCs in vitro, and the resulting events sometimes occurred within genes and had the capacity to alter target gene expression. Transgenic mice harboring a human L1-EGFP reporter exhibited neuronal retrotransposition events consistent with L1 mobilization during both embryonic and adult neurogenesis. These observations led to the hypothesis that cells of the neuronal lineage can accommodate L1 retrotransposition and that L1 activity may contribute to genomic and functional diversity among individual neurons (119).

In 2009, Coufal et al. (24) extended the above rodent studies to humans, demonstrating that the L1-EGFP reporter can retrotranspose in NPCs isolated from the human fetal brain and in NPCs derived from human embryonic stem cells (hESCs). Endogenous L1 copies were shown to be hypomethylated in the fetal brain relative to skin, suggesting that endogenous L1s could be transcriptionally active in the human brain and that this process was regulated by Sox2 and MeCP2. To estimate endogenous L1 copy-number variation (CNV) in human tissue, Coufal et al. (24) developed a multiplex TaqMan quantitative polymerase chain reaction (qPCR) assay and applied this approach to a wide range of human organs and brain subregions from several postmortem donors. This assay suggested that, on average, the adult hippocampus contained approximately 80 more L1 ORF2 copies per cell than heart or liver, with substantial variability between individuals. The hippocampus also exhibited elevated L1 CNV compared with other brain regions, a result explained by the presence of the subgranular zone, a major neurogenic niche, in the hippocampus. Although lacking sequence-based characterization of endogenous retrotransposition events from the brain, these L1 CNV data provided tantalizing evidence that L1 mobilization caused extensive somatic mosaicism in the human brain.

These experiments were a fundamental advance beyond the results of Muotri et al. (99) because (*a*) active retrotransposons are far more common in the mouse genome than in the human genome (>3,000 versus ~100 RC-L1s, respectively) (7, 16, 46, 102) and (*b*) the L1-EGFP reporter may not have accurately recapitulated endogenous L1 activation rates because of differential epigenetic suppression of an engineered human L1 transgene in mouse cells. Observation of L1 CNV in human tissues therefore excluded the conclusion that somatic L1 retrotransposition was an artifact of the L1-EGFP system, or a phenomenon restricted to rodents, and demonstrated that L1 was likely active across a broad spectrum of mammalian neurons.

L1 Regulation in the Neuronal Lineage

The finding that neuronal cell types are permissive for L1 retrotransposition raises the question of how L1 circumvents suppression in these cells. Muotri et al. (99) found that Sox2 interacts with the L1 promoter and represses L1 expression in rat neural stem cells (NSCs) and upon differentiation into NPCs, a decrease in Sox2 levels permits L1 transcriptional activation and presumably retrotransposition. Subsequent work by Coufal et al. (24) in human fetal NSCs and hESC-derived NPCs corroborated this result, indicating that L1 regulation by Sox2 is conserved in rodents and humans. Kuwabara et al. (68) found that the L1 promoter, as well as the L1 ORF2 sequence, contains overlapping Sox2 and T-cell factor/lymphoid enhancer factor (TCF/LEF) binding sites (Sox/LEF). Sox2, a negative regulator of neuronal differentiation, was demonstrated to repress promoter activity from these sites, whereas Wnt3a and β-catenin signaling increased L1 promoter activity. Moreover, the transcription factor NeuroD1, which promotes neuronal differentiation in adult hippocampal neural progenitors, is also regulated by Sox2 and Wnt/β-catenin signaling through Sox/LEF sites in its promoter (68). These results are consistent with derepression of L1 transcription concurrent with commitment of NPCs to the neuronal lineage and, strikingly, Sox2 regulation of the core neurogenesis pathway incorporating NeuroD1 is inseparable from Sox2 regulation of L1. If we also consider the expression and function of Sox2 in male gametes, it is plausible that establishment of the Sox2 regulatory program acts to limit L1 mobilization in the germline while not inhibiting retrotransposition during neurogenesis. Whether the latter property of Sox2 affects evolutionary fitness is unknown.

DNA methylation is another critical component of L1 repression in somatic cells. The X-linked DNA methyl-binding protein MeCP2 has been demonstrated to associate with the L1 promoter and repress L1 expression in cultured cells (137). Coufal et al. (24) investigated MeCP2-mediated L1 regulation in the brain and found higher levels of MeCP2 association with the L1 promoter in NSCs than in hESC-derived NPCs, suggesting a role for MeCP2 in modulating L1 activity during neuronal development. Indeed, using the L1-EGFP reporter, Muotri et al. (100) demonstrated that MeCP2 represses L1 promoter activity in a methylation-dependent manner. Notably, L1 regulation was specific to MeCP2, as perturbation of the DNA methyl-binding protein MDB1 did not affect L1 expression. Studies of MeCP2 knockout mice harboring the L1-EGFP reporter revealed an elevated rate of neuronal L1 retrotransposition in vivo in the absence of MeCP2. Endogenous L1s also appeared to undergo elevated rates of retrotransposition in the absence of

MeCP2, as a single-cell genomic qPCR assay revealed an increase in L1 copy number in MeCP2 knockout mouse neuroepithelial cells, but not fibroblasts, compared with control animals.

Germline and de novo MeCP2 mutations in humans cause Rett Syndrome (RTT), a condition characterized by profound neurodevelopmental abnormality. The disease affects ∼1/10,000 females but is rarely seen in males, as mutations in the X-linked MeCP2 are hemizygous lethal (3). Given the importance of MeCP2 in L1 suppression, L1 may be transcriptionally more active in RTT neurons, and thus RTT neurons may harbor a far higher somatic L1 mobilization rate than seen for healthy controls. How increased L1 activity is related to the neurological symptoms of RTT is unknown, but it is possible that elevated levels of L1 transcription could interfere with normal cellular processes or that an increased rate of retrotransposition events into neuronally expressed genes could have deleterious effects on neuronal function.

To study the relationship between MeCP2 deficiency, RTT, and L1 mobilization, Muotri et al. (100) derived NPCs from induced pluripotent stem cells (iPSCs) previously reprogrammed from RTT patient fibroblasts. They found that MeCP2 mutant NPCs accommodated significantly more retrotransposition of the L1-EGFP reporter than NPCs derived from control iPSCs. Furthermore, qPCR analysis showed a significantly higher L1 copy number in postmortem human brain tissue from RTT patients compared with controls. These experiments conclusively indicated that MeCP2 regulates L1 expression and retrotransposition activity in the mammalian brain. However, as for the L1 CNV data reported by Coufal et al. (24), these results must ultimately be corroborated by the genomic mapping of endogenous L1 insertions to be certain that L1 CNV is associated with L1 copies integrated into the genome and not with the accumulation of extrachromosomal L1 DNA via a largely uncharacterized mechanism (51, 59, 124). It is also important to note that, despite obvious differences in how well neurological phenotypes can be assessed in humans and animals, conditional restoration of MeCP2 function in MeCP2 mutant mice appears to ameliorate neurological dysfunction (49). Thus, although the role of MeCP2 in L1 regulation is clear, it remains unknown whether elevated L1 activity contributes to RTT etiology.

Regulation of L1 retrotransposition in neuronal cell types may not be limited to transcriptional and epigenetic control. Ataxia telangiectasia mutated (ATM) is a serine/threonine kinase that functions as a sensor of DNA damage (117). As its name reflects, ATM is mutated in the autosomal recessive disorder ataxia telangiectasia, which is characterized by progressive neuronal degeneration, variable immunodeficiency, ocular telangiectasias, and cancer susceptibility (2). Coufal et al. (23) demonstrated in 2011 that a human L1-EGFP can retrotranspose with increased efficiency in the brains of ATM-deficient mice as well as with human NPCs derived from hESCs in which ATM expression had been knocked down by RNAi. Furthermore, endogenous L1 CNV detection by qPCR revealed higher levels of the L1 ORF2 sequence in hippocampal samples from ataxia telangiectasia patients compared with hippocampal samples from normal matched controls. Experiments in non-neuronal cultured cell types (HeLa and HCT116 cells) revealed that L1 may generate more or possibly longer insertions in the absence of ATM, consistent with ATM recognition of the L1 TPRT intermediate as DNA damage, and resultant abrogation or truncation of the nascent L1 insertion. Thus, in normal cells, ATM is predicted to limit L1 retrotransposition, whereas in ataxia telangiectasia patients, loss of ATM function may allow elevated rates of retrotransposition or longer L1 insertions. Mapping of increased retrotransposition events using sequence-based approaches from the brains of ataxia telangiectasia patients relative to controls would provide definitive evidence for a role of ATM in regulating neuronal retrotransposition. Furthermore, mapping and characterization of retrotransposition events in ATM-deficient neurons may provide clues about the relationship, if there is one, between increased L1 retrotransposition and the progressive neurodegeneration observed in ataxia telangiectasia patients.

The studies discussed above have begun to uncover the mechanisms responsible for L1 regulation in the mammalian brain. It is very likely that additional host factors affect neuronal retrotransposition. Considering that L1 is usually repressed in somatic tissues, the appropriate line of inquiry may focus on factors that have previously been demonstrated to regulate retroelement activity in cells that occasionally accommodate retrotransposition, such as the early embryo (25, 37, 131). For example, the epigenetic regulator TRIM28/KAP1 has previously been demonstrated to regulate LTR retrotransposons in mouse embryonic stem cells (114). Furthermore, deletion of TRIM28/KAP1 in the forebrain of adult mice leads to stress-related behavioral abnormalities in learning and memory (58). It would therefore be interesting to determine whether TRIM28/KAP1 has a role in regulating retrotransposons in the mouse brain and whether neurological abnormalities associated with forebrain-specific TRIM28/KAP1 deletion are related to increased transcription and mobilization of retrotransposons. Similarly, other epigenetic effectors, such as the histone lysine methyltransferase SETDB1 (84), the histone deacetylase 1 (HDAC1) (112), the polycomb repressive complexes PRC1 and PRC2 (72), and the lysine-specific demethylase KDM1A/LSD1 (78), have been demonstrated to affect LTR retrotransposon activity in mouse embryonic stem cells. Whether these factors also modulate retrotransposon activity in the brain presents an interesting line of future inquiry.

MAPPING DE NOVO RETROTRANSPOSON INSERTIONS: CRITERIA AND DOCUMENTED EXAMPLES

Engineered L1s undergoing retrotransposition in cultured neuronal cell types and in the brains of transgenic animals, coupled with qPCR detection of L1 CNV in the human brain, constituted compelling yet incomplete evidence for endogenous L1 retrotransposition in mammalian neurons. In the following sections, we review recent studies in which deep-sequencing technologies and high-throughput analysis have been employed to identify and characterize somatic retrotransposon insertions, providing critical proof of bona fide retrotransposition in neurons. When considering such studies, however, it is important to bear in mind the unique challenges associated with mapping somatic retrotransposition events. New somatic insertions must be identified among the hundreds of thousands of copies already residing in the genome. Furthermore, somatic retrotransposon insertions are expected to be present in only a subset of cells and may even be unique to an individual cell. Determination of the extent of such somatic mosaicism requires analysis of single cell genomes, which can be achieved by whole-genome amplification (WGA). However, WGA may in turn introduce artifacts, such as chimeric sequences. In all cases, it is imperative that rigorous standards are upheld when calling and validating somatic transposon insertions from large sequencing data sets. Fortunately, retrotransposition events are usually accompanied by certain structural hallmarks that can be used to discern true insertions from other forms of genomic rearrangements or artifacts that may, at first pass, mimic bona fide transposable element activity.

L1-mediated retrotransposition, which includes mobilization of L1 as well as the nonautonomous retrotransposons *Alu* and SVA, accounts for all transposable element activity in humans. The ideal characterization of a de novo somatic L1-mediated retrotransposition event would comprise the following (**Figure 4**):

1. Mapping of L1-genome junctions at both the 5′ and 3′ flanks of the insertion, to single-nucleotide resolution.
2. Identification of structural features consistent with mobilization by retrotransposition: target-site duplications, poly-A tail, and 5′ truncation.
3. Insertion at a nucleotide motif resembling the loose L1 EN cleavage consensus site 5′-TTTT/A-3′.

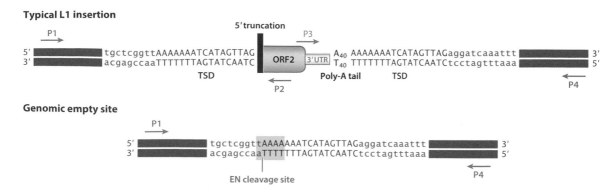

Typical L1 insertion

Genomic empty site

Figure 4

A fully-characterized typical L1 (long interspersed element 1) insertion. Hallmarks of retrotransposition include 5′ truncation (*red*), a poly-A tail (*purple*), and target-site duplications (TSDs) (*black*). Surrounding genomic DNA not contributing to the TSD is represented in blue. Below, the genomic empty site is depicted; the canonical L1 endonuclease cleavage motif (5′-TTTT/A-3′) is emphasized in green. Typical validation primers are depicted as gray arrows: P1 and P4 distinguish empty versus filled sites and can be used to confirm the empty site in control tissues. P1 and P2 amplify the 5′ L1-genome junction; P3 and P4 amplify the 3′ L1-genome junction. Abbreviation: EN, endonuclease.

4. Validation of both 5′ and 3′ L1-genome junctions by PCR and capillary sequencing.
5. Confirmation that the insertion is absent from matched control tissue.

Although it is preferable that all reported somatic insertions fulfill these specifications, certain technical and biological constraints make this task difficult or impossible to achieve in some cases. L1 insertions are frequently associated with deletions of target-site genomic DNA, as well as internal deletions and rearrangements of the L1 sequence itself (40, 41, 90, 126, 127) (**Figure 2**). Therefore, one end of an insertion may be easily identified from sequencing reads, but mapping the other end may be considerably more difficult. In the case of large deletions of genomic DNA or chromosomal translocations associated with retrotransposition events, mapping both L1-genome junctions of a new insertion may be virtually impossible. L1-mediated retrotransposition often occurs into highly repetitive genomic regions, which can confound mapping strategies and make PCR-based validation of insertions challenging. PCR validation can also be difficult for the 3′ ends of insertions because of the presence of the poly-A tail, as Baillie et al. (6) found. Transductions of 3′ flanking genomic DNA and, less frequently, 5′ flanking genomic DNA can occur during L1 mobilization (**Figure 3**). A transduced sequence presents an additional challenge to mapping strategies, but, excitingly, 3′ transductions provide a powerful tool to identify highly active progenitor elements that give rise to de novo insertions (79). Another point to consider is that not all retrotransposition events generate target-site duplications; a substantial percentage of insertions are blunt and can be associated with deletions of target-site DNA (40, 41). Indeed, as noted above, ENi L1 insertions typically lack TSDs, are frequently associated with L1 3′ truncations and deletion of target-site DNA, and are not expected to occur at the L1 EN consensus cleavage site (95, 96).

With the above limitations in mind, what constitutes sufficient evidence for a somatic insertion? Calling insertions based on discordant paired-end reads without detailed resolution of at least one L1-genome junction, because of a gap in the read pairs, is not ideal. Considering the prevalence of retroelement sequences in the genome, it is possible that such read pairs could indicate other types of genomic rearrangements or indeed could simply be DNA chimeras produced as an artifact of sequencing. We argue that for each putative insertion, at least one L1-genome junction should be identified at single-nucleotide resolution, from which the presence of an L1 EN cleavage consensus site can be discerned. Validation by PCR and capillary sequencing, although the strongest evidence to verify a putative insertion, cannot practically be applied to every insertion from data sets numbering in the thousands. Therefore, a reasonable expectation is that a random subset of insertions should be chosen for PCR validation so that a false-positive rate can be determined. Validation PCRs performed in parallel on matched control tissues are also necessary to conclusively confirm that the insertion represents a somatic rather than germline event. These criteria may appear prescriptive. However, the spatiotemporal boundaries of reported somatic L1 activity are expanding rapidly and, critically, incorporate human diseases in which somatic L1 mobilization could be considered as an etiological factor targeted for clinical intervention. In this setting, stringent requirements for reporting new L1 insertions are arguably both necessary and appropriate.

Mapping Neuronal L1 Retrotransposition Events by Retrotransposon Capture Sequencing

The first study to successfully map and characterize bona fide L1 insertions from human tissue employed a high-throughput approach for detection of endogenous retrotransposition events termed retrotransposon capture sequencing (RC-seq) (6) (**Figure 5a**). In RC-seq, genomic DNA is captured using custom arrays targeting the termini of full-length L1, *Alu*, and SVA consensus sequences, and then is deeply sequenced to generate paired-end reads spanning retrotransposon-genome junctions. These read pairs are then mapped to the reference genome to identify loci containing known and novel insertions, following the premise that L1 insertion heterogeneity in the brain can be overcome through targeted L1 sequencing.

To identify brain-specific somatic L1 insertions, Baillie et al. (6) used genomic DNA extracted from five different brain regions of three elderly postmortem donors without pathological signs of neurological disease. Quantitative PCR detection of L1 CNV confirmed elevated L1 copy number in the hippocampus relative to other brain regions in two donors, consistent with previous observations (24). RC-seq was then performed on genomic DNA from the hippocampus (highest L1 copy number) and caudate nucleus (lowest L1 copy number) from all three individuals. Putative insertions present in more than one individual or brain region, existing catalogs of retroelement polymorphisms or RC-seq previously performed on an exclusive cohort of pooled genomic DNA extracted from human blood donors (as other tissues from the brain donors were not available), were designated as germline. Strikingly, only 8.4% of *Alu* insertions and 1.9% of L1 insertions

---

Figure 5

Strategies for sequencing somatic L1 (long interspersed element 1) insertions. (*a*) Retrotransposon capture sequencing (RC-seq). RC-seq selectively enriches randomly fragmented genomic DNA for 5′ and 3′ L1-genome junctions. Illumina adapters are ligated to enriched libraries and fragments are deeply sequenced. Paired-end reads are tiled across L1-genome junctions. The library insert size used by Baillie et al. (6) allowed a gap between paired-end reads; subsequent iterations of the RC-seq protocol require overlapping read pairs and therefore allow single-nucleotide resolution of L1-genome junctions (118). (*b*) Whole-genome sequencing (WGS) without L1 enrichment. WGS is carried out on fragmented genomic DNA. Putative L1 insertions are identified as paired-end reads, wherein one end aligns with the reference genome, and the other aligns with the retroelement sequence. Genomic DNA is shown in green; the 5′ termini and 3′ termini of L1 insertions are represented in light purple and blue, respectively. L1 target-site duplications are represented as black triangles. (*c*) L1 insertion profiling (L1-IP). L1-IP employs a hemi-specific PCR (polymerase chain reaction) scheme to amplify the 3′ flanking regions of L1 insertions. Asymmetric PCR with an L1-specific primer targeting active subfamily L1s (*blue arrow*) is followed by hemi-specific PCR reaction using degenerate primers (*gray arrows*) with linker sequences (*orange lines*). A second round of PCR with a second L1-specific primer (*dashed blue arrow*) introduces Illumina sequencing adapters (*purple lines*) to facilitate deep sequencing and mapping of putative insertions. Abbreviation: UTR, untranslated region.

a Retrotransposon capture sequencing

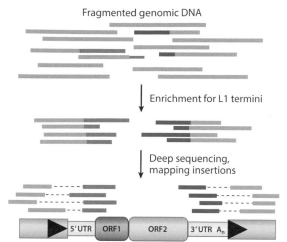

Fragmented genomic DNA

Enrichment for L1 termini

Deep sequencing,
mapping insertions

5′UTR | ORF1 | ORF2 | 3′UTR A$_n$

b Whole-genome paired-end sequencing

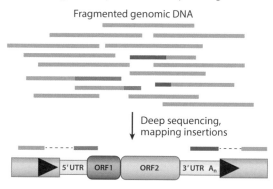

Fragmented genomic DNA

Deep sequencing,
mapping insertions

5′UTR | ORF1 | ORF2 | 3′UTR A$_n$

c L1 insertion profiling

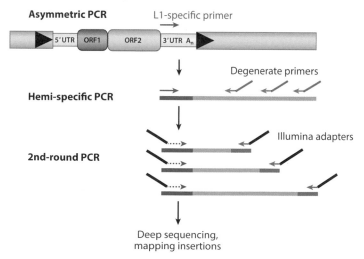

Asymmetric PCR

L1-specific primer

5′UTR | ORF1 | ORF2 | 3′UTR A$_n$

Degenerate primers

Hemi-specific PCR

Illumina adapters

2nd-round PCR

Deep sequencing,
mapping insertions

detected from the brain were determined to be germline insertions, suggesting a vast number of potential brain-specific somatic insertions: 7,743 L1, 13,692 *Alu*, and 1,350 SVA insertions. The majority of the 29 selected insertions could be validated by PCR targeting the L1 5′ end-insertion site junction, consistent with confident identification of new retrotransposon insertions as laid out in **Figure 4**. For three insertions (one L1 and two *Alu*), both the 5′ and 3′ L1-insertion site junctions could be discerned from RC-seq reads. Among these, one L1 and one *Alu* insertion were flanked by TSDs, indicative of retrotransposition by canonical TPRT. Consistent with retrotransposon evolution and mobilization in the germline, more than 80% of somatic L1 and *Alu* insertions arose from the most recently active L1 and *Alu* subfamilies. However, the lack of a nonbrain control tissue, such as liver, meant that only insertions present in one brain region and not the other were termed somatic, effectively excluding from reporting any somatic insertions present in multiple brain regions.

L1 and *Alu* insertions in the brain occurred into genes with a higher frequency than known germline insertions. This observation supported the notion that intragenic insertions are generally deleterious, and therefore those occurring in the germline would quickly be lost through negative selection, whereas in the brain there is no opportunity for this selection to occur (130). However, the frequency of intragenic insertions in the brain was also higher than random expectation, and, indeed, L1 insertions occurred preferentially into genes highly expressed in the brain. Genes related to neurogenesis and synaptic function were shown to be favored for L1 insertion by gene ontology analysis, even when the higher average length of genes active in the brain was accounted for. Taken together, these results suggest that L1 mobilization in the brain occurs more readily into actively expressed and euchromatinized genes. These loci favor L1 EN target-site nicking, and, similarly, L1 integration here is more likely to impact host gene expression (52), as compared with a gene not expressed in the brain. Considering the numerous examples of how L1 insertions can impact gene function and regulation, it is tempting to speculate that mosaicism generated by somatic retrotransposition in the brain can alter the functional output of individual neurons, and perhaps contribute to neuronal plasticity.

Baillie et al. (6) detected somatic L1 retrotransposition in the brains of individuals who were of advanced age (~92 years on average) but otherwise presented no neuropathological abnormalities. If L1 retrotransposition is a normal part of mammalian brain physiology, what, if any, relationship is there between L1 retrotransposition and human neurological disease? As described above, Muotri et al. (100) approached this question with their discovery of increased L1 retrotransposition in mice deficient in MeCP2, which is mutated in RTT. Coufal et al. (23) similarly uncovered evidence for increased neuronal retrotransposition in ataxia telangiectasia patients. Future studies will no doubt shed light on the relationship between increased neuronal retroelement expression, retrotransposition, and neuronal dysfunction in RTT and ataxia telangiectasia.

In a very recent study, Bundo et al. (17) sought a link between increased L1 copy number and major mental health disorders, including schizophrenia. Schizophrenia is a psychiatric disorder characterized by psychosis and includes delusions, hallucinations, disorganized speech and behavior, impaired cognition, and altered emotional reactivity (98, 106). Schizophrenia rarely, if ever, exhibits Mendelian inheritance, and many genomic loci and copy-number variants, a prominent example being 22q11 deletion, have been associated with the disease (98). Bundo et al. (17) employed quantitative PCR and demonstrated an increase in L1 ORF2 sequences in neurons from the prefrontal cortex of schizophrenia patients compared with control patients. This result was recapitulated in mice and macaques exposed to treatments mimicking environmental risk factors for perturbed early neuronal development. In addition, iPSC-derived neurons from schizophrenia patients with the 22q11 deletion exhibited an elevated L1 copy number in the L1 CNV assay pioneered by Coufal et al. (24). Strikingly, however, whole-genome sequencing (WGS) of the brain and liver tissue of schizophrenia patients, performed without enrichment for L1 sequences (**Figure 5b**), revealed no significant increase in the number of brain-specific L1 insertions in schizophrenia patients compared with controls. The authors report that brain-specific L1s had a significantly higher tendency to retrotranspose into neuronal- and mental disorder–associated genes in schizophrenia patient brains than in control patients. However, examination of their WGS data set reveals that nearly half of the putative L1 insertions arose from older L1 subfamilies not currently active in the human genome, rather than from young, highly active elements, which is not consistent with de novo L1 retrotransposition. Furthermore, insertions were detected based only on discordant paired-end reads and therefore were not characterized at single-nucleotide resolution, and validation by PCR and capillary sequencing (**Figure 4**) was not performed. In sum, the relationship, if one exists, between L1 activity in the brain and the pathogenesis of schizophrenia requires extensive additional study before definitive conclusions can be drawn.

Detection of an increased L1 copy number by qPCR, in the absence of a corresponding increase in mapped L1 insertions in the brain, suggests that phenomena other than de novo L1 retrotransposition may contribute to L1 CNV in disease. Intriguingly, a recent study reported a massive increase in L1 cDNA in HIV-infected cells compared with control cells (59). A report by Han & Shao (51) has also demonstrated that TPRT events can undergo alternate pathways of resolution, leading to the generation of extrachromosomal circular DNA products bearing L1 sequence. Thus, one may speculate that under conditions of cellular stress, L1 reverse transcription could take place ectopically, perhaps by using an alternate cellular nucleic acid as a primer. Likewise, initiated TPRT events could resolve without integration of a new L1 copy into the genome. To further speculate, it is possible that cellular abnormalities associated with schizophrenia, or perhaps the drugs used in its treatment, could trigger aberrant L1 reverse transcription initiation or outcomes, thus accounting for an increase in the L1 copy number without a corresponding increase in genomic L1 integration. Indeed, a recent report used L1 CNV qPCR and detected a startling 255 L1 copy/cell increase in wild-type mouse embryonic stem cells and an ~860 copy/cell increase in *Dicer*-deficient mouse embryonic stem cells over 20 passages (19). As the mutational load exacted by hundreds of genomic insertions would likely be intolerable by embryonic stem cells, the generation of extrachromosomal L1 DNA may provide an alternate explanation for these results.

Similarly, large-scale genomic CNV between control tissues and brain, between brain subregions, and between individual neurons, is another potential confounding factor in the assessment of L1 CNV in disease. Indeed, recent reports employing WGA strategies have detected CNVs in individual neurons and other cell types (42, 86), although further work is required to definitively address whether these events represent technical artifacts of WGA. This phenomenon could result in the uneven deletion or duplication of resident L1 insertions, leading to differences in relative differences in the L1 copy number not arising from retrotransposition. Therefore, as stressed above, it is critical that qPCR-based L1 CNV assays are corroborated by the mapping of insertions bearing hallmarks of retrotransposition by TPRT.

THE EXTENT OF NEURONAL MOSAICISM: SINGLE-CELL APPROACHES

Detection of bona fide somatic L1 insertions from bulk human brain tissue, complemented by earlier observations, provides definitive evidence that the neuronal genome is a somatic mosaic due to L1 mobilization. However, these analyses do not directly address the key question of how complex that mosaicism is, i.e., how frequently does L1 mobilize in the neuronal lineage? Low read depth for brain-specific L1 insertions relative to germline insertions uncovered by Baillie et al. (6) suggests that each insertion is present in only a subset of neurons, but does not constitute a quantitative measure of the prevalence of a given L1 insertion in the brain. It is possible that, owing to L1 mobilization in the latter phases of neurogenesis, each individual neuron is genetically unique with regard to its cohort of loci containing somatic L1 insertions. It is also possible that L1 insertions arising during embryogenesis could be present in large subpopulations of neurons throughout the brain. Indeed, both of these scenarios could be true, painting a complex picture of L1-driven neuronal mosaicism in the adult brain.

The extent to which individual neurons share or differ in L1 content can be resolved by single cell–based approaches. Analyzing the genome of single cells presents several technical challenges, as WGA is required to generate sufficient material for analysis. The individual cell in question is consumed during the amplification process, and no starting material remains from which to validate results obtained from amplified DNA. Single-cell analysis experiments must therefore be carefully designed and subject to rigorous quality-control measures. Data from single-cell analysis must also be interpreted with technical caveats, such as WGA bias and the likely presence of molecular artifacts, in mind (70).

In the first study to successfully perform single-neuron analysis of L1 insertions, Evrony et al. (32) isolated single nuclei from the caudate nucleus and pyramidal nuclei from the frontal cortex of three individuals by fluorescence-activated cell sorting (FACS) and performed multiple displacement amplification (MDA) to obtain 15–20 µg of DNA from each of 300 single nuclei. Putative L1 insertions were identified using a modified version of the method of Ewing & Kazazian (33), which entails hemi-nested PCR amplification of L1-genome junctions, using primers annealing specifically to the 3′ end of human-specific L1s in combination with degenerate primers to anneal to genomic flanking DNA (**Figure 5c**). For this study, the method of Ewing et al. (33) was adapted for high-throughput, multiplexed sequencing and, along with the MDA, required an estimated equivalent of ~60 cycles of PCR for library preparation. In the amplified genomes of individual neurons, greater than 80% of known reference insertions could be detected on average and for single-copy X-linked insertions in nuclei from a male donor, this figure was approximately 75%. For detection of somatic L1 insertions from individual neurons, the authors used a sensitivity threshold at which only 50% of known reference insertions can be detected. The frequency of unique L1 insertions was estimated at less than one per neuron, with 82% of neurons harboring no unique insertions. PCR validation of L1 3′ end-genome junctions (**Figure 4**) yielded results consistent with a bona fide L1 retrotransposition event for only 5 out of 81 putative insertions, suggesting a high rate of false positives due to MDA chimeras and technical artifacts. Among the five validated insertions, one could be fully characterized and bore the hallmarks of L1 retrotransposition; owing to a 5′ transduction, this insertion could be traced to a progenitor element on chromosome 8. This insertion was present in two cortical nuclei and could be amplified from bulk samples. The other four insertions could be characterized only at their 3′ L1-genome junction and appeared to be specific to the individual neuron in which they were detected.

The work of Evrony et al. (32) represented a major methodological advance for studies of L1 and for single-cell genomics in general. Evrony et al. (32) identified and validated the first somatic full-length L1 insertion found in the brain, emphatically demonstrating that L1-mediated mosaicism can occur in neurons. Notably, the estimates of L1 mobilization frequency produced by Evrony et al. (32) were more than two orders of magnitude lower than the qPCR-based estimates of Coufal et al. (24) (0.6 versus 80 per neuron, respectively). The reasons for this disparity remain to be ascertained. However, even if the former figure is provisionally accepted and extrapolated, 6×10^{10} somatic L1 insertions would be present in each human brain, a figure contrasting with the aforementioned estimate of 4×10^8 polymorphic retrotransposon insertions in the entire human population and, in our opinion, at odds with Evrony et al.'s (32) description of somatic

L1 mobilization as rare. It is also important to note that the figures reported by Coufal et al. (24) specified the hippocampus as a hotspot of L1 retrotransposition, whereas Evrony et al. (32) studied cortical neurons. Moreover, chimera formation during MDA is an established source of artifacts (70). To exclude events representing localized chimeras between existing L1 insertions and nearby genomic DNA, Evrony et al. (32) only considered somatic L1 insertions greater than 20 kb from known L1 reference genome and polymorphic insertions. Finally, the true-positive rate from the $3'$ L1-genome junction PCR is critical to estimating the somatic L1 mobilization rate with this approach. It is possible that MDA amplification is biased in favor of chimeric sequences, which could lead to a higher read count for chimeras relative to bona fide L1 insertions, thus skewing the pool of potential insertions selected for PCR validation in favor of false positives. Therefore, prioritizing insertions for validation on the basis of high read count could artificially elevate the false-positive rate. In sum, although the identification by Evrony et al. (32) of a full-length L1 insertion bearing the hallmarks of TPRT was a landmark achievement, substantial additional work is still required to accurately quantify somatic L1 mobilization in the human brain.

NEURONAL RETROTRANSPOSITION IN *DROSOPHILA*

Retrotransposition in neurons, particularly in brain regions critical for learning and memory, was very recently revealed to be not exclusive to mammals. Perrat et al. (108) uncovered evidence for increased transposable element expression in the *Drosophila* mushroom body, a brain structure required for olfactory memory. Specifically, transposon transcripts were highly expressed in $\alpha\beta$ neurons, which are proposed to be required for storage and recall of memories, as contrasted to $\alpha'\beta'$ or γ neurons, which are proposed to be required for acquisition and stabilization of memories. Transposable element transcripts expressed in $\alpha\beta$ neurons include LTR and non-LTR retrotransposons, as well as DNA transposons. To identify somatic mobilization events, Perrat et al. (108) employed WGS on genomic DNA extracted from $\alpha\beta$ neurons, other brain tissue, and embryos genetically identical to the flies used in the neuron analysis. They reported 215 insertions present in $\alpha\beta$ neurons but absent from embryos and the *Drosophila* reference genome, and 200 such insertions in other brain tissue, 19 of which overlapped with the set identified in $\alpha\beta$ neurons. Putative insertions were identified by discordant paired-end reads, which do not allow single-nucleotide resolution of retroelement-genome junctions or identification of the structural hallmarks of transposable element integration (**Figure 4**). Furthermore, sequencing depth was ~3.1 times for a sample of 5,000 pooled $\alpha\beta$ neurons, requiring extrapolation to assess the per-neuron transposition rate. Future validation of putative insertions by PCR and capillary sequencing, perhaps combined with single-neuron analysis, are required to definitively assess how frequently endogenous transposable element insertions occur in the *Drosophila* brain.

Evidence for transposon activity in the *Drosophila* brain was also recently uncovered by Li et al. (74), who found that some transposable elements exhibited age-dependent transcriptional upregulation. The authors used a *gypsy-TRAP* reporter system, in which insertions of the LTR retrotransposon *gypsy* into an engineered reporter construct activate GFP expression in the affected cell, to demonstrate retrotransposition in mushroom body neurons of aged flies but not their younger counterparts. The *gypsy*-TRAP system specifically detects *gypsy* retrotransposition events occurring into a transgenic preferred target locus and therefore does not provide information about the activity of other transposable elements or the full spectrum of genomic insertion sites targeted by transposons. Nevertheless, this experiment clearly demonstrates that neuronal retrotransposition occurs in *Drosophila* in an age-dependent manner. This result raises the question of whether transposable element insertions also accumulate in the mammalian brain as a function of advancing age.

Both of the above studies also investigated how transposable elements are regulated in the *Drosophila* brain. Prompted by the observation that the translocated *Stellate* locus *STE12DOR*, which is usually repressed by piRNAs (PIWI-interacting RNAs), was upregulated in αβ neurons relative to other brain tissues, Perrat et al. (108) investigated whether the piRNA pathway is involved in differential transposon regulation among neuronal types in the *Drosophila* brain. Indeed, they found that αβ neurons do not express the PIWI clade proteins Aub or Ago3, consistent with higher transposon expression compared with PIWI-expressing α′β′ and γ neurons. Furthermore, analysis of RNA isolated from the heads of PIWI-mutant, piRNA-defective flies revealed elevated expression levels of some transposons. Consistent with previous reports (39), Perrat et al. (108) also found that heads from endo-siRNA (small interfering RNA)-defective *ago2* and *dcr-2* mutant flies contained elevated levels of certain transposon transcripts. Taken together, these results suggest that the endo-siRNA pathway and the piRNA pathway, which was previously demonstrated to repress transposons in the *Drosophila* germline, may both contribute to transposon regulation in the *Drosophila* brain.

In parallel, Li et al. (74) investigated the functional consequences of transposon derepression in the *Drosophila* brain by employing *ago2* mutant flies in which endo-siRNA-mediated transposon repression is released in somatic tissues. In *ago2* mutant heads, R2 and gypsy transposon transcripts were elevated relative to controls, and transcript levels from *ago2* mutant young flies were comparable to those observed in wild-type aged flies. Elevated *gypsy* transcript levels were accompanied by the detection of de novo gypsy insertions in the *ovo* locus, an established preferred *gypsy* integration target. To investigate a functional link between elevated transposon activity and age-dependent neuronal impairment, *ago2* mutant flies were subjected to learning and memory tests. Young *ago2* mutant flies exhibited memory defects that could be rescued by *ago2* transgene expression and that worsened with increasing age. *Ago2* mutant flies also had shorter life spans. Disruption in neurons of *loki*, the *Drosophila* Chk2 ortholog that mediates DNA damage-induced apoptosis, increased life span in wild-type flies, and also partially relieved age-dependent memory loss. Although links between increased transposon activity, memory loss, and mortality uncovered by this study are correlative and await more direct mechanistic studies of the consequences of transposon activity on neuronal function and comprehensive mapping of transposon insertions in the *Drosophila* brain, the evidence for a functional consequence of neuronal transposition is nevertheless intriguing.

REMAINING QUESTIONS AND FUTURE DIRECTIONS

Perhaps the most interesting question regarding neuronal retrotransposition, and also the most difficult to answer, is whether these events play a role in normal brain function. In order to elucidate a function for retrotransposition in the brain, it will be important to accomplish a thorough characterization of the rate, developmental timing, and genomic impact of neuronal retrotransposition. From this information, testable hypotheses about the function of neuronal retrotransposition may emerge. Below, we highlight fundamental questions that should be resolved in the near future.

What Is the Rate of Neuronal Retrotransposon Insertions?

On the basis of a quantitative PCR assay, Coufal et al. (24) estimated that the adult human hippocampus contains approximately 80 new L1 copies per cell, suggesting a strikingly high rate of somatic retrotransposition in the brain. Baillie et al. (6) identified 7,743 neuronal L1 insertions from RC-seq analysis of bulk DNA, yielding a conservative estimate of 0.04 unique insertions per neuron. Evrony et al. (32) employed single-cell analysis and arrived at a figure of <0.6 unique somatic insertions per neuron; when taking into account only insertions that could be validated by PCR, this figure drops to 0.04 unique insertions per neuron. There are several potential reasons for the discrepancy between estimates based on qPCR and those based on numbers of mapped insertions. As discussed above, qPCR assays detect relative L1 DNA content but do not discriminate between bona fide genomic L1 insertions and other sources of increased L1 DNA, such as genomic instability not arising from retrotransposition and perhaps the accumulation of extrachromosomal L1 reverse transcription products (59, 124). It is therefore possible that qPCR-based assays could produce an overestimate of L1 retrotransposition. However, estimating the actual rate of L1 retrotransposition using insertions detected in bulk DNA by the method of Baillie et al. (6) is fraught with uncertainty because detection is probably not at saturation and insertions present in only one or a small subset of neurons are likely to evade detection. Indeed, it is impossible to determine how many insertions are missed in this method. Analysis of single cells will ultimately produce the most reliable quantification of neuronal retrotransposition. Evrony et al. (32) have taken the first step toward this end, and as methods for WGA from single cells improve and strategies for detection become more sensitive, an accurate estimate of the rate of neuronal retrotransposition will ultimately emerge.

What Is the Developmental Timing and Cell-Type Specificity of Neuronal Retrotransposition?

The developmental timing and cell-type specificity of neuronal retrotransposition are important subjects for future investigation. Neuronal retrotransposition events may accumulate during a specific developmental stage, such as embryonic development, or the generation of neuronal somatic mosaicism may be an ongoing process throughout the life of an organism. It is important to determine whether retrotransposition occurs mainly in dividing neuronal precursor cells, as previous studies have suggested (24, 99), or whether new events can occur in fully differentiated neurons. The answers to such questions will contribute to a complete picture of retrotransposition-derived neuronal mosaicism. Furthermore, this knowledge will direct speculation as to whether retrotransposition generates pre-existing neuronal diversity that can be exploited during learning and memory formation, or whether new neuronal retrotransposition events occur in response to external stimuli. Indeed, Muotri et al. (101) found that voluntary exercise is correlated with an increase in hippocampal retrotransposition in mice, suggesting that external stimuli may indeed lead to increased retrotransposition and therefore neuronal genomic diversity.

What Are the Functional Consequences of Neuronal Retrotransposition?

Muotri et al. (99) demonstrated that an engineered L1 insertion could alter gene expression and cell fate in rat NPCs cultured in vitro. Future studies should focus on how such alterations to target gene expression occur on a mechanistic level. Numerous studies have put forth mechanisms by which L1 insertions can disrupt transcript integrity and expression levels of target genes, and L1 insertions are also associated with epigenetic alterations to target DNA. It will be interesting to determine the prevalence of these various alterations in neuronal retrotransposition, and whether these changes alter the expression and integrity of target genes. Addressing these questions on a per-neuron basis presents a formidable technical challenge, as current techniques do not allow analysis of genomic DNA and RNA expression from the same cell.

How Does Neuronal Retrotransposon Activity Relate to Human Disease?

Understanding a potential role for neuronal retrotransposition in normal physiology will be aided by examining the relationship between retrotransposition and brain disorders and aging. The observations that L1 is highly expressed and retrotransposition occurs with high frequency in apparently normal human and rodent brains indicate that a certain level of L1 activity is tolerable and may even have a physiological function. However, demonstrated correlations between L1 upregulation and RTT in humans and dysregulation of transposable elements by *ago2* deficiency and memory impairment in flies suggest that too much neuronal transposition may have negative consequences (74, 100). Similarly, numerous cellular factors, including the APOBEC3 cytidine deaminases (13, 14, 18, 64, 97, 103, 123), the Aicardi-Goutieres syndrome gene products Trex1 (124) and SAMHD1 (138), and the Mov10 putative helicase (4, 43, 75), have been demonstrated to restrict L1 retrotransposition in vitro. It would be interesting to determine whether deficiency in such factors in humans or rodents in vivo leads to an increase in neuronal retrotransposition and whether this activity has any impact on brain function.

Given recent reports demonstrating somatic L1 mobilization in cancer (57, 71, 118, 121) and the emergence of the brain as a major site of somatic retrotransposition in mammals, the logical follow-up question is whether L1 mobilization occurs in brain tumors. Strikingly, examination of 5 medulloblastoma and 5 glioblastoma genomes by Iskow et al. (57) and 19 glioblastoma genomes by Lee et al. (71) revealed no tumor-specific retroelement insertions. One explanation for this result is that retrotransposition in the brain appears to occur primarily in the neuronal lineage (99) (**Figure 3b**) and may not, therefore, be prevalent in tumors arising from the glial cell lineage (76). Alternatively, if retrotransposition is indeed a normal or even necessary factor in brain physiology, host mechanisms may direct new insertions to specific regions of the neuronal genome, precluding potentially oncogenic retrotransposition events.

To conclude, the surprising discovery in recent years that somatic retrotransposition can occur in the metazoan brain provides fuel for speculation regarding the functional impact of transposable elements upon both normal and abnormal physiology. A thorough characterization of the rate, timing, consequences, and regulation of neuronal retrotransposition is, however, required before a functional role for retrotransposition in the brain can be conclusively discerned. In any case, that genetic mosaicism associated with the mobilization of transposable elements is now a major focus of research is somewhat ironic, considering that the original characterization of transposition in maize by McClintock (85) nearly 60 years ago described a somatic phenomenon. There, a phenotypic outcome was clear. Here, the functional consequences of L1 activity in neurons may prove comparatively elusive and yet, arguably, equally important for our understanding of the genetic basis for life.

SUMMARY POINTS

1. Recent studies have demonstrated that mobile elements are active and generate somatic mosaicism in the mammalian and *Drosophila* brain.

2. Certain regulatory factors, including the methyl-binding protein MeCP2 and the master cell cycle regulator ATM, play a role in limiting retrotransposon activity in the mammalian brain. Deficiencies in such factors suggest a correlation between deregulated neuronal retrotransposition and neurological disease.

3. Advances in sequencing technology have allowed mapping and characterization of somatic retrotransposon insertions. However, sequencing data must be carefully analyzed to distinguish bona fide retrotransposon insertions from other genomic rearrangements or technical artifacts.

FUTURE ISSUES

1. What are the rate, developmental timing, and cell-type specificity of neuronal retrotransposition?

2. How are elevated levels of retroelement expression and mobilization observed in neurological diseases, such as RTT, related to disease etiology?

3. Does retrotransposition in the brain, and the resultant somatic mosaicism, play a beneficial or vital role in normal brain physiology?

DISCLOSURE STATEMENT

The authors are not aware of any affiliations, memberships, funding, or financial holdings that might be perceived as affecting the objectivity of this review.

ACKNOWLEDGMENTS

We thank Jose Garcia-Perez and Adam Ewing for reading the manuscript and constructive comments, and apologize to colleagues whose work could not be cited because of space limitations. G.J.F. acknowledges the support of an Australian NHMRC Career Development Fellowship (GNT1045237), NHMRC Project Grants GNT1042449, GNT1045991, GNT1052303, GNT1067983, and GNT1068789 and the European Union's Seventh Framework Program (FP7/2007-2013) under grant agreement No. 259743 underpinning the MODHEP consortium.

LITERATURE CITED

1. Alisch RS, Garcia-Perez JL, Muotri AR, Gage FH, Moran JV. 2006. Unconventional translation of mammalian LINE-1 retrotransposons. *Genes Dev.* 20:210–24

2. Ambrose M, Gatti RA. 2013. Pathogenesis of ataxia-telangiectasia: the next generation of ATM functions. *Blood* 121:4036–45

3. Amir RE, Van den Veyver IB, Wan M, Tran CQ, Francke U, Zoghbi HY. 1999. Rett syndrome is caused by mutations in X-linked MECP2, encoding methyl-CpG-binding protein 2. *Nat. Genet.* 23:185–88

4. Arjan-Odedra S, Swanson CM, Sherer NM, Wolinsky SM, Malim MH. 2012. Endogenous MOV10 inhibits the retrotransposition of endogenous retroelements but not the replication of exogenous retroviruses. *Retrovirology* 9:53

5. Athanikar JN, Badge RM, Moran JV. 2004. A YY1-binding site is required for accurate human LINE-1 transcription initiation. *Nucleic Acids Res.* 32:3846–55

6. Baillie JK, Barnett MW, Upton KR, Gerhardt DJ, Richmond TA, et al. 2011. Somatic retrotransposition alters the genetic landscape of the human brain. *Nature* 479:534–37

7. Beck CR, Collier P, Macfarlane C, Malig M, Kidd JM, et al. 2010. LINE-1 retrotransposition activity in human genomes. *Cell* 141:1159–70

8. Beck CR, Garcia-Perez JL, Badge RM, Moran JV. 2010. LINE-1 elements in structural variation and disease. *Annu. Rev. Genomics Hum. Genet.* 12:187–215

9. Becker KG, Swergold GD, Ozato K, Thayer RE. 1993. Binding of the ubiquitous nuclear transcription factor YY1 to a *cis* regulatory sequence in the human LINE-1 transposable element. *Hum. Mol. Genet.* 2:1697–702

10. Belancio VP, Hedges DJ, Deininger P. 2008. Mammalian non-LTR retrotransposons: for better or worse, in sickness and in health. *Genome Res.* 18:343–58

11. Belancio VP, Roy-Engel AM, Deininger P. 2008. The impact of multiple splice sites in human L1 elements. *Gene* 411:38–45

12. Bestor TH, Bourc'his D. 2004. Transposon silencing and imprint establishment in mammalian germ cells. *Cold Spring Harb. Symp. Quant. Biol.* 69:381–87

13. Bogerd HP, Wiegand HL, Doehle BP, Lueders KK, Cullen BR. 2006. APOBEC3A and APOBEC3B are potent inhibitors of LTR-retrotransposon function in human cells. *Nucleic Acids Res.* 34:89–95

14. Bogerd HP, Wiegand HL, Hulme AE, Garcia-Perez JL, O'Shea KS, et al. 2006. Cellular inhibitors of long interspersed element 1 and Alu retrotransposition. *Proc. Natl. Acad. Sci. USA* 103:8780–85

15. Brouha B, Meischl C, Ostertag E, de Boer M, Zhang Y, et al. 2002. Evidence consistent with human L1 retrotransposition in maternal meiosis I. *Am. J. Hum. Genet.* 71:327–36

16. Brouha B, Schustak J, Badge RM, Lutz-Prigge S, Farley AH, et al. 2003. Hot L1s account for the bulk of retrotransposition in the human population. *Proc. Natl. Acad. Sci. USA* 100:5280–85

17. Bundo M, Toyoshima M, Okada Y, Akamatsu W, Ueda J, et al. 2014. Increased L1 retrotransposition in the neuronal genome in schizophrenia. *Neuron* 81:306–13

18. Chen H, Lilley CE, Yu Q, Lee DV, Chou J, et al. 2006. APOBEC3A is a potent inhibitor of adeno-associated virus and retrotransposons. *Curr. Biol.* 16:480–85

19. Ciaudo C, Jay F, Okamoto I, Chen CJ, Sarazin A, et al. 2013. RNAi-dependent and independent control of LINE1 accumulation and mobility in mouse embryonic stem cells. *PLoS Genet.* 9:e1003791

20. Cordaux R, Batzer MA. 2009. The impact of retrotransposons on human genome evolution. *Nat. Rev. Genet.* 10:691–703

21. Cost GJ, Boeke JD. 1998. Targeting of human retrotransposon integration is directed by the specificity of the L1 endonuclease for regions of unusual DNA structure. *Biochemistry* 37:18081–93

22. Cost GJ, Golding A, Schlissel MS, Boeke JD. 2001. Target DNA chromatinization modulates nicking by L1 endonuclease. *Nucleic Acids Res.* 29:573–77

23. Coufal NG, Garcia-Perez JL, Peng GE, Marchetto MC, Muotri AR, et al. 2011. Ataxia telangiectasia mutated (ATM) modulates long interspersed element-1 (L1) retrotransposition in human neural stem cells. *Proc. Natl. Acad. Sci. USA* 108:20382–87

24. Coufal NG, Garcia-Perez JL, Peng GE, Yeo GW, Mu Y, et al. 2009. L1 retrotransposition in human neural progenitor cells. *Nature* 460:1127–31

25. de Boer M, van Leeuwen K, Geissler J, Weemaes CM, van den Berg TK, et al. 2014. Primary immunodeficiency caused by an exonized retroposed gene copy inserted in the *CYBB* gene. *Hum. Mutat.* 35:486–96

26. de Koning AP, Gu W, Castoe TA, Batzer MA, Pollock DD. 2011. Repetitive elements may comprise over two-thirds of the human genome. *PLoS Genet.* 7:e1002384

27. Dmitriev SE, Andreev DE, Terenin IM, Olovnikov IA, Prassolov VS, et al. 2007. Efficient translation initiation directed by the 900-nucleotide-long and GC-rich 5′ untranslated region of the human retrotransposon LINE-1 mRNA is strictly cap dependent rather than internal ribosome entry site mediated. *Mol. Cell. Biol.* 27:4685–97

28. Dombroski BA, Mathias SL, Nanthakumar E, Scott AF, Kazazian HH Jr. 1991. Isolation of an active human transposable element. *Science* 254:1805–8

29. Doucet AJ, Hulme AE, Sahinovic E, Kulpa DA, Moldovan JB, et al. 2010. Characterization of LINE-1 ribonucleoprotein particles. *PLoS Genet.* 6:e1001150

30. Ergun S, Buschmann C, Heukeshoven J, Dammann K, Schnieders F, et al. 2004. Cell type–specific expression of LINE-1 open reading frames 1 and 2 in fetal and adult human tissues. *J. Biol. Chem.* 279:27753–63

31. Esnault C, Maestre J, Heidmann T. 2000. Human LINE retrotransposons generate processed pseudo-genes. *Nat. Genet.* 24:363–67
32. Evrony GD, Cai X, Lee E, Hills LB, Elhosary PC, et al. 2012. Single-neuron sequencing analysis of L1 retrotransposition and somatic mutation in the human brain. *Cell* 151:483–96
33. Ewing AD, Kazazian HH Jr. 2010. High-throughput sequencing reveals extensive variation in human-specific L1 content in individual human genomes. *Genome Res.* 20:1262–70
34. Faulkner GJ. 2011. Retrotransposons: mobile and mutagenic from conception to death. *FEBS Lett.* 585:1589–94
35. Faulkner GJ, Kimura Y, Daub CO, Wani S, Plessy C, et al. 2009. The regulated retrotransposon tran-scriptome of mammalian cells. *Nat. Genet.* 41:563–71
36. Feng Q, Moran JV, Kazazian HH Jr, Boeke JD. 1996. Human L1 retrotransposon encodes a conserved endonuclease required for retrotransposition. *Cell* 87:905–16
37. Garcia-Perez JL, Marchetto MC, Muotri AR, Coufal NG, Gage FH, et al. 2007. LINE-1 retrotranspo-sition in human embryonic stem cells. *Hum. Mol. Genet.* 16:1569–77
38. Garcia-Perez JL, Morell M, Scheys JO, Kulpa DA, Morell S, et al. 2010. Epigenetic silencing of engi-neered L1 retrotransposition events in human embryonic carcinoma cells. *Nature* 466:769–73
39. Ghildiyal M, Seitz H, Horwich MD, Li C, Du T, et al. 2008. Endogenous siRNAs derived from trans-posons and mRNAs in *Drosophila* somatic cells. *Science* 320:1077–81
40. Gilbert N, Lutz S, Morrish TA, Moran JV. 2005. Multiple fates of L1 retrotransposition intermediates in cultured human cells. *Mol. Cell. Biol.* 25:7780–95
41. Gilbert N, Lutz-Prigge S, Moran JV. 2002. Genomic deletions created upon LINE-1 retrotransposition. *Cell* 110:315–25
42. Gole J, Gore A, Richards A, Chiu YJ, Fung HL, et al. 2013. Massively parallel polymerase cloning and genome sequencing of single cells using nanoliter microwells. *Nat. Biotechnol.* 31:1126–32
43. Goodier JL, Cheung LE, Kazazian HH Jr. 2012. MOV10 RNA helicase is a potent inhibitor of retro-transposition in cells. *PLoS Genet.* 8:e1002941
44. Goodier JL, Cheung LE, Kazazian HH Jr. 2013. Mapping the LINE1 ORF1 protein interactome reveals associated inhibitors of human retrotransposition. *Nucleic Acids Res.* 41:7401–19
45. Goodier JL, Kazazian HH Jr. 2008. Retrotransposons revisited: the restraint and rehabilitation of para-sites. *Cell* 135:23–35
46. Goodier JL, Ostertag EM, Du K, Kazazian HH Jr. 2001. A novel active L1 retrotransposon subfamily in the mouse. *Genome Res.* 11:1677–85
47. Goodier JL, Ostertag EM, Kazazian HH Jr. 2000. Transduction of 3′-flanking sequences is common in L1 retrotransposition. *Hum. Mol. Genet.* 9:653–57
48. Grimaldi G, Skowronski J, Singer MF. 1984. Defining the beginning and end of KpnI family segments. *EMBO J.* 3:1753–59
49. Guy J, Gan J, Selfridge J, Cobb S, Bird A. 2007. Reversal of neurological defects in a mouse model of Rett syndrome. *Science* 315:1143–47
50. Hall LL, Carone DM, Gomez AV, Kolpa HJ, Byron M, et al. 2014. Stable C0T-1 repeat RNA is abundant and is associated with euchromatic interphase chromosomes. *Cell* 156:907–19
51. Han JS, Shao S. 2012. Circular retrotransposition products generated by a LINE retrotransposon. *Nucleic Acids Res.* 40:10866–77
52. Han JS, Szak ST, Boeke JD. 2004. Transcriptional disruption by the L1 retrotransposon and implications for mammalian transcriptomes. *Nature* 429:268–74
53. Hancks DC, Kazazian HH Jr. 2012. Active human retrotransposons: variation and disease. *Curr. Opin. Genet. Dev.* 22:191–203
54. Heras SR, Macias S, Plass M, Fernandez N, Cano D, et al. 2013. The Microprocessor controls the activity of mammalian retrotransposons. *Nat. Struct. Mol. Biol.* 20:1173–81
55. Hohjoh H, Singer MF. 1996. Cytoplasmic ribonucleoprotein complexes containing human LINE-1 protein and RNA. *EMBO J.* 15:630–39
56. Iida S, Morita Y, Choi JD, Park KI, Hoshino A. 2004. Genetics and epigenetics in flower pigmentation associated with transposable elements in morning glories. *Adv. Biophys.* 38:141–59

57. Iskow RC, McCabe MT, Mills RE, Torene S, Pittard WS, et al. 2010. Natural mutagenesis of human genomes by endogenous retrotransposons. *Cell* 141:1253–61

58. Jakobsson J, Cordero MI, Bisaz R, Groner AC, Busskamp V, et al. 2008. KAP1-mediated epigenetic repression in the forebrain modulates behavioral vulnerability to stress. *Neuron* 60:818–31

59. Jones RB, Song H, Xu Y, Garrison KE, Buzdin AA, et al. 2013. LINE-1 retrotransposable element DNA accumulates in HIV-1-infected cells. *J. Virol.* 87:13307–20

60. Jurka J. 1997. Sequence patterns indicate an enzymatic involvement in integration of mammalian retroposons. *Proc. Natl. Acad. Sci. USA* 94:1872–77

61. Kano H, Godoy I, Courtney C, Vetter MR, Gerton GL, et al. 2009. L1 retrotransposition occurs mainly in embryogenesis and creates somatic mosaicism. *Genes Dev.* 23:1303–12

62. Kazazian HH Jr, Wong C, Youssoufian H, Scott AF, Phillips DG, Antonarakis SE. 1988. Haemophilia A resulting from de novo insertion of L1 sequences represents a novel mechanism for mutation in man. *Nature* 332:164–66

63. Khazina E, Weichenrieder O. 2009. Non-LTR retrotransposons encode noncanonical RRM domains in their first open reading frame. *Proc. Natl. Acad. Sci. USA* 106:731–36

64. Kinomoto M, Kanno T, Shimura M, Ishizaka Y, Kojima A, et al. 2007. All APOBEC3 family proteins differentially inhibit LINE-1 retrotransposition. *Nucleic Acids Res.* 35:2955–64

65. Kubo S, Seleme MC, Soifer HS, Perez JL, Moran JV, et al. 2006. L1 retrotransposition in nondividing and primary human somatic cells. *Proc. Natl. Acad. Sci. USA* 103:8036–41

66. Kulpa DA, Moran JV. 2005. Ribonucleoprotein particle formation is necessary but not sufficient for LINE-1 retrotransposition. *Hum. Mol. Genet.* 14:3237–48

67. Kulpa DA, Moran JV. 2006. *cis*-Preferential LINE-1 reverse transcriptase activity in ribonucleoprotein particles. *Nat. Struct. Mol. Biol.* 13:655–60

68. Kuwabara T, Hsieh J, Muotri A, Yeo G, Warashina M, et al. 2009. Wnt-mediated activation of NeuroD1 and retro-elements during adult neurogenesis. *Nat. Neurosci.* 12:1097–105

69. Lander ES, Linton LM, Birren B, Nusbaum C, Zody MC, et al. 2001. Initial sequencing and analysis of the human genome. *Nature* 409:860–921

70. Lasken RS, Stockwell TB. 2007. Mechanism of chimera formation during the multiple displacement amplification reaction. *BMC Biotechnol.* 7:19

71. Lee E, Iskow R, Yang L, Gokcumen O, Haseley P, et al. 2012. Landscape of somatic retrotransposition in human cancers. *Science* 337:967–71

72. Leeb M, Pasini D, Novatchkova M, Jaritz M, Helin K, Wutz A. 2010. Polycomb complexes act redundantly to repress genomic repeats and genes. *Genes Dev.* 24:265–76

73. Levin HL, Moran JV. 2011. Dynamic interactions between transposable elements and their hosts. *Nat. Rev. Genet.* 12:615–27

74. Li W, Prazak L, Chatterjee N, Gruninger S, Krug L, et al. 2013. Activation of transposable elements during aging and neuronal decline in *Drosophila*. *Nat. Neurosci.* 16:529–31

75. Li X, Zhang J, Jia R, Cheng V, Xu X, et al. 2013. The MOV10 helicase inhibits LINE-1 mobility. *J. Biol. Chem.* 288:21148–60

76. Liu C, Sage JC, Miller MR, Verhaak RG, Hippenmeyer S, et al. 2011. Mosaic analysis with double markers reveals tumor cell of origin in glioma. *Cell* 146:209–21

77. Luan DD, Korman MH, Jakubczak JL, Eickbush TH. 1993. Reverse transcription of R2Bm RNA is primed by a nick at the chromosomal target site: a mechanism for non-LTR retrotransposition. *Cell* 72:595–605

78. Macfarlan TS, Gifford WD, Agarwal S, Driscoll S, Lettieri K, et al. 2011. Endogenous retroviruses and neighboring genes are coordinately repressed by LSD1/KDM1A. *Genes Dev.* 25:594–607

79. Macfarlane CM, Collier P, Rahbari R, Beck CR, Wagstaff JF, et al. 2013. Transduction-specific ATLAS reveals a cohort of highly active L1 retrotransposons in human populations. *Hum. Mutat.* 34:974–85

80. Martin SL. 1991. Ribonucleoprotein particles with LINE-1 RNA in mouse embryonal carcinoma cells. *Mol. Cell. Biol.* 11:4804–7

81. Martin SL, Bushman FD. 2001. Nucleic acid chaperone activity of the ORF1 protein from the mouse LINE-1 retrotransposon. *Mol. Cell. Biol.* 21:467–75

82. Mathias SL, Scott AF, Kazazian HH Jr, Boeke JD, Gabriel A. 1991. Reverse transcriptase encoded by a human transposable element. *Science* 254:1808–10

83. Matlik K, Redik K, Speek M. 2006. L1 antisense promoter drives tissue-specific transcription of human genes. *J. Biomed. Biotechnol.* 2006:71753

84. Matsui T, Leung D, Miyashita H, Maksakova IA, Miyachi H, et al. 2010. Proviral silencing in embryonic stem cells requires the histone methyltransferase ESET. *Nature* 464:927–31

85. McClintock B. 1950. The origin and behavior of mutable loci in maize. *Proc. Natl. Acad. Sci. USA* 36:344–55

86. McConnell MJ, Lindberg MR, Brennand KJ, Piper JC, Voet T, et al. 2013. Mosaic copy number variation in human neurons. *Science* 342:632–37

87. McKay R. 1997. Stem cells in the central nervous system. *Science* 276:66–71

88. Miki Y, Nishisho I, Horii A, Miyoshi Y, Utsunomiya J, et al. 1992. Disruption of the *APC* gene by a retrotransposal insertion of L1 sequence in a colon cancer. *Cancer Res.* 52:643–45

89. Minakami R, Kurose K, Etoh K, Furuhata Y, Hattori M, Sakaki Y. 1992. Identification of an internal *cis*-element essential for the human L1 transcription and a nuclear factor(s) binding to the element. *Nucleic Acids Res.* 20:3139–45

90. Mine M, Chen JM, Brivet M, Desguerre I, Marchant D, et al. 2007. A large genomic deletion in the *PDHX* gene caused by the retrotranspositional insertion of a full-length LINE-1 element. *Hum. Mutat.* 28:137–42

91. Monot C, Kuciak M, Viollet S, Mir AA, Gabus C, et al. 2013. The specificity and flexibility of L1 reverse transcription priming at imperfect T-tracts. *PLoS Genet.* 9:e1003499

92. Moran JV, DeBerardinis RJ, Kazazian HH Jr. 1999. Exon shuffling by L1 retrotransposition. *Science* 283:1530–34

93. Moran JV, Holmes SE, Naas TP, DeBerardinis RJ, Boeke JD, Kazazian HH Jr. 1996. High frequency retrotransposition in cultured mammalian cells. *Cell* 87:917–27

94. Morgan HD, Sutherland HG, Martin DI, Whitelaw E. 1999. Epigenetic inheritance at the agouti locus in the mouse. *Nat. Genet.* 23:314–18

95. Morrish TA, Garcia-Perez JL, Stamato TD, Taccioli GE, Sekiguchi J, Moran JV. 2007. Endonuclease-independent LINE-1 retrotransposition at mammalian telomeres. *Nature* 446:208–12

96. Morrish TA, Gilbert N, Myers JS, Vincent BJ, Stamato TD, et al. 2002. DNA repair mediated by endonuclease-independent LINE-1 retrotransposition. *Nat. Genet.* 31:159–65

97. Muckenfuss H, Hamdorf M, Held U, Perkovic M, Lower J, et al. 2006. APOBEC3 proteins inhibit human LINE-1 retrotransposition. *J. Biol. Chem.* 281:22161–72

98. Mulle JG. 2012. Schizophrenia genetics: progress, at last. *Curr. Opin. Genet. Dev.* 22:238–44

99. Muotri AR, Chu VT, Marchetto MC, Deng W, Moran JV, Gage FH. 2005. Somatic mosaicism in neuronal precursor cells mediated by L1 retrotransposition. *Nature* 435:903–10

100. Muotri AR, Marchetto MC, Coufal NG, Oefner R, Yeo G, et al. 2010. L1 retrotransposition in neurons is modulated by MeCP2. *Nature* 468:443–46

101. Muotri AR, Zhao C, Marchetto MC, Gage FH. 2009. Environmental influence on L1 retrotransposons in the adult hippocampus. *Hippocampus* 19:1002–7

102. Naas TP, DeBerardinis RJ, Moran JV, Ostertag EM, Kingsmore SF, et al. 1998. An actively retrotransposing, novel subfamily of mouse L1 elements. *EMBO J.* 17:590–97

103. Niewiadomska AM, Tian C, Tan L, Wang T, Sarkis PT, Yu XF. 2007. Differential inhibition of long interspersed element 1 by APOBEC3 does not correlate with high-molecular-mass-complex formation or P-body association. *J. Virol.* 81:9577–83

104. Nigumann P, Redik K, Matlik K, Speek M. 2002. Many human genes are transcribed from the antisense promoter of L1 retrotransposon. *Genomics* 79:628–34

105. Ostertag EM, DeBerardinis RJ, Goodier JL, Zhang Y, Yang N, et al. 2002. A mouse model of human L1 retrotransposition. *Nat. Genet.* 32:655–60

106. Owen MJ, Williams HJ, O'Donovan MC. 2009. Schizophrenia genetics: advancing on two fronts. *Curr. Opin. Genet. Dev.* 19:266–70

107. Perepelitsa-Belancio V, Deininger P. 2003. RNA truncation by premature polyadenylation attenuates human mobile element activity. *Nat. Genet.* 35:363–66

108. Perrat PN, DasGupta S, Wang J, Theurkauf W, Weng Z, et al. 2013. Transposition-driven genomic heterogeneity in the *Drosophila* brain. *Science* 340:91–95

109. Pickeral OK, Makalowski W, Boguski MS, Boeke JD. 2000. Frequent human genomic DNA transduction driven by LINE-1 retrotransposition. *Genome Res.* 10:411–15

110. Ray DA, Batzer MA. 2011. Reading TE leaves: new approaches to the identification of transposable element insertions. *Genome Res.* 21:813–20

111. Rebollo R, Karimi MM, Bilenky M, Gagnier L, Miceli-Royer K, et al. 2011. Retrotransposon-induced heterochromatin spreading in the mouse revealed by insertional polymorphisms. *PLoS Genet.* 7:e1002301

112. Reichmann J, Crichton JH, Madej MJ, Taggart M, Gautier P, et al. 2012. Microarray analysis of LTR retrotransposon silencing identifies Hdac1 as a regulator of retrotransposon expression in mouse embryonic stem cells. *PLoS Comput. Biol.* 8:e1002486

113. Richardson SR, Narvaiza I, Planegger RA, Weitzman MD, Moran JV. 2014. APOBEC3A deaminates transiently exposed single-strand DNA during LINE-1 retrotransposition. *eLife* 3:e02008

114. Rowe HM, Jakobsson J, Mesnard D, Rougemont J, Reynard S, et al. 2010. KAP1 controls endogenous retroviruses in embryonic stem cells. *Nature* 463:237–40

115. Scott AF, Schmeckpeper BJ, Abdelrazik M, Comey CT, O'Hara B, et al. 1987. Origin of the human L1 elements: proposed progenitor genes deduced from a consensus DNA sequence. *Genomics* 1:113–25

116. Sen SK, Huang CT, Han K, Batzer MA. 2007. Endonuclease-independent insertion provides an alternative pathway for L1 retrotransposition in the human genome. *Nucleic Acids Res.* 35:3741–51

117. Shiloh Y. 2001. ATM (ataxia telangiectasia mutated): expanding roles in the DNA damage response and cellular homeostasis. *Biochem. Soc. Trans.* 29:661–66

118. Shukla R, Upton KR, Munoz-Lopez M, Gerhardt DJ, Fisher ME, et al. 2013. Endogenous retrotransposition activates oncogenic pathways in hepatocellular carcinoma. *Cell* 153:101–11

119. Singer T, McConnell MJ, Marchetto MC, Coufal NG, Gage FH. 2010. LINE-1 retrotransposons: mediators of somatic variation in neuronal genomes? *Trends Neurosci.* 33:345–54

120. Slotkin RK, Martienssen R. 2007. Transposable elements and the epigenetic regulation of the genome. *Nat. Rev. Genet.* 8:272–85

121. Solyom S, Ewing AD, Rahrmann EP, Doucet T, Nelson HH, et al. 2012. Extensive somatic L1 retrotransposition in colorectal tumors. *Genome Res.* 22:2328–38

122. Speek M. 2001. Antisense promoter of human L1 retrotransposon drives transcription of adjacent cellular genes. *Mol. Cell. Biol.* 21:1973–85

123. Stenglein MD, Harris RS. 2006. APOBEC3B and APOBEC3F inhibit L1 retrotransposition by a DNA deamination-independent mechanism. *J. Biol. Chem.* 281:16837–41

124. Stetson DB, Ko JS, Heidmann T, Medzhitov R. 2008. Trex1 prevents cell-intrinsic initiation of autoimmunity. *Cell* 134:587–98

125. Swergold GD. 1990. Identification, characterization, and cell specificity of a human LINE-1 promoter. *Mol. Cell. Biol.* 10:6718–29

126. Symer DE, Connelly C, Szak ST, Caputo EM, Cost GJ, et al. 2002. Human L1 retrotransposition is associated with genetic instability in vivo. *Cell* 110:327–38

127. Takasu M, Hayashi R, Maruya E, Ota M, Imura K, et al. 2007. Deletion of entire HLA-A gene accompanied by an insertion of a retrotransposon. *Tissue Antigens* 70:144–50

128. Taylor MS, Lacava J, Mita P, Molloy KR, Huang CR, et al. 2013. Affinity proteomics reveals human host factors implicated in discrete stages of LINE-1 retrotransposition. *Cell* 155:1034–48

129. Tchenio T, Casella JF, Heidmann T. 2000. Members of the SRY family regulate the human LINE retrotransposons. *Nucleic Acids Res.* 28:411–15

130. Upton KR, Baillie JK, Faulkner GJ. 2011. Is somatic retrotransposition a parasitic or symbiotic phenomenon? *Mob. Genet. Elements* 1:279–82

131. van den Hurk JA, Meij IC, Seleme MC, Kano H, Nikopoulos K, et al. 2007. L1 retrotransposition can occur early in human embryonic development. *Hum. Mol. Genet.* 16:1587–92

132. van den Hurk JA, van de Pol DJ, Wissinger B, van Driel MA, Hoefsloot LH, et al. 2003. Novel types of mutation in the choroideremia (CHM) gene: a full-length L1 insertion and an intronic mutation activating a cryptic exon. *Hum. Genet.* 113:268–75

133. Wei W, Gilbert N, Ooi SL, Lawler JF, Ostertag EM, et al. 2001. Human L1 retrotransposition: *cis* preference versus *trans* complementation. *Mol. Cell. Biol.* 21:1429–39

134. Wheelan SJ, Aizawa Y, Han JS, Boeke JD. 2005. Gene-breaking: a new paradigm for human retrotransposon-mediated gene evolution. *Genome Res.* 15:1073–78

135. Yang N, Kazazian HH Jr. 2006. L1 retrotransposition is suppressed by endogenously encoded small interfering RNAs in human cultured cells. *Nat. Struct. Mol. Biol.* 13:763–71

136. Yang N, Zhang L, Zhang Y, Kazazian HH Jr. 2003. An important role for RUNX3 in human L1 transcription and retrotransposition. *Nucleic Acids Res.* 31:4929–40

137. Yu F, Zingler N, Schumann G, Stratling WH. 2001. Methyl-CpG-binding protein 2 represses LINE-1 expression and retrotransposition but not Alu transcription. *Nucleic Acids Res.* 29:4493–501

138. Zhao K, Du J, Han X, Goodier JL, Li P, et al. 2013. Modulation of LINE-1 and Alu/SVA retrotransposition by Aicardi-Goutieres syndrome-related SAMHD1. *Cell Rep.* 4:1108–15

Factors Underlying Restricted Crossover Localization in Barley Meiosis

James D. Higgins,[1] Kim Osman,[2] Gareth H. Jones,[2] and F. Chris H. Franklin[2,*]

[1] School of Biological Sciences, University of Leicester, Leicester LE1 7RH, United Kingdom; email: jh555@leicester.ac.uk

[2] School of Biosciences, University of Birmingham, Edgbaston, Birmingham B15 2TT, United Kingdom; email: k.osman@bham.ac.uk, garethjones125@btinternet.com, F.C.H.Franklin@bham.ac.uk

Annu. Rev. Genet. 2014. 48:29–47

First published online as a Review in Advance on August 1, 2014

The *Annual Review of Genetics* is online at genet.annualreviews.org

This article's doi: 10.1146/annurev-genet-120213-092509

*Corresponding author

Keywords

recombination, chiasma, poaceae, cereals, chromosome synapsis

Abstract

Meiotic recombination results in the formation of cytological structures known as chiasmata at the sites of genetic crossovers (COs). The formation of at least one chiasma/CO between homologous chromosome pairs is essential for accurate chromosome segregation at the first meiotic division as well as for generating genetic variation. Although DNA double-strand breaks, which initiate recombination, are widely distributed along the chromosomes, this is not necessarily reflected in the chiasma distribution. In many species there is a tendency for chiasmata to be distributed in favored regions along the chromosomes, whereas in others, such as barley and some other grasses, chiasma localization is extremely pronounced. Localization of chiasma to the distal regions of barley chromosomes restricts the genetic variation available to breeders. Studies reviewed herein are beginning to provide an explanation for chiasma localization in barley. Moreover, they suggest a potential route to manipulating chiasma distribution that could be of value to plant breeders.

INTRODUCTION

A Brief Overview of Meiosis

The formation of genetic crossovers (COs) is essential for the accurate segregation of chromosomes during meiosis in most sexually reproducing eukaryotes. Moreover, it provides a source of genetic variation between the generations that has long been exploited by both plant and animal breeders. During prophase I of meiosis, physical links between homologous chromosomes (homologs), referred to cytologically as chiasmata, arise at the sites of COs through homologous recombination (46, 48). These enable the homologs to correctly orientate on the meiotic spindle equator at metaphase I. Homolog disjunction at the first meiotic division is directly followed by a second division that segregates the sister chromatids to form four haploid gametes. In mutants defective for chiasma formation, homologs segregate randomly at the first meiotic division, leading to the formation of aneuploid gametes. Although meiosis has been investigated for more than a century, significant progress toward understanding the underlying molecular processes has been comparatively recent. Studies in budding yeast have been instrumental in this respect and have provided the basis for functional analyses in a range of organisms, including plants. These studies have revealed that many aspects of meiotic recombination are highly conserved. Nevertheless, there are also intriguing differences.

Meiotic recombination is initiated by the programmed formation of numerous DNA double-strand breaks (DSBs) catalyzed by the topoisomerase-like protein SPO11 (50). DSBs preferentially occur in short DNA regions termed recombination hot spots that are distributed along the chromosomes but with significantly reduced frequency in the centromeric and telomeric regions (80). In budding yeast, hot spots are associated with regions of low nucleosome density predominantly located at gene promoters (81). In mammals, DSBs are directed away from gene promoters to intergenic sequence motifs through the activity of PRDM9, a rapidly evolving zinc-finger protein containing a histone 3 lysine 4 methyltransferase (8, 73, 82, 89). In *Arabidopsis thaliana*, hot spots overlap gene promoters and are linked with a number of chromatin features. These include the presence of histone H2A.Z nucleosomes at the +1 position, low nucleosome density, low DNA methylation, and H3K4 trimethylation (21). Repair of the DSBs is controlled to ensure that each pair of homologs receives at least one obligate CO (referred to as CO assurance), which is essential for accurate segregation. In budding yeast, approximately 50% of the DSBs are repaired to form 80–90 COs per cell, with the remainder being repaired as non-COs. However, in plants and animals only approximately 5% of DSBs are repaired as COs, whereas the vast majority are repaired via a non-CO pathway (78). The important, yet poorly understood, phenomenon termed CO interference ensures that multiple COs do not occur in adjacent regions along the chromosomes (11, 49). As a result of these control mechanisms, the numerical distribution of most COs is strikingly non-random. In addition, a small proportion of COs, estimated to be approximately 15% in *A. thaliana*, arise via a non-interference pathway (39, 71, 78).

Studies have established that recombination is closely coordinated with the extensive remodeling of the homologous chromosomes that characterizes prophase I of meiosis (52). At leptotene, the first substage of prophase I, a linear protein axis is formed along each homolog. This organizes the sister chromatids that make up each homolog into linear looped arrays of chromatin conjoined at the loop bases. Some of the proteins that make up the axis, for example, Hop1-Red1 in budding yeast and the corresponding proteins ASY1-ASY3 in *Arabidopsis* and PAIR2-PAIR3 in rice, play a key role by influencing DSB repair through the creation of a repair template bias in favor of using one of the non-sister chromatids, thus promoting interhomolog recombination (18, 31, 43, 76, 93, 95). As prophase I progresses from leptotene to zygotene, the homolog pairs become increasingly aligned. During zygotene, the paired homologs undergo synapsis through the formation of the synaptonemal complex (SC) (79). The SC has a tripartite structure consisting of the aligned linear axes linked by overlapping transverse filaments that lie perpendicular to the axes, bringing them in close apposition at a distance of 100 nm; at the pachytene stage, SC formation is complete. Recombination is ongoing throughout prophase I. Importantly, in most species, chromosome pairing, synapsis, and recombination progression are interdependent (99). When the homologs have recombined to form COs during diplotene the SC breaks down. At diakinesis the homolog pairs appear cytologically as condensed bivalent structures linked by one or more chiasmata. Subsequently, at metaphase I the bivalents align on the equator before undergoing the first meiotic division.

It might be assumed that, considering the constraints of the obligate CO and interference, the position of COs along the chromosomes could be quite variable from cell to cell. Although positional variation does occur, it is clear that distribution is somehow influenced such that COs tend to arise in favored chromosomal regions. In some species, localized distribution is highly pronounced (24, 46, 48). This is particularly true for members of the grass family, including cereal crops, and has important implications for plant breeding (29, 55, 65). Although the phenomenon of CO/chiasma localization has been known for many years from genetic and cytogenetic studies, an understanding of the basis for this restriction has been lacking. However, the development of molecular cytogenetic tools for the analysis of plant meiosis is beginning to provide a route toward unraveling this question. Here, we review the factors that may account for localized CO formation in the grasses, with particular reference to recent insights into how CO distribution in barley (*Hordeum vulgare*), a key member of the Poaceae, is influenced.

The Phenomenon of Crossover Localization

The construction of genetic maps in species such as *Drosophila* and maize during the early part of the twentieth century was predicated on the assumption that genetic COs were more or less evenly distributed across chromosomes so that the CO frequency between any two genetic markers could be regarded as an indication of the physical distance separating them (28, 90). Almost contemporaneously, cytogenetic observations were increasingly suggesting that COs, visualized as chiasmata, were in many instances far from even in their distribution along chromosomes (24).

These cytogenetic observations led to the development of the concept of chiasma/CO localization, whereby chiasmata may be, depending on species, preferentially and non-randomly restricted to certain chromosome regions (24). Two forms of pronounced localization were recognized: distal localization, in which chiasmata are restricted to the terminal regions of chromosomes, which are usually remote from centromeric regions, and proximal localization, in which chiasmata are restricted to regions bordering centromeres (48). Despite some early reservations that distal chiasmata may have originated in more central chromosome regions and migrated by a process called terminalization to distal regions, it is now widely accepted that chiasmata do not terminalize and hence their locations reflect their real sites of origin (91, 92).

Examples of such localization have been recorded across a diverse range of animal and plant species (48). In several cases of pronounced chiasma localization, restriction of chiasmata to certain regions is associated with restricted synapsis of homologous chromosomes. Probably the earliest demonstration of this association was in the plant genus *Fritillaria* (75). The snake's head fritillary (*Fritillaria meleagris*) exhibits localization of chiasmata in pollen mother cells (PMCs) to proximal chromosome regions bordering the centromeres, which in this species occupy mid-chromosome locations. A study of prophase I chromosomes in this species revealed that chromosome pairing at the light-microscopic level is also restricted to proximal regions. At this time the existence of the SC

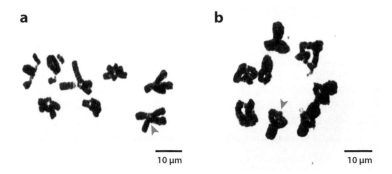

a

b

10 μm

10 μm

Figure 1

Squash preparations of *Allium* chromosomes at metaphase I of meiosis stained with orcein. (*a*) *Allium fistulosum*; note stellate-shaped bivalents due to the extreme proximal position of the chiasmata (*arrowhead*). (*b*) *Allium cepa* chiasmata are distal (*arrowhead*) or interstitial. Reproduced with permission from Reference 2 (NRC Res. Press License: 3318821336122).

(see above) was not known. In later studies, involving electron-microscopic analysis, an association between chiasma localization and restricted synapsis was confirmed. For example, in male meiosis of the large marsh grasshopper *Stethophyma grossum* extreme proximal chiasma localization in eight of the eleven chromosome pairs is associated with restricted synapsis to proximal regions of the same chromosomes (16, 32, 94). A similar association of chiasma localization, this time distal, and restricted synapsis is found in male meiosis of the rhabdocoel planarian worm *Mesostoma ehrenbergii* (77). Thus, an association between restricted synapsis and chiasma localization has been observed in both plant and animal kingdoms. Nevertheless, in most recorded cases of chiasma localization, chromosome synapsis is complete. An instructive example is the case of the closely related onion species *Allium cepa* and *Allium fistulosum* (**Figure 1**). *A. fistulosum* exhibits proximal chiasma localization (**Figure 1a**), with more than 90% occurring within the proximal 25% of the SC length, whereas *A. cepa* has distal to unrestricted chiasmata (**Figure 1b**) (2). In both species, chromosome synapsis is complete, and, furthermore, there is no discernible difference in the initiation or progression of synapsis between the two species.

It is worth noting that many species of plants and animals, although not exhibiting pronounced localization of chiasmata, nevertheless do exhibit some degree of localization so that when chiasma positions are carefully measured and mapped it is evident that some regions have elevated chiasma frequencies compared with others, for example, in the Orthopteran insects *Schistocerca gregaria* (34) and *Chorthippus brunneus* (58), which exhibit polarity in chiasma distribution from the telomeres to the centromeric regions. Subtelomeric and distal chiasmata are also strongly predominant in the human male, with proximal chiasmata rare other than on acrocentric chromosomes (44). A final point of interest is that chiasma localization within a species can vary between the sexes. For example, chiasmata in males of the newt *Triturus helveticus* show a marked distal localization, whereas in females, chiasmata are interstitial (97). Similarly, when female meiosis was finally analyzed in *F. meleagris*, chiasmata were frequently found to occupy interstitial positions rather than the highly proximal position observed in male meiosis (33).

Chiasma Localization in Members of the Grass Family

The grass family (Poaceae) comprises more than 10,000 species classified into subfamilies that each comprise a number of tribes, among which are all the major cereal species, forage grasses, and many minor grains (e.g., millet) as well as other economically important crops (e.g., sugar cane) (51). Given their major importance as sources of human and animal nutrition, these species have been the subject of extensive genetic analysis, particularly in relation to crop improvement. Aside from their practical importance, some of the key members of the Poaceae are particularly well suited to cytogenetic analysis because they possess very large chromosomes. Thus, since the early part of the twentieth century, cytogenetic studies in cereals have made important contributions to our understanding of chromosome behavior during meiosis.

The grass tribe Triticeae includes the cereal crops wheat (*Triticum aestivum*), rye (*Secale cereale* L.), oat (*Avena sativa*), and barley (*H. vulgare* L.). A striking feature of these species is that CO formation along the chromosomes is non-uniform. Allohexaploid bread wheat (*T. aestivum*) (2n = 6x = 42) is the product of hybridizations between progenitor species carrying the AA, BB, and DD genomes (51). A common feature of the Poaceae in general, reflected in wheat, is the dominance of distally located chiasmata. Chiasma counts by Lukaszewski (63) on wheat metaphase I chromosome spreads revealed that among 40 to 41 paired arms per PMC, 34 to 35 had terminal chiasmata, 1 to 2 had interstitial chiasmata, and none had proximal chiasmata. COs are restricted to the distal half of each arm, where they increase exponentially within proximity to the telomeres (63). An analysis of inverted wheat chromosome arms 2BS and 4AL revealed that COs are biased to particular chromosomal segments independent of their locations along the chromosomes (66). The positions of chiasmata moved with the chromosome segments from distal to interstitial regions.

Rye (*Secale cereale* L.) is also a member of the wheat family and has been extensively studied at the cytological level because it possesses a small number of very large chromosomes (2n = 2x = 14) (64). The distal bias of chiasmata is so skewed that homologous chromosome associations at metaphase I are often referred to as end-to-end (46, 47). However, using Giemsa-banding, Jones (47) revealed that the distal chiasmata were actually adjacent and proximal in location to the terminal heterochromatic repeats that are present in rye chromosomes. Analysis of a line in which the long arm of chromosome 1 was nearly entirely inverted revealed a shift to proximal chiasma formation on the inverted arm rather than the normal distal localization (64). The author therefore reasoned that the recombination frequency along a chromosome is position independent and segment specific. In addition, instances of synapsis being limited to the chiasma-proficient chromosomal regions were observed, but it is not clear why full synapsis was not achieved. Analysis of the behavior of a rye deletion chromosome in a wheat addition line also highlights the link between chromosomal segment and chiasma formation. Deletion of all but the proximal 30% of the long arm of rye chromosome 5 resulted in loss of COs in the remaining region (74). However, chromosome pairing and synapsis in the deleted arm appeared normal. This finding is consistent with those in barley (see below), which show that recombination initiation is not deficient in the interstitial/proximal regions but rather that these initiation events are unlikely to progress to form COs.

In the diploid oat (*Avena strigosa*) and tetraploid oat (*A. barbata*), a skewed distal bias of chiasmata was also observed, although the frequency was not quantified (57). In twelve cultivated hexaploid oat (*A. sativa*) varieties, an over-representation of terminal chiasmata was observed (6). Despite the low frequency of interstitial chiasmata, there was considerable variation among varieties, suggesting that they could differ at the genetic recombination level.

Maize (*Zea mays* L.) possesses ten large (2n = 2x = 20) chromosomes (2,300 Mb genome) characterized by large, heterochromatic DNA knobs (15). The bulk of the maize genome is composed of highly repetitive transposable elements (TEs) (69). Analyses of CO frequency and distribution have largely been carried out using recombination nodule (RN)-based assays. RNs are electron-dense protein complexes detected using electron microscopy that are closely correlated with COs and lie at sites where chiasmata will later form (3, 17). A detailed analysis of RNs in maize revealed a high number in the distal regions that dramatically declined with increased proximity to the centromeres. The gradient of RN distribution correlated with SC length. SCs that were longer (SC1) had a weak gradient, whereas shorter SCs (SC9 and SC10) had very steep gradients (3). These data were in accordance with chiasma counts performed concurrently. In addition, Falque et al. (30) [using an antibody against the MutL homolog MLH1, which, as a heterodimer with MLH3, marks the sites of interfering COs (19, 45, 67)] revealed that ~85% of the RNs belonged to the interference-sensitive pathway and the remaining ~15% to the interference-independent pathway. The number of residual non-interfering COs also correlated with SC length.

Brachypodium distachyon is a temperate grass species closely related to the cereal crops (barley, wheat, and rye) as well as to forage grasses such as ryegrass (27). However, in contrast to the cereals, it has a relatively small genome for a grass species (~355 Mb) containing ten small diploid (2n = 2x = 10) chromosomes (10). A cytological analysis of meiotic metaphase I chromosome spreads revealed that the majority of nuclei possess five ring bivalents, indicating at least one CO in each chromosome arm (16 out of 20), and the remaining four nuclei contained four rings and one rod bivalent per cell (27). Moreover, chiasmata were not strictly localized to any particular region of the chromosomes and were observed in proximal, distal, and interstitial positions.

Although chromosome size may be a major contributory factor in determining CO number and position, evidence from rice suggests that meiotic genes that control important steps in the process also have a substantial influence. Compared with the large chromosomes of the cereal crops, rice (*Oryza sativa*) has a larger number of relatively small chromosomes (2n = 2x = 24) and a smaller genome (~420 Mb) (14). However, during meiosis, wild-type metaphase I chromosome spreads revealed that bivalents consisted of a mixture of rods and rings with chiasma location biased toward the ends of chromosomes (60, 96). The chiasma frequency and distribution was dramatically altered in a mutant of the rice synaptonemal complex transverse filament protein ZEP1 (96). Although the analysis lacked immunolocalization of MLH1 and MLH3 or marker-based recombination assays, the metaphase I bivalents appeared to contain greater numbers of chiasmata that were observed in proximal, distal, and interstitial chromosomal regions (96).

CROSSOVER LOCALIZATION IN BARLEY: IDENTIFYING THE CONTRIBUTORY FACTORS

Barley (*H. vulgare* L.) is a diploid member of the Triticeae and the fourth most abundant cereal after wheat, maize, and rice. It has a haploid genome size of 5,100 Mb comprising seven chromosomes encoding a putative 53,220 genes, of which approximately 50% have been identified with high confidence (68). Genetic analysis of recombination frequencies in barley mapping populations has revealed that, similar to that found in barley's near relatives, the distribution of CO events is not uniform, with a strong bias toward distal chromosomal regions (55). This has been confirmed by cytological analysis, including chiasma counts and immunolocalization of MLH1 on pachytene chromosome spreads (**Figure 2a,b**). CO formation is strongly suppressed in the centromeric and pericentromeric chromosomal regions, which data indicate correspond to almost 50% of the physical map (56, 68). The heterogeneity of CO formation along the chromosomes is mirrored by the gene distribution, which shows a strong enrichment in the distal regions. Nevertheless, studies indicate that approximately 30% of the genes lie outside the recombinogenic distal DNA. It is suggested that this may be a significant barrier for plant breeders, as it has the potential to limit available genetic variability, creating difficulties for both gene introgression through linkage drag and map-based cloning.

Despite the long awareness of the strong bias toward distal localization of COs in the cereals, identifying the factors responsible for this has until recently proved difficult. However, in recent

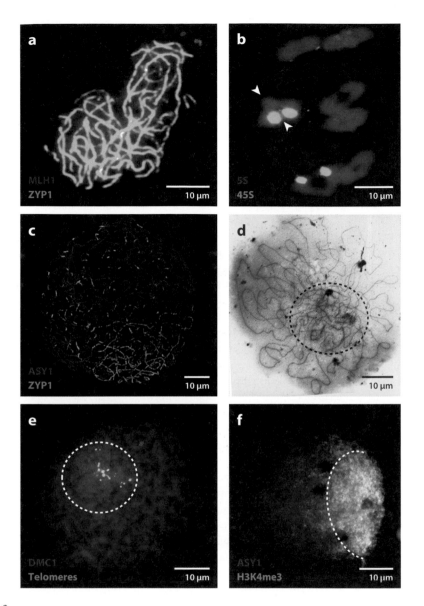

Figure 2

Barley chromosomes reveal a spatiotemporal progression of meiotic events that initiate at the distal ends and consequently form crossovers (COs) in these regions. (*a*) During pachytene, immunolocalization of MLH1 foci (*red*), which mark the future sites of interfering COs, are observed at the ends of the synaptonemal complexes marked by the transverse filament protein ZYP1 (*green*). (*b*) At metaphase I, chiasmata are detected at the distal ends of the chromosomes (*white arrowheads*), highlighted by the location of fluorescence in situ hybridization ribosomal DNA probes 5S (*red*) and 45S (*green*). (*c*) Structured illumination microscopy of an early zygotene nucleus showing extensive polymerization of ZYP1 (*green*) in a specific region of the nucleus with small stretches throughout the nucleus. The chromosome axes are marked by ASY1 (*red*). (*d*) An electron micrograph of silver-stained chromosomes at zygotene showing the initiation of synapsis in the distal regions (*highlighted by the dotted circle*). (*e*) The strand-exchange protein DMC1 initially localizes in the vicinity of the telomeres (*green*; *highlighted by the white dotted circle*) in a leptotene nucleus. (*f*) The chromosome axis protein ASY1 (*red*) initially localizes and extends in the euchromatic hyperabundant regions marked by H3K4me3 (histone 3 lysine 4 trimethylation) (*green*; *highlighted by white dotted line*). The chromosomes are counterstained with DAPI (*blue*).

years the application of molecular cytological approaches based around immunocytochemistry using antibodies that recognize meiotic proteins has led to substantial progress in the understanding of meiosis in the model plant *A. thaliana* (5, 19, 70). By combining immunolocalization studies with 5'-bromo-2'-deoxyuridine (BrdU) labeling of the DNA during meiotic S phase (4), it has been possible to establish an accurate chronology of progress through meiosis. Fortuitously, many of the antibodies raised against the *Arabidopsis* meiotic proteins also recognize the corresponding proteins in barley, thus permitting a more detailed analysis of barley than was hitherto possible. These studies are beginning to reveal the factors that contribute to the skewed distribution of COs in barley and likely other members of the Poaceae.

Spatiotemporal Asymmetry of Chromosome Axis Formation and Synapsis During Prophase I of Meiosis

An earlier study based on sequential sampling of developing spikelets estimated that meiosis in barley occupied 39.4 h (9). A more recent analysis applying a modified version of the BrdU labeling method developed for *Arabidopsis* indicated a figure of 43 h, which is not too different from the earlier study (40). The majority of this time is accounted for by prophase I, with the two division stages completed in approximately 3 h. As with many species, early prophase I in barley is characterized by the appearance of the telomere bouquet, one of the early landmarks in the meiotic pathway (40, 84, 87). The bouquet arises through the attachment and clustering of the telomeres in a restricted region on the nuclear envelope (87). As a result, physical contacts between the subtelomeric/distal regions of chromosomes are promoted. This is thought to lead to stable association and pairing of homologous chromosomes at these distal sites prior to interstitial and proximal regions of the chromosomes. This is supported by cytological analyses of chromosome spread preparations from meiocytes in a range of species (38, 88). Formation of the telomere bouquet is completed by late G2, some 8 h post–S phase. Immunolocalization of the chromosome axis-associated protein ASY1 (40) reveals that appearance of the telomere bouquet coincides with the elaboration of a linear chromosome axis initiating in the subtelomeric/distal regions of the chromosomes. The linearization of the axis continues into interstitial/proximal regions, such that by 13 h post–S phase axis formation is complete. Synapsis of the homologous chromosomes denoting the onset of zygotene can be monitored using electron microscopy or an antibody raised against the SC transverse filament protein ZYP1 (41) (**Figure 2c,d**). ZYP1 is first observed approximately 25 h after S phase, again initiating in the subtelomeric/distal regions. As these signals extend, additional interstitial synapsis initiation sites are observed, which also extend and coalesce to form a complete synaptonemal complex at pachytene. The SC persists until diplotene at 39 h post–S phase, at which point the chromosomes desynapse. The immunolocalization studies are in accord with earlier analyses regarding chromosome synapsis in the Poaceae based on electron microscopy. Thus, it is clear that barley chromosomes complete synapsis at pachytene. Hence, the skewed spatial distribution of chiasmata cannot be due to synapsis being restricted to a limited region of the homologous chromosomes as in male meiosis of the grasshopper *Stethophyma grossum* (16, 32, 94).

An interesting observation from the analysis of synapsis in barley is that there is a nearly fourfold excess of synapsis initiation sites relative to COs/chiasmata (55 v ∼ 17). In budding yeast, it appears that each synapsis initiation site corresponds to a future CO site (36). However, this is clearly not the case in barley. Indeed, it seems likely that barley may reflect the norm for multicellular eukaryotes given that observations in *Arabidopsis* suggest a similar excess of synapsis initiation sites over COs (39). One could suppose that limiting synapsis initiation to sites of recombination intermediates destined to form COs in barley and other species with large chromosomes would compromise the ability to undergo efficient synapsis. Conversely, high levels of CO formation could lead to insurmountable problems due to chromosome entanglements during meiotic prophase I. Thus, it may be that CO control in most species has been adjusted such that synapsis is initiated efficiently, but CO formation is limited to a relatively low level. That said, recent studies in *Arabidopsis* have shown that mutation of the *FANCM* gene can result in a substantial elevation of CO formation without any deleterious impact on chromosome stability (23, 54). It remains to be determined whether this elevation occurs in plants with large chromosomes.

Recombination Pathway Progression

The chronology of the early recombination pathway in barley has been investigated by immunolocalization studies in conjunction with a meiotic time course (40). Immunolocalization of γH2AX, the phosphorylated form of the histone 2 variant H2AX, which is widely used as a proxy for DSB formation, and the recombinases RAD51 and DMC1, which are required for strand exchange, has revealed that recombination is initiated in the same spatially biased manner as chromosome axis formation and synapsis. At approximately 4 h post–S phase, approximately 200 γH2AX foci are detected in the subtelomeric/distal regions coincident with the appearance of the axis marker ASY1. The number of foci then progressively increases, concurrent with their appearance in the interstitial and proximal regions of the chromosomes. Maximum numbers are reached at approximately 13 h post–S phase when approximately 450 foci per nucleus are detectable and chromosome axis formation is complete. Immunolocalization of RAD51 and DMC1 foci follows a similar spatial and numerical pattern, although they are initially detected at approximately 10 h post–S phase (**Figure 2e**). Thus, overall recombination initiation and progression appear to occur in a spatiotemporal wave across the nucleus, with events in distal regions preceding those in proximal regions by up to 3 h. Immunolocalization of the MutS homolog MSH4, which is thought to stabilize progenitor Holliday junction intermediates, also follows this spatiotemporal distribution. Nevertheless, progression to form COs in interstitial/proximal DNA is clearly rare given that chiasmata are generally not found in these regions. Moreover, immunolocalization with an anti-MLH3 antibody, which localizes to CO sites in pachytene, confirms this distribution, suggesting that the initiating DSBs in interstitial/proximal regions are repaired prior to double Holliday junction (dHj) formation by an alternative repair pathway, possibly via synthesis-dependent strand annealing or using the sister chromatid as the repair template (85).

Meiotic Progression and Crossover Formation Are Correlated with Chromatin Organization

It is well established that the chromosomes in eukaryotes show linear differentiation into regions of euchromatin and heterochromatin. In barley and some other members of the Poaceae with similarly large chromosomes, the euchromatin-rich DNA is distributed along the distal regions of the chromosomes, whereas the heterochromatic DNA is localized to the centromeric region. Additional heterochromatic DNA may also occur at interstitial sites or in the telomeric region, as is the case in rye. Immunolocalization using antibodies that recognize the histone modifications K3K9me3, H3K27me3, H3K4me3 (**Figure 2f**), and H4K16ac, which are associated with transcriptionally active, gene-rich DNA, reveals a high degree of enrichment in the distal regions (35, 40). Heterochromatic marks, such as H3K9me2, H3K27me2, and H4K20me1, are abundant along the entire length of the chromosomes, save for the distal regions where they are depleted. These observations suggest a link between euchromatin and CO formation. This association is also supported by the studies in rye, outlined earlier, which indicate that CO formation is excluded from heterochromatic DNA.

Effect of Temperature on Chiasma Frequency and Distribution

Further insight into the underlying control of CO distribution in barley has come from temperature-shift experiments. That meiosis is sensitive to elevated temperature has been known for many years. Unsurprisingly, extreme temperature during meiosis leads to a complete disruption of the process (62). However, studies in *Tradescantia bracteata* and *Uvularia perfoliata* revealed that although high temperatures (>35°C) resulted in a rapid decrease in chiasma formation, plants exposed to a range of temperatures below this level exhibit a progressive shift of chiasma distribution with an increased frequency of interstitial COs (26). Similarly, in barley, exposure to 35°C also resulted in complete meiotic failure, whereas comparison of the chiasma distribution in plants exposed to 22°C and 30°C during meiosis revealed a significant decrease in distal COs and a coincident increase in interstitial events at the higher temperature (40). Interestingly, this effect was not uniform across all the chromosomes, hinting at a possible influence of chromosome structure. The shift in distribution at 30°C was also accompanied by a slight reduction in the mean chiasma frequency per PMC from 14.8 to 13.5. It would seem reasonable to suppose that the reduction in COs and change in chiasma distribution are in some way linked, and further studies have been conducted to try to address this.

Effect of Temperature on Meiotic Recombination

As PMCs progress through prophase I, extensive chromosome remodeling occurs, which evidence suggests is closely coupled with meiotic recombination and vice versa. Thus, mutations affecting one of these processes can have profound effects on the other. This fact is illustrated in a range of *Arabidopsis* and rice mutants. For example, mutation of the *Arabidopsis* chromosome axis protein ASY3 or the corresponding rice protein PAIR3 leads to defects in recombination that result in a reduction of COs (31, 98). In some instances, mutations in components of the recombination machinery, such as the *MutS* genes *MSH4* and *MSH5*, lead to a delay in meiotic progression of several hours (39, 42). Also, a mutant allele of the cell cycle control gene *RETINOBLASTOMA* (*RBR*) affects meiotic progression, resulting in a defect in synapsis and CO formation (20). An *Arabidopsis arp6* mutant that is defective in deposition of H2A.Z into nucleosomes at DSB sites during meiosis phenocopies H2A.Z localization at higher temperatures and exhibits a small but significant reduction in COs (21). Hence, the reduction in COs in barley at 30°C could be accounted for by one or more factors.

Immunolocalization studies conducted on chromosome spread preparations from PMCs isolated from plants held at 30°C compared with those from plants at 22°C suggest that there are no obvious defects in the chromosome axes and that overall levels of DSBs and early recombination intermediates are not significantly different. However, localization of the SC transverse filament protein ZYP1 revealed a defect in installation of the SC. Formation was delayed and accumulations of ZYP1 protein, possibly corresponding to polycomplexes, were apparent (40). It is known from studies in budding yeast that the corresponding Zip1 protein acts at CO-designated recombination intermediates to impose the CO fate. Assuming ZYP1 fulfills a similar role in barley, it seems likely that the loading defect may result in an occasional failure in CO imposition. Presumably, the magnitude of the defect increases as the temperature rises above 30°C given that it seems that no COs are formed at 35°C and the chromosomes are completely asynaptic. Moreover, in barley ZYP1 RNAi knockdown lines in which SC formation is absent or reduced, there is a corresponding reduction in CO formation (7). Although the impact on SC formation at 30°C likely explains the slight reduction in COs at this temperature, additional defects in other components of the meiotic machinery or meiotic progression (see below) cannot be ruled out on the basis of existing evidence.

Effect of Temperature on Chromatin and Meiotic Progression

Although the reduction in CO frequency at 30°C is likely related to the SC defect, the change in chiasma distribution may be due to other factors. Because most COs in barley are associated with euchromatin-rich chromosomal regions, one explanation for the change in CO distribution could be a shift in the pattern of histone modifications at the elevated temperature. There is strong evidence from a variety of species that meiotic recombination is influenced by the chromatin landscape. For example, studies in mammals and budding yeast have demonstrated that recombination hot spots are associated with trimethylation of the lysine 4 residue in histone 3 (8, 13, 73, 82, 89). Histone acetylation has also been shown to influence recombination in both budding yeast and *Arabidopsis* (72, 83). In the former, deletion of the *SIR2* gene, which encodes a histone deacetylase, changes the genomic distribution of recombination. In *Arabidopsis*, hyperacetylation arising through overexpression of a histone acetylase altered the frequency and distribution of chiasmata.

However, in barley PMCs there are no obvious changes in the distribution of the euchromatic and heterochromatic marks at 30°C compared to 22°C (40). Although subtle short-range changes cannot be excluded, there is as yet no evidence to indicate that the change in distribution is directly driven by altered chromatin marks. Nevertheless, the global chromatin organization in barley does appear to have a significant part to play in influencing chiasma distribution. Comparison of meiotic time courses at the two temperatures reveals that the duration is approximately 43 h in both instances. This suggests that meiotic progression in barley is buffered against fluctuations in temperature. Although the duration of meiosis does not appear to be affected by a moderate shift in temperature, dual labeling of the chromosomal DNA during meiotic S phase with BrdU and 5-ethynyl-2′-deoxyuridine (EdU) reveals a significant effect on replication. It is well established that euchromatin-rich chromosomal segments are replicated earlier than heterochromatin-rich regions (61, 86). BrdU-EdU dual labeling has revealed that in barley at 22°C, the distal DNA is replicated within 4 h of the initiation of S phase; interstitial DNA is replicated within 6 h. However, replication of the heterochromatin-rich proximal DNA is not completed until 13 h. Shifting the temperature to 30°C reduces the length of S phase. This shift does not seem to have an influence on replication of the distal euchromatic DNA, but replication of the interstitial and proximal DNA is completed by 9 h, some 4 h earlier than at 22°C. It is proposed that the increase in temperature makes the heterochromatic DNA more accessible to the replication machinery, possibly through reducing the occupancy of the histone H2A.Z, which acts as a thermosensor. As mentioned earlier, aside from the effect on SC formation, the temperature shift does not compromise the ability to form the chromosome axes or reduce the overall number of recombination initiation events. That said, there is a clear effect on the localization of the meiotic proteins during the early stages of meiosis. In particular, the strong bias toward initiation of both axis formation and recombination in the distal regions of the chromosomes in the vicinity of the telomere bouquet seen at 22°C is less pronounced at 30°C. As a result, elaboration of the chromosome axes and initiation of recombination take place in distal and interstitial regions at a similar time, although a degree of bias still exists. As a consequence, DSBs that occur in interstitial/proximal DNA are repaired more frequently via the CO pathway, such that the 25:1 ratio of distal to interstitial chiasmata observed at 22°C is reduced to 11:1 at 30°C. Thus, it seems that elevated temperature tends to synchronize early meiotic events along the chromosomes and that this increases the probability that a DSB occurring in interstitial DNA may be repaired as a CO rather than as a non-CO. Why this should be the case remains to be established; nevertheless, there are a number of factors that may influence events.

Can Crossover Frequency and Distribution in Barley Be Explained by Crossover Interference?

The distribution of meiotic COs is highly controlled such that each pair of homologous chromosomes receives a minimum of one obligate CO, with most additional COs subject to interference (see above). As a result, COs are well spaced along the chromosomes, with often only a single CO per chromosome arm. Studies in budding yeast indicate that the fate of individual DSBs to be repaired as either a CO or a non-CO is taken early in prophase I, implying that CO interference is established at this point (12). Because there are many commonalities between meiotic control in budding yeast and plants, it seems likely that CO designation also occurs in early prophase I in plants. This relationship could suggest that in barley under normal conditions there would be a strong bias for CO designation at distal sites given that recombination initiates in this region 2–3 h before interstitial/proximal sites. Thus, interference would be established at distal sites, thus disfavoring interstitial/proximal DSBs from progressing to form COs. Although this hypothesis may seem attractive, there are a number of observations that suggest the explanation may lie elsewhere.

Immunolocalization of MLH3 foci along barley chromosomes at pachytene has revealed that the mean interfocus distances range from 29.2% to 44.35% of arm length for chromosomes 2H and 3H, with a minimum distance of 6.1% (85). Overall, nearly 40% of the MLH3 foci were separated by less than 20% of arm length. These data were analyzed using the CODA (crossover distribution analyzer) gamma distribution method to quantify the strength of interference (*nu*), whereby a value of $nu = 1$ indicates no interference, >1 indicates positive interference, and <1 indicates negative interference (37). This analysis gave values for *nu* of 1.44 and 1.58 for 2H and 3H, respectively. These figures are substantially lower than interference calculations in some other species, including tomato, where *nu* values of 7.9 and 6.9 for chromosomes 1 and 2, respectively, were recorded (59). Although this could be interpreted as indicating that interference in barley is relatively weak in these chromosomal regions, some caution in direct comparisons may be required because of the nature of meiotic progression in barley and the proposed role of the chromosome axis in mediating interference. The gamma distribution method is based on the relative separation of MLH3 foci along the chromosome axes when synapsis is complete at pachytene. However, the differential timing of events along barley chromosomes could allow CO designation to take place at recombination intermediates along a distal chromosome segment before axis elaboration has been completed in the interstitial/proximal region. If so, then it could be argued that interference should be measured in the context of the degree of axis formation at the time of CO designation rather than when axis formation has proceeded to completion. In this context, it is arguable that CO interference in barley may actually be stronger than currently estimated. Counter to this argument, mutation of the axial element protein SYCP3 in mouse does not affect interference between MLH1 foci (25). However, the cohesin complex, which is a key component and organizer of the chromosome axes, was present in these mice; hence, it is unclear whether axis function in relation to any role in mediating interference was compromised.

Although interfocus separation between MLH3 foci can be relatively short, the majority of foci are nevertheless separated by $>70\%$ of the total chromosome length because most COs are restricted to the distal regions of the chromosomes. Because CO interference is known to operate across the centromeric region (22), interference over the interstitial/proximal regions is likely stronger than that in distal regions. Indeed, Phillips et al. (85) calculated *nu* across the centromeric regions for 2H and 3H to be 5.86 and 6.42, respectively. This raises the question of whether interference is imposed differentially along the chromosomes or, alternatively, the repair of interstitial/proximal recombination intermediates is mediated via a non-CO route. Although no categorical answer yet exists, studies indicate that dynamic changes in the chromatin environment may be a significant influence.

On the basis of data from a range of species, it appears that during mitosis and meiotic prophase I, chromosomes undergo a programmed set of cycles of chromatin expansion and contraction (53). These observations have led to the development of the "mechanical basis of chromosome function" model, which, in relation to meiosis, proposes a functional inter-relationship between the chromatin cycles and the four key transitions in the meiotic pathway, namely DSB formation, single-end invasion, second-end capture, and dHj resolution, that lead to CO formation. It is proposed that the four transitions are coordinated by three rounds of mechanical stress and relaxation generated during prophase I by the chromatin cycles, with each transition coincident with a phase of chromatin expansion. Studies confirm that the chromosome cycles are also conserved in barley (40). Analyzing these in conjunction with meiotic progression has led to an interesting observation concerning the spatial differentiation across the barley chromosomes. The 2–3 h time difference between meiotic transitions, such as axis formation and DSB formation, in distal versus interstitial/proximal regions monitored using immunocytochemistry to detect meiotic proteins, reveals that although the transitions in distal regions occur in synchrony with a chromatin expansion phase, interstitial events occur during periods of contraction. If it is a requirement that imposition of a CO fate on a designated recombination intermediate is coincident with a chromatin expansion phase, then it would explain why interstitial/proximal intermediates are not repaired via a CO route. There is no direct experimental evidence to confirm this supposition. Nevertheless, one of the effects of exposing barley PMCs to 30°C is that the spatial differentiation in meiotic progression observed at 22°C is much less pronounced. As a result, the timing of the meiotic transitions in the interstitial regions tends to be more coincident with those in the distal DNA and in phase with periods of chromatin expansion. Importantly, this is accompanied by a significant increase in CO formation at interstitial sites.

CONCLUSION

The phenomenon of chiasma localization has been recognized for many years, in some instances since the early days of cytogenetics. It is clear that it occurs widely throughout the different eukaryotic kingdoms. Examples of both distal and proximal localization have been observed. In some cases, extreme localization is directly associated with limited or incomplete chromosome synapsis. There is now compelling evidence to indicate that in many organisms SC formation is essential to ensure that a CO fate is imposed on CO-designated recombination intermediates. Hence, by restricting the degree of synapsis, recombination is concomitantly limited. However, in many species, including the cereals, chiasma localization is not linked to limited SC formation. An explanation has therefore proved less tractable. Elucidating the basis for chiasma localization, particularly in the cereals, is potentially important from a plant breeding viewpoint. The studies outlined above provide a tantalizing indication that the global organization of the chromatin in relation to timing of replication is a key influence. Moreover, they suggest some scope for modifying the chiasma distribution that may provide a simple basis for manipulating recombination through temperature.

An additional, intriguing question, which has previously been raised, is whether the chromosome architecture in the cereals determines CO position or whether crossing over has been a major factor in the evolution of genome organization. Akhunov et al. (1) have shown that wheat loci derived by duplication were most frequently located in distal, high-recombination chromosome regions, whereas ancestral loci were evenly distributed between the proximal two-thirds of the chromosomes and the distal third. These authors suggest that recombination has played a central role in the evolution of the wheat genome structure and that gradients of recombination rates along chromosome arms promote more rapid rates of genome evolution in distal, high-recombination regions than in proximal, low-recombination regions. Similarly to wheat, it has been argued that meiotic recombination in maize has been one of the main factors of maize genome evolution and the two may be intimately linked. Meiotic drive, the subversion of meiosis so that particular genes are preferentially transmitted to the progeny, appears to affect heterochromatin knob chromosomal position and size. Hence, it is likely that meiotic recombination influences genome organization and possibly vice versa, but further study is required to resolve the issue.

SUMMARY POINTS

1. The formation of COs, which are cytologically manifested as chiasmata at metaphase I of meiosis, is carefully regulated to ensure a minimum of at least one obligate CO between homologous chromosome pairs (bivalents). CO interference ensures that additional COs along a bivalent are widely spaced.

2. Although most species studied show a tendency for COs/chiasmata to be localized in favored chromosomal regions, in some species this localization is highly pronounced.

3. A number of important members of the grass family (including cereals) such as barley and forage grasses, exhibit CO localization, which effectively limits COs to the distal regions of the chromosomes. This presents a potential barrier for plant breeders.

4. In some species, CO localization is associated with restricted chromosome synapsis. However, this restriction is not the case in barley and other cereals.

5. Immunocytochemistry in conjunction with fluorescence microscopy and super-resolution microscopy using a panel of antibodies against meiotic proteins have revealed that spatiotemporal asymmetry of meiotic chromosome remodeling and recombination progression underlie CO/chiasma localization in barley. Studies show that chromosome axis formation, chromosome pairing, and synapsis and recombination are initiated in the distal chromosome regions 2–3 h in advance of the corresponding events in proximal DNA. As a consequence, a proportion of recombination events in the distal regions progress to form COs, whereas virtually all those in the proximal regions are repaired without CO formation.

6. Studies indicate that late replication of the heterochromatic DNA (relative to euchromatic DNA), which is enriched in the proximal regions of the barley chromosomes, is an important factor in establishing the asymmetry of meiotic progression.

7. The application of a moderate temperature pulse during meiosis has been found to alter chiasma distribution, leading to a greater proportion of interstitial/proximal COs. It seems that the differential timing of replication between the euchromatic and heterochromatic DNA is less marked, such that recombination is initiated more synchronously along the chromosomes. This provides a potential route for plant breeders to manipulate recombination.

FUTURE ISSUES

1. Although an elevation of temperature to 30°C during meiosis in barley leads to an increase in interstitial/proximal COs, this does not appear to affect all the chromosomes to the same degree. One hypothesis to test is whether the effect of temperature is governed by the organization of the individual chromosomes. Hence, it is of interest to determine whether factors such as chromosome size and the proportion and distribution of heterochromatin underlie this variation. The effect of different temperatures could be explored, or similarly, the effect of modification of chromatin through chemical treatments. For example, application of trichostatin A to modify histone acetylation has been shown to affect chiasma distribution in *Arabidopsis* (83). Whether or not a heat-pulse strategy could be used to modify chiasma distribution in other cereals could also be investigated.

2. The fact that a modest increase in temperature reduces CO frequency is also significant, as it could contribute to yield reduction in areas affected by climate change. It will be interesting to determine whether accessions can be identified that are resilient to elevated temperatures. At present, the mechanistic basis of the temperature susceptibility remains to be determined. Work on barley has identified a problem with SC formation, but whether this effect is due to an impact on the SC proteins themselves, to synapsis initiation, or to remodeling of the chromosome axis during zygotene has yet to be established.

3. The reason DSBs that form in the interstitial/proximal regions do not progress to form COs remains to be discovered. The observation that there is a time delay in DSB formation relative to distal regions that alters the relationship between the repair processes and the conserved chromatin cycles appears significant. Clearly, this influences how the breaks are repaired. It is conceivable that CO interference may be involved. However, studies in other species with large chromosomes, such as grasshoppers, indicate that interference is dissipated over a region of 30% of a chromosome arm. Also, the MLH3 interfocus distance observed in barley itself can be less than 20% of arm length. Hence, more interstitial COs may be anticipated. One possibility is that by the time interstitial and proximal DSBs are undergoing repair, the bias toward interhomolog repair mediated by the chromosome axis proteins is lifted such that a switch to using the sister chromatid as the repair template occurs, precluding additional COs from forming. It is also possible that the local chromatin environment directs repair down a non-CO route.

DISCLOSURE STATEMENT

The authors are not aware of any affiliations, memberships, funding, or financial holdings that might be perceived as affecting the objectivity of this review.

ACKNOWLEDGMENTS

We thank all members of the meiosis community whose work has directly or indirectly contributed to this review. We apologize to anyone whose work has not been included due to space limitation. We thank the Biotechnology and Biological Sciences Research Council for support (Grant BB/F019351/1).

LITERATURE CITED

1. Akhunov ED, Akhunova AR, Linkiewicz AM, Dubcovsky J, Hummel D, et al. 2003. Synteny perturbations between wheat homoeologous chromosomes caused by locus duplications and deletions correlate with recombination rates. *Proc. Natl. Acad. Sci. USA* 100:10836–41

2. Albini SM, Jones GH. 1988. Synaptonemal complex spreading in *Allium cepa* and *Allium fistulosum*. II. Pachytene observations: the SC karyotype and the correspondence of late recombination nodules and chiasmata. *Genome* 30:399–410

3. Anderson LK, Doyle GG, Brigham B, Carter J, Hooker KD, et al. 2003. High-resolution crossover maps for each bivalent of *Zea mays* using recombination nodules. *Genetics* 165:849–65

4. Armstrong SJ, Franklin FCH, Jones GH. 2003. A meiotic time-course for *Arabidopsis thaliana*. *Sex. Plant Reprod.* 16:141–49

5. Armstrong SJ, Jones GH. 2003. Meiotic cytology and chromosome behaviour in wild-type *Arabidopsis thaliana*. *J. Exp. Bot.* 54:1–10

6. Baptista-Giacomelli FR, Pagliarini MS, de Almeida JL. 2000. Meiotic behavior in several Brazilian oat cultivars (*Avena sativa* L.). *Cytologia (Tokyo)* 65:371–78

7. Barakate A, Higgins JD, Vivera S, Stephens J, Perry RM, et al. 2014. The synaptonemal complex protein ZYP1 is required for imposition of meiotic crossovers in barley. *Plant Cell* 26:729–40

8. Baudat F, Buard J, Grey C, Fledel-Alon A, Ober C, et al. 2010. PRDM9 is a major determinant of meiotic recombination hotspots in humans and mice. *Science* 327:836–40

9. Bennett MD, Finch RA. 1971. Duration of meiosis in barley. *Genet. Res.* 17:209–14

10. Bennett MD, Leitch IJ. 2005. Nuclear DNA amounts in angiosperms: progress, problems and prospects. *Ann. Bot.* 95:45–90

11. Berchowitz LE, Copenhaver GP. 2010. Genetic interference: Don't stand so close to me. *Curr. Genomics* 11:91–102

12. Bishop DK, Zickler D. 2004. Early decision: meiotic crossover interference prior to stable strand exchange and synapsis. *Cell* 117:9–15

13. Borde V, Robine N, Lin W, Bonfils S, Geli V, Nicolas A. 2009. Histone H3 lysine 4 trimethylation marks meiotic recombination initiation sites. *EMBO J.* 28:99–111

14. Bowers JE, Arias MA, Asher R, Avise JA, Ball RT, et al. 2005. Comparative physical mapping links conservation of microsynteny to chromosome structure and recombination in grasses. *Proc. Natl. Acad. Sci. USA* 102:13206–11

15. Buckler ES, Phelps-Durr TL, Buckler CSK, Dawe RK, Doebley JF, Holtsford TP. 1999. Meiotic drive of chromosomal knobs reshaped the maize genome. *Genetics* 153:415–26

16. Calvente A, Viera A, Page J, Parra MT, Gomez R, et al. 2005. DNA double-strand breaks and homology search: inferences from a species with incomplete pairing and synapsis. *J. Cell Sci.* 118:2957–63

17. Carpenter ATC. 1975. Electron microscopy of meiosis in *Drosophila melanogaster* females. II. The recombination nodule: a recombination-associated structure at pachytene? *Proc. Natl. Acad. Sci. USA* 72:3186–89

18. Caryl AP, Armstrong SJ, Jones GH, Franklin FCH. 2000. A homologue of the yeast *HOP1* gene is inactivated in the *Arabidopsis* meiotic mutant *asy1*. *Chromosoma* 109:62–71

19. Chelysheva L, Grandont L, Vrielynck N, le Guin S, Mercier R, Grelon M. 2010. An easy protocol for studying chromatin and recombination protein dynamics during *Arabidopsis thaliana* meiosis: immunodetection of cohesins, histones and MLH1. *Cytogenet. Genome Res.* 129:143–53

20. Chen Z, Higgins JD, Hui JTL, Li J, Franklin FCH, Berger F. 2011. Retinoblastoma protein is essential for early meiotic events in *Arabidopsis*. *EMBO J.* 30:744–55

21. Choi K, Zhao X, Kelly KA, Venn O, Higgins JD, et al. 2013. *Arabidopsis* meiotic crossover hot spots overlap with H2A.Z nucleosomes at gene promoters. *Nat. Genet.* 45:1327–36

22. Colombo PC, Jones GH. 1997. Chiasma interference is blind to centromeres. *Heredity* 79:214–27

23. Crismani W, Girard C, Froger N, Pradillo M, Santos JL, et al. 2012. FANCM limits meiotic crossovers. *Science* 336:1588–90

24. Darlington CD. 1931. Meiosis. *Biol. Rev. Biol. Proc. Camb. Philos. Soc.* 6:221–64

25. de Boer E, Dietrich AJ, Hoog C, Stam P, Heyting C. 2007. Meiotic interference among MLH1 foci requires neither an intact axial element structure nor full synapsis. *J. Cell Sci.* 120:731–36

26. Dowrick G. 1957. The influence of temperature on meiosis. *Heredity* 11:37–49

27. Draper J, Mur LAJ, Jenkins G, Ghosh-Biswas GC, Bablak P, et al. 2001. *Brachypodium distachyon*. A new model system for functional genomics in grasses. *Plant Physiol.* 127:1539–55

28. Emerson RA, Beadle GW, Fraser AC. 1935. A summary of linkage studies in maize. *Cornell Univ. Agric. Stn. Mem.* 180:1–83

29. Erayman M, Sandhu D, Sidhu D, Dilbirligi M, Baenziger PS, Gill KS. 2004. Demarcating the gene-rich regions of the wheat genome. *Nucleic Acids Res.* 32:3546–65

30. Falque M, Anderson LK, Stack SM, Gauthier F, Martin OC. 2009. Two types of meiotic crossovers coexist in maize. *Plant Cell* 21:3915–25

31. Ferdous M, Higgins JD, Osman K, Lambing C, Roitinger E, et al. 2012. Inter-homolog crossing-over and synapsis in *Arabidopsis* meiosis are dependent on the chromosome axis protein AtASY3. *PLoS Genet.* 8:e1002507

32. Fletcher HL. 1978. Localized chiasmata due to partial pairing: 3D reconstruction of synaptonemal complexes in male *Stethophyma grossum*. *Chromosoma* 65:247–69

33. Fogwill M. 1958. Differences in crossing-over and chromosome size in the sex cells of *Lilium* and *Fritillaria*. *Chromosoma* 9:493–504

34. Fox DP. 1973. Control of chiasma distribution in locust, *Schistocerca gregaria* (Forskal). *Chromosoma* 43:289–328

35. Fuchs J, Demidov D, Houben A, Schubert I. 2006. Chromosomal histone modification patterns: from conservation to diversity. *Trends Plant Sci.* 11:199–208

36. Fung JC, Rockmill B, Odell M, Roeder GS. 2004. Imposition of crossover interference through the nonrandom distribution of synapsis initiation complexes. *Cell* 116:795–802

37. Gauthier F, Martin OC, Falque M. CODA (crossover distribution analyzer): quantitative characterization of crossover position patterns along chromosomes. *BMC Bioinform.* 12:27

38. Harper L, Golubovskaya I, Cande WZ. 2004. A bouquet of chromosomes. *J. Cell Sci.* 117:4025–32

39. Higgins JD, Armstrong SJ, Franklin FCH, Jones GH. 2004. The *Arabidopsis* MutS homolog AtMSH4 functions at an early step in recombination: evidence for two classes of recombination in *Arabidopsis*. *Genes Dev.* 18:2557–70

40. Higgins JD, Perry RM, Barakat A, Ramsay L, Waugh R, et al. 2012. Spatiotemporal asymmetry of the meiotic program underlies the predominantly distal distribution of meiotic crossovers in barley. *Plant Cell* 24:4096–109

41. Higgins JD, Sanchez-Moran E, Armstrong SJ, Jones GH, Franklin FCH. 2005. The *Arabidopsis* synaptonemal complex protein ZYP1 is required for chromosome synapsis and normal fidelity of crossing over. *Genes Dev.* 19:2488–500

42. Higgins JD, Vignard J, Mercier R, Pugh AG, Franklin FCH, Jones GH. 2008. AtMSH5 partners AtMSH4 in the class I meiotic crossover pathway in *Arabidopsis thaliana*, but is not required for synapsis. *Plant J.* 55:28–39

43. Hollingsworth NM, Byers B. 1989. *Hop1*: a yeast meiotic pairing gene. *Genetics* 121:445–62

44. Hultén M. 1974. Chiasma distribution at diakinesis in the normal human male. *Hereditas* 76:55–78

45. Jackson N, Sanchez-Moran E, Buckling E, Armstrong SJ, Jones GH, Franklin FCH. 2006. Reduced meiotic crossovers and delayed prophase I progression in AtMLH3-deficient *Arabidopsis*. *EMBO J.* 25:1315–23

46. Jones GH. 1987. Chiasmata. In *Meiosis*, ed. PB Moens, pp. 213–44. Waltham, MA: Acad. Press

47. Jones GH. 1978. Giemsa C-banding of rye meiotic chromosomes and nature of terminal chiasmata. *Chromosoma* 66:45–57

48. Jones GH. 1984. The control of chiasma distribution. *SEB Symp.* 38:293–320

49. Jones GH, Franklin FC. 2006. Meiotic crossing-over: obligation and interference. *Cell* 126:246–48

50. Keeney S, Giroux CN, Kleckner N. 1997. Meiosis-specific DNA double-strand breaks are catalyzed by Spo11, a member of a widely conserved protein family. *Cell* 88:375–84

51. Kellogg EA. 1998. Relationships of cereal crops and other grasses. *Proc. Natl. Acad. Sci. USA* 95:2005–10

52. Kleckner N. 2006. Chiasma formation: chromatin/axis interplay and the role(s) of the synaptonemal complex. *Chromosoma* 115:175–94

53. Kleckner N, Zickler D, Jones GH, Dekker J, Padmore R, et al. 2004. A mechanical basis for chromosome function. *Proc. Natl. Acad. Sci. USA* 101:12592–97

54. Knoll A, Higgins JD, Seeliger K, Reha SJ, Dangel NJ, et al. 2012. The Fanconi anemia ortholog FANCM ensures ordered homologous recombination in both somatic and meiotic cells in *Arabidopsis*. *Plant Cell* 24:1448–64

55. Kunzel G, Korzun L, Meister A. 2000. Cytologically integrated physical restriction fragment length polymorphism maps for the barley genome based on translocation breakpoints. *Genetics* 154:397–412

56. Kunzel G, Waugh R. 2002. Integration of microsatellite markers into the translocation-based physical RFLP map of barley chromosome 3H. *Theor. Appl. Genet.* 105:660–65

57. Ladizinsky G. 2012. *Studies in Oat Evolution: A Man's Life with* Avena (*Springer Briefs in Agriculture*). New York: Springer

58. Laurie DA, Jones GH. 1981. Interindividual variation in chiasma distribution in *Chorthippus brunneus* (*Orthoptera*, *Acrididae*). *Heredity* 47:409–16

59. Lhuissier FGP, Offenberg HH, Wittich PE, Vischer NOE, Heyting C. 2007. The mismatch repair protein MLH1 marks a subset of strongly interfering crossovers in tomato. *Plant Cell* 19:862–76

60. Li X, Chang Y, Xin X, Zhu C, Li X, et al. 2013. Replication protein A2c coupled with replication protein A1c regulates crossover formation during meiosis in rice. *Plant Cell* 25:3885–99

61. Lima de Faria A, Jaworska H. 1972. Relation between chromosome size gradient and sequence of DNA replication in rye. *Hereditas* 70:39–57

62. Loidl J. 1989. Effects of elevated temperature on meiotic chromosome synapsis in *Allium ursinum*. *Chromosoma* 97:449–58

63. Lukaszewski AJ. 1992. A comparison of physical distribution of recombination in chromosome 1R in diploid rye and in hexaploid triticale. *Theor. Appl. Genet.* 83:1048–53

64. Lukaszewski AJ. 2008. Unexpected behavior of an inverted rye chromosome arm in wheat. *Chromosoma* 117:569–78

65. Lukaszewski AJ, Curtis CA. 1993. Physical distribution of recombination in B-genome chromosomes of tetraploid wheat. *Theor. Appl. Genet.* 86:121–27

66. Lukaszewski AJ, Kopecky D, Linc G. 2012. Inversions of chromosome arms 4AL and 2BS in wheat invert the patterns of chiasma distribution. *Chromosoma* 121:201–8

67. Marcon E, Moens P. 2003. MLH1p and MLH3p localize to precociously induced chiasmata of okadaic-acid-treated mouse spermatocytes. *Genetics* 165:2283–87

68. Mayer KFX, Waugh R, Langridge P, Close TJ, Wise RP, et al. 2012. A physical, genetic and functional sequence assembly of the barley genome. *Nature* 491:711–16

69. McClintock B. 1948. Mutable loci in maize. *Carnegie Inst. Wash. Year Book* 47:155–69

70. Mercier R, Grelon M. 2008. Meiosis in plants: ten years of gene discovery. *Cytogenet. Genome Res.* 120:281–90

71. Mercier R, Jolivet S, Vezon D, Huppe E, Chelysheva L, et al. 2005. Two meiotic crossover classes cohabit in *Arabidopsis*: One is dependent on MER3, whereas the other one is not. *Curr. Biol.* 15:692–701

72. Mieczkowski PA, Dominska M, Buck MJ, Gerton JL, Lieb JD, Petes TD. 2006. Global analysis of the relationship between the binding of the Bas1p transcription factor and meiosis-specific double-strand DNA breaks in *Saccharomyces cerevisiae*. *Mol. Cell. Biol.* 26:1014–27

73. Myers S, Bowden R, Tumian A, Bontrop RE, Freeman C, et al. 2010. Drive against hotspot motifs in primates implicates the *PRDM9* gene in meiotic recombination. *Science* 327:876–79

74. Naranjo T, Valenzuela NT, Perera E. 2010. Chiasma frequency is region specific and chromosome conformation dependent in a rye chromosome added to wheat. *Cytogenet. Genome Res.* 129:133–42

75. Newton WCF, Darlington CD. 1930. *Fritillaria meleagris* chiasma formation and distribution. *J. Genet.* 22:1–14

76. Nonomura KI, Nakano M, Eiguchi M, Suzuki T, Kurata N. 2006. PAIR2 is essential for homologous chromosome synapsis in rice meiosis I. *J. Cell Sci.* 119:217–25

77. Oakley HA, Jones GH. 1982. Meiosis in *Mesostoma ehrenbergii ehrenbergii* (Turbellaria, Rhabdocoela). I. Chromosoma pairing, synaptonemal complexes and chiasma localization in spermatogenesis. *Chromosoma* 85:311–22

78. Osman K, Higgins JD, Sanchez-Moran E, Armstrong SJ, Franklin FCH. 2011. Pathways to meiotic recombination in *Arabidopsis thaliana*. *New Phytol.* 190:523–44

79. Page SL, Hawley RS. 2004. The genetics and molecular biology of the synaptonemal complex. *Annu. Rev. Cell Dev. Biol.* 20:525–58

80. Paigen K, Petkov P. 2010. Mammalian recombination hot spots: properties, control and evolution. *Nat. Rev. Genet.* 11:221–33

81. Pan J, Sasaki M, Kniewel R, Murakami H, Blitzblau HG, et al. 2011. A hierarchical combination of factors shapes the genome-wide topography of yeast meiotic recombination initiation. *Cell* 144:719–31

82. Parvanov ED, Petkov PM, Paigen K. 2010. *Prdm9* controls activation of mammalian recombination hotspots. *Science* 327:835

83. Perrella G, Consiglio MF, Aiese-Cigliano R, Cremona G, Sanchez-Moran E, et al. 2010. Histone hyper-acetylation affects meiotic recombination and chromosome segregation in *Arabidopsis*. *Plant J.* 62:796–806

84. Phillips D, Nibau C, Wnetrzak J, Jenkins G. 2012. High resolution analysis of meiotic chromosome structure and behaviour in barley (*Hordeum vulgare* L.). *PLoS ONE* 7:e39539

85. Phillips D, Wnetrzak J, Nibau C, Barakate A, Ramsay L, et al. 2013. Quantitative high resolution mapping of HvMLH3 foci in barley pachytene nuclei reveals a strong distal bias and weak interference. *J. Exp. Bot.* 64:2139–54

86. Pryor A, Faulkner K, Rhoades MM, Peacock WJ. 1980. Asynchronous replication of heterochromatin in maize. *Proc. Natl. Acad. Sci. USA* 77:6705–9

87. Ronceret A, Pawlowski WP. 2010. Chromosome dynamics in meiotic prophase I in plants. *Cytogenet. Genome Res.* 129:173–83

88. Scherthan H. 2001. A bouquet makes ends meet. *Nat. Rev. Mol. Cell Biol.* 2:621–27

89. Smagulova F, Gregoretti IV, Brick K, Khil P, Camerini-Otero RD, Petukhova GV. 2011. Genome-wide analysis reveals novel molecular features of mouse recombination hotspots. *Nature* 472:375–78

90. Sturtevant AH. 1913. The linear arrangement of six sex-linked factors in *Drosophila*, as shown by their mode of association. *J. Exp. Zool.* 14:43–59

91. Tease C. 1978. Cytological detection of crossing-over in BrdU substituted meiotic chromosomes using the fluorescent plus giemsa technique. *Nature* 272:823–24

92. Tease C, Jones GH. 1978. Analysis of exchanges in differentially stained meiotic chromosomes of *Locusta migrotoria* after BrdU substitution and FPG staining. I. Crossover exchanges in monochiasmate bivalents. *Chromosoma* 69:163–78

93. Thompson EA, Roeder GS. 1989. Expression and DNA sequence of *RED1*, a gene required for meiosis I chromosome segregation in yeast. *Mol. Gen. Genet.* 218:293–301

94. Wallace BMN, Jones GH. 1978. Incomplete chromosome pairing and its relation to chiasma localization in *Stethophyma grossum* spermocytes. *Heredity* 40:385–96

95. Wang KJ, Wang M, Tang D, Shen Y, Qin BX, et al. 2011. PAIR3, an axis-associated protein, is essential for the recruitment of recombination elements onto meiotic chromosomes in rice. *Mol. Biol. Cell* 22:12–19

96. Wang M, Wang KJ, Tang D, Wei CX, Li M, et al. 2010. The central element protein ZEP1 of the synaptonemal complex regulates the number of crossovers during meiosis in rice. *Plant Cell* 22:417–30

97. Watson JD, Callan HG. 1963. The form of bivalent chromosomes in newt oocytes at first metaphase of meiosis. *Q. J. Microsc. Sci.* 104:281–95

98. Yuan WY, Li XW, Chang YX, Wen RY, Chen GX, et al. 2009. Mutation of the rice gene PAIR3 results in lack of bivalent formation in meiosis. *Plant J.* 59:303–15

99. Zickler D, Kleckner N. 1999. Meiotic chromosomes: integrating structure and function. *Annu. Rev. Genet.* 33:603–754

pENCODE: A Plant Encyclopedia of DNA Elements

Amanda K. Lane,[1] Chad E. Niederhuth,[1] Lexiang Ji,[1,2] and Robert J. Schmitz[1,2]

[1]Department of Genetics, University of Georgia, Athens, Georgia 30602; email: schmitz@uga.edu

[2]Institute of Bioinformatics, University of Georgia, Athens, Georgia 30602

Annu. Rev. Genet. 2014. 48:49–70

First published online as a Review in Advance on August 15, 2014

The *Annual Review of Genetics* is online at genet.annualreviews.org

This article's doi: 10.1146/annurev-genet-120213-092443

Keywords

DNA elements, comparative epigenomics, epigenetics

Abstract

ENCODE projects exist for many eukaryotes, including humans, but as of yet no defined project exists for plants. A plant ENCODE would be invaluable to the research community and could be more readily produced than its metazoan equivalents by capitalizing on the preexisting infrastructure provided from similar projects. Collecting and normalizing plant epigenomic data for a range of species will facilitate hypothesis generation, cross-species comparisons, annotation of genomes, and an understanding of epigenomic functions throughout plant evolution. Here, we discuss the need for such a project, outline the challenges it faces, and suggest ways forward to build a plant ENCODE.

INTRODUCTION

ENCODE:
ENCyclopedia of
DNA Elements;
**http://www.genome.
gov/10005107**

**Chromatin
modifications:**
covalent modifications,
such as DNA
methylation and
histone modifications,
to DNA and histones

DNA elements:
DNA sequences that
inherently provide
sequence specificity to
diverse biological
processes through
interactions with
proteins and/or RNAs

Epigenomics: the
study of genome-wide
maps of chromatin
modifications, RNAs,
protein:DNA
interactions, and
chromatin accessibility

International efforts are underway to advance plant sciences, with the goal of addressing concerns about bioenergy, food security, and climate change. One of the most significant contributions to these efforts is the recent and continuing production of high-quality plant genome sequences. The first plant genome sequenced was from *Arabidopsis thaliana* in 2000, and this provided the first comprehensive view of the genomic landscape of a plant (3). It revealed the presence of more than 25,000 genes and plant-specific gene families not found in animal or bacterial genomes. It also provided the infrastructure to support the daunting task of determining the function and the biological process to which each of these genes belongs. Since that time, more than 30 high-quality plant genomes have been published for a wide range of both model and crop species. The availability of these genome sequences is enabling useful annotations, such as gene identification, QTL (quantitative trait loci) mapping, and marker-assisted introgression of favorable alleles in crops, to name just a few examples. Furthermore, large-scale resequencing projects have been initiated on the basis of the availability of these genome assemblies, which aim to catalog within-species sequence variation, to facilitate genome-wide association mapping, and to enable comparative genomic studies between species.

A major omission from these current endeavors is the presence of a comparative epigenomic plant resource. In conjunction with the advances in sequencing throughput and the ease with which we are acquiring large volumes of data, a serious discussion about a coordinated effort by the international plant sciences community to initiate a plant ENCODE (pENCODE) project is warranted. The goal of such a project would be to coordinate the ongoing work in individual laboratories across the globe; to focus community efforts on a set of high priorities; and to standardize sample/data preparation, acquisition, and dissemination. ENCODE projects exist for human (38) as well as other major model organisms, such as mice, flies, and worms (52, 85, 118). One of the major goals of ENCODE projects is to build upon reference genomes by trying to understand how DNA sequence information is translated into different cell types, tissues, organs, and ultimately entire organisms. One of the findings from the human ENCODE project that is of direct interest to plant scientists are the epigenomic maps that were determined for cell lineages that represent different developmental states. The integration of transcription-factor binding sites, RNA expression states, DNase I hypersensitivity sites, and chromatin modification maps revealed enormous complexity in translating sequence to phenotype. Fortunately, this vast sea of sequence information can now be broken down into smaller more manageable domains as a result of the ENCODE project. Another major finding from the human ENCODE project that is highly relevant to the plant science community was the identification of large numbers of trait-associated sequence variants localized to regulatory DNA elements (84). These ENCODE projects have not only generated genome-wide maps of sequence variation, RNAs (both coding and noncoding), chromatin modifications, protein:DNA interactions, and inter/intrachromosomal interactions, but have also developed the protocols required to generate these data, the software required to analyze them, and the genome browsers required to visualize them (5, 8, 11, 20, 23, 33–35, 40, 41, 51, 57, 60, 61, 66, 72, 79, 89–92, 103, 109, 118, 120, 130, 137). Therefore, pENCODE could take full advantage of this existing infrastructure and dedicate most of its resources to sample selection, preparation, and analysis. Furthermore, it could provide the driving force for organization and standardization within the community.

To organize an international community of plant scientists with overlapping goals to decode plant genomes, the Epigenomics of Plants International Consortium (EPIC; **https://www.plant-epigenome.org/**) was formed in 2008 (39). EPIC has successfully built a community of scientists (to join the EPIC community, register here: **https://www.plant-epigenome.org/user/register**), developed a core mission and specific focus areas, and facilitated the exchange of ideas in public forums at international conferences, and it could serve as the coordinating body for pENCODE. One of the key features of pENCODE is that plants provide an ideal organism to study how the environment interacts with the genome to coordinate phenotypic changes. Plant species do not contain a nervous system but instead take advantage of a complex transcriptional regulatory code to execute many of the same responses that animals experience. This is partly exemplified by the massive expansion of transcription factor (TF) families present in plant genomes. Plant genomes also offer an excellent system to understand how genomes manage newly duplicated sequences, such as genes, chromosomes, and/or genomes. Clearly, this is a major mechanism that plant species have adopted in their evolution as compared with most major animal model systems, and understanding how and which pathways are affected after duplication events could be facilitated by pENCODE. Another major advantage of a pENCODE project would be the ability to translate novel findings to the field. Already, major efforts are underway to understand how the epigenome is reprogrammed in hybrids and in response to environmental stress conditions. A more complete understanding of how DNA sequence information in plant genomes is translated into phenotypic changes is foundational to rapidly generating novel cultivars that could be introduced into the field. With all of the benefits that would be afforded by a pENCODE project, the next major step for this community is to secure international support to fund the execution of the outlined goals. Although such efforts come at a substantial cost, the funds necessary are not near the amount required for the original human ENCODE and modENCODE (model ENCODE projects for *Caenorhabditis elegans* and *Drosophila melanogaster*) projects, largely because the cost to acquire the data is now much lower and many of the analysis and visualization tools already exist. For example, there is a major effort already underway, referred to as EPIC-CoGe (Comparative Genomics; **http://genomevolution.org/CoGe/** and **http://genomevolution.org/r/9360**), that is storing and publicly disseminating published data sets (**http://www.iplantcollaborative.org/**). EPIC-CoGe leverages the Powered by iPlant Program for computational and data management scalability (**http://www.iplantcollaborative.org/** and **http://genomevolution.org/r/bi0u**). Resources such as this that make the data accessible to the individual investigators are essential to the success of the scientific community to realize the full potential of the published information. However, these resources are not geared toward standardizing data sets generated from different groups to make them comparable.

Given the significant cyberinfrastructural support associated with CoGe, efforts are being made to reanalyze and distribute published sequencing data sets from the raw data. This standardization of data is one of the most important features of community-wide ENCODE-like projects, and requirements for releasing raw experimental data have resulted in standards such as MIAME (minimum information about a microarray experiment), BAM [a binary file of a SAM (sequence alignment map) file], etc. (15, 73). This practice is important for laboratories that want to analyze publicly available data that are produced by different groups because the processed data sets can all be run through the exact same workflow. With the standardization of data generation and analysis, the greater community can reliably and repeatedly use the data produced over long periods of time. Finally, after determining that the data are of high quality, it is essential that this information is publicly released in a timely fashion to promote advancements in plant sciences by individual laboratories. These data release policies could follow the standards agreed upon by scientists as outlined in the Fort Lauderdale agreement on Sharing Data from Large-Scale Biological Research Projects (**http://www.genome.gov/27528022**).

Some will ask whether there is a need for an internationally coordinated pENCODE. In fact, mini-ENCODE-like projects are operating from individual laboratories and loosely formed international consortiums. This is the case for *Arabidopsis*, for which there exist genome-wide maps of histone modifications (12, 74, 140, 141), histone variants (25, 124, 143), RNAs (1, 42, 45, 50, 58, 77), DNA methylation (24, 77), nucleosome occupancy (22), chromatin accessibility (139), and chromosomal interactions (86). Additionally, the 1,001 *Arabidopsis* Genomes Project is cataloging genetic variants and building the infrastructure to execute genome-wide association studies using natural accessions that were isolated from throughout the Northern Hemisphere (18, 47, 80, 95, 112). Similar communities exist for rice (62), maize (21, 56, 83, 133), brassica (59), and soybean (70), and are beginning to surface for other plant species. However, several species with assembled genomes have not developed such collaborative support. Although these data are incredibly useful for each of these communities, there is no standard for sample collection, which makes it challenging to accurately perform comparative epigenomics between species. An internationally coordinated effort will reduce overlap in developing methods and acquiring data between individual laboratories, which would serve to increase the efficiency of releasing deliverables to the public. It would also provide standardization to the processing of these data sets. With data rapidly being deposited in the public domain, advances in plant sciences would be accelerated.

One reason genome resequencing projects have successfully launched is because diverse collections of accessions or cultivars exist, making sample identification obvious, although there are currently no standardized practices for gDNA isolation, library preparation, or data analysis. For pENCODE, a consensus needs to be reached to determine the samples from which epigenomic data are collected. In addition to a genotype(s) for each species, specific tissues, cell types, developmental time points, and environmental treatments need to be selected. Therefore, identifying samples that have broad support from the community is much more challenging than selecting genotypes for genome resequencing projects because of the possible variation in data selection. Steps are required to reach this consensus. First, current data must be collected, which is already being done by other projects. Next, the consensus for missing data and for data processing must be determined. Finally, reprocessing of existing data and filling in missing gaps will provide the final tools needed by the community.

Furthermore, sample preparation is much more challenging than simply isolating genomic DNA for genome resequencing. For this community-wide effort to be successful, it would be beneficial to make certain that these data are comparable across plant species. Here lies another challenge. Most laboratories have experts working with a single plant species and with a specific developmental or environmental process. Ideally, to be able to compare developmental or environmental programs between species, the identical developmental stage must be matched or treatment administered. In some cases, it is technically challenging to determine what the comparable stage of development means for diverse plant species. Regardless, standards can be reached between many different laboratories for collection of samples from different developmental states and upon different environmental treatments by focusing on those most readily accessible and that coordinate with multiple existing efforts. Normalizing acceptable data quality to an average of the realistic output of these protocols is simpler than selecting the data to be collected. Additionally, the quality of each data set is dependent on the type of data being generated. For example, RNA-seq data sets may have a standardized library preparation protocol: A minimum number of sequenced reads and all raw data are processed the same way. Other data types, such as whole-genome bisulfite sequencing (WGBS) and MethylC-seq, have their own set of requirements. Similarly, MethylC-seq data sets require not only minimum read depths and data processing through the same analysis pipeline for identifying methylated cytosines and determining methylation levels (115) but also a minimum conversion rate of unmethylated cytosines by the sodium bisulfite reaction.

In this review, we discuss the need for pENCODE, the challenges a project like this poses, and the benefits this project could have for advancing our understanding of plant sciences. Additionally, we discuss the needs for standardization of sample collection, sample preparation, and data

Comparative epigenomics: within- and between-species comparisons of epigenome maps that may or may not include DNA sequence variation

processing, including tools for analysis pipelines, visualization, and dissemination. Data-driven, discovery-based research projects are hypothesis-generating factories. Given the collegiality within the plant sciences community, a concerted effort to execute a successful pENCODE project would have long-lasting effects on plant sciences.

THE DISTINCTION BETWEEN EPIGENOMICS AND EPIGENETICS

Here, we make the case for the need for an epigenomic resource rather than an epigenetic one. Because of the widely used nature of the terms epigenetics and epigenomics in the relevant literature, it is important to be clear about our use of them. The key differences being that epigenetics requires demonstration of heritability of phenotypes in addition to an absence of differences in DNA sequence, whereas the study of epigenomics is broadly used to encompass all factors that interact with DNA and contain the possibility of affecting gene regulation, such as chromatin modifications, DNA methylation, RNAs (coding and noncoding), etc. Originally, epigenomics referred to chromatin modifications throughout the genome (17), but the term has been expanded into a more recent definition, which also includes RNAs, TF binding, nucleosome positioning, and chromosomal interactions (13).

Although the topic of epigenomics may appear broad, it can be utilized at great length to create maps of genomic features. Maps such as these are useful for hypothesis generation of readily testable, genome-wide studies, which can be rapidly completed because of the existence of these same genomic resources. For example, these epigenomic maps allow for the search for true epigenetic phenomena at wide scale rather than by a singular gene approach. Having these data located at a central hub with compatible formatting greatly increases the ease of hypothesis generation and testing. Simply put, laboratories do not need to reinvent the wheel for each analysis.

An excellent example of the benefits of a multipronged, genome-wide approach to studying a developmental program is a project by Zhong et al. (142) that elucidated the molecular events that lead to ripening in tomato through a combination of WGBS, RNA-seq, and ChIP (chromatin immunoprecipitation)-seq. They used these high-throughput methods on samples from various mutants and at various developmental time points and were able to create a list of 292 candidate genes. Utilizing an antibody for RIN (RIPENING INHIBITOR), a MADS-box TF that directly regulates fruit-ripening genes, the authors performed ChIP-seq. Combining these results with expression data from fruits that were either wild type or homozygous for a *rin* loss-of-function mutation, they were able to curate their list of 292 candidate genes, which included all 16 genes already associated with fruit ripening.

Many projects result in large numbers of candidate genes that have to be further narrowed or randomly selected for additional hypothesis testing. From this perspective, a list of 292 is small and testable, providing numerous hypotheses that only became available through combining high-throughput technologies. Furthermore, there are now developmental time-course data for gene expression and methylation patterns that can be mined for future work, which does not necessarily need to relate to fruit ripening specifically. Other projects can use this epigenomic map of the tomato genome to determine lists of candidate genes for their points of interest as well as for comparative epigenomic studies. These data also support testing of the 292 possible fruit-ripening genes without having to spend the money or time to repeat or add additional data sets. These kinds of projects readily stem from pENCODE.

EPIGENOMIC DATA TYPES

The success of pENCODE relies heavily on the individual building blocks that, when combined, unveil the epigenome. The epigenome of a cell describes the activity of a genome, and the building blocks represent distinct data types (13). What are some of the epigenomic data sets that should be acquired to create these genome-wide maps? Described below are the most common techniques (113) used to generate different types of epigenomic maps, along with their advantages and disadvantages.

ChIP-seq

Chromatin immunoprecipitation combined with deep sequencing (ChIP-seq) is regarded as the standard technique to identify genome-wide distributions of DNA-bound factors and histone tail modifications (64). Specific antibodies are used to immunoprecipitate proteins or histones with specific tail modifications of interest and the cross-linked chromatin, which is subsequently sequenced to identify genomic regions associated with the protein or histone tail modification of interest.

Pros: This sequencing technique requires low sequencing depth and typically fewer than 20 million reads to detect these protein:DNA interactions.

Cons: This technique is specifically used for anchoring known sequences to a reference genome, so it is only applicable to published plant genome assemblies, requires significant input of starting chromatin, is inherently low throughput, and relies heavily on the availability and quality of the antibody. Often overexpression or manipulation of higher target TF protein levels is required for successful chromatin immunoprecipitation.

DNase-seq, FAIRE-seq, and ATAC-seq

As complementary methods to ChIP-seq, formaldehyde-assisted isolation of regulatory elements with sequencing (FAIRE-seq) (53, 54), DNase I hypersensitive sequencing (DNase-seq) (29), and assay for transposase-accessible chromatin sequencing (ATAC-seq) (16) are able to identify the vast majority of putative bound sites in nucleosome-depleted regions at a genome scale. More specifically, FAIRE-seq is based on formaldehyde cross-linking followed with sonication and phenol-chloroform extraction, and is capable of detecting potential regulatory regions. DNase-seq depends on the genome-wide distributions of DNase I hypersensitive (DH) sites. DNase-seq not only sensitively identifies *cis*-regulatory DNA but also provides information for motif and protein occupancy for *trans*-acting factors, which bind to the aforementioned *cis*-regulatory DNA sequences. ATAC-seq is dependent on an adapter-loaded transposase system that performs tagmention (fragmentation of gDNA and addition of an adapter in a single step) of open chromatin. Such predictions can ultimately be verified through follow-up experiments.

Pros: Can identify DNA footprints to base-pair resolution, which can be combined with known DNA binding motifs for placement of DNA:protein interactions. These methods are also powerful in that they can uncover completely novel binding motifs not detected by other methods.

Cons: These techniques require a reference genome for alignment of sequencing reads and refinement of cross-linking and/or DNase I digestion times for optimal results.

Hi-C-seq and ChIA-PET-seq

Neither ChIP-seq nor other complementary techniques can capture chromatin interactions, which has led to the development of new technologies, such as Hi-C sequencing (75) and chromatin interaction analysis with paired-end tag sequencing (ChIA-PET) (46). Both require cross-linking between DNA and proteins in the initial step. The former technique requires samples to be gathered after enzymatic digestion, whereas the latter technique relies on immunoprecipitation using a specific antibody to a protein of interest.

Pros: Reveals inter- and intrachromosomal interactions, which are useful for accurate association of DNA elements to genes.

Cons: These techniques are best suited for cell-type specific samples, otherwise complications quickly arise when trying to detect these chromosomal interactions. Hi-C also requires very high sequencing depth, which scales with genome size when compared to other seq assays, such as ChIP-seq and RNA-seq.

MNase-seq

ChIP and other techniques cannot determine nucleosome occupancy, regional accessibility, or stability. To address this, micrococcal nuclease (MNase) coupled with sequencing (MNase-seq) (114) can be used to anchor the locations of nucleosomes based on the boundary sequences of linker DNA that are released from chromatin according to nucleosome accessibility, occupancy, and stability. MNase-seq relies on the activity of an enzyme, which releases DNA sequences from chromatin in a time-dependent manner.

Pros: This technique requires lower input quantities, and the length of digestion can be adjusted to discern different features of nucleosomes, such as occupancy, stability, and accessibility.

Cons: The length of the digestion must be carefully monitored, as overdigestion occurs within minutes.

MethylC-seq

Cytosine methylation is a covalent base modification that can be surveyed genome wide using WGBS (24, 77), which is regarded as the gold-standard method to detect DNA methylation levels at single-base resolution. The principle of this technique is to couple the sodium bisulfite conversion reaction, which converts unmethylated cytosines to uracil and ultimately to thymine after PCR (polymerase chain reaction) amplification, with high-throughput sequencing.

Pros: Can detect single-base resolution DNA methylation states of any cytosine with high precision and requires much lower input material compared with most high-throughput sequencing techniques.

Cons: Requires high coverage sequencing compared with other techniques described in this section (although reduced representation methods do exist) and requires sufficient chemical conversion rates of unmethylated cytosines to uracils by the sodium bisulfite reaction.

RNA-seq

Transcription in the genome of both coding and noncoding sequences can be measured using RNA-seq (87). There exists a multitude of RNA-seq approaches, including cDNA-seq, strand-specific RNA-seq (77), polyA RNA-seq, ribosomal RNA depletion RNA-seq, and small RNA-seq.

Pros: Requires incredibly low amounts of starting material and can even work from single cell samples. Lower read numbers can still be used to obtain sufficient information to evaluate RNA abundances, as the genome size does not generally affect the total RNA in the cell. Instead, this is generally a reflection of expressed gene number.

Cons: It is generally more difficult to compare RNA-seq data between different laboratories, as most data producers rely on different RNA enrichment and library construction methods. Moreover, ribosomal depletion methods increase the number of uninformative reads per sample, as the depletion methods are not as efficient as polyA selection for enriching transcripts.

The Benefits of a pENCODE Project

Although there is currently no official pENCODE, there are a number of groups that have been generating high-throughput epigenomic data sets in a wide range of plant species. So far, the most abundant data sets in existence are RNA-seq, which is mostly due to the ease with which this experiment can be performed. Genome-wide, single-base resolution DNA methylation data exist for a number of plant species, including *Arabidopsis* (24, 77), maize (36, 49, 101), soybean (110, 119), rice (135), sorghum (94), *Brachypodium distachyon* (127), *Amborella* (7), and tomato (142). Additionally, a limited number of ChIP-seq maps for histone modifications and TFs are available in *Arabidopsis*, rice, and maize, but in other plant species ChIP-seq maps are more limited. The rarest data sets currently available for plant genomes include nucleosome positions, DNase-seq maps (138, 139), and chromosomal interaction maps (86) that are mainly only available in a single accession of a reference plant species.

A major goal of pENCODE would be to facilitate decoding the manner in which plant genomes are expressed. This goal directly builds upon the success of sequencing de novo plant genomes, which have been invaluable for annotating the gene content, gene structure, locations of genes, and intergenic space as well as other structural features such as centromeres and telomeres. One of our next major challenges is to understand how sequence information is translated into expression variation. With this knowledge, the link between genetic variation and phenotypic variation can be advanced for a large number of plant traits being studied across the globe. For example, the plant science community has had a number of successes using quantitative genetic approaches to identify favorable alleles in crop species. Although identification of QTL is relatively straight forward through either linkage or genome-wide association mapping techniques, the actual identification of the causal variant(s) is still incredibly challenging (134). Similar to studies in human populations in which great strides are being made at predicting causal variants using ENCODE data (67), numerous genetic variants linked with the trait of interest are found outside of coding sequences. In many cases, having epigenomic maps would facilitate a more rapid identification of these causal variants by providing an additional layer of information, especially in species that have large genomes. To enable hypothesis testing of predicted causal variants, mutant strains provide a vital resource, and fortunately there are already numerous projects aimed at creating large mutant populations of diverse plant species using T-DNA, transposon tagging, or TILLING mutagenesis (6, 14, 19, 26, 27, 30, 69, 93, 100, 104, 107, 123, 131). These mutant populations, in combination with results from pENCODE, will be invaluable for identifying trait-associated sequence variants.

Many of the techniques previously described generate data that lead to an emerging picture of the genomic landscape of a cell at a specific developmental stage or upon a specific environmental treatment similar to the pictures that arise from sequence variation detected by resequencing projects. Just as patterns emerge for sequence variants (59, 129) that can inform us about the evolutionary history of the sequence, such as rates and locations of synonymous versus nonsynonymous base substitutions, patterns emerge from comparisons of epigenomic data sets. For example, it is well known that a pattern of enrichment of the histone modification lysine 4 trimethylation on histone 3 (H3K4me3) often clearly demarcates the transcriptional start site of an expressed locus (140), whereas H3K9me2 is found at loci that are actively silenced (12) (**Figure 1**). Essentially, all epigenomic techniques described in this review generate data that have been linked to a mechanism or process. The power of genomics is the ability to rapidly create high-resolution maps of the genome, which leads to the generation of hypotheses that can be tested for specific genes or regions of interest.

For an example of how epigenomic maps could accelerate the identification of a causal variant imagine the following scenario: There is a trait of interest associated with an allele present in a plant species with a very large genome, but the region of interest is more than 100 kb. Fortunately, this particular trait is governed by an impact on expression variation, so now the search begins for the sequence change that leads to this variation. Unfortunately, genomic DNA sequencing of this region fails to detect such a variant and now requires a rare recombinant to fine map this causative

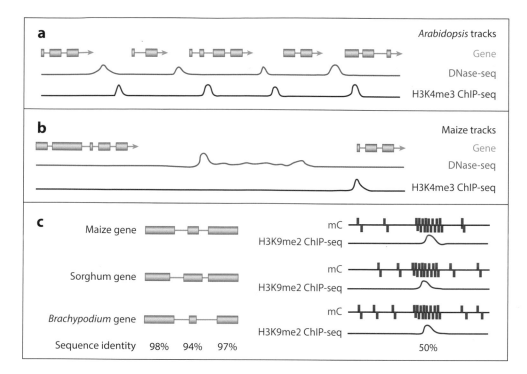

Figure 1

This model is a simplified version of the data that would be uncovered through a comparative epigenomics browser. (*a*) Shorter intergenic space in a smaller, more compact genome, such as *Arabidopsis*, allows for location of DNA elements without the need for several data sets. The area in which these elements can be located is restricted. Here, this is modeled by peaks for DNA elements in H3K4m3 ChIP-seq (*purple*) and DNase-seq (*orange*) data sets. H3K4me3 is associated with transcriptional start sites, and DNase-seq is associated with promoter regions. They are located between each gene model (*green*), and either data set would clearly define them. (*b*) Larger genomes, such as maize, can have much larger intergenic spaces, as depicted here. These region lengths can make locating DNA elements more difficult because data sets may not have a single clear peak. However, multiple data sets locating points of consistency can lead to clearer recognition of these DNA elements. (*c*) When comparing related species, important conserved elements, such as genes (*green*), can be easily annotated through sequence identity (*black*; *below both halves of the figure*) as a percent of the sequence conserved across species. A model is shown on the left of the figure. However, there are cases in which sequence conservation is not enough to identify important elements, especially in short sequences. A model is shown on the right, which could occur in a promoter region. In this example, even though there is low sequence identity at the nucleotide level, a combination of conserved methylation data (mC; *pink*) and H3K9me2 ChIP-seq data (*purple*) is used to accurately identify an important genomic region.

allele. After years of person hours, it is finally recognized that this particular allele is under the control of small RNAs associated with repressed loci, as opposed to sequence variation between the differentially expressed alleles. This particular scenario is incredibly challenging to solve but is not unheard of for some of the causal variants identified by research laboratories across the globe. Moreover, the pursuit of this specific example would have benefited greatly from epigenomic maps. If these genome-wide maps for DNA methylation, histone modifications, nucleosome occupancy, small RNAs, RNAs, and chromosomal interactions existed, the identification of this causal variant would be greatly accelerated. In reality, the scenario described above is not hypothetical; it describes the countless years of effort and the many approaches used to clone and understand the mechanistic action of an allelic state associated with the paramutation properties of the *B* locus in maize (4, 96–98, 121, 122).

The ability to generate genome-wide maps of epigenomes was not possible ten years ago but is today, and the generation of these maps will undoubtedly advance research within the plant sciences. Many examples exist in which these maps have accelerated the identification of long-range enhancers in plant species (81, 108, 125). Essentially, generating these maps improves our ability to decode genomes by unveiling features that are not readily apparent from the underlying sequence information alone. These maps will also facilitate annotation of novel genes, refinement of current gene annotations, and potentially uncover locations of transposon and repeat sequences, which are prevalent in plant genomes. With new assemblies for plant genomes rapidly appearing in the public domain, it is often assumed by most researchers that these are highly polished assemblies and annotations, but in most cases the available assemblies represent drafts. They are fantastic resources that will expedite research, but they still require refinement. Epigenomic maps are not only important for identification of novel causal variants but they are also powerful for annotating genomes. Genomes are most commonly annotated using sequence and transcript-based methods to identify gene structures such as untranslated regions, exons, etc. The production of high quality epigenomic maps could rapidly refine annotations by revealing transcriptional start sites, gene-body DNA methylation (associated with expressed loci), small RNAs, and repressive DNA methylation associated with repeats, transposons, and some genic regions (94). Genome assemblies and annotations are taken for granted, but although draft genomes and annotations are valuable, it is important to consider the continued pursuit of decoding these genomes until the genome and annotation are at the highest possible resolution. There is no doubt that the generation of epigenomic maps will result in more accurate annotations of their respective genomes.

CATALOGING NONCODING ELEMENTS IN PLANT GENOMES

In plants and animals, chromatin domains, defined by DNA methylation, sets of modified histones, and nucleosome positioning, play a role in gene expression. Work performed in *Arabidopsis*, rice, and maize is the primary source of chromatin modification data in plant species and has laid the groundwork for expanding this course of study into different plant species.

Knowledge provided by studying chromatin domains is not limited to the patterns and functions of the domains themselves. Most of these data have been useful in predicting gene regulatory regions. For example, mapping DH sites, which correspond to open chromatin domains, has provided genome-wide information about TFBSs (transcription factor binding sites; **http://www.plantregulome.org/**) and RNA polymerase II binding sites (138, 139). Furthermore, specific chromatin modifications are correlated to different genomic sections. For example, in *Arabidopsis*, eight different chromosome modifications have been mapped together and, in concert, indicate four chromatin states that occur preferentially around specific genomic features, including active genes, repressed genes, silent repeat elements, and intergenic regions (105). These patterns can also be used to predictively annotate genomes for these elements. This tool becomes even more powerful if conservation is included across species. When annotating genomes, information from related species can be utilized through application of sequence conservation as an annotation assistant. Situations may arise, however, in which there is a lack of sequence conservation, yet a small regulatory element is present in multiple species. Sequence conservation alone can overlook these small elements because they are simply too short. In these instances, alternative data sets can be used to locate such repeated elements by comparing similar patterns across species. Therein lies the power of a comparison of chromatin domains (**Figure 1**).

Many mechanistic questions remain as to how these patterns of histone modifications and chromatin domains function to alter gene expression, but there are also missing patterns. Most genome-wide studies examining patterns of chromatin domains in plants compare a type of chromatin modification (H3K9me2 or H3K27me3) with sequence structures (such as transposons and repetitive sequences) and DNA methylation or small RNAs. In the past few years, there has been an increase of comparisons across chromatin modification types, which has revealed not only that correlations exist between different chromatin domains and DNA methylation/gene sequences but also that there are combinatorial effects of chromatin domains on gene expression (105). No one epigenomic state has patterns completely independent of all other chromatin states and thus some regulatory mechanisms will emerge when this is studied between species and more inclusively.

Epigenomics approaches can be easily applied to plant species that are not traditionally considered good genetic systems, such as fruit tree crops, which have long generation times. Furthermore, random mutagenesis is not readily useful in many of these same plant species. Therefore, application of epigenomics to create a list of candidates to study specific developmental or environmental questions can bypass some of the issues that arise when studying plant species that are not as amenable to genetics (i.e., generation time, space, number of offspring, and transformability). Fortunately, for those species that are transformable, genome targeting technologies such as CRISPRs (clustered regularly interspaced short palindromic repeats) are promising methods for targeted mutagenesis, which will be vital for testing hypotheses with regard to these interesting candidate gene lists that were identified from epigenomics approaches (10, 44, 88, 116).

This additional ease also translates to plants with large genomes. For these plants, the additional intergenic space makes it more difficult to locate potential DNA elements that define transcriptional programming. In smaller plant genomes, such as *Arabidopsis*, it has been shown, for example, that DNase-seq can readily identify the majority of regions occupied by TFs (139; **http://www.plantregulome.org**). For example, one study found DH sites were associated 94.9% and 89.7% with two well-known TFBSs through comparing DNase-seq data and ChIP-seq data (139). In this genome, DNase-seq alone becomes a powerful tool to locate promoter regions. However, almost 45% of the DH sites were within 1-kb upstream of genes, which is indicative of the much more compact genome and high gene density in *Arabidopsis* as compared with other plant species such as rice, which has a value of 27% (138). The short intergenic spaces in compact plant genomes make location of DNA elements simpler than in these larger genomes. For example, locating DNA elements in plants with larger genomes, such as maize, is much more difficult, as these DNA elements can occur tens to hundreds of kilobases away from their corresponding gene (**Figure 1**). Furthermore, these types of questions could be examined within and across species given the correct tools and data organization. In addition to significant differences between plants and animals concerning gene regulation through chromatin domains, there are known differences in other epigenomic factors, like DNA methylation between plant species such as rice, maize, *B. distachyon*, and *Arabidopsis* (126, 127).

COMPARATIVE AND POPULATION EPIGENOMICS

Comparative epigenomics is the use of epigenomic maps to identify similarities and differences in epigenomes within and between species. Just as comparative genomics has proven to be a powerful tool, giving deep insight into the evolution and functional elements of the genome, comparative epigenomics can provide a broad understanding of epigenomic features, leading to the formation of new hypotheses. The two approaches are in fact complementary, as data from one can be used to inform the other. Between-species comparisons can give insight into the evolution of the epigenome and the different ways the same epigenomic tool kit is used by different species.

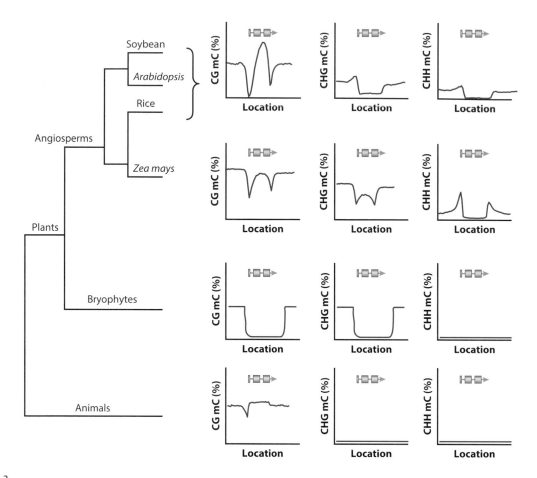

Figure 2

Methylation patterns within gene bodies vary even among closely related species. On the left is an approximate phylogeny illustrating the relationships between the shown species. On the right are graphical representations of the average gene-body methylation pattern for each species broken down by methylation context (CG, CHG, and CHH, where H = A, C, or T). The pink lines indicate methylation levels (*y axis*) across a gene (*shown in green; location on the x axis*). Plants and animals vary drastically across each methylation context, with animals, such as puffer fish, lacking CHG and CHH in the gene bodies. *Selaginella moellendorffii* and *Physcomitrella patens* have a distinct lack of methylation in the gene body. Angiosperms again diverge with maize. They show different patterns from rice, soybean, and *Arabidopsis*. These defined differences highlight the need for and the unexpected results generated from comparative epigenomic studies.

Within-species comparisons reveal the breadth of epigenomic variation and the tools to link this information to phenotypes.

Between-species comparisons of DNA methylation have already been done for the few species whose methylomes have been sequenced (43, 127, 136). These studies provide insight into the evolutionary past of DNA methylation while showing key differences that have developed over time (**Figure 2**). Two studies compared methylation not only in plants but across eukaryotes (43, 136). This work showed that gene-body methylation is highly conserved, as it is associated with genes expressed at moderate levels and basal to the divergence of plants and animals. There the similarities end. Methylation in plants occurs within all three sequence contexts, whereas animals have primarily CG methylation except in the brain and embryonic stem cells (76, 78). Although silencing of transposons by DNA methylation is found in plants, fungi, and vertebrates, it appears to be absent in invertebrates. Thus, transposon silencing appears to have shifted in mechanism in different lineages. Examination of the angiosperms *Arabidopsis*, rice, and poplar showed very similar patterns, indicating conservation of DNA methylation in these plant species (43, 127, 136). Comparing methylation between *B. distachyon*, rice, and *Arabidopsis*, further evidence was found for the conservation of gene-body methylation between orthologs in angiosperms. More striking differences were discovered between the angiosperms and the land plants *Selaginella moellendorffii* and *Physcomitrella patens*, which diverged early from the angiosperms (136). Here, methylation of both genes and regions around transcriptional start sites appears to be absent. Comparing deeper evolutionary divergences, various green algae species show that CG and CHG methylation is very ancient in plants (43, 136).

These studies show that comparative epigenomics is possible and is informative about the evolutionary history of species and the usage of the epigenome. Feng et al. (43) and Zemach et al. (136) both linked their results to phylogenetic analysis of the enzymes involved in DNA methylation and subsequently reflected their results back to genetic explanations of some of the differences observed. A limiting factor in these studies has been the lack of epigenomic data from a wide range of plant species, masking potential differences and even subtle similarities in the usage of the epigenome. Furthermore, the species commonly studied, such as *Arabidopsis*, often have small genomes, are diploid, and have relatively low amounts of repetitive DNA. Many of our most economically important species have very different genomic content and as a result may possess important differences in how the epigenome is used. A recent example can be found in maize, where CHH methylation was enriched in regions upstream of the genes, which were dubbed CHH islands (49). Within soybean, which is an ancient polyploid, there is a clear preferential methylation of orthologs from one of the ancestral genomes versus the other (110). A pENCODE project could begin to address many of these major questions in the plant sciences by providing additional epigenomic data sets from a variety of species.

Within-species comparisons enable the discovery of natural epigenomic variants, such as differentially methylated regions (DMRs) and single methylation polymorphisms (SMPs). This type of study can advance our basic understanding of epigenomic variation, including the rate at which such variants arise. An example can be found in *Arabidopsis*, where two studies of DNA methylation across generations of a mutation accumulation line made it possible to calculate the rate at which SMPs arise, showing it to be several-fold higher than the rate of genetic mutations (9, 111). There have been an increasing number of studies examining DNA methylation in natural populations of *Arabidopsis* (112, 132), maize (36, 37, 101), and soybean (110). These reveal widespread epigenomic variation. Although many methylation variants identified in these studies segregate with parental genotypes, a significant number do not and may be true epigenetic variants (101, 110, 112). Such approaches will help us further understand the extent at which true epigenetic variants exist within natural populations.

By treating epigenomic features as phenotypes, it will be possible to use association or QTL mapping to identify genetic variation underlying methylation variants or methylQTL (36, 110, 112). The power of such approaches has already shown that natural variants in the CMT2 DNA methyltransferase underlie natural methylation variation in *Arabidopsis* populations and their adaptation to temperature (117). These approaches can be further strengthened by application to experimental populations such as epigenetic recombinant inbred lines (epiRILs), which are largely isogenic but differ in their methylation content (63, 102). Combined with work on natural populations, the association mapping and QTL analysis previously discussed can be used to further link phenotypic variation to epigenomic variation (71, 106). In fact, epigenomic variants could be used in lieu of traditionally used genetic markers, as was recently demonstrated for mapping the basis of complex traits that are associated with heritable epigenetic variants in

Arabidopsis (28). This approach, however, will require the discovery of new epigenomic variants across many populations, a task that pENCODE could begin to address.

FUTURE CHALLENGES AND DIRECTIONS

Numerous challenges exist to establish pENCODE. Fortunately, many of these challenges can be overcome by international coordination of the plant research community. Unlike the human ENCODE project, plant species do not have readily obtainable cell lines available for most cell types because of their inherent differentiation properties. Therefore, most epigenomic data sets require the generation of maps from tissues/organs that contain multiple cell types. This fact does not pose an issue in terms of annotating genomes using epigenomic data, but it will confound analysis of developmental and environmentally treated samples for obvious reasons. Fortunately, the plant community is ahead of their animal counterparts in their ability to isolate specific cell types in vivo for species that are readily transformable (31, 32), but this is a cumbersome process, especially for plant species that require years to generate stable transgenics. Additional challenges exist for assays such as DNase-seq that require high-input material, but technologies to reduce input material for assays such as ATAC-seq, nano-ChIP-seq (2), MethylC-seq, etc. are constantly being improved, primarily because of the interest in surveying low-input sample material.

In addition to determining the samples for pENCODE, it will be necessary to select the plant species to be included. These species will likely be selected on the basis of the quality of reference genomes available and the ability to survey specific cell types, but should also include a wide range of species from across the plant kingdom.

As discussed above, for such a project to succeed the plant sciences community will not only need to come to a consensus through ready communication and an organized venue regarding the samples and plant species to be surveyed but also agree upon the protocols used in sample preparation, sequencing library construction, analysis pipelines, quality metrics, and visualization methods of disseminated data sets.

Digital Reconstruction of an Expression Atlas

The future is bright for decoding plant genomes because of the rapid advances in sequencing throughput and because of the existing infrastructure that is required to execute such a goal. In the future, as new technologies permit, it may be possible to generate high-resolution digital reconstructions of plants at all stages of development. This goal is limited for most sequencing-based techniques at this time but is feasible for a transcriptome map, given that sequencing libraries can be generated from a single cell (128). The rate-limiting step is sample collection, as methods need to be developed to section a plant at high resolution and at the same time preserve and collect the sectioned tissue. Of course, the ultimate resolution requires reconstruction of an expression atlas from each individual cell or at least cell type, but again the challenge for the plant sciences community is the replicable extraction and isolation of these specific cells, which is complicated because of the existence of plant cell walls, as downstream methods to lyse cells and create sequencing libraries already exist.

Scalable Cyberinfrastructure

In order to rapidly process, integrate, analyze, and disseminate the data generated by pENCODE, a scalable computational platform is required. The iPlant Collaborative is the first large national investment by the National Science Foundation to develop these resources for life science research and has developed a panoply of resources to enable scalable computing, data management, and distributed virtual organizations (55). This cyberinfrastructure has been the computational foundation for EPIC-CoGe and has the framework in place to permit pENCODE researchers to integrate and share their data processing and analysis pipelines, develop virtual communities, and create additional pENCODE bioinformatic platforms. Unifying these computational applications on a common infrastructure allows each resource to more easily interoperate with one another and allows researchers to more easily move their data and analyses among these systems to accelerate scientific discovery.

Synthetic Biology

Techniques such as INTACT have been developed in plants to allow for the collection of data from specific in vivo cell types. These methods and their applicability make plants a useful system to study specific cell types in living tissues, which is not feasible in most animal ENCODE projects. Surveying in vivo epigenomic states results in data sets that have boundless possibilities for hypothesis generation, but testing these hypotheses can be cumbersome. Methods need to be developed to test the significance of identified epigenomic features on resulting gene expression patterns. These methods should take advantage of advancements in synthetic biology. Technologies are available to rapidly generate DNA sequences that can in turn be assayed for their effects on gene expression states, as has been nicely demonstrated in mice species (99). For this to succeed, plant transient assays, such as the STAY GREEN reporter system (82) and high-throughput yeast-1-hybrid systems (48), will need to be used to test hypotheses generated from these genome-wide maps. These synthetic approaches are excellent ways to rapidly test hypotheses and further reduce a genome-wide list of candidates to a validated, more-refined set that can be experimentally confirmed in planta.

Epigenome Engineering

Although pENCODE will assist with hypothesis generation and testing, it will also supply a necessary resource for testing and preparing epigenome engineering techniques by providing resources to adapt techniques from one species to another. Besides disrupting the epigenome through the use of pharmacological variation, which has poorly understood effects, and capitalizing on already present natural variants, methods are being developed to perform directed epigenomic reprogramming of specific genes or regions of the genome. Thus far, these methods have focused upon altering histone or DNA methylation (65, 68). Methods for this directed approach include using zinc finger nucleases, transcription activator–like effectors, and the CRISPR-Cas system present in bacteria, which all locate specific short sequences and can target methylation-altering proteins, such as methyltransferases or DNA demethylases. In order to bring these pieces together and adapt their use to multiple species, a database of testable hypotheses would be invaluable.

CONCLUSION

This work describes the benefits and the need for pENCODE. It is clear that this effort will result in significant deliverables to the plant sciences community, but we should not underestimate the unknown. One of the most exciting aspects of the discovery-based research approach associated with pENCODE is the potential for paradigm shifting results that could possibly emerge from creating these epigenomic maps.

SUMMARY POINTS

1. The need and ability to utilize pENCODE already exist throughout the plant science community. This is supported by the presence of several species-based collaborative efforts to unlock the story of plant epigenomes.

2. Current epigenomic data collection methods are becoming financially feasible, and new technologies are continuously developed for the analysis of these data and to create more cost-effective methods for its collection.

3. The major goal of pENCODE would be to facilitate the decoding of the manner in which plant genomes are expressed, which is analogous to the nature in which plant genomes are assembled and dissected for gene structure and genomic patterns.

4. Epigenomic maps can facilitate numerous forms of hypothesis generation and testing, including the development of conservative gene-candidate lists and prediction of the location of causal variants.

5. Comparative and population epigenomics made possible through the existence of pENCODE will shed light upon natural variation in epigenomic markers, such as DMRs, and chromatin domains. They will create the ability to discover novel epigenomic patterns across plant species and facilitate learning about plant evolutionary history.

FUTURE ISSUES

1. How do we develop a list of qualities for seq data sets that will create uniformity across the field?

2. How do we set developmental time points and environmental assay procedures that will be applicable and comparable across all plant species?

3. How do we as a community collect and coordinate all existing efforts to organize and annotate existing epigenomic plant data?

4. How do we take advantage of the ability to profile epigenomes of specific cell types using the INTACT system?

DISCLOSURE STATEMENT

The authors are not aware of any affiliations, memberships, funding, or financial holdings that might be perceived as affecting the objectivity of this review.

ACKNOWLEDGMENTS

We would like to thank Joseph Ecker, Rick Myers, Vicki Chandler, Jeremy Schmutz, Doris Wagner, Eric Lyons, Christine Queitsch, Scott Jackson, Alessandra Oddone, Ryan Lister, Keiko Torii, Rhiannon McCrae and Brian Gregory for helpful discussions and suggestions on this timely topic. Funding from the University of Georgia Graduate Recruitment Opportunities Assistantship to A.K.L. and funding from the University of Georgia Research Foundation, the National Institutes of Health (R00GM100000), and the National Science Foundation (IOS-1339194) to R.J.S. supported this work.

LITERATURE CITED

1. Addoquaye C, Eshoo T, Bartel D, Axtell M. 2008. Endogenous siRNA and miRNA targets identified by sequencing of the *Arabidopsis* degradome. *Curr. Biol.* 18:758–62

2. Adli M, Bernstein BE. 2011. Whole-genome chromatin profiling from limited numbers of cells using nano-ChIP-seq. *Nat. Protoc.* 6:1656–68

3. *Arabidopsis* Genome Initiat. 2000. Analysis of the genome sequence of the flowering plant *Arabidopsis thaliana*. *Nature* 408:796–815

4. Alleman M, Sidorenko L, McGinnis K, Seshadri V, Dorweiler JE, et al. 2006. An RNA-dependent RNA polymerase is required for paramutation in maize. *Nature* 442:295–98

5. Allen MA, Hillier LW, Waterston RH, Blumenthal T. 2011. A global analysis of *C. elegans trans*-splicing. *Genome Res.* 21:255–64

6. Alonso JM, Stepanova AN, Leisse TJ, Kim CJ, Chen H, et al. 2003. Genome-wide insertional mutagenesis of *Arabidopsis thaliana*. *Science* 301:653–57

7. *Amborella* Genome Proj. 2013. The *Amborella* genome and the evolution of flowering plants. *Science* 342:1241089

8. Arvey A, Tempera I, Tsai K, Chen HS, Tikhmyanova N, et al. 2012. An atlas of the Epstein-Barr virus transcriptome and epigenome reveals host-virus regulatory interactions. *Cell Host Microbe* 12:233–45

9. Becker C, Hagmann J, Muller J, Koenig D, Stegle O, et al. 2011. Spontaneous epigenetic variation in the *Arabidopsis thaliana* methylome. *Nature* 480:245–49

10. Belhaj K, Chaparro-Garcia A, Kamoun S, Nekrasov V. 2013. Plant genome editing made easy: targeted mutagenesis in model and crop plants using the CRISPR/Cas system. *Plant Methods* 9:39

11. Berezikov E, Robine N, Samsonova A, Westholm JO, Naqvi A, et al. 2011. Deep annotation of *Drosophila melanogaster* microRNAs yields insights into their processing, modification, and emergence. *Genome Res.* 21:203–15

12. Bernatavichute Y, Zhang X, Cokus S, Pellegrini M, Jacobsen S, Dilkes B. 2008. Genome-wide association of histone H3 lysine nine methylation with CHG DNA methylation in *Arabidopsis thaliana*. *PLOS ONE* 3:e3156

13. Bernstein BE, Stamatoyannopoulos JA, Costello JF, Ren B, Milosavljevic A, et al. 2010. The NIH roadmap epigenomics mapping consortium. *Nat. Biotechnol.* 28:1045–48

14. Bragg JN, Wu J, Gordon SP, Guttman ME, Thilmony R, et al. 2012. Generation and characterization of the Western Regional Research Center *Brachypodium* T-DNA insertional mutant collection. *PLOS ONE* 7:e41916

15. Brazma A, Hingamp P, Quackenbush J, Sherlock G, Spellman P, et al. 2001. Minimum information about a microarray experiment (MIAME): toward standards for microarray data. *Nat. Genet.* 29:365–71

16. Buenrostro JD, Giresi PG, Zaba LC, Chang HY, Greenleaf WJ. 2013. Transposition of native chromatin for fast and sensitive epigenomic profiling of open chromatin, DNA-binding proteins and nucleosome position. *Nat. Methods* 10:1213–18

17. Callinan PA, Feinberg AP. 2006. The emerging science of epigenomics. *Hum. Mol. Genet.* 15(Spec. No. 1):R95–101

18. Cao J, Schneeberger K, Ossowski S, Gunther T, Bender S, et al. 2011. Whole-genome sequencing of multiple *Arabidopsis thaliana* populations. *Nat. Genet.* 43:956–63

19. Carelli M, Calderini O, Panara F, Porceddu A, Losini I, et al. 2013. Reverse genetics in *Medicago truncatula* using a TILLING mutant collection. *Methods Mol. Biol.* 1069:101–18

20. Cherbas L, Willingham A, Zhang D, Yang L, Zou Y, et al. 2011. The transcriptional diversity of 25 *Drosophila* cell lines. *Genome Res.* 21:301–14

21. Chia JM, Song C, Bradbury PJ, Costich D, de Leon N, et al. 2012. Maize HapMap2 identifies extant variation from a genome in flux. *Nat. Genet.* 44:803–7

22. Chodavarapu RK, Feng S, Bernatavichute YV, Chen PY, Stroud H, et al. 2010. Relationship between nucleosome positioning and DNA methylation. *Nature* 466:388–92

23. Chung WJ, Agius P, Westholm JO, Chen M, Okamura K, et al. 2011. Computational and experimental identification of mirtrons in *Drosophila melanogaster* and *Caenorhabditis elegans*. *Genome Res.* 21:286–300

24. Cokus SJ, Feng S, Zhang X, Chen Z, Merriman B, et al. 2008. Shotgun bisulphite sequencing of the *Arabidopsis* genome reveals DNA methylation patterning. *Nature* 452:215–19

25. Coleman-Derr D, Zilberman D. 2012. Deposition of histone variant H2A.Z within gene bodies regulates responsive genes. *PLOS Genet.* 8:e1002988

26. Cooper JL, Henikoff S, Comai L, Till BJ. 2013. TILLING and ecotilling for rice. *Methods Mol. Biol.* 956:39–56

27. Cooper JL, Till BJ, Laport RG, Darlow MC, Kleffner JM, et al. 2008. TILLING to detect induced mutations in soybean. *BMC Plant Biol.* 8:9

28. Cortijo S, Wardenaar R, Colome-Tatche M, Gilly A, Etcheverry M, et al. 2014. Mapping the epigenetic basis of complex traits. *Science* 343:1145–48

29. Crawford GE, Holt IE, Whittle J, Webb BD, Tai D, et al. 2006. Genome-wide mapping of DNase hypersensitive sites using massively parallel signature sequencing (MPSS). *Genome Res.* 16:123–31

30. Dalmais M, Antelme S, Ho-Yue-Kuang S, Wang Y, Darracq O, et al. 2013. A TILLING platform for functional genomics in *Brachypodium distachyon*. *PLOS ONE* 8:e65503

31. Deal RB, Henikoff S. 2010. A simple method for gene expression and chromatin profiling of individual cell types within a tissue. *Dev. Cell* 18:1030–40

32. Deal RB, Henikoff S. 2011. The INTACT method for cell type–specific gene expression and chromatin profiling in *Arabidopsis thaliana*. *Nat. Protoc.* 6:56–68

33. Dixon JR, Selvaraj S, Yue F, Kim A, Li Y, et al. 2012. Topological domains in mammalian genomes identified by analysis of chromatin interactions. *Nature* 485:376–80

34. Djebali S, Davis CA, Merkel A, Dobin A, Lassmann T, et al. 2012. Landscape of transcription in human cells. *Nature* 489:101–8

35. Eaton ML, Prinz JA, MacAlpine HK, Tretyakov G, Kharchenko PV, MacAlpine DM. 2011. Chromatin signatures of the *Drosophila* replication program. *Genome Res.* 21:164–74

36. Eichten SR, Briskine R, Song J, Li Q, Swanson-Wagner R, et al. 2013. Epigenetic and genetic influences on DNA methylation variation in maize populations. *Plant Cell* 25:2783–97

37. Eichten SR, Swanson-Wagner RA, Schnable JC, Waters AJ, Hermanson PJ, et al. 2011. Heritable epigenetic variation among maize inbreds. *PLOS Genet.* 7:e1002372

38. ENCODE Proj. Consort. EP, Bernstein BE, Birney E, Dunham I, Green ED, et al. 2012. An integrated encyclopedia of DNA elements in the human genome. *Nature* 489:57–74

39. EPIC Plan. Consort. 2012. Reading the second code: mapping epigenomes to understand plant growth, development, and adaptation to the environment. *Plant Cell* 24:2257–61

40. Ercan S, Lubling Y, Segal E, Lieb JD. 2011. High nucleosome occupancy is encoded at X-linked gene promoters in *C. elegans*. *Genome Res.* 21:237–44

41. Ezkurdia I, del Pozo A, Frankish A, Rodriguez JM, Harrow J, et al. 2012. Comparative proteomics reveals a significant bias toward alternative protein isoforms with conserved structure and function. *Mol. Biol. Evol.* 29:2265–83

42. Fahlgren N, Howell MD, Kasschau KD, Chapman EJ, Sullivan CM, et al. 2007. High-throughput sequencing of *Arabidopsis* microRNAs: evidence for frequent birth and death of *MIRNA* genes. *PLOS ONE* 2:e219

43. Feng S, Cokus SJ, Zhang X, Chen PY, Bostick M, et al. 2010. Conservation and divergence of methylation patterning in plants and animals. *Proc. Natl. Acad. Sci. USA* 107:8689–94

44. Feng Z, Zhang B, Ding W, Liu X, Yang DL, et al. 2013. Efficient genome editing in plants using a CRISPR/Cas system. *Cell Res.* 23:1229–32

45. Filichkin SA, Priest HD, Givan SA, Shen R, Bryant DW, et al. 2010. Genome-wide mapping of alternative splicing in *Arabidopsis thaliana*. *Genome Res.* 20:45–58

46. Fullwood MJ, Liu MH, Pan YF, Liu J, Xu H, et al. 2009. An oestrogen-receptor-α-bound human chromatin interactome. *Nature* 462:58–64

47. Gan X, Stegle O, Behr J, Steffen JG, Drewe P, et al. 2011. Multiple reference genomes and transcriptomes for *Arabidopsis thaliana*. *Nature* 477:419–23

48. Gaudinier A, Zhang L, Reece-Hoyes JS, Taylor-Teeples M, Pu L, et al. 2011. Enhanced Y1H assays for *Arabidopsis*. *Nat. Methods* 8:1053–55

49. Gent JI, Ellis NA, Guo L, Harkess AE, Yao Y, et al. 2013. CHH islands: de novo DNA methylation in near-gene chromatin regulation in maize. *Genome Res.* 23:628–37

50. German MA, Pillay M, Jeong DH, Hetawal A, Luo S, et al. 2008. Global identification of microRNA-target RNA pairs by parallel analysis of RNA ends. *Nat. Biotechnol.* 26:941–46

51. Gerstein MB, Kundaje A, Hariharan M, Landt SG, Yan KK, et al. 2012. Architecture of the human regulatory network derived from ENCODE data. *Nature* 489:91–100

52. Gerstein MB, Lu ZJ, Van Nostrand EL, Cheng C, Arshinoff BI, et al. 2010. Integrative analysis of the *Caenorhabditis elegans* genome by the modENCODE project. *Science* 330:1775–87

53. Giresi PG, Kim J, McDaniell RM, Iyer VR, Lieb JD. 2007. FAIRE (formaldehyde-assisted isolation of regulatory elements) isolates active regulatory elements from human chromatin. *Genome Res.* 17:877–85

54. Giresi PG, Lieb JD. 2009. Isolation of active regulatory elements from eukaryotic chromatin using FAIRE (formaldehyde assisted isolation of regulatory elements). *Methods* 48:233–39

55. Goff SA, Vaughn M, McKay S, Lyons E, Stapleton AE, et al. 2011. The iPlant Collaborative: cyberinfrastructure for plant biology. *Front. Plant Sci.* 2:34

56. Gore MA, Chia JM, Elshire RJ, Sun Q, Ersoz ES, et al. 2009. A first-generation haplotype map of maize. *Science* 326:1115–17

57. Graveley BR, Brooks AN, Carlson JW, Duff MO, Landolin JM, et al. 2011. The developmental transcriptome of *Drosophila melanogaster*. *Nature* 471:473–79

58. Gregory BD, O'Malley RC, Lister R, Urich MA, Tonti-Filippini J, et al. 2008. A link between RNA metabolism and silencing affecting *Arabidopsis* development. *Dev. Cell* 14:854–66

59. Haudry A, Platts AE, Vello E, Hoen DR, Leclercq M, et al. 2013. An atlas of over 90,000 conserved noncoding sequences provides insight into crucifer regulatory regions. *Nat. Genet.* 45:891–98

60. Hoffman MM, Ernst J, Wilder SP, Kundaje A, Harris RS, et al. 2013. Integrative annotation of chromatin elements from ENCODE data. *Nucleic Acids Res.* 41:827–41

61. Hoskins RA, Landolin JM, Brown JB, Sandler JE, Takahashi H, et al. 2011. Genome-wide analysis of promoter architecture in *Drosophila melanogaster*. *Genome Res.* 21:182–92

62. Jacquemin J, Bhatia D, Singh K, Wing RA. 2013. The International Oryza Map Alignment Project: development of a genus-wide comparative genomics platform to help solve the 9 billion–people question. *Curr. Opin. Plant Biol.* 16:147–56

63. Johannes F, Porcher E, Teixeira FK, Saliba-Colombani V, Simon M, et al. 2009. Assessing the impact of transgenerational epigenetic variation on complex traits. *PLOS Genet.* 5:e1000530

64. Johnson DS, Mortazavi A, Myers RM, Wold B. 2007. Genome-wide mapping of in vivo protein-DNA interactions. *Science* 316:1497–502

65. Johnson LM, Du J, Hale CJ, Bischof S, Feng S, et al. 2014. SRA- and SET-domain-containing proteins link RNA polymerase V occupancy to DNA methylation. *Nature* 507:124–28

66. Kharchenko PV, Alekseyenko AA, Schwartz YB, Minoda A, Riddle NC, et al. 2011. Comprehensive analysis of the chromatin landscape in *Drosophila melanogaster*. *Nature* 471:480–85

67. Kircher M, Witten DM, Jain P, O'Roak BJ, Cooper GM, Shendure J. 2014. A general framework for estimating the relative pathogenicity of human genetic variants. *Nat. Genet.* 46:310–15

68. Konermann S, Brigham MD, Trevino AE, Hsu PD, Heidenreich M, et al. 2013. Optical control of mammalian endogenous transcription and epigenetic states. *Nature* 500:472–76

69. Kumar AP, Boualem A, Bhattacharya A, Parikh S, Desai N, et al. 2013. SMART: sunflower mutant population and reverse genetic tool for crop improvement. *BMC Plant Biol.* 13:38

70. Lam HM, Xu X, Liu X, Chen W, Yang G, et al. 2010. Resequencing of 31 wild and cultivated soybean genomes identifies patterns of genetic diversity and selection. *Nat. Genet.* 42:1053–59

71. Latzel V, Allan E, Bortolini Silveira A, Colot V, Fischer M, Bossdorf O. 2013. Epigenetic diversity increases the productivity and stability of plant populations. *Nat. Commun.* 4:2875

72. Lee BK, Iyer VR. 2012. Genome-wide studies of CCCTC-binding factor (CTCF) and cohesin provide insight into chromatin structure and regulation. *J. Biol. Chem.* 287:30906–13

73. Li H, Handsaker B, Wysoker A, Fennell T, Ruan J, et al. 2009. The sequence alignment/map format and SAM tools. *Bioinformatics* 25:2078–79

74. Li X, Wang X, He K, Ma Y, Su N, et al. 2008. High-resolution mapping of epigenetic modifications of the rice genome uncovers interplay between DNA methylation, histone methylation, and gene expression. *Plant Cell* 20:259–76

75. Lieberman-Aiden E, van Berkum NL, Williams L, Imakaev M, Ragoczy T, et al. 2009. Comprehensive mapping of long-range interactions reveals folding principles of the human genome. *Science* 326:289–93

76. Lister R, Mukamel EA, Nery JR, Urich M, Puddifoot CA, et al. 2013. Global epigenomic reconfiguration during mammalian brain development. *Science* 341:1237905

77. Lister R, O'Malley RC, Tonti-Filippini J, Gregory BD, Berry CC, et al. 2008. Highly integrated single-base resolution maps of the epigenome in *Arabidopsis*. *Cell* 133:523–36

78. Lister R, Pelizzola M, Dowen RH, Hawkins RD, Hon G, et al. 2009. Human DNA methylomes at base resolution show widespread epigenomic differences. *Nature* 462:315–22

79. Liu T, Rechtsteiner A, Egelhofer TA, Vielle A, Latorre I, et al. 2011. Broad chromosomal domains of histone modification patterns in *C. elegans*. *Genome Res.* 21:227–36

80. Long Q, Rabanal FA, Meng D, Huber CD, Farlow A, et al. 2013. Massive genomic variation and strong selection in *Arabidopsis thaliana* lines from Sweden. *Nat. Genet.* 45:884–90

81. Louwers M, Bader R, Haring M, van Driel R, de Laat W, Stam M. 2009. Tissue- and expression level–specific chromatin looping at maize *b1* epialleles. *Plant Cell* 21:832–42

82. Ma S, Shah S, Bohnert HJ, Snyder M, Dinesh-Kumar SP. 2013. Incorporating motif analysis into gene co-expression networks reveals novel modular expression pattern and new signaling pathways. *PLOS Genet.* 9:e1003840

83. Makarevitch I, Eichten SR, Briskine R, Waters AJ, Danilevskaya ON, et al. 2013. Genomic distribution of maize facultative heterochromatin marked by trimethylation of H3K27. *Plant Cell* 25:780–93

84. Maurano MT, Humbert R, Rynes E, Thurman RE, Haugen E, et al. 2012. Systematic localization of common disease-associated variation in regulatory DNA. *Science* 337:1190–95

85. modENCODE Consort., Roy S, Ernst J, Kharchenko PV, Kheradpour P, et al.. 2010. Identification of functional elements and regulatory circuits by *Drosophila* modENCODE. *Science* 330:1787–97

86. Moissiard G, Cokus SJ, Cary J, Feng S, Billi AC, et al. 2012. MORC family ATPases required for heterochromatin condensation and gene silencing. *Science* 336:1448–51

87. Mortazavi A, Williams B, Mccue K, Schaeffer L, Wold B. 2008. Mapping and quantifying mammalian transcriptomes by RNA-Seq. *Nat. Methods* 5:621–28

88. Nekrasov V, Staskawicz B, Weigel D, Jones JD, Kamoun S. 2013. Targeted mutagenesis in the model plant *Nicotiana benthamiana* using Cas9 RNA-guided endonuclease. *Nat. Biotechnol.* 31:691–93

89. Neph S, Vierstra J, Stergachis AB, Reynolds AP, Haugen E, et al. 2012. An expansive human regulatory lexicon encoded in transcription factor footprints. *Nature* 489:83–90

90. Nielsen CB, Younesy H, O'Geen H, Xu X, Jackson AR, et al. 2012. Spark: a navigational paradigm for genomic data exploration. *Genome Res.* 22:2262–69

91. Niu W, Lu ZJ, Zhong M, Sarov M, Murray JI, et al. 2011. Diverse transcription factor binding features revealed by genome-wide ChIP-seq in *C. elegans*. *Genome Res.* 21:245–54

92. Nordman J, Li S, Eng T, Macalpine D, Orr-Weaver TL. 2011. Developmental control of the DNA replication and transcription programs. *Genome Res.* 21:175–81

93. Okabe Y, Asamizu E, Saito T, Matsukura C, Ariizumi T, et al. 2011. Tomato TILLING technology: development of a reverse genetics tool for the efficient isolation of mutants from Micro-Tom mutant libraries. *Plant Cell Physiol.* 52:1994–2005

94. Olson A, Klein RR, Dugas DV, Lu Z, Regulski M, et al. 2014. Expanding and vetting sorghum bicolor gene annotations through transcriptome and methylome sequencing. *Plant Genome* doi: 10.3835/plantgenome2013.08.0025

95. Ossowski S, Schneeberger K, Clark R, Lanz C, Warthmann N, Weigel D. 2008. Sequencing of natural strains of *Arabidopsis thaliana* with short reads. *Genome Res.* 18:2024–33

96. Patterson GI, Harris LJ, Walbot V, Chandler VL. 1991. Genetic analysis of *B-Peru*, a regulatory gene in maize. *Genetics* 127:205–20

97. Patterson GI, Kubo KM, Shroyer T, Chandler VL. 1995. Sequences required for paramutation of the maize *b* gene map to a region containing the promoter and upstream sequences. *Genetics* 140:1389–406

98. Patterson GI, Thorpe CJ, Chandler VL. 1993. Paramutation, an allelic interaction, is associated with a stable and heritable reduction of transcription of the maize *b* regulatory gene. *Genetics* 135:881–94

99. Patwardhan RP, Hiatt JB, Witten DM, Kim MJ, Smith RP, et al. 2012. Massively parallel functional dissection of mammalian enhancers in vivo. *Nat. Biotechnol.* 30:265–70

100. Rawat N, Sehgal SK, Joshi A, Rothe N, Wilson DL, et al. 2012. A diploid wheat TILLING resource for wheat functional genomics. *BMC Plant Biol.* 12:205

101. Regulski M, Lu Z, Kendall J, Donoghue MT, Reinders J, et al. 2013. The maize methylome influences mRNA splice sites and reveals widespread paramutation-like switches guided by small RNA. *Genome Res.* 231651–62

102. Reinders J, Wulff BB, Mirouze M, Mari-Ordonez A, Dapp M, et al. 2009. Compromised stability of DNA methylation and transposon immobilization in mosaic *Arabidopsis* epigenomes. *Genes Dev.* 23:939–50

103. Riddle NC, Minoda A, Kharchenko PV, Alekseyenko AA, Schwartz YB, et al. 2011. Plasticity in patterns of histone modifications and chromosomal proteins in *Drosophila* heterochromatin. *Genome Res.* 21:147–63

104. Rogers C, Wen J, Chen R, Oldroyd G. 2009. Deletion-based reverse genetics in *Medicago truncatula*. *Plant Physiol.* 151:1077–86

105. Roudier F, Ahmed I, Berard C, Sarazin A, Mary-Huard T, et al. 2011. Integrative epigenomic mapping defines four main chromatin states in *Arabidopsis*. *EMBO J.* 30:1928–38

106. Roux F, Colome-Tatche M, Edelist C, Wardenaar R, Guerche P, et al. 2011. Genome-wide epigenetic perturbation jump-starts patterns of heritable variation found in nature. *Genetics* 188:1015–17

107. Sallaud C, Gay C, Larmande P, Bes M, Piffanelli P, et al. 2004. High throughput T-DNA insertion mutagenesis in rice: a first step towards in silico reverse genetics. *Plant J.* 39:450–64

108. Salvi S, Sponza G, Morgante M, Tomes D, Niu X, et al. 2007. Conserved noncoding genomic sequences associated with a flowering-time quantitative trait locus in maize. *Proc. Natl. Acad. Sci. USA* 104:11376–81

109. Sanyal A, Lajoie BR, Jain G, Dekker J. 2012. The long-range interaction landscape of gene promoters. *Nature* 489:109–13

110. Schmitz RJ, He Y, Valdes-Lopez O, Khan SM, Joshi T, et al. 2013. Epigenome-wide inheritance of cytosine methylation variants in a recombinant inbred population. *Genome Res.* 23:1663–74

111. Schmitz RJ, Schultz MD, Lewsey MG, O'Malley RC, Urich MA, et al. 2011. Transgenerational epigenetic instability is a source of novel methylation variants. *Science* 334:369–73

112. Schmitz RJ, Schultz MD, Urich MA, Nery JR, Pelizzola M, et al. 2013. Patterns of population epigenomic diversity. *Nature* 495:193–98

113. Schmitz RJ, Zhang X. 2011. High-throughput approaches for plant epigenomic studies. *Curr. Opin. Plant Biol.* 14:130–36

114. Schones DE, Cui K, Cuddapah S, Roh TY, Barski A, et al. 2008. Dynamic regulation of nucleosome positioning in the human genome. *Cell* 132:887–98

115. Schultz MD, Schmitz RJ, Ecker JR. 2012. "Leveling" the playing field for analyses of single-base resolution DNA methylomes. *Trends Genet.* 28:583–85

116. Shan Q, Wang Y, Li J, Zhang Y, Chen K, et al. 2013. Targeted genome modification of crop plants using a CRISPR-Cas system. *Nat. Biotechnol.* 31:686–88

117. Shen X, Forsberg S, Petterson M, Sheng Z, Carlborg O. 2013. Natural CMT2 variation is associated with genome-wide methylation changes and temperature adaptation. *arXiv* arXiv:1310.4522 [q-bio.PE]

118. Shen Y, Yue F, McCleary DF, Ye Z, Edsall L, et al. 2012. A map of the *cis*-regulatory sequences in the mouse genome. *Nature* 488:116–20

119. Song QX, Lu X, Li QT, Chen H, Hu XY, et al. 2013. Genome-wide analysis of DNA methylation in soybean. *Mol. Plant* 6:1961–74

120. Spencer WC, Zeller G, Watson JD, Henz SR, Watkins KL, et al. 2011. A spatial and temporal map of *C. elegans* gene expression. *Genome Res.* 21:325–41

121. Stam M, Belele C, Dorweiler JE, Chandler VL. 2002. Differential chromatin structure within a tandem array 100 kb upstream of the maize *b1* locus is associated with paramutation. *Genes Dev.* 16:1906–18

122. Stam M, Belele C, Ramakrishna W, Dorweiler JE, Bennetzen JL, Chandler VL. 2002. The regulatory regions required for B′ paramutation and expression are located far upstream of the maize *b1* transcribed sequences. *Genetics* 162:917–30

123. Stephenson P, Baker D, Girin T, Perez A, Amoah S, et al. 2010. A rich TILLING resource for studying gene function in *Brassica rapa*. *BMC Plant Biol.* 10:62

124. Stroud H, Otero S, Desvoyes B, Ramirez-Parra E, Jacobsen SE, Gutierrez C. 2012. Genome-wide analysis of histone H3.1 and H3.3 variants in *Arabidopsis thaliana*. *Proc. Natl. Acad. Sci. USA* 109:5370–75

125. Studer A, Zhao Q, Ross-Ibarra J, Doebley J. 2011. Identification of a functional transposon insertion in the maize domestication gene *tb1*. *Nat. Genet.* 43:1160–63

126. Takuno S, Gaut BS. 2012. Body-methylated genes in *Arabidopsis thaliana* are functionally important and evolve slowly. *Mol. Biol. Evol.* 29:219–27

127. Takuno S, Gaut BS. 2013. Gene body methylation is conserved between plant orthologs and is of evolutionary consequence. *Proc. Natl. Acad. Sci. USA* 110:1797–802

128. Tang F, Barbacioru C, Wang Y, Nordman E, Lee C, et al. 2009. mRNA-Seq whole-transcriptome analysis of a single cell. *Nat. Methods* 6:377–82

129. Thomas BC, Rapaka L, Lyons E, Pedersen B, Freeling M. 2007. *Arabidopsis* intragenomic conserved noncoding sequence. *Proc. Natl. Acad. Sci. USA* 104:3348–53

130. Thurman RE, Rynes E, Humbert R, Vierstra J, Maurano MT, et al. 2012. The accessible chromatin landscape of the human genome. *Nature* 489:75–82

131. Uauy C, Paraiso F, Colasuonno P, Tran RK, Tsai H, et al. 2009. A modified TILLING approach to detect induced mutations in tetraploid and hexaploid wheat. *BMC Plant Biol.* 9:115

132. Vaughn MW, Tanurdzić M, Lippman Z, Jiang H, Carrasquillo R, et al. 2007. Epigenetic natural variation in *Arabidopsis thaliana*. *PLOS Biol.* 5:e174

133. Wang X, Elling AA, Li X, Li N, Peng Z, et al. 2009. Genome-wide and organ-specific landscapes of epigenetic modifications and their relationships to mRNA and small RNA transcriptomes in maize. *Plant Cell* 21:1053–69

134. Weigel D, Nordborg M. 2005. Natural variation in *Arabidopsis*. How do we find the causal genes? *Plant Physiol.* 138:567–68

135. Zemach A, Kim MY, Silva P, Rodrigues JA, Dotson B, et al. 2010. Local DNA hypomethylation activates genes in rice endosperm. *Proc. Natl. Acad. Sci. USA* 107:18729–34

136. Zemach A, McDaniel IE, Silva P, Zilberman D. 2010. Genome-wide evolutionary analysis of eukaryotic DNA methylation. *Science* 328:916–19

137. Zentner GE, Scacheri PC. 2012. The chromatin fingerprint of gene enhancer elements. *J. Biol. Chem.* 287:30888–96

138. Zhang W, Wu Y, Schnable JC, Zeng Z, Freeling M, et al. 2012. High-resolution mapping of open chromatin in the rice genome. *Genome Res.* 22:151–62

139. Zhang W, Zhang T, Wu Y, Jiang J. 2012. Genome-wide identification of regulatory DNA elements and protein-binding footprints using signatures of open chromatin in *Arabidopsis*. *Plant Cell* 24:2719–31

140. Zhang X, Bernatavichute YV, Cokus S, Pellegrini M, Jacobsen SE. 2009. Genome-wide analysis of mono-, di- and trimethylation of histone H3 lysine 4 in *Arabidopsis thaliana*. *Genome Biol.* 10:R62

141. Zhang X, Clarenz O, Cokus S, Bernatavichute YV, Pellegrini M, et al. 2007. Whole-genome analysis of histone H3 lysine 27 trimethylation in *Arabidopsis*. *PLOS Biol.* 5:e129

142. Zhong S, Fei Z, Chen YR, Zheng Y, Huang M, et al. 2013. Single-base resolution methylomes of tomato fruit development reveal epigenome modifications associated with ripening. *Nat. Biotechnol.* 31:154–59

143. Zilberman D, Coleman-Derr D, Ballinger T, Henikoff S. 2008. Histone H2A.Z and DNA methylation are mutually antagonistic chromatin marks. *Nature* 456:125–29

Archaeal DNA Replication

Lori M. Kelman[1] and Zvi Kelman[2]

[1]Program in Biotechnology, Montgomery College, Germantown, Maryland 20876;
email: lori.kelman@montgomerycollege.edu

[2]National Institute of Standards and Technology and Institute for Bioscience and Biotechnology
Research, Rockville, Maryland 20850; email: zkelman@umd.edu

Annu. Rev. Genet. 2014. 48:71–97

The *Annual Review of Genetics* is online at
genet.annualreviews.org

This article's doi:
10.1146/annurev-genet-120213-092148

Keywords

Archaea, initiation of DNA replication, protein-DNA interactions,
protein-protein interactions, regulation of DNA replication, protein
structure

Abstract

DNA replication is essential for all life forms. Although the process is fundamentally conserved in the three domains of life, bioinformatic, biochemical,
structural, and genetic studies have demonstrated that the process and the
proteins involved in archaeal DNA replication are more similar to those in
eukaryal DNA replication than in bacterial DNA replication, but have some
archaeal-specific features. The archaeal replication system, however, is not
monolithic, and there are some differences in the replication process between
different species. In this review, the current knowledge of the mechanisms
governing DNA replication in Archaea is summarized. The general features
of the replication process as well as some of the differences are discussed.

INTRODUCTION

In 1977, Carl Woese and colleagues recognized that the rRNA sequences and other physiological features of a group of prokaryotes were different enough from those in Bacteria to classify them as a separate domain: the Archaebacteria (now Archaea) (177). The archaeal domain is currently divided into six different phyla (or kingdoms): Aigarchaeota, Crenarchaeota, Euryarchaeota, Korarchaeota, Nanoarchaeota, and Thaumarchaeota (50).

Early bioinformatic studies using complete archaeal genome sequences suggested that although Archaea contain circular chromosomes, as do Bacteria, the proteins and complexes that participate in DNA replication are more closely related to those of Eukarya than to those of Bacteria (**Table 1**). Since then, phylogenetic, biochemical, structural, and genetic studies have demonstrated the relationship between the archaeal and eukaryal DNA replication systems (47, 50, 60, 61, 71, 127, 143).

Here, the current state of knowledge on the mechanism of DNA replication in Archaea, the similarities and differences between species, and the properties of individual proteins and complexes are discussed. Because of space limitations, only representative references are provided, and when appropriate the reader is referred to other reviews.

DNA replication takes place during the S phase of the cell cycle and is tightly coordinated with other cell-cycle events and cell division. This review concentrates on replication, and the reader is referred to several reviews for information on other aspects of the archaeal cell cycle (103, 113, 153). Topoisomerases play a major role in DNA replication by untwisting the DNA in front of and behind the moving replication fork. Topoisomerases are not covered here but have been reviewed (21, 42).

OVERVIEW OF DNA REPLICATION

Chromosomal DNA replication is an essential process that ensures the accurate and timely duplication of the genetic information. It is a complex and highly regulated process that is functionally and structurally conserved in all life forms (127). In all organisms, the process, which occurs during S phase, begins at a specific region on the chromosome called the origin of replication. Origin binding proteins (OBPs) bind to the origin and locally unwind it (35) and, together with the helicase loader, assemble the two replicative helicases around the DNA to form the prereplication complex (pre-RC) (**Figure 1**). Additional proteins bind to the pre-RC to form the preinitiation complex (pre-IC), in which the DNA duplex is unwound and single-stranded (ss) DNA binding protein (SSB) coats the ssDNA. DNA primase, DNA polymerase, and the rest of the replication machinery are recruited to the SSB-ssDNA nucleofilament at the origin. Upon activation, the two helicases begin to unwind the duplex in opposite directions using energy derived from nucleoside triphosphate hydrolysis. The moving helicases form the two replication forks to initiate bidirectional DNA synthesis. The ssDNA exposed behind the helicase is coated with SSB. Owing to the antiparallel nature of DNA and the unidirectionality of DNA polymerase, the leading strand is synthesized continuously, whereas the lagging strand is copied discontinuously as a series of Okazaki fragments (**Figure 2**, **Figure 3a**). A number of enzymes, including a nuclease, a DNA polymerase, and DNA ligase, process the Okazaki fragments to form the mature duplex DNA (**Figure 3b**). At the end of the replication process, the two replication forks (on a circular chromosome, or adjacent forks on a linear chromosome) collide and DNA synthesis terminates.

ORIGIN OF REPLICATION

In all organisms, DNA replication starts at specific sites on the chromosome known as origins of replication. All origins contain similar characteristics, including one or more A/T-rich regions

Table 1 A summary of DNA replication features in the three domains of life[a]

Attribute	Bacteria	Eukarya	Archaea[b]	
			Euryarchaea	Crenarchaea
Chromosome	Circular	Linear	Circular	Circular
Replication origin	Single	Multiple	Single or Multiple	Single or Multiple
Prereplication complex (pre-RC)				
Origin recognition	DnaA (1)	ORC (6)	Cdc6[c] (≥1)	Cdc6 (≥1) WhiP[d] (1)
Helicase[e]	DnaB (1)	MCM (6)	MCM (≥1)	MCM (1)
Helicase loader[e]	DnaC (1)	Cdc6 (1) Cdt1 (1)	Cdc6 (≥1)	Cdc6 (≥1)
Preinitiation complex (pre-IC)				
Cdc45	-	Cdc45 (1)	Cdc45 (1)	Cdc45 (1)
GINS	-	GINS (4)	GINS (1–2)	GINS (1–2)
Single-stranded DNA binding protein (SSB)	SSB (1)	RPA (3)	RPA (1–3)	SSB (1)[f]
Elongation complex				
Primase	DnaG (1)	Polα/primase (4)	Primase (2)	Primase (2)
Sliding clamp	β-subunit (1)	PCNA (1)	PCNA (1)	PCNA (3)
Clamp loader	τ-complex (5)	RFC (5)	RFC (2)	RFC (2)
DNA polymerase	PolC (3)	PolB[g] (1)	PolB/PolD (1[h]/2)	PolB (2–3)
Okazaki fragment maturation				
Primer removal	PolI (1)	Fen1 (1)	Fen1 (1)	Fen1 (1)
Gap filling	PolI (1)	PolB[i] (1)	PolB/PolD (1[h]/2)	PolB (2–3)
DNA ligase	NAD⁺-dependent (1)	ATP-dependent (1)	ATP-dependent[j] (1)	ATP-dependent (1)

[a]Bacteria-like features and proteins are in red, eukaryal-like features and proteins are in blue, and archaeal-specific proteins are in green. The range of homologs identified in different species is shown in parentheses.

[b]Two representative kingdoms are shown.

[c]The genomes of species belonging to *Methanococcales* and *Methanopyrales* do not contain genes encoding for Cdc6 homologs.

[d]An archaeal homolog of the eukaryotic Cdt1 protein, WhiP, was identified in some species.

[e]In bacteria, the helicase and helicase loader are not considered to be part of the pre-RC but rather the pre-IC. Because this paper is about archaea, these proteins are included under pre-RC.

[f]The crenarchaeal SSB shares some features with the bacterial SSB and other features with the eukaryotic replication protein A (RPA).

[g]All three replicative DNA polymerases in eukarya (Polα, Polδ, and Polε) belong to family B.

[h]In some archaeal species, PolB is not essential for cell viability.

[i]Polδ, the lagging-strand DNA polymerase, participates in Okazaki fragment maturation.

[j]Some archaeal DNA ligases use NAD⁺ as a cofactor.

Abbreviations: MCM, minichromosome maintenance; ORC, origin recognition complex; PCNA, proliferating cell nuclear antigen; RFC, replication factor C.

referred to as duplex unwinding elements (DUE), and specific sequences that facilitate the binding of OBPs. Upon binding to the origin, the proteins melt the duplex DNA to form the initial replication bubble (**Figure 1**) (15).

Origins of replication in Archaea were first identified using in silico skew analysis (74, 105). This method made use of an observation in Bacteria that strand-specific biases in nucleotide, oligomer, and codon frequencies could be identified along the chromosome, with an abrupt change in the bias at the origin of replication and termination regions.

The origin of *Pyrococcus abyssi* was the first to be mapped in vivo using pulsed-field gel electrophoresis and two-dimensional gel analysis (118, 126). As predicted by skew analysis (105),

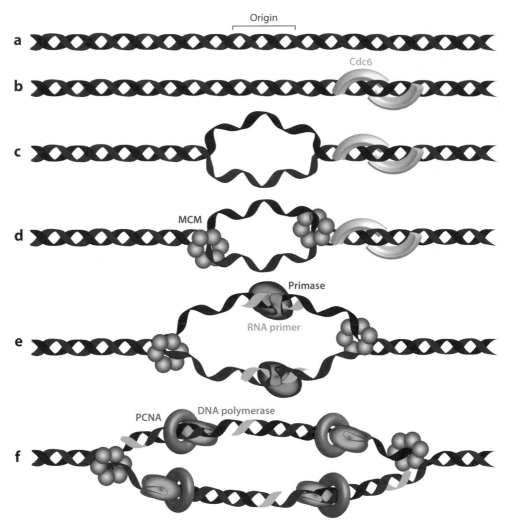

Figure 1

Initiation of DNA replication. The (*a*) origin of replication is (*b*) recognized by Cdc6 (*yellow*). (*c*) The binding of Cdc6 to the origin region forms the initial replication bubble. (*d*) Cdc6 participates in the assembly of the two minichromosome maintenance (MCM) helicases (*green*) around the DNA to form the two replication forks. (*e*) The helicases move in opposite directions, enlarging the bubble. Primase (*pink*) is recruited to the DNA and synthesizes the RNA primers (*light green*). (*f*) The rest of the replication machinery assembles and initiates bidirectional DNA synthesis. For simplicity, the single-stranded DNA binding protein is omitted and only the DNA polymerase (*orange*) and proliferating cell nuclear antigen (PCNA) (*blue*) are shown.

P. abyssi contains a single origin of replication (referred to as oriC) located in an intergenic region of the chromosome. Bidirectional DNA synthesis initiates from the origin and terminates in a region of the chromosome opposite the origin (119, 126).

The presence of a single origin in a circular archaeal genome was not surprising, because bacterial circular chromosomes contain a single origin. However, skew analysis failed to identify a clear origin of replication in some archaeal species, and the use of more refined in silico algorithms suggested the presence of multiple origins in certain species (184). The presence of multiple

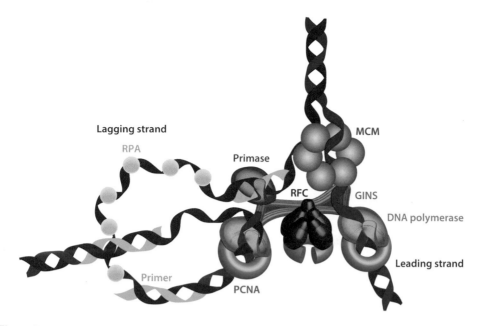

Figure 2

A schematic representation of the archaeal replisome. MCM (minichromosome maintenance) is green, DNA polymerase is orange, primase is pink, RFC (replication factor C) is purple, GINS is gray, RPA (replication protein A) is yellow, and PCNA (proliferating cell nuclear antigen) is blue. Primers on the lagging strand are shown in light green.

origins could confound skew analysis and explain the failure to identify an origin. Studies using two-dimensional gel analysis, marker frequency analysis, and replication initiation point (RIP) mapping of two *Sulfolobus* species revealed that these organisms contain three origins of replication (72, 106, 149). Subsequent studies showed that other archaeal species also contain multiple origins.

Archaeal origins have characteristics similar to origins of replication from other organisms (15). They are located in intergenic regions, are rich in A and T nucleotides, contain one or more DUEs, and contain binding sites for OBPs [referred to as origin recognition boxes (ORBs)]. Many origins are located in regions of the chromosome that encodes DNA replication proteins. Many archaeal origins are located upstream from genes that encode the archaeal homolog of the eukaryotic Cdc6 protein. The Cdc6 proteins are the archaeal OBPs and were shown to bind to ORBs. The proximity of the origin of replication to the OBPs is also common in Bacteria (89). Several hypotheses have been proposed to explain the proximity of the gene that encodes the OBP to the origin, such as the ability of the proteins to associate with the origin as soon as they are synthesized or to reduce the probability of loss of the gene that encodes the OBP due to genomic rearrangement (71, 103).

In several archaeal species with multiple origins of replication, the origin is located in the vicinity of the homolog of another eukaryotic replication protein, Cdt1. This protein binds to the origin of replication in its vicinity (148, 154). Several origins are not located upstream of a known initiator protein (138), and the proteins that bind these origins have not yet been identified.

The origins of replication from most Archaea studied contain clear ORBs. ORBs are inverted repeats located on both sides of the DUE region(s) and were shown to be the binding site for the Cdc6 proteins (36, 45, 174). The ORBs from different species share sequence similarity with

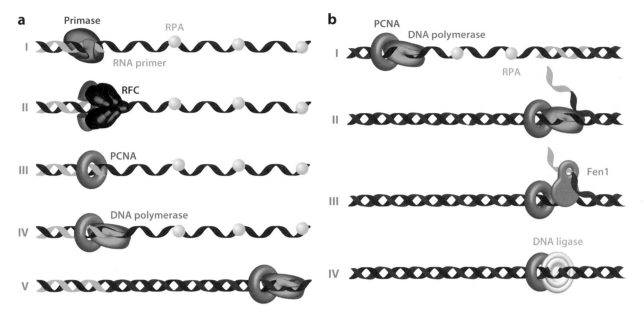

Figure 3

Okazaki fragment synthesis and maturation. (*a*) Okazaki fragment synthesis. I. Primase (*pink*) synthesizes an RNA primer (*light green*) on RPA (replication protein A) (*yellow*)-coated single-stranded DNA. II. This primer is recognized by RFC (replication factor C) (*purple*). III. RFC assembles PCNA (proliferating cell nuclear antigen) (*blue*) around the primer. IV. The DNA polymerase (*orange*) associates with PCNA. V. This association initiates processive DNA synthesis. (*b*) Okazaki fragment maturation. I. The lagging-strand DNA polymerase (*orange*) elongates the Okazaki fragment with the assistance of PCNA (*blue*). II. When the DNA polymerase reaches the previous Okazaki fragment, its strand displacement activity forms a flap structure. III. The flap is removed by the activity of Fen1 (flap endonuclease 1) (*teal*) and PCNA. IV. The nick remaining on the DNA is sealed by DNA ligase (*yellow*) and PCNA.

a consensus sequence referred to as the mini-ORB, followed, in some origins, by a string of G nucleotides (a G-string) to form the intact ORB (10, 27). It was shown that the mini-ORB is sufficient to bind Cdc6 proteins and that Cdc6 from one organism can bind an ORB from another species (149). Some origins do not have a clear ORB or mini-ORB consensus sequence, although they contain other inverted repeats (138). It is not clear whether Cdc6, Cdt1, or other species-specific initiator proteins can bind those repeats.

Interestingly, in some species the origins of replication could be deleted with no loss of viability (53). However, in the absence of a canonical origin, the cellular recombination machinery appears to be required for cell viability. The mechanism of initiation in this case is not clear. It is also not known how widespread this phenomenon is among the Archaea.

PREREPLICATION COMPLEX

The pre-RC is the protein complex that assembles at the origin of replication during the G1 phase of the cell cycle and is responsible for the regulation of the initiation process. In Eukarya, the pre-RC is composed of four main components: the six-subunit origin recognition complex (ORC), Cdc6 and Cdt1 proteins, and the six-subunit minichromosome maintenance (MCM) helicase. In Archaea, pre-RC formation includes the assembly of the Cdc6 protein and the MCM helicase. In some Archaea, a Cdt1 homolog (WhiP) or other, not yet identified, proteins may also be part of the pre-RC.

Cdc6

The OBP in Bacteria is the DnaA protein, which binds the DnaA box (the bacterial ORB) at the origin of replication. In Eukarya, the OBP is the hexameric ORC complex composed of the Orc1–6 proteins. ORC associates with the origin and recruits additional components of the pre-RC, such as the MCM helicase and the Cdc6 protein, to participate in MCM assembly at the origin. The eukaryotic Orc1–5 proteins, Cdc6, and the bacterial DnaA protein all belong to the initiator clade of the AAA+ family of ATPases (27), and Orc1–5 and Cdc6 were proposed to be paralogs (110).

In nearly all Archaea, at least one clear homolog of the eukaryotic Orc or Cdc6 proteins has been identified, and most contain two or three genes encoding for the protein (143). Owing to similarity to both the eukaryotic Orc and Cdc6 proteins, the archaeal enzyme has been called Cdc6, Orc, Cdc6/Orc, and Cdc6/Orc1. In this review, the protein is referred to as Cdc6. Although the Cdc6 protein is essential for cell viability in most species studied (154), it appears to be dispensable in others (T. Santangelo & Z. Kelman, unpublished results).

Because of their similarity to both the ORC complex and the Cdc6 protein, the archaeal enzymes were suggested to be involved in both origin recognition and helicase loading (**Table 1**; **Figure 1**). The role of Cdc6 protein in origin recognition has been established. The first suggestion of the association of Cdc6 with the origin of replication came from chromatin immunoprecipitation (ChIP) analysis with *P. abyssi* (118). Subsequent in vivo and in vitro studies illustrated that although the proteins can bind both ssDNA and double-stranded (ds) DNA, they show clear preference for the ORB dsDNA (10, 27).

Although the archaeal Cdc6 proteins bind origins, the mechanism by which they initiate DNA replication is not clear. The bacterial DnaA protein unwinds origin duplex DNA to facilitate the assembly of the replicative helicase DnaB on the exposed ssDNA (27). In contrast, binding of the eukaryotic ORC to the origin does not induce unwinding. Furthermore, there is accumulating evidence to suggest that the eukaryotic MCM assembles at the origin around dsDNA and not around ssDNA, as suggested for Bacteria (27). The similarities between the archaeal Cdc6 and the eukaryotic ORC and Cdc6 proteins and the lack of clear data on the ability of the archaeal Cdc6 protein to melt the origin may suggest that, as in Eukarya, during pre-RC formation the MCM helicase is assembled on dsDNA and origin melting occurs during pre-IC formation.

Eukaryotic Cdc6 plays an essential role in the assembly of the MCM helicase at the origin, and thus it was suggested that the archaeal Cdc6 plays a similar role. Direct interaction between Cdc6 and MCM has been reported in several archaeal species, and the interaction was capable of regulating helicase activity (152). Studies showed that the interaction between Cdc6 and MCM inhibited helicase activity (158), which is likely due to the dissociation of the MCM hexamers in the presence of Cdc6 (79, 159). The inhibitory effect of Cdc6 on helicase activity is reminiscent of the observation made in bacteria, where binding of the helicase loader DnaC to the helicase DnaB inhibits helicase activity (172). Thus, this observation may support the hypothesis that the archaeal Cdc6 functions as a helicase loader. *Thermoplasma acidophilum* Cdc6, however, stimulates the activity of the MCM helicase (51, 52), which may suggest a different mechanism of helicase loading in that species. Nevertheless, the mechanism of helicase assembly at the archaeal origin and the role of Cdc6 protein in the process are not known, although several mechanisms have been proposed (151).

Cdc6, DnaA, and Orc1-5 belong to the AAA+ superfamily of ATPases (27, 34). The structural organization of the Cdc6 protein is similar to that of the bacterial DnaA proteins in which the N-terminal AAA+ catalytic unit is followed by a DNA binding domain. However, although in DnaA the DNA binding domain is a helix-turn-helix (HTH), the Cdc6 protein contains a winged-helix domain (WHD) (27). Biochemical studies demonstrated that an intact WHD is required for

ORB binding (10, 27, 35). The WHD also participates in the interaction between the Cdc6 and MCM proteins (30, 66).

In addition to all known motifs present in other members of the AAA+ family of ATPases, an α-helix inserted into the canonical AAA+ fold was also revealed by the structure of Cdc6 and DnaA. This helical insertion is found in all proteins belonging to the initiator clade of the AAA+ family and is referred to as the initiator specific motif (ISM) (27).

The three-dimensional structure of Cdc6 bound to the ORB has been determined (36, 45). In addition to extensive contact between the WHD and the DNA, the ISM also contacts the DNA. However, although ORB sequences are conserved and in vivo and in vitro studies showed the specificity of Cdc6 binding to the ORBs, only a few sequence-specific contacts were observed in the co-crystals (36, 45). The mechanism by which binding specificity is achieved is not yet known and may involve the helicase or other proteins (10, 27, 174).

Cdc6 is expressed during the G1 phase and degraded after initiation (107, 149). However, the regulation of expression and of subsequent degradation is not understood. In Eukarya, phosphorylation of Cdc6 following the initiation process marks the enzyme for degradation. A similar mechanism may regulate Cdc6 in Archaea. A number of archaeal protein kinases are cell-cycle regulated (107), and in vivo phosphorylation of Cdc6 protein has been reported (144).

Alternatively, the archaeal Cdc6 proteins can autophosphorylate in vitro using the γ-phosphate of ATP (46). Studies suggest that autophosphorylation may be regulated by MCM binding to Cdc6 and by the association of Cdc6 with DNA (46, 66). Although the role of autophosphorylation in Cdc6 function is not known, it may play a regulatory role during pre-RC formation (47). It will be important to determine whether Cdc6 can autophosphorylate in vivo as a first step to establish whether the process has a physiological role.

In bacteria, the activity of DnaA is regulated by a mechanism referred to as regulatory inactivation of DnaA (RIDA), which ensures that the chromosome replicates only once per cell cycle (98). RIDA is dependent upon interaction between the bacterial processivity factor of DNA polymerase III (PolIII) holoenzyme (the β-subunit), DnaA, and a homologous-to-DnaA (Hda) protein. Interaction between Hda and DnaA stimulates ATP hydrolysis by DnaA, resulting in an ADP-bound DnaA protein that cannot bind to the origin of replication. However, Hda must associate with the β-subunit on DNA to stimulate ATP hydrolysis by DnaA. Proliferating cell nuclear antigen (PCNA) is the archaeal functional homolog of the β-subunit (80). In several species, the Cdc6 protein contains a PCNA-interacting protein (PIP) motif required for interactions with PCNA, and PCNA and Cdc6 can form a complex in vivo (102) and in vitro (3). It was therefore proposed that regulation of Cdc6 function in Archaea might resemble the bacterial RIDA process (100). However, no homolog of Hda has been identified in Archaea. It is possible that another unidentified protein is the archaeal functional homolog of Hda, or that an Hda-like protein is not needed in Archaea as Cdc6 can interact directly with PCNA.

Other mechanisms may also regulate Cdc6 function. Archaeal Cdc6 proteins crystallized as monomers with ADP tightly bound at the active site, and denaturation of Cdc6 was required for removal of the bound ADP. As it is thought that the Cdc6 protein is active in the ATP-bound form, several hypotheses for the role of the tight ADP binding for regulating the enzyme have been proposed (10).

Winged-Helix Initiator Protein

Cdt1 is a eukaryotic protein that participates in the initiation of DNA replication as a part of the pre-RC (27). Structurally, the middle and the C-terminal portions of the eukaryotic Cdt1 protein have a WHD (97). When the three origins of replication in *Sulfolobus solfataricus* were identified, it was noted that only two origins are located upstream to a gene encoding Cdc6 (106). However, the protein encoded by the gene upstream of the third *S. solfataricus* origin also contains two WHD, although at the N and C termini of the molecule. This suggested that the protein might be the archaeal homolog of Cdt1 (148). Because amino acid similarity between the archaeal and eukaryal enzymes is low, the archaeal protein is called WhiP (winged-helix initiator protein). It was found that the WhiP protein could bind to the origin of replication near its gene (148). To date, WhiP homologs have been identified only in a small subset of species belonging to the crenarchaeota branch. However, as many archaeal genomes contain proteins with a single WHD that is similar to that found in WhiP, these proteins may play a role in the initiation process in those species.

Minichromosome Maintenance Helicase

The MCM proteins are the replicative helicases in Archaea responsible for the separation of the dsDNA in front of the replication fork. The archaeal MCM proteins have been extensively studied using structural, biochemical, and genetic tools (11, 16, 28, 93, 123, 152, 164).

In Eukarya, MCM is a complex of six different but related proteins (Mcm2–7). Most archaeal species contain a single gene that encodes MCM. To date, in archaeal species with multiple MCM genes, only one is essential for viability (59, 133). The single MCM homolog from the archaeon *Methanothermobacter thermautotrophicus* was the first to be biochemically characterized (23, 79, 156), but others have since been studied. As with other helicases, the MCM utilizes energy from ATP hydrolysis to translocate along one strand of the DNA and displace the other. The enzyme binds to and translocates along ssDNA and dsDNA in the 3′ to 5′ direction, and in vitro the enzyme is processive on its own and able to displace several hundred bases (16, 28, 93, 152, 164). However, on dsDNA the enzyme probably interacts with one strand, resulting in the observed 3′ to 5′ directionality. Biochemical and biophysical studies with mutated enzymes suggested that ssDNA and dsDNA are located in the central hole of the hexameric ring. During DNA unwinding, the helicase may encounter proteins on DNA (such as histones) or short RNA molecules, forming R-loops. Archaeal MCM can displace proteins from DNA and unwind DNA-RNA hybrids (161, 162).

Early studies suggested that archaeal MCM proteins form dodecameric structures in solution as judged by size-exclusion chromatography and sedimentation methods (23, 79, 156). Further support for dodecameric structures came from the three-dimensional structure of the N-terminal portion of the enzyme (40, 104) and the full-length protein (91). Electron microscope reconstruction studies, however, indicated that the enzyme could form other structures, including hexamers, heptamers, octamers, filaments, and open rings (28, 152, 164).

Although the ability of MCM to form both hexamer and head-to-head dodecameric structures was first reported with an archaeal enzyme, it was later shown for the eukaryotic MCM (145) and the bacterial DnaB (166). It is thought that the replicative helicases are assembled as dodecameric rings at the origin. It is not clear whether the two hexamers associate during elongation or are separated during DNA synthesis and move away from each other (151, 168). However, in vitro studies indicate that the archaeal enzymes are active as single hexamers (160), suggesting that upon loading the two rings work independently.

Structurally, archaeal MCM proteins can be divided into three main parts: an N-terminal part, a catalytic region, and a C-terminal HTH domain (40, 152, 164). The N-terminal part of MCM forms hexamers and dodecamers on its own (40, 67, 92). This part of the molecule contains a C_4-type zinc finger and a β-hairpin motif shown to be involved in MCM interaction with DNA (78, 141). The structure also revealed the presence of a long loop shown to play a role in communication between the N-terminal DNA binding region and the catalytic part of the molecule (6, 150). The central part of MCM contains the catalytic domains and all the conserved motifs found in other

AAA+ family members (28, 152, 164). The structure also revealed the presence of three β-hairpins in the catalytic part; all are needed for helicase activity (17, 63, 122). The C-terminal part of MCM was suggested to fold into an HTH domain (2), although the domain was not seen clearly in any of the structures of the molecule. Biochemical studies suggested that the domain might play a regulatory role, as removal resulted in increased helicase activity in vitro (7).

PREINITIATION COMPLEX

During S phase of the cell cycle, the pre-RC recruits additional factors, including Cdc45 and GINS, to the origin to establish the pre-IC. Following pre-IC formation, the dsDNA is unwound and SSB associates with the exposed ssDNA. The replisome is assembled at the origin and DNA replication initiates.

Single-Stranded DNA Binding Protein

SSB plays an essential role in DNA replication by coating ssDNA behind the helicase to prevent reannealing and protect the ssDNA from attack by nucleases and chemical modification (32). In Bacteria and Eukarya, SSB also stimulates the activity of DNA polymerases by generating a uniform substrate without secondary structure (142). SSBs were also shown to play an important role in coordinating DNA synthesis on the lagging strand in Bacteria and Eukarya (182, 183), but this activity has not yet been reported in archaea.

In all domains of life, SSB binds to ssDNA via an oligonucleotide/oligosaccharide binding (OB) fold (32). In bacteria, the SSB protein forms homotetramers in which each monomer contains a single OB fold at the N terminus and a flexible acidic C terminus, which is needed for interaction with other proteins (82).

In Eukarya, the SSB is the heterotrimeric replication protein A (RPA) and comprises the RPA70, RPA32, and RPA14 proteins (142). RPA70 contains four OB folds, whereas RPA32 and RPA14 each contain one OB fold; however, only three OB folds from RPA70 and the OB fold of RPA32 are involved in ssDNA binding. The RPA70 protein also contains a C_4-type zinc-finger motif and RPA32 contains a WHD, and both are found near the C-terminus. These motifs, however, do not participate in DNA binding but rather in complex stability and protein-protein interactions (142).

The single RPA homolog from *Methanocaldococcus jannaschii* was the first archaeal SSB to be characterized (70). Other archaeal SSB proteins show significant variation in structure, domain organization, and subunit composition. In most Aigarchaea and Crenarchaea, the SSBs are similar to those in Bacteria in that they have a single OB fold at the N terminus followed by a flexible, acidic C terminus, which is not required for DNA binding (171), whereas the genomes of Korarchaea and Thaumarchaea contain homologs of both SSB and RPA (143). The structure of the Crenarchaea OB fold, however, is more similar to that of the eukaryotic RPA70 than to that of the bacterial SSB (84).

The SSB proteins from the other archaeal lineages are diverse in structure and domain organization, although overall they are similar to the eukaryotic RPA. Whereas some organisms encode a single RPA with multiple OB folds (70), others contain multiple RPA homologs (88). It was found that some organisms with multiple RPAs form trimeric complexes similar to the eukaryotic RPA complex (88). Genetic studies also showed that in some cases when multiple RPAs are present, only a subset is essential for cell viability (163). In addition to the OB fold, several archaeal RPA proteins also contain a zinc-finger motif (147, 163).

In silico approaches identified SSBs in most archaeal species either by similarity to the eukaryotic or bacterial SSB or by searches for putative OB folds. However, to date, no canonical

SSB has been detected in one crenarchaeal order, Thermoproteales. Screening of a *Thermoproteus tenax* extract for proteins capable of binding to ssDNA revealed a new type of SSB [referred to as ThermoDBP (Thermoproteales-specific DNA binding protein)] (136). The protein was shown to form a homodimer with a novel DNA binding fold at the N terminus and a leucine zipper at the C terminus that mediates dimerization (136).

The Cdc45-MCM-GINS Complex

In Eukarya, the MCM helicase is not active on its own but is activated by association with two accessory factors: the GINS complex and the Cdc45 protein. The resulting complex is referred to as the CMG (Cdc45-MCM-GINS) complex and is thought to function as the replicative helicase (129). However, in contrast to the eukaryotic Mcm2–7 complex, in vitro studies showed that the archaeal MCM is active on its own without association with other proteins. Nevertheless, a CMG complex may also be present in Archaea and may play a role in vivo.

The GINS complex. In Eukarya, the GINS complex is a ring-shaped heterotetramer of four related proteins, Sld5, Psf1, Psf2, and Psf3 (GINS is derived from the Japanese go-ichi-ni-san, meaning 5-1-2-3), that plays an essential role in establishment and maintenance of replication forks (65, 108). In addition to interacting with MCM and Cdc45, the GINS complex also interacts with the DNA polymerase α (Polα)-primase complex that synthesizes primers on the lagging strand and with the leading-strand DNA polymerase Polε (65, 108). It is thought that the GINS complex may be the eukaryotic functional homolog of the bacterial τ-subunit that couples the helicase, primase, and polymerases at the replication fork and thus plays an important scaffolding role in replisome formation and in coordinating leading- and lagging-strand synthesis (64).

The first archaeal GINS homologs were identified using in silico approaches that detected their weak similarity to the eukaryotic proteins (112). This study was followed by the isolation and characterization of the GINS complex from *S. solfataricus* and other species (9, 114). Although some species contain a single homolog, designated GINS15 (also called GINS51) for its similarity to the eukaryotic Psf1 and Sld5 proteins, other species contain two homologs, a GINS15 and another referred to as GINS23 for its similarities to the eukaryotic Psf2 and Psf3 proteins (108, 110). However, it has been shown that in all cases studied, the GINS proteins form tetrameric complexes, either homotetramers of GINS15 (128) or a heterotetramer of two subunits of GINS15 and two of GINS23 (114). As is the case in Eukarya, the archaeal GINS proteins are essential for cell viability (155).

Supporting evidence for GINS's function as a scaffolding complex comes from the observation that the archaeal complex interacts with a number of replisome components, including primase, MCM, Cdc45, DNA polymerase D (PolD), and PCNA (102, 109, 114, 140). The genes encoding GINS15 and GINS23 are often in an operon with the genes encoding PCNA and/or the small subunit of primase (PriS) and MCM, respectively (14, 108).

Structural studies revealed that each of the two GINS subunits is composed of two distinct domains: a large A domain and a smaller B domain. The order of the two domains, however, is different in the different subunits. In the GINS15 protein the A domain is at the N terminus and the B domain is at the C terminus (AB type), whereas the orientation is reversed in GINS23 (BA type) (65).

The structure of the *Thermococcus kodakarensis* GINS revealed that two GINS15 proteins form a top layer and two GINS23 proteins form a bottom layer with an overall structure that resembles a trapezoid with a narrow cavity in the center (130). The structure suggests that the B domain of GINS15 does not participate in tetramer formation (130). It is possible that the B domain is needed for GINS interaction with other components of the CMG complex or with other replication proteins.

The structure also revealed similarities between the B domain and the C-terminal part of PriS (65, 167). This C-terminal domain, however, is not present in PriS from all species. Because the gene encoding GINS15 in several archaeal species is in an operon with the gene encoding PriS (14, 108), it has been suggested that the C-terminal domain of PriS may have been acquired from the GINS15 protein via tandem duplication (167).

Cdc45. The first suggestion that Archaea might contain a Cdc45 homolog came from a study that identified a nuclease in *T. kodakarensis* associated with the GINS complex (101, 102). A structural prediction analysis suggested that the protein, referred to as GAN (GINS-associated nuclease), has limited similarity to the bacterial RecJ and the eukaryotic Cdc45 proteins (90, 101). Detailed in silico analyses have shown that all archaeal species contain a protein with similarity to Cdc45 and the bacterial RecJ (111). Although the *T. kodakarensis* protein possesses nuclease activity, this is not the case for all archaeal Cdc45 proteins. When the GINS complex was purified from *S. solfataricus*, a protein with similarity to the RecJ DNA binding domain copurified with the complex (114). This protein does not include the RecJ nuclease domain and does not possess nuclease activity. Similarly, the eukaryotic Cdc45 protein does not contain key residues for nuclease activity and no nuclease activity could be detected in vitro with the eukaryotic enzyme (J. Hurwitz & Z. Kelman, unpublished results). Therefore, the function of the archaeal Cdc45 in DNA replication probably does not depend on its nuclease domain. Nevertheless, it is likely that the archaeal and eukaryal Cdc45 evolved from the bacterial RecJ (110), and the archaeal Cdc45 and the bacterial RecJ have similar three-dimensional structures (S. Nair & Z. Kelman, unpublished results).

The role of Cdc45 in archaeal DNA replication is not clear. The protein interacts with GINS and PolD, which may suggest a role at the replication fork. However, although the eukaryotic Cdc45 is essential for cell viability (115), the genes encoding the archaeal enzyme can be deleted from the chromosome without a major effect on cell growth (T. Santangelo & Z. Kelman, unpublished results). This may suggest that the protein is not essential for DNA replication or that another factor can replace it. Future studies are needed to determine the role, if any, for the archaeal Cdc45 protein in vivo.

ELONGATION COMPLEX

Following MCM activation, the replication bubble is formed, and the other components of the replisome are recruited to the DNA to establish the two replication forks and initiate bidirectional DNA synthesis.

Primase

DNA polymerases are incapable of initiating DNA synthesis de novo and require a 3′-hydroxyl-primed template in order to elongate DNA chains. DNA primases synthesize short RNA primers on template DNA, which are subsequently extended by DNA polymerase (**Figure 2**; **Figure 3***a*) (44). In bacteria, DNA primase consists of a single subunit, the DnaG protein, which associates with both the DnaB helicase and PolIII. It synthesizes a 10–12 oligoribonucleotide primer that is transferred to the bacterial clamp loader. In Eukarya, DNA primase is a heterodimer containing a catalytic p48 subunit in tight association with a regulatory p58 subunit. In vivo, the heterodimer is found in a complex with two other proteins, the Polα catalytic subunit (p180) and the B subunit (p70), to form the Polα-primase complex (44). The primase component synthesizes 10–15 oligoribonucleotide primers that are elongated by the Polα component, generating covalently linked RNA-DNA oligonucleotide chains of 30–40 nucleotides (44). These pre-Okazaki fragments are recognized by the clamp loader.

The archaeal primase is a two-subunit complex; a small subunit p41 (PriS) contains the catalytic activity, and a large subunit p46 (PriL) regulates primase activity and contains an iron-sulfur domain. Both subunits were shown to be essential for cell viability (155). Nanoarchaeota genomes encode a shorter form of primase in which the iron-sulfur domain of PriL is fused to the catalytic domain of PriS (143).

The first report on the biochemical properties of the archaeal primase came from studies on the small subunit from *M. jannaschii* (31), followed by studies on the dimeric enzyme from other species (60). No homologs of the eukaryotic Polα or B subunit have been identified in archaeal genomes (96). In contrast to DNA primases from Bacteria and Eukarya that can utilize only ribonucleotides, in vitro studies with the archaeal catalytic subunit PriS and the two-subunit complex PriS-PriL demonstrated that the archaeal primase is capable of initiating oligonucleotide chains de novo from either ribonucleotide triphosphates (rNTPs) or deoxyribonucleotide triphosphates (dNTPs) (60, 96).

Although the primase can utilize both ribo- and deoxynucleotides in vitro, several different assays, including RNA unmasking, RNA labeling, and RIP mapping, strongly suggest that primers are composed of oligoribonucleotides in vivo (119, 149). The much higher cellular levels of rNTPs in comparison to dNTPs (89) may explain the discrepancies between primase activity in vitro and primer synthesis in vivo.

Archaeal primases also have strand displacement, gap-filling, terminal transferase, and pyrophosphatase activity (60), but the roles of those activities are not clear. In vitro studies also showed that the enzymes are capable of generating nucleotide adducts in the presence of rNTPs, dNTPs, and small molecules with $^-$OH and $^-$NH groups. The reaction results in the formation of r/dNMP-O-R and r/dNMP-N-R (19, 20). It is not clear whether the enzyme makes similar compounds in vivo and whether such products have a physiological role.

The three-dimensional structure of PriS from several species has been determined, revealing a two-domain structure (95, 137): a large primase domain (62), which contains the catalytic part, and a small domain that appears to have a species-specific fold. The catalytic domain also contains a zinc-finger motif. Although the role of zinc binding is not clear, the proximity of this motif to the active site may suggest a role in catalysis. DnaG also contains a zinc binding domain that participates in ssDNA binding (25), and therefore it is possible that zinc binding has a similar role in the archaeal primase.

The structures of the N-terminal part of PriL alone and a complex of the N-terminal part of PriL and PriS have been determined (95). The N-terminal part of PriL interacts with PriS and also forms an arm that connects the catalytic subunit to the DNA binding domain located at the C terminus of PriL (116). The C-terminal part of PriL contains an iron-sulfur domain (87). The structure and role of the iron-sulfur domain is not yet known, but this domain may play a role in maintaining the correct three-dimensional structure of the C-terminal domain (137).

Polymerase Accessory Proteins

The replicative polymerase on its own has very low processivity. High processivity is conferred by a ring-shaped processivity factor (sliding clamp) that encircles DNA and acts to tether the polymerase catalytic unit to the template for processive DNA synthesis (**Figure 2**; **Figure 3a**) (58, 94, 131). The sliding clamps are stable rings and thus cannot assemble themselves around DNA but must be loaded onto DNA by a clamp loader complex (69, 181). The clamp loader recognizes the 3′ end of the single strand–duplex (primer-template) junction and utilizes ATP hydrolysis to assemble the clamp around the primer. The clamp encircles the primer and then binds the polymerase for rapid and processive DNA synthesis. Upon completion of an Okazaki fragment, the polymerase dissociates from the clamp, leaving it assembled around the duplex DNA. The clamps left on the DNA interact with proteins needed for Okazaki fragment maturation, chromatin remodeling, and other cellular processes (170).

Replication factor C. In eukarya, the clamp loader is the pentameric replication factor C (RFC) complex composed of one large subunit (Rfc1) and four small subunits (Rfc2-5), all belonging to the AAA+ family of ATPases (69, 181). Most archaea contain two homologs of RFC (22). One is similar to the small subunits of the eukaryotic RFC and thus is referred to as RFC small (RFCS), and the other is similar to the large subunit of the eukaryotic RFC and is named RFC large (RFCL). Like the eukaryotic RFC complex, the archaeal complex is pentameric, with four subunits of RFCS associated with one subunit of RFCL (58, 127).

RFC must be bound to ATP in order to associate with and open the PCNA ring. Following the assembly of PCNA around the primer, the interaction of RFC with the DNA and PCNA stimulates its ATPase activity. Upon ATP hydrolysis, the affinity of RFC to DNA and PCNA is substantially reduced, resulting in the dissociation of RFC from the DNA and PCNA (58, 127).

In Bacteria and Eukarya, primase remains associated with the primer following primer synthesis. It was shown that primases and the clamp loaders interact with SSB, and competition between the clamp loader and primase for the interaction with SSB plays an important role in the handoff of the primer from primase to the clamp loader (182, 183). A similar RPA-dependent handoff mechanism may also exist in archaea, although it has not yet been reported. Studies have shown, however, that the archaeal primase interacts directly with RFC (109, 178). It is thus possible that in Archaea the handoff of the primers from primase to RFC does not require the involvement of RPA.

As mentioned above, PCNA remains on the DNA after the completion of an Okazaki fragment. However, during replication there are more Okazaki fragments than PCNA trimers within the cell, and so the clamps must be recycled. It was shown that the eukaryotic RFC can actively remove PCNA rings from DNA (180) and a similar unloading activity was proposed for archaeal RFC (77).

Biochemical, biophysical, structural, and molecular modeling studies with RFC have shed light on the mechanism by which RFC assembles the PCNA around DNA (68, 124). The data suggest that during the assembly process, RFC, in the ATP-bound form, forms a right-handed spiral on the top of the clamp with a pitch that is congruent with the helical geometry of duplex DNA. Binding to the clamp results in clamp opening with an out-of-plane configuration resembling a right-handed spring washer (69), enabling assembly around the DNA.

Proliferating cell nuclear antigen. PCNA is a trimeric ring that encircles dsDNA and can slide bidirectionally along it. All activities described for the PCNA proteins require them to encircle the DNA duplex; no biochemical function for PCNA separate from DNA has been reported. PCNA was first reported as a processivity factor for replicative DNA polymerases, but subsequent studies have established that PCNAs also associate with, and modulate the activity of, many other proteins involved in nucleic acid metabolic transactions and cell-cycle regulation (131, 170). It was proposed that the movement of PCNA along dsDNA might function as a moving platform for enzymes that participate in DNA metabolic processes but have low affinity for DNA (76).

Most of the proteins that interact with PCNA do so via a PIP motif (173). The PIP motif interacts with the loop that connects the two domains in each PCNA monomer [referred to as the interdomain connecting loop (IDCL)] (49). Biochemical and structural analysis elucidated the requirement for an intact PIP motif for interaction with PCNA (60).

The three-dimensional structures of PCNA proteins from several archaeal species were determined (131), as were PCNA complexes with interacting enzymes, including DNA ligase, flap endonuclease 1 (Fen1), and DNA polymerase B (PolB) (60). Although the crenarchaeota PCNA protein forms heterotrimers while PCNA from the other kingdoms form homotrimers, the ring structures are very similar (131), and they are also similar to the bacterial β-subunit and the eukaryotic PCNA (58, 80).

The structure revealed that three monomers interact in a head-to-tail manner to form a trimeric ring sufficiently large to accommodate dsDNA. Each monomer is composed of two structurally similar domains, so the trimer has a pseudo six-fold symmetry (75). The two domains in each monomer are connected by the IDCL (49).

Although all PCNA proteins are acidic (80), the charge distribution on the ring surface is not symmetrical. The outer surface is negatively charged, whereas there is a net positive electrostatic potential in the central cavity where the dsDNA is located (131). The positive charge in the central cavity was shown to be required for interaction with the DNA, and it was suggested that the negatively charged surface might prevent nonspecific interaction with DNA. In contrast to the structures of other PCNA proteins, the structures of the halophilic PCNA proteins revealed very few positively charged residues within the central cavity (125, 176), which may suggest a different mechanism of interaction for halophile PCNA with DNA (125).

When the structures of the sliding clamps were initially determined, it was hypothesized that the DNA passes perpendicularly through the central hole, allowing bidirectional sliding along dsDNA (180). However, structural studies of PCNA-DNA complexes from several archaeal species suggest that the DNA has a substantial tilt while passing through the central hole of the ring (121).

Several mechanisms regulate the activity of eukaryotic PCNA. One involves the binding of small PIP-containing regulatory proteins to PCNA, which in turn prevents other proteins from binding and thus regulates (inhibits) the effect of PCNA on their activity (41). Archaea may use similar mechanisms in which binding of small proteins to PCNA regulates its interaction with other enzymes. Studies have shown that a small protein from *T. kodakarensis*, referred to as Thermococcales inhibitor of PCNA (TIP), binds to PCNA and prevents its interaction with DNA polymerase and other enzymes. However, the mechanism of binding of TIP to PCNA does not involve a canonical PIP motif (99).

Another mechanism by which the eukaryotic PCNA is regulated involves modification by small proteins, such as ubiquitin and SUMO. These proteins, when binding PCNA, modulate PCNA interaction with other enzymes (33). Although small modifier proteins have also been identified in Archaea (120), it remains to be determined whether they modify PCNA and whether the modification plays a role in regulating PCNA function.

A third mechanism used by Eukarya to regulate PCNA activity is phosphorylation of the interacting enzymes, which results in the modulation of their interactions with PCNA (56). Protein phosphorylation is common in Archaea (83), and in vivo phosphorylation of PCNA and several replication proteins has been reported (144). Therefore, it is possible that phosphorylation may regulate the activity of the archaeal PCNA.

DNA Polymerase

DNA replication is achieved by a DNA-dependent DNA polymerase that uses primed ssDNA as a template to synthesize the complementary strand. Bacterial PolIII replicates both the leading and lagging strands (127). In Eukarya, two different polymerases, Polε and Polδ, replicate the leading and lagging strands, respectively (127). Two different DNA polymerases, PolB and PolD, have been implicated in archaeal DNA replication.

B-type polymerases. The archaeal PolB DNA polymerases have been extensively studied, mainly for their importance in polymerase chain reaction. A large number of enzymes have been identified, purified, and biochemically characterized (60, 110, 139). At least one PolB homolog has been identified in every archaeal genome. In addition to polymerase activity, these enzymes possess a 3′ to 5′ exonuclease proofreading activity. Polymerases are not processive; processivity is achieved by PolB interaction with PCNA.

For extended processivity, bacterial and eukaryal replicative polymerases require SSB. In vitro studies showed, however, that at least in some archaea, RPA inhibited rather than stimulated the primer extension activity of PolB (81), and direct interaction between RPA and PolB has also been reported (81, 102). The presence of PCNA relieved, but did not eliminate, the inhibitory effect of RPA. Why RPA inhibits PolB and what the role of the inhibition is in vivo is not clear. It is possible that when a polymerase is a part of the replisome, RPA does not inhibit its activity. Alternatively, PolB may not be the replicative enzyme in Archaea (29, 155) but instead is involved in DNA repair and/or recombination. If this is the case, then inhibition by RPA may insure that PolB will replicate only short DNA fragments, as expected for a repair enzyme. It is also possible that PolB replicates only the leading strand, and therefore its ssDNA template is not coated with RPA. Future studies are needed to determine how common the inhibition of PolB by RPA is, as well as the physiological role of that inhibition.

The three-dimensional structures of several archaeal PolB enzymes have been determined and revealed similar topology to other DNA polymerases, including the palm, finger, thumb, and exonuclease domains (60, 73). One feature of the archaeal PolB not found in other family members is an N-terminal domain involved in specific interactions with uracil and hypoxanthine on the template strand (24). Copying of damaged DNA could result in irreversible mutations. Therefore, the ability of a polymerase to recognize uracil or hypoxanthine, followed by replication stalling, may enable the cell to repair the DNA damage prior to replication (24, 60).

D-type polymerases. The genomes of all Archaea except Crenarchaea contain the archaeal-specific PolD in addition to PolB. PolD was originally isolated from *Pyrococcus furiosus* by screening for DNA polymerase activity in cell extract (169). PolD is a dimeric enzyme, and the large subunit (DP2) contains polymerase catalytic activity while the small subunit (DP1) contains the 3′ to 5′ exonuclease activity. The DP2 protein does not display any sequence similarity with other protein families (110), whereas DP1 shares amino acid sequence similarity with several of the small, noncatalytic subunits of the eukaryotic Polα, Polδ, and Polε (1). Although each subunit alone has low activity, the dimeric PolD complex exhibits polymerase and exonuclease activities (60).

The three-dimensional structures of the N-terminal parts of DP1 and DP2 have been solved (117, 179). The structure of the N-terminal part of DP1 confirmed the bioinformatic prediction that it shares similarity with the small subunits of Polα and Polε (179). Although high-resolution structures of the full-length proteins have not yet been reported, bioinformatic predictions provide insight into the structure and domain organization, including the presence of an OB fold in DP1 and a zinc-finger motif in DP2 (110). Mutational analyses of conserved residues, regions, and motifs have shed light on the regions required for activity, on interactions between DP1 and DP2, and on interactions with DNA (157).

Which is the replicative polymerase in Archaea? The three replicative polymerases in Eukarya, Polα, Polδ, and Polε, all belong to family B, and therefore it was presumed that members of this family also replicate the archaeal genomes. Crenarchaea genomes encode only members of PolB, and therefore PolB must be the replicative enzyme in this kingdom. In other archaeal branches, however, both PolB and PolD are present. It is not clear whether both polymerases are involved in chromosomal replication, as in Eukarya, or only one of the two is, as in Bacteria, with the other polymerase playing a different cellular role. Genetic studies with PolB from several species suggested that both PolB and PolD are required for cell viability (13). In other organisms, however, PolB is dispensable for cell growth (29, 155). These contradictory observations need further evaluation. They may suggest, however, that at least in some species, PolD replicates both the leading and lagging strands.

Aphidicolin is an inhibitor of the eukaryotic replicative polymerases and thus was thought to inhibit all members of the family B enzymes. PolD, however, is not sensitive to aphidicolin (48, 169). Therefore, sensitivity to aphidicolin was suggested to serve as an additional tool to determine whether the archaeal PolB is essential for cell growth. The picture, however, is not that clear. In vitro studies have shown that although aphidicolin inhibits PolB from some species (48, 169), it has no effect on PolB activity in others (81). In vivo studies did not provide much clarification. Although aphidicolin was shown to inhibit the growth of *Halobacterium halobium*, there was no effect on the growth of the crenarchaea *Sulfolobus acidocaldarius* (43). The *S. acidocaldarius* genome contains only PolB, again demonstrating that some archaeal PolB enzymes are not sensitive to the drug. Taking the in vivo and in vitro evidence together, it is clear that aphidicolin sensitivity cannot be used as a tool to determine whether PolB is required for chromosomal replication.

Additional studies are needed to determine whether both PolB and PolD are a part of the replisome or whether only PolD is involved in chromosomal replication and PolB has another function. It is possible that although both polymerases are replicative enzymes, in the absence of PolB, PolD can substitute for PolB activity. Although in Eukarya Polε is the leading-strand polymerase (127), it was shown that cells harboring mutant Polε are viable and that Polδ can replicate both DNA strands (85). It is possible that in archaea, PolD replicates the lagging strand and PolB copies the leading strand. The role for PolD in lagging-strand synthesis may be supported by the observation that PolD exhibits a strand displacement activity (54) that is required for Okazaki fragment maturation. In addition, PolD can utilize RNA primers more efficiently than PolB (55), whereas processivity of PolB, in the presence of PCNA, is greater than that of PolD (132). Similar observations were made in Eukarya, where the leading-strand polymerase is more processive than the lagging-strand enzyme (12). In addition, the inhibition of PolB by RPA may suggest that it is not replicating the lagging strand because the lagging-strand polymerase probably needs to displace RPA.

OKAZAKI FRAGMENT MATURATION

During chromosomal replication, the leading strand is replicated as a continuous strand, whereas the lagging strand is replicated discontinuously as a series of Okazaki fragments (**Figure 3**). The RNA primers that initiate each Okazaki fragment must be removed, the gap in the DNA must be filled, and the newly synthesized Okazaki fragment must be ligated to the previous fragment to create the mature dsDNA. This task is achieved by the concerted activity of several enzymes, including DNA polymerase, Fen1, and DNA ligase (**Figure 3b**) (5). The lagging-strand polymerase, upon reaching the previous Okazaki fragment, displaces the primer and provides the substrate for Fen1. Following Fen1 removal of the primer, the lagging-strand DNA polymerase fills the gap and DNA ligase joins the two adjacent Okazaki fragments.

Flap Endonuclease 1

Fen1 is a structure-specific nuclease that plays an important role in Okazaki fragment maturation on the lagging strand (4). The lagging-strand DNA polymerase, using its strand displacement activity, removes the RNA primer and forms a branched DNA molecule (flap structure). The 5′-flap endonuclease activity of Fen1 removes the primer (**Figure 3b**).

The three-dimensional structures of Fen1 from several archaeal species and co-complexes with DNA have been determined (60). The structures reveal that the protein catalytic domain is similar to other nucleases. In addition, Fen1 contains a helical clamp that encircles the 5′-flap and interacts with the upstream and downstream duplex DNA. Upon binding to the substrate, the protein undergoes a conformational change that closes the clamp around the single-stranded flap. In Eukarya, it was shown that the preferred substrate of Fen1 contains an unpaired 3′ nucleotide (3′-flap) overlapping with a variable length region of 5′ ssDNA (5′-flap), referred to as an overlap-flap (39). The structure of the archaeal Fen1 with DNA suggested that an overlap-flap may also be the preferred substrate for the archaeal enzyme (18).

As is the case with the eukaryotic Fen1, the archaeal enzyme is not essential for cell viability (155). In Eukarya, Exonuclease I (ExoI), a homolog of Fen1 with similar catalytic activity, may replace Fen1 activity in the deletion strain (4). However, in Archaea there are no clear homologs of ExoI or other known nucleases with Fen1-like activity. Thus, it is not clear which protein(s) replaces Fen1 activity in the deleted strain, although it was suggested that the archaeal Cdc45 might substitute for Fen1 in the deleted strain (101). Support for this idea comes from the observation that although Cdc45 or Fen1 can be readily deleted from *T. kodakarensis* cells, the double mutation is lethal (T. Santangelo & Z. Kelman, unpublished results). Alternatively, RNase H, which degrades RNA in a RNA-DNA hybrid, may be responsible for primer removal in the Fen1 deleted strain. Future studies are needed, however, to identify the mechanism by which Okazaki fragments mature in the absence of Fen1.

DNA Ligase

DNA ligase plays an essential role in chromosomal DNA replication by joining adjacent Okazaki fragments on the lagging strand. DNA ligases can be divided into two groups according to the cofactor required for activity: ATP or NAD^+. NAD^+-dependent ligases are found predominantly in Bacteria, whereas eukaryotic genomes contain ATP-dependent ligases (110).

Genes that encode DNA ligase have been identified in all archaeal genomes and were shown to be essential for cell viability. The first archaeal DNA ligase was identified in the genome of the thermophilic archaeon *Desulfolobus ambivalens* (86). The first characterized archaeal ligase from *M. thermautotrophicus* demonstrated a strict requirement for ATP as a cofactor (165), but subsequent studies identified a small subset of archaeal ligases that are NAD^+-dependent (185), whereas others can utilize both cofactors (60).

The structures of several archaeal DNA ligases were determined (121, 135) and together with the structure of the eukaryotic DNA ligase bound to DNA (134) revealed a three-domain organization consisting of a DNA binding domain (DBD), a catalytic nucleotidyltransferase domain, and a C-terminal OB fold domain. In the absence of DNA substrate, the enzyme has an open, extended form. Upon DNA binding, the enzyme forms a closed structure around the ds-DNA with both the DBD and OB domains binding to the minor groove on both sides of the nick, thus positioning the catalytic domain for catalysis. It was suggested that binding to PCNA facilitates the transition from extended to closed structure (57).

The Role of Proliferating Cell Nuclear Antigen in Okazaki Fragment Maturation

The enzymes required for Okazaki fragment maturation, including DNA polymerase, Fen1, and DNA ligase, interact with PCNA via a PIP motif, and interaction with PCNA modulates their enzymatic activities (60, 170). These interactions suggest that PCNA coordinates the Okazaki fragment maturation process by sequentially recruiting factors to the lagging strand.

In crenarchaea, PCNA is a heterotrimer and each subunit of PCNA has specificity for binding to DNA polymerase, Fen1, or DNA ligase (8). This specificity is likely to have arisen from differences in the structure of the IDCL among the three different PCNA proteins (175), enabling them to discriminate between small differences in the PIP motifs of the different interacting enzymes. The simultaneous binding of all three proteins to PCNA provides an efficient coupling of the three enzyme activities. Upon flap formation by a lagging-strand polymerase, Fen1 cleaves the flap, the gap is filled by the polymerase, and the Okazaki fragments are joined by DNA ligase.

PCNA proteins from all other archaeal species, however, form homotrimeric rings (131). Therefore, it is not clear whether the role of PCNA in regulating Okazaki fragment maturation in these species is similar to that in Crenarchaea. Although DNA polymerase, Fen1, and DNA ligase all interact with PCNA, it is not clear whether they can bind simultaneously.

TERMINATION OF DNA REPLICATION AND PRODUCT RESOLUTION

At the end of chromosomal replication, converging replication forks must complete replication and the replisomes must dissociate from the DNA. It is essential that converging forks not pass each other and continue replication, as this results in over-replication of part of the chromosome. In Bacteria, termination occurs in a specific chromosomal region referred to as the termination region (*ter*). The specific *ter*-binding protein Tus participates in the termination process by regulating replisome movement, ensuring precise replication of the chromosome (146).

Limited information is available on the mechanism of replication termination in Archaea. The circular nature of archaeal chromosomes may suggest a mechanism similar to bacterial termination, but a study using *S. solfataricus* suggested that termination occurs by random collision of the two replication forks and not at a specific site (37). In addition, the presence of *ter*-like sequences or Tus-like proteins has not yet been reported in Archaea.

Bacterial chromosomes contain *dif* (deletion induced filamentation) regions near *ter*, where concatemers formed during chromosome segregation are resolved by the Xer site-specific recombinases (XerC and XerD) in conjunction with FtsK (146). Only a few studies on archaeal chromosome resolution have been published. Archaeal genomes contain a *dif*-like sequence and Xer homologs (26, 37), but the location of the *dif* sequence in the chromosome is different in different species. In some species the *dif* region is located in the termination zone (26), but it is not near the termination region in others (37). The results to date, although limited, suggest that Archaea may use a bacterial-like mechanism for chromosome resolution.

CONCLUDING REMARKS

Much progress has been made in understanding archaeal DNA replication, but there remains much to be elucidated. Many factors have been identified and characterized, and the three-dimensional structures of most proteins have been determined. New genetic tools (38) have identified new proteins that may be involved in replication. These genetic tools, together with fluorescence and single-molecule approaches, should lead to new and exciting studies on the replication process in vivo.

Future studies will concentrate on poorly understood aspects of the replication process. The mechanism that regulates the initiation process is unknown, and the coordination between initiation and other cell-cycle events is yet to be explored. Similarly, the study of termination and chromosome segregation is in its infancy and more research is needed.

The development of an in vitro replication system is one of the goals of future research. Replication in bacteria, viruses, and phages was characterized after in vitro replication systems were developed, and it is anticipated that such a system will similarly serve as a catalyst for the understanding of archaeal replication.

DISCLOSURE STATEMENT

The authors are not aware of any affiliations, memberships, funding, or financial holdings that might be perceived as affecting the objectivity of this review.

ACKNOWLEDGMENTS

We thank Drs. Jerard Hurwitz, Satish Nair, and Thomas Santangelo for sharing data prior to publication. We thank Dr. Debra Weinstein for help with figures. We wish to apologize to colleagues whose primary work was not cited because of space limitations.

LITERATURE CITED

1. Aravind L, Koonin EV. 1998. Phosphoesterase domains associated with DNA polymerases of diverse origins. *Nucleic Acids Res.* 26:3746–52
2. Aravind L, Koonin EV. 1999. DNA-binding proteins and evolution of transcription regulation in the archaea. *Nucleic Acids Res.* 27:4658–70
3. Arora J, Goswami K, Saha S. 2014. Characterization of the replication initiator Orc1/Cdc6 from the archaeon *Picrophilus torridus*. *J. Bacteriol.* 196:276–86
4. Balakrishnan L, Bambara RA. 2013. Flap endonuclease 1. *Annu. Rev. Biochem.* 82:119–38
5. Balakrishnan L, Bambara RA. 2013. Okazaki fragment metabolism. *Cold Spring Harb. Perspect. Biol.* 5:a010173
6. Barry ER, Lovett JE, Costa A, Lea SM, Bell SD. 2009. Intersubunit allosteric communication mediated by a conserved loop in the MCM helicase. *Proc. Natl. Acad. Sci. USA* 106:1051–56
7. Barry ER, McGeoch AT, Kelman Z, Bell SD. 2007. Archaeal MCM has separable processivity, substrate choice and helicase domains. *Nucleic Acids Res.* 35:988–98
8. Beattie TR, Bell SD. 2012. Coordination of multiple enzyme activities by a single PCNA in archaeal Okazaki fragment maturation. *EMBO J.* 31:1556–67
9. Bell SD. 2011. DNA replication: archaeal oriGINS. *BMC Biol.* 9:36
10. Bell SD. 2012. Archaeal orc1/cdc6 proteins. *Subcell. Biochem.* 62:59–69
11. Bell SD, Botchan MR. 2013. The minichromosome maintenance replicative helicase. *Cold Spring Harb. Perspect. Biol.* 5:a012807
12. Bermudez VP, Farina A, Raghavan V, Tappin I, Hurwitz J. 2011. Studies on human DNA polymerase ε and GINS complex and their role in DNA replication. *J. Biol. Chem.* 286:28963–77
13. Berquist BR, DasSarma P, DasSarma S. 2007. Essential and non-essential DNA replication genes in the model halophilic Archaeon, *Halobacterium* sp. NRC-1. *BMC Genet.* 8:31
14. Berthon J, Cortez D, Forterre P. 2008. Genomic context analysis in Archaea suggests previously unrecognized links between DNA replication and translation. *Genome Biol.* 9:R71
15. Boulikas T. 1996. Common structural features of replication origins in all life forms. *J. Cell Biochem.* 60:297–316
16. Brewster AS, Chen XS. 2010. Insights into the MCM functional mechanism: lessons learned from the archaeal MCM complex. *Crit. Rev. Biochem. Mol. Biol.* 45:243–56
17. Brewster AS, Slaymaker IM, Afif SA, Chen XS. 2010. Mutational analysis of an archaeal minichromosome maintenance protein exterior hairpin reveals critical residues for helicase activity and DNA binding. *BMC Mol. Biol.* 11:62
18. Chapados BR, Hosfield DJ, Han S, Qiu J, Yelent B, et al. 2004. Structural basis for FEN-1 substrate specificity and PCNA-mediated activation in DNA replication and repair. *Cell* 116:39–50

19. Chemnitz Galal W, Pan M, Giulian G, Yuan W, Li S, et al. 2012. Formation of dAMP-glycerol and dAMP-Tris derivatives by the *Thermococcus kodakaraensis* DNA primase. *J. Biol. Chem.* 287:16220–29

20. Chemnitz Galal W, Pan M, Kelman Z, Hurwitz J. 2012. Characterization of the DNA primase complex isolated from the archaeon, *Thermococcus kodakaraensis*. *J. Biol. Chem.* 287:16209–19

21. Chen SH, Chan N-L, Hsieh T-S. 2013. New mechanistic and functional insights into DNA topoisomerases. *Annu. Rev. Biochem.* 82:139–70

22. Chia N, Cann I, Olsen GJ. 2010. Evolution of DNA replication protein complexes in eukaryotes and archaea. *PLoS ONE* 5:e10866

23. Chong JP, Hayashi MK, Simon MN, Xu RM, Stillman B. 2000. A double-hexamer archaeal minichromosome maintenance protein is an ATP-dependent DNA helicase. *Proc. Natl. Acad. Sci. USA* 97:1530–35

24. Connolly BA. 2009. Recognition of deaminated bases by archaeal family-B DNA polymerases. *Biochem. Soc. Trans.* 37:65–68

25. Corn JE, Pease PJ, Hura GL, Berger JM. 2005. Crosstalk between primase subunits can act to regulate primer synthesis in trans. *Mol. Cell* 20:391–401

26. Cortez D, Quevillon-Cheruel S, Gribaldo S, Desnoues N, Sezonov G, et al. 2010. Evidence for a Xer/*dif* system for chromosome resolution in archaea. *PLoS Genet.* 6:e1001166

27. Costa A, Hood IV, Berger JM. 2013. Mechanisms for initiating cellular DNA replication. *Annu. Rev. Biochem.* 82:25–54

28. Costa A, Onesti S. 2009. Structural biology of MCM helicases. *Crit. Rev. Biochem. Mol. Biol.* 44:326–42

29. Čuboňová L, Richardson T, Burkhart BW, Kelman Z, Reeve JN, et al. 2013. Archaeal DNA polymerase D but not DNA polymerase B is required for genome replication in *Thermococcus kodakarensis*. *J. Bacteriol.* 195:2322–28

30. De Felice M, Esposito L, Pucci B, De Falco M, Manco G, et al. 2004. Modular organization of a Cdc6-like protein from the crenarchaeon *Sulfolobus solfataricus*. *Biochem. J.* 381:645–53

31. Desogus G, Onesti S, Brick P, Rossi M, Pisani FM. 1999. Identification and characterization of a DNA primase from the hyperthermophilic archaeon *Methanococcus jannaschii*. *Nucleic Acids Res.* 27:4444–50

32. Dickey TH, Altschuler SE, Wuttke DS. 2013. Single-stranded DNA-binding proteins: multiple domains for multiple functions. *Structure* 21:1074–84

33. Dieckman LM, Freudenthal BD, Washington MT. 2012. PCNA structure and function: insights from structures of PCNA complexes and post-translationally modified PCNA. *Subcell. Biochem.* 62:281–99

34. Duderstadt KE, Berger JM. 2008. AAA+ ATPases in the initiation of DNA replication. *Crit. Rev. Biochem. Mol. Biol.* 43:163–87

35. Duderstadt KE, Berger JM. 2012. A structural framework for replication origin opening by AAA+ initiation factors. *Curr. Opin. Struct. Biol.* 23:144–53

36. Dueber EL, Corn JE, Bell SD, Berger JM. 2007. Replication origin recognition and deformation by a heterodimeric archaeal Orc1 complex. *Science* 317:1210–13

37. Duggin IG, Dubarry N, Bell SD. 2011. Replication termination and chromosome dimer resolution in the archaeon *Sulfolobus solfataricus*. *EMBO J.* 30:145–53

38. Farkas JA, Picking JW, Santangelo TJ. 2013. Genetic techniques for the archaea. *Annu. Rev. Genet.* 47:539–61

39. Finger LD, Atack JM, Tsutakawa S, Classen S, Tainer J, et al. 2012. The wonders of flap endonucleases: structure, function, mechanism and regulation. *Subcell. Biochem.* 62:301–26

40. Fletcher RJ, Bishop BE, Leon RP, Sclafani RA, Ogata CM, Chen XS. 2003. The structure and function of MCM from archaeal *M. thermoautotrophicum*. *Nat. Struct. Biol.* 10:160–67

41. Flores-Rozas H, Kelman Z, Dean FB, Pan ZQ, Harper JW, et al. 1994. Cdk-interacting protein 1 directly binds with proliferating cell nuclear antigen and inhibits DNA replication catalyzed by the DNA polymerase δ holoenzyme. *Proc. Natl. Acad. Sci. USA* 91:8655–59

42. Forterre P. 2012. Introduction and historical perspective. In *DNA Topoisomerases and Cancer*, ed. Y Pommier, pp. 1–52. New York: Humana Press

43. Forterre P, Elie C, Kohiyama M. 1984. Aphidicolin inhibits growth and DNA synthesis in halophilic archaebacteria. *J. Bacteriol.* 159:800–2

44. Frick DN, Richardson CC. 2001. DNA primases. *Annu. Rev. Biochem.* 70:39–80

45. Gaudier M, Schuwirth BS, Westcott SL, Wigley DB. 2007. Structural basis of DNA replication origin recognition by an ORC protein. *Science* 317:1213–16

46. Grabowski B, Kelman Z. 2001. Autophosphorylation of the archaeal Cdc6 homologues is regulated by DNA. *J. Bacteriol.* 183:5459–64

47. Grabowski B, Kelman Z. 2003. Archaeal DNA replication: eukaryal proteins in a bacterial context. *Annu. Rev. Microbiol.* 57:487–516

48. Greenough L, Menin JF, Desai NS, Kelman Z, Gardner AF. 2014. Characterization of Family D DNA polymerase from *Thermococcus* sp. 9°N. *Extremophiles* 18:653–64

49. Gulbis JM, Kelman Z, Hurwitz J, O'Donnell M, Kuriyan J. 1996. Structure of the C-terminal region of p21$^{WAF1/CIP1}$ complexed with human PCNA. *Cell* 87:297–306

50. Guy L, Saw JH, Ettema TJG. 2014. The archaeal legacy of eukaryotes: a phylogenomic perspective. *Cold Spring Harb. Perspect. Biol.* In press

51. Haugland GT, Sakakibara N, Pey AL, Rollor CR, Birkeland N-K, Kelman Z. 2008. *Thermoplasma acidophilum* Cdc6 protein stimulates MCM helicase activity by regulating its ATPase activity. *Nucleic Acids Res.* 36:5602–9

52. Haugland GT, Shin J-H, Birkeland NK, Kelman Z. 2006. Stimulation of MCM helicase activity by a Cdc6 protein in the archaeon *Thermoplasma acidophilum*. *Nucleic Acids Res.* 34:6337–44

53. Hawkins M, Malla S, Blythe MJ, Nieduszynski CA, Allers T. 2013. Accelerated growth in the absence of DNA replication origins. *Nature* 503:544–47

54. Henneke G. 2012. In vitro reconstitution of RNA primer removal in Archaea reveals the existence of two pathways. *Biochem. J.* 447:271–80

55. Henneke G, Flament D, Hubscher U, Querellou J, Raffin JP. 2005. The hyperthermophilic euryarchaeota *Pyrococcus abyssi* likely requires the two DNA polymerases D and B for DNA replication. *J. Mol. Biol.* 350:53–64

56. Henneke G, Koundrioukoff S, Hubscher U. 2003. Phosphorylation of human Fen1 by cyclin-dependent kinase modulates its role in replication fork regulation. *Oncogene* 22:4301–13

57. Howes TR, Tomkinson AE. 2012. DNA ligase I, the replicative DNA ligase. *Subcell. Biochem.* 62:327–41

58. Indiani C, O'Donnell M. 2006. The replication clamp-loading machine at work in the three domains of life. *Nat. Rev. Mol. Cell Biol.* 7:751–61

59. Ishino S, Fujino S, Tomita H, Ogino H, Takao K, et al. 2011. Biochemical and genetical analyses of the three MCM genes from the hyperthermophilic archaeon, *Thermococcus kodakarensis*. *Genes Cells* 16:1176–89

60. Ishino Y, Ishino S. 2012. Rapid progress of DNA replication studies in Archaea, the third domain of life. *Sci. China Life Sci.* 55:386–403

61. Ishino S, Kelman LM, Kelman Z, Ishino Y. 2013. The archaeal DNA replication machinery: past, present and future. *Genes Genet. Syst.* 88:315–19

62. Iyer LM, Koonin EV, Leipe DD, Aravind L. 2005. Origin and evolution of the archaeo-eukaryotic primase superfamily and related palm-domain proteins: structural insights and new members. *Nucleic Acids Res.* 33:3875–96

63. Jenkinson ER, Chong JP. 2006. Minichromosome maintenance helicase activity is controlled by N- and C-terminal motifs and requires the ATPase domain helix-2 insert. *Proc. Natl. Acad. Sci. USA* 103:7613–18

64. Johnson A, O'Donnell M. 2005. Cellular DNA replicases: components and dynamics at the replication fork. *Annu. Rev. Biochem.* 74:283–315

65. Kamada K. 2012. The GINS complex: structure and function. *Subcell. Biochem.* 62:135–56

66. Kasiviswanathan R, Shin J-H, Kelman Z. 2005. Interactions between the archaeal Cdc6 and MCM proteins modulate their biochemical properties. *Nucleic Acids Res.* 33:4940–50

67. Kasiviswanathan R, Shin J-H, Melamud E, Kelman Z. 2004. Biochemical characterization of the *Methanothermobacter thermautotrophicus* minichromosome maintenance (MCM) helicase N-terminal domains. *J. Biol. Chem.* 279:28358–66

68. Kelch BA, Makino DL, O'Donnell M, Kuriyan J. 2011. How a DNA polymerase clamp loader opens a sliding clamp. *Science* 334:1675–80

69. Kelch BA, Makino DL, O'Donnell M, Kuriyan J. 2012. Clamp loader ATPases and the evolution of DNA replication machinery. *BMC Biol.* 10:34

70. Kelly TJ, Simancek P, Brush GS. 1998. Identification and characterization of a single-stranded DNA-binding protein from the archaeon *Methanococcus jannaschii*. *Proc. Natl. Acad. Sci. USA* 95:14634–39

71. Kelman LM, Kelman Z. 2003. Archaea: an archetype for replication initiation studies? *Mol. Microbiol.* 48:605–15

72. Kelman LM, Kelman Z. 2004. Multiple origins of replication in archaea. *Trends Microbiol.* 12:399–401

73. Kelman Z. 2000. DNA replication in the third domain (of life). *Curr. Protein Pept. Sci.* 1:139–54

74. Kelman Z. 2000. The replication origin of archaea is finally revealed. *Trends Biochem. Sci.* 25:521–23

75. Kelman Z, Finkelstein J, O'Donnell M. 1995. Why have six-fold symmetry? *Curr. Biol.* 5:1239–42

76. Kelman Z, Hurwitz J. 1998. Protein-PCNA interactions: a DNA-scanning mechanism? *Trends Biochem. Sci.* 23:236–38

77. Kelman Z, Hurwitz J. 2000. A unique organization of the protein subunits of the DNA polymerase clamp loader in the archaeon *Methanobacterium thermoautotrophicum* ΔH. *J. Biol. Chem.* 275:7327–36

78. Kelman Z, Hurwitz J. 2003. Structural lessons in DNA replication from the third domain of life. *Nat. Struct. Biol.* 10:148–50

79. Kelman Z, Lee JK, Hurwitz J. 1999. The single minichromosome maintenance protein of *Methanobacterium thermoautotrophicum* ΔH contains DNA helicase activity. *Proc. Natl. Acad. Sci. USA* 96:14783–88

80. Kelman Z, O'Donnell M. 1995. Structural and functional similarities of prokaryotic and eukaryotic DNA polymerase sliding clamps. *Nucleic Acids Res.* 23:3613–20

81. Kelman Z, Pietrokovski S, Hurwitz J. 1999. Isolation and characterization of a split B-type DNA polymerase from the archaeon *Methanobacterium thermoautotrophicum* ΔH. *J. Biol. Chem.* 274:28751–61

82. Kelman Z, Yuzhakov A, Andjelkovic J, O'Donnell M. 1998. Devoted to the lagging strand: the χ subunit of DNA polymerase III holoenzyme contacts SSB to promote processive elongation and sliding clamp assembly. *EMBO J.* 17:2436–49

83. Kennelly PJ. 2003. Archaeal protein kinases and protein phosphatases: insights from genomics and biochemistry. *Biochem. J.* 370:373–89

84. Kerr ID, Wadsworth RI, Cubeddu L, Blankenfeldt W, Naismith JH, White MF. 2003. Insights into ssDNA recognition by the OB fold from a structural and thermodynamic study of *Sulfolobus* SSB protein. *EMBO J.* 22:2561–70

85. Kesti T, Flick K, Keranen S, Syvaoja JE, Wittenberg C. 1999. DNA polymerase ε catalytic domains are dispensable for DNA replication, DNA repair, and cell viability. *Mol. Cell* 3:679–85

86. Kletzin A. 1992. Molecular characterisation of a DNA ligase gene of the extremely thermophilic archaeon *Desulfurolobus ambivalens* shows close phylogenetic relationship to eukaryotic ligases. *Nucleic Acids Res.* 20:5389–96

87. Klinge S, Hirst J, Maman JD, Krude T, Pellegrini L. 2007. An iron-sulfur domain of the eukaryotic primase is essential for RNA primer synthesis. *Nat. Struct. Mol. Biol.* 14:875–77

88. Komori K, Ishino Y. 2001. Replication protein A in *Pyrococcus furiosus* is involved in homologous DNA recombination. *J. Biol. Chem.* 276:25654–60

89. Kornberg A, Baker TA. 1992. *DNA Replication*. New York: W.H. Freeman. 931 pp.

90. Krastanova I, Sannino V, Amenitsch H, Gileadi O, Pisani FM, Onesti S. 2012. Structural and functional insights into the DNA replication factor Cdc45 reveal an evolutionary relationship to the DHH family of phosphoesterases. *J. Biol. Chem.* 287:4121–28

91. Krueger S, Shin J-H, Curtis JE, Rubinson KA, Kelman Z. 2014. The solution structure of full-length dodecameric MCM by SANS and molecular modeling. *Proteins*. In press

92. Krueger S, Shin J-H, Raghunandan S, Curtis JE, Kelman Z. 2011. Atomistic ensemble modeling and small-angle neutron scattering of intrinsically disordered protein complexes: applied to minichromosome maintenance protein. *Biophys. J.* 101:2999–3007

93. Krupovic M, Gribaldo S, Bamford DH, Forterre P. 2010. The evolutionary history of archaeal MCM helicases: a case study of vertical evolution combined with hitch-hiking of mobile genetic elements. *Mol. Biol. Evol.* 27:2716–32

94. Langston LD, Indiani C, O'Donnell M. 2009. Whither the replisome: emerging perspectives on the dynamic nature of the DNA replication machinery. *Cell Cycle* 8:2686–91

95. Lao-Sirieix SH, Nookala RK, Roversi P, Bell SD, Pellegrini L. 2005. Structure of the heterodimeric core primase. *Nat. Struct. Mol. Biol.* 12:1137–44

96. Lao-Sirieix SH, Pellegrini L, Bell SD. 2005. The promiscuous primase. *Trends Genet.* 21:568–72

97. Lee C, Hong B, Choi JM, Kim Y, Watanabe S, et al. 2004. Structural basis for inhibition of the replication licensing factor Cdt1 by geminin. *Nature* 430:913–17

98. Leonard AC, Grimwade JE. 2011. Regulation of DnaA assembly and activity: taking directions from the genome. *Annu. Rev. Microbiol.* 65:19–35

99. Li Z, Huang RY, Yopp DC, Hileman TH, Santangelo TJ, et al. 2014. A novel mechanism for regulating the activity of proliferating cell nuclear antigen by a small protein. *Nucleic Acids Res.* 42:5776–89

100. Li Z, Kelman LM, Kelman Z. 2013. *Thermococcus kodakarensis* DNA replication. *Biochem. Soc. Trans.* 41:332–38

101. Li Z, Pan M, Santangelo TJ, Chemnitz W, Yuan W, et al. 2011. A novel DNA nuclease is stimulated by association with the GINS complex. *Nucleic Acids Res.* 39:6114–23

102. Li Z, Santangelo TJ, Čuboňová L, Reeve JN, Kelman Z. 2010. Affinity purification of an archaeal DNA replication protein network. *MBio* 1:e00221–10

103. Lindas AC, Bernander R. 2013. The cell cycle of archaea. *Nat. Rev. Microbiol.* 11:627–38

104. Liu W, Pucci B, Rossi M, Pisani FM, Ladenstein R. 2008. Structural analysis of the *Sulfolobus solfataricus* MCM protein N-terminal domain. *Nucleic Acids Res.* 36:3235–43

105. Lopez P, Philippe H, Myllykallio H, Forterre P. 1999. Identification of putative chromosomal origins of replication in Archaea. *Mol. Microbiol.* 32:883–86

106. Lundgren M, Andersson A, Chen L, Nilsson P, Bernander R. 2004. Three replication origins in *Sulfolobus* species: synchronous initiation of chromosome replication and asynchronous termination. *Proc. Natl. Acad. Sci. USA* 101:7046–51

107. Lundgren M, Bernander R. 2007. Genome-wide transcription map of an archaeal cell cycle. *Proc. Natl. Acad. Sci. USA* 104:2939–44

108. MacNeill SA. 2010. Structure and function of the GINS complex, a key component of the eukaryotic replisome. *Biochem. J.* 425:489–500

109. MacNeill SA. 2011. Protein-protein interactions in the archaeal core replisome. *Biochem. Soc. Trans.* 39:163–68

110. Makarova KS, Koonin EV. 2013. Archaeology of eukaryotic DNA replication. *Cold Spring Harb. Perspect. Biol.* 5:a012963

111. Makarova KS, Koonin EV, Kelman Z. 2012. The CMG (CDC45/RecJ, MCM, GINS) complex is a conserved component of the DNA replication system in all archaea and eukaryotes. *Biol. Direct* 7:7

112. Makarova KS, Wolf YI, Mekhedov SL, Mirkin BG, Koonin EV. 2005. Ancestral paralogs and pseudoparalogs and their role in the emergence of the eukaryotic cell. *Nucleic Acids Res.* 33:4626–38

113. Makarova KS, Yutin N, Bell SD, Koonin EV. 2010. Evolution of diverse cell division and vesicle formation systems in Archaea. *Nat. Rev. Microbiol.* 8:731–41

114. Marinsek N, Barry ER, Makarova KS, Dionne I, Koonin EV, Bell SD. 2006. GINS, a central nexus in the archaeal DNA replication fork. *EMBO Rep.* 7:539–45

115. Masai H, Matsumoto S, You Z, Yoshizawa-Sugata N, Oda M. 2010. Eukaryotic chromosome DNA replication: where, when, and how? *Annu. Rev. Biochem.* 79:89–130

116. Matsui E, Nishio M, Yokoyama H, Harata K, Darnis S, Matsui I. 2003. Distinct domain functions regulating de novo DNA synthesis of thermostable DNA primase from hyperthermophile *Pyrococcus horikoshii*. *Biochemistry* 42:14968–76

117. Matsui I, Urushibata Y, Shen Y, Matsui E, Yokoyama H. 2011. Novel structure of an N-terminal domain that is crucial for the dimeric assembly and DNA-binding of an archaeal DNA polymerase D large subunit from *Pyrococcus horikoshii*. *FEBS Lett.* 585:452–58

118. Matsunaga F, Forterre P, Ishino Y, Myllykallio H. 2001. In vivo interactions of archaeal Cdc6/Orc1 and minichromosome maintenance protein with the replication origin. *Proc. Natl. Acad. Sci. USA* 98:11152–57

119. Matsunaga F, Norais C, Forterre P, Myllykallio H. 2003. Identification of short "eukaryotic" Okazaki fragments synthesized from a prokaryotic replication origin. *EMBO Rep.* 4:154–58

120. Maupin-Furlow JA. 2013. Ubiquitin-like proteins and their roles in archaea. *Trends Microbiol.* 21:31–38

121. Mayanagi K, Kiyonari S, Saito M, Shirai T, Ishino Y, Morikawa K. 2009. Mechanism of replication machinery assembly as revealed by the DNA ligase-PCNA-DNA complex architecture. *Proc. Natl. Acad. Sci. USA* 106:4647–52

122. McGeoch AT, Trakselis MA, Laskey RA, Bell SD. 2005. Organization of the archaeal MCM complex on DNA and implications for the helicase mechanism. *Nat. Struct. Mol. Biol.* 12:756–62

123. Medagli B, Onesti S. 2013. Structure and mechanism of hexameric helicases. *Adv. Exp. Med. Biol.* 767:75–95

124. Miyata T, Oyama T, Mayanagi K, Ishino S, Ishino Y, Morikawa K. 2004. The clamp-loading complex for processive DNA replication. *Nat. Struct. Mol. Biol.* 11:632–36

125. Morgunova E, Gray FC, MacNeill SA, Ladenstein R. 2009. Structural insights into the adaptation of proliferating cell nuclear antigen (PCNA) from *Haloferax volcanii* to a high-salt environment. *Acta Crystallogr. D* 65:1081–88

126. Myllykallio H, Lopez P, Lopez-Garcia P, Heilig R, Saurin W, et al. 2000. Bacterial mode of replication with eukaryotic-like machinery in a hyperthermophilic archaeon. *Science* 288:2212–15

127. O'Donnell M, Langston L, Stillman B. 2013. Principles and concepts of DNA replication in bacteria, archaea, and eukarya. *Cold Spring Harb. Perspect. Biol.* 5:a010108

128. Ogino H, Ishino S, Mayanagi K, Haugland GT, Birkeland NK, et al. 2011. The GINS complex from the thermophilic archaeon, *Thermoplasma acidophilum* may function as a homotetramer in DNA replication. *Extremophiles* 15:529–39

129. Onesti S, MacNeill SA. 2013. Structure and evolutionary origins of the CMG complex. *Chromosoma* 122:47–53

130. Oyama T, Ishino S, Fujino S, Ogino H, Shirai T, et al. 2011. Architectures of archaeal GINS complexes, essential DNA replication initiation factors. *BMC Biol.* 9:28

131. Pan M, Kelman LM, Kelman Z. 2011. The archaeal PCNA proteins. *Biochem. Soc. Trans.* 39:20–24

132. Pan M, Santangelo TJ, Čuboňová L, Li Z, Metangmo H, et al. 2013. *Thermococcus kodakarensis* has two functional PCNA homologues but only one is required for viability. *Extremophiles* 17:453–61

133. Pan M, Santangelo TJ, Li Z, Reeve JN, Kelman Z. 2011. *Thermococcus kodakarensis* encodes three MCM homologs but only one is essential. *Nucleic Acids Res.* 39:9671–80

134. Pascal JM, O'Brien PJ, Tomkinson AE, Ellenberger T. 2004. Human DNA ligase I completely encircles and partially unwinds nicked DNA. *Nature* 432:473–78

135. Pascal JM, Tsodikov OV, Hura GL, Song W, Cotner EA, et al. 2006. A flexible interface between DNA ligase and PCNA supports conformational switching and efficient ligation of DNA. *Mol. Cell* 24:279–91

136. Paytubi S, McMahon SA, Graham S, Liu H, Botting CH, et al. 2012. Displacement of the canonical single-stranded DNA-binding protein in the Thermoproteales. *Proc. Natl. Acad. Sci. USA* 109:E398–405

137. Pellegrini L. 2012. The Pol α-primase complex. *Subcell. Biochem.* 62:157–69

138. Pelve EA, Lindas AC, Knoppel A, Mira A, Bernander R. 2012. Four chromosome replication origins in the archaeon *Pyrobaculum calidifontis*. *Mol. Microbiol.* 85:986–95

139. Perler FB, Kumar S, Kong H. 1996. Thermostable DNA polymerases. *Adv. Protein Chem.* 48:377–435

140. Pluchon PF, Fouqueau T, Creze C, Laurent S, Briffotaux J, et al. 2013. An extended network of genomic maintenance in the archaeon *Pyrococcus abyssi* highlights unexpected associations between eucaryotic homologs. *PLoS ONE* 8:e79707

141. Poplawski A, Grabowski B, Long SE, Kelman Z. 2001. The zinc finger domain of the archaeal minichromosome maintenance protein is required for helicase activity. *J. Biol. Chem.* 276:49371–77

142. Prakash A, Borgstahl GE. 2012. The structure and function of replication protein A in DNA replication. *Subcell. Biochem.* 62:171–96

143. Raymann K, Forterre P, Brochier-Armanet C, Gribaldo S. 2014. Global phylogenomic analysis disentangles the complex evolutionary history of DNA replication in Archaea. *Genome Biol. Evol.* 6:192–212

144. Reimann J, Esser D, Orell A, Amman F, Pham TK, et al. 2013. Archaeal signal transduction: impact of protein phosphatase deletions on cell size, motility and energy metabolism in *Sulfolobus acidocaldarius*. *Mol. Cell. Proteomics* 12:3908–23

145. Remus D, Beuron F, Tolun G, Griffith JD, Morris EP, Diffley JF. 2009. Concerted loading of Mcm2-7 double hexamers around DNA during DNA replication origin licensing. *Cell* 139:719–30

146. Reyes-Lamothe R, Nicolas E, Sherratt DJ. 2012. Chromosome replication and segregation in bacteria. *Annu. Rev. Genet.* 46:121–43

147. Robbins JB, McKinney MC, Guzman CE, Sriratana B, Fitz-Gibbon S, et al. 2005. The euryarchaeota, nature's medium for engineering of single-stranded DNA-binding proteins. *J. Biol. Chem.* 280:15325–39

148. Robinson NP, Bell SD. 2007. Extrachromosomal element capture and the evolution of multiple replication origins in archaeal chromosomes. *Proc. Natl. Acad. Sci. USA* 104:5806–11

149. Robinson NP, Dionne I, Lundgren M, Marsh VL, Bernander R, Bell SD. 2004. Identification of two origins of replication in the single chromosome of the archaeon *Sulfolobus solfataricus*. *Cell* 116:25–38

150. Sakakibara N, Kasiviswanathan R, Melamud E, Han M, Schwarz FP, Kelman Z. 2008. Coupling of DNA binding and helicase activity is mediated by a conserved loop in the MCM protein. *Nucleic Acids Res.* 36:1309–20

151. Sakakibara N, Kelman LM, Kelman Z. 2009. How is the archaeal MCM helicase assembled at the origin? Possible mechanisms. *Biochem. Soc. Trans.* 37:7–11

152. Sakakibara N, Kelman LM, Kelman Z. 2009. Unwinding the structure and function of the archaeal MCM helicase. *Mol. Microbiol.* 72:286–96

153. Samson RY, Bell SD. 2011. Cell cycles and cell division in the archaea. *Curr. Opin. Microbiol.* 14:350–56

154. Samson RY, Xu Y, Gadelha C, Stone TA, Faqiri JN, et al. 2013. Specificity and function of archaeal DNA replication initiator proteins. *Cell Rep.* 3:485–96

155. Sarmiento F, Mrazek J, Whitman WB. 2013. Genome-scale analysis of gene function in the hydrogenotrophic methanogenic archaeon *Methanococcus maripaludis*. *Proc. Natl. Acad. Sci. USA* 110:4726–31

156. Shechter DF, Ying CY, Gautier J. 2000. The intrinsic DNA helicase activity of *Methanobacterium thermoautotrophicum* ΔH minichromosome maintenance protein. *J. Biol. Chem.* 275:15049–59

157. Shen Y, Tang XF, Matsui E, Matsui I. 2004. Subunit interaction and regulation of activity through terminal domains of the family D DNA polymerase from *Pyrococcus horikoshii*. *Biochem. Soc. Trans.* 32:245–49

158. Shin J-H, Grabowski B, Kasiviswanathan R, Bell SD, Kelman Z. 2003. Regulation of minichromosome maintenance helicase activity by Cdc6. *J. Biol. Chem.* 278:38059–67

159. Shin J-H, Heo GY, Kelman Z. 2008. The *Methanothermobacter thermautotrophicus* Cdc6-2 protein, the putative helicase loader, dissociates the minichromosome maintenance helicase. *J. Bacteriol.* 190:4091–94

160. Shin J-H, Heo G-Y, Kelman Z. 2009. The *Methanothermobacter thermautotrophicus* MCM helicase is active as a hexameric ring. *J. Biol. Chem.* 284:540–46

161. Shin J-H, Kelman Z. 2006. The replicative helicases of bacteria, archaea and eukarya can unwind RNA-DNA hybrid substrates. *J. Biol. Chem.* 281:26914–21

162. Shin J-H, Santangelo TJ, Xie Y, Reeve JN, Kelman Z. 2007. Archaeal minichromosome maintenance (MCM) helicase can unwind DNA bound by archaeal histones and transcription factors. *J. Biol. Chem.* 282:4908–15

163. Skowyra A, MacNeill SA. 2012. Identification of essential and non-essential single-stranded DNA-binding proteins in a model archaeal organism. *Nucleic Acids Res.* 40:1077–90

164. Slaymaker IM, Chen XS. 2012. MCM structure and mechanics: what we have learned from archaeal MCM. *Subcell. Biochem.* 62:89–111

165. Sriskanda V, Kelman Z, Hurwitz J, Shuman S. 2000. Characterization of an ATP-dependent DNA ligase from the thermophilic archaeon *Methanobacterium thermoautotrophicum*. *Nucleic Acids Res.* 28:2221–28

166. Stelter M, Gutsche I, Kapp U, Bazin A, Bajic G, et al. 2012. Architecture of a dodecameric bacterial replicative helicase. *Structure* 20:554–64

167. Swiatek A, MacNeill SA. 2010. The archaeo-eukaryotic GINS proteins and the archaeal primase catalytic subunit PriS share a common domain. *Biol. Direct* 5:17

168. Takahashi TS, Wigley DB, Walter JC. 2005. Pumps, paradoxes and ploughshares: mechanism of the MCM2-7 DNA helicase. *Trends Biochem. Sci.* 30:437–44

169. Uemori T, Sato Y, Kato I, Doi H, Ishino Y. 1997. A novel DNA polymerase in the hyperthermophilic archaeon, *Pyrococcus furiosus*: gene cloning, expression, and characterization. *Genes Cells* 2:499–512

170. Vivona JB, Kelman Z. 2003. The diverse spectrum of sliding clamp interacting proteins. *FEBS Lett.* 546:167–72

171. Wadsworth RI, White MF. 2001. Identification and properties of the crenarchaeal single-stranded DNA binding protein from *Sulfolobus solfataricus*. *Nucleic Acids Res.* 29:914–20

172. Wahle E, Lasken RS, Kornberg A. 1989. The dnaB-dnaC replication protein complex of *Escherichia coli*. II. Role of the complex in mobilizing dnaB functions. *J. Biol. Chem.* 264:2469–75

173. Warbrick E, Heatherington W, Lane DP, Glover DM. 1998. PCNA binding proteins in *Drosophila melanogaster*: the analysis of a conserved PCNA binding domain. *Nucleic Acids Res.* 26:3925–32

174. Wigley DB. 2009. ORC proteins: marking the start. *Curr. Opin. Struct. Biol.* 19:72–78

175. Williams GJ, Johnson K, Rudolf J, McMahon SA, Carter L, et al. 2006. Structure of the heterotrimeric PCNA from *Sulfolobus solfataricus*. *Acta Crystallogr. Sect. F* 62:944–48

176. Winter JA, Christofi P, Morroll S, Bunting KA. 2009. The crystal structure of *Haloferax volcanii* proliferating cell nuclear antigen reveals unique surface charge characteristics due to halophilic adaptation. *BMC Struct. Biol.* 9:55

177. Woese CR, Fox GE. 1977. Phylogenetic structure of the prokaryotic domain: the primary kingdoms. *Proc. Natl. Acad. Sci. USA* 74:5088–90

178. Wu K, Lai X, Guo X, Hu J, Xiang X, Huang L. 2007. Interplay between primase and replication factor C in the hyperthermophilic archaeon *Sulfolobus solfataricus*. *Mol. Microbiol.* 63:826–37

179. Yamasaki K, Urushibata Y, Yamasaki T, Arisaka F, Matsui I. 2010. Solution structure of the N-terminal domain of the archaeal D-family DNA polymerase small subunit reveals evolutionary relationship to eukaryotic B-family polymerases. *FEBS Lett.* 584:3370–75

180. Yao N, Turner J, Kelman Z, Stukenberg PT, Dean F, et al. 1996. Clamp loading, unloading and intrinsic stability of the PCNA, β and gp45 sliding clamps of human, *E. coli* and T4 replicases. *Genes Cells* 1:101–13

181. Yao NY, O'Donnell M. 2012. The RFC clamp loader: structure and function. *Subcell. Biochem.* 62:259–79

182. Yuzhakov A, Kelman Z, Hurwitz J, O'Donnell M. 1999. Multiple competition reactions for RPA order the assembly of the DNA polymerase δ holoenzyme. *EMBO J.* 18:6189–99

183. Yuzhakov A, Kelman Z, O'Donnell M. 1999. Trading places on DNA: a three-point switch underlies primer handoff from primase to the replicative DNA polymerase. *Cell* 96:153–63

184. Zhang R, Zhang CT. 2003. Multiple replication origins of the archaeon *Halobacterium* species NRC-1. *Biochem. Biophys. Res. Commun.* 302:728–34

185. Zhao A, Gray FC, MacNeill SA. 2006. ATP- and NAD$^+$-dependent DNA ligases share an essential function in the halophilic archaeon *Haloferax volcanii*. *Mol. Microbiol.* 59:743–52

Molecular Genetic Dissection of Quantitative Trait Loci Regulating Rice Grain Size

Jianru Zuo and Jiayang Li

State Key Laboratory of Plant Genomics and National Plant Gene Research Center, Institute of Genetics and Developmental Biology, Chinese Academy of Sciences, Beijing 100101, China; email: jyli@genetics.ac.cn

Annu. Rev. Genet. 2014. 48:99–118

First published online as a Review in Advance on August 18, 2014

The *Annual Review of Genetics* is online at genet.annualreviews.org

This article's doi: 10.1146/annurev-genet-120213-092138

Keywords

grain size, grain weight, quantitative trait loci, domestication, rice, *Oryza sativa* L.

Abstract

Grain size is one of the most important factors determining rice yield. As a quantitative trait, grain size is predominantly and tightly controlled by genetic factors. Several quantitative trait loci (QTLs) for grain size have been molecularly identified and characterized. These QTLs may act in independent genetic pathways and, along with other identified genes for grain size, are mainly involved in the signaling pathways mediated by proteasomal degradation, phytohormones, and G proteins to regulate cell proliferation and cell elongation. Many of these QTLs and genes have been strongly selected for enhanced rice productivity during domestication and breeding. These findings have paved new ways for understanding the molecular basis of grain size and have substantial implications for genetic improvement of crops.

INTRODUCTION

Global food security is challenged by multiple factors, including continuously increasing population, reduced arable land, global climate change, and, more recently, the demands for the production of biofuels (83). During the past half century, the green revolution, featuring the use of semidwarf genes in rice and wheat (62, 70, 76), has greatly improved yields of these two major crops. Similarly important, the rice yield has also increased since the 1970s via the use of hybrids in China and Southeastern Asia. However, a recent study revealed that yields of maize, rice, wheat, and soybean in 24–39% of growing areas stagnated, collapsed, or never improved (65). In particular, more than 78%, 37%, and 81% of rice growing areas in China, India, and Indonesia (the top three producers of rice), respectively, show yield stagnation from 1961 to 2008 (65), illustrating both the challenge and the potential of increasing crop yield in the coming decades.

Rice is one of the most important food crops worldwide, and more than half of the global population uses rice as the main food source. Rice is also an excellent model species in plant biology, especially for studies on monocotyledonous plants, because of its small genome size and completed genome sequence as well as efficient genetic transformation technology and vast genetic resources. Grain yield of rice is directly determined by four major components, namely the number of panicles or effective tillers per plant, the number of grains per panicle, grain weight, and the ratio of filled grains (68, 101). These first three traits, along with many other related agronomically important traits, have been extensively studied in recent decades and tremendous progress has been made in almost every aspect of rice biology (17, 25, 54, 68, 95, 96, 101, 107, 110, 112). The genetics-based approach, combined with the state-of-the-art functional genomic technologies, has been able to identify and functionally characterize hundreds of agronomically important genes and partially elucidate the underpinning regulatory mechanisms. Remarkably, the molecular characterization of a number of grain yield–related quantitative trait loci (QTLs) in recent years has shed great light on the regulatory mechanisms of the complex agronomic traits and offered new promises for the next generation of super rice (17, 54, 96, 101, 107, 112).

The major consumable products of rice are grains or seeds, which contain starches, lipids, proteins, and mineral nutrients for humans. Typically, a seed mainly consists of the embryo and the endosperm, which are enclosed by the seed coat. Similar to grains of other cereal crops, rice grain is structurally and anatomically different from seeds of dicotyledonous plants, such as *Arabidopsis* (**Figure 1**). In *Arabidopsis*, the embryo occupies the majority space of a mature seed and most nutrients are stored in the cotyledons. In rice, however, the endosperm occupies most of the space in a mature seed and contains most of the storage compounds, mainly as starches, with relatively smaller amounts of storage proteins and lipids and trace amounts of other substances (**Figure 1**). Therefore, rice endosperm is a direct determinant of grain weight and is also the major food source for humans.

Grain weight is mainly determined by grain size (volume) and the degree of grain filling (plumpness) (68, 101). During grain filling, ovary and endosperm cells rapidly divide and expand, accompanied by the transport of nutrients, mainly carbohydrates, from the photosynthetic tissues (source) to the endosperm (sink), and the eventual accumulation of the storage compounds in the sink. Endosperm development in rice has been extensively studied and comprehensively reviewed (25, 61, 110). Grain size, another factor determining grain weight, is specified by its three dimensional structures: grain length (GL), grain width (GW), and grain thickness (GT). Grain size or shape is also an important quality trait of rice grains because of the preferences of consumers in different geographical locations around the world. In addition, grain size is one of the most frequently selected traits during domestication and breeding, thus providing an excellent model for evolution studies (52, 55, 80).

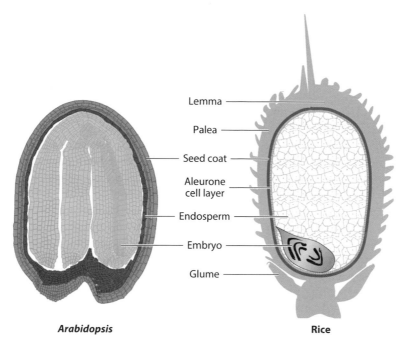

Lemma

Palea

Seed coat

Aleurone cell layer

Endosperm

Embryo

Glume

Arabidopsis

Rice

Figure 1

Longitudinal sections of mature seeds of *Arabidopsis* and rice. Only main structures are shown. Not drawn to scale.

Seed size is predominantly determined by genetic factors, whereas grain filling is controlled by both genetic and environmental factors (68). Yet grain size is also regulated by active interactions between the growth of the maternal integument and the development of the endosperm (13), illustrating a complex network that regulates grain size. In the two model plant species *Arabidopsis* and rice, systematic studies on seed size are still underway, and comprehensive understanding on this complex developmental process is still lacking. Similar to many other agronomic traits, QTLs play a vital role in regulating grain size in rice. In this review, we summarize recent progresses in molecular identification of several important QTLs for rice grain size, their molecular mechanisms in regulating grain size, and their variations during domestication and breeding when applicable. Relevant studies in *Arabidopsis*, maize, and wheat are discussed. We also examine the implications and potentials of these discoveries in molecular breeding.

GENETIC AND MOLECULAR ANALYSES OF QUANTITATIVE TRAIT LOCI FOR GRAIN SIZE

In rice, numerous studies have been conducted to genetically map QTLs for grain yield traits, and thousands of QTLs have been detected during the past several decades. Grain size and shape are important determinants of grain yield and grain quality, which are usually controlled by QTLs. More than 400 QTLs that control grain size and shape have been detected by using various mapping populations (14, 17, 51, 54, 101, 107; see **http://www.gramene.org/qtl** for more information) (**Figure 2**). Several major QTLs for grain size have been molecularly characterized, and their regulatory roles in determining grain size or weight have also been explored (**Table 1**).

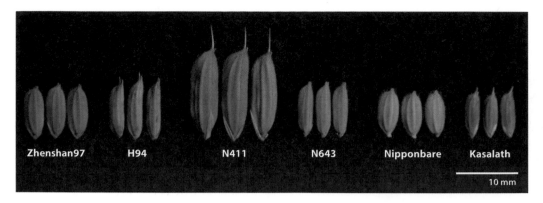

Figure 2

Rice grains produced from various cultivars with different grain sizes. Zhenshan97 and H94 were used for mapping *GS5* (*GRAIN SIZE 5*); N411 and N643 were used for mapping *qGL3* (*GRAIN LENGTH 3*); and Nipponbare and Kasalath were used for mapping *qSW5* (*SEED WIDTH 5*) and *TGW6* (*THOUSAND-GRAIN WIDTH 5*). The photo was provided by Professor Hongsheng Zhang.

GRAIN SIZE 3

GS3 (*GRAIN SIZE 3*) is a major QTL that controls both grain weight and grain length, with minor effects on grain width and thickness, and was the first QTL molecularly characterized to regulate grain size. *GS3* was identified from the progeny derived from a cross between Minghui 63 (large grain) and Chuan 7 (small grain), which show remarkable difference in grain length and weight (9). *GS3*, identified by map-based cloning, encodes a putative transmembrane protein containing a plant-specific organ size regulation (OSR) domain, a tumor necrosis factor receptor/nerve growth factor receptor (TNFR/NGFR) family cysteine-rich domain, and a von Willebrand factor type C (VWFC) domain (9, 49). Four *GS3* alleles were identified in an analysis of 82 accessions. Similar

Table 1   A summary of major quantitative trait loci (QTLs) controlling grain size

QTLs	Regulated traits	Functional annotation	Nature of the beneficial alleles	Domestication and selection	References
GS3	GL, GW	G-protein γ subunit	Loss-of-function	Yes	(9, 49)
GS5	GW, GF	Serine carboxypeptidase	Enhanced expression by polymorphisms in the promoter	Yes	(42)
qGL3/qGL3.1	GL, GF	Protein phosphatase	Reduced enzymatic activity by substitution	No	(16, 64, 109)
GW2	GW, GF	RING-type E3 ligase	Loss-of-function	No	(75)
GW5/qSW5	GW	Polyubiquitin-interacting protein	Loss-of-function	Yes	(74, 97)
GW8	GW	Transcription factor	Enhanced expression by polymorphisms in the promoter	Yes	(94)
GIF1	GF	Invertase	Loss-of-function	Yes	(92)
TGW6	GS, GF	IAA-glucose hydrolase	Loss-of-function	No	(23)

Abbreviations: GF, grain filling, GL, grain length; GS, grain size; GW, grain width.

to Minghui 63 (*GS3-3*), all 11 long-grain cultivars carry a nonsense mutation that causes the formation of a polypeptide lacking all major functional domains mentioned above. Zhenshan 97 (*GS3-1*) and Nipponbare (*GS3-2*) represent the wild type with all the predicted domains and show intermediate grain size. The *GS3-4* alleles, represented by Chuan 7 and several variants that carry various mutations in exon 5, encode truncated proteins lacking the TNFR/NGFR and VWFC domains and show the strongest phenotype in inhibiting grain length (49, 82). These observations, along with transgenic studies, revealed that *GS3* is a negative regulator of grain size, and the OSR domain functions as a negative regulatory motif whose activity is repressed by the TNFR/NGFR and VWFC domains (9, 49, 82). At the cellular level, *GS3* modulates seed length by regulating the cell number in the upper epidermis of the glume, with marginal effects on the cell size, indicating that *GS3* plays an important role in regulating cell division (82).

DENSE AND ERECT PANICLE1 (*DEP1*)/*qPE9-1* was characterized as a major QTL regulating panicle architecture. A dominant mutant allele *dep1* causes an increased number of grains per panicle and a reduction in grain size (19, 111). Interestingly, DEP1 contains an OSR-like domain, which remains intact in the *dep1* allele, implying a possible functional similarity between GS3 and DEP1/qPE9-1 as negative regulators of cell proliferation. More intriguingly, both GS3 and DEP1/qPE9-1 share considerable homology with an atypical heterotrimeric G protein γ subunit (AGG3) in *Arabidopsis* that also contains an OSR-like domain, a TNFR/NGFR domain, and a VWFC domain (5, 39). *AGG3* is involved in the regulation of guard cell K^+-channel and organ size (5, 39). In contrast to that of *GS3* and *DEP1*/*qPE9-1*, however, *AGG3* positively regulates organ size, including seed size (39, 67). Moreover, different results of subcellular localization of DEP1/qPE9-1 and AGG3 were reported (19, 39, 111), further complicating the underpinning regulatory mechanisms of this class of proteins for organ size. Nevertheless, given that AGG3 is a functional heterotrimeric G-protein component (5), these studies highlight the G protein–mediated signaling as a conserved mechanism in regulating organ size in higher plants (see below).

GRAIN SIZE 5

Using a double haploid (DH) population derived from a cross between Zhenshan 97 (wide grain) and H94 (slender grain) (**Figure 2**), the QTL GS5 (*GRAIN SIZE 5*) was detected. *GS5* positively regulates grain width, filling, and weight by promoting cell division and, to a lesser extent, cell elongation of the palea and lemma. *GS5*, identified by positional cloning, encodes a putative serine carboxypeptidase and acts as a positive regulator of a subset of the G1-to-S transition genes of the cell cycle. Transgenic studies and association analysis revealed that the *GS5*-promoted larger grain size is caused by polymorphisms in the *GS5* promoter, which may result in different expression levels of *GS5* (42).

GRAIN LENGTH 3

qGL3/*qGL3.1* (*GRAIN LENGTH 3*), a major QTL that contributes significantly to the three-dimensional size of grains (length, width, and thickness), grain weight, and filling, was identified by three independent studies (16, 64, 109) (see **Figure 2**). *qGL3*/*qGL3.1* encodes a putative serine/threonine protein phosphatase containing a Kelch-like repeat domain (OsPPKL1). Although the allelic QTLs were identified in different genetic mapping populations, an identical mutation, Asp-to-Glu transition at residue 364, was found in the conserved AVLDT motif of the second Kelch domain, which caused the parental cultivars carrying this mutation to produce heavier grains, indicating that Asp[364] plays a critical role for the phosphatase activity of qGL3/qGL3.1 (16, 64, 109). Consistently, the wild-type qGL3/qGL3.1 protein shows higher phosphatase activity than the mutated proteins on its substrate, Cyclin-T1;3, and the knockdown of the *Cyclin-T1;3* expression resulted in shorter grains (64). The reduced phosphatase activity may alter the progression of the cell cycle, thereby causing the increased cell number in the outer glume and, consequently, longer grains (64, 109).

The rice genome contains two *qGL3/qGL3.1*-like genes, *OsPPKL2* and *OsPPKL3*. Overexpression of *OsPPKL1* (*qGL3/qGL3.1*) and *OsPPKL3* produced short grains, whereas the overexpression of *OsPPKL2* resulted in long grains. Consistently, the T-DNA insertion mutants in *OSPPKL3* and *OSPPKL2* showed longer and shorter grains, respectively, indicating that these three homologous proteins play distinctive roles in regulating grain size (109). The *qgl3/qgl3.1* is a rare allele that has not been selected during breeding, thus offering great potential in rice breeding (109). As demonstrated by field trails, the introgression of *qgl3/qgl3.1* into various varieties significantly enhanced grain yield (64, 109).

GRAIN WIDTH 2

The *GW2* (*GRAIN WIDTH 2*) QTL was detected by using the progeny of a cross between WY3 (large grain) and Fengaizhan-1 (FAZ1; small grain). *GW2*, also identified by positional cloning, encodes a novel RING-type protein with the E3 ubiquitin ligase activity. The loss-of-function mutation in *GW2* results in the increased cell number at the outer parenchyma cell layer and, consequently, wider spikelet hull. However, the *gw2* allele does not affect the cell number of endosperm cells but rather causes larger endosperm cells, suggesting that *GW2* regulates development of the spikelet and endosperm by distinctive mechanisms. Meanwhile, *gw2* also enhances the grain milk filling rate and increases the rate of dry matter accumulation, resulting in the increased grain weight without detectable effects on grain quality. Together, these traits lead to a higher grain yield, demonstrating that *GW2* negatively regulates grain size and weight (75).

Notably, a single nucleotide polymorphism (SNP) in the promoter of *TaGW2*, a homolog of *GW2* in wheat (*Triticum aestivum* L.), is tightly associated with grain width and grain weight, and this SNP has been strongly and positively selected during breeding (78). However, the knockdown of *TaGW2* expression by RNAi in wheat resulted in decreased grain size and weight, a phenotype opposite to that of rice *gw2* (3). This phenotypic difference might be caused by different subcellular localization patterns of GW2 (cytoplasm) and TaGW2 (cytoplasm and nucleolus) when transiently expressed in onion epidermal cells and tobacco leaf epidermal cells, respectively (3, 75). Alternatively, GW2 and TaGW2 may have distinctive substrate specificities (see below).

GRAIN WIDTH 5/SEED WIDTH 5

GW5/qSW5 (*GRAIN WIDTH 5/SEED WIDTH 5*) was independently identified as a major QTL that controls grain width and weight by two groups in different recombinant inbred lines (RILs) generated from crosses between Asominori/Nipponbare (wide grains) and IR24/Kasalath (slender grains) (74, 97) (**Figure 2**). The wide-grain phenotype of Nipponbare is attributed to the increased cell number in the outer glumes (74). *GW5/qSW5* encodes a novel nuclear protein that is localized in the nucleus and physically interacts with polyubiquitin, suggesting that GW5 may act in the proteasomal degradation pathway to regulate grain size (97), a role similar to that of *GW2* (see below). Both the Nipponbare and Asominori alleles carry an identical large deletion in *GW5/qSW5*, which is tightly associated with the grain width phenotype in all 146 examined rice cultivars, suggesting a strong artificial selection during breeding. Therefore, *GW5/qSW5* may negatively regulate grain width through the proteasomal degradation pathway (97).

Although both *GW2* and *GW5/qSW5* negatively regulate grain width, presumably in the proteasomal degradation pathway, plants pyramiding *gw2* and *gw5* showed an enhanced phenotype of grain width compared with those carrying one of the two major QTLs, suggesting that these two loci function in independent pathways to regulate grain traits (106).

GRAIN WIDTH 8

The QTL *GW8* (*GRAIN WIDTH 8*) positively regulates grain width and weight via the promotion of cell proliferation (94). *GW8* encodes the transcription factor OsSPL16 that positively regulates the expression of several genes involved in the G1-to-S transition, a regulatory role similar to that of *GS5* (42, 94). A higher expression level of *GW8* promotes cell division and grain filling, thereby increasing grain width and yield. Mutations in the *GW8* promoter, such as in Basmati varieties, cause the formation of more slender grains, which were likely an important target during breeding. Interestingly, a *gs3 gw8* double mutant (NIL-*gw8 gs3* in the HJX74 background) shows an additive phenotype with a more slender grain than its parents, indicating that *GS3* and *GW8* genetically act in independent pathways. Practically, this fact also offers a strategy to simultaneously improve grain quality and yield in breeding (94).

GRAIN INCOMPLETE FILLING 1

As mentioned above, grain size is mainly determined by grain filling and the size of the spikelet hull. During grain filling, carbohydrates synthesized in the photosynthetic organs (source) are transported into grains (sink), which directly determines grain weight. Whereas most QTLs of grain size characterized thus far are directly involved in the regulation of the size of the spikelet hull, the *GRAIN INCOMPLETE FILLING 1* (*GIF1*) gene and the QTL *THOUSAND-GRAIN WEIGHT 6* (*TGW6*) were shown to regulate grain filling, source ability, and sink size (23, 92). *GIF1* encodes a cell wall invertase that is required for carbon partitioning during early grain filling. The *gif1* mutation slows down grain filling, resulting in the reduction of seed weight. Analysis of the *GIF1* sequences from a panel of cultivated rice varieties and wild rice (*Oryza rufipogon*) revealed the presence of the domestication signature in the *GIF1* promoter region. Consistent with this, genetic mapping with introgression lines revealed that the wild rice *GIF1* is responsible for grain weight reduction (92).

THOUSAND-GRAIN WEIGHT 6

During endosperm development, the primary endosperm nucleus undergoes nuclear division to produce multinucleate cells, followed by cellularization. The timing of the transition from the syncytial phase to cellularization is critical for endosperm development and eventually for rice grain size (61, 110). Using a mapping population generated between Nipponbare (wider grains) and Kasalath (slender grains) (**Figure 2**), the QTL *TGW6* (*THOUSAND-GRAIN WEIGHT 6*) was identified by positional cloning, which revealed an important mechanism regulating grain filling during endosperm development. The *TGW6* allele in the Nipponbare cultivar encodes a novel protein with indole-3-acetic acid (IAA)-glucose hydrolase activity and positively regulates free IAA levels in grains. However, the Kasalath *tgw6* allele, carrying a premature stop codon caused by a frame-shift mutation, had a remarkably reduced level of free IAA in NIL^{tgw6}. The increased IAA level in Nipponbare delays the transition from the syncytial to the cellular phase during early endosperm development, which causes the limited cell number and reduced grain length, as evident by the observation that the number of endosperm cell layers in Nipponbare (*TGW6* allele) is significantly reduced compared with that of NIL^{tgw6}. Moreover, the *tgw6* allele

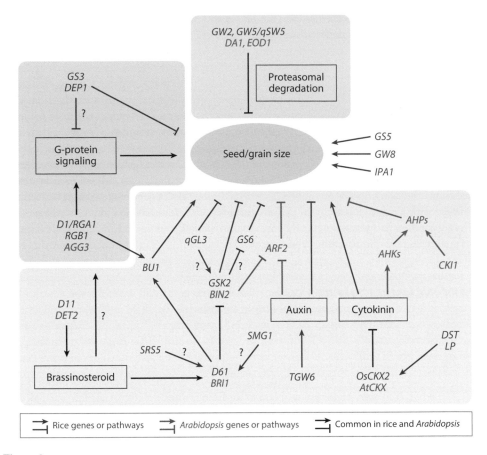

Figure 3

A schematic representation illustrating major regulatory genes that control seed size. These genes are involved in the regulation of cell division and cell expansion, presumably via different signaling pathways (indicated by shaded areas) or via unknown pathways (*GS5*, *GW8*, and *IPA1*). *qGL3* is also known as *qGL3.1*. For conciseness, only components that are known to be involved in the control of seed size are shown in each signaling pathway, and homologous genes of maize and wheat are not shown.

may also be indirectly involved in promoting the translocation rates from source organs to the sink, thereby simultaneously increasing source ability and the size of the sink (23).

Analysis of 14 wild rice lines (*O. rufipogon*) and 69 varieties from the world rice core collection revealed that the Kasalath-type *tgw6* allele was only present in an *O. rufipogon* line and in four cultivars from Indochina, indicating that the Kasalath-type *tgw6* was not a target during rice domestication (23), a case similar to that of *qGL3/qGL3.1* (*OsPPKL1*) and *GW2* (75, 109).

MOLECULAR MECHANISMS REGULATING GRAIN SIZE

The analyses of several major QTLs and a number of other genes for grain size have provided important clues about the molecular mechanisms that regulate this key agronomic trait. Currently available evidence suggests that grain size is controlled by multiple signaling pathways involving ubiquitination-mediated proteasomal degradation, phytohormones, and G-protein signaling pathways (**Figure 3**).

The Proteasomal Degradation Pathway

The characterization of two major QTLs, *GW2* and *GW5/qSW5*, which encode a RING-type E3 ubiquitin ligase and a nuclear protein that interacts with polyubiquitin, respectively, highlights the critical role of proteasomal degradation in regulation of grain size (75, 97). In both cases, the increased cell number was found in the outer glumes of the loss-of-function alleles, indicating that both loci negatively regulate cell proliferation. The regulation of the cell cycle by two ubiquitin ligases, the anaphase-promoting complex/cyclosome (APC/C) and the Skp1/Cullin/F-box (SCF), is a highly conserved mechanism in higher eukaryotes (59, 63). In rice, the APC/C complex was recently found to directly target and degrade MOC1, a key transcription factor controlling tilling, in a cell cycle–dependent manner (44, 102). Therefore, it is reasonable to assume that GW2 and GW5/qSW5 are involved in the degradation of key regulators of the cell cycle (75, 97).

Do *GW2* and *GW5/qSW5* function in the same genetic pathway? A *gw2 gw5* double mutant showed an additive or enhanced phenotype of grain size, suggesting that these two loci function independently (106) or that these two proteins are present in the same complex. In favor of the first possibility, GW2 and GW5/qSW5 were found to be localized in the cytoplasm (75) and the nucleus (97), respectively, when transiently expressed in onion epidermal cells. Given that these two proteins are localized in planta in a similar manner to their fusion proteins that are tagged with green fluorescent protein (GFP) in a heterologous expression system, currently available evidence does not exclude the possibility that GW2 is also localized in the nucleus as its wheat homolog TaGW2 (3). An additional concern is that GW2, a regulator of the cell cycle, should be at least partly localized in the nucleus. Therefore, it is also plausible that GW2 and GW5/qSW5 act in the same pathway as components in a protein complex. In addition to GW2 and GW5/qSW5, several other rice proteins related to proteasomal degradation have been implied to play a role in regulating grain size (6, 35, 37).

Proteasomal degradation–mediated seed development appears to be a conserved mechanism. In *Arabidopsis*, the mutant protein encoded by the *da1-1* allele (DA means big or large in Chinese) imposes a negative effect on DA1, a ubiquitin receptor, and a DA1-related protein (DAR1), which function redundantly to negatively regulate organ size, including seed size, thereby causing the formation of enlarged seeds (43). Moreover, mutations in *ENHANCER OF da1-1* (*EOD1*), encoding an E3 ubiquitin ligase, synergistically enhance the seed size phenotype of *da1-1*, illustrating the importance of the proteasomal degradation pathway in the regulation of seed size. Intriguingly, mutations in *Arabidopsis DA2*, a homolog of the rice *GW2* gene encoding a RING-type E3 ubiquitin E3 ligase (43.1% identity outside of the RING domain), cause the formation of enlarged seeds, whereas *Arabidopsis* plants overexpressing *DA2* or rice *GW2* produce smaller seeds than the wild type, demonstrating a similar mechanism operating in *Arabidopsis* and rice in the regulation of seed size. DA2 physically interacts with DA1 and, consistently, *DA2* acts synergistically with *DA1* to negatively regulate seed size (99). The identification of the substrates of GW2 and DA2 will shed more light on the molecular mechanism of proteasomal degradation–regulated seed size.

Phytohormones

Plant hormones have profound effects on seed development and are directly or indirectly involved in the control of seed size. Although all phytohormones are more or less involved in the regulation of seed development, recent studies highlight the importance of cytokinin, brassinosteroid (BR), and, to a lesser extent, auxin in the control of grain or seed size. Emerging evidence also uncovers a regulatory mechanism on seed size control possibly mediated by the interaction of BR and G-protein signaling.

Auxin

Auxin is involved in almost every aspect of plant growth and development. As discussed above, *TGW6*, a major QTL that controls rice grain weight and grain filling, encodes an IAA-glucose hydrolase and plays an important role in the regulation of auxin homeostasis during endosperm development (23). The maize (*Zea mays*) *defective endosperm-B18* (*de18*) mutant has 40% less dry mass than wild-type seeds, which is associated with a reduced IAA level in *de18* endosperms (88) This reduced level is caused by mutations in the maize *YUCCA1* (*ZmYUC1*) gene that encodes a seed-specific flavin monooxygenase involved in tryptophan-dependent IAA biosynthesis (4). The application of a synthetic auxin naphthalene-acetic acid to developing seeds partially rescues the reduced grain weight phenotype of *de18* (88).

In *Arabidopsis*, mutations in *MEGAINTEGUMENTA* (*MNT*) cause excessive cell division in the integuments, resulting in the formation of enlarged seed coats and dramatically increased seed size. *MNT* is allelic to *AUXIN RESPONSE FACTOR 2* (*ARF2*), encoding a transcription factor that negatively regulates auxin signaling (71). Although *mnt/arf2* shares some phenotypic similarity with *da1-1*, *MNT/ARF2* and *DA1* act in different genetic pathways (43).

Cytokinin

In *Arabidopsis*, enlarged seeds are produced by the reduced cytokinin level or by the impaired cytokinin signaling, including mutations in the receptor *AHK* genes, the histidine phosphotransfer protein (*AHP*) gene, and a histidine kinase gene, *CKI1* (7, 22, 66, 98). In these cases, although the enlarged seeds are mainly attributed to the increased size of the embryo, the endospermal and/or maternal genotypes have a major effect on the control of seed size, which is directly regulated by the cytokinin level and cytokinin signaling through integrating epigenetic and genetic cues (36, 66).

In rice, a major QTL, *Grain number 1a* (*Gn1a*), encodes a cytokinin oxidase/dehydrogenase (OsCKX2) that irreversibly catalyzes degradation of cytokinin and is preferentially expressed in the inflorescence meristem and flowers (1). Mutations in *Gn1a/OsCKX2* (1) and *DROUGHT AND SALT TOLERANCE* (*DST*), which encode a zinc finger transcription factor directly and positively regulate the expression of *Gn1a/OsCKX2* (40), cause the increased accumulation of cytokinin in the inflorescence meristem and, consequently, the increased grain number. In many cases, the increased number of seeds or grains is closely associated with the reduced size or vice versa, presumably owing to the availability of the fixed carbon in the source and the efficiency of transport to the sink (19, 75, 109). However, grain size is also positively correlated with the grain number in some cases. The *DST*[reg1] mutant allele causes the increased cytokinin level in the inflorescence meristem, resulting in the significant increase of both the grain number and grain weight (40). An additional example is the rice *larger panicle* (*lp*) mutant, in which the reduced expression of *OsCKX2* correlates with the increase in both grain number and grain weight (37), suggesting that cytokinin may positively regulate these two traits.

In both *DST*[reg1] and *lp* mutants, the reduced expression of *OsCKX2* and other *OsCKX* genes, which causes an elevated cytokinin level in the inflorescence meristem (37, 40), results in an increase in grain size. This phenotype, however, is opposite to that observed in *Arabidopsis*, in which overexpression of *AtCKX1* and *AtCKX3* results in a dramatic increase in seed weight (98). This phenotypic difference between rice and *Arabidopsis* may be attributed to different mechanisms of cytokinin in regulating reproductive development in these two species. This notion is supported by a recent study on the wheat *TaCKX6-D1* gene (108). Five haplotypes of *TaCKX6-D1* (*a* through *e*) were identified and an 18-bp deletion in intron 2 was found in haplotype *a*, which caused a reduced expression of *TaCKX6-D1*. Association analysis revealed that this deletion was tightly associated with grain weight, a role similar to that in rice, albeit it remains unknown whether grain number is also affected by this deletion. Importantly, a significant domestication signature at the *TaCKX6-D1* locus was found in Chinese wheat germplasms, suggesting that this trait has been positively selected for during breeding programs (108).

Brassinosteroid

The regulatory mechanism of BR on seed development has been vigorously investigated in recent years in *Arabidopsis*, rice, and a number of other species (27). The rice *dwarf 11 (d11)* mutant shows a typical BR-deficient phenotype, characterized as dwarfism, with erect leaves and small round seeds (84). D11 encodes a cytochrome P450 (CYP724B1) that shares considerable homology with several enzymes involved in BR biosynthesis. Consistent with this, the application of BR restores the *d11* mutant phenotype, suggesting that D11 plays a role in BR biosynthesis (84). Similarly, the *Arabidopsis* BR-deficient mutant *deetiolated2 (det2)* causes a reduction in seed size, and BR was found to directly regulate a subset of genes known to be involved in the regulation of seed size (26). In the rice BR-signaling pathway, mutations in the BR receptor gene *D61/BRASSINOSTEROID INSENSITIVE1 (OsBRI1)* and overexpression of *GSK2*, a negative regulator of BR signaling, cause the formation of small grains (57, 84, 87). Mutations in the *SMALL AND ROUND SEED 5 (SRS5)* gene, encoding an α tubulin, impair cell elongation, and a *d61 srs5* double mutant produces seeds smaller than both parents (72). However, the mechanism of *SRS5*-regulated BR signaling remains unclear. *BRASSINOSTEROID UPREGULATED1 (BU1)* was identified as a BR-inducible gene, encoding a putative helix-loop-helix transcription factor that likely acts downstream of *D61/OsBRI1* as a positive regulator of the BR response. Overexpression of *BU1* induces a typical BR phenotype in rice, including the formation of enlarged grains (85).

In a number of genetic screens for rice grain size mutants, several mutants were identified and their wild-type alleles were molecularly characterized. In most cases (with one exception; see below), the identified genes were found to positively regulate grain size through BR signaling–mediated cell division and cell expansion (8, 28, 58). However, the *DWARF AND LOW-TILLERING (DLT)/D62/GS6* gene, encoding a GRAS (refers to GAI, RGA, SCR) family transcription factor that positively regulates BR signaling and gibberellic acid (GA) metabolism (41, 86, 87), is a negative regulator of grain size (79), a phenotype apparently different from other loss-of-function mutations that affect the positive regulators of the BR response. Notably, whereas DLT/D62/GS6 is directly phosphorylated by GSK2, a protein kinase negatively regulating BR signaling, the overexpression of *GSK2* produces small and round seeds similar to other BR mutants (87). Considering that the increased grain size phenotype is specific to the *gs6* allele with marginal effects on grain size by the *dlt* and *d62* alleles (41, 79) and that *DLT/D62/GS6* appears to be strongly selected for in *japonica* cultivars during breeding (79), it is likely that the *gs6* allele regulates grain size through an unknown mechanism independent of BR signaling.

It is worthwhile to point out that most, if not all, BR mutants with a seed size phenotype show severe growth defects, thus raising the possibility that the altered seed size may be a secondary effect of the related mutations. This concern was partly relieved by a recent study on *qGL3/qGL3.1/OsPPKL1*, which encodes a protein phosphatase and specifically affects grain length (16, 64, 109). *OsPPKL1* belongs to a small gene family in which *OsPPKL1* and *OsPPKL3* fall into the same subgroup, whereas *OsPPKL2* and *BSU1 (BRI1 SUPPRESSOR 1)*, an *Arabidopsis* homolog, are in another subgroup (109). BSU1 and its homolog BSL (BSU1-like) directly dephosphorylates the GSK3-like kinase BIN2 (BR-insensitive 2) (32) or antagonizes the BIN2 activity on a downstream component (56) to positively regulate BR signaling. Analogously, *OsPPKL2* may function similarly to BSU1 and BSL to positively regulate BR signaling, whereas *OsPPKL1* and *OsPPKL3* may play an opposite role in BR signaling (109). In this regard, the loss-of-function mutations in *qGL3/qGL3.1/OsPPKL1* may specifically control grain size by modulating BR signaling. Interestingly, *BIN2* was found to directly phosphorylate ARF2, a negative regulator of auxin signaling and seed size (71), and hence was proposed to link BR and auxin signaling pathways (91). However, the possible role of *BIN2* in regulating *Arabidopsis* seed size remains unknown.

G-Protein Signaling

The small and round grain phenotypes have also been noticed in several other rice dwarf mutants, including *d1*, which shows a similar phenotype to *d11* (2, 11). The *d1* mutant displays reduced sensitivity to both GA and BR (2, 11, 60, 93) and is characterized by defects in cell proliferation (24). *D1* was characterized to encode the α subunit of G protein (RGA1 for rice G-protein subunit α) (2, 11), a component of the heterotrimeric G protein, which plays a fundamental role in regulating multiple signaling pathways in plants (30, 89). Similar to that of *d1*, knockdown of the β-subunit (*RGB1*) expression also causes a reduction in grain size (90). However, the loss-of-function mutations in the putative γ subunit gene *GS3* increase grain size (9, 49). Conversely, overexpression of the OSR domain of *GS3* or a gain-of-function mutation in another γ subunit gene *DEP1* (the *dep1* allele), which maintains the intactness of the inhibitory OSR domain, results in the formation of smaller grains than the wild type (19, 49). It appears that the rice γ subunits GS3 and DEP1 function distinctively from their *Arabidopsis* partner AGG3 (39) and the α and β subunits of rice G protein (RGA1 and RGB1) (2, 11, 90). One explanation is that rice may use G-protein signaling mechanisms distinctive from *Arabidopsis* (89), which, however, cannot explain the phenotype of the *d1* mutant and *RGB1*-knockdown plants. Considering that AGG3 genetically and physically interacts with the *Arabidopsis* G-protein β-subunit AGB1 (5, 39), a second possibility would be that GS3 and DEP1 may interact with specific cofactors to negatively regulate RGB1 or to regulate a pathway independent of canonical G-protein signaling.

Although emerging evidence obtained from both rice and *Arabidopsis* strongly suggests the involvement of G-protein signaling in the control of seed size, several questions have yet to be answered. Is G-protein signaling specifically involved in the control of seed size, and if this is indeed the case, is there cross talk between G-protein and BR-signaling pathways? Whereas *D1/RGA1* positively regulates the BR response (60, 93), the BR-induced expression of *BU1* is partly dependent on *D1/RGA1* (85), strongly suggesting that G protein is involved in BR signaling in rice, including the regulation of grain size. Perhaps the most challenging question is the fundamental differences between the rice γ subunits with their *Arabidopsis* partners and the rice α and β subunits, a case similar to that of the functional divergence of cytokinin in these two model species.

Interactions of Different Pathways

Currently available evidence, albeit limited, suggests the control of grain size is mainly involved in three major pathways, which are mediated by proteasomal degradation, phytohormones (especially BR), and G protein, respectively (**Figure 3**). With the identification of several QTLs controlling grain size, as highlighted in **Figure 3**, it is extremely interesting to learn the interactions among these loci and eventually among various signaling pathways. Unfortunately, limited information is available. As both GW2 and GW5/qSW5 encode proteins involved in proteasomal degradation (74, 75, 97), it is tempting to speculate that these two loci act in a same pathway. However, plants carrying *gw2 gw5* and *gs3 gw8* alleles show additive phenotypes compared with their parental lines, ruling out the possibility that these loci act in a linear pathway (94, 106). In a more striking case, N411, a *japonica* accession carrying five known positive grain-size loci (*qgl3/qgl3.1*, *gw2*, *gs3*, *gw5/qsw5*, and *GS5*), shows a giant-like grain shape (1,000-grain weight over 70 g for N411 versus approximately ~20–30 g for most *japonica* varieties; see **Figure 2**), strongly suggesting that at least some of these five loci function in different genetic pathways. Although these observations support a model of the additive effects or a nonlinear module of these loci, they cannot exclude the possibility of mutual regulation in some of these loci. In an analysis of a panel of 180 cultivars, a mutual masking effect on seed length and seed width was observed between *GW5/qSW5* and *GS3*, correlated to the increased expression level of *GS3* in *gw5/qsw5*, suggesting possible genetic interactions between these two loci (105).

Whereas the negative regulation of *GW5/qSW5* on *GS3* is likely an indirect effect, a direct and positive regulation on *DEP1* by *IPA1/WFP* was recently reported (46). Both *IPA1/WFP* and *DEP1* are mainly involved in the regulation of branching with minor effects on grain weight (19, 29). The enhanced expression of *IPA1/WFP* in the *ipa1* allele causes an increase in grain weight, and a *dep1* gain-of-function mutation results in a reduction of grain weight. However, in contrast to that of the branching phenotype, RIL of *ipa1 dep1* shows similar grain weight to RIL-*ipa1 DEP1* (Z. Lu & J. Li, unpublished data), suggesting that *IPA1* may act independently of *DEP1* to regulate grain weight.

The interactive signaling is not only restricted to those regulating a given trait but also occurs in pathways regulating related or different traits. In seed plants, small-seeded species usually produce more seeds for a given amount of energy than large-seeded species (55). This rule also fits for various rice varieties. For example, negative correlation of grain number and grain weight has often been observed (19, 34, 75, 94, 109), although both the grain number and grain weight are increased in some cases (29, 37, 40). Therefore, the coordination of grain weight and grain quality is a major issue in breeding, as the increased grain weight often causes reduction in grain quality (68, 94, 100, 107). In a more extended view, a beneficial locus is sometimes tightly linked with undesirable traits or even controlled by the same gene. In the latter case, an example is that the increased expression of *GW8* enhances productivity but compromises in grain quality (94). Such circumstances, usually caused by the pleiotropic effects of a given locus, should be a major challenge in future breeding programs. Practically, the negative effects or undesirable traits of a locus can be fixed in some cases, as recently demonstrated by pyramiding *gw8* and *gs3*, two genetically unlinked loci, into an elite variety, after which both grain yield and grain quality are improved (94).

EVOLUTION AND DOMESTICATION OF GRAIN-SIZE TRAITS

Domestication is an excellent model for studying evolutionary processes since the Charles Darwin era. Recently, the study of crop domestication has greatly benefited from QTL mapping, genome-wide association studies (GWASs), and whole-genome resequencing studies (15, 18, 33, 47, 52, 103). The genus *Oryza* is considered to have originated approximately 130 mya and to have been domesticated from wild rice (*O. rufipogon*) thousands of years ago (31). The most widely grown Asian cultivated rice (*O. sativa*) has two major subspecies, *japonica* and *indica*, which were domesticated from *O. rufipogon* and crosses between *japonica* and local wild rice, respectively, in southern China (18). It has been estimated that there are more than 120,000 varieties of rice around the world (31). During domestication, many agronomically important traits have been strongly selected, including tiller number, panicle architecture, mating type, shattering, seed coat color, grain number, grain size, and dormancy (69, 80). Similar to other traits, major QTLs for grain size have been major targets of domestication due to their importance in determining yield (**Table 1**).

The best-studied example of the domesticated QTL related to grain size is *GS3*, which has been subjected to strong positive selection during breeding (9, 49, 81, 104). Multiple *GS3* alleles have been identified (45, 49, 81, 82), and the most important allele is the *GS3-3* or A allele, characterized as a C-to-A substitution in exon 2 that converts a Cys residue into a premature stop codon, which results in the loss of all functional domains (9, 10, 81). The A or *GS3-3* allele was first identified in Minghui 63, a long-grain cultivar, and in the same position, the short-grain variety Chuan 7 carries a C nucleotide (C or *GS3-4* allele) (9, 49). Several independent association studies on large collections of wild accessions and cultivated varieties indicated that the A allele, which is likely to originate from a *japonica*-like ancestor and subsequently flow into the *indica* varieties by introgression, was tightly associated with the long-grain phenotype (10, 81, 104). By analyzing 322 wild accessions and 235 accessions of *O. sativa*, the A allele was found in 34% of *O. sativa*, with a lower frequency (4%) in wild rice (*O. rufipogon/Oryza nivara/Oryza spontanea*). Surprisingly, no significant differences were found in average seed length between wild rice and the cultivated varieties, although significant differences in seed width and seed weight were indeed observed between *O. sativa* and *O. rufipogon*. Moreover, no substantial differences in grain size were observed between the A alleles and C alleles of *O. rufipogon*, whereas the *O. sativa* accessions carrying the A or C alleles showed significant variations in grain size, strongly suggesting that the *GS3*-regulated grain size is specific to the genetic background of *O. sativa* (81).

A small deletion in the promoter of the *gw8* allele results in the substantially reduced expression of *GW8* in all developmental stages, which is attributed to the wider and heavier grains produced from plants carrying a *GW8* allele. By examining a panel of 115 modern cultivars, landraces, and wild progenitors, the *GW8* allele is tightly associated with the high-yield *indica* cultivars, suggesting that this allelic variation in the promoter region is selected during breeding (94). In contrast to *GW8*, *GW5/qSW5* negatively regulates grain width and grain weight (74, 97). A large deletion (~1.2 Kb) in *GW5/qSW5*, which causes a loss-of-function mutation, is highly correlated to the grain width phenotype in three examined collections consisting of 142, 46, and 127 cultivars or varieties (45, 74, 97). *GS5*, a minor QTL for seed width, is a positive regulator of grain size and is physically located close to *GW5/qSW5* (~2 Mb). Various expression levels of *GS5*, attributed to polymorphisms in the promoter region in various accessions, are positively correlated to grain width (42). In a comparative study of four domesticated genes related to grain size, it was estimated that *GS3* and *GW5/qSW5* have exerted major effects on grain size, correlated to their strong selection during breeding, whereas *GS5* and perhaps *GW2* (see below) show a relatively weaker effect (45).

No allelic variations were detected in the *GW2* locus in several independent studies (81, 104, 105), although a recent report suggested that *GW2* might be somehow selected for in some *japonica* varieties (45). Nevertheless, the orthologs of rice *GW2* in wheat and maize have been subjected to positive selection during breeding. A rice *GW2*-like gene was identified in chromosome 6A of the wheat genome, designated as *TaGW2-6A*, and two haplotypes, Hap-6A-A and Hap-6A-G, were characterized in the promoter region of *TaGW2-6A*. The Hap-6A-A haplotype causes a reduced expression level of *TaGW2-6A*, which is tightly associated with wider grain and heavier grain weight in a panel of 265 Chinese wheat varieties and shows an increased frequency in recently released varieties. Intriguingly, among European varieties released from 1899 to 1999, the Hap-6A-A haplotype showed a strong bias in geographic distributions, with it mainly found in southern European varieties but rarely found in northern Europe (78). In maize, a linkage analysis located *ZmGW2-CHR4* within a QTL for kernel weight. Association analysis of a diverse panel of 121 inbred lines identified a SNP in the promoter region of *ZmGW2-CHR4*, which is highly related with kernel width and kernel weight (38). The variations found in the promoter regions of *TaGW2-6A* and *ZmGW2-CHR4* are distinctive from those found in rice *GW2*, which carries a 1-bp deletion in the coding region (75), indicating that different regulatory mechanisms and domestication patterns are very likely involved among these three crops.

Similar to *GW2*, *qGL3/qGL3.1* and *TGW6* do not show apparent signs to be selected for during domestication and breeding (23, 64, 109), likely owing to the rareness of the alleles of these three loci, which offer great potential for future breeding programs.

CHALLENGES AND FUTURE PERSPECTIVES

During the past decade, we have witnessed rapid progress in the elucidation of the regulatory mechanisms of grain size. These achievements are highlighted by the identification and characterization of a large number of genes involved in the regulation of grain size, in particular several QTL genes that act as key regulators of grain size. However, we are also facing challenges to a full understanding of the regulatory mechanisms of seed size in rice and other species. As summarized in **Figure 3**, our current knowledge on this critical agronomic trait is rather fragmented, relying on several seemingly independent pathways full of gaps and far beyond understanding in a view of systems biology. All of these highlighted signaling pathways, mainly involving G protein, ubiquitination-mediated proteasomal degradation, and BR, are involved in the regulation of multiple biological processes. Therefore, an additional key question to be addressed is how these signaling pathways specifically regulate grain size, which can be partly answered by filling up the major gaps in each pathway and by characterizing the targets of key regulators.

Obviously, it is essential to identify and characterize additional components of the related signaling pathways. This task can be partly accomplished by the forward genetic approach combined with the reverse genetic approach, as the majority of QTLs for rice grain size remain to be molecularly characterized. An immediate goal of this task is to molecularly characterize more than 400 QTLs for grain size and grain shape. Because of functional redundancy, some of these loci or genes are difficult to identify using the forward genetic approach. In this regard, as demonstrated by a number of recent studies (12, 20, 21, 48, 103), the analysis of rice genome variations during evolution, domestication, and artificial selection, combined with GWASs, should allow fast and efficient identification of important trait-associated loci and alleles that are difficult to identify using genetic approaches. Moreover, systems biology approaches, especially the analyses of transcriptomes, proteasomes, and metabolomes, should enable the identification and construction of the networks underlying rice grain development (50, 77, 110). The use of targeting-induced local lesions in genomes (TILLING) and recently developed genome-editing technologies (53, 73) will greatly facilitate the functional characterization of genes of interest. Taken together, these advances should allow the eventual pyramiding of the beneficial alleles of grain size with other agronomically important traits in a new generation of super rice varieties, which will ideally produce extra-large grains similar to the N411 accession that carries at least five known positive grain size loci (*qgl3*, *gw2*, *gs3*, *GW5/qSW5*, and *GS5*) (109) and that has also significantly improved grain quality.

Finally, the progress summarized in this review focuses on rice grain size, which, together with grain filling, largely determines grain weight. Grain weight, especially grain filling, are eventually dependent on the availability of carbohydrates produced in the source organs and subsequent transport of the photosynthetic products into grains (the sink). Yet regulation of the latter two processes remains poorly understood and is the most important challenge for genetic improvement of major crops, including rice.

DISCLOSURE STATEMENT

The authors are not aware of any affiliations, memberships, funding, or financial holdings that might be perceived as affecting the objectivity of this review.

ACKNOWLEDGMENTS

We thank Dr. Jinye Mu and Ms. Juli Peng for their assistance in editing the literature; Ms. Juli Peng for preparing **Figure 1**; Professor Hongsheng Zhang (Nanjing Agricultural University) for providing photos for **Figure 2**; and Professor Yunhai Li (Institute of Genetics and Developmental Biology, Chinese Academy of Sciences) and Professor Hongsheng Zhang for critical reading of the manuscript. We apologize to the colleagues whose work is not covered in this review because of limited space. This work was supported by grants from National Natural Science Foundation of China (91335204, 91217302) and State Key Laboratory of Plant Genomics.

LITERATURE CITED

1. Ashikari M, Sakakibara H, Lin SY, Yamamoto T, Takashi T, et al. 2005. Cytokinin oxidase regulates rice grain production. *Science* 309:741–45

2. Ashikari M, Wu J, Yano M, Sasaki T, Yoshimura A. 1999. Rice gibberellin-insensitive dwarf mutant gene *Dwarf 1* encodes the α-subunit of GTP-binding protein. *Proc. Natl. Acad. Sci. USA* 96:10284–89

3. Bednarek J, Boulaflous A, Girousse C, Ravel C, Tassy C, et al. 2012. Down-regulation of the *TaGW2* gene by RNA interference results in decreased grain size and weight in wheat. *J. Exp. Bot.* 63:5945–55

4. Bernardi J, Lanubile A, Li Q-B, Kumar D, Kladnik A, et al. 2012. Impaired auxin biosynthesis in the *defective endosperm18* mutant is due to mutational loss of expression in the *ZmYuc1* gene encoding endosperm-specific YUCCA1 protein in maize. *Plant Physiol.* 160:1318–28

5. Chakravorty D, Trusov Y, Zhang W, Acharya BR, Sheahan MB, et al. 2011. An atypical heterotrimeric G-protein γ-subunit is involved in guard cell K⁺-channel regulation and morphological development in *Arabidopsis thaliana*. *Plant J.* 67:840–51

6. Chen Y, Xu Y, Luo W, Li W, Chen N, et al. 2013. The F-box protein OsFBK12 targets OsSAMS1 for degradation and affects pleiotropic phenotypes, including leaf senescence, in rice. *Plant Physiol.* 163:1673–85

7. Deng Y, Dong H, Mu J, Ren B, Zheng B, et al. 2010. *Arabidopsis* histidine kinase CKI1 acts upstream of HISTIDINE PHOSPHOTRANSFER PROTEINS to regulate female gametophyte development and vegetative growth. *Plant Cell* 22:1232–48

8. Duan P, Rao Y, Zeng D, Yang Y, Xu R, et al. 2014. *SMALL GRAIN 1*, which encodes a mitogen-activated protein kinase kinase 4, influences grain size in rice. *Plant J.* 77:547–57

9. Fan C, Xing Y, Mao H, Lu T, Han B, et al. 2006. GS3, a major QTL for grain length and weight and minor QTL for grain width and thickness in rice, encodes a putative transmembrane protein. *Theor. Appl. Genet.* 112:1164–71

10. Fan C, Yu S, Wang C, Xing Y. 2009. A causal C-A mutation in the second exon of *GS3* highly associated with rice grain length and validated as a functional marker. *Theor. Appl. Genet.* 118:465–72

11. Fujisawa Y, Kato T, Ohki S, Ishikawa A, Kitano H, et al. 1999. Suppression of the heterotrimeric G protein causes abnormal morphology, including dwarfism, in rice. *Proc. Natl. Acad. Sci. USA* 96:7575–80

12. Gao Z, Zhao S, He W, Guo L, Peng Y, et al. 2013. Dissecting yield-associated loci in super hybrid rice by resequencing recombinant inbred lines and improving parental genome sequences. *Proc. Natl. Acad. Sci. USA* 110:14492–97

13. Haig D. 2013. Kin conflict in seed development: an interdependent but fractious collective. *Annu. Rev. Cell Dev. Biol.* 29:189–211

14. Hao W, Lin H. 2010. Toward understanding genetic mechanisms of complex traits in rice. *J. Genet. Genomics* 37:653–66

15. He Z, Zhai W, Wen H, Tang T, Wang Y, et al. 2011. Two evolutionary histories in the genome of rice: the roles of domestication genes. *PLoS Genet.* 7:e1002100

16. Hu Z, He H, Zhang S, Sun F, Xin X, et al. 2012. A Kelch motif-containing serine/threonine protein phosphatase determines the large grain QTL trait in rice. *J. Integr. Plant Biol.* 54:979–90

17. Huang R, Jiang L, Zheng J, Wang T, Wang H, et al. 2013. Genetic bases of rice grain shape: so many genes, so little known. *Trends Plant Sci.* 18:218–26

18. Huang X, Kurata N, Wei X, Wang Z, Wang A, et al. 2012. A map of rice genome variation reveals the origin of cultivated rice. *Nature* 490:497–501

19. Huang X, Qian Q, Liu Z, Sun H, He S, et al. 2009. Natural variation at the *DEP1* locus enhances grain yield in rice. *Nat. Genet.* 41:494–97

20. Huang X, Wei X, Sang T, Zhao Q, Feng Q, et al. 2010. Genome-wide association studies of 14 agronomic traits in rice landraces. *Nat. Genet.* 42:961–67

21. Huang X, Zhao Y, Wei X, Li C, Wang A, et al. 2012. Genome-wide association study of flowering time and grain yield traits in a worldwide collection of rice germplasm. *Nat. Genet.* 44:32–39

22. Hutchison CE, Li J, Argueso C, Gonzalez M, Lee E, et al. 2006. The *Arabidopsis* histidine phosphotransfer proteins are redundant positive regulators of cytokinin signaling. *Plant Cell* 18:3073–87

23. Ishimaru K, Hirotsu N, Madoka Y, Murakami N, Hara N, et al. 2013. Loss of function of the IAA-glucose hydrolase gene *TGW6* enhances rice grain weight and increases yield. *Nat. Genet.* 45:707–11

24. Izawa Y, Takayanagi Y, Inaba N, Abe Y, Minami M, et al. 2010. Function and expression pattern of the α subunit of the heterotrimeric G protein in rice. *Plant Cell Physiol.* 51:271–81

25. James MG, Denyer K, Myers AM. 2003. Starch synthesis in the cereal endosperm. *Curr. Opin. Plant Biol.* 6:215–22

26. Jiang W, Huang H, Hu Y, Zhu S, Wang Z, Lin W. 2013. Brassinosteroid regulates seed size and shape in *Arabidopsis*. *Plant Physiol.* 162:1965–77

27. Jiang W, Lin W. 2013. Brassinosteroid functions in *Arabidopsis* seed development. *Plant Signal. Behav.* 8:e25928

28. Jiang Y, Bao L, Jeong S, Kim SK, Xu C, et al. 2012. XIAO is involved in the control of organ size by contributing to the regulation of signaling and homeostasis of brassinosteroids and cell cycling in rice. *Plant J.* 70:398–408

29. Jiao Y, Wang Y, Xue D, Wang J, Yan M, et al. 2010. Regulation of *OsSPL14* by OsmiR156 defines ideal plant architecture in rice. *Nat. Genet.* 42:541–44

30. Jones AM, Assmann SM. 2004. Plants: the latest model system for G-protein research. *EMBO Rep.* 5:572–78

31. Khush GS. 1997. Origin, dispersal, cultivation and variation of rice. *Plant Mol. Biol.* 35:25–34

32. Kim T-W, Guan S, Sun Y, Deng Z, Tang W, et al. 2009. Brassinosteroid signal transduction from cell-surface receptor kinases to nuclear transcription factors. *Nat. Cell Biol.* 11:1254–60

33. Lenser T, Theissen G. 2013. Molecular mechanisms involved in convergent crop domestication. *Trends Plant Sci.* 18:704–14

34. Li F, Liu W, Tang J, Chen J, Tong H, et al. 2010. Rice DENSE AND ERECT PANICLE 2 is essential for determining panicle outgrowth and elongation. *Cell Res.* 20:838–49

35. Li J, Chu H, Zhang Y, Mou T, Wu C, et al. 2012. The rice *HGW* gene encodes a ubiquitin-associated (UBA) domain protein that regulates heading date and grain weight. *PLoS ONE* 7:e34231

36. Li J, Nie X, Tan JLH, Berger F. 2013. Integration of epigenetic and genetic controls of seed size by cytokinin in *Arabidopsis*. *Proc. Natl. Acad. Sci. USA* 110:15479–84

37. Li M, Tang D, Wang K, Wu X, Lu L, et al. 2011. Mutations in the F-box gene *LARGER PANICLE* improve the panicle architecture and enhance the grain yield in rice. *Plant Biotechnol. J.* 9:1002–13

38. Li Q, Li L, Yang X, Warburton M, Bai G, et al. 2010. Relationship, evolutionary fate and function of two maize co-orthologs of rice GW2 associated with kernel size and weight. *BMC Plant Biol.* 10:143

39. Li S, Liu Y, Zheng L, Chen L, Li N, et al. 2012. The plant-specific G protein γ subunit AGG3 influences organ size and shape in *Arabidopsis thaliana*. *New Phytol.* 194:690–703

40. Li S, Zhao B, Yuan D, Duan M, Qian Q, et al. 2013. Rice zinc finger protein DST enhances grain production through controlling *Gn1a/OsCKX2* expression. *Proc. Natl. Acad. Sci. USA* 110:3167–72

41. Li W, Wu J, Weng S, Zhang Y, Zhang D, Shi C. 2010. Identification and characterization of *dwarf 62*, a loss-of-function mutation in *DLT/OsGRAS-32* affecting gibberellin metabolism in rice. *Planta* 232:1383–96

42. Li Y, Fan C, Xing Y, Jiang Y, Luo L, et al. 2011. Natural variation in *GS5* plays an important role in regulating grain size and yield in rice. *Nat. Genet.* 43:1266–69

43. Li Y, Zheng L, Corke F, Smith C, Bevan MW. 2008. Control of final seed and organ size by the *DA1* gene family in *Arabidopsis thaliana*. *Genes Dev.* 22:1331–36

44. Lin Q, Wang D, Dong H, Gu S, Cheng Z, et al. 2012. Rice APC/C^{TE} controls tillering by mediating the degradation of MONOCULM 1. *Nat. Commun.* 3:752–59

45. Lu L, Shao D, Qiu X, Sun L, Yan W, et al. 2013. Natural variation and artificial selection in four genes determine grain shape in rice. *New Phytol.* 200:1269–80

46. Lu Z, Yu H, Xiong G, Wang J, Jiao Y, et al. 2013. Genome-wide binding analysis of the transcription activator IDEAL PLANT ARCHITECTURE1 reveals a complex network regulating rice plant architecture. *Plant Cell* 25:3743–59

47. Luo J, Liu H, Zhou T, Gu B, Huang X, et al. 2013. *An-1* encodes a basic helix-loop-helix protein that regulates awn development, grain size, and grain number in rice. *Plant Cell* 25:3360–76

48. Lyu J, Zhang S, Dong Y, He W, Zhang J, et al. 2013. Analysis of elite variety tag SNPs reveals an important allele in upland rice. *Nat. Commun.* 4:2138–46

49. Mao H, Sun S, Yao J, Wang C, Yu S, et al. 2010. Linking differential domain functions of the GS3 protein to natural variation of grain size in rice. *Proc. Natl. Acad. Sci. USA* 107:19579–84

50. Matsuda F, Okazaki Y, Oikawa A, Kusano M, Nakabayashi R, et al. 2012. Dissection of genotype-phenotype associations in rice grains using metabolome quantitative trait loci analysis. *Plant J.* 70:624–36

51. McCough SR, Doerge RW. 1995. QTL mapping in rice. *Trends Genet.* 11:482–87

52. Meyer RS, Purugganan MD. 2013. Evolution of crop species: genetics of domestication and diversification. *Nat. Rev. Genet.* 14:840–52

53. Miao J, Guo D, Zhang J, Huang Q, Qin G, et al. 2013. Targeted mutagenesis in rice using CRISPR-Cas system. *Cell Res.* 10:1233–36

54. Miura K, Ashikari M, Matsuoka M. 2011. The role of QTLs in the breeding of high-yielding rice. *Trends Plant Sci.* 16:319–26

55. Moles AT, Ackerly DD, Webb CO, Tweddle JC, Dickie JB, Westoby M. 2005. A brief history of seed size. *Science* 307:576–80

56. Mora-García S, Vert G, Yin Y, Caño-Delgado A, Cheong H, Chory J. 2004. Nuclear protein phosphatases with Kelch-repeat domains modulate the response to brassinosteroids in *Arabidopsis*. *Genes Dev.* 18:448–60

57. Morinaka Y, Sakamoto T, Inukai Y, Agetsuma M, Kitano H, et al. 2006. Morphological alteration caused by brassinosteroid insensitivity increases the biomass and grain production of rice. *Plant Physiol.* 141:924–31

58. Nakagawa H, Tanaka A, Tanabata T, Ohtake M, Fujioka S, et al. 2012. *Short Grain1* decreases organ elongation and brassinosteroid response in rice. *Plant Physiol.* 158:1208–19

59. Nakayama KI, Nakayama K. 2006. Ubiquitin ligases: cell-cycle control and cancer. *Nat. Rev. Cancer* 6:369–81

60. Oki K, Inaba N, Kitagawa K, Fujioka S, Kitano H, et al. 2009. Function of the α subunit of rice heterotrimeric G protein in brassinosteroid signaling. *Plant Cell Physiol.* 50:161–72

61. Olsen OA, Linnestad C, Nichols SE. 1999. Developmental biology of the cereal endosperm. *Trends Plant Sci.* 4:253–57

62. Peng J, Richards DE, Hartley NM, Murphy GP, Devos KM, et al. 1999. "Green revolution" genes encode mutant gibberellin response modulators. *Nature* 400:256–61

63. Pines J. 2011. Cubism and the cell cycle: the many faces of the APC/C. *Nat. Rev. Mol. Cell Biol.* 12:427–38

64. Qi P, Lin Y, Song X, Shen J, Huang W, et al. 2012. The novel quantitative trait locus *GL3.1* controls rice grain size and yield by regulating Cyclin-T1;3. *Cell Res.* 22:1666–80

65. Ray DK, Ramankutty N, Mueller ND, West PC, Foley JA. 2012. Recent patterns of crop yield growth and stagnation. *Nat. Commun.* 3:1293–94

66. Riefler M, Novak O, Strnad M, Schmülling T. 2006. *Arabidopsis* cytokinin receptor mutants reveal functions in shoot growth, leaf senescence, seed size, germination, root development, and cytokinin metabolism. *Plant Cell* 18:40–54

67. Roy Choudhury S, Riesselman AJ, Pandey S. 2014. Constitutive or seed-specific overexpression of *Arabidopsis* G-protein γ subunit 3 (AGG3) results in increased seed and oil production and improved stress tolerance in *Camelina sativa*. *Plant Biotechnol. J.* 12:49–59

68. Sakamoto T, Matsuoka M. 2008. Identifying and exploiting grain yield genes in rice. *Curr. Opin. Plant Biol.* 11:209–14

69. Sang T, Ge S. 2007. The puzzle of rice domestication. *J. Integr. Plant Biol.* 49:760–68

70. Sasaki A, Ashikari M, Ueguchi-Tanaka M, Itoh H, Nishimura A, et al. 2002. Green revolution: a mutant gibberellin-synthesis gene in rice. *Nature* 416:701–2

71. Schruff MC, Spielman M, Tiwari S, Adams S, Fenby N, Scott RJ. 2006. The *AUXIN RESPONSE FACTOR 2* gene of *Arabidopsis* links auxin signaling, cell division, and the size of seeds and other organs. *Development* 133:251–61

72. Segami S, Kono I, Ando T, Yano M, Kitano H, et al. 2012. *Small and round seed 5* gene encodes alpha-tubulin regulating seed cell elongation in rice. *Rice* 5:4

73. Shan Q, Wang Y, Li J, Zhang Y, Chen K, et al. 2013. Targeted genome modification of crop plants using a CRISPR-Cas system. *Nat. Biotech.* 31:686–88

74. Shomura A, Izawa T, Ebana K, Ebitani T, Kanegae H, et al. 2008. Deletion in a gene associated with grain size increased yields during rice domestication. *Nat. Genet.* 40:1023–28

75. Song X, Huang W, Shi M, Zhu M, Lin H. 2007. A QTL for rice grain width and weight encodes a previously unknown RING-type E3 ubiquitin ligase. *Nat. Genet.* 39:623–30

76. Spielmeyer W, Ellis MH, Chandler PM. 2002. Semidwarf (*sd-1*), "green revolution" rice, contains a defective gibberellin 20-oxidase gene. *Proc. Natl. Acad. Sci. USA* 99:9043–48

77. Sreenivasulu N, Wobus U. 2013. Seed-development programs: a systems biology–based comparison between dicots and monocots. *Annu. Rev. Plant Biol.* 64:189–217

78. Su Z, Hao C, Wang L, Dong Y, Zhang X. 2011. Identification and development of a functional marker of *TaGW2* associated with grain weight in bread wheat (*Triticum aestivum* L.). *Theor. Appl. Genet.* 122:211–23

79. Sun L, Li X, Fu Y, Zhu Z, Tan L, et al. 2013. *GS6*, a member of the GRAS gene family, negatively regulates grain size in rice. *J. Integr. Plant Biol.* 55:938–49

80. Sweeney M, McCouch S. 2007. The complex history of the domestication of rice. *Ann. Bot.* 100:951–57

81. Takano-Kai N, Jiang H, Kubo T, Sweeney M, Matsumoto T, et al. 2009. Evolutionary history of *GS3*, a gene conferring grain length in rice. *Genetics* 182:1323–34

82. Takano-Kai N, Jiang H, Powell A, McCouch S, Takamure I, et al. 2013. Multiple and independent origins of short seeded alleles of *GS3* in rice. *Breed. Sci.* 63:77–85

83. Takeda S, Matsuoka M. 2008. Genetic approaches to crop improvement: responding to environmental and population changes. *Nat. Rev. Genet.* 9:444–57

84. Tanabe S, Ashikari M, Fujioka S, Takatsuto S, Yoshida S, et al. 2005. A novel cytochrome P450 is implicated in brassinosteroid biosynthesis via the characterization of a rice dwarf mutant, *dwarf11*, with reduced seed length. *Plant Cell* 17:776–90

85. Tanaka A, Nakagawa H, Tomita C, Shimatani Z, Ohtake M, et al. 2009. *BRASSINOSTEROID UPREG-ULATED1*, encoding a helix-loop-helix protein, is a novel gene involved in brassinosteroid signaling and controls bending of the lamina joint in rice. *Plant Physiol.* 151:669–80

86. Tong H, Jin Y, Liu W, Li F, Fang J, et al. 2009. DWARF AND LOW-TILLERING, a new member of the GRAS family, plays positive roles in brassinosteroid signaling in rice. *Plant J.* 58:803–16

87. Tong H, Liu L, Jin Y, Du L, Yin Y, et al. 2012. DWARF AND LOW-TILLERING acts as a direct downstream target of a GSK3/SHAGGY-like kinase to mediate brassinosteroid responses in rice. *Plant Cell* 24:2562–77

88. Torti G, Manzocchi L, Salamini F. 1986. Free and bound indole-acetic acid is low in the endosperm of the maize mutant defective *endosperm-B18*. *Theor. Appl. Genet.* 72:602–5

89. Urano D, Chen J-G, Botella JR, Jones AM. 2013. Heterotrimeric G protein signalling in the plant kingdom. *Open. Biol.* 3:120186

90. Utsunomiya Y, Samejima C, Takayanagi Y, Izawa Y, Yoshida T, et al. 2011. Suppression of the rice heterotrimeric G protein β-subunit gene, *RGB1*, causes dwarfism and browning of internodes and lamina joint regions. *Plant J.* 67:907–16

91. Vert G, Walcher CL, Chory J, Nemhauser JL. 2008. Integration of auxin and brassinosteroid pathways by Auxin Response Factor 2. *Proc. Natl. Acad. Sci. USA* 105:9829–34

92. Wang E, Wang J, Zhu X, Hao W, Wang L, et al. 2008. Control of rice grain-filling and yield by a gene with a potential signature of domestication. *Nat. Genet.* 40:1370–74

93. Wang L, Xu Y, Ma Q, Li D, Xu Z, Chong K. 2006. Heterotrimeric G protein α subunit is involved in rice brassinosteroid response. *Cell Res.* 16:916–22

94. Wang S, Wu K, Yuan Q, Liu X, Liu Z, et al. 2012. Control of grain size, shape and quality by *OsSPL16* in rice. *Nat. Genet.* 44:950–54

95. Wang Y, Li J. 2008. Molecular basis of plant architecture. *Annu. Rev. Plant Biol.* 59:253–79

96. Wang Y, Li J. 2011. Branching in rice. *Curr. Opin. Plant Biol.* 14:94–99

97. Weng J, Gu S, Wan X, Gao H, Guo T, et al. 2008. Isolation and initial characterization of *GW5*, a major QTL associated with rice grain width and weight. *Cell Res.* 18:1199–209

98. Werner T, Motyka V, Laucou V, Smets R, Van Onckelen H, Schmülling T. 2003. Cytokinin-deficient transgenic *Arabidopsis* plants show multiple developmental alterations indicating opposite functions of cytokinins in the regulation of shoot and root meristem activity. *Plant Cell* 15:2532–50

99. Xia T, Li N, Dumenil J, Li J, Kamenski A, et al. 2013. The ubiquitin receptor DA1 interacts with the E3 ubiquitin ligase DA2 to regulate seed and organ size in *Arabidopsis*. *Plant Cell* 25:3347–59

100. Xie X, Song M-H, Jin F, Ahn S-N, Suh J-P, et al. 2006. Fine mapping of a grain weight quantitative trait locus on rice chromosome 8 using near-isogenic lines derived from a cross between *Oryza sativa* and *Oryza rufipogon*. *Theor. Appl. Genet.* 113:885–94

101. Xing Y, Zhang Q. 2010. Genetic and molecular bases of rice yield. *Annu. Rev. Plant Biol.* 61:421–42

102. Xu C, Wang Y, Yu Y, Duan J, Liao Z, et al. 2012. Degradation of MONOCULM 1 by APC/C[TAD1] regulates rice tillering. *Nat. Commun.* 3:750–58

103. Xu X, Liu X, Ge S, Jensen JD, Hu F, et al. 2012. Resequencing 50 accessions of cultivated and wild rice yields markers for identifying agronomically important genes. *Nat. Biotechnol.* 30:105–11

104. Yan C, Yan S, Yang Y, Zeng X, Fang Y, et al. 2009. Development of gene-tagged markers for quantitative trait loci underlying rice yield components. *Euphytica* 169:215–26

105. Yan S, Zou G, Li S, Wang H, Liu H, et al. 2011. Seed size is determined by the combinations of the genes controlling different seed characteristics in rice. *Theor. Appl. Genet.* 123:1173–81

106. Ying J, Gao J, Shan J, Zhu M, Shi M, Lin H. 2012. Dissecting the genetic basis of extremely large grain shape in rice cultivar "JZ1560." *J. Genet. Genomics* 39:325–33

107. Yu Y, Wing RA, Li J. 2013. Grain quality. In *Genetics and Genomics of Rice*, ed. Q Zhang, RA Wing, pp. 237–54. New York: Springer

108. Zhang L, Zhao Y, Gao L, Zhao G, Zhou R, et al. 2012. *TaCKX6-D1*, the ortholog of rice *OsCKX2*, is associated with grain weight in hexaploid wheat. *New Phytol.* 195:574–84

109. Zhang X, Wang J, Huang J, Lan H, Wang C, et al. 2012. Rare allele of *OsPPKL1* associated with grain length causes extra-large grain and a significant yield increase in rice. *Proc. Natl. Acad. Sci. USA* 109:21534–39

110. Zhou S, Yin L, Xue H. 2013. Functional genomics based understanding of rice endosperm development. *Curr. Opin. Plant Biol.* 16:236–46

111. Zhou Y, Zhu J, Li Z, Yi C, Liu J, et al. 2009. Deletion in a quantitative trait gene *qPE9-1* associated with panicle erectness improves plant architecture during rice domestication. *Genetics* 183:315–24

112. Zuo J, Li J. 2014. Molecular dissection of complex agronomic traits of rice: a team effort by Chinese scientists in recent years. *Nat. Sci. Rev.* 1:253–76

Exploring Developmental and Physiological Functions of Fatty Acid and Lipid Variants Through Worm and Fly Genetics

Huanhu Zhu and Min Han

Howard Hughes Medical Institute and Department of Molecular, Cellular, and Developmental Biology, University of Colorado, Boulder, Colorado 80309; email: huanhu.zhu@colorado.edu

Annu. Rev. Genet. 2014. 48:119–48

First published online as a Review in Advance on August 25, 2014

The *Annual Review of Genetics* is online at genet.annualreviews.org

This article's doi: 10.1146/annurev-genet-041814-095928

Keywords

metabolism, lipid biology, *Caenorhabditis elegans*, *Drosophila*, cell signaling, human, disease

Abstract

Lipids are more than biomolecules for energy storage and membrane structure. With ample structural variation, lipids critically participate in nearly all aspects of cellular function. Lipid homeostasis and metabolism are closely related to major human diseases and health problems. However, lipid functional studies have been significantly underdeveloped, partly because of the difficulty in applying genetics and common molecular approaches to tackle the complexity associated with lipid biosynthesis, metabolism, and function. In the past decade, a number of laboratories began to analyze the roles of lipid metabolism in development and other physiological functions using animal models and combining genetics, genomics, and biochemical approaches. These pioneering efforts have not only provided valuable insights regarding lipid functions in vivo but have also established feasible methodology for future studies. Here, we review a subset of these studies using *Caenorhabditis elegans* and *Drosophila melanogaster*.

INTRODUCTION: THE STUDY OF LIPID VARIANT FUNCTIONS IS A NEW RESEARCH FRONTIER

Lipids are best known for their roles as major components of the biomembrane and as energy storage molecules. The fact that lipid molecules are highly variable in structure, with hundreds or even thousands of different species, indicates that they perform far more functions than just serving in simple structural roles. Indeed, decades of research have indicated that lipids participate in a broad range of cellular functions, including signaling events that regulate animal growth, development, and behaviors. Although we presently have a relatively better understanding of the structures and chemical properties of various lipids, our knowledge about the biological functions of many of these variations is quite poor. In particular, studies using genetic model organisms such as flies and worms have resulted in a plethora of breakthrough findings to establish important paradigms regarding the functions of proteins and noncoding RNAs in regulating cellular processes and animal development/behaviors, but functional studies of lipid variants lag far behind. For example, the lipid composition in specific tissues and at specific stages of these model organisms had not been characterized until some limited analyses were begun in recent years (6, 54, 90). Our knowledge of the impact of a healthy membrane lipid composition on physiological functions, and the mechanisms by which a healthy composition is achieved, is very limited.

There are multiple reasons for inadequate functional studies of lipid variants using genetic model organisms. Although the competing attraction from the exciting protein and RNA research fields in past decades and the lack of awareness of diverse functions associated with lipids are likely significant contributors to the problem, technical difficulty also presents important obstacles to lipid functional studies. Unlike the situation with proteins and RNAs, there is usually no linear relationship between genomic sequence variation and the lipid structural variation, making it difficult to specifically ablate the function of a specific lipid by genetic means. For example, the biosynthesis of even a simple fatty acid (FA) involves the functions of multiple enzymes that rarely limit their roles to one specific FA species. Furthermore, the destination of a given FA is multiple and complex, as it can be elongated or shortened and/or be desaturated or hydroxylated to give rise to different FAs or FA derivatives, be incorporated into complex lipids such as triglycerides, glycerophospholipids (GPLs), sphingolipids, and sterols, or be used in post-translational modification of proteins through FA acylation. Therefore, phenotypes caused by a single genetic variant in FA biosynthesis are commonly insufficient to indicate the function of a specific FA or its derivative. Another obstacle of lipid functional studies is the technical limitation beyond genetics. For example, a specific protein can be easily detected by antibody-based immunohistochemistry and immunofluorescence, or by following transgenic proteins with short peptide tags or fluorescent markers. Similarly, a specific RNA can be detected by in situ hybridization, reverse transcription polymerase chain reaction, or microarray analysis. These techniques, unfortunately, cannot be directly applied to lipids. By contrast, lipids have traditionally been labeled or analyzed by chemical dyes or thin-layer chromatography (5, 42), but these methods usually are not precise, reliable, or convenient. Furthermore, powerful assays to characterize protein-protein, protein-DNA, or protein-RNA interactions [e.g., yeast two-hybridization and immunoprecipitation (IP) and chromatin IP followed by high-throughput analysis] are also generally not available for lipid analysis. It may also be fair to state that lipid profiling using mass spectrometry is significantly more difficult than profiling expression levels of proteins and RNAs, despite the recent development of shotgun analysis methodology (49, 61).

Despite the obstacles mentioned above, functional analysis of lipids has attracted the attention of many scientists in recent years, including those who have historically focused on the study of protein or RNA functions using genetic model organisms. The sense of urgency to advance the lipid research field was incited by information from several sources. First, cell biology and biochemical studies using cultured cells or yeast have accumulated exciting information indicating specific roles of certain lipid molecules in various cellular processes, such as cell division, endocytosis, exocytosis, and apoptosis as well as many other cell signaling events (e.g., 36, 132). Second, human disease studies have indicated the association of lipid metabolism and lipids with the pathology of major diseases, including diabetes, cancers, and cardiovascular diseases, and the problem of obesity (4, 68). Finally, genetic screens and genome-wide RNAi screens in model organisms, such as the nematode *Caenorhabditis elegans* (referred to as *C. elegans* hereafter) and the fruit fly *Drosophila melanogaster* (referred to as *Drosophila* hereafter), have also identified the roles of many lipid biosynthesis/metabolism enzymes in development, behavior, stress responses, and aging (e.g., 5, 51, 53). The advances from these different fields inevitably triggered the merge of disciplines aiming to understand lipid functions at a more advanced level. At the same time, the development of new techniques has enabled scientists to do lipid-related analyses in a more accurate and efficient way. The combination of gas chromatography, high-performance liquid chromatography, nuclear magnetic resonance spectrometry, and, especially, mass spectrometry permits the more precise identification and quantification of a broad range of lipid molecules (e.g., 20). Fluorescent-conjugated lipid analogs can be used to mark the potential subcellular localization of related lipids (147). Chemical synthesis of lipids may also facilitate the verification of the proposed function of its endogenous counterpart by supplementation (129). IP methods for the detection of acylated proteins could potentially facilitate mechanistic studies of lipid-protein interactions (85). All of these advances render lipid biology as an emerging research frontier that may continue to attract the efforts of scientists, including geneticists.

Currently, lipid-related studies using genetic model organisms could be roughly divided into three areas. First, extensive effort and progress have been made in understanding the regulations of lipid biosynthesis (anabolism and catabolism) by signal transduction and transcription factor–mediated regulatory pathways under various physiological conditions. Because the primary targets are mainly proteins and nucleic acids in regulatory pathways, the application of genetic tools is relatively more straightforward. The second area involves the analysis of the impact of specific lipid metabolic pathways on the global lipid composition. Polyunsaturated FAs (PUFAs), GPLs, sphingolipids, and cholesterol have been shown to have a role in the lipid homeostasis regulation network (7, 14, 31, 181). The third area involves studies aiming to understand the specific functions associated with individual lipid molecules/classes and their impact on development and other physiological functions. This review mainly focuses on this third area, even though it is diverse in its scope and relatively less developed, because studies using model organisms are vitally important for dissecting physiological functions. The progress of other aspects of lipid metabolism–related studies, especially research on the impact of lipid metabolism and transportation on obesity and age-related programs, can be found in several excellent recent reviews (12, 63, 96, 182). Here, we discuss advances in understanding the physiological roles of several lipid classes from *C. elegans* and *Drosophila* genetic studies.

PHYSIOLOGICAL FUNCTIONS ASSOCIATED WITH FATTY ACID VARIANTS

FAs are highly variable in their structures, with hundreds of different species (**Figure 1**). The degree of saturation (number of double bonds) and chain length (number of carbons) are two well-known variations that are under the regulation of FA desaturases and elongases. In addition, there are FAs with odd-number or branched-carbon chains, as well as the common even-number and straight-carbon chains. Although a functional importance to maintain these variants is expected,

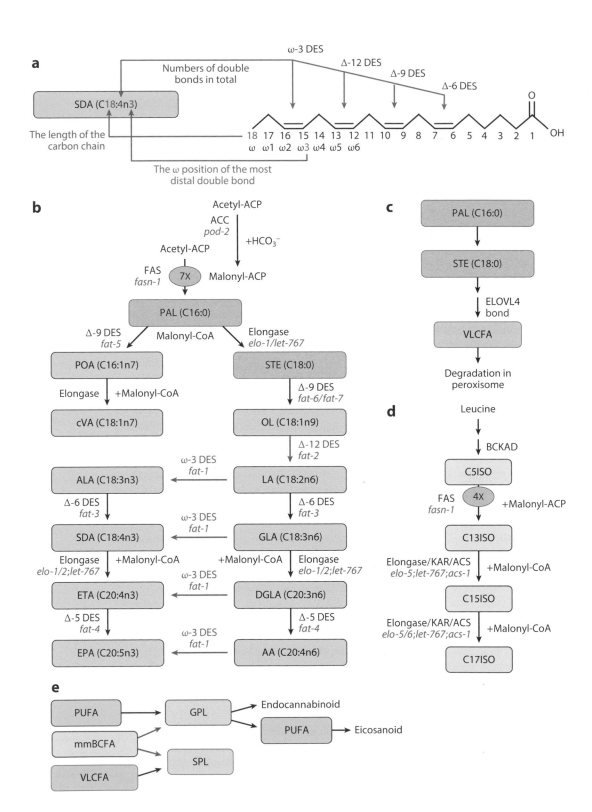

our knowledge about their functional specificities is poor, despite sporadic reports showing the association of certain human diseases with certain FA deficiencies (124, 196). Perhaps the most well-known FAs are ω-3 PUFAs. These FAs are enriched in fish oil, which has consequently become an extremely popular dietary supplement worldwide due to numerous claimed health benefits, including the prevention of mental illness and cardiovascular diseases (64). Although some of these claims were suggested by correlation data from population and disease studies (64), controlled population studies have failed to show a strong benefit of fish oil supplementation for people with a normal diet (151). Importantly, mechanistic studies of the claimed benefits of high levels of ω-3 FAs over other PUFAs are lacking. This issue alone points to the need to carry out functional analyses of FA variants and their homeostases by studying model organisms.

Functional Specificities Associated with Various Unsaturated Fatty Acids

FAs are biosynthesized from acyl carrier proteins (ACP), specifically acetyl-ACP and malonyl-ACP, by acetyl-CoA carboxylase (ACC) and FA synthase (FAS) in the mitochondria (**Figure 1b**) (104). After palmitic acid (C16:0) has been synthesized, the majority is transported from the mitochondria to the endoplasmic reticulum (ER) to be further elongated to long-chain FAs (77). During this process, a series of desaturases can introduce double bonds to the carbon chain to produce monounsaturated FAs (MUFAs) and PUFAs. In mammals, PUFAs are considered essential because mammals lack the related Δ-12 desaturases and ω-3 desaturases necessary to produce ω-6 and ω-3 FAs, respectively (**Figure 1a,b**). Mammals fed on ω-6- and ω-3-free food present multiple severe developmental defects and die early (55). PUFAs can be incorporated into membrane lipids or be further metabolized into important signaling lipids, such as eicosanoids or endocannabinoids (**Figure 1e**).

Unlike mammals, the nematode *C. elegans* has Δ-12 desaturase and ω-3 desaturase enzymes that permit biosynthesis of all PUFAs endogenously (171) (**Figure 1b**). Because the common laboratory food of *C. elegans*, *E. coli* OP50, does not produce PUFAs (171), the dependence of *C. elegans* on endogenous PUFA biosynthesis facilitates genetic studies of the physiological functions of certain PUFAs in specific tissues. By combining gas chromatography and FA supplementation on mutants, many genes involved in MUFA and PUFA synthesis in *C. elegans* have been discovered (186) (**Figure 1b**). Defects in MUFA/PUFA biosynthesis from single or multiple genetic mutations have been reported to cause various developmental phenotypes, including slow growth, small body size, molting defects, and reduced brood size (71, 91, 186).

Figure 1

Metabolic pathways of various fatty acids (FAs). (*a*) A diagram using stearidonic acid (SDA) as an example shows the nomenclature of common FAs. (*b–d*) Simplified pathways of FA biosynthesis, including lipid metabolites in rectangles [saturated FAs in gray, unsaturated FAs in red, VLCFAs (very-long-chain fatty acids) in orange, mmBCFAs (monomethyl branched-chain fatty acids) in light blue], catalytic steps (*arrow*), enzymes (black indicates that the step/enzyme is also conserved in mammals, and blue indicates that the step/enzyme is unique in *Caenorhabditis elegans*), in corresponding genes of *C. elegans* or *Drosophila* (*C. elegans* in blue and *Drosophila* in red). (*b*) Unsaturated FA biosynthetic pathway in *C. elegans*. (*c*) VLCFA biosynthetic pathway in *Drosophila*. (*d*) mmBCFA biosynthetic pathway in *C. elegans*. (*e*) A diagram showing how various FAs can be incorporated into glycerophospholipids (GPLs) and sphingolipids (SPLs), or become precursors of eicosanoids and endocannabinoids. Abbreviations: AA, arachidonic acid; ACC, acetyl-CoA carboxylase; ACP, acyl carrier proteins; ACS, acyl-CoA synthetase; ALA, α-linolenic acid; BCKAD, branched-chain α-keto acid dehydrogenase complex; cVA, *cis*-vaccenic acid; C5ISO, isovaleric acid; C13ISO, 11-methyldodecanoic acid; C15ISO, 13-methyltetradecanoic acid; C17ISO, 15-methylhexadecanoic acid; DES, desaturase; DGLA, dihomo-γ-linolenic acid; EPA, eicosapentaenoic acid; ETA, eicosatetraenoic acid; FAS, fatty acid synthase; GLA, γ-linolenic acid; KAR, 3-ketoacyl-CoA reductase; LA, linolenic acid; OL, oleic acid; PAL, palmitic acid; POA, palmitoleic acid; PUFA, polyunsaturated fatty acid; SDA, stearidonic acid; STE, stearic acid.

More importantly, the establishment of the MUFA/PUFA synthetic pathways in *C. elegans* enables the analyses of the functional relationships between PUFAs and known cellular signaling/regulatory pathways for specific physiological functions, especially in stress response. For example, Δ-6 FAs, such as stearidonic acid (C18:4n3; SDA) or γ-linolenic acid (C18:3n6; GLA), were found to be essential for p38 MAPK (mitogen-associated protein kinase)-dependent innate immunity against pathogen attack (131). The mRNA levels of multiple genes in the p38 pathway are dependent on the intestinal FAT-3 (Δ-6 desaturase) activity or its products (SDA and GLA) rather than on downstream PUFAs (131). In another interesting study, ω-6 PUFAs, but not ω-3 PUFAs, were indicated to induce autophagy and extend life span under dietary restrictive conditions (133).

Another set of PUFA-related studies in *C. elegans* is also potentially of great significance for human health. It was observed that the lack of long-chain ω-3 and ω-6 PUFAs can cause chemotaxis and avoidance defects in *C. elegans* (81, 105). Neuronal restoration of a related enzyme (Δ-6 desaturase encoded by *fat-3*) can rescue the defect, suggesting that it is not a generic FA metabolic problem. Those defects could be rescued by supplementation of long-chain ω-3 and ω-6 PUFA to *fat-3*(−) adults, suggesting that the defects are not due to developmental miscues (81, 105). Further analysis suggests that the defect is caused by a decrease in synaptic vesicles at the recycling step (117). A study done in mammals suggests that a Parkinson's disease–related protein α-synuclein may be involved in PUFA-dependent synaptic vesicle recycling, which is also clathrin dependent (9). The detailed mechanism by which long-chain ω-3 and ω-6 PUFAs regulate the synaptic vesicle recycling, however, is still unknown.

Because *C. elegans* can also obtain major PUFAs through diet in addition to endogenous biosynthesis, the physiological effect of an excess supplement can also be studied (133, 187). An intriguing example involves dihomo-γ-linolenic acid (C20:3n6; DGLA) that is synthesized endogenously at the level of ~4% of total FAs in *C. elegans* (186). Double mutants of *fat-1*(−);*fat-4*(−), which lack ω-3 and Δ-5 desaturase activities, result in accumulation of DGLA to 22% of total FA and show almost wild-type fertility, whereas 22% DGLA accumulation from an exogenous supplement results in complete sterility (187). Interestingly, this sterility was later shown to be efficiently suppressed by several mutations that disrupt the insulin receptor TGF-β or the peroxisomal β-oxidation pathway without a significant decrease in the DGLA level. This finding suggests a signaling role of DGLA in germline development that can be modified by changing a downstream activity (189). Why the high level of exogenous, but not endogenous, DGLA disrupts germline development and which downstream pathway senses this level of DGLA are interesting questions to address in the future.

Compared with *C. elegans*, *Drosophila* has a limited capacity for PUFA biosynthesis (180). *Drosophila* has been reported to lack Δ-5 and Δ-6 desaturases and thus cannot synthesize the C20 and C22 PUFAs (161). Those PUFAs are also reported to be nonessential for *Drosophila* development (60, 161). Among the desaturases for MUFA and PUFA biosynthesis, the Δ-9 desaturases have been extensively studied in flies for their functions in sex pheromone biosynthesis and related courtship behaviors (26, 40). One class of these pheromones is synthesized from saturated FA substrates, usually after undergoing one or two desaturations, elongated to generate 23–29 carbon very-long-chain unsaturated FAs, decarboxylated, and finally secreted in hydrocarbon (monoene or diene) forms (40). The specific desaturases and elongases responsible for synthesizing these very-long-chain unsaturated FAs have been identified (22, 23). Another unsaturated FA derivative known to be involved in mating behavior is *cis*-vaccenyl acetate (cVA). Found in semen, cVA helps male flies avoid recently mated nonvirgin females during the courtship process (33). The pheromone protein LUSH that binds to cVA, the pheromone receptor Or67d, and the cofactor SNMP has been identified and analyzed for this fascinating function (10, 97, 100). How cVA activates its receptor, however, is still under debate (47, 100). Additionally, the cVA pheromone system has also been used in some learning and behavior studies in *Drosophila* (84, 111), possibly because of the convenience of FA supplementation.

The Roles of Very-Long-Chain Fatty Acids Are Mostly Undefined

The length variation of FAs is quite impressive, ranging from short-chain FAs (≤6 carbons) to very-long-chain FAs (VLCFAs) (from 22 to 40 carbons). The length differences in FA side chains are expected to impact the functions or metabolism of higher-order lipids, but nearly nothing is known in this regard. One of the few sporadic studies related to this came from human genetic analysis of an early onset Stargardt-like macular dystrophy, resulting from a heterozygous mutation in the *ELOVL4* gene, which encodes the FA elongase of VLCFAs with >26 carbons (196). Genetic analyses also indicate that epidermal Elovl4 is essential for survival and development in mice (122, 178). Despite extensive studies that provide plausible hypotheses (178), we still do not know how missing the presumed VLCFA products of *ELOVL4* causes macular degeneration in the dominant condition and neonatal death in the homozygous condition (179).

Interestingly, in *Drosophila*, the VLCFA is also reported to be essential. Ablation of oenocyte expression of genes involved in VLCFA synthesis causes larval lethality at the L2/L3 stage after an away-from-food behavior, possibly because of hypoxia resulting from the water tightness defect of spiracles combined with liquid media that may then leak into the tracheal trunk (138). Interestingly, an epidermal permeability barrier function was also reported to be defective in mouse *Elovl4* homozygous mutants, raising a question regarding the potential connection between the fly and mouse VLCFA functions. Proper VLCFA level is also reported to be important for *Drosophila* fertility (19, 170). Both the low VLCFA level caused by mutating a *Drosophila* VLCFA elongase (Bond) and the high VLCFA level caused by mutating *Drosophila* peroxins such as PEX2, PEX10, or PEX12 cause similar spermatogenesis defects, including spermatocyte cytokinesis failure (19, 170) (**Figure 1c**).

Although some of these studies tested the changed level of VLCFAs in vivo (19, 80), it is still not known which VLCFA species, or its metabolite, is responsible for this function or how it is achieved. A VLCFA-containing sphingolipid is a potential candidate for this function and is found in *Drosophila* and *C. elegans* (24, 145) (**Figure 1e**). For example, double mutations of acyltransferases ACS-20 and ACS-22 in *C. elegans*, which potentially transfer VLCFAs to sphingomyelin, are reported to cause a cuticle structure defect (80). Tissue-specific knockdown of sphingolipid biosynthetic genes and lipid profiling in *Drosophila* and *C. elegans* may help to advance the understanding of the problem (see section Sphingolipids Have Diverse Functions in Regulating Cellular and Developmental Events).

Most of the functions of PUFAs or VLCFAs are likely to be executed by higher-order lipids, but yet little is known about the linkage between specific FAs with these higher-order lipids regarding these specific functions. Studies on PUFA-derived eicosanoids and endocannabinoids and monomethyl branched-chain FA (mmBCFA)-derived sphingolipids (reviewed below) have made significant progress in this regard (**Figure 1e**).

Roles of Eicosanoids and Endocannabinoids in Germline Development and Aging

In addition to the functions of PUFAs themselves, PUFA derivatives such as eicosanoids and endocannabinoids are important signaling lipids with numerous physiological functions in mammals (142, 166). *C. elegans* also produces a series of eicosanoids (32, 180). Shown through a series of

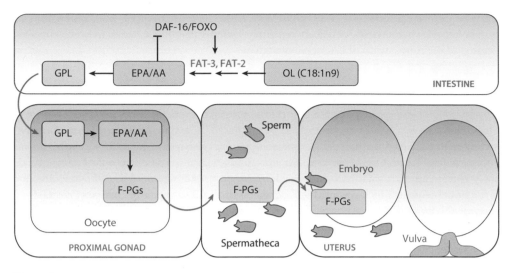

Figure 2

Eicosanoids guide the target movement of sperm in *Caenorhabditis elegans*. A graphical illustration of how eicosanoid F-series prostaglandins (F-PGs) synthesized in the gonad target *C. elegans* sperm to the spermatheca for egg fertilization. DAF-16/FOXO promotes intestinal polyunsaturated fatty acid (PUFA) biosynthesis through increased expression of *fat*-family genes. Glycerophospholipid (GPL)-containing yolk transfers to oocytes to synthesize F-PGs that promote sperm guidance. PUFAs negatively regulate DAF-16 to promote oocyte yolk endocytosis and PG synthesis (32). The green arrows indicate intercellular transportation of lipid molecules such as GPLs and F-PGs. Abbreviations: AA, arachidonic acid; EPA, eicosapentaenoic acid; OL, oleic acid.

elegant studies, sperm guidance during fertilization was first found to depend on synthesis of C18 or C20 PUFAs by the Δ-6 and Δ-12 desaturases FAT-2 and FAT-3, respectively (95). Follow-up studies showed that these PUFAs, including arachidonic acid (AA) and eicosapentaenoic acid (EPA), are synthesized in the intestine and transported to oocytes to be converted into F-series prostaglandins (a class of eicosanoid) to guide the sperm under the regulation of the insulin and TGF-β pathways (32, 69, 95) (**Figure 2**). Another interesting study indicated the role of eicosanoid biosynthesis in the reoxygenation response after anoxia. The O_2-dependent enzyme EGL-9 hydroxylates the hypoxia inducible factor (HIF) to reduce the HIF level in the nucleus and thereby restore the level of its negative target cyp-13A12, a cytochrome P450 oxygenase that catalyzes the synthesis of eicosanoids from PUFAs. Eicosanoids are responsible for the acute oxygen response shift from apoxia (114).

In *Drosophila* the prostaglandin pathway has been found to play important roles in egg follicle maturation (174). One such role involves regulating the actin skeletal organization, possibly through interacting with the actin-bundling protein Fascin and causing actin remodeling (52, 174). Because loss of the same enzyme involved in prostaglandin synthesis causes similar ovary defects in mice (107), and because a high level of both prostaglandin and Fascin correlated with malignancy of cancers (52), further studies in *Drosophila* on the subject may greatly advance our understanding of prostaglandin functions in mammals.

Endocannabinoids, the active compounds in marijuana, are conserved lipid signals that regulate multiple biological processes in a variety of organisms (30). Found in *C. elegans* by mass spectrometry (103), endocannabinoids have been indicated to play regulatory roles in axon regeneration and TOR (target of rapamycin)-dependent life-span extension (113, 139). These studies have identified *C. elegans* orthologs involved in endocannabinoid biosynthesis and metabolism, as well as in the downstream signaling pathway. However, identifying the endocannabinoid receptors should make *C. elegans* an even better in vivo model to study endocannabinoid-related functions in the future.

Monomethyl Branched-Chain Fatty Acids Critically Impact Worm Development

mmBCFAs are saturated FAs with a methyl group at the ω-1 (iso-) or ω-2 (anteiso-) position (**Figure 1a**). mmBCFAs are abundant in some bacteria species and present in fungi, plants, and animals. In humans, mmBCFAs have been detected in the skin, vernix caseosa, harderian and sebaceous glands, hair, and brain (88). Humans obtain long-chain mmBCFAs through intake of dairy products (146) and by endogenous biosynthesis (78). Branched-chain amino acids (leucine, isoleucine, and valine) are precursors of mmBCFA biosynthesis in eukaryotes (82), but some specific enzymatic steps of biosynthesis have not been characterized in mammals. Genetics in *C. elegans*, which has high levels of endogenous mmBCFAs (24), identified several key enzymes that function in mmBCFA biosynthesis (35, 88, 89) (**Figure 1d**). Furthermore, a feedback regulatory pathway involving *C. elegans* sterol regulatory element-binding protein SBP-1 and transcriptional cofactor CBP-1 has been indicated in mmBCFA homeostasis (89).

Although the physiological functions of mmBCFAs in mammals are essentially unknown, studies in *C. elegans* have provided valuable insights regarding the functions of mmBCFAs. The essential roles of mmBCFAs in development were established by observing growth arrest, egg-laying defects, and gonad degeneration after blocking mmBCFA biosynthesis (35, 88, 89). These phenotypes can be fully rescued by dietary supplementation of C15 or C17 mmBCFAs (in iso or ante-iso form) but not by any straight-chain FAs (35, 88, 89). Further genetic and biochemical analyses have indicated that most of these physiological roles of mmBCFAs were executed by higher-order lipids, such as sphingolipids and phospholipids (PLs), containing mmBCFAs (24, 88–90, 197).

Worms hatched from mmBCFA-depleted eggs uniformly arrest at L1 (89). Although this arrest is similar to the food-deprivation-induced L1 arrest that is regulated by the insulin and IGF-like (IIS) pathway (8, 101), mmBCFA function is independent of the IIS pathway (89, 197). Analysis of mutations that suppressed the L1 arrest reveals three aspects of mmBCFA function and metabolism (159, 184, 197). First, mmBCFA-derived branched-chain sphingolipids positively regulate the TORC1 signaling pathway in controlling postembryonic growth and development (197) (see section Advances in Understanding the Roles of Glycosphingolipids in Promoting Cell Signaling, Membrane Polarity, and Development). Second, the functional peroxisome and, likely, a novel mechanism are responsible for degradation of mmBCFAs (184). Finally, a P-4 type ATPase aminophospholipid flippase (TAT-2) antagonizes the function or transportation of an mmBCFA-containing lipid (possibly glucosylceramide) (159, 197). These studies highlight the functional significance of a single class of FAs.

FUNCTIONAL STUDIES OF GLYCEROPHOSPHOLIPIDS IN FLIES AND WORMS: LIMITED BUT PROGRESSING

GPLs, often referred to as PLs, mainly consist of sn-1,2-diacylglycerols and a phosphate residue in position sn-3 linked to various functional moieties. GPLs are key components of all biological membranes as well as second messengers and enzyme activators, and are involved in many signaling pathways, cellular physiological functions, and human diseases (39, 99, 121, 144, 177). We only discuss studies covering several major categories (**Figure 3a**).

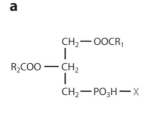

X group	Structure	GPL name
Cho	HO-CH₂CH₂N⁺(CH₃)₃	PC
Etn	HO-CH₂CH₂NH₂	PE
Ser	HO-CH₂C(NH₂)HCOOH	PS
Ins	(inositol ring structure)	PI

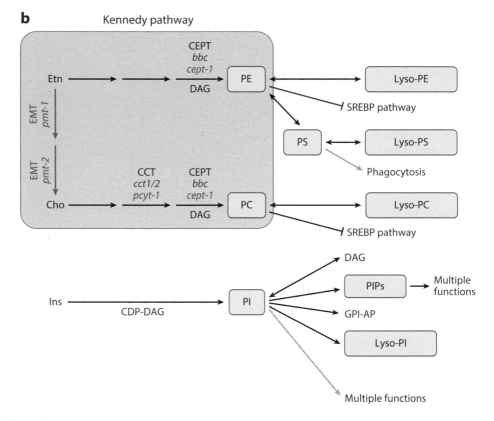

Figure 3

Metabolism of glycerophospholipids (GPLs). (*a*) Basic molecular structure of GPLs (*left*), with X representing the unique head group of each GPL class, listed in the table (*right*). (*b*) Simplified pathways of GPL biosynthesis, including lipid metabolites (GPLs in blue rectangle), catalytic steps (*arrows*), and enzymes (black capitalized text indicates the step/enzyme is also conserved in mammals, blue capitalized text indicates the step/enzyme is unique in *Caenorhabditis elegans*), in *C. elegans* and/or *Drosophila* corresponding genes (*C. elegans* in blue and *Drosophila* in red). Downstream targets are regulated positively (*gray arrows*) or negatively (*bars*). Abbreviations: CCT, cholinephosphate cytidylyl transferase; CDP-DAG, cytidine diphosphate diacylglycerol; CEPT, choline/ethanolamine-phospho-transferase; Cho, choline; DAG, diacylglycerol; EMT, ethanolamine methyltransferase; Etn, ethanolamine; GPI-AP, glycosylphosphatidylinositol anchored proteins; Ins, inositol; PC, phosphatidylcholine; PE, phosphatidylethanolamine; PI, phosphatidylinositol; PIPs, phosphoinositides; PS, phosphatidylserine; Ser, serine; SREBP, sterol regulatory element-binding protein.

Functional Studies on Phosphatidylcholine and Phosphatidylethanolamine

Phosphatidylcholine (PC) and phosphatidylethanolamine (PE) are the most abundant GPLs in animals and two of the major building blocks of bilayer membranes (concentrated at the outer and inner leaflet, respectively) **(Figure 3b)**. Studies in mammals have indicated that PC is involved in intracellular triglyceride storage, ER cholesterol homeostasis, and multiple ER stress pathways (99). Functional analyses of PC and PE in invertebrates are very limited; only de novo PC biosynthesis-related genes have been found in *C. elegans* (13, 41, 135) and *Drosophila* (57, 173) **(Figure 3b)**. In *C. elegans*, along with reported general developmental phenotypes like larval arrest and sterility in PC biosynthesis mutants, a recent study reported that defective PC biosynthesis leads to the upregulation of the SREBP signaling pathway, which transcriptionally promotes FA and lipid biogenesis (including PC biosynthesis) and lipid droplet accumulation (181). In other words, like the well-known role of cholesterol in mammals, the level of PC may serve as an SREBP signaling sensor in lipid feedback regulation in *C. elegans* (181). In *Drosophila*, such a negative connection between PC synthesis/level and lipid droplet expansion has also been established, as a PC de novo pathway enzyme CTP:phosphocholine cytidylyltransferase (CCT) is activated by targeting it to lipid droplets (93). Because the connection between PC biosynthesis/level and lipid droplet accumulation/expansion has also been observed in mammals, these studies in model organisms have uncovered a conserved alternative SREBP pathway–involved regulation of lipid homeostasis (93, 181). It remains an interesting question whether the SREBP signaling pathway is regulated by lower PC levels via the change of membrane curvature (181) or via changes in endocytosis, given that de novo synthesized PC is also reported to affect endocytosis and the downstream Notch and EGFR signaling pathways (188).

In *Drosophila*, PE was also suggested to be a negative regulator of the SREBP feedback regulatory pathway (31). This PE-regulated SREBP signaling was suggested to function in the maintenance of normal cardiac physiology, as mutating ethanolamine kinase of the de novo PE synthesis pathway causes a hyperactive SREBP signal-dependent tachycardia and defects in cardiac relaxation (108). PE-related studies in *C. elegans* are generally lacking. In one report, a P-4 type ATPase aminophospholipid flippase, TAT-5, was indicated to play a role in maintaining the asymmetric distribution of PE at the plasma membrane of *C. elegans* embryos, potentially by regulating the release of extracellular vesicles in a *rab-11*- and ESCRT-dependent manner. Disruption of *tat-5* causes large-scale budding of vesicles from the plasma membrane, compromising cell adhesion and leading to defects in cell shape (190).

Role of Phosphatidylserine in Apoptotic Cell Clearance

Phosphatidylserine (PS) has a relatively low abundance on mammalian cell membranes (176) but has significant functions in many cellular and physiological processes. PS is usually distributed along the inner leaflet of plasma membranes, and this asymmetrical distribution can be changed upon specific physiological events, such as activation of platelets or sperm in mammals (175). The best-known function of PS is to expose an "eat me" signal at the plasma membrane outer leaflet of apoptotic cells to be recognized by a PS receptor expressed at the surface of phagocytic cells (38). PS also targets and/or modulates activities of signaling proteins at certain membranes via their phospholipid-binding domains (175). In *C. elegans*, significant progress has been made in understanding the role of PS in the engulfment of apoptotic cells during phagocytosis. A series of studies reported the roles of PS receptors PSR-1 and CED-1, the flippase TAT-1 (which maintains the PS asymmetric localization and prevents PS externalization until the initiation of apoptosis), and the cleaved CED-8 protein (which triggers PS externalization upon apoptosis) (21, 27, 169,

185). Interestingly, in *Drosophila*, a potential PS receptor, dPSR, has been shown to be nonessential for apoptosis. Furthermore, dPSR is expressed in the nucleus instead of at the cell membrane and its overexpression inhibits normal apoptosis (94). Given the high similarity between the *C. elegans* PSR-1 and *Drosophila* dPSR sequences, the discrepancy between their functions in phagocytosis is intriguing. Because the *Drosophila* CED-1 homolog *draper* has also been shown to function as a PS receptor (123), it is thus possible that *draper* is the only known functional PS receptor in *Drosophila*.

Roles of Phosphatidylinositol in Multiple Signaling Processes

Phosphatidylinositol (PI) is not as abundant as the other PLs mentioned above, but PI and its derivatives are probably the most extensively studied for cellular regulatory functions (**Figure 3***b*). PI can be metabolized to PUFA-derived signaling molecules like prostaglandin and endocannabinoid (144) (see section Roles of Eicosanoids and Endocannabinoids in Germline Development and Aging). PI can be cleaved by phospholipase C to generate DAG, a signaling lipid with multiple roles of its own (16, 144). PI serves as a substrate for glycosylphosphatidylinositol (GPI) anchors, which can be post-translationally added to the C terminus of many eukaryotic proteins and anchor them to the outer leaflet of the cell membrane (141). The most well-studied phosphorylated metabolites of PI are phosphoinositides [PIPs; including PI3P, PI4P, PI(4,5)P2, and PI(3,4,5)P3]. Phosphoinositides can bind and activate a large number of signaling proteins for important regulatory functions, such as relaying signaling activity in the IIS and TORC1 pathways, and their activities are linked to various human diseases, including cancer and type two diabetes (34). In *C. elegans* and *Drosophila*, in addition to the extensively studied PI3K-PTEN pathway, functional analyses of many other phosphoinositides have been reported. For example, disrupting dephosphorylation of PI3P affects the endocytosis-dependent Wnt signaling pathway (165). Normal function of the IMPase TTX-7, the enzyme that catabolizes inositol-monophosphate to inositol, has been reported to localize to both the pre- and postsynaptic proteins within the neuritis of RIA interneurons and regulate sensory behaviors in the mature nervous system (172). In *Drosophila*, the specific PI3P pool, which is maintained by the activities of the phosphoinositide phosphatase dMTM and the class II PI3-kinase dPi3K68D, has been shown to be important for integrin localization, integrin-mediated adhesion, and related muscle remodeling (150). The PI4P pool, maintained by the activity of a kinase (Sst4) and a phosphatase (Sac1), was indicated to regulate the activity of the Hedgehog signaling pathway (193). In a recent report, the GPI transamidase was shown to be essential for the polarized sorting of integral membrane proteins at the *trans*-Golgi network in *Drosophila* photoreceptor cells (156). Furthermore, neuronal PI(4,5)P2 was reported to recruit and activate the presynaptic Wiscott-Aldrich syndrome protein to repress the activity of Moesin in regulating growth of neuromuscular junctions (86).

Model organisms are expected to play prominent roles in tackling the diverse and complex physiological functions associated with PI metabolites. Fundamental questions include: How is a specific PI metabolite generated in vivo; what is the biochemical nature of its function; when and where (cell type and subcellular location) during animal development does a given metabolite execute its function; how is the asymmetric temporal and spatial distribution of the metabolite achieved; and what are the downstream proteins regulated by specific PI metabolites for specific functions? PI-related studies in worms and flies usually do provide valuable information toward answering those questions. For example, the asymmetric distribution of PIP2 in embryonic cells has been shown to be generated by the activity of PAR-2, PAR-3, and CSNK-1 (casein kinase 1-γ), which modulate the polarized function of a PIP2-synthesizing enzyme, PPK-1. This PIP2 asymmetry was proposed to guide the asymmetric localization of the Gα-coupled receptors GPR-1/2, leading to posterior spindle displacement (136). Another study also identified the role of the phosphatidylinositol transfer protein PITP-1 in transporting PI to neuronal synapses, where the PI metabolite DAG functions to regulate synaptic transmission and related behavior response (76). Temporally controlled reversible phosphorylation of PI3P by PI3P kinase PIKI-1 and VPS-34 along with PI3P phosphatase MTM-1 was shown to regulate the PI3P cycling pattern on phagosomes and thus ensure the efficient degradation of apoptotic cells in *C. elegans* (112). Phosphoinositides are also known to target proteins or vesicle cargos to certain locations for their functions (121, 144). In *C. elegans*, PI(4,5)P2 on membrane vesicles was reported to bind to kinesin for transport along microtubules (87). Similarly in *Drosophila*, PI(4,5)P2 has been shown to target the adaptor protein dMyD88 to plasma membranes to activate the Toll-like receptor pathway for the innate immune response (116). In another study, synthesis of PI(4,5)P2 at the growing end of the cysts was shown to recruit the exocyst to promote targeted membrane delivery and polarization of the elongating cysts during spermatogenesis (37). Because PI and its metabolites are involved in multiple signaling pathways that affect animal development and other physiological functions, the analysis of temporal- and spatial-specific functions of PI will likely continue to be a major theme in future studies using model organisms.

Functional Impact of Fatty Acid Side Chains in Phospholipids

A typical phospholipid (PL) has two FA side chains attached to the sn-1 and sn-2 positions of a glycerophosphate backbone, but only a limited number of recent studies have begun to address two major questions regarding the impact of having different FA side chains: (*a*) What are the functional specificities associated with FA variants on a given PL? (*b*) Does the incorporation of specific FAs into a specific PL significantly affect the level of that PL, which in turn affects the PL composition in particular tissues? The composition of PLs on cellular membranes of different tissues is likely to be distinctly different and may have a strong influence on various cellular functions (39, 177).

Studies on the first question have been limited but informative. For example, it has been found that docosahexaenoyl residues are enriched in the acyl chains of PS in the human brain, and this enrichment appears to be critically important for the nervous system (172, 175). Sometimes this specific acyl chain is needed for generating related GPL derivatives, such as certain prostaglandins and endocannabinoids (see section Roles of Eicosanoids and Endocannabinoids in Germline Development and Aging). These specific manipulations of the acyl chain are usually performed by the Lands cycle, which includes lipases to remove a certain acyl chain (phospholipase A2) and a lysophospholipid acyltransferase (164). In *C. elegans*, mutation of the sn-2 lysoPI acyltransferase MBOA-7 has been shown to dramatically reduce the abundance of AA and EPA in PI, which is associated with a series of developmental defects, including early larval arrest and egg-laying defects (102). Another study showed that the phospholipase A1 (IPLA-1) and sn-1 lysoPI acyltransferases (ACL-8–10) incorporate stearic acid (C18:0) into the sn-1 position of PI, and this activity is important for asymmetric cell-fate determination and orientation of dividing seam cells (75). The activities of *Drosophila* lysophospholipid acyltransferases, such as Oysgedart, Nessy, and Farjavit, which prefer transferring unsaturated FAs, are also important for multiple steps in germ-cell development (168). Considering the potential significance of GPL remodeling for membrane diversity involved in various cellular processes and human diseases (163), research using model organisms in this relatively new area may need to expand.

A recent study in *C. elegans* presents an important advance in addressing the second question (90). Incorporation of a specific FA variant, C17ISO, into two different classes of PLs in a specific tissue (somatic gonad) depends on different acyl-coA synthetases (ACSs). These PLs are then presumably transported to oocytes and zygotes to support membrane functions. When ACS-1 is eliminated in whole animals or only in the somatic gonad, incorporation of C17ISO into PE and related PLs is blocked, leading to an imbalanced PL composition in the zygote and dramatic phenotypes in the embryos where membrane dynamics, including cytokinesis, are disrupted. The effect of PL composition change likely acts through IP3 signaling that occurs at the cell and ER membrane because hyperactivity of IP3 signaling can suppress the phenotypes. This study indicates that channeling of a particular FA to different PLs may be critically regulated by ACS enzymes, a large family of enzymes in both worms and mammals (90, 167). It also suggests that PL remodeling may play key roles in dictating the composition of PLs in specific cells during development.

A study in *Drosophila* has indicated an alternative way to remodel the acyl chain of GPLs (115). Tafazzin, a cardiolipin remodeling enzyme, was identified as a highly efficient reversible acyl-transferase that shuttles fatty acyl chains between PC and cardiolipin in a manner that is independent of acyl-CoA, phospholipase, and acyltransferase (115). This finding potentially uncovered a new mechanism for specific FA composition regulation in GPLs and might be beneficial for understanding of related human disease (115).

SPHINGOLIPIDS HAVE DIVERSE FUNCTIONS IN REGULATING CELLULAR AND DEVELOPMENTAL EVENTS

The name of sphingolipids comes from the Greek word sphinx, denoting the enigmatic character of this lipid when it was first identified in brain tissue (127). Now sphingolipids have been shown to be a class of the most complex and bioactive lipids in all eukaryotes and in some bacteria species, and their functions in various cellular/developmental processes and human diseases have been extensively studied (45, 127). The backbone of a sphingolipid consists of an aliphatic amino alcohol called the sphingoid base. Various modifications on the backbone generate several subclasses of sphingolipids. Here, we group sphingolipids into simple sphingolipids (sphingoid bases, ceramides, and their phosphorylated products), glycosphingolipids (GSLs), and sphingomyelin (**Figure 4a**). We limit our discussion to the first two groups.

Simple Sphingolipids as Signaling Molecules

Simple sphingolipids are synthesized de novo or metabolized from existing sphingolipids in animals (**Figure 4b**). Many simple sphingolipids are important signaling molecules and are essential for

$\longrightarrow$

Figure 4

Metabolism of sphingolipids (SPLs). (*a*) A typical molecular structure of SPLs (d18:1, C22:0:1 for this example), with X representing the unique head group of each SPL class, listed in the table (*right*). (*b*) A simplified pathway of SPL biosynthesis and metabolism in *Caenorhabditis elegans* and *Drosophila*, including lipid metabolites in rectangles [VLCFA (very-long-chain fatty acid) in orange, simple SPLs in light green, SM/CPE in blue, and GSLs in dark green], catalytic steps (*arrow*) and required enzymes (*black*), and corresponding genes (*C. elegans* in blue and *Drosophila* in red). Abbreviations: CDase, ceramidase; Cer, ceramide; Cer-1-P, ceramide-1-phosphate; CerK, ceramide kinase; CerS, ceramide synthase; CGT, ceramide galactosyltransferase; Cho, choline; CPE, ceramide phosphoethanoamine; DHC, dihydroceramide; Etn, ethanolamine; Gal, galactose; GalCer, galactosylceramide; GCS, glucosylceramide synthase; GlcCer, glucosylceramide; GlcNAc, N-acetylglucosamine; GlcNAc-T, N-acetylglucosamine transferase; Glu, glucose; GSL, glycosphingolipid; H, hydrogen; LCA, long-chain aldehyde; LCS, lactosylceramide synthase; Man, mannosyl; mmBCFA, monomethyl branched-chain fatty acid; SFA, saturated fatty acid; SK, sphingosine kinase; SM, sphingomyelin; SMase, sphingomyelinase; SMS, sphingomyelin synthase; SPA, sphinganine; SPH, sphingosine; SPH-1-P, sphingosine-1-phosphate; SPT, serine-palmitoyltransferase; S1PL, sphingosine-1-phosphate lyase.

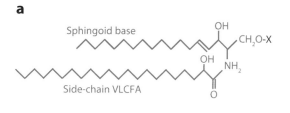

X group	SPL name
H	Cer
Phosphate	Cer-1-P
Etn	CPE
Cho	SM
Glc	GlcCer
Gal	GalCer

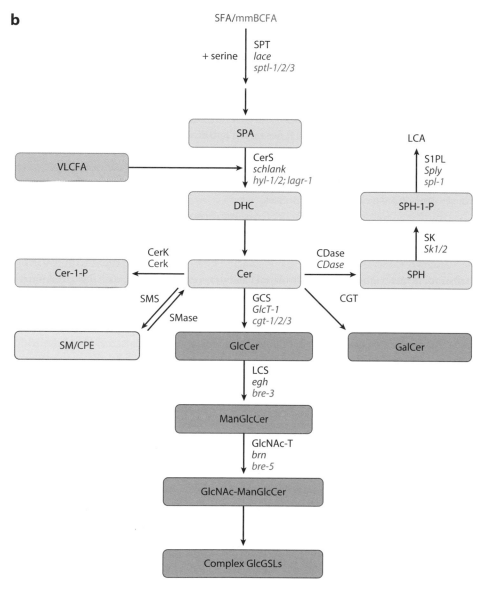

animal development. For example, extensive studies have indicated that sphingosine-1-phosphate (S1P) plays important roles in multiple cellular processes, including tumorigenesis, angiogenesis, and lymphocyte-trafficking processes (43). Ceramide-1-phosphate is also involved in cell proliferation and the inflammatory process (70). Ceramide itself plays roles in apoptosis, stress response, insulin resistance, tumor progression, and neurodegeneration (130).

In *Drosophila*, most genes involved in the metabolism of simple sphingolipids have been discovered. Overexpression of ceramidase (CDase) was reported to protect the degeneration of photoreceptor cells from apoptosis caused by arrestin or phospholipase C mutations via facilitating endocytosis and membrane turnover (1, 2). This effect is probably due to a reduced level of proapoptotic ceramide in the CDase mutant, given that a mutation in ceramide kinase (Cerk) causes the opposite effect (28). Upregulation of S1P by mutations of S1P lyase (Sply) and a lipid phosphatase (Wunen) or by overexpression of a serine palmitoyltransferase (Lace, SPT2) was found to suppress muscle dystrophy in a Dystrophin mutant, a model for Duchenne muscular dystrophy in humans (137). This result points to the importance of S1P signaling in the prevention of age-dependent muscle wasting. Null alleles of some sphingosine kinase genes such as Sk2 are viable, with various developmental phenotypes (66). Null alleles of several enzymes involved in simple sphingolipid de novo synthesis and metabolic pathways, such as *Lace* (serine palmitoyltransferase), *Schlank* [ceramide synthase (CerS)], and *Sply* (S1P lyase), are lethal, whereas partial loss-of-function alleles of these genes display various developmental phenotypes (3, 7, 67), indicating that maintaining normal sphingolipid biosynthesis/homeostasis is essential in multiple developmental processes. It is currently unclear whether and how the structural specificity of these sphingolipids influences their roles in these developmental processes.

It is worth mentioning that genes in the *Drosophila* sphingolipid metabolic pathways commonly have fewer functional homologs than their mammalian or *C. elegans* counterparts. The impact of this feature on the genetic analysis of sphingolipid function is twofold. On one hand, it makes it much easier to eliminate the total enzyme activity by using single mutations or RNAi. On the other hand, as these enzymes have pleiotropic and often essential functions, the severe developmental defects of the mutants would prevent the study of other nonessential functions or essential functions at later stages. Moreover, the "one enzyme for all" situation reduces the opportunity to analyze specific aspects of enzymatic function because homologs of these enzymes in other organisms often have specific substrate preferences and expression patterns. However, with the advantage of the FLP-FRT and GAL4-UAS system to create tissue-specific homozygous mutant clones (11, 192), *Drosophila* could be an ideal system to study the cellular/tissue function of a recessive lethal gene in lipid pathways. For example, a null allele of *lace/SPT2* is homozygous lethal (3), but by using clonal analysis, it was found that *lace*(−) clones overproliferate by affecting downstream Notch and Wingless signaling pathways (155). Such a strategy may be applied to study the function of Schlank, the sole homolog of CerS in *Drosophila*, as these null alleles are larval lethal and cause a dramatic decrease in fat storage by negatively affecting the dSREBP pathway (7). Mosaic analysis could facilitate the study aiming to understand where and how dSREBP is regulated by Schlank in vivo, as a similar question would be difficult to address in other organisms with multiple homologs.

In *C. elegans*, genes for the sphingolipid de novo pathway, such as SPT (serine-palmitoyltransferase) (*sptl-1,2,3*), CerS (*hyl-1,2* and *lagr-1*), and S1P lyase (*spl-1*), are also essential for development and viability (125, 126, 159) (**Figure 4b**). Functional studies of ceramide molecules have shown their roles in stress response, including anoxia and radiation-induced apoptosis (29, 126). S1P is reported to be necessary, by a muscarinic signaling pathway, at presynaptic terminals to promote neurotransmitter release (18).

Advances in Understanding the Roles of Glycosphingolipids in Promoting Cell Signaling, Membrane Polarity, and Development

After transfer to the Golgi apparatus, ceramide can receive a hexasaccharide on the 1-hydroxy residue to form two types of mono-glycosylceramides (cerebroside): galactosylceramide (GalCer) and glucosylceramide (GlcCer) (**Figure 4b**). GalCer can be further modified to become sulfatide. Both GalCer and sulfatide are important for myelin function and stability but are not essential for animal viability (25). GlcCer can be further modified to generate a large group of GlcCer-derived GSLs (higher-order GlcGSLs) (58), which have been shown to be essential for mammalian development as early as the embryonic stage (20, 45). GlcGSL metabolism has also been shown to impact human health, as mutations that disrupt the degradation of GlcGSL (as well as sphingomyelin and ceramide) cause a series of lipid storage–related diseases (153).

In *Drosophila*, GlcCer and its derived higher-order sphingolipids have also been shown to be essential for development as null alleles of related synthases [the mannosyl GlcCer transferase Egh and the GlcNAc-transferase Brn are lethal (157, 183)]. Egh and Brn were also reported to play regulatory roles in the Notch signaling pathway (48). Notch ligand activity was suggested to be modulated by a higher-order GlcGSL synthesized by the N-acetylgalactosamine transferase α4GT1. Because α4GT1 is nonessential, the lethality caused by mutations in earlier steps of GlcGSL synthesis is likely due to a different function associated with these lipids (59), indicating a possible role in activating the EGFR pathway (143).

GlcT1, the *Drosophila* glucosylceramide synthase catalyzing the first step of GlcGSL biosynthesis, is essential for development and also plays a role in regulating fat storage in the fat body (92). This latter function was not observed in Egh or Brn mutants, suggesting that the lipid storage defect associated with GlcT1 is possibly due to deficiency in GlcCer itself rather than in a higher-order GlcGSL (92). It would be interesting to determine whether the GlcT1 mutant phenotype is mechanistically related to the defects caused by CerS mutations that disrupt SREBP-dependent FA synthesis (7) (see section Simple Sphingolipids as Signaling Molecules).

In *C. elegans*, GlcCer is the first and simplest GlcGSL synthesized by three glucosylceramide synthases (CGT-1–3) (118). As in flies, GlcCer can also be further glycosylated to yield higher-order GlcGSLs in worms, but these higher-order GlcGSLs are not essential for development (50). Nevertheless, these lipids impact development and other physiological functions. For example, higher-order GlcGSLs in worms have also been shown to regulate Notch signaling (83). In addition, some higher-order GlcGSLs have been shown to be important for *Bacillus thuringiensis* toxin sensitivity (50). GlcCer appears to be the only essential GlcGSL, as worms lacking it arrest at an early larval stage (50, 197). Because synthesis of GlcCer in the intestine, but not in other tissues, is sufficient to meet all major developmental requirements for GlcCer (118), GlcCer likely acts in the intestine for its essential role in worms.

The essential functions of GlcCer have been further analyzed regarding two different downstream activities in *C. elegans*. First, GlcCer, along with clathrin and the AP-1 complex, was found to play a key role in establishing apical polarity in the intestine (194, 195). Second, like its mmBCFA precursors (see section Monomethyl Branched-Chain Fatty Acids Critically Impact Worm Development), branched-chain GlcCer (the predominant GlcCer in worms) is required for the initiation of postembryonic development of newly hatched larvae (118, 197). Further analysis indicated that GlcCer executes this developmental role by promoting the TORC1 signaling pathway in the intestine. In particular, constitutive TORC1 activity by genetic mutation of a TORC1 repressor or by transgenes that constitutively activate TORC1 in the intestine was able to suppress the postembryonic developmental arrest caused by mmBCFAs and GlcCer deficiency (197). In other words, the only essential role of GlcCer identified thus far is to activate the TORC1

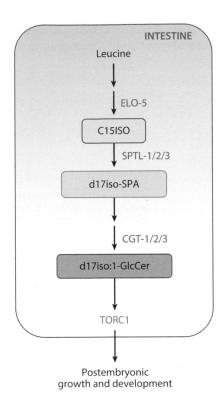

INTESTINE

Leucine

↓

↓ ELO-5

C15ISO

↓ SPTL-1/2/3

d17iso-SPA

↓

↓ CGT-1/2/3

d17iso:1-GlcCer

↓

TORC1

↓

Postembryonic
growth and development

Figure 5

Intestinal d17iso:1-GlcCer promotes TORC1 activity and developmental growth in *Caenorhabditis elegans*. A model for the regulation of postembryonic growth and development by monomethyl branched-chain fatty acids (mmBCFAs) and d17iso:1-glucosylceramide (d17iso:1-GlcCer), including lipid metabolites in rectangles (mmBCFAs in light blue and sphingolipids in green). Intestinal d17iso:1-GlcCer biosynthesis, from leucine and intermediate metabolite C15iso FA (C15ISO), positively regulates target of rapamycin complex 1 (TORC1) to modulate postembryonic development of *C. elegans*.

signaling pathway and thereby regulate animal growth and development in *C. elegans* (197) (**Figure 5**). It would be interesting to determine whether the roles of GlcCer in endocytosis/ polarity and TORC1 activation are directly related and how GlcCer carries out these functions.

The Unique Structure of Sphingolipids in *Drosophila* and *Caenorhabditis elegans*

Although overall the sphingolipid metabolic pathways are mostly conserved between *Drosophila*, *C. elegans*, and mammals, the core structures of major sphingolipids in *Drosophila* and *C. elegans* are different from their mammalian counterparts. Knowledge of these structural differences is specifically important to avoid incorrect assumptions that would hinder exogenous lipid supplement experiments and mass-spectrometer analysis.

First, the sphingoid bases vary between organisms. In mammals, the major sphingoid bases are d18:1 sphingosine and d20:1 sphingosine (20). In *Drosophila*, they are d14:1 sphingosine and d16:1 sphingosine (67). A recent report also indicated that d14:2$^{\Delta4,6}$ and d16:2$^{\Delta4,6}$ sphingadienines are relatively abundant in wild-type flies (54). In *C. elegans*, the major sphingoid base is iso-branched d17:1 (15-methyl-d16:1) sphingosine (24). Under the mmBCFA-deficient condition, d16:1 sphingosine is synthesized instead but does not substitute for the normal function of d17iso:1-glucosylceramide in vivo (35, 197). Sphingolipids in different organisms likely perform similar functions, but the structural variations point to differences in their biosynthesis pathways. Therefore, some conclusions in previous sphingolipid studies in flies or worms that were based on mammalian sphingolipid structure may not be accurate.

Second, glycosylation varies between these organisms. In mammals, the second hexose added to GlcCer is galactose, whereas it is usually mannose in *C. elegans* and *Drosophila* (24, 160). However, in *Drosophila*, ectopic expression of mammalian UDP-galactose glucosylceramide transferase was found to rescue the developmental defect caused by mutating the UDP-mannose glucosylceramide transferase, where endogenous mannosylceramide was substituted by lactosylceramide (183).

Last, the composition of the fatty acyl chain attached to the sphingoid base is different in flies and worms than in mammals (24, 54), and these differences may be related to the physiological functions (54). Specifically, HYL-1 and HYL-2 are two ceramide synthases in *C. elegans* that have different ceramide substrate specificities (varying carbon-chain length) and independent functions. HYL-2 mutants showed sensitivity to apoxia, which could not be rescued by ectopic expression of HYL-1 driven by the *hyl-2* promoter, indicating that the functional difference between these two enzymes is not due to the difference in expression patterns but rather to the difference in substrate specificity. Mass spectrometer analyses suggest that HYL-2 preferably transfers shorter FA chains to sphingoid bases, whereas HYL-1 prefers the longer FA chain as substrate. Therefore, sphingolipids of complex structures with shorter side-chain FAs are likely responsible for the apoxia resistance (126). This hypothesis could be further tested by asking whether blocking VLCFA synthases [ACS-20 and ACS-22 (80)] could provide a protective role under the apoxia condition in *C. elegans*.

FUNCTIONS OF CHOLESTEROL AND ITS DERIVATIVES

Cholesterol and its derivatives, such as vitamin D3, steroid hormones, and bile acids, have a variety of important physiological functions related to many aspects of human health and diseases such as Tangier disease, Niemann-Pick disease type C, and atherosclerosis (120). General lipid synthesis is also regulated by the cholesterol-dependent SREBP pathway in mammals (15). In *Drosophila* and *C. elegans*, cholesterol is also essential for development and other physiological functions (74). For example, hormone derivatives of cholesterol, such as dafachronic acid in *C. elegans* and ecdysone in *Drosophila*, were found to be ligands of nuclear hormone receptors that were previously identified as orphan receptors to regulate related signaling pathways in both mammals and the invertebrate organisms (129, 191). Biosynthesis, metabolism, and transportation of cholesterol-derived hormones are facilitated and regulated by multiple enzymatic and regulatory pathways (62, 72, 73, 140, 152). In *Drosophila*, a series of studies demonstrated that cholesterol binds Hedgehog and regulates its transportation and assembly to affect downstream signaling (44, 106).

Unlike mammals, where cholesterol can be synthesized endogenously, invertebrates like *Drosophila* and *C. elegans* are cholesterol auxotrophs, lacking key enzymes in a homologous de novo pathway (17, 98). Furthermore, both *C. elegans* and *Drosophila* seem to require only a low level of cholesterol for their development. In *C. elegans*, a relatively high level of cholesterol appears to be required only for functions in certain specific stages, such as gonad development and molting (during the transition from one larval stage to another) (98, 119, 162). Similarly, *Drosophila* can survive and modulate development with low amounts of cholesterol and *Drosophila* S2 cells can survive under cholesterol-depleted conditions (17, 56). This partial independence from cholesterol may imply that other lipids can substitute for part of the functions associated with cholesterol or its derivatives, raising an intriguing possibility that these model organisms may be good systems to study the fundamental function of cholesterol and its functionally related lipids for specific physiological functions. For example, one cholesterol function is to help the formation of lipid rafts, potentially fluctuating assemblies of GSLs, cholesterol, and proteins that can serve as a platform for membrane signaling and trafficking (109). It is possible that the lipid raft-related function of cholesterol can be substituted, albeit not likely with the same effectiveness, by GSLs in most *C. elegans* and *Drosophila* cells. In another study, mutations in the *C. elegans* aging-related gene *clk-1* causing a slow defecation phenotype could be rescued by decreasing cholesterol supplementation or by mutations in *tat-2*, an ortholog of mammalian phospholipid flippase (ATP8B1) that is important for bile acid secretion and cholesterol uptake (110). Given the reported suppression effect of *tat-2* mutants on mmBCFA and sphingolipid biosynthesis defects (159), the cholesterol-related role of TAT-2 also raises an intriguing connection between sphingolipid and cholesterol homeostasis.

The mevalonate pathway synthesizes important metabolites, including cholesterol, co-enzyme Q10, and heme A, as well as functions in protein prenylation in mammals. The cholesterol auxotrophic nature of *C. elegans* also makes it a good model to study roles of metabolites of the mevalonate pathway in processes other than cholesterol biosynthesis (148). For example, statins are a well-known inhibitor of HMG-CoA reductase in the mevalonate pathway and have been widely used by human patients to lower cholesterol and for other health benefits. Statins also cause multiple adverse effects in human patients, possibly due to a lack of other useful metabolites as a result of inhibition of HMG-CoA reductase (46). In *C. elegans*, statins also cause cholesterol-independent defects like growth arrest and lethality, possibly due to decreased protein prenylation and induction of the ER unfolded protein response (UPR) (128). By using forward genetics, gain-of-function mutations of a leucine zipper transcription factor ATFS-1 have been found to suppress these cholesterol-independent effects of statins, possibly through constitutive activation of the UPR in mitochondria (149). In *Drosophila*, the mevalonate pathway metabolite isoprenoid is also indicated to be required for protein prenylation to guide germ-cell migration in embryos (154).

CHALLENGES AND FUTURE DIRECTIONS FOR LIPID RESEARCH IN MODEL ORGANISMS

One major obstacle to studying the functional specificities associated with various lipids and lipid composition in specific cellular locations is the complexity of lipid structures and related biosynthetic and metabolic processes. Usually when we talk about a lipid, it is actually a group of lipids sharing the same core structure (for example, glucosylceramide consists of a class of lipids with different sphingoid base chains or fatty acyl side chains.). Structural differences may influence the function of a given lipid or its metabolism. Eliminating a certain biosynthetic/metabolic enzyme activity may alter the levels of many lipid variants in this class and sometimes may also change the levels of lipids outside of this class. For example, double mutants of two Δ-9 desaturases, FAT-6 and FAT-7, cause a systematic profile change of unsaturated FAs (14). This complication may prevent us from making a simple conclusion solely on the basis of genetic analysis of this enzyme. The flip side of this issue is that two or more enzyme homologs that share roles at a particular lipid biosynthesis or metabolic step may have differential catalytic efficiency on different lipid variants in the same class. For example, HYL-2 rather than HYL-1 is important for *C. elegans* survival upon apoxia (126). Therefore, the physiological defects associated with mutating a particular enzyme may only reflect the roles of a subgroup of the lipids of the same class. Therefore, to make greater advances in this field, researchers need to effectively apply all of the sophisticated genetic tools developed in the past decades to tackle genetic redundancy and pleiotropism associated with individual lipid metabolism enzymes. Moreover, to be able to accurately interpret genetic results and make significant connections between lipids and functions, genetic approaches must be accompanied with biochemical methods that analyze the levels and composition of lipids under specific physiological conditions. Although many lipid researchers using genetic model organisms have begun to analyze lipids by biochemical methods, including mass spectrometry, these efforts need to be significantly extended.

Another challenging question to answer is how lipids interact with proteins. Lipids often achieve their functions in cell signaling with the help of lipid-binding proteins. For examples, GPLs bind the PH domain–containing protein phospholipase C and regulate its activity (79), and S1P binds its receptor to deliver the signal (43). Systematic assays and bioinformatic methods to investigate interactions between lipid molecules such as phosphoinositides and their binding proteins were recently reviewed (158). A promising method to label the distal end of a certain FA with biotin and search for acylated protein by IP technology has also been established because most lipids contain FA chains (85). However, efficient methods to identify nonconjugated binding between lipids and proteins, especially for those membrane lipids of complex structures, remain to be developed.

Another major obstacle is the lack of a convenient way to label a specific lipid class in vivo; this is even more difficult for an individual lipid. One broadly used method is to employ fluorescently conjugated lipids as supplements to mimic the distribution of endogenous lipids (134, 194). Although it is relatively convenient, several drawbacks of this method may indicate that conclusions drawn from this method need further confirmation. First, commercially available lipids that are synthesized chemically or purified from mammal tissues and plants, especially those lipids of complex structure such as ceramide/GSLs, are usually structurally different from their endogenous counterparts in model organisms. Second, the exogenously supplied lipids may not function like endogenously synthesized lipids (133, 187). Third, uptake and incorporation of exogenous lipids of complex structure into certain tissues are usually difficult (e.g., they may be metabolized in the intestine or absorbed inefficiently) in a living organism.

Another approach is to label lipids directly with antibodies. Developing an antibody library against major lipids would be very helpful, especially if the antibodies could be generated in a way that they have high specificities to lipids of specific structure (for example, an antibody that recognizes only d18:1 sphingosine but not d16:1, d18:0 sphingosine, or d18:1 sphingosine-1-P). One development in this regard is using a fluorescent/tag conjugated-peptide domain, which is known to specifically bind to a certain class of GSLs as the tracing marker, to observe subcellular localization and movement of sphingolipids under physiological conditions in flies and worms (65, 198). Depending on the success in establishing specific binding between lipids and proteins, this method has the potential to be extremely helpful in the future. Given the complexity of lipid molecules, it is highly unlikely that we can generate antibodies for most of the individual lipids. Therefore, with the limitation in reorganizing lipids in specific cellular and subcellular locations, high-throughput lipid profiling using mass spectrometry (20) is still the major way to recognize individual lipids or profile a class of lipids in animals. Combined with high-performance liquid chromatography analyses, the result is usually sufficient to identify individual lipids, distinct from stereoisomers, from a crude extraction (20, 172).

CONCLUSION AND REMARKS

The lipid work covered in this review represents recent functional studies of lipid variants done in *C. elegans* and *Drosophila*. Clearly, this field is relatively young and the understanding of physiological roles associated with this gigantic family of molecules has just begun. The field will continue to grow in the next decade and produce a plethora of important breakthrough findings, contributing to our understanding of basic science and important human health problems. Model organisms will continue to be at the forefront of this research, and with the development of more sophisticated methods, insights gained from model organisms will be applicable to human biology.

DISCLOSURE STATEMENT

The authors are not aware of any affiliations, memberships, funding, or financial holdings that might be perceived as affecting the objectivity of this review.

ACKNOWLEDGMENTS

We thank Aileen Sewell for insightful comments and editing of the manuscript. Research done in the authors' laboratory was supported by HHMI and NIH.

LITERATURE CITED

1. Acharya U, Mowen MB, Nagashima K, Acharya JK. 2004. Ceramidase expression facilitates membrane turnover and endocytosis of rhodopsin in photoreceptors. *Proc. Natl. Acad. Sci. USA* 101:1922–26
2. Acharya U, Patel S, Koundakjian E, Nagashima K, Han X, Acharya JK. 2003. Modulating sphingolipid biosynthetic pathway rescues photoreceptor degeneration. *Science* 299:1740–43
3. Adachi-Yamada T, Gotoh T, Sugimura I, Tateno M, Nishida Y, et al. 1999. De novo synthesis of sphingolipids is required for cell survival by down-regulating c-Jun N-terminal kinase in *Drosophila* imaginal discs. *Mol. Cell. Biol.* 19:7276–86
4. Arsenault BJ, Boekholdt SM, Kastelein JJ. 2011. Lipid parameters for measuring risk of cardiovascular disease. *Nat. Rev. Cardiol.* 8:197–206
5. Ashrafi K, Chang FY, Watts JL, Fraser AG, Kamath RS, et al. 2003. Genome-wide RNAi analysis of *Caenorhabditis elegans* fat regulatory genes. *Nature* 421:268–72
6. Atilla-Gokcumen GE, Muro E, Relat-Goberna J, Sasse S, Bedigian A, et al. 2014. Dividing cells regulate their lipid composition and localization. *Cell* 156:428–39
7. Bauer R, Voelzmann A, Breiden B, Schepers U, Farwanah H, et al. 2009. Schlank, a member of the ceramide synthase family controls growth and body fat in *Drosophila*. *EMBO J.* 28:3706–16
8. Baugh LR, Sternberg PW. 2006. DAF-16/FOXO regulates transcription of cki-1/Cip/Kip and repression of lin-4 during *C. elegans* L1 arrest. *Curr. Biol.* 16:780–85
9. Ben Gedalya T, Loeb V, Israeli E, Altschuler Y, Selkoe DJ, Sharon R. 2009. α-Synuclein and polyunsaturated fatty acids promote clathrin-mediated endocytosis and synaptic vesicle recycling. *Traffic* 10:218–34
10. Benton R, Vannice KS, Vosshall LB. 2007. An essential role for a CD36-related receptor in pheromone detection in *Drosophila*. *Nature* 450:289–93
11. Brand AH, Perrimon N. 1993. Targeted gene expression as a means of altering cell fates and generating dominant phenotypes. *Development* 118:401–15
12. Branicky R, Desjardins D, Liu JL, Hekimi S. 2010. Lipid transport and signaling in *Caenorhabditis elegans*. *Dev. Dyn.* 239:1365–77
13. Brendza KM, Haakenson W, Cahoon RE, Hicks LM, Palavalli LH, et al. 2007. Phosphoethanolamine N-methyltransferase (PMT-1) catalyses the first reaction of a new pathway for phosphocholine biosynthesis in *Caenorhabditis elegans*. *Biochem. J.* 404:439–48
14. Brock TJ, Browse J, Watts JL. 2007. Fatty acid desaturation and the regulation of adiposity in *Caenorhabditis elegans*. *Genetics* 176:865–75
15. Brown MS, Goldstein JL. 1997. The SREBP pathway: regulation of cholesterol metabolism by proteolysis of a membrane-bound transcription factor. *Cell* 89:331–40
16. Carrasco S, Merida I. 2007. Diacylglycerol, when simplicity becomes complex. *Trends Biochem. Sci.* 32:27–36
17. Carvalho M, Schwudke D, Sampaio JL, Palm W, Riezman I, et al. 2010. Survival strategies of a sterol auxotroph. *Development* 137:3675–85

18. Chan JP, Hu Z, Sieburth D. 2012. Recruitment of sphingosine kinase to presynaptic terminals by a conserved muscarinic signaling pathway promotes neurotransmitter release. *Genes Dev.* 26:1070–85

19. Chen H, Liu Z, Huang X. 2010. *Drosophila* models of peroxisomal biogenesis disorder: peroxins are required for spermatogenesis and very-long-chain fatty acid metabolism. *Hum. Mol. Genet.* 19:494–505

20. Chen Y, Liu Y, Sullards MC, Merrill AH Jr. 2010. An introduction to sphingolipid metabolism and analysis by new technologies. *Neuromol. Med.* 12:306–19

21. Chen YZ, Mapes J, Lee ES, Skeen-Gaar RR, Xue D. 2013. Caspase-mediated activation of *Caenorhabditis elegans* CED-8 promotes apoptosis and phosphatidylserine externalization. *Nat. Commun.* 4:2726

22. Chertemps T, Duportets L, Labeur C, Ueda R, Takahashi K, et al. 2007. A female-biased expressed elongase involved in long-chain hydrocarbon biosynthesis and courtship behavior in *Drosophila melanogaster*. *Proc. Natl. Acad. Sci. USA* 104:4273–78

23. Chertemps T, Duportets L, Labeur C, Ueyama M, Wicker-Thomas C. 2006. A female-specific desaturase gene responsible for diene hydrocarbon biosynthesis and courtship behaviour in *Drosophila melanogaster*. *Insect Mol. Biol.* 15:465–73

24. Chitwood DJ, Lusby WR, Thompson MJ, Kochansky JP, Howarth OW. 1995. The glycosylceramides of the nematode *Caenorhabditis elegans* contain an unusual, branched-chain sphingoid base. *Lipids* 30:567–73

25. Coetzee T, Fujita N, Dupree J, Shi R, Blight A, et al. 1996. Myelination in the absence of galactocerebroside and sulfatide: normal structure with abnormal function and regional instability. *Cell* 86:209–19

26. Dallerac R, Labeur C, Jallon JM, Knipple DC, Roelofs WL, Wicker-Thomas C. 2000. A Δ9 desaturase gene with a different substrate specificity is responsible for the cuticular diene hydrocarbon polymorphism in *Drosophila melanogaster*. *Proc. Natl. Acad. Sci. USA* 97:9449–54

27. Darland-Ransom M, Wang X, Sun CL, Mapes J, Gengyo-Ando K, et al. 2008. Role of *C. elegans* TAT-1 protein in maintaining plasma membrane phosphatidylserine asymmetry. *Science* 320:528–31

28. Dasgupta U, Bamba T, Chiantia S, Karim P, Tayoun AN, et al. 2009. Ceramide kinase regulates phospholipase C and phosphatidylinositol 4, 5, bisphosphate in phototransduction. *Proc. Natl. Acad. Sci. USA* 106:20063–68

29. Deng X, Yin X, Allan R, Lu DD, Maurer CW, et al. 2008. Ceramide biogenesis is required for radiation-induced apoptosis in the germ line of *C. elegans*. *Science* 322:110–15

30. Di Marzo V, Bifulco M, De Petrocellis L. 2004. The endocannabinoid system and its therapeutic exploitation. *Nat. Rev. Drug Discov.* 3:771–84

31. Dobrosotskaya IY, Seegmiller AC, Brown MS, Goldstein JL, Rawson RB. 2002. Regulation of SREBP processing and membrane lipid production by phospholipids in *Drosophila*. *Science* 296:879–83

32. Edmonds JW, Prasain JK, Dorand D, Yang Y, Hoang HD, et al. 2010. Insulin/FOXO signaling regulates ovarian prostaglandins critical for reproduction. *Dev. Cell* 19:858–71

33. Ejima A, Smith BP, Lucas C, van der Goes van Naters W, Miller CJ, et al. 2007. Generalization of courtship learning in *Drosophila* is mediated by *cis*-vaccenyl acetate. *Curr. Biol.* 17:599–605

34. Engelman JA, Luo J, Cantley LC. 2006. The evolution of phosphatidylinositol 3-kinases as regulators of growth and metabolism. *Nat. Rev. Genet.* 7:606–19

35. Entchev EV, Schwudke D, Zagoriy V, Matyash V, Bogdanova A, et al. 2008. LET-767 is required for the production of branched chain and long chain fatty acids in *Caenorhabditis elegans*. *J. Biol. Chem.* 283:17550–60

36. Ewers H, Helenius A. 2011. Lipid-mediated endocytosis. *Cold Spring Harb. Perspect. Biol.* 3:a004721

37. Fabian L, Wei HC, Rollins J, Noguchi T, Blankenship JT, et al. 2010. Phosphatidylinositol 4,5-bisphosphate directs spermatid cell polarity and exocyst localization in *Drosophila*. *Mol. Biol. Cell* 21:1546–55

38. Fadok VA, Voelker DR, Campbell PA, Cohen JJ, Bratton DL, Henson PM. 1992. Exposure of phosphatidylserine on the surface of apoptotic lymphocytes triggers specific recognition and removal by macrophages. *J. Immunol.* 148:2207–16

39. Fagone P, Jackowski S. 2009. Membrane phospholipid synthesis and endoplasmic reticulum function. *J. Lipid Res.* 50(Suppl.):S311–16

40. Ferveur JF. 2005. Cuticular hydrocarbons: their evolution and roles in *Drosophila* pheromonal communication. *Behav. Genet.* 35:279–95

41. Friesen JA, Liu MF, Kent C. 2001. Cloning and characterization of a lipid-activated CTP:phosphocholine cytidylyltransferase from *Caenorhabditis elegans*: identification of a 21-residue segment critical for lipid activation. *Biochim. Biophys. Acta* 1533:86–98

42. Fuchs B, Suss R, Teuber K, Eibisch M, Schiller J. 2011. Lipid analysis by thin-layer chromatography: a review of the current state. *J. Chromatogr. A* 1218:2754–74

43. Fyrst H, Saba JD. 2010. An update on sphingosine-1-phosphate and other sphingolipid mediators. *Nat. Chem. Biol.* 6:489–97

44. Gallet A, Rodriguez R, Ruel L, Therond PP. 2003. Cholesterol modification of hedgehog is required for trafficking and movement, revealing an asymmetric cellular response to hedgehog. *Dev. Cell* 4:191–204

45. Gault CR, Obeid LM, Hannun YA. 2010. An overview of sphingolipid metabolism: from synthesis to breakdown. *Adv. Exp. Med. Biol.* 688:1–23

46. Golomb BA, Evans MA. 2008. Statin adverse effects: a review of the literature and evidence for a mitochondrial mechanism. *Am. J. Cardiovasc. Drugs* 8:373–418

47. Gomez-Diaz C, Reina JH, Cambillau C, Benton R. 2013. Ligands for pheromone-sensing neurons are not conformationally activated odorant binding proteins. *PLOS Biol.* 11:e1001546

48. Goode S, Melnick M, Chou TB, Perrimon N. 1996. The neurogenic genes egghead and brainiac define a novel signaling pathway essential for epithelial morphogenesis during *Drosophila* oogenesis. *Development* 122:3863–79

49. Griffiths WJ, Wang Y. 2009. Mass spectrometry: from proteomics to metabolomics and lipidomics. *Chem. Soc. Rev.* 38:1882–96

50. Griffitts JS, Haslam SM, Yang T, Garczynski SF, Mulloy B, et al. 2005. Glycolipids as receptors for *Bacillus thuringiensis* crystal toxin. *Science* 307:922–25

51. Griffitts JS, Whitacre JL, Stevens DE, Aroian RV. 2001. Bt toxin resistance from loss of a putative carbohydrate-modifying enzyme. *Science* 293:860–64

52. Groen CM, Spracklen AJ, Fagan TN, Tootle TL. 2012. *Drosophila* Fascin is a novel downstream target of prostaglandin signaling during actin remodeling. *Mol. Biol. Cell* 23:4567–78

53. Gronke S, Mildner A, Fellert S, Tennagels N, Petry S, et al. 2005. Brummer lipase is an evolutionary conserved fat storage regulator in *Drosophila*. *Cell Metab.* 1:323–30

54. Guan XL, Cestra G, Shui G, Kuhrs A, Schittenhelm RB, et al. 2013. Biochemical membrane lipidomics during *Drosophila* development. *Dev. Cell* 24:98–111

55. Guillou H, Zadravec D, Martin PG, Jacobsson A. 2010. The key roles of elongases and desaturases in mammalian fatty acid metabolism: insights from transgenic mice. *Prog. Lipid Res.* 49:186–99

56. Gupta GD, Swetha MG, Kumari S, Lakshminarayan R, Dey G, Mayor S. 2009. Analysis of endocytic pathways in *Drosophila* cells reveals a conserved role for GBF1 in internalization via GEECs. *PLOS ONE* 4:e6768

57. Gupta T, Schupbach T. 2003. Cct1, a phosphatidylcholine biosynthesis enzyme, is required for *Drosophila* oogenesis and ovarian morphogenesis. *Development* 130:6075–87

58. Hakomori S. 2000. Traveling for the glycosphingolipid path. *Glycoconj. J.* 17:627–47

59. Hamel S, Fantini J, Schweisguth F. 2010. Notch ligand activity is modulated by glycosphingolipid membrane composition in *Drosophila melanogaster*. *J. Cell Biol.* 188:581–94

60. Hammad LA, Cooper BS, Fisher NP, Montooth KL, Karty JA. 2011. Profiling and quantification of *Drosophila melanogaster* lipids using liquid chromatography/mass spectrometry. *Rapid Commun. Mass Spectrom.* 25:2959–68

61. Han X, Gross RW. 2005. Shotgun lipidomics: electrospray ionization mass spectrometric analysis and quantitation of cellular lipidomes directly from crude extracts of biological samples. *Mass Spectrom. Rev.* 24:367–412

62. Hannich JT, Entchev EV, Mende F, Boytchev H, Martin R, et al. 2009. Methylation of the sterol nucleus by STRM-1 regulates dauer larva formation in *Caenorhabditis elegans*. *Dev. Cell* 16:833–43

63. Hansen M, Flatt T, Aguilaniu H. 2013. Reproduction, fat metabolism, and life span: What is the connection? *Cell Metab.* 17:10–19

64. Harris WS, Mozaffarian D, Lefevre M, Toner CD, Colombo J, et al. 2009. Towards establishing dietary reference intakes for eicosapentaenoic and docosahexaenoic acids. *J. Nutr.* 139:804S–19

65. Hebbar S, Lee E, Manna M, Steinert S, Kumar GS, et al. 2008. A fluorescent sphingolipid binding domain peptide probe interacts with sphingolipids and cholesterol-dependent raft domains. *J. Lipid Res.* 49:1077–89

66. Herr DR, Fyrst H, Creason MB, Phan VH, Saba JD, Harris GL. 2004. Characterization of the *Drosophila* sphingosine kinases and requirement for Sk2 in normal reproductive function. *J. Biol. Chem.* 279:12685–94

67. Herr DR, Fyrst H, Phan V, Heinecke K, Georges R, et al. 2003. *Sply* regulation of sphingolipid signaling molecules is essential for *Drosophila* development. *Development* 130:2443–53

68. Hla T, Dannenberg AJ. 2012. Sphingolipid signaling in metabolic disorders. *Cell Metab.* 16:420–34

69. Hoang HD, Prasain JK, Dorand D, Miller MA. 2013. A heterogeneous mixture of F-series prostaglandins promotes sperm guidance in the *Caenorhabditis elegans* reproductive tract. *PLOS Genet.* 9:e1003271

70. Hoeferlin LA, Wijesinghe DS, Chalfant CE. 2013. The role of ceramide-1-phosphate in biological functions. *Handb. Exp. Pharmacol.* 215:153–66

71. Horikawa M, Nomura T, Hashimoto T, Sakamoto K. 2008. Elongation and desaturation of fatty acids are critical in growth, lipid metabolism and ontogeny of *Caenorhabditis elegans*. *J. Biochem.* 144:149–58

72. Horner MA, Pardee K, Liu S, King-Jones K, Lajoie G, et al. 2009. The *Drosophila* DHR96 nuclear receptor binds cholesterol and regulates cholesterol homeostasis. *Genes Dev.* 23:2711–16

73. Huang X, Warren JT, Buchanan J, Gilbert LI, Scott MP. 2007. *Drosophila Niemann-Pick Type C-2* genes control sterol homeostasis and steroid biosynthesis: a model of human neurodegenerative disease. *Development* 134:3733–42

74. Huber TB, Schermer B, Muller RU, Hohne M, Bartram M, et al. 2006. Podocin and MEC-2 bind cholesterol to regulate the activity of associated ion channels. *Proc. Natl. Acad. Sci. USA* 103:17079–86

75. Imae R, Inoue T, Kimura M, Kanamori T, Tomioka NH, et al. 2010. Intracellular phospholipase A1 and acyltransferase, which are involved in *Caenorhabditis elegans* stem cell divisions, determine the sn-1 fatty acyl chain of phosphatidylinositol. *Mol. Biol. Cell* 21:3114–24

76. Iwata R, Oda S, Kunitomo H, Iino Y. 2011. Roles for class IIA phosphatidylinositol transfer protein in neurotransmission and behavioral plasticity at the sensory neuron synapses of *Caenorhabditis elegans*. *Proc. Natl. Acad. Sci. USA* 108:7589–94

77. Jakobsson A, Westerberg R, Jacobsson A. 2006. Fatty acid elongases in mammals: their regulation and roles in metabolism. *Prog. Lipid Res.* 45:237–49

78. Jones LN, Rivett DE. 1997. The role of 18-methyleicosanoic acid in the structure and formation of mammalian hair fibres. *Micron* 28:469–85

79. Kadamur G, Ross EM. 2013. Mammalian phospholipase C. *Annu. Rev. Physiol.* 75:127–54

80. Kage-Nakadai E, Kobuna H, Kimura M, Gengyo-Ando K, Inoue T, et al. 2010. Two very long chain fatty acid acyl-CoA synthetase genes, acs-20 and acs-22, have roles in the cuticle surface barrier in *Caenorhabditis elegans*. *PLOS ONE* 5:e8857

81. Kahn-Kirby AH, Dantzker JL, Apicella AJ, Schafer WR, Browse J, et al. 2004. Specific polyunsaturated fatty acids drive TRPV-dependent sensory signaling in vivo. *Cell* 119:889–900

82. Kaneda T, Smith EJ. 1980. Relationship of primer specificity of fatty acid de novo synthetase to fatty acid composition in 10 species of bacteria and yeasts. *Can. J. Microbiol.* 26:893–98

83. Katic I, Vallier LG, Greenwald I. 2005. New positive regulators of lin-12 activity in *Caenorhabditis elegans* include the BRE-5/Brainiac glycosphingolipid biosynthesis enzyme. *Genetics* 171:1605–15

84. Keleman K, Vrontou E, Kruttner S, Yu JY, Kurtovic-Kozaric A, Dickson BJ. 2012. Dopamine neurons modulate pheromone responses in *Drosophila* courtship learning. *Nature* 489:145–49

85. Kho Y, Kim SC, Jiang C, Barma D, Kwon SW, et al. 2004. A tagging-via-substrate technology for detection and proteomics of farnesylated proteins. *Proc. Natl. Acad. Sci. USA* 101:12479–84

86. Khuong TM, Habets RL, Slabbaert JR, Verstreken P. 2010. WASP is activated by phosphatidylinositol-4,5-bisphosphate to restrict synapse growth in a pathway parallel to bone morphogenetic protein signaling. *Proc. Natl. Acad. Sci. USA* 107:17379–84

87. Klopfenstein DR, Tomishige M, Stuurman N, Vale RD. 2002. Role of phosphatidylinositol(4,5)bisphosphate organization in membrane transport by the Unc104 kinesin motor. *Cell* 109:347–58

88. Kniazeva M, Crawford QT, Seiber M, Wang CY, Han M. 2004. Monomethyl branched-chain fatty acids play an essential role in *Caenorhabditis elegans* development. *PLOS Biol.* 2:E257

89. Kniazeva M, Euler T, Han M. 2008. A branched-chain fatty acid is involved in post-embryonic growth control in parallel to the insulin receptor pathway and its biosynthesis is feedback-regulated in *C. elegans*. *Genes Dev.* 22:2102–10

90. Kniazeva M, Shen H, Euler T, Wang C, Han M. 2012. Regulation of maternal phospholipid composition and IP3-dependent embryonic membrane dynamics by a specific fatty acid metabolic event in *C. elegans*. *Genes Dev.* 26:554–66

91. Kniazeva M, Sieber M, McCauley S, Zhang K, Watts JL, Han M. 2003. Suppression of the ELO-2 FA elongation activity results in alterations of the fatty acid composition and multiple physiological defects, including abnormal ultradian rhythms, in *Caenorhabditis elegans*. *Genetics* 163:159–69

92. Kohyama-Koganeya A, Nabetani T, Miura M, Hirabayashi Y. 2011. Glucosylceramide synthase in the fat body controls energy metabolism in *Drosophila*. *J. Lipid Res.* 52:1392–99

93. Krahmer N, Guo Y, Wilfling F, Hilger M, Lingrell S, et al. 2011. Phosphatidylcholine synthesis for lipid droplet expansion is mediated by localized activation of CTP:phosphocholine cytidylyltransferase. *Cell Metab.* 14:504–15

94. Krieser RJ, Moore FE, Dresnek D, Pellock BJ, Patel R, et al. 2007. The *Drosophila* homolog of the putative phosphatidylserine receptor functions to inhibit apoptosis. *Development* 134:2407–14

95. Kubagawa HM, Watts JL, Corrigan C, Edmonds JW, Sztul E, et al. 2006. Oocyte signals derived from polyunsaturated fatty acids control sperm recruitment in vivo. *Nat. Cell Biol.* 8:1143–48

96. Kuhnlein RP. 2011. The contribution of the *Drosophila* model to lipid droplet research. *Prog. Lipid Res.* 50:348–56

97. Kurtovic A, Widmer A, Dickson BJ. 2007. A single class of olfactory neurons mediates behavioural responses to a *Drosophila* sex pheromone. *Nature* 446:542–46

98. Kurzchalia TV, Ward S. 2003. Why do worms need cholesterol? *Nat. Cell Biol.* 5:684–88

99. Lagace TA, Ridgway ND. 2013. The role of phospholipids in the biological activity and structure of the endoplasmic reticulum. *Biochim. Biophys. Acta* 1833:2499–510

100. Laughlin JD, Ha TS, Jones DN, Smith DP. 2008. Activation of pheromone-sensitive neurons is mediated by conformational activation of pheromone-binding protein. *Cell* 133:1255–65

101. Lee BH, Ashrafi K. 2008. A TRPV channel modulates *C. elegans* neurosecretion, larval starvation survival, and adult lifespan. *PLOS Genet.* 4:e1000213

102. Lee HC, Inoue T, Imae R, Kono N, Shirae S, et al. 2008. *Caenorhabditis elegans* mboa-7, a member of the MBOAT family, is required for selective incorporation of polyunsaturated fatty acids into phosphatidylinositol. *Mol. Biol. Cell* 19:1174–84

103. Lehtonen M, Reisner K, Auriola S, Wong G, Callaway JC. 2008. Mass-spectrometric identification of anandamide and 2-arachidonoylglycerol in nematodes. *Chem. Biodivers.* 5:2431–41

104. Leibundgut M, Maier T, Jenni S, Ban N. 2008. The multienzyme architecture of eukaryotic fatty acid synthases. *Curr. Opin. Struct. Biol.* 18:714–25

105. Lesa GM, Palfreyman M, Hall DH, Clandinin MT, Rudolph C, et al. 2003. Long chain polyunsaturated fatty acids are required for efficient neurotransmission in *C. elegans*. *J. Cell Sci.* 116:4965–75

106. Li Y, Zhang H, Litingtung Y, Chiang C. 2006. Cholesterol modification restricts the spread of Shh gradient in the limb bud. *Proc. Natl. Acad. Sci. USA* 103:6548–53

107. Lim H, Paria BC, Das SK, Dinchuk JE, Langenbach R, et al. 1997. Multiple female reproductive failures in cyclooxygenase 2–deficient mice. *Cell* 91:197–208

108. Lim HY, Wang W, Wessells RJ, Ocorr K, Bodmer R. 2011. Phospholipid homeostasis regulates lipid metabolism and cardiac function through SREBP signaling in *Drosophila*. *Genes Dev.* 25:189–200

109. Lingwood D, Simons K. 2010. Lipid rafts as a membrane-organizing principle. *Science* 327:46–50

110. Liu JL, Desjardins D, Branicky R, Agellon LB, Hekimi S. 2012. Mitochondrial oxidative stress alters a pathway in *Caenorhabditis elegans* strongly resembling that of bile acid biosynthesis and secretion in vertebrates. *PLOS Genet.* 8:e1002553

111. Liu W, Liang X, Gong J, Yang Z, Zhang YH, et al. 2011. Social regulation of aggression by pheromonal activation of Or65a olfactory neurons in *Drosophila*. *Nat. Neurosci.* 14:896–902

112. Lu N, Shen Q, Mahoney TR, Neukomm LJ, Wang Y, Zhou Z. 2012. Two PI 3-kinases and one PI 3-phosphatase together establish the cyclic waves of phagosomal PtdIns(3)P critical for the degradation of apoptotic cells. *PLOS Biol.* 10:e1001245

113. Lucanic M, Held JM, Vantipalli MC, Klang IM, Graham JB, et al. 2011. N-acylethanolamine signalling mediates the effect of diet on lifespan in *Caenorhabditis elegans*. *Nature* 473:226–29

114. Ma DK, Rothe M, Zheng S, Bhatla N, Pender CL, et al. 2013. Cytochrome P450 drives a HIF-regulated behavioral response to reoxygenation by *C. elegans*. *Science* 341:554–58

115. Malhotra A, Xu Y, Ren M, Schlame M. 2009. Formation of molecular species of mitochondrial cardiolipin. 1. A novel transacylation mechanism to shuttle fatty acids between sn-1 and sn-2 positions of multiple phospholipid species. *Biochim. Biophys. Acta* 1791:314–20

116. Marek LR, Kagan JC. 2012. Phosphoinositide binding by the Toll adaptor dMyD88 controls antibacterial responses in *Drosophila*. *Immunity* 36:612–22

117. Marza E, Long T, Saiardi A, Sumakovic M, Eimer S, et al. 2008. Polyunsaturated fatty acids influence synaptojanin localization to regulate synaptic vesicle recycling. *Mol. Biol. Cell* 19:833–42

118. Marza E, Simonsen KT, Faergeman NJ, Lesa GM. 2009. Expression of ceramide glucosyltransferases, which are essential for glycosphingolipid synthesis, is only required in a small subset of *C. elegans* cells. *J. Cell Sci.* 122:822–33

119. Matyash V, Geier C, Henske A, Mukherjee S, Hirsh D, et al. 2001. Distribution and transport of cholesterol in *Caenorhabditis elegans*. *Mol. Biol. Cell* 12:1725–36

120. Maxfield FR, Tabas I. 2005. Role of cholesterol and lipid organization in disease. *Nature* 438:612–21

121. Mayinger P. 2012. Phosphoinositides and vesicular membrane traffic. *Biochim. Biophys. Acta* 1821:1104–13

122. McMahon A, Butovich IA, Kedzierski W. 2011. Epidermal expression of an Elovl4 transgene rescues neonatal lethality of homozygous Stargardt disease-3 mice. *J. Lipid Res.* 52:1128–38

123. McPhee CK, Logan MA, Freeman MR, Baehrecke EH. 2010. Activation of autophagy during cell death requires the engulfment receptor Draper. *Nature* 465:1093–96

124. Meloni I, Muscettola M, Raynaud M, Longo I, Bruttini M, et al. 2002. FACL4, encoding fatty acid-CoA ligase 4, is mutated in nonspecific X-linked mental retardation. *Nat. Genet.* 30:436–40

125. Mendel J, Heinecke K, Fyrst H, Saba JD. 2003. Sphingosine phosphate lyase expression is essential for normal development in *Caenorhabditis elegans*. *J. Biol. Chem.* 278:22341–49

126. Menuz V, Howell KS, Gentina S, Epstein S, Riezman I, et al. 2009. Protection of *C. elegans* from anoxia by HYL-2 ceramide synthase. *Science* 324:381–84

127. Merrill AH Jr, Schmelz EM, Dillehay DL, Spiegel S, Shayman JA, et al. 1997. Sphingolipids—the enigmatic lipid class: biochemistry, physiology, and pathophysiology. *Toxicol. Appl. Pharmacol.* 142:208–25

128. Morck C, Olsen L, Kurth C, Persson A, Storm NJ, et al. 2009. Statins inhibit protein lipidation and induce the unfolded protein response in the non-sterol producing nematode *Caenorhabditis elegans*. *Proc. Natl. Acad. Sci. USA* 106:18285–90

129. Motola DL, Cummins CL, Rottiers V, Sharma KK, Li T, et al. 2006. Identification of ligands for DAF-12 that govern dauer formation and reproduction in *C. elegans*. *Cell* 124:1209–23

130. Mullen TD, Hannun YA, Obeid LM. 2012. Ceramide synthases at the centre of sphingolipid metabolism and biology. *Biochem. J.* 441:789–802

131. Nandakumar M, Tan MW. 2008. Gamma-linolenic and stearidonic acids are required for basal immunity in *Caenorhabditis elegans* through their effects on p38 MAP kinase activity. *PLOS Genet.* 4:e1000273

132. Nezis IP, Sagona AP, Schink KO, Stenmark H. 2010. Divide and ProsPer: the emerging role of PtdIns3P in cytokinesis. *Trends Cell Biol.* 20:642–49

133. O'Rourke EJ, Kuballa P, Xavier R, Ruvkun G. 2013. Omega-6 polyunsaturated fatty acids extend life span through the activation of autophagy. *Genes Dev.* 27:429–40

134. Pagano RE, Martin OC, Kang HC, Haugland RP. 1991. A novel fluorescent ceramide analogue for studying membrane traffic in animal cells: accumulation at the Golgi apparatus results in altered spectral properties of the sphingolipid precursor. *J. Cell Biol.* 113:1267–79

135. Palavalli LH, Brendza KM, Haakenson W, Cahoon RE, McLaird M, et al. 2006. Defining the role of phosphomethylethanolamine N-methyltransferase from *Caenorhabditis elegans* in phosphocholine biosynthesis by biochemical and kinetic analysis. *Biochemistry* 45:6056–65

136. Panbianco C, Weinkove D, Zanin E, Jones D, Divecha N, et al. 2008. A casein kinase 1 and PAR proteins regulate asymmetry of a PIP(2) synthesis enzyme for asymmetric spindle positioning. *Dev. Cell* 15:198–208

137. Pantoja M, Fischer KA, Ieronimakis N, Reyes M, Ruohola-Baker H. 2013. Genetic elevation of sphingosine 1–phosphate suppresses dystrophic muscle phenotypes in *Drosophila*. *Development* 140:136–46

138. Parvy JP, Napal L, Rubin T, Poidevin M, Perrin L, et al. 2012. *Drosophila melanogaster* acetyl-CoA-carboxylase sustains a fatty acid–dependent remote signal to waterproof the respiratory system. *PLOS Genet.* 8:e1002925

139. Pastuhov SI, Fujiki K, Nix P, Kanao S, Bastiani M, et al. 2012. Endocannabinoid-Goα signalling inhibits axon regeneration in *Caenorhabditis elegans* by antagonizing Gqα-PKC-JNK signalling. *Nat. Commun.* 3:1136

140. Patel DS, Fang LL, Svy DK, Ruvkun G, Li W. 2008. Genetic identification of HSD-1, a conserved steroidogenic enzyme that directs larval development in *Caenorhabditis elegans*. *Development* 135:2239–49

141. Paulick MG, Bertozzi CR. 2008. The glycosylphosphatidylinositol anchor: a complex membrane-anchoring structure for proteins. *Biochemistry* 47:6991–7000

142. Piomelli D. 2003. The molecular logic of endocannabinoid signalling. *Nat. Rev. Neurosci.* 4:873–84

143. Pizette S, Rabouille C, Cohen SM, Therond P. 2009. Glycosphingolipids control the extracellular gradient of the *Drosophila* EGFR ligand Gurken. *Development* 136:551–61

144. Poccia D, Larijani B. 2009. Phosphatidylinositol metabolism and membrane fusion. *Biochem. J.* 418:233–46

145. Poulos A. 1995. Very long chain fatty acids in higher animals: a review. *Lipids* 30:1–14

146. Ran-Ressler RR, Sim D, O'Donnell-Megaro AM, Bauman DE, Barbano DM, Brenna JT. 2011. Branched chain fatty acid content of United States retail cow's milk and implications for dietary intake. *Lipids* 46:569–76

147. Rasmussen JA, Hermetter A. 2008. Chemical synthesis of fluorescent glycero- and sphingolipids. *Prog. Lipid Res.* 47:436–60

148. Rauthan M, Pilon M. 2011. The mevalonate pathway in *C. elegans*. *Lipids Health Dis.* 10:243

149. Rauthan M, Ranji P, Aguilera Pradenas N, Pitot C, Pilon M. 2013. The mitochondrial unfolded protein response activator ATFS-1 protects cells from inhibition of the mevalonate pathway. *Proc. Natl. Acad. Sci. USA* 110:5981–86

150. Ribeiro I, Yuan L, Tanentzapf G, Dowling JJ, Kiger A. 2011. Phosphoinositide regulation of integrin trafficking required for muscle attachment and maintenance. *PLOS Genet.* 7:e1001295

151. Roncaglioni MC, Tombesi M, Avanzini F, Barlera S, Caimi V, et al. 2013. n-3 fatty acids in patients with multiple cardiovascular risk factors. *N. Engl. J. Med.* 368:1800–8

152. Roth GE, Gierl MS, Vollborn L, Meise M, Lintermann R, Korge G. 2004. The *Drosophila* gene *Start1*: a putative cholesterol transporter and key regulator of ecdysteroid synthesis. *Proc. Natl. Acad. Sci. USA* 101:1601–6

153. Sabourdy F, Kedjouar B, Sorli SC, Colie S, Milhas D, et al. 2008. Functions of sphingolipid metabolism in mammals: lessons from genetic defects. *Biochim. Biophys. Acta* 1781:145–83

154. Santos AC, Lehmann R. 2004. Isoprenoids control germ cell migration downstream of HMGCoA reductase. *Dev. Cell* 6:283–93

155. Sasamura T, Matsuno K, Fortini ME. 2013. Disruption of *Drosophila melanogaster* lipid metabolism genes causes tissue overgrowth associated with altered developmental signaling. *PLOS Genet.* 9:e1003917

156. Satoh T, Inagaki T, Liu Z, Watanabe R, Satoh AK. 2013. GPI biosynthesis is essential for rhodopsin sorting at the trans-Golgi network in *Drosophila* photoreceptors. *Development* 140:385–94

157. Schwientek T, Keck B, Levery SB, Jensen MA, Pedersen JW, et al. 2002. The *Drosophila* gene *brainiac* encodes a glycosyltransferase putatively involved in glycosphingolipid synthesis. *J. Biol. Chem.* 277:32421–29

158. Scott JL, Musselman CA, Adu-Gyamfi E, Kutateladze TG, Stahelin RV. 2012. Emerging methodologies to investigate lipid-protein interactions. *Integr. Biol.* 4:247–58

159. Seamen E, Blanchette JM, Han M. 2009. P-type ATPase TAT-2 negatively regulates monomethyl branched-chain fatty acid mediated function in post-embryonic growth and development in *C. elegans*. *PLOS Genet.* 5:e1000589

160. Seppo A, Moreland M, Schweingruber H, Tiemeyer M. 2000. Zwitterionic and acidic glycosphingolipids of the *Drosophila melanogaster* embryo. *Eur. J. Biochem.* 267:3549–58

161. Shen LR, Lai CQ, Feng X, Parnell LD, Wan JB, et al. 2010. *Drosophila* lacks C20 and C22 PUFAs. *J. Lipid Res.* 51:2985–92

162. Shim YH, Chun JH, Lee EY, Paik YK. 2002. Role of cholesterol in germ-line development of *Caenorhabditis elegans*. *Mol. Reprod. Dev.* 61:358–66

163. Shindou H, Hishikawa D, Harayama T, Yuki K, Shimizu T. 2009. Recent progress on acyl CoA:lysophospholipid acyltransferase research. *J. Lipid Res.* 50(Suppl.):S46–51

164. Shindou H, Shimizu T. 2009. Acyl-CoA:lysophospholipid acyltransferases. *J. Biol. Chem.* 284:1–5

165. Silhankova M, Port F, Harterink M, Basler K, Korswagen HC. 2010. Wnt signalling requires MTM-6 and MTM-9 myotubularin lipid-phosphatase function in Wnt-producing cells. *EMBO J.* 29:4094–105

166. Smith WL. 1989. The eicosanoids and their biochemical mechanisms of action. *Biochem. J.* 259:315–24

167. Soupene E, Kuypers FA. 2008. Mammalian long-chain acyl-CoA synthetases. *Exp. Biol. Med. (Maywood)* 233:507–21

168. Steinhauer J, Gijon MA, Riekhof WR, Voelker DR, Murphy RC, Treisman JE. 2009. *Drosophila* lysophospholipid acyltransferases are specifically required for germ cell development. *Mol. Biol. Cell* 20:5224–35

169. Suzuki J, Denning DP, Imanishi E, Horvitz HR, Nagata S. 2013. Xk-related protein 8 and CED-8 promote phosphatidylserine exposure in apoptotic cells. *Science* 341:403–6

170. Szafer-Glusman E, Giansanti MG, Nishihama R, Bolival B, Pringle J, et al. 2008. A role for very-long-chain fatty acids in furrow ingression during cytokinesis in *Drosophila* spermatocytes. *Curr. Biol.* 18:1426–31

171. Tanaka T, Ikita K, Ashida T, Motoyama Y, Yamaguchi Y, Satouchi K. 1996. Effects of growth temperature on the fatty acid composition of the free-living nematode *Caenorhabditis elegans*. *Lipids* 31:1173–78

172. Tanizawa Y, Kuhara A, Inada H, Kodama E, Mizuno T, Mori I. 2006. Inositol monophosphatase regulates localization of synaptic components and behavior in the mature nervous system of *C. elegans*. *Genes Dev.* 20:3296–310

173. Tilley DM, Evans CR, Larson TM, Edwards KA, Friesen JA. 2008. Identification and characterization of the nuclear isoform of *Drosophila melanogaster* CTP:phosphocholine cytidylyltransferase. *Biochemistry* 47:11838–46

174. Tootle TL, Spradling AC. 2008. *Drosophila* Pxt: a cyclooxygenase-like facilitator of follicle maturation. *Development* 135:839–47

175. Vance DE. 2013. Physiological roles of phosphatidylethanolamine N-methyltransferase. *Biochim. Biophys. Acta* 1831:626–32

176. Vance JE. 2008. Phosphatidylserine and phosphatidylethanolamine in mammalian cells: two metabolically related aminophospholipids. *J. Lipid Res.* 49:1377–87

177. Vance JE, Tasseva G. 2013. Formation and function of phosphatidylserine and phosphatidylethanolamine in mammalian cells. *Biochim. Biophys. Acta* 1831:543–54

178. Vasireddy V, Uchida Y, Salem N Jr., Kim SY, Mandal MN, et al. 2007. Loss of functional ELOVL4 depletes very long-chain fatty acids (> or = C28) and the unique omega-O-acylceramides in skin leading to neonatal death. *Hum. Mol. Genet.* 16:471–82

179. Vasireddy V, Wong P, Ayyagari R. 2010. Genetics and molecular pathology of Stargardt-like macular degeneration. *Prog. Retin. Eye Res.* 29:191–207

180. Vrablik TL, Watts JL. 2013. Polyunsaturated fatty acid derived signaling in reproduction and development: insights from *Caenorhabditis elegans* and *Drosophila melanogaster*. *Mol. Reprod. Dev.* 80:244–59

181. Walker AK, Jacobs RL, Watts JL, Rottiers V, Jiang K, et al. 2011. A conserved SREBP-1/phosphatidylcholine feedback circuit regulates lipogenesis in metazoans. *Cell* 147:840–52

182. Walther TC, Farese RV Jr. 2012. Lipid droplets and cellular lipid metabolism. *Annu. Rev. Biochem.* 81:687–714

183. Wandall HH, Pizette S, Pedersen JW, Eichert H, Levery SB, et al. 2005. Egghead and brainiac are essential for glycosphingolipid biosynthesis in vivo. *J. Biol. Chem.* 280:4858–63

184. Wang R, Kniazeva M, Han M. 2013. Peroxisome protein transportation affects metabolism of branched-chain fatty acids that critically impact growth and development of *C. elegans*. *PLOS ONE* 8:e76270

185. Wang X, Wu YC, Fadok VA, Lee MC, Gengyo-Ando K, et al. 2003. Cell corpse engulfment mediated by *C. elegans* phosphatidylserine receptor through CED-5 and CED-12. *Science* 302:1563–66

186. Watts JL, Browse J. 2002. Genetic dissection of polyunsaturated fatty acid synthesis in *Caenorhabditis elegans*. *Proc. Natl. Acad. Sci. USA* 99:5854–59

187. Watts JL, Browse J. 2006. Dietary manipulation implicates lipid signaling in the regulation of germ cell maintenance in *C. elegans*. *Dev. Biol.* 292:381–92

188. Weber U, Eroglu C, Mlodzik M. 2003. Phospholipid membrane composition affects EGF receptor and Notch signaling through effects on endocytosis during *Drosophila* development. *Dev. Cell* 5:559–70

189. Webster CM, Deline ML, Watts JL. 2013. Stress response pathways protect germ cells from omega-6 polyunsaturated fatty acid–mediated toxicity in *Caenorhabditis elegans*. *Dev. Biol* 373:14–25

190. Wehman AM, Poggioli C, Schweinsberg P, Grant BD, Nance J. 2011. The P4-ATPase TAT-5 inhibits the budding of extracellular vesicles in *C. elegans* embryos. *Curr. Biol.* 21:1951–59

191. Wollam J, Magner DB, Magomedova L, Rass E, Shen Y, et al. 2012. A novel 3-hydroxysteroid dehydrogenase that regulates reproductive development and longevity. *PLOS Biol.* 10:e1001305

192. Xu T, Rubin GM. 1993. Analysis of genetic mosaics in developing and adult *Drosophila* tissues. *Development* 117:1223–37

193. Yavari A, Nagaraj R, Owusu-Ansah E, Folick A, Ngo K, et al. 2010. Role of lipid metabolism in smoothened derepression in hedgehog signaling. *Dev. Cell* 19:54–65

194. Zhang H, Abraham N, Khan LA, Hall DH, Fleming JT, Gobel V. 2011. Apicobasal domain identities of expanding tubular membranes depend on glycosphingolipid biosynthesis. *Nat. Cell Biol.* 13:1189–201

195. Zhang H, Kim A, Abraham N, Khan LA, Hall DH, et al. 2012. Clathrin and AP-1 regulate apical polarity and lumen formation during *C. elegans* tubulogenesis. *Development* 139:2071–83

196. Zhang K, Kniazeva M, Han M, Li W, Yu Z, et al. 2001. A 5-bp deletion in ELOVL4 is associated with two related forms of autosomal dominant macular dystrophy. *Nat. Genet.* 27:89–93

197. Zhu H, Shen H, Sewell AK, Kniazeva M, Han M. 2013. A novel sphingolipid-TORC1 pathway critically promotes postembryonic development in *Caenorhabditis elegans*. *Elife* 2:e00429

198. Zullig S, Neukomm LJ, Jovanovic M, Charette SJ, Lyssenko NN, et al. 2007. Aminophospholipid translocase TAT-1 promotes phosphatidylserine exposure during *C. elegans* apoptosis. *Curr. Biol.* 17:994–99

Quality Control and Infiltration of Translation by Amino Acids Outside of the Genetic Code

Tammy Bullwinkle,[1] Beth Lazazzera,[2] and Michael Ibba[1,3]

[1]Department of Microbiology, Ohio State University, Columbus, Ohio 43210

[2]Department of Microbiology, Immunology and Molecular Genetics, University of California, Los Angeles, California 90095

[3]Ohio State Biochemistry Program and Center for RNA Biology, Ohio State University, Columbus, Ohio 43210; email: ibba.1@osu.edu

Annu. Rev. Genet. 2014. 48:149–66

First published online as a Review in Advance on August 28, 2014

The *Annual Review of Genetics* is online at genet.annualreviews.org

This article's doi: 10.1146/annurev-genet-120213-092101

Keywords

protein synthesis, tRNA, nonproteinogenic, quality control, aminoacyl-tRNA synthetase

Abstract

Translation of the genome into functional proteins is critical for cellular life. Accurate protein synthesis relies on proper decoding of mRNAs by the ribosome using aminoacyl-tRNAs. During aminoacyl-tRNA synthesis, stringent substrate discrimination and rigorous product proofreading ensure tRNAs are paired with the correct amino acid, as defined by the rules of the genetic code. What has remained far less clear is the extent to which amino acids that are not part of the genetic code might also threaten translational accuracy. Here, we review the broad range of nonproteinogenic, or nonprotein, amino acids that can naturally accumulate under different conditions, the ability of the translation quality control machinery to deal with such substrates, and their potential impact on the integrity of the genetic code and cellular viability.

INTRODUCTION

Transfer RNA (tRNA): alanine tRNA or tRNA^{Ala} denotes tRNA specific for alanine

Aminoacyl-tRNA synthetases (aaRSs): denoted by their three-letter amino acid designation, e.g., AlaRS for alanyl-tRNA synthetase

NPA: nonproteinogenic amino acid

The genetic code defines the rules by which information stored in the nucleic acid sequences of genes is translated into the corresponding amino acid sequences in proteins. How accurately the genetic code is translated depends mainly on two steps in protein synthesis: precise decoding of mRNAs and accurate synthesis of aminoacyl-tRNAs (aa–transfer RNAs). aa-tRNAs are made by aminoacyl-tRNA synthetases (aaRSs), which match specific amino acids with the corresponding tRNAs as defined by the genetic code. As described in more detail below, tRNA aminoacylation is the subject of extensive quality control (QC), and a reduction in the accuracy of this step in protein synthesis can have catastrophic effects on the cell. Such effects are well illustrated in a landmark study that linked QC by alanyl-tRNA synthetase (AlaRS) to protein-folding defects and neurodegeneration in mice (52). In a related study in mammalian cells, inactivation of the valyl-tRNA synthetase (ValRS) QC mechanism that discriminates valine (Val) from threonine (Thr) disrupted cell morphology and led to membrane blebbing, activation of caspase-3, and apoptosis (67). A notable finding of the latter study was that the phenotypes observed upon disruption of QC were exacerbated dramatically by the introduction of α-aminobutyrate into the growth media. α-Aminobutyrate is a naturally occurring metabolite that can potentially act as a substrate for protein synthesis despite the fact that it is not encoded within the genetic code; i.e., it is a nonproteinogenic, or nonprotein, amino acid (NPA; see sidebar, Genetically Encoded Nonproteinogenic and Nonnatural Amino Acids). One of the broader implications of this study is that mistranslation of Val codons with the NPA α-aminobutyrate is more disruptive to the synthesis of functional proteins than misincorporation of the genetic code amino acid Thr. Although other examples of the harmfulness of NPAs are well documented (e.g., 8), their use in translation can also be beneficial, for example, to maintaining growth during amino acid starvation (79).

Although NPAs can dramatically reduce the accuracy of protein synthesis, the majority of studies to date on QC and mistranslation have focused on errors within the confines of the genetic code, i.e., the substitution of one canonical amino acid for another canonical amino acid. The ability of the QC machinery to recognize and proofread amino acids outside the genetic code, the natural occurrence of different NPAs, and how translation with nonproteinogenic substrates ultimately impacts cell growth have received comparatively little attention and are the focus of this review (**Figure 1**).

GENETICALLY ENCODED NONPROTEINOGENIC AND NONNATURAL AMINO ACIDS

The genetic code contains 22 amino acids: the 20 canonical amino acids found in all organisms plus selenocysteine and pyrrolysine, which are encoded in only some genomes. In addition to these genetically encoded protein amino acids (GPAs), protein synthesis is able to use a vast range of other natural and synthetic substrates. Nonproteinogenic, or nonprotein, amino acids (NPAs) is a term used to refer to naturally occurring amino acids that are not part of the genetic code but nevertheless can act as substrates for protein synthesis. Examples of NPAs include D–amino acids and other precursors and products of GPA metabolism. The other category of amino acids that can be used as substrates for translation is the synthetic nonnatural amino acids (NNAs). Although translation can naturally utilize a modest range of NNAs, genetic engineering has allowed synthetic expansion of the genetic code to accommodate a wide range of potentially useful NNAs (reviewed in 69).

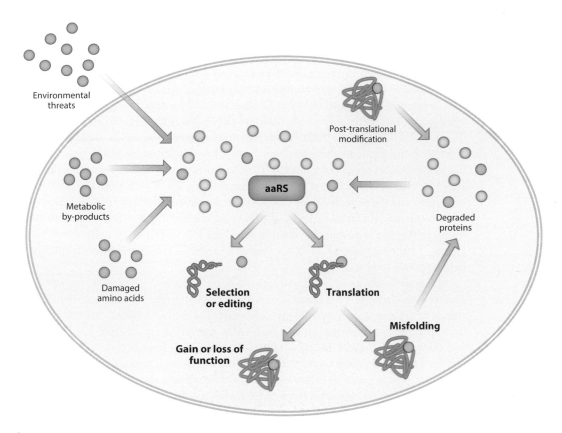

Figure 1

The central role of aminoacyl-tRNA synthetase (aaRS) enzymes in determining the fate of nonproteinogenic amino acids (NPAs; *red circles*) in the cell. Sources of NPA in the cellular pool of translation substrates are shown as gray arrows, and the outcomes of aaRS recognition of these compounds are shown as blue arrows. If an NPA is recognized by an aaRS, then it is either edited or used in protein synthesis. NPAs that are capable of making their way into the proteome can have various effects on protein functions in the cell, many of which are still unknown. NPAs in the proteome can also be recycled back into the amino acid pool following protein degradation. Genetically encoded cognate amino acids are shown as blue circles.

AMINO ACID QUALITY CONTROL

The potential utilization of NPAs during translation depends in part on their ability to be recognized by aaRSs. The aaRS family comprises a total of 23 enzymes, various combinations of which are found in different organisms (14). Aminoacylation of tRNA by aaRSs is a two-step reaction, consisting of ATP-dependent amino acid activation followed by ligation of amino acid to the 3′ end of tRNA, forming an aminoacyl ester bond (40). During amino acid activation, approximately half of the aaRSs display a level of specificity of 3,000:1 or greater for cognate versus noncognate proteinogenic amino acids, which is similar to overall error rates typically observed during protein synthesis. For the other aaRSs, the existence of closely related near-cognate substrates precludes a high level of cognate amino acid discrimination, and for QC these enzymes depend on an additional proofreading function called editing (109) (**Figure 1**). Editing is categorized as pre- or post-transfer, depending on whether products of the first or second step of the aminoacylation reaction are hydrolyzed, respectively (37, 57, 111).

AaRS editing contributes to QC as part of a double-sieve mechanism (30, 31). The first sieve is the active site of the aaRS, which determines the specificity for the cognate substrate. The second sieve is an editing or proofreading activity to clear either activated near-cognate amino acids or mischarged tRNAs. For example, isoleucyl-tRNA synthetase (IleRS) activates Ile only ~200-fold more efficiently than the noncognate amino acid Val, which differs by a single methyl group (92). To maintain translation fidelity, IleRS employs a form of proofreading that results in hydrolysis of misactivated Val-AMP (pre-transfer editing) and misacylated Val-tRNA^{Ile} (post-transfer editing), preventing mischarged tRNA from being released for translation (5, 23, 27). Similarly, the editing activities of the enzyme phenylalanyl-tRNA synthetase (PheRS) prevent the delivery of Tyr (tyrosine)-tRNA^{Phe} to the ribosome and protect against the mistranslation of Phe codons as Tyr (58). Editing activities have also been described for ValRS, leucine-tRNA synthetase (LeuRS), methionine-tRNA synthetase (MetRS), ThrRS, AlaRS, proline-tRNA synthetase (ProRS), lysine-tRNA synthetase (LysRS), and serine-tRNA synthetase (SerRS) in a broad range of organisms, consistent with the widespread use of aaRS editing to maintain translational fidelity (109 and references therein). Finally, if mischarged tRNAs are released before editing has occurred, in some instances they are subsequently edited in *trans* either by aaRSs or by freestanding *trans*-editing factors. This resampling of aa-tRNA synthesis errors in *trans* provides an additional QC step before protein synthesis.

Although the ability of aaRSs to discriminate and proofread noncognate, genetically encoded protein amino acids (GPAs) is well documented, the importance of their role in preventing infiltration of the genetic code by NPAs is considerably less well understood. Several aaRSs have been demonstrated to efficiently proofread both GPAs and NPAs; for example, LeuRS edits norvaline and Ile (36), and PheRS edits *meta*-Tyr (*m*-Tyr) and *para*-Tyr (*p*-Tyr) (49) (see **Table 1**). The ability of aaRS QC to target substrates outside the genetic code with adequate substrate specificity to prevent protein synthesis errors suggests that NPAs pose a threat to accurate translation in the cell (22). This idea is supported by previous studies showing that NPAs present major challenges for QC when supplied exogenously to cells, and under some conditions this is even more problematic than near-cognate GPA replacements (e.g., 35, 67). The known naturally occurring NPAs include D-enantiomers, which are used for cell wall biosynthesis but are excluded from protein synthesis by broad specificity D–amino acid deacylase *trans*-editing factors, and a range of naturally occurring metabolites (**Table 1**; see below for details). A typical NPA metabolite is *m*-Tyr, a product of Phe oxidation that is partially excluded from translation by the PheRS editing pathway in some organisms but not others (49). In *Escherichia coli*, the editing of *m*-Tyr-tRNA^{Phe} produced after cells are exposed to reactive oxygen species (ROS) constitutes a major component of the cell's oxidative stress response, underscoring the potential threat to cellular viability posed by NPAs (15). These findings illustrate that in order to fully appreciate the potential impact of NPAs on cell growth, it is necessary to examine conditions under which they are expected to accumulate to significant levels and how this might affect protein synthesis and cellular physiology.

MODIFIED AMINO ACID POOLS

The accumulation of NPAs in the cell can occur in a number of ways. One is via the degradation of proteins containing residues that have undergone post-translational modification. Post-translational modifications involve various chemical changes, such as acylation, methylation, phosphorylation, and oxidation. Degradation of modified proteins introduces NPAs into the pool of free amino acids in the cell, potentially making them available as substrates for aaRS enzymes and, subsequently, use in protein synthesis. It is not entirely clear how drastically free NPA pools are altered by these modified amino acids and whether they serve as substrates for translation or affect the need for aaRS proofreading. For example, phosphorylated amino acids are not substrates for protein synthesis, and substantial manipulation of the translation machinery is needed

Table 1 Naturally occurring nonproteinogenic amino acids that are substrates for aminoacyl-tRNA synthetases (aaRSs)

Source	Nonprotein amino acid	Structure	aaRS	Edited by aaRSs	Cotranslationally inserted into proteins	References
Post-translational oxidation	Leucine hydroxides (e.g., γ-hydroxyleucine)		LeuRS	Yes	Yes	(28, 33, 83)
Post-translational oxidation	Valine hydroxides (e.g., 4-hydroxyvaline)		NA	NA	Yes	(33, 83)
Post-translational oxidation	*Trans*-4-hydroxyproline		ProRS	Yes	Yes	(13)
Post-translational oxidation	Hydroxylysine (5-hydroxy-L-lysine)		LysRSII	Yes	NA	(2)
Oxidation of free amino acids; post-translational oxidation	3,4-dihydroxyphenylalanine (L-dopa)		PheRS	Yes	Yes	(64, 83, 105)
Plant product; post-translational oxidation	*Meta*-tyrosine		PheRS TyrRS	Yes	Yes	(15, 16, 35, 49)
Plant product	Canavanine		ArgRS	No	Yes	(86)
Plant product	Azetidine-2-carboxylic acid		ProRS	No	Yes	(78, 94, 101)
Plant product	Mimosine		PheRS	Yes	NA	(34)

(Continued)

Table 1 (*Continued*)

Source	Nonprotein amino acid	Structure	aaRS	Edited by aaRSs	Cotranslationally inserted into proteins	References
Plant product	2-Amino-4-methylhex-4-enoic acid (AMHA)		PheRS	Yes	NA	(3)
Plant product	β-N-methylamino-L-alanine (BMAA)		SerRS	NA	Yes	(24)
Biosynthesis by-products	Homocysteine		Several	Yes	NA	(45, 46)
Biosynthesis by-products	Homoserine		LysRSII	Yes	NA	(42)
Biosynthesis by-products	Ornithine		LysRSII	Yes	NA	(2, 43)
Biosynthesis by-products	α-Aminobutyrate		ValRS, LysRSII	Yes	Yes	(2, 66)
Biosynthesis by-products	Norleucine		MetRS	Yes	Yes	(11, 60)
Biosynthesis by-products	Norvaline		LeuRS, MetRS	Yes	Yes	(4, 48)
Biosynthesis by-products	β-Methylnorleucine		IleRS	NA	Yes	(65, 100)

Abbreviations: ArgRS, arginine-tRNA synthetase; IleRS, isoleucyl-tRNA synthetase; LeuRS, leucine-tRNA synthetase; LysRS, lysine-tRNA synthetase; MetRS, methionine-tRNA synthetase; NA, not available; PheRS, phenylalanyl-tRNA synthetase; ProRS, proline-tRNA synthetase; SerRS, serine-tRNA synthetase; TyrRS, tyrosine-tRNA synthetase; ValRS, valine-tRNA synthetase.

to accommodate them (74). The other potential sources of NPAs in the cellular metabolite pool are chemical modification of free GPAs and enzymatic synthesis, for example, of amino acid biosynthesis intermediates. Examples of the different routes of NPA accumulation that have been demonstrated to generate substrates used in protein synthesis are discussed in more detail below.

Oxidation of Amino Acids

Carefully controlled oxidation-reduction reactions within proteins have critical physiological roles in the cell, such as regulating the function of enzymes containing active cysteine (Cys) residues. However, unregulated oxidation resulting from the presence of elevated levels of ROS can both damage proteins and transform GPAs into NPAs. Protein modification and oxidative damage are well characterized in eukaryotic systems and have been found to be strongly correlated with aging and disease (99). More recently, oxidation of free amino acids prior to translation has also been shown to be a source of damaged residues in proteins (35). In both cases, the generated oxidized amino acids, whether they contribute directly or indirectly to the cellular NPA pool, can have a significant impact on the accuracy of translation and the overall functionality of the proteome.

In addition to the sulfur-containing amino acids Met and Cys, there are several other amino acids prone to oxidation, in particular the aromatic amino acids (98). Phenylalanine can undergo hydroxyl radical attack, which places a hydroxyl group at either the *ortho* or *meta* position of the carbon ring, generating *o*-Tyr and *m*-Tyr, respectively. *m*-Tyr is activated efficiently by the catalytic site of PheRS, and in *E. coli*, post-transfer editing by PheRS is required for growth in the presence of this NPA (15). In the absence of PheRS editing, *m*-Tyr is readily misincorporated at Phe codons, which limits, and can even inhibit, cell growth at elevated NPA levels. The primary role of *E. coli* PheRS proofreading, at least under laboratory conditions, seems to be to protect the cell from *m*-Tyr misincorporation, as similar experiments with the noncognate GPA *p*-Tyr do not have any discernible effects on growth. In yeast, by contrast, PheRS editing directly prevents *p*-Tyr incorporation but provides little protection against *m*-Tyr. The variability in the roles of the same pathway in different organisms, protecting the proteome from NPA incorporation in *E. coli* and GPA misincorporation in yeast, illustrates how the QC machinery can adapt to the demands posed by different growth environments and cellular physiologies.

Hydroxyl radicals can also generate other amino acids in the cell; for example, attack on Tyr results in addition of a hydroxyl group at the *meta* position producing the NPA L-3,4-dihydroxyphenylalanine (L-DOPA). L-DOPA is a precursor for dopamine and is used as a treatment for Parkinson's disease. The potential effects of L-DOPA misincorporation into proteins as a secondary outcome of therapeutic use of this NPA are of great interest. In vitro experiments show that L-DOPA is a substrate for PheRS and also that it can be incorporated into mammalian proteomes via TyrRS, an aaRS that lacks editing mechanisms to correct aminoacylation errors (64, 70, 83). *E. coli* and *Bacillus subtilis* TyrRS have also been shown to use L-DOPA as a substrate (16). The location within proteins at which L-DOPA is misincorporated determines the impact of this NPA on the cell, as only the replacement of Tyr residues that are not exposed to solvent leads to protein unfolding and aggregation (25, 26).

Aliphatic amino acids can also be targets for oxidation, leading to formation of noncognate, nonprotein substrates for aaRSs and the translation machinery. Oxidized GPAs, such as Leu hydroxide and Val hydroxide, have been shown to be incorporated into nascent proteins in mammalian cells (33, 83). In addition to oxidation of the host cell amino acid pool, these nonprotein amino acids are also produced post-translationally in plants, providing a dietary source of NPAs (82). Proline hydroxides (e.g., *trans*-3-hydroxyproline and *trans*-4-hydroxyproline) are also found in plants, where they are used for cell wall biosynthesis and may constitute as much as 16% of the residues in potato lectin (1). In animals, post-translational hydroxylation of Pro residues contributes to the role of collagen in cell stability and regulates the activity of the hypoxia-inducible factor, which is central to oxygen homeostasis (93). *E. coli* ProRS is able to edit the Pro analog *trans*-4-hydroxy-proline, but the QC machinery can be overwhelmed at high intracellular concentrations of this NPA, leading to efficient incorporation into recombinant proteins (9, 13).

De Novo Synthesis of Nonprotein Amino Acids

In addition to the accumulation of damaged amino acids in the cell, there are numerous examples of GPA mimics that are produced in large quantities by plants and microbes as part of what appear to be defense mechanisms, with toxic effects on protein synthesis in neighboring species. Such NPAs, in particular, plant-produced amino acids, have been studied mainly for their toxic effects on humans through consumption or in some cases their neurological benefits, as with theanine found in green tea (reviewed in 81). The targets of extracellular NPAs are often biosynthetic pathways, where they mimic intermediates, resulting in direct disruption of these critical processes. However, in other cases these NPAs also have high similarity to GPAs and therefore are recognized by aaRSs and used as substrates for translation.

Nonprotein Amino Acid–Based Defense Mechanisms in Plants and Microbes

Two NPAs produced in plants, L-canavanine and azetidine-2-carboxylic acid, have been well characterized for their ability to be used for protein synthesis by some organisms but not others. L-Canavanine, a mimic of arginine (Arg), is produced by a subfamily of Leguminosae and was shown to be incorporated into the proteins of insect larvae in place of Arg (32, 85). Ingestion of canavanine by humans becomes toxic at high levels and is being investigated in combination with Arg starvation for its potential anticancer properties (107). Differences exist in the ability of ArgRS from various organisms to use canavanine as a substrate and correlate with the presence of this amino acid in the organisms' environments, indicating the selection pressures such NPAs place on translation and the development of QC mechanisms, in this case increased enzyme specificity for the GPA Arg. The jack bean plant, which produces high levels of canavanine, and some species of insects such as the brucid beetle, have highly selective ArgRSs that do not efficiently use canavanine as a substrate and so provide a mechanism of resistance to the NPA in these organisms (41, 62, 87). Azetidine-2-carboxylate is produced in large amounts by several plant species, including sugar beets (*Beta vulgaris*) (78). This NPA is effectively misincorporated into proteins at Pro codons by several plant species as well as animals, where it has been shown to lead to neurodegeneration and autoimmune disorders (89, 90). Similar to L-canavanine and ArgRS, the ProRSs of azetidine-2-carboxylate-producing plant species are able to discriminate effectively against the NPA. Azetidine-2-carboxylate is toxic to *E. coli*, where it has been shown to elude post-transfer editing by ProRS (9).

Several plant NPAs chemically similar to Phe have been investigated for their ability to target PheRSs of other organisms. *Aesculus californica* seeds produce AMHA (2-amino-4-methylhex-4-enoic acid), which is a poor substrate for the host PheRS but is efficiently activated by other plant PheRSs. Whether AMHA is excluded during transfer to tRNA^{Phe} or is edited by the proofreading activity of PheRS is unknown (3). Mimosine (β-3-hydroxy-4-pyridone) is an aromatic amino acid analog found in high amounts in the genus *Mimosa* and in *Leucaena leucocephala*, and is toxic to animals, *E. coli*, and mung bean seedlings (96). This NPA is known to have cytotoxic effects outside of protein synthesis, and it is interesting to note that mimosine is activated by mung bean PheRS as well as PheRSs from mimosine-producing plants (34, 96). However, transfer of mimosine to tRNA^{Phe} has not been observed, and the NPA is not incorporated into proteins (96).

These observations suggest that PheRS pre-transfer and/or post-transfer QC mechanisms protect the proteomes of producer and target plants from mimosine incorporation, and that differences in mimosine sensitivity are unrelated to the potential impact of the NPA on protein synthesis. m-Tyr, a product of phenyalanine oxidation discussed above, is produced at high concentrations by fescue grasses and is toxic to other competing plant species (8). L-DOPA, another hydroxylation product, is made by the macuna plants, and is incorporated into mammalian cells in place of Tyr, indicating its use as an NPA by TyrRS of nonproducing species (83, 84). BMAA (β-N-methylamino-L-alanine) is an NPA that mimics Ser and is produced by cyanobacteria (genus *Nostoc*), which include free-living organisms and root symbionts of the cycad palms native to Guam (20). BMAA is found in contaminated seafood and drinking water, and can accumulate in seeds of *Cycas circinalis* as well as in the animals that feed on such plants. This NPA is associated with neurological diseases of the South Pacific that resemble amyotrophic lateral sclerosis (ALS), Parkinson's, and dementia (6). BMAA is found in proteins in the place of Ser and upon accumulation results in protein misfolding and aggregation, which is particularly problematic in nerve cells (24, 108).

The vast number of NPAs produced by plants, many of which target protein synthesis, presents the question of what evolutionary roles these toxins play. NPAs present an effective mechanism to limit unwanted competitors in a particular niche as a form of defense against competing plants, insect herbivores, or microbial pathogens that live in the surrounding rhizosphere. However, the selective pressure cytotoxic NPAs exert on aaRS specificity and proofreading in the producer suggests that to remain effective against targeted species, the development of widespread resistance is limited. More broadly, as with other antibiotics (21), the fact that sensitive organisms have persisted in niches where NPAs are synthesized also suggest other as yet unknown, perhaps positive, roles beyond interfering with protein synthesis.

PRIMARY METABOLITES AND NONPROTEIN AMINO ACIDS

The complexity and diversity of cellular metabolism lead to various by-products and intermediates, many of which are, or resemble, amino acids. Owing to the high number of NPAs and other metabolites in the cell, the translational machinery is constantly faced with an immense pool of compounds to select from, in addition to the coded amino acids normally found in proteins. In most cases, a low concentration of these intermediates and high aaRS enzyme specificity prevent any threat to protein synthesis. However there are several examples where metabolites that fall outside of the coded set of amino acids used in translation have been shown to function as efficient aaRS substrates and in some cases have been seen to be incorporated into the protein. Different editing mechanisms are used to limit the incorporation of these metabolites in protein biosynthesis, including aaRS-dependent pre-transfer editing of homocysteine (Hcy) and post-transfer editing of norvaline, and editing by an independent editing domain in the case of D–amino acids. Metabolic imbalances caused by mutations or changes in the organisms' environment can lead to significant upshifts in the levels of these metabolites, which in the case of Hcy is associated with significant toxicity. However, many questions remain regarding the full range of conditions that can lead to accumulation of these metabolites and the complete list of cellular mechanisms to prevent these NPAs from entering protein synthesis and eliciting cellular toxicity.

Pre-Transfer Editing of Metabolites: Homocysteine, Homoserine, and Ornithine

Several biosynthetic intermediates (Hcy, homoserine, and ornithine) are activated by aaRSs and then subsequently hydrolyzed in a tRNA-independent, pre-transfer editing reaction. In each case, a cyclized lactone or lactam is formed, which in the case of Hcy-thiolactone is a highly toxic molecule that requires an additional enzyme for further breakdown. How frequently the levels of homoserine or ornithine exceed the levels of the cognate amino acid for the relevant aaRSs (LysRS, ValRS, and IleRS in the case of homoserine and LysRS in the case of ornithine) is unknown, but for Hcy, the end product of Hcy editing, Hcy-thiolactone, can accumulate to significant levels in humans (46).

Hcy is the precursor of Met and differs from Met by one methyl group (19, 77). As a consequence, Hcy is readily activated by MetRS and, to a lesser extent, by IleRS, LeuRS, ValRS, and LysRS (46). Hcy-AMP is subsequently hydrolyzed in the aaRS active site through an intramolecular reaction between the activated carboxyl group and the thiolate to yield Hcy-thiolactone and AMP (46). Hcy-thiolactone is itself a toxic molecule that reacts with the amino side chain of lysines in proteins, resulting in protein N-homocysteinylation (77). In the cases in which N-homocysteinylation has been studied, it typically has resulted in protein inactivation and commonly caused the protein to aggregate (44, 73). Thus, it is not surprising that at least some organisms have a lactonase (PON1 in humans) that can degrade Hcy-thiolactone (45). In fact, PON1 lactonase activity was inversely correlated with serum protein N-homocysteinylation in humans (76).

As part of the biosynthetic pathway for Met, from bacteria to mammals, and also for cysteine in mammals, it would seem unlikely a priori that Hcy would accumulate to sufficient levels to compete with the cognate amino acids of the aaRSs. However, several genetic disorders and dietary deficiencies lead to increased levels of Hcy and Hcy-thiolactone in humans, a condition called hyperhomocysteinemia, which is a risk factor for a number of diseases, including cardiovascular disease (45, 47). The conversion of Hcy to Met is a folate- and vitamin B_{12}–dependent reaction. Human mutations in the gene for methylenetetrahydrofolate reductase, the enzyme that generates the folate substrate, or dietary deficiency in either folate or vitamin B_{12} lead to hyperhomocysteinemia (103). Similarly, the conversion of Hcy to cysteine is a vitamin B_6–dependent reaction, and genetic defects in cystathionine beta-synthetase, the first step in the cystenine biosynthesis pathway, and dietary deficiencies in vitamin B_6 lead to hyperhomocysteinemia, or high Cys in the blood (53).

Although defects in Hcy processing are well known in humans, it is less readily apparent whether other organisms encounter elevated levels of Hcy. For the bacterial species that can synthesize their own vitamins, Hcy should not readily accumulate. However, most bacterial species utilize a LuxS pathway to degrade S-adenosylhomocysteine (SAH), a by-product of reactions involving S-adenosylmethionine (SAM). LuxS catalyzes the second step in the pathway that generates Hcy and 4,5-dihydroxy-2,3-pentanedione (DPD) (75). DPD spontaneously cyclizes to form the AI-2 quorum-sensing molecule, which regulates processes such as biofilm formation and pathogenesis (29, 91). When bacteria produce high levels of AI-2, are the levels of Hcy high enough that MetRS and other aaRS are able to transform Hcy to Hcy-thiolactone? Although mammals and certain vertebrates and nematodes have homologs of the human PON enzymes that can degrade Hcy-thiolactone (7), it is unclear whether bacteria have a mechanism to degrade to do the same. Bacterial lactonases have been identified that degrade acyl-homoserine lactones, a class of quorum-sensing molecules of bacteria (54). Mammalian PON enzymes have also been found to degrade bacterial acyl-homoserine lactones, indicating the broad specificity of this class of enzymes (71). Bacterial genes recently identified as having >30% identity to mammalian PON1 demonstrate the phylogenetic relationship between the mammalian PON enzymes and the bacterial lactonases that degrade acyl-homoserine lactones (7). This finding raises the possibility that bacterial PONs and other lactonases may have a second function in degrading toxic Hcy-thiolactone, in addition to a quorum-quenching role.

Post-Transfer Editing of Metabolites from Branched-Chain Amino Acid Biosynthesis

By-products of the branched-chain amino acid pathways that have been shown to be substrates for aaRSs include norleucine, norvaline, and β-methylnorleucine. These have all been shown to be used in protein synthesis, particularly in the case of over-produced recombinant proteins in *E. coli*. Norleucine has been shown to be incorporated in place of Met in many different proteins in *E. coli* (60). Norvaline is incorporated at Leu codons in human hemoglobin expressed in *E. coli* (4), and β-methylnorleucine is misincorporated in overexpressed hirudin (65). These NPAs are charged onto tRNAs by aaRSs, including norvaline by LeuRS, norleucine by MetRS, and β-methylnorleucine by IleRS, but are then removed by a post-transfer editing reaction that hydrolyzes the aminoacylated tRNA to release the free amino acid and tRNA (102). Therefore, misincorporation must occur at concentrations at which these NPAs saturate the editing mechanism. Norvaline and norleucine have been shown to accumulate in cells when there is an imbalance in branched-chain amino acid biosynthesis, particularly when there is a downshift in oxygen concentration (10, 97). However, much remains to be learned about the environmental growth conditions that affect the accumulation of these NPAs.

A Free-Standing Editing Domain and D–Amino Acids

Another class of NPAs is D–amino acids, which are by-products of the synthesis of the corresponding L–amino acids. D–amino acids are highly toxic if incorporated into cellular proteins; however, many cell types do not exclude D–amino acids but rather use them for specific cellular functions and even actively produce D–amino acids (112). Bacteria specifically produce D–amino acids for incorporation into peptidoglycan and into nonribosomally synthesized antimicrobial peptides (17). Interestingly, several bacterial species have recently been shown to secrete micromolar concentrations of D–amino acids during entry into the stationary phase (for reviews, see 17, 39). In *Vibrio cholerae* and *B. subtilis*, for example, these stationary-phase secreted D–amino acids are not the canonical D-Ala and D-Glu found in peptidoglycan but instead are D-Leu, D-Met, and D-Val for *V. cholerae* (51) and D-Leu, D-Met, and D-Tyr for *B. subtilis* (50). Although the full range of function for these secreted D–amino acids is unknown, they can be incorporated into peptidoglycan, resulting in a change in structure and function. In *V. cholerae*, for instance, incorporation of D-Met and D-Leu into peptidoglycan increased osmotic stress resistance 20-fold (51).

For *B. subtilis* and possibly other organisms, the increased production of D–amino acids during the stationary phase poses a problem, as it can compromise protein synthesis. Addition of D–amino acids [D-Leu, D-Tyr, and D-Trp (tryptophan)] to *B. subtilis* cells resulted not only in incorporation of these amino acids into peptidoglycan but also in inhibited growth (50, 51, 55). This growth inhibition appears to be due to the incorporation of these D–amino acids into proteins. The addition to cells of the D–amino acid along with the specific L-enantiomer prevented growth inhibition and did not affect incorporation of the D–amino acid into peptidoglycan (55). Interestingly, the *B. subtilis* strain used in these studies has a loss-of-function mutation in the *dtd* gene that encodes D-tyrosyl-tRNA deacylase, an editing enzyme with broad specificity for D-aminoacylated tRNAs. Replacing the mutant *dtd* allele with a wild-type allele of the gene resulted in a strain that was now resistant to D–amino acids with regard to growth inhibition (55). Although it has not been formally shown that these D–amino acids are incorporated into proteins, the data clearly support the idea that these D–amino acids can be charged onto tRNAs.

D-tyrosyl-tRNA deacylases are widespread and represent the major mechanism cells use to edit tRNAs that are mischarged with D–amino acids (112). However, for particular D–amino acids that are not effectively selected against by the tRNA synthetases, the D-tyrosyl-tRNA deacylase may not be sufficient to resist the surge of D–amino acids produced by bacteria on entry into stationary phase. TyrRS is able to aminoacylate tRNA^{Tyr} with either D- or L-Tyr (16, 95). In *B. subtilis*, there are two TyrRS encoding genes, *tyrS* and *tyrZ*. TyrS is the major TyrRS used by the cell during vegetative growth (38), and like TyrRS in other organisms, TyrS is able to use D-Tyr as a substrate (16; R. Williams-Wagner, M. Raina, M. Ibba, and T. Henkin, personal communication). TyrZ, by contrast, exhibits a slower rate of aminoacylation than TyrS but is 20-fold more selective for L-Tyr than for D-Tyr (R. Williams-Wagner, M. Raina, M. Ibba, and T. Henkin, personal communication). Although the function of TyrZ in *B. subtilis* is not known, its ability to select against D-Tyr strongly suggests that this aaRS may be important during stationary phase, when the processes of biofilm formation and sporulation are induced and levels of D-Tyr produced by the cell dramatically increase. The observation that *B. subtilis* encodes a TyrRS that is able to effectively discriminate D-Tyr presents the question of why TyrZ is not the major TyrRS of the cell. Is the slower aminoacylation rate of TyrZ insufficient to support a fast growth rate?

CONCLUSIONS AND FUTURE OBJECTIVES

In general, cells go to great lengths to minimize errors during protein synthesis, employing extensive QC mechanisms to ensure that the genetic code is translated with high fidelity (113). The role of aminoacylation QC in limiting mistranslation of the genetic code and the diverse outcomes, both positive and negative, of coding errors have been studied extensively and are the subject of several recent reviews (72, 80). Although the general assumption is that there are no specific hot spots for mistranslation, there are proteins (called intrinsically disordered proteins) that are more prone to aggregation, such as α-synuclein and tau proteins (106). However, despite the widespread use of QC and proofreading, it is not always safe to assume that aminoacylation errors lead to protein unfolding and aggregation and limited growth and viability. *E. coli* can tolerate up to 10% mistranslation of some codons (88), microbes such as *Mycoplasma mobile* have lost aaRS QC mechanisms leading to widespread mistranslation (56, 110), and the yeast *Candida albicans* uses misaminoacylation to generate morphological diversity and increase competiveness (63). Perhaps one of the most striking examples is the mischarging of up to 1% of all tRNAs, with Met in eukaryotic cells exposed to viruses, Toll-like receptor ligands, or chemically induced oxidative stress (68). The common feature of all these examples is that the reported errors involve the replacement during aminoacylation of one GPA by another that is often closely related in structure. Consequently, the resulting errors in mRNA decoding are confined to amino acids within the genetic code and frequently lead to conservative substitutions. This is not the case with NPAs, which by definition are outside the genetic code and whose use during protein synthesis is often considerably more detrimental than the equivalent GPA substitutions. For example, substitutions with norleucine increase sensitivity to oxidative stress, and mistranslation with α-aminobutyrate or *m*-Tyr is considerably more problematic for the cell than are the corresponding GPA errors (15, 61, 67). As a better understanding of the challenges posed to translation by NPAs emerges, it is likely that a broader appreciation will develop for their role in both healthy and stressed cells.

The observation that NPAs sometimes pose a greater threat than misincorporated GPAs to the synthesis of a functional proteome is consistent with the evolution of aaRS QC to recognize and deal with such threats. For some aaRSs, such as TyrZ and PheRS, the NPA targeted by QC is known, and in other instances divergent evolution suggests as yet undefined challenges to protein synthesis. Yeast, for example, encodes two GlyRS (glycine-tRNA synthetase) enzymes, the second of which (GlyRS2) is dispensable for growth under normal conditions but is expressed under certain stress conditions (heat, oxidation, pH) in which NPAs might accumulate (18, 104). Similarly, *E. coli* encodes a second LysRS, LysU, which is synthesized as part of the heat shock response and exhibits differences in amino acid specificity compared with the corresponding housekeeping enzyme LysS (12). In another case, ThrRS editing activity against the GPA Ser is altered under oxidative stress conditions via modification of a critical Cys residue. These examples suggest that more NPAs that impact protein synthesis remain to be discovered. Recent developments in analytical approaches have considerably facilitated the ability to characterize and quantify cellular metabolite pools and are likely to dramatically change our future appreciation of the changing balance of GPAs and NPAs under different growth conditions, as well as our understanding of how this influences protein synthesis.

SUMMARY POINTS

1. Studies of translation QC have typically focused primarily on errors within the genetic code.

2. AaRS QC pathways can proofread nonproteinogenic amino acids outside of the genetic code.

3. Proofreading and editing of nonproteinogenic amino acids are essential for normal cell growth.

4. Nonproteinogenic amino acids can accumulate to potentially toxic levels in cells.

5. Growth under stress conditions can significantly increase nonproteinogenic amino acid synthesis.

6. QC of nonprotein amino acids is more important than proofreading of genetic code amino acids under some growth conditions.

7. Continuing advances in metabolomics are starting to reveal the full range of NPAs that accumulate in different cells under varied growth conditions.

DISCLOSURE STATEMENT

The authors are not aware of any affiliations, memberships, funding, or financial holdings that might be perceived as affecting the objectivity of this review.

ACKNOWLEDGMENTS

We thank T. Henkin, M. Raina, and R. Williams-Wagner for sharing unpublished data. Work in the authors' labs on this topic was supported by grants MCB 1052344 (to M.I.) and 1052493 (to B.L.) from the National Science Foundation.

LITERATURE CITED

1. Allen AK, Neuberger A. 1973. The purification and properties of the lectin from potato tubers, a hydroxyproline-containing glycoprotein. *Biochem. J.* 135:307–14

2. Ambrogelly A, O'Donoghue P, Soll D, Moses S. 2010. A bacterial ortholog of class II lysyl-tRNA synthetase activates lysine. *FEBS Lett.* 584:3055–60

3. Anderson JW, Fowden L. 1970. Properties and substrate specificities of the phenylalanyl-transfer-ribonucleic acid synthetases of *Aesculus* species. *Biochem. J.* 119:677–90

4. Apostol I, Levine J, Lippincott J, Leach J, Hess E, et al. 1997. Incorporation of norvaline at leucine positions in recombinant human hemoglobin expressed in *Escherichia coli*. *J. Biol. Chem.* 272:28980–88

5. Baldwin AN, Berg P. 1966. Transfer ribonucleic acid–induced hydrolysis of valyladenylate bound to isoleucyl ribonucleic acid synthetase. *J. Biol. Chem.* 241:839–45

6. Banack SA, Cox PA. 2003. Biomagnification of cycad neurotoxins in flying foxes: implications for ALS-PDC in Guam. *Neurology* 61:387–89

7. Bar-Rogovsky H, Hugenmatter A, Tawfik DS. 2013. The evolutionary origins of detoxifying enzymes: The mammalian serum paraoxonases (PONs) relate to bacterial homoserine lactonases. *J. Biol. Chem.* 288:23914–27

8. Bertin C, Weston LA, Huang T, Jander G, Owens T, et al. 2007. Grass roots chemistry: meta-tyrosine, an herbicidal nonprotein amino acid. *Proc. Natl. Acad. Sci. USA* 104:16964–69

9. Beuning PJ, Musier-Forsyth K. 2000. Hydrolytic editing by a class II aminoacyl-tRNA synthetase. *Proc. Natl. Acad. Sci. USA* 97:8916–20

10. Biermann M, Linnemann J, Knüpfer U, Vollstädt S, Bardl B, et al. 2013. Trace element associated reduction of norleucine and norvaline accumulation during oxygen limitation in a recombinant *Escherichia coli* fermentation. *Microb. Cell Fact.* 12:116

11. Bogosian G, Violand BN, Dorward-King EJ, Workman WE, Jung PE, Kane JF. 1989. Biosynthesis and incorporation into protein of norleucine by *Escherichia coli*. *J. Biol. Chem.* 264:531–39

12. Brevet A, Chen J, Leveque F, Blanquet S, Plateau P. 1995. Comparison of the enzymatic properties of the two *Escherichia coli* lysyl-tRNA synthetase species. *J. Biol. Chem.* 270:14439–44

13. Buechter DD, Paolella DN, Leslie BS, Brown MS, Mehos KA, Gruskin EA. 2003. Co-translational incorporation of *trans*-4-hydroxyproline into recombinant proteins in bacteria. *J. Biol. Chem.* 278:645–50

14. Bullwinkle TJ, Ibba M. 2014. Emergence and evolution. *Top. Curr. Chem.* 344:43–88

15. Bullwinkle TJ, Reynolds NM, Raina M, Moghal A, Matsa E, et al. 2014. Oxidation of cellular amino acid pools leads to cytotoxic mistranslation of the genetic code. *eLife* 3:e02501

16. Calendar R, Berg P. 1966. The catalytic properties of tyrosyl ribonucleic synthetases from *Escherichia coli* and *Bacillus subtilis*. *Biochemistry* 5:1690–95

17. Cava F, Lam H, de Pedro MA, Waldor MK. 2011. Emerging knowledge of regulatory roles of D-amino acids in bacteria. *Cell. Mol. Life Sci.* 68:817–31

18. Chen SJ, Wu YH, Huang HY, Wang CC. 2012. *Saccharomyces cerevisiae* possesses a stress-inducible glycyl-tRNA synthetase gene. *PLOS ONE* 7:e33363

19. Cohen GN, Saint-Girons I. 1987. Biosynthesis of threonine, lysine, and methionine. In Escherichia coli *and* Salmonella typhimurium, ed. FC Neidhardt, pp. 429–44. Washington, DC: American Soc. Microbiol.

20. Cox PA, Banack SA, Murch SJ. 2003. Biomagnification of cyanobacterial neurotoxins and neurodegenerative disease among the Chamorro people of Guam. *Proc. Natl. Acad. Sci. USA* 100:13380–83

21. Davies J. 2013. Specialized microbial metabolites: functions and origins. *J. Antibiot. (Tokyo)* 66:361–64

22. Döring V, Mootz HD, Nangle LA, Hendrickson TL, De Crécy-Lagard V, et al. 2001. Enlarging the amino acid set of *Escherichia coli* by infiltration of the valine coding pathway. *Science* 292:501–4

23. Dulic M, Cvetesic N, Perona JJ, Gruic-Sovulj I. 2010. Partitioning of tRNA-dependent editing between pre- and post-transfer pathways in class I aminoacyl-tRNA synthetases. *J. Biol. Chem.* 285:23799–809

24. Dunlop RA, Cox PA, Banack SA, Rodgers KJ. 2013. The non-protein amino acid BMAA is misincorporated into human proteins in place of L-serine causing protein misfolding and aggregation. *PLOS ONE* 8:e75376

25. Dunlop RA, Dean RT, Rodgers KJ. 2008. The impact of specific oxidized amino acids on protein turnover in J774 cells. *Biochem. J.* 410:131–40

26. Dunlop RA, Rodgers KJ, Dean RT. 2002. Recent developments in the intracellular degradation of oxidized proteins. *Free Radic. Biol. Med.* 33:894–906

27. Eldred EW, Schimmel PR. 1972. Rapid deacylation by isoleucyl transfer ribonucleic synthetase of isoleucine-specific transfer ribonucleic acid aminoacylated with valine. *J. Biol. Chem.* 247:2961–64

28. Englisch S, Englisch U, von der Haar F, Cramer F. 1986. The proofreading of hydroxy analogues of leucine and isoleucine by leucyl-tRNA synthetases from *E. coli* and yeast. *Nucleic Acids Res.* 14:7529–39

29. Federle MJ. 2009. Autoinducer-2-based chemical communication in bacteria: complexities of interspecies signaling. *Contrib. Microbiol.* 16:18–32

30. Fersht AR. 1977. Editing mechanisms in protein synthesis. Rejection of valine by the isoleucyl-tRNA synthetase. *Biochemistry* 16:1025–30

31. Fersht AR, Dingwall C. 1979. Evidence for the double-sieve editing mechanism in protein synthesis. Steric exclusion of isoleucine by valyl-tRNA synthetases. *Biochemistry* 18:2627–31

32. Fowden L, Lea PJ, Bell EA. 1979. The nonprotein amino acids of plants. *Adv. Enzymol. Relat. Areas Mol. Biol.* 50:117–75

33. Fu SL, Dean RT. 1997. Structural characterization of the products of hydroxyl-radical damage to leucine and their detection on proteins. *Biochem. J.* 324(Pt. 1):41–48

34. Gabius HJ, von der Haar F, Cramer F. 1983. Evolutionary aspects of accuracy of phenylalanyl-tRNA synthetase. A comparative study with enzymes from *Escherichia coli, Saccharomyces cerevisiae, Neurospora crassa*, and turkey liver using phenylalanine analogues. *Biochemistry* 22:2331–39

35. Gurer-Orhan H, Ercal N, Mare S, Pennathur S, Orhan H, Heinecke JW. 2006. Misincorporation of free m-tyrosine into cellular proteins: a potential cytotoxic mechanism for oxidized amino acids. *Biochem. J.* 395:277–84

36. Hellmann RA, Martinis SA. 2009. Defects in transient tRNA translocation bypass tRNA synthetase quality control mechanisms. *J. Biol. Chem.* 284:11478–84

37. Hendrickson TL, Schimmel P. 2003. Transfer RNA–dependent amino acid discrimination by aminoacyl-tRNA synthetases. In *Translation Mechanisms*, ed. J Lapointe, L Brakier-Gingras, pp. 34–64. New York: Kluwer Acad./Plenum Publ.

38. Henkin TM, Glass BL, Grundy FJ. 1992. Analysis of the *Bacillus subtilis tyrS* gene: conservation of a regulatory sequence in multiple tRNA synthetase genes. *J. Bacteriol.* 174:1299–306

39. Horcajo P, de Pedro MA, Cava F. 2012. Peptidoglycan plasticity in bacteria: stress-induced peptidoglycan editing by noncanonical D-amino acids. *Microb. Drug Resist.* 18:306–13

40. Ibba M, Söll D. 2000. Aminoacyl-tRNA synthesis. *Annu. Rev. Biochem.* 69:617–50

41. Igloi GL, Schiefermayr E. 2009. Amino acid discrimination by arginyl-tRNA synthetases as revealed by an examination of natural specificity variants. *FEBS J.* 276:1307–18

42. Jakubowski H. 1997. Aminoacyl thioester chemistry of class II aminoacyl-tRNA synthetases. *Biochemistry* 36:11077–85

43. Jakubowski H. 1999. Misacylation of tRNALys with noncognate amino acids by lysyl-tRNA synthetase. *Biochemistry* 38:8088–93

44. Jakubowski H. 1999. Protein homocysteinylation: possible mechanism underlying pathological consequences of elevated homocysteine levels. *FASEB J.* 13:2277–83

45. Jakubowski H. 2000. Homocysteine thiolactone: metabolic origin and protein homocysteinylation in humans. *J. Nutr.* 130:377S–81

46. Jakubowski H. 2011. Quality control in tRNA charging: editing of homocysteine. *Acta Biochim. Pol.* 58:149–63

47. Jakubowski H, Perla-Kajan J, Finnell RH, Cabrera RM, Wang H, et al. 2009. Genetic or nutritional disorders in homocysteine or folate metabolism increase protein N-homocysteinylation in mice. *FASEB J.* 23:1721–27

48. Kiick KL, Weberskirch R, Tirrell DA. 2001. Identification of an expanded set of translationally active methionine analogues in *Escherichia coli*. *FEBS Lett.* 502:25–30

49. Klipcan L, Moor N, Kessler N, Safro MG. 2009. Eukaryotic cytosolic and mitochondrial phenylalanyl-tRNA synthetases catalyze the charging of tRNA with the meta-tyrosine. *Proc. Natl. Acad. Sci. USA* 106:11045–48

50. Kolodkin-Gal I, Romero D, Cao S, Clardy J, Kolter R, Losick R. 2010. D-amino acids trigger biofilm disassembly. *Science* 328:627–29

51. Lam H, Oh DC, Cava F, Takacs CN, Clardy J, et al. 2009. D-amino acids govern stationary phase cell wall remodeling in bacteria. *Science* 325:1552–55

52. Lee JW, Beebe K, Nangle LA, Jang J, Longo-Guess CM, et al. 2006. Editing-defective tRNA synthetase causes protein misfolding and neurodegeneration. *Nature* 443:50–55

53. Lee SJ, Lee DH, Yoo HW, Koo SK, Park ES, et al. 2005. Identification and functional analysis of cystathionine beta-synthase gene mutations in patients with homocystinuria. *J. Hum. Genet.* 50:648–54

54. Lee SJ, Park SY, Lee JJ, Yum DY, Koo BT, Lee JK. 2002. Genes encoding the N-acyl homoserine lactone-degrading enzyme are widespread in many subspecies of *Bacillus thuringiensis*. *Appl. Environ. Microbiol.* 68:3919–24

55. Leiman SA, May JM, Lebar MD, Kahne D, Kolter R, Losick R. 2013. D-amino acids indirectly inhibit biofilm formation in *Bacillus subtilis* by interfering with protein synthesis. *J. Bacteriol.* 195:5391–95

56. Li L, Boniecki MT, Jaffe JD, Imai BS, Yau PM, et al. 2011. Naturally occurring aminoacyl-tRNA synthetases editing-domain mutations that cause mistranslation in *Mycoplasma* parasites. *Proc. Natl. Acad. Sci. USA* 108:9378–83

57. Ling J, Reynolds N, Ibba M. 2009. Aminoacyl-tRNA synthesis and translational quality control. *Annu. Rev. Microbiol.* 63:61–78

58. Ling J, So BR, Yadavalli SS, Roy H, Shoji S, et al. 2009. Resampling and editing of mischarged tRNA prior to translation elongation. *Mol. Cell* 33:654–60

59. Ling J, Soll D. 2010. Severe oxidative stress induces protein mistranslation through impairment of an aminoacyl-tRNA synthetase editing site. *Proc. Natl. Acad. Sci. USA* 107:4028–33

60. Lu HS, Tsai LB, Kenney WC, Lai PH. 1988. Identification of unusual replacement of methionine by norleucine in recombinant interleukin-2 produced by *E. coli*. *Biochem. Biophys. Res. Commun.* 156:807–13

61. Luo S, Levine RL. 2009. Methionine in proteins defends against oxidative stress. *FASEB J.* 23:464–72

62. Malinow MR, Bardana EJ Jr, Pirofsky B, Craig S, McLaughlin P. 1982. Systemic lupus erythematosus-like syndrome in monkeys fed alfalfa sprouts: role of a nonprotein amino acid. *Science* 216:415–17

63. Miranda I, Silva-Dias A, Rocha R, Teixeira-Santos R, Coelho C, et al. 2013. *Candida albicans* CUG mistranslation is a mechanism to create cell surface variation. *mBio* 4:e00285–13

64. Moor N, Klipcan L, Safro MG. 2011. Bacterial and eukaryotic phenylalanyl-tRNA synthetases catalyze misaminoacylation of tRNA(Phe) with 3,4-dihydroxy-L-phenylalanine. *Chem. Biol.* 18:1221–29

65. Muramatsu R, Negishi T, Mimoto T, Miura A, Misawa S, Hayashi H. 2002. Existence of beta-methylnorleucine in recombinant hirudin produced by *Escherichia coli*. *J. Biotechnol.* 93:131–42

66. Nangle LA, De Crécy-Lagard V, Döring V, Schimmel P. 2002. Genetic code ambiguity. Cell viability related to the severity of editing defects in mutant tRNA synthetases. *J. Biol. Chem.* 277:45729–33

67. Nangle LA, Motta CM, Schimmel P. 2006. Global effects of mistranslation from an editing defect in mammalian cells. *Chem. Biol.* 13:1091–100

68. Netzer N, Goodenbour JM, David A, Dittmar KA, Jones RB, et al. 2009. Innate immune and chemically triggered oxidative stress modifies translational fidelity. *Nature* 462:522–26

69. O'Donoghue P, Ling J, Wang YS, Soll D. 2013. Upgrading protein synthesis for synthetic biology. *Nat. Chem. Biol.* 9:594–98

70. Ozawa K, Headlam MJ, Mouradov D, Watt SJ, Beck JL, et al. 2005. Translational incorporation of L-3,4-dihydroxyphenylalanine into proteins. *FEBS J.* 272:3162–71

71. Ozer EA, Pezzulo A, Shih DM, Chun C, Furlong C, et al. 2005. Human and murine paraoxonase 1 are host modulators of *Pseudomonas aeruginosa* quorum-sensing. *FEMS Microbiol. Lett.* 253:29–37

72. Pan T. 2013. Adaptive translation as a mechanism of stress response and adaptation. *Annu. Rev. Genet.* 47:121–37

73. Paoli P, Sbrana F, Tiribilli B, Caselli A, Pantera B, et al. 2010. Protein N-homocysteinylation induces the formation of toxic amyloid-like protofibrils. *J. Mol. Biol.* 400:889–907

74. Park HS, Hohn MJ, Umehara T, Guo LT, Osborne EM, et al. 2011. Expanding the genetic code of *Escherichia coli* with phosphoserine. *Science* 333:1151–54

75. Pei D, Zhu J. 2004. Mechanism of action of S-ribosylhomocysteinase (LuxS). *Curr. Opin. Chem. Biol.* 8:492–97

76. Perla-Kajan J, Jakubowski H. 2010. Paraoxonase 1 protects against protein N-homocysteinylation in humans. *FASEB J.* 24:931–36

77. Perla-Kajan J, Twardowski T, Jakubowski H. 2007. Mechanisms of homocysteine toxicity in humans. *Amino Acids* 32:561–72

78. Peterson PJ, Fowden L. 1965. Purification, properties and comparative specificities of the enzyme prolyl-transfer ribonucleic acid synthetase from *Phaseolus aureus* and *Polygonatum multiflorum*. *Biochem. J.* 97:112–24

79. Pezo V, Metzgar D, Hendrickson TL, Waas WF, Hazebrouck S, et al. 2004. Artificially ambiguous genetic code confers growth yield advantage. *Proc. Natl. Acad. Sci. USA* 101:8593–97

80. Reynolds NM, Lazazzera BA, Ibba M. 2010. Cellular mechanisms that control mistranslation. *Nat. Rev. Microbiol.* 8:849–56

81. Rodgers KJ. 2014. Non-protein amino acids and neurodegeneration: the enemy within. *Exp. Neurol.* 253C:192–96

82. Rodgers KJ, Shiozawa N. 2008. Misincorporation of amino acid analogues into proteins by biosynthesis. *Int. J. Biochem. Cell Biol.* 40:1452–66

83. Rodgers KJ, Wang H, Fu S, Dean RT. 2002. Biosynthetic incorporation of oxidized amino acids into proteins and their cellular proteolysis. *Free Radic. Biol. Med.* 32:766–75

84. Rodgers KJ, Hume PM, Dunlop RA, Dean RT. 2004. Biosynthesis and turnover of DOPA-containing proteins by human cells. *Free Radic. Biol. Med.* 37:1756–64

85. Rosenthal GA. 2001. L-Canavanine: a higher plant insecticidal allelochemical. *Amino Acids* 21:319–30

86. Rosenthal GA, Dahlman DL. 1986. L-Canavanine and protein synthesis in the tobacco hornworm *Manduca sexta*. *Proc. Natl. Acad. Sci. USA* 83:14–18

87. Rosenthal GA, Dahlman DL, Janzen DH. 1976. A novel means for dealing with L-canavanine, a toxic metabolite. *Science* 192:256–58

88. Ruan B, Palioura S, Sabina J, Marvin-Guy L, Kochhar S, et al. 2008. Quality control despite mistranslation caused by an ambiguous genetic code. *Proc. Natl. Acad. Sci. USA* 105:16502–7

89. Rubenstein E. 2000. Biologic effects of and clinical disorders caused by nonprotein amino acids. *Medicine (Baltimore)* 79:80–89

90. Rubenstein E, McLaughlin T, Winant RC, Sanchez A, Eckart M, et al. 2009. Azetidine-2-carboxylic acid in the food chain. *Phytochemistry* 70:100–4

91. Rui F, Marques JC, Miller ST, Maycock CD, Xavier KB, Ventura MR. 2012. Stereochemical diversity of AI-2 analogs modulates quorum sensing in *Vibrio harveyi* and *Escherichia coli*. *Bioorg. Med. Chem.* 20:249–56

92. Schmidt E, Schimmel P. 1994. Mutational isolation of a sieve for editing in a transfer RNA synthetase. *Science* 264:265–67

93. Schofield CJ, Ratcliffe PJ. 2004. Oxygen sensing by HIF hydroxylases. *Nat. Rev. Mol. Cell Biol.* 5:343–54

94. Schwartz TW. 1988. Effect of amino acid analogs on the processing of the pancreatic polypeptide precursor in primary cell cultures. *J. Biol. Chem.* 263:11504–10

95. Sheoran A, Sharma G, First EA. 2008. Activation of D-tyrosine by *Bacillus stearothermophilus* tyrosyl-tRNA synthetase: 1. Pre-steady-state kinetic analysis reveals the mechanistic basis for the recognition of D-tyrosine. *J. Biol. Chem.* 283:12960–70

96. Smith IK, Fowden L. 1968. Studies on specificities of phenylalanyl- and tyrosyl-sRNA synthetases from plants. *Phytochemistry* 7:1065–75

97. Soini J, Falschlehner C, Liedert C, Bernhardt J, Vuoristo J, Neubauer P. 2008. Norvaline is accumulated after a down-shift of oxygen in *Escherichia coli* W3110. *Microb. Cell Fact.* 7:30

98. Stadtman ER. 1993. Oxidation of free amino acids and amino acid residues in proteins by radiolysis and by metal-catalyzed reactions. *Annu. Rev. Biochem.* 62:797–821

99. Stadtman ER. 2006. Protein oxidation and aging. *Free Radic. Res.* 40:1250–58

100. Sugiura M, Kisumi M, Chibata I. 1981. Beta-methylnorleucine, an antimetabolite produced by *Serratia marcescens*. *J. Antibiot. (Tokyo)* 34:1278–82

101. Takeuchi T, Rosenbloom J, Prockop DJ. 1969. Biosynthesis of abnormal collagens with amino acid analogues. II. Inability of cartilage cells to extrude collagen polypeptides containing L-azetidine-2-carboxylic acid or *cis*-4-fluoro-L-proline. *Biochim. Biophys. Acta* 175:156–64

102. Tang Y, Tirrell DA. 2002. Attenuation of the editing activity of the *Escherichia coli* leucyl-tRNA synthetase allows incorporation of novel amino acids into proteins in vivo. *Biochemistry* 41:10635–45

103. Trimmer EE. 2013. Methylenetetrahydrofolate reductase: biochemical characterization and medical significance. *Curr. Pharm. Des.* 19:2574–93

104. Turner RJ, Lovato M, Schimmel P. 2000. One of two genes encoding glycyl-tRNA synthetase in *Saccharomyces cerevisiae* provides mitochondrial and cytoplasmic functions. *J. Biol. Chem.* 275:27681–88

105. Umeda A, Thibodeaux GN, Zhu J, Lee Y, Zhang ZJ. 2009. Site-specific protein cross-linking with genetically incorporated 3,4-dihydroxy-L-phenylalanine. *ChemBioChem* 10:1302–4

106. Uversky VN, Oldfield CJ, Dunker AK. 2008. Intrinsically disordered proteins in human diseases: introducing the D2 concept. *Annu. Rev. Biophys.* 37:215–46

107. Vynnytska-Myronovska B, Bobak Y, Garbe Y, Dittfeld C, Stasyk O, Kunz-Schughart LA. 2012. Single amino acid arginine starvation efficiently sensitizes cancer cells to canavanine treatment and irradiation. *Int. J. Cancer* 130:2164–75

108. Xie X, Basile M, Mash DC. 2013. Cerebral uptake and protein incorporation of cyanobacterial toxin beta-N-methylamino-L-alanine. *Neuroreport* 24:779–84

109. Yadavalli SS, Ibba M. 2012. Quality control in aminoacyl-tRNA synthesis its role in translational fidelity. *Adv. Protein Chem. Struct. Biol.* 86:1–43

110. Yadavalli SS, Ibba M. 2012. Selection of tRNA charging quality control mechanisms that increase mistranslation of the genetic code. *Nucleic Acids Res.* 41:1104–12

111. Yadavalli SS, Musier-Forsyth K, Ibba M. 2008. The return of pretransfer editing in protein synthesis. *Proc. Natl. Acad. Sci. USA* 105:19031–32

112. Yang H, Zheng G, Peng X, Qiang B, Yuan J. 2003. D-Amino acids and D-Tyr-tRNA(Tyr) deacylase: stereospecificity of the translation machine revisited. *FEBS Lett.* 552:95–98

113. Zaher HS, Green R. 2009. Fidelity at the molecular level: lessons from protein synthesis. *Cell* 136:746–62

Vulnerabilities on the Lagging-Strand Template: Opportunities for Mobile Elements

Ashwana D. Fricker and Joseph E. Peters

Department of Microbiology, Cornell University, Ithaca, New York 14853;
email: jep48@cornell.edu

Annu. Rev. Genet. 2014. 48:167–86

First published online as a Review in Advance on
September 3, 2014

The *Annual Review of Genetics* is online at
genet.annualreviews.org

This article's doi:
10.1146/annurev-genet-120213-092046

Keywords

transposition, DNA replication, replication fork stalls, replication fork
collapse, lagging-strand template, Red recombination, HUH elements,
group II intron

Abstract

Mobile genetic elements have the ability to move between positions in a
genome. Some of these elements are capable of targeting one of the template
strands during DNA replication. Examples found in bacteria include
(*a*) Red recombination mediated by bacteriophage λ, (*b*) integration of group
II mobile introns that reverse splice and reverse transcribe into DNA,
(*c*) HUH endonuclease elements that move as single-stranded DNA, and
(*d*) Tn7, a DNA cut-and-paste transposon that uses a target-site-selecting
protein to target transposition into certain forms of DNA replication. In all
of these examples, the lagging-strand template appears to be targeted using
a variety of features specific to this strand. These features appear especially
available in certain situations, such as when replication forks stall or collapse.
In this review, we address the idea that features specific to the lagging-strand
template represent vulnerabilities that are capitalized on by mobile genetic
elements.

INTRODUCTION

After DNA replication, each daughter cell receives an original and a copied strand of parental DNA. Although this view might suggest that the two daughter chromosomes are equal in all ways, in reality, processing events related to the direction of DNA replication can impose differences in the mutation potential of the two strands. Depending on the direction of DNA replication, each stretch of nascent DNA an individual cell receives is derived from either the leading-strand template or the lagging-strand template. It is known that repair biases lead to a general skew toward G > C in one strand of the chromosome in bacteria and archaea (22) and that there are other processing differences between the strands (see below). In this review, we focus on the finding that the lagging-strand template is more vulnerable to mobile DNA elements than is the leading-strand template. We focus on bacteria because there are exceptional numbers of mobile elements accumulated by horizontal transfer in bacteria and because the regions of the genome derived from the leading- and lagging-strand templates are easier to unambiguously define. However, these findings likely apply to all three domains of life.

We suggest that some processing events found more frequently on the lagging-strand template make this strand more accessible or vulnerable, and therefore serve as an opportunity for mobile DNA elements seeking to insert into the host genome. Furthermore, we propose that the lagging-strand template is especially vulnerable in regions where DNA replication is perturbed. We also address the idea that hosts may have evolved systems to protect the vulnerable lagging-strand template.

Differences Between the Template Strands

In all bacteria analyzed to date, DNA replication proceeds bi-directionally from one origin of replication per chromosome (*oriC*). Each replication fork complex (called a replisome) is responsible for replicating approximately half of the chromosome, and each of these regions is called a replichore. DNA polymerases in the replisome are held to the DNA by association with a protein ring called the sliding-clamp processivity factor (or sliding clamp). Another protein complex (collectively called tau in bacteria) couples the polymerases together across the strands (**Figure 1a**). DNA polymerases replicate DNA in a $5' \rightarrow 3'$ direction. Owing to the antiparallel nature of the complementary strands, the polymerase associated with the leading-strand template moves continuously in the same direction as the larger replisome, whereas DNA replication on the lagging-strand template progresses away from the replisome. To provide the $3'$ OH needed for DNA polymerase, new RNA primers are produced by primase every 1–2 kb in bacteria. The resulting DNA fragments on the lagging-strand template found prior to completion are called Okazaki fragments. The transient single-stranded DNA (ssDNA) in the lagging-strand template occurring between priming events is coated with ssDNA binding protein (SSB).

In *Escherichia coli*, and likely all bacteria, two coordinated forces move the replisome: the action of the DNA polymerase on the leading-strand template and the action of the DnaB helicase (the enzyme responsible for separating the DNA strands) on the lagging-strand template (42). Primase (DnaG) is known to closely associate with DnaB. The process of initiating a new Okazaki fragment involves making a new RNA primer, loading a sliding clamp at this primer, and signaling DNA polymerase to switch from the last sliding clamp to the new one, leaving a sliding clamp free on the DNA for other interactions (reviewed in 15). Therefore, there are at least two regions of ssDNA on the lagging-strand template during replication: one between the Okazaki fragments that shrinks as the polymerase advances and one between the helicase and Okazaki fragment that grows in

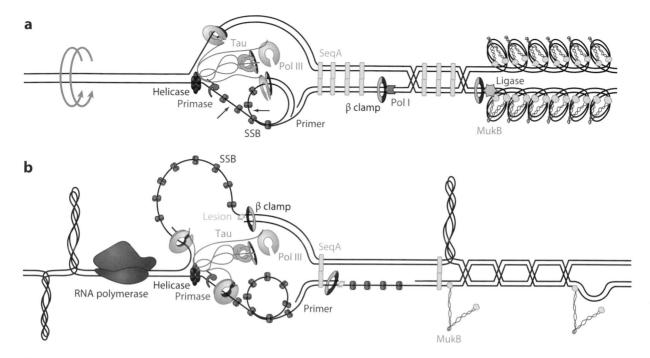

Figure 1

Normal and damaged DNA replication forks. The three-polymerase model, in which the polymerases on the leading- (*top*) and lagging- (*bottom*) strand template are coupled by tau (*green tentacles*), is shown. Pol III (DNA polymerase III) (*orange split ring*) is held to each strand by the sliding clamp (*blue/purple homodimeric circles*). The helicase (*dark-red hexamer*) tracks along the lagging-strand template and opens the two parental strands. Each Okazaki fragment is started with an RNA primer (*red line*) generated by primase (*brown comma*). The ssDNA (single-stranded DNA) on the lagging-strand is loaded with SSB (single-stranded DNA binding protein) (*light-purple homotetramer*). (*a*) Normal DNA replication fork. Okazaki fragments are processed by the progressive action of RNA primer removal by Pol 1 (DNA polymerase I) (*blue chevron*) and repair of the remaining nick by ligase (*pink star*). After replication, the hemimethylated duplex DNA has been suggested to be tethered together by SeqA (*light-green rods*) until MukB (*teal ball-and-chains*) forms a secondary structure. The two arrows indicate regions of ssDNA during normal DNA replication. (*b*) Damaged DNA replication fork. DNA replication forks can stall or collapse due to a number of processes. Many of these are shown, but any one could stall or collapse a DNA replication fork. Supercoiling, large-protein complexes, such as RNA polymerase (*dark-blue crab claw*), or lesions (*yellow blob*) can lead to replication fork stalling. The ssDNA on both strands is loaded with SSB. After replication, a failure in the coupling protein SeqA or in the secondary structure forming protein MukB may lead to altered supercoiling, delayed precatenane removal, and other issues with daughter-strand segregation.

size until a new primer is started (**Figure 1**, arrows). There is compelling evidence suggesting that three DNA polymerases are active in a replisome, leaving the possibility that two Okazaki fragments may be replicated in a single replisome, adding an additional stretch of ssDNA (28, 51, 85). Multiple nonexclusive models have been suggested for how replication of a new Okazaki fragment is signaled. These basically differ on whether a second gap-filling activity is needed to complete each fragment (47, 121). Multiple processing events need to occur in order to mature the lagging-strand template into a continuous DNA strand. In *E. coli*, DNA polymerase I (Pol I) fills in any remaining ssDNA gaps and, using its flap endonuclease (57) and/or $5' \rightarrow 3'$ exonuclease (55) activity, may displace and remove the RNA primer. After processing by Pol I, the 5' PO$_4$ and 3' OH remain until DNA ligase seals the nick (53). Ligase may associate with the sliding clamp previously occupied by DNA Pol I to help identify its substrates (53).

Exogenous and Endogenous Forces Complicate DNA Replication

Damaged DNA and physical or topological blockages in the template strands perturb the orderly coordination of events described above. Many types of lesions can block DNA polymerase on the leading-strand or lagging-strand template, resulting in significant gaps that require special repair processes, such as homologous recombination and lesion repair systems, and the recruitment of various DNA polymerases (**Figure 1b**). The entire replisome can be stalled at cross-links that occur across the DNA strands at collisions with protein complexes (such as RNA polymerase) or in regions of high supercoiling (14, 82, 83). In many of these cases, a stalled replication fork needs to be restarted or completely re-assembled using special machinery involving the Pri proteins and associated factors (31, 91). Multiple pathways for restarting DNA replication forks have been reconstituted in vitro, where, following reloading of the DnaB helicase, other replisome components can be coordinated. In some bacteria with circular chromosomes, replication forks are also actively stalled at specific DNA sites in the chromosome (called *ter* sites), where a *trans*-acting protein binds to the site, inhibiting the progression of the replicative DNA helicase (reviewed in 16). In other bacteria, replication may terminate when the two forks collide or by some unknown mechanism. It remains unclear how replication fork stalling specifically affects processing events on the lagging-strand template, but this template appears to be more vulnerable to mobile elements in these situations (see below).

MOBILE DNA ELEMENTS THAT TAKE ADVANTAGE OF ATTRIBUTES OF THE LAGGING-STRAND TEMPLATE

The following examples present cases in which mobile elements appear to take advantage of aspects found predominantly on the lagging-strand template, including stretches of ssDNA, free 3′ OH ends that can prime replication, and interactions found with specific proteins enriched on the lagging-strand template. The lagging-strand template may be especially accessible in cases in which a replisome encounters DNA damage and then stalls or collapses, and at sites where DNA replication actively terminates. We refer to this accessibility of the lagging-strand template as a vulnerability in the host-mediated process of DNA replication, given that the insertion of mobile elements is more likely to be detrimental than beneficial to the host organism. As addressed below, it can be argued that both the mobile element and the host may derive some benefit from greater accessibility during periods of stress.

λ Red Recombination and Other Phage Systems

Host-mediated recombination is extremely important to the cell for restarting stalled or collapsed replication forks. Homologous recombination systems that are unrelated to those found in the bacterial host can be found in some viruses of bacteria called bacteriophages (phages). It is still under debate as to why phages would have recombination systems, although multiple ideas have been proposed. Strains without λ Red recombination grow poorly, which may be attributed to the formation of concatemers during λ replication (94). Alternatively, λ Red recombination may play an important role in the highly mosaic nature of the bacteriophage (61). By being able to prime interphage recombination using very small patches of homology, there are more opportunities to stitch together new phage variants. However, phage-mediated recombination events are not needed for integration into the chromosome (2). The best-studied system for phage-mediated recombination is the Red recombination system from bacteriophage λ, although other phage systems seem to show the same properties (84, 100, 106) (see below).

The λ Red recombination system involves two proteins, Exo (or Redα) and Bet (or Redβ). The λ Red system has been reviewed in detail recently (32, 69) and is reviewed briefly here. The Bet recombinase is a single-strand annealing protein that binds to ssDNA, scans DNAs in the cell for a homologous sequence, and anneals the bound DNA to a homologous or mostly homologous sequence, allowing it to be present in progeny. Host-encoded genetic recombination generally utilizes a different type of protein, the well-studied RecA recombinase, in bacteria (reviewed in 56). Exo is an exonuclease that is responsible for processing duplex DNAs to ssDNA that can be bound by Bet without an accessory activity. A third protein that is found with the λ Red system is Gam, which protects DNAs with blunt or nearly blunt ends from the action of RecBCD (and other host exonucleases). Without Gam, RecBCD aggressively digests λ DNA because it lacks *chi* sites that normally convert this complex from degrading to actively loading the RecA recombinase (68). In fact, work with *gam* mutants identified the *chi* sequence by the ability of this sequence to protect the phage from RecBCD (70).

An interesting aspect of λ Red recombination is that there is an apparent dependency on DNA replication and an enhanced recombination with the lagging-strand template, indicating that discontinuously replicated DNA is more active for Red-mediated recombination. The bias for the lagging-strand template is revealed from work with recombination with ssDNA oligonucleotides. To monitor recombination into the chromosome or plasmid, successful recombination of oligonucleotides containing mismatches can be screened or directly selected. In experiments utilizing oligonucleotides, an early observation indicated that recombination was more efficient when it was complementary to the lagging-strand template than when it was complementary to the leading-strand template (17, 48). One model to explain a bias for recombination on the lagging-strand template holds that, unlike the RecA recombination system, which has the capacity to survey duplex DNA for homologous sequences, the phage recombinase must have access to DNA that is already single stranded.

Given that DNA replication is an essential process, it is difficult to test whether DNA replication is required for Red recombination. However, a great many experiments are highly consistent with the idea that ssDNA oligonucleotides recombine into transient ssDNA regions at the replication fork. For example, a decrease in recombination efficiency is found with oligonucleotides that lack a 3′ OH, suggesting that DNA replication must be primed from this end (34). Furthermore, a mutant allele of DnaG primase, which extends the length of Okazaki fragments and therefore expands the regions of ssDNA, increases the frequency of Red recombination (45). Important hints about the processes of Red recombination also come from the extent of oligonucleotide processing after recombination, which can be monitored using oligonucleotides containing mismatches to the host target sequence. After Red recombination has integrated the oligonucleotide, sequencing can determine which regions of the oligonucleotide were recombined into the host genome (49, 66). In addition to observing that oligonucleotides are frequently processed when they are recombined into the host, it was also found that the exonuclease activities of the normal DNA polymerases used in *E. coli* replication, DNA polymerase I and III, appeared to be responsible for these processing events (49). Neither exonucleases involved in DNA processing nor a DNA polymerase that is induced with DNA damage, Pol II, appeared to be involved in processing the oligonucleotides, supporting the idea that Red recombination occurs during normal DNA replication.

The effect of host-encoded methyl-directed mismatch repair (MMR) on the Red recombination process is also consistent with a process that occurs during DNA replication. A requirement for monitoring Red recombination is that at least one mismatch is present on the oligonucleotide used in the experiment. Red recombination occurs at a 100-fold increased frequency in *E. coli* hosts that are deficient in MMR (7). The *E. coli* MMR system is responsible for identifying and correcting mismatches that occur following DNA replication (reviewed in 24, 38). Upon recognition of the mismatched bases, MutS recruits MutL and MutH to nick one strand, which is removed and replaced. One important feature of the MMR system involves an ability of the system to recognize the newly replicated strand (which presumably contains a polymerase mistake). In *E. coli* and related bacteria, this recognition process involves DNA methylation. A separate enzyme, DNA adenine methyltransferase (Dam), is responsible for methylation of the A site at GATC/CTAG sequences. The methylase tracks behind the replication fork, leaving the new GATC sequence transiently unmethylated. This so-called hemimethylated state at GATC/CTAG sequences where one A is methylated and the other is not is recognized by the MutH protein. MutH then nicks the new (unmethylated) DNA strand. A helicase removes the strand with the nick along with the incorrect base, and subsequent replication replaces the missing strand. In bacteria, archaea, and eukaryotes that lack the Dam methylation system, a different and incompletely understood process is involved in identifying the newly replicated strand. Although the type of mismatch can affect the overall effect of MMR deficiency on the final recombination process, the loss of MMR does not change the bias for the lagging-strand template (48). In practice, high-efficiency recombination can occur in wild-type cells by using oligonucleotides with consecutive mismatches that prohibit MutS binding, oligonucleotides containing chemically modified bases that subvert the MMR system, or oligonucleotides with mismatches that are not recognized by the MMR system (7, 93, 108, 118).

There is reason to believe that the phage-encoded recombination systems take advantage of a vulnerability that is already present during DNA replication. In support of this, oligonucleotides can recombine in bacteria that have no obvious homologs to the phage systems. At a very low frequency, RecA-independent oligonucleotide recombination can be monitored in various members of gamma-proteobacteria with the same bias for the lagging-strand template. These systems show many of the same hallmarks of the λ Red system, in that the concentration of the oligonucleotide and the ability of certain DNA sequences to anneal more stably are important for obtaining higher frequencies of recombination (93, 101). However, there is an interesting difference between phage-mediated oligonucleotide recombination and oligonucleotide recombination associated with the host, concerning the length of the oligonucleotide used for recombination. In phage-mediated oligonucleotide recombination, the Red recombination frequency of smaller oligonucleotides (minimally 23 bp) increases exponentially as the length increases to 40 bp but plateaus as it increases to 70 bp (93). However, in the host oligonucleotide recombination systems, shorter oligonucleotides (~20 bp in length) were integrated with approximately the same efficiency as oligonucleotides of up to 70 bp in length in *E. coli* and 120 bp in *Pseudomonas syringae* (101). The observation that, with Red-independent oligonucleotide recombination, smaller oligonucleotides (shorter than 20 bp) can recombine with a low frequency that does not increase with the length of the oligonucleotide has been used to suggest that the naked oligonucleotide simply binds to regions of ssDNA at the replication fork (101). Conversely, the observation that the efficiency of Red recombination increases with progressively longer oligonucleotides has been used to support the importance of Bet-binding (93). Work with a phage system derived from *Mycobacteria* suggests that these same attributes found in the proteobacteria are also found in highly diverged bacteria; the mycobacterial phage Che9c encodes the exonuclease gp60 and the recombinase gp61, which carry out the same functions as Exo and Bet, respectively (106).

Work with λ Red and other phage recombination systems suggests that regions of transiently exposed ssDNA are available for Bet on the lagging-strand template but not generally on the leading-strand template. A single oligonucleotide containing sequences flanking a region up to 45 kb can recombine with the DNA, causing a deletion of the region (107). This finding suggests that annealed oligonucleotides are stable and that sequences found in an oligonucleotide can bridge regions that are well separated during active replication (107). Generally, phage recombination systems only have high efficiency in the hosts from which they were originally isolated (100, 106). This is consistent with the idea that these systems do not passively seek out ssDNA but instead rely on interactions between the phage proteins and host proteins to facilitate the entry process into the replisome. It remains unclear which mechanisms afford the replisome protection from phage recombination, but work in this area could synergize with work in other systems that target the lagging-strand template.

Group II Mobile Introns

Group II mobile introns are another type of mobile element that appears able to capitalize on features on the lagging-strand template (reviewed in 46). Full integration of these elements can be associated with host DNA replication. Group II mobile introns move as RNA after self-splicing out of an mRNA. The element itself is a catalytic intron RNA (ribozyme) that carries out two transesterification reactions for excision from an mRNA as a lariat structure. Group II mobile introns additionally use an intron-encoded protein (IEP) that acts as a maturase to help the mRNA fold into the correct structure and a reverse transcriptase activity to make a DNA copy of the RNA element after integration (62). The IEP can also have endonuclease activity, for reasons described below. As the word intron implies, group II introns use the same chemistry as introns that splice out of mRNA in eukaryotes (and are likely relatives of introns found in mRNA). After excision, the element integrates into DNA by using one of two pathways that differ by the targets that are recognized. In one pathway, called retrohoming, the element inserts into a single conserved site that base pairs with the RNA (and typically is recognized by the IEP). A second targeting pathway that occurs at a much lower frequency, called retrotransposition, involves a broader array of sites that have imperfect matches to the site used in retrohoming.

Integration events in DNA involve reverse splicing, where the mobile RNA is joined to the target DNA (87). This process is facilitated by the ability of the IEP to unwind the DNA strands to allow the element to search for homology for target recognition (98). After the single-stranded RNA (ssRNA) is integrated into one DNA strand, the subsequent steps can be variable, depending on the element and the host (67, 123). However, in all cases, a 3′ OH must be available to initiate reverse transcription to make a DNA copy of the ssRNA element that has been integrated (122, 123). In some cases, the IEP also contains an endonuclease activity that is capable of breaking the other strand (92). This ability to cleave the second strand for priming reverse transcription is often used in the retrohoming reaction (9). However, the process of retrotransposition that allows the movement of the element to other sites often requires another mechanism for initiating reverse transcription (36). Retrotransposition favors integration into DNA replication forks, and depending on the element and host, can show a bias to the lagging-strand template found during DNA replication (35, 36).

One well-studied group II mobile intron is the Ll.LtrB intron found in the relaxase gene *ltrB*, which is involved in conjugal plasmid transfer in *Lactococcus lactis* (63). Ll.LtrB preferentially retrohomes into the *ltrB* gene using its IEP, LtrA. In vivo, intron movement can be monitored using a system in which Ll.LtrB insertion events in the chromosome can be detected by direct antibiotic selection (8). Using this system, retrohoming occurs at a high frequency (approximately half the substrates that are available result in integration). The importance of the endonuclease activity of LtrA to retrohoming was assessed using a triple LtrA mutant, Y529A-R531A-T533A (LtrA[YRT]), which has normal maturase and DNA binding ability but lacks endonuclease activity (92). The LtrA[YRT] protein showed an ~100-fold drop in retrohoming compared with the wild-type protein, indicating that retrohoming is highly dependent on the endonuclease activity of LtrA (8, 36). However, in the process of retrotransposition, where a broader variety of sites is recognized, movement does not appear to be as strongly affected. Although retrotransposition with wild-type LtrA only occurred at a frequency of 0.02%, this translated into an unexpectedly modest drop (~40%) in the frequency of retrotransposition with LtrA^{YRT} (36). Interestingly, retrotransposition was also strongly biased into the lagging-strand template, as was particularly obvious in a unidirectionally replicating plasmid in the host, where 32 out of 33 events were in the lagging-strand template (35). Multiple factors could account for a bias for this strand: a requirement for ssDNA over double-stranded DNA (dsDNA), a need for more frequent priming via 3′ OH ends, or interactions with lagging-strand-dependent host replication factors. Using an in vitro system, retrotransposition into ssDNA was found to be preferred over retrotransposition into dsDNA, a bias that was markedly more evident in some DNA sequences. Analysis of natural retrotransposition events found in sequenced genomes also indicated a bias to the lagging-strand template (36). An explanation for why retrotransposition, but not retrohoming, is biased to lagging-strand replication comes from the knowledge that LtrA-mediated dsDNA unwinding and endonuclease activities are dependent upon binding of specific target sequences (65, 98). The contacts that are important for unwinding and second-strand nicking would therefore be unlikely to be found in sites used for retrotransposition. Thus, these integration events could be expected to be dependent on the ssDNA nature of the lagging-strand template and the availability of 3′ OH from Okazaki fragments to prime reverse transcription (36). Although other factors cannot be ruled out, these results suggest there is a strong bias for the lagging-strand template because it provides a higher availability of ssDNA targets. This would remove the need for unwinding a dsDNA template and may provide more opportunities for priming on the lagging-strand template that involves discontinuous replication.

Together, these data suggest that ssDNA and Okazaki fragments on the lagging-strand template provide an opportunity for group II elements to utilize a wider variety of insertion sites during retrotransposition, suggesting that these elements may specifically target these regions of the DNA because they are naturally available or accessible. It is hypothesized that more ancestral group II mobile elements, those without endonuclease domains, may have used ssDNA preferentially and biased retrotransposition into the lagging strand. Elements that acquired IEPs with a C-terminal DNA-binding domain could increase DNA unwinding ability (29), allowing for insertions into dsDNA, and thereby providing a gateway for evolving more stringent target-site selection. There is also evidence that insertion during DNA replication and into the lagging-strand template may reduce or eliminate, via priming from Okazaki fragments, the need for cleavage of the second strand. This is consistent with the finding that RmInt1, a group II mobile intron from *Sinorhizobium meliloti* that naturally lacks the C-terminal endonuclease domain, has a distinct preference for the lagging-strand template (60). The acquisition of an endonuclease domain could be seen as another advance that would obviate the need for insertion into actively replicating DNA, which would provide a primer for reverse transcription. This pathway may have been of paramount importance for early group II mobile elements and may still be important to many simpler elements found today.

HUH Endonuclease Elements

Another class of mobile elements that takes advantage of aspects of the lagging-strand template are the HUH (where H is a histidine and U is a hydrophobic residue) endonuclease elements. They are distinct from other classes of mobile elements in that they move as ssDNA. As described below, the movement of these elements is highly linked to the availability of ssDNA, both for excision and for integration. Two well-studied IS*200/605* family members that represent this class are IS*608* and IS*Dra2*.

IS*608* was originally discovered in the pathogen *Helicobacter pylori* and is particularly well studied because it is able to mobilize in a heterologous *E. coli* system (41). Using this *E. coli* system, it has been shown that IS*608* integration occurs at many positions but is always immediately downstream of a tetranucleotide sequence (5′-TTAC) (41). IS*608* naturally carries two genes: *tnpA*, which encodes a transposase, and *tnpB*, which is not required for transposition and is of unknown function (41). The TnpA transposase is a member of a large family of proteins that includes a conserved HUH domain to coordinate metal ions involved in catalysis (3). Members of the HUH family are found in all domains of life and appear to have adapted to many cellular processes involving cleavage and ligation at specific sequences in ssDNA (3).

TnpA contains the catalytic HUH motif, and work with this protein provides insight into the function of other HUH family proteins that have additional domains. Reconstitution of the transposition pathway in vitro indicated that TnpA catalyzes excision from ssDNA substrates (but not from dsDNA substrates) flanked by the terminal left and right ends (30). This ssDNA requirement inherently biases transposition to replication on the lagging-strand template. Consistent with this idea, a bioinformatics analysis of the natural occurrence of this family of elements indicated that there is a strong bias for elements in the lagging-strand template (104). To address strand differences in vivo, excision of the transposon from either the leading- or lagging-strand template of a plasmid was monitored (104). After overnight growth, a high degree of excision was found to occur only when the element resided in the lagging-strand template and was hardly detectable when situated in the leading-strand template (104). Furthermore, the frequency of excision correlated to what is known about the length of Okazaki fragments: Transposition frequencies were highest in a 0.3-kb element but decreased as the length was increased to 4 kb (104). This suggests the IS*608* element may need to reside entirely in one ssDNA region found on the lagging-strand template. Additional studies related IS*608* movement to DNA replication; transposition was dependent on replication in experiments that used strains containing temperature-sensitive mutations in essential DNA replication genes (104). Mutants that produce less DnaG primase, which increases the length of stretches of ssDNA on the lagging-strand template, had higher excision frequencies (104). This phenotype can be rescued by overexpression of DnaG, which consistently resulted in fewer excision events (104). Moreover, the excision frequency with less DnaG as compared with the wild type was greater in longer synthetic constructs than in shorter constructs (104).

Examining where transposition events occurred also provided interesting information about the presumed availability of ssDNA in cells. Insertion events that occurred into mobile plasmids and the chromosome were biased in an orientation expected for insertions into the lagging-strand template (104). Chromosomal insertions showed hot spots in highly transcribed rRNA operons, presumably because replication forks overtake RNA polymerase in these regions (104). In cases in which there is a rear-end collision between the faster moving DNA polymerase and the RNA polymerase, DNA replication is not believed to be stalled (81), but this event could impart other effects that can be capitalized on by these elements. The replication termination system of *E. coli* could be used as a tool to relate transposition and replication fork pausing (104). In *E. coli*, the progression of DNA replication forks is inhibited at specific sites in the DNA, called *ter* sites, through the action of a *trans*-acting protein, Tus (reviewed in 71). Interestingly, a unique insertion hot spot was observed proximal to an active *ter* site in a plasmid-based system and was dependent on the Tus protein (104). These results indicate that ssDNA in the lagging-strand template is the preferred transposition target and that access to this structure may increase when replication forks stall.

Similar results have been obtained for IS*Dra2*, another member of the IS*200/605* family from *Deinococcus radiodurans* (104). As was found with IS*608*, sequenced transposition events across the chromosome show a bias with the direction of DNA replication (104). Additionally, in a transposition assay based on the ability of excision events to activate a gene (where presumably orientation would not matter), events were biased to the lagging-strand template (104). The bias toward ssDNA found in other instances could also be shown with *D. radiodurans* as a function of its ability to withstand very high levels of radiation (75). When exposed to such extreme conditions, the chromosome experiences high levels of fragmentation, and extended sections of ssDNA are used to rebuild the chromosome. It was found that under these conditions, transposition is stimulated (75), and as expected, any bias to the leading or lagging strand is lost (104). This is consistent with the idea that ssDNA is the most important feature of the lagging-strand template for transposition and not necessarily some of the other aspects specific to DNA replication. Many transposons seem to regulate transposition with the physiologic state or stressed state of the cell (see below). It is possible that sensitivity of the IS*Dra2* element to ssDNA could also be an adaptive mechanism to stimulate transposition in response to host stress.

These results suggest that the ssDNA found on the lagging-strand template during DNA replication provides an opportunity for the very existence of HUH endonuclease elements. Work with these elements suggests that ssDNA may also be a signal of cell stress from extreme DNA damage stimulating the movement of this class of element. These studies also support the notion that not all lagging-strand templates are the same. Places where DNA replication forks stall, overtake RNA polymerase, or are subject to frequent priming by DnaG may lead to differential opportunities for movement of HUH endonuclease elements involving the lagging-strand template.

TnsE-Mediated Tn7 Transposition

Transposons are discrete elements that can move within a genome. Transposons that move as ssDNA appear to require the lagging-strand template during replication in a donor or target DNA (see above). There are examples of DNA transposons that move as dsDNA that are also sensitive to DNA replication. Transposon Tn7 provides an interesting example in which the lagging-strand template is actively targeted for transposition (78). Tn7 transposes via a dsDNA intermediate in a process known as cut-and-paste transposition, in which the element is excised from one site and inserted into another position in the cell (50, 74, 79). The process of Tn7 transposition requires multiple proteins that act on nonidentical *cis*-acting left and right ends that flank the element. The transposase that removes and rejoins the element is composed of two proteins, TnsA and TnsB (TnsAB), which work with a regulator protein, TnsC. Transposition targets are identified by one of two dedicated targeting proteins, TnsD or TnsE. The TnsD protein specifically targets a unique sequence found in bacterial chromosomes. The ability to target the lagging-strand template involves one of the Tn7-encoded proteins, TnsE, which is able to specifically recognize components found on the lagging-strand template during replication.

TnsE-mediated transposition was initially of interest because of its ability to specifically direct transposition into mobile plasmids called conjugal plasmids, which are capable of moving between bacteria (117). When a conjugal plasmid is present in the strain, transposition is stimulated >100-fold and the vast majority of these insertions are targeted into the mobile plasmids, despite the fact that they make up ~1% of the DNA in the cell (78, 117). Analysis of these insertions indicated that there is no sequence specificity with the TnsE-mediated pathway (117). However, a striking orientation bias of insertions is found in these conjugal plasmids (117). During conjugation, the relaxase nicks the origin of transfer (*oriT*) of the conjugal plasmid, and one strand of the plasmid DNA is transferred into the recipient cell, initiating DNA processing events (115). Host-mediated DNA replication synthesizes the complementary strands in the donor and recipient cell during the process of transfer. In the donor cell, a continuous process akin to processing on a leading-strand template occurs that is initiated by the liberated 3′ OH end. However, in the recipient cell, DNA replication is continuously reprimed in a process more similar to events found on the lagging-strand template. In early work with TnsE (117), it was unclear which specific molecular target was recognized and how it could drive the orientation bias with transposition. Targeting was dependent on active conjugation: Conjugal plasmids were activated as transposition targets in recipient cells, but the plasmid-encoded or host-encoded proteins involved in the process were not identified.

The essential role of host-mediated processes on the lagging-strand template was indicated by experiments in strains without conjugal plasmids (78). In these strains, TnsE-mediated transposition events occur at a very low frequency into the chromosome, with a regional bias centered on DNA replication termination sites. Strikingly, the rare TnsE-mediated insertion events in the chromosome occurred only in a single orientation across each replichore (78). The strict orientation bias indicated that a replication process from the host was the preferred target for TnsE-mediated transposition. This finding, in addition to the previous observation that TnsABC+E transposition targets events in conjugal plasmids in recipient cells, indicates that the lagging-strand template provides a DNA structure and/or protein complex that is recognized by TnsE-mediated transposition.

Additional experiments revealed that two components found during DNA replication on the lagging-strand template are essential for recognition by TnsE. Although TnsE generally has a strong affinity for DNA, competition experiments revealed that its preferred binding substrates are structures that contain a 3' recessed end. Supporting the role of this structure, it was found that TnsE gain-of-activity mutant proteins show an enhanced ability to interact with 3' recessed-end structures (78). These structures are abundant during replication, an observation that reinforces the idea that TnsABC+E transposition targets the lagging-strand template. Further investigation revealed that there was also an essential protein component for TnsE-mediated transposition, the sliding-clamp processivity factor (73). As noted above, the sliding-clamp proteins are deployed with each new priming cycle on the lagging-strand template, and sliding clamps left behind appear to be important for recruiting proteins that are responsible for actions that mature the lagging-strand template, including RNA primer removal, ligation of Okazaki fragments, and mismatch repair (53, 80). Proteins that interact with the sliding clamp have a conserved motif that facilitates part of this interaction (11). Interestingly, a putative clamp-interacting motif was identified in TnsE and was also found to be conserved across homologs of the TnsE protein (73). Interaction between TnsE and the sliding clamp was confirmed biochemically and genetically (73). Mutations in TnsE that perturbed the TnsE–sliding clamp interaction also either abolished or significantly decreased TnsE-mediated transposition but did not affect transposition targeted in other pathways (73). The interaction between TnsE and the sliding clamp was found to be weak, a possible adaptation that might mitigate some of the consequences of interacting with an essential component in the cell (73).

The TnsABC+E transposition system was also reconstituted in an in vitro system and showed TnsABC+E-dependent transposition if the target plasmid had a 20-bp gap (73). Of note, plasmids were not used as targets for TnsABC+E-dependent transposition if they were only nicked. When the sliding-clamp protein was preloaded onto this substrate, insertion events were strongly biased to the same orientation as found with in vivo TnsABC+E transposition (73). In this assay, the sliding clamps are loaded onto the DNA substrate in a single orientation and are believed to preferentially reside at gaps (27). Consistent with this idea was the finding that a specific interaction with sliding clamps on the lagging-strand template directed insertions in a single orientation (73). Although the in vitro reaction seems to recapitulate the minimum requirements for TnsE-mediated transposition, there are likely to be other components in the system. The in vitro reaction required gain-of-activity mutants that have ~1,000-fold higher transposition frequencies than wild-type proteins (73). These mutants showed the same bias in transposition as the wild type (78).

It is unclear whether these mutants simply amplify a low signal or whether they are compensating for an unknown component that normally must be present for transposition in vivo.

TnsE-mediated transposition not only targets the lagging-strand template, but these events are also strongly biased to regions of natural terminators of DNA replication (77, 78). As explained above with IS608, transposition is stimulated at sites where DNA replication terminates. A similar increase in the frequency of transposition in response to interfering with DNA replication is also observed in other elements, as explained below. Perturbing DNA replication forks also appears to make the lagging-strand template more vulnerable to TnsE-mediated transposition. These together suggest that forks become more vulnerable under periods of replication stress. Enhanced targeting of TnsE-mediated transposition events to conjugal plasmids could also be due to a loss of some type of protection found with normal DNA replication forks. One possibility is that unknown features may protect replication forks where both the leading- and lagging-strand templates are processed in a coordinated fashion. The protection may be lost during conjugation in which only one strand is replicated in each cell. This result would also be consistent with the finding that the filamentous bacteriophage M13, which replicates in a process where the replication of both strands is not spatially coordinated, is also a target for TnsE-mediated Tn7 transposition (21).

TnsABC+E transposition is also aggressively stimulated by replication events associated with DNA double-strand break (DSB) repair (77, 97). Repair of DSBs in bacteria usually involves initiating DNA replication with one of the broken ends using homologous recombination (reviewed in 44, 56). In *E. coli*, RecBCD exonuclease loads onto a DSB, degrading both strands of dsDNA until it encounters a specific DNA sequence called a *chi* site. *Chi* sites are recognized only in one orientation and are overrepresented in the chromosome in such a way that would be expected to quickly facilitate the reestablishment of DNA replication forks progressing in the normal direction toward the terminus region. After RecBCD engages a *chi* site, it then degrades from the 5′ end while actively loading RecA onto the 3′ end of extended ssDNA. RecA-coated ssDNA can invade a sister copy of the chromosome, forming a structure called a D-loop, in which one of the strands of the duplex DNA is displaced by the incoming ssDNA. The D-loop structure is recognized by the Pri proteins and associated factors that assemble and initiate a DNA replication fork from the 3′ end. Subsequent work with the system ruled out specific interactions with proteins involved in replication restart needed for repair and instead showed that replication initiated at the break was very likely a target (97). Of further interest, the insertion events primarily occurred at hot spots that were dependent on a regional DSB at an origin-proximal position. Hot spots for TnsE-mediated insertion were not found during normal DNA replication. These results suggest that not all DNA replication is the same with regard to Tn7 transposition because normal DNA replication events initiated at *oriC* did not result in highly focused hot spots for Tn7, but DNA replication initiated for DNA repair resulted in highly active hot spots that attracted most of the transposition events. This would suggest that Tn7 might be sensitive to differences in the replisome that stem from how they were originally initiated. It is also possible that proteins expressed during DNA damage alter the replisome, something that has been suggested in other work (37, 52, 59), and other changes in the cell may alter the ability of the replisome to proceed.

TnsE-mediated transposition appears to be specifically adapted to target transposition into the lagging-strand template, especially in regions where replication forks tend to stall or during atypical DNA replication, such as that found during conjugation or replication-mediated DSB repair. Like Tn7 and IS608, there are examples in which other very different types of DNA transposons preferentially transpose into replication forks that are actively stopped. For example, IS903 insertion in *E. coli* shows strong biases toward places where DNA replication terminates and shows an orientation dependency during conjugation (33, 102). A bias for transposition to a region of replication termination can also be found in some Firmicutes with Tn917. This bias may also involve an interaction with the sliding clamp because the Tn917 transposase contains a putative sliding-clamp-interacting motif (26, 96) (also see below). A major difference between Tn7 and the other elements discussed above is that Tn7 appears to choose to actively target the lagging-strand template without an obvious need for a feature (i.e., ssDNA and/or a 3′ OH end) found on this strand for cut-and-paste transposition.

IS THE LAGGING-STRAND TEMPLATE VULNERABLE, AND ARE MOLECULAR SYSTEMS IN PLACE TO REDUCE THIS VULNERABILITY?

Features of the lagging-strand template provide an opportunity for the mobilization of a variety of genetic elements. Other processes could also be considered vulnerabilities specific to the lagging-strand template. For example, the frequency of mutation differs depending on the placement of *lacZ* alleles on the leading- versus lagging-strand template (20). Constitutive expression of the bacterial DNA damage response genes (the SOS response) can magnify this effect, possibly due to the increased levels of activated RecA found during the SOS response overwhelming SSB or to the fact that the discontinuous nature of replication on the lagging-strand template allows more opportunities for SOS-induced DNA polymerases to pirate free 3′ OH ends. Hairpins formed from inverted repeats are normally not energetically favorable in dsDNA but may be favored when a stretch of ssDNA is available on the lagging-strand template, making it vulnerable to processing by enzymes that cleave hairpins like SbcCD (18). In yeast, triplet repeats, which are able to form hairpins, are more unstable on the lagging-strand template than on the leading-strand template (23), something that may relate to the abundance of ssDNA on this strand and the ability of these sequences to be expanded or deleted during Okazaki-fragment processing (40). Although stretches of ssDNA, free 3′ OH ends, and free sliding-clamp proteins may occasionally be found on the leading-strand template (119), they are more common on the lagging-strand template by the nature of discontinuous replication. Therefore, it seems fair to consider whether these features represent potential vulnerabilities for the host. This characterization seems especially appropriate given that these features are preferentially targeted by mobile elements in certain atypical situations when replication is perturbed, actively terminated, or initiated as a result of DSB repair, or during replication initiated by other genetic elements. In the following sections, we address the idea that there may be specific benefits for mobile elements to target features on the lagging-strand template and that there may be molecular systems in hosts that help protect these features.

Is It Advantageous for Mobile Elements to Respond to DNA Replication?

Many transposons have been shown to upregulate their movement during DNA replication. This is an important advantage for transposons that use cut-and-paste transposition in that the DSB created in donor DNA at the site left by the transposon can then be repaired by homologous recombination from the sister chromosome. Because the sister chromosome would still have a copy of the transposon, this process would also re-establish the element at the site that it vacated. A second benefit would be the ability to test a new insertion site in only one daughter cell: If insertions occur into an essential gene, only the daughter cell with the chromosome that received the insertion would be lost. Specific molecular systems have been identified for upregulating transposition after replication (86, 120) and an association of transposition with replication has been shown with some eukaryotic elements (88, 99). In a variety of other cases, eukaryotic transposases have been found to interact with the sliding clamp or to contain putative sliding-clamp-interacting motifs

(19, 73, 103, 112, 113). Association with the sliding clamp could act as a way to coordinate other interactions on DNA (103). Alternatively, association with sliding clamps could provide a mechanism to regulate transposition. This stems from the fact that sliding clamps are only found loaded on DNA during DNA replication and during certain DNA repair events. Therefore, transposons that require an interaction with sliding clamps on DNA would be active only during these times and at these places. In the case of bacteria, if DNA replication associated with conjugal plasmid transfer and bacteriophage replication is readily used by mobile genetic elements, these elements would also have the advantage of facilitated horizontal transfer, as is found with TnsE-mediated Tn7 transposition.

Is Replication Fork Stress an Important Indicator of Host Stress?

Genetic elements have been suggested to monitor the growth state and stress level of their host. Perhaps the most classic example is the activation of the lytic cycle of bacteriophage λ following SOS induction. In this process, ssDNA-bound RecA acts as the SOS signaling molecule, inducing cleavage of the λ repressor, thereby initiating the lysis program (10, 25). This feature of induced mobility in response to host stress is also observed by other mobile elements. Molecular systems exist that allow transposons to eavesdrop on the metabolic state of the host, permitting them to increase the frequency of transposition in cells that are subject to nutrient stress (4–6, 12, 95, 102, 105, 114). DNA replication is also acutely sensitive to many natural processes on the DNA (64), and the elongation phase of DNA replication is actively regulated in response to nutrient stress with a variety of molecular mechanisms (13, 58, 76, 90). Therefore, genetic elements that move more frequently when DNA replication is perturbed also end up mobilizing more frequently under the same stress conditions that affect replication.

If replication found during DNA DSB repair is differentially recognized by some genetic elements, there could be consequences on the frequency of transposition in natural populations. Unlike bacterial growth in the laboratory, the majority of natural environments are nutrient limited, and it is generally accepted that there are protracted periods of slow or no growth (43, 89). Under these conditions, DNA replication associated with the duplication of cells is limited; however, DNA replication associated with DNA repair may take on a more dominant role. Given these proposed environmental DNA replication patterns, transposition rates measured in the laboratory using actively growing cells may be misleading for understanding the mobility of genetic elements that are sensitive to the type of DNA replication. In addition, natural transformation also provides an opportunity for repair-initiated replication when ssDNA fragments are integrated into the genome (54, 109). Therefore, if replication associated with DNA repair is a more general target found with some genetic elements, it could also be an indicator of cell stress in multiple distinct ways.

Do Cells Have Distinct Mechanisms to Allow Protection?

If the lagging-strand template is vulnerable, it seems reasonable to ask whether systems are in place to protect the lagging-strand template. Technically, any system that helps to ensure the stability and orderly progress of replication forks would also limit access to the types of genetic elements discussed above (**Figure 1**). Ahead of the replication fork, gyrase is responsible for removing positive supercoils. Behind the replication fork, precatenanes can accumulate and must be unlinked by Topo IV for the sister chromosomes to eventually be segregated (111). Unlinking of chromosomes via Topo IV appears to be coordinated with the bridging activity of SeqA between (and/or within) sister chromosomes in a process that is also affected by the eventual condensation of

DNA by the MukBEF system (39, 72). The eventual segregation of the two sister chromosomes also appears to be a highly regulated process (110, 116). Organizing processes controlling supercoiling and strand separation, coupling DNA replication on both template strands, and orchestrating primer removal, gap filling, and ligation before the chromosomes are unlinked and segregated may limit many negative outcomes, including access to mobile genetic elements. The coordination of these features might also provide a steric barrier to proteins and nucleic acids from horizontally acquired genetic elements. Although it would be hard to implicate any part of replication and segregation systems as processes that evolved specifically to protect the chromosome from genetic elements, it will be interesting to know whether specific molecular systems in these elements have evolved to disrupt aspects of replication as a mechanism to get access to the lagging-strand template. In a practical sense, any mechanisms that we uncover in mobile genetic elements may find use as tools to help manipulate bacterial chromosomes.

DISCLOSURE STATEMENT

The authors are not aware of any affiliations, memberships, funding, or financial holdings that might be perceived as affecting the objectivity of this review.

ACKNOWLEDGMENTS

We thank Marlene Belfort, Michael Chandler, Don Court, Adam Parks, and the members of the Peters lab for comments on the manuscript. Work in the Peters lab is funded by the National Science Foundation (MCB-1244227).

LITERATURE CITED

1. Bidnenko V, Ehrlich SD, Michel B. 2002. Replication fork collapse at replication terminator sequences. *EMBO J.* 21(14):3898–907
2. Bobay LM, Touchon M, Rocha EPC. 2013. Manipulating or superseding host recombination functions: a dilemma that shapes phage evolvability. *PLOS Genet.* 9(9):e1003825
3. Chandler M, de la Cruz F, Dyda F, Hickman AB, Moncalian G, Ton-Hoang B. 2013. Breaking and joining single-stranded DNA: the HUH endonuclease superfamily. *Nat. Rev. Microbiol.* 11(8):525–38
4. Claverys JP, Prudhomme M, Martin B. 2006. Induction of competence regulons as a general response to stress in Gram-positive bacteria. *Annu. Rev. Microbiol.* 60(1):451–75
5. Coros AM, Twiss E, Tavakoli NP, Derbyshire KM. 2005. Genetic evidence that GTP is required for transposition of IS*903* and Tn*552* in *Escherichia coli*. *J. Bacteriol.* 187(13):4598–606
6. Coros CJ, Piazza CL, Chalamcharla VR, Smith D, Belfort M. 2009. Global regulators orchestrate group II intron retromobility. *Mol. Cell* 34(2):250–56
7. Costantino N, Court DL. 2003. Enhanced levels of λ Red-mediated recombinants in mismatch repair mutants. *Proc. Natl. Acad. Sci. USA* 100(26):15748–53
8. Cousineau B, Lawrence S, Smith D, Belfort M. 2000. Retrotransposition of a bacterial group II intron. *Nature* 404(6781):1018–21
9. Cousineau B, Smith D, Lawrence-Cavanagh S, Mueller JE, Yang J, et al. 1998. Retrohoming of a bacterial group II intron: mobility via complete reverse splicing, independent of homologous DNA recombination. *Cell* 94(4):451–62
10. Craig NL, Roberts JW. 1980. *E. coli* RecA protein-directed cleavage of phage λ repressor requires polynucleotide. *Nature* 283(5742):26–30
11. Dalrymple BP, Kongsuwan K, Wijffels G, Dixon NE, Jennings PA. 2001. A universal protein-protein interaction motif in the eubacterial DNA replication and repair systems. *Proc. Natl. Acad. Sci. USA* 98(20):11627–32

12. DeBoy RT, Craig NL. 2000. Target site selection by Tn7:*att*Tn7 transcription and target activity. *J Bacteriol.* 182(11):3310–13

13. DeNapoli J, Tehranchi AK, Wang JD. 2013. Dose-dependent reduction of replication elongation rate by (p)ppGpp in *Escherichia coli* and *Bacillus subtilis*. *Mol. Microbiol.* 88(1):93–104

14. De Septenville AL, Duigou S, Boubakri H, Michel B. 2012. Replication fork reversal after replication-transcription collision. *PLOS Genet.* 8(4):e1002622

15. Duderstadt KE, Reyes-Lamothe R, van Oijen AM, Sherratt DJ. 2014. Replication-fork dynamics. *Cold Spring Harb. Perspect. Biol.* 6(1):a010157

16. Duggin IG, Wake RG, Bell SD, Hill TM. 2008. The replication fork trap and termination of chromosome replication. *Mol. Microbiol.* 70(6):1323–33

17. Ellis HM, Yu D, DiTizio T, Court DL. 2001. High efficiency mutagenesis, repair, and engineering of chromosomal DNA using single-stranded oligonucleotides. *Proc. Natl. Acad. Sci. USA* 98(12):6742–46

18. Eykelenboom JK, Blackwood JK, Okely E, Leach DRF. 2008. SbcCD causes a double-strand break at a DNA palindrome in the *Escherichia coli* chromosome. *Mol. Cell* 29(5):644–51

19. Feschotte C, Mouchès C. 2000. Evidence that a family of miniature inverted-repeat transposable elements (MITES) from the *Arabidopsis thaliana* genome has arisen from a *pogo*-like DNA transposon. *Mol. Biol. Evol.* 17(5):730–37

20. Fijalkowska IJ, Jonczyk P, Tkaczyk MM, Bialoskorska M, Schaaper RM. 1998. Unequal fidelity of leading strand and lagging strand DNA replication on the *Escherichia coli* chromosome. *Proc. Natl. Acad. Sci. USA* 95(17):10020–25

21. Finn JA, Parks AR, Peters JE. 2007. Transposon Tn7 directs transposition into the genome of filamentous bacteriophage M13 using the element-encoded TnsE protein. *J. Bacteriol.* 189(24):9122–25

22. Frank AC, Lobry JR. 1999. Asymmetric substitution patterns: a review of possible underlying mutational or selective mechanisms. *Gene* 238(1):65–77

23. Freudenreich CH, Stavenhagen JB, Zakian VA. 1997. Stability of a CTG/CAG trinucleotide repeat in yeast is dependent on its orientation in the genome. *Mol. Cell. Biol.* 17(4):2090–98

24. Fukui K. 2010. DNA mismatch repair in eukaryotes and bacteria. *J. Nucleic Acids* 27:260512

25. Galkin VE, Yu X, Bielnicki J, Ndjonka D, Bell CE, Egelman EH. 2009. Cleavage of bacteriophage λ cI repressor involves the RecA C-terminal domain. *J. Mol. Biol.* 385(3):779–87

26. Garsin DA, Urbach J, Huguet-Tapia JC, Peters JE, Ausubel FM. 2004. Construction of an *Enterococcus faecalis* Tn*917*-mediated-gene-disruption library offers insight into Tn*917* insertion patterns. *J. Bacteriol.* 186(21):7280–89

27. Georgescu RE, Kim SS, Yurieva O, Kuriyan J, Kong XP, O'Donnell M. 2008. Structure of a sliding clamp on DNA. *Cell* 132(1):43–54

28. Georgescu RE, Kurth I, O'Donnell ME. 2012. Single-molecule studies reveal the function of a third polymerase in the replisome. *Nat. Struct. Mol. Biol.* 19(1):113–16

29. Guo H, Zimmerly S, Perlman PS, Lambowitz AM. 1997. Group II intron endonucleases use both RNA and protein subunits for recognition of specific sequences in double-stranded DNA. *EMBO J.* 16(22):6835–48

30. Guynet C, Hickman AB, Barabas O, Dyda F, Chandler M, Ton-Hoang B. 2008. In vitro reconstitution of a single-stranded transposition mechanism of IS*608*. *Mol. Cell* 29(3):302–12

31. Heller RC, Marians KJ. 2006. Replication fork reactivation downstream of a blocked nascent leading strand. *Nature* 439(7076):557–62

32. Hillyar CRT. 2012. Genetic recombination in bacteriophage lambda. *Biosci. Horiz.* 5:hzs001

33. Hu WY, Derbyshire KM. 1998. Target choice and orientation preference of the insertion sequence IS*903*. *J. Bacteriol.* 180(12):3039–48

34. Huen MSY, Li X, Lu LY, Watt RM, Liu DP, Huang JD. 2006. The involvement of replication in single stranded oligonucleotide-mediated gene repair. *Nucleic Acids Res.* 34(21):6183–94

35. Ichiyanagi K, Beauregard A, Belfort M. 2003. A bacterial group II intron favors retrotransposition into plasmid targets. *Proc. Natl. Acad. Sci. USA* 100(26):15742–47

36. Ichiyanagi K, Beauregard A, Lawrence S, Smith D, Cousineau B, Belfort M. 2002. Retrotransposition of the Ll.LtrB group II intron proceeds predominantly via reverse splicing into DNA targets. *Mol. Microbiol.* 46(5):1259–72

37. Indiani C, Patel M, Goodman MF, O'Donnell ME. 2013. RecA acts as a switch to regulate polymerase occupancy in a moving replication fork. *Proc. Natl. Acad. Sci. USA* 110(14):5410–15

38. Iyer RR, Pluciennik A, Burdett V, Modrich PL. 2006. DNA mismatch repair: functions and mechanisms. *Chem. Rev.* 106(2):302–23

39. Joshi MC, Magnan D, Montminy TP, Lies M, Stepankiw N, Bates D. 2013. Regulation of sister chromosome cohesion by the replication fork tracking protein SeqA. *PLOS Genet.* 9(8):e1003673

40. Kantartzis A, Williams GM, Balakrishnan L, Roberts RL, Surtees JA, Bambara RA. 2012. Msh2-Msh3 interferes with Okazaki fragment processing to promote trinucleotide repeat expansions. *Cell Rep.* 2(2):216–22

41. Kersulyte D, Velapatiño B, Dailide G, Mukhopadhyay AK, Ito Y, et al. 2002. Transposable element ISH*p608* of *Helicobacter pylori*: nonrandom geographic distribution, functional organization, and insertion specificity. *J. Bacteriol.* 184(4):992–1002

42. Kim S, Dallmann HG, McHenry CS, Marians KJ. 1996. Coupling of a replicative polymerase and helicase: a τ-DnaB interaction mediates rapid replication fork movement. *Cell* 84(4):643–50

43. Kolter R, Siegele DA, Tormo A. 1993. The stationary phase of the bacterial life cycle. *Annu. Rev. Microbiol.* 47:855–74

44. Kowalczykowski SC. 2000. Initiation of genetic recombination and recombination-dependent replication. *Trends Biochem. Sci.* 25(4):156–65

45. Lajoie MJ, Gregg CJ, Mosberg JA, Washington GC, Church GM. 2012. Manipulating replisome dynamics to enhance lambda Red-mediated multiplex genome engineering. *Nucleic Acids Res.* 40(22):e170

46. Lambowitz AM, Zimmerly S. 2010. Group II introns: mobile ribozymes that invade DNA. *Cold Spring Harb. Perspect. Biol.* 3(8):a003616

47. Leu FP, Georgescu R, O'Donnell M. 2003. Mechanism of the *E. coli* tau processivity switch during lagging-strand synthesis. *Mol. Cell* 11(2):315–27

48. Li X, Costantino N, Lu L, Liu D, Watt RM, et al. 2003. Identification of factors influencing strand bias in oligonucleotide-mediated recombination in *Escherichia coli*. *Nucleic Acids Res.* 31(22):6674–87

49. Li X, Thomason LC, Sawitzke JA, Costantino N, Court DL. 2013. Bacterial DNA polymerases participate in oligonucleotide recombination. *Mol. Microbiol.* 88(5):906–20

50. Li Z, Craig NL, Peters JE. 2013. Transposon Tn7. In *Bacterial Integrative Mobile Genetic Elements*, ed. AP Roberts, P Mullany, pp. 1–32. Austin, TX: Landes Biosci.

51. Lia G, Michel B, Allemand JF. 2012. Polymerase exchange during Okazaki fragment synthesis observed in living cells. *Science* 335(6066):328–31

52. Lia G, Rigato A, Long E, Chagneau C, Le Masson M, et al. 2013. RecA-promoted, RecFOR-independent progressive disassembly of replisomes stalled by helicase inactivation. *Mol. Cell* 49(3):547–57

53. Lopez de Saro FJ, O'Donnell M. 2001. Interaction of the β sliding clamp with MutS, ligase, and DNA polymerase I. *Proc. Natl. Acad. Sci. USA* 98(15):8376–80

54. Lorenz MG, Wackernagel W. 1994. Bacterial gene transfer by natural genetic transformation in the environment. *Microbiol. Rev.* 58(3):563–602

55. Lundquist RC, Olivera BM. 1982. Transient generation of displaced single-stranded DNA during nick translation. *Cell* 31(1):53–60

56. Lusetti SL, Cox MM. 2002. The bacterial RecA protein and the recombinational DNA repair of stalled replication forks. *Annu. Rev. Biochem.* 71:71–100

57. Lyamichev V, Brow MA, Dahlberg JE. 1993. Structure-specific endonucleolytic cleavage of nucleic acids by eubacterial DNA polymerases. *Science* 260(5109):778–83

58. Maciąg M, Kochanowska M, Łyżeń R, Węgrzyn G, Szalewska-Pałasz A. 2010. ppGpp inhibits the activity of *Escherichia coli* DnaG primase. *Plasmid* 63(1):61–67

59. Maliszewska-Tkaczyk M, Jonczyk P, Bialoskorska M, Schaaper RM, Fijalkowska IJ. 2000. SOS mutator activity: unequal mutagenesis on leading and lagging strands. *Proc. Natl. Acad. Sci. USA* 97(23):12678–83

60. Martínez-Abarca F, Barrientos-Durán A, Fernández-López M, Toro N. 2004. The RmInt1 group II intron has two different retrohoming pathways for mobility using predominantly the nascent lagging strand at DNA replication forks for priming. *Nucleic Acids Res.* 32(9):2880–88

61. Martinsohn JT, Radman M, Petit MA. 2008. The λ Red proteins promote efficient recombination between diverged sequences: implications for bacteriophage genome mosaicism. *PLOS Genet.* 4(5):e1000065

62. Matsuura M, Saldanha R, Ma H, Wank H, Yang J, et al. 1997. A bacterial group II intron encoding reverse transcriptase, maturase, and DNA endonuclease activities: biochemical demonstration of maturase activity and insertion of new genetic information within the intron. *Genes Dev.* 11(21):2910–24

63. Mills DA, McKay LL, Dunny GM. 1996. Splicing of a group II intron involved in the conjugative transfer of pRS01 in lactococci. *J. Bacteriol.* 178(12):3531–38

64. Mirkin EV, Mirkin SM. 2007. Replication fork stalling at natural impediments. *Microbiol. Mol. Biol. Rev.* 71(1):13–35

65. Mohr G, Smith D, Belfort M, Lambowitz AM. 2000. Rules for DNA target-site recognition by a lactococcal group II intron enable retargeting of the intron to specific DNA sequences. *Genes Dev.* 14(5):559–73

66. Mosberg JA, Lajoie MJ, Church GM. 2010. Lambda Red recombineering in *Escherichia coli* occurs through a fully single-stranded intermediate. *Genetics* 186(3):791–99

67. Muñoz-Adelantado E, San Filippo J, Martínez-Abarca F, García-Rodríguez FM, Lambowitz AM, Toro N. 2003. Mobility of the *Sinorhizobium meliloti* group II intron RmInt1 occurs by reverse splicing into DNA, but requires an unknown reverse transcriptase priming mechanism. *J. Mol. Biol.* 327(5):931–43

68. Murphy KC. 1991. Lambda Gam protein inhibits the helicase and *chi*-stimulated recombination activities of *Escherichia coli* RecBCD enzyme. *J. Bacteriol.* 173(18):5808–21

69. Murphy KC. 2012. Phage recombinases and their applications. In *Advances in Virus Research*, ed. M Łobocka, W Szybalski, pp. 367–414. Waltham, MA: Acad. Press

70. Myers RS, Stahl FW. 1994. Chi and the RecBCD enzyme of *Escherichia coli. Annu. Rev. Genet.* 28:49–70

71. Neylon C, Kralicek AV, Hill TM, Dixon NE. 2005. Replication termination in *Escherichia coli*: structure and antihelicase activity of the Tus-*ter* complex. *Microbiol. Mol. Biol. Rev.* 69(3):501–26

72. Nicolas E, Upton AL, Uphoff S, Henry O, Badrinarayanan A, Sherratt D. 2014. The SMC complex MukBEF recruits topoisomerase IV to the origin of replication region in live *Escherichia coli. mBio* 5(1):e01001-13

73. Parks AR, Li Z, Shi Q, Owens RM, Jin MM, Peters JE. 2009. Transposition into replicating DNA occurs through interaction with the processivity factor. *Cell* 138(4):685–95

74. Parks AR, Peters JE. 2009. Tn7 elements: engendering diversity from chromosomes to episomes. *Plasmid* 61(1):1–14

75. Pasternak C, Ton-Hoang B, Coste G, Bailone A, Chandler M, Sommer S. 2010. Irradiation-induced *Deinococcus radiodurans* genome fragmentation triggers transposition of a single resident insertion sequence. *PLOS Genet.* 6(1):e1000799

76. Persky NS, Ferullo DJ, Cooper DL, Moore HR, Lovett ST. 2009. The ObgE/CgtA GTPase influences the stringent response to amino acid starvation in *Escherichia coli. Mol. Microbiol.* 73(2):253–66

77. Peters JE, Craig NL. 2000. Tn7 transposes proximal to DNA double-strand breaks and into regions where chromosomal DNA replication terminates. *Mol. Cell* 6(3):573–82

78. Peters JE, Craig NL. 2001. Tn7 recognizes transposition target structures associated with DNA replication using the DNA-binding protein TnsE. *Genes Dev.* 15(6):737–47

79. Peters JE, Craig NL. 2001. Tn7: smarter than we thought. *Nat. Rev. Mol. Cell Biol.* 2(11):806–14

80. Pluciennik A, Burdett V, Lukianova O, O'Donnell M, Modrich P. 2009. Involvement of the β clamp in methyl-directed mismatch repair in vitro. *J. Biol. Chem.* 284(47):32782–91

81. Pomerantz RT, O'Donnell M. 2008. The replisome uses mRNA as a primer after colliding with RNA polymerase. *Nature* 456(7223):762–66

82. Possoz C, Filipe SR, Grainge I, Sherratt DJ. 2006. Tracking of controlled *Escherichia coli* replication fork stalling and restart at repressor-bound DNA in vivo. *EMBO J.* 25(11):2596–604

83. Postow L, Ullsperger C, Keller RW, Bustamante C, Vologodskii AV, Cozzarelli NR. 2001. Positive torsional strain causes the formation of a four-way junction at replication forks. *J. Biol. Chem.* 276(4):2790–96

84. Poteete AR, Fenton AC. 1993. Efficient double-strand break-stimulated recombination promoted by the general recombination systems of phages λ and p22. *Genetics* 134(4):1013–21

85. Reyes-Lamothe R, Sherratt DJ, Leake MC. 2010. Stoichiometry and architecture of active DNA replication machinery in *Escherichia coli. Science* 328(5977):498–501

86. Roberts D, Hoopes BC, McClure WR, Kleckner N. 1985. IS*10* transposition is regulated by DNA adenine methylation. *Cell* 43(1):117–30

87. Roitzsch M, Pyle AM. 2009. The linear form of a group II intron catalyzes efficient autocatalytic reverse splicing, establishing a potential for mobility. *RNA* 15(3):473–82

88. Ros F, Kunze R. 2001. Regulation of activator/dissociation transposition by replication and DNA methylation. *Genetics* 157(4):1723–33

89. Roszak DB, Colwell RR. 1987. Survival strategies of bacteria in the natural environment. *Microbiol. Rev.* 51(3):365–79

90. Rymer RU, Solorio FA, Tehranchi AK, Chu C, Corn JE, et al. 2012. Binding mechanism of metal·NTP substrates and stringent-response alarmones to bacterial DnaG-type primases. *Structure* 20(9):1478–89

91. Sandler SJ. 2000. Multiple genetic pathways for restarting DNA replication forks in *Escherichia coli* K-12. *Genetics* 155(2):487–97

92. San Filippo J, Lambowitz AM. 2002. Characterization of the C-terminal DNA-binding/DNA endonuclease region of a group II intron-encoded protein. *J. Mol. Biol.* 324(5):933–51

93. Sawitzke JA, Costantino N, Li X, Thomason LC, Bubunenko M, et al. 2011. Probing cellular processes with oligo-mediated recombination and using the knowledge gained to optimize recombineering. *J. Mol. Biol.* 407(1):45–59

94. Segawa T, Tomizawa J. 1971. Formation of concatemers of lambda phage DNA in a recombination-deficient system. *Mol. Gen. Genet.* 111(3):197–201

95. Sharpe PL, Craig NL. 1998. Host proteins can stimulate Tn7 transposition: a novel role for the ribosomal protein L29 and the acyl carrier protein. *EMBO J.* 17(19):5822–31

96. Shi Q, Huguet-Tapia JC, Peters JE. 2009. Tn*917* targets the region where DNA replication terminates in *Bacillus subtilis*, highlighting a difference in chromosome processing in the Firmicutes. *J. Bacteriol.* 191(24):7623–27

97. Shi Q, Parks AR, Potter BD, Safir IJ, Luo Y, et al. 2008. DNA damage differentially activates regional chromosomal loci for Tn7 transposition in *Escherichia coli*. *Genetics* 179(3):1237–50

98. Singh NN, Lambowitz AM. 2001. Interaction of a group II intron ribonucleoprotein endonuclease with its DNA target site investigated by DNA footprinting and modification interference. *J. Mol. Biol.* 309(2):361–86

99. Spradling AC, Bellen HJ, Hoskins RA. 2011. *Drosophila* P elements preferentially transpose to replication origins. *Proc. Natl. Acad. Sci. USA* 108(38):15948–53

100. Swingle B, Bao Z, Markel E, Chambers A, Cartinhour S. 2010. Recombineering using RecTE from *Pseudomonas syringae*. *Appl. Environ. Microbiol.* 76(15):4960–68

101. Swingle B, Markel E, Costantino N, Bubunenko MG, Cartinhour S, Court DL. 2010. Oligonucleotide recombination in Gram-negative bacteria. *Mol. Microbiol.* 75(1):138–48

102. Swingle B, O'Carroll M, Haniford D, Derbyshire KM. 2004. The effect of host-encoded nucleoid proteins on transposition: H-NS influences targeting of both IS*903* and Tn*10*. *Mol. Microbiol.* 52(4):1055–67

103. Taylor MS, LaCava J, Mita P, Molloy KR, Huang CRL, et al. 2013. Affinity proteomics reveals human host factors implicated in discrete stages of LINE-1 retrotransposition. *Cell* 155(5):1034–48

104. Ton-Hoang B, Pasternak C, Siguier P, Guynet C, Hickman AB, et al. 2010. Single-stranded DNA transposition is coupled to host replication. *Cell* 142(3):398–408

105. Twiss E, Coros AM, Tavakoli NP, Derbyshire KM. 2005. Transposition is modulated by a diverse set of host factors in *Escherichia coli* and is stimulated by nutritional stress: host factors and transposition. *Mol. Microbiol.* 57(6):1593–607

106. Van Kessel JC, Hatfull GF. 2007. Recombineering in *Mycobacterium tuberculosis*. *Nat. Methods* 4(2):147–52

107. Wang HH, Isaacs FJ, Carr PA, Sun ZZ, Xu G, et al. 2009. Programming cells by multiplex genome engineering and accelerated evolution. *Nature* 460(7257):894–98

108. Wang HH, Xu G, Vonner AJ, Church G. 2011. Modified bases enable high-efficiency oligonucleotide-mediated allelic replacement via mismatch repair evasion. *Nucleic Acids Res.* 39(16):7336–47

109. Wang JD, Sanders GM, Grossman AD. 2007. Nutritional control of elongation of DNA replication by (p)ppGpp. *Cell* 128(5):865–75

110. Wang X, Liu X, Possoz C, Sherratt DJ. 2006. The two *Escherichia coli* chromosome arms locate to separate cell halves. *Genes Dev.* 20(13):1727–31

111. Wang X, Reyes-Lamothe R, Sherratt DJ. 2008. Modulation of *Escherichia coli* sister chromosome cohesion by topoisomerase IV. *Genes Dev.* 22(17):2426–33

112. Warbrick E. 2000. The puzzle of PCNA's many partners. *BioEssays* 22(11):997–1006

113. Warbrick E, Lane DP, Glover DM, Heatherington W. 1998. PCNA binding proteins in *Drosophila melanogaster*: the analysis of a conserved PCNA binding domain. *Nucleic Acids Res.* 26(17):3925–32

114. Wardle SJ, O'Carroll M, Derbyshire KM, Haniford DB. 2005. The global regulator H-NS acts directly on the transpososome to promote Tn*10* transposition. *Genes Dev.* 19(18):2224–35

115. Waters VL, Guiney DG. 1993. Processes at the nick region link conjugation, t-DNA transfer and rolling circle replication. *Mol. Microbiol.* 9(6):1123–30

116. White MA, Eykelenboom JK, Lopez-Vernaza MA, Wilson E, Leach DRF. 2008. Non-random segregation of sister chromosomes in *Escherichia coli*. *Nature* 455(7217):1248–50

117. Wolkow CA, DeBoy RT, Craig NL. 1996. Conjugating plasmids are preferred targets for Tn*7*. *Genes Dev.* 10(17):2145–57

118. Yang Y, Sharan SK. 2003. A simple two-step, "hit and fix" method to generate subtle mutations in BACs using short denatured PCR fragments. *Nucleic Acids Res.* 31(15):e80

119. Yeeles JTP, Marians KJ. 2013. Dynamics of leading-strand lesion skipping by the replisome. *Mol. Cell* 52(6):855–65

120. Yin JCP, Krebs MP, Reznikoff WS. 1988. Effect of DAM methylation on Tn*5* transposition. *J. Mol. Biol.* 199(1):35–45

121. Yuan Q, McHenry CS. 2013. Cycling of the *E. coli* lagging strand polymerase is triggered exclusively by the availability of a new primer at the replication fork. *Nucleic Acids Res.* 42(3):1747–56

122. Zhong J, Lambowitz AM. 2003. Group II intron mobility using nascent strands at DNA replication forks to prime reverse transcription. *EMBO J.* 22(17):4555–65

123. Zimmerly S, Guo H, Perlman PS, Lambowitz AM. 1995. Group II intron mobility occurs by target DNA-primed reverse transcription. *Cell* 82(4):545–54

Self-Organization of Meiotic Recombination Initiation: General Principles and Molecular Pathways

Scott Keeney,[1,2] Julian Lange,[2]
and Neeman Mohibullah[1,2]

[1]Howard Hughes Medical Institute, Memorial Sloan Kettering Cancer Center, New York, NY 10065; email: s-keeney@ski.mskcc.org

[2]Molecular Biology Program, Memorial Sloan Kettering Cancer Center, New York, NY 10065

Annu. Rev. Genet. 2014. 48:187–214

The *Annual Review of Genetics* is online at genet.annualreviews.org

This article's doi:
10.1146/annurev-genet-120213-092304

Keywords

Spo11, DNA double-strand breaks, cell cycle, ATM, DNA replication

Abstract

Recombination in meiosis is a fascinating case study for the coordination of chromosomal duplication, repair, and segregation with each other and with progression through a cell-division cycle. Meiotic recombination initiates with formation of developmentally programmed DNA double-strand breaks (DSBs) at many places across the genome. DSBs are important for successful meiosis but are also dangerous lesions that can mutate or kill, so cells ensure that DSBs are made only at the right times, places, and amounts. This review examines the complex web of pathways that accomplish this control. We explore how chromosome breakage is integrated with meiotic progression and how feedback mechanisms spatially pattern DSB formation and make it homeostatic, robust, and error correcting. Common regulatory themes recur in different organisms or in different contexts in the same organism. We review this evolutionary and mechanistic conservation but also highlight where control modules have diverged. The framework that emerges helps explain how meiotic chromosomes behave as a self-organizing system.

INTRODUCTION

Meiosis is the specialized cell division that generates gametes in sexually reproducing organisms. It appends two rounds of chromosome segregation to one round of DNA replication, thereby achieving the necessary genome reduction prior to gamete fusion, which restores proper ploidy (121) (**Figure 1a**). The second meiotic division is like mitosis in that it separates centromeres of sister chromatids, but the first meiotic division is different: It separates homologous maternal and paternal chromosomes. Meiosis I poses unique challenges because homologous chromosomes need not share any special spatial relationship before meiosis, unlike sister chromatids, which are born alongside one another when DNA is replicated. To segregate accurately, homologous chromosomes must find one another, pair up, and form temporary physical connections that stabilize them on the metaphase I spindle.

In most species, the physical connections are formed by reciprocal exchange of chromosome arms via homologous recombination in conjunction with sister chromatid cohesion (121). Recombination also fosters genetic diversification by breaking up linkage groups. In many taxa, including fungi, plants, and mammals, it promotes chromosome pairing by providing a mechanism for

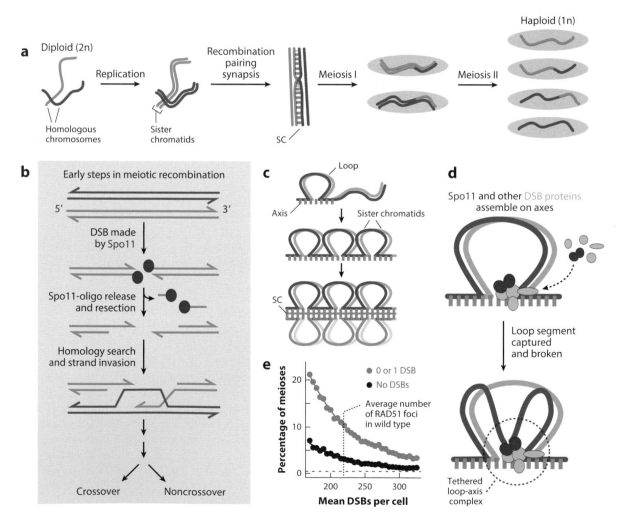

identifying DNA sequence homology (14). Recombination failure often leads to meiotic arrest or chromosome segregation failure, with dire consequences for fertility (113, 121).

Recombination initiates with DNA double-strand breaks (DSBs), which cells inflict on their own genomes (39, 75) (**Figure 1b**). Major steps in the recombination pathway are best defined in the budding yeast *Saccharomyces cerevisiae*, but conservation of key players indicates that many of the events proceed in similar fashion in different species (39). DSBs are formed by Spo11, a conserved topoisomerase relative that cleaves DNA via a covalent protein-DNA intermediate (12, 76). This intermediate is then endonucleolytically cleaved to release Spo11 attached to a short oligonucleotide (oligo) (76, 101, 114). Because Spo11-oligo complexes are a quantitative by-product of DSB formation—each DSB results in the generation of two SPO11-oligo complexes—they have proven useful in quantifying total DSB levels, even in organisms like mice, where direct molecular detection of DSBs is thus far not possible (e.g., 81).

Endonucleolytic release of Spo11-oligo complexes frees DSB ends so that the 5′ strand termini can be exonucleolytically resected to yield 3′ single-stranded tails. These tails invade intact homologous duplexes in reactions dependent on strand-exchange proteins related to bacterial RecA (Rad51 and, in some species, its meiosis-specific paralog Dmc1), ultimately giving rise to recombinant products (63). The repair of any given DSB can result in either the reciprocal exchange of chromosome arms flanking the break (a crossover) or no exchange of flanking arms (a noncrossover) (**Figure 1b**). The crossovers help link homologous chromosomes on the metaphase I spindle, but all interhomolog recombination events (including those leading to noncrossovers) promote pairing in those organisms that rely on recombination for this process.

←

Figure 1

Chromosome behaviors in meiosis. (*a*) Stages in the progression through meiosis. Homologous chromosomes undergo replication to form pairs of sister chromatids, which are held together by cohesin (not shown). Replicated chromosomes initiate recombination, then pair and become closely juxtaposed along their lengths. The aligned configuration is transiently stabilized by the synaptonemal complex (SC). A subset of recombination events are resolved by reciprocal exchange of chromosome arms (crossing over), which, in conjunction with sister chromatid cohesion, provide physical connections between homologous chromosomes that allow them to be aligned on the spindle at metaphase of the first meiotic division. Meiosis I separates homologous chromosomes, then Meiosis II separates sister centromeres. The two divisions yield progeny with half the genetic content of the parent and with new combinations of parental alleles in some of the meiotic products because of crossing over. (*b*) Overview of early steps in meiotic recombination. Recombination is initiated by a double-strand break (DSB) catalyzed by the topoisomerase-like protein Spo11. The DNA cleavage reaction leaves a pair of Spo11 molecules attached to the 5′ DNA ends on either side of the DSB. Endonucleolytic cleavage releases Spo11 attached to a short oligonucleotide, and further exonucleolytic resection generates 3′ single-stranded tails that are bound by strand-exchange proteins (Dmc1 and/or Rad51; not shown). Once a homologous DNA duplex is located, the single-stranded DNA tail invades the intact homologous duplex. Further DNA transactions (not pictured) give rise to mature recombination products in which homologs have (crossover) or have not (noncrossover) exchanged arms. (*c*) Loop-axis structure of meiotic chromosomes. Early in prophase of Meiosis I, each pair of sister chromatids develops a proteinaceous axis with chromatin extending out in loops. As prophase proceeds, the axes elongate and axes of homologous chromosomes are brought together and joined via the zipper-like SC. (*d*) Model for integration of DSB formation with loop-axis organization of chromosomes. Recombination occurs in spatial proximity to axes, but DSBs usually form in DNA thought to be in the chromatin loops. To reconcile this apparent paradox, it has been proposed that most of the DSB-forming machinery—including Spo11 itself—assembles on chromosome axes and then captures and breaks a DNA segment from a nearby loop, forming a tethered loop-axis complex (15, 79, 124). (*e*) Monte Carlo simulations were used to evaluate whether randomly distributing DSBs among chromosomes could support the efficient pairing and synapsis seen in normal mouse spermatocytes [<0.5% of cells with an unsynapsed pair of autosomes (e.g., 71)]. We varied the mean number of DSBs per cell and mimicked natural cell-to-cell fluctuation estimated from variability in numbers of RAD51 foci (219.2 ± 69.8, mean $\pm$ SD; from 36). For each value of the mean, we simulated a population of 10,000 cells with that mean and with a coefficient of variation of 30%. For each simulated cell, DSBs were then randomly distributed among the 19 pairs of autosomes, with each chromosome weighted in proportion to its axis length (from 45). The figure plots the fraction of simulated cells in which at least one chromosome pair had no DSBs (*black points*) or had one or no DSBs (*gray*), as such chromosomes would probably fail to synapse (71). The results indicate that a random distribution could not provide the very low failure rate seen in normal cells (*horizontal dashed line*), even if DSB numbers were much higher than the wild-type average.

Recombination is closely integrated with the development of meiosis-specific higher-order chromosome structures (79). Early in prophase I, sister chromatids develop a proteinaceous axis (the axial element), with chromatin extending out in loops (**Figure 1c**). As chromosomes pair, their axes align and are held together to form the zipper-like synaptonemal complex (SC). The tripartite SC comprises the juxtaposed chromosome axes plus central region components, including transverse filaments (coiled-coil proteins spanning the gap between axes). The SC has as-yet poorly understood roles in promoting completion of recombination and may also foster the exchange of chromosome axes at crossover sites, modulate sister chromatid cohesion, and/or sense and help resolve instances where nonhomologous chromosomes have become topologically intertwined (known as interlocks) (79). Direct cytological visualization shows that recombination protein complexes reside on chromosome axes, but molecular studies in yeast place the most frequently cleaved DNA sequences on chromatin loops; this paradox has led to the proposal that DSBs are formed within loop segments that become transiently tethered to chromosome axes (15, 79) (**Figure 1d**). Because Spo11 and its accessory factors are also enriched on chromosome axes, it is thought that DSB machinery assembled on axes captures and breaks loop segments (1, 79, 102, 124, 150). In fungi, plants, and mammals, DSBs form at approximately the same time as axes are forming, and recombination is completed within the context of the SC (39, 63, 79).

The positive roles DSBs play in promoting normal meiotic chromosome behavior come with risk because errors in DSB repair can lead to mutation, cell death, aneuploid gametes, and/or infertility (59, 113, 138). The potentially lethal nature of these lesions puts a premium on the cell's ability to control the timing, number, and location of DSBs to foster their essential functions and minimize deleterious effects. Here, we review recent discoveries that illuminate the molecular underpinnings of this control.

RECOMBINATION INITIATION IS A ROBUST, SELF-ORGANIZING PROCESS

DSBs are more likely to occur in some genomic regions than in others (10, 72, 85). Depictions of this nonrandom distribution often focus on hot spots, the small regions (typically ~150–250-base-pairs wide in budding yeast) where DSBs occur most often. However, hot spots are only one aspect of the DSB distribution because essentially every base pair in the genome is a potential substrate for Spo11, with cleavage probability varying over orders of magnitude (123). The shape of this probability distribution, i.e., the DSB landscape, is molded by the combinatorial action of many factors (chromosomal proteins and the DNA they bind) that interact hierarchically over different size scales (72, 79, 85, 123, 161). The number of potential break sites is thus enormous, so each cell ends up with a different array of DSB positions that are chosen on the fly as meiosis proceeds. Likewise, the exact final number of DSBs is also not genetically predetermined and varies substantially from cell to cell (e.g., with a coefficient of variation of >30% in mouse) (33, 36).

Despite these stochastic aspects, DSB formation is robust, homeostatic, and error correcting (27, 48, 71, 160). The geneticist's stock-in-trade is analysis of how things go wrong in mutants, but of course the implicit starting point for most studies is the fact that things generally go right in wild type. For example, the great majority of cells achieve a sufficient number of DSBs that are distributed appropriately so that each chromosome pair has enough for at least one crossover to form (33, 67, 100, 132). Moreover, in organisms with recombination-promoted pairing, sufficient DSBs nearly always form to support that process as well (55, 71, 159). Simple modeling suggests that randomly distributing DSBs among and along chromosomes would not achieve such a high success rate (**Figure 1*e***), implying the existence of mechanisms that control DSB number and distribution to ensure proper chromosome behavior.

Although there is substantial cell-to-cell variation of DSB numbers within species, greater differences are seen when comparing between organisms. Relatively few DSBs occur in species that can pair chromosomes without recombination, such as *Drosophila melanogaster* and *Caenorhabditis elegans* (~20–30 per cell on average) (65, 99, 115, 135). In contrast, more DSBs tend to be generated in organisms that rely on recombination for efficient pairing, such as *S. cerevisiae* (average of ~150–200 per cell), plants (e.g., ~200–300 in *Arabidopsis thaliana* and >1,500 per cell in lily), or mammals (e.g., ~200–300 per cell in mouse) (25, 31, 36, 123, 127, 158, 166). These apparent species-specific set points, without genetic (or epigenetic) predetermination of precise numbers, imply that DSB regulation is self-organizing and homeostatic, which also provides potential for robustness and error correction (27, 71, 160).

DSB formation is a suicide reaction for Spo11 because the endonucleolytic release pathway leaves the protein's active-site tyrosine residue covalently linked to DNA (**Figure 1*b***). In principle, this feature could be a means of controlling total DSB numbers, but in fact most Spo11 protein molecules never make a break (81, 114) and Spo11 and other proteins essential for DSB formation remain abundant on chromatin after most DSBs have formed (75). These features suggest that mechanisms that control DSB numbers work in part by restraining Spo11 activity.

Understanding of these DSB-regulating mechanisms has grown substantially in recent years. Studies in several species have uncovered elements that integrate DSB formation with progression through meiosis and that coordinate DSB formation with other chromosomal events (e.g., DNA replication). Recent studies have also revealed an intersecting network of negative feedback circuits that work locally along chromosomes to fine-tune the control of DSBs. In its broad framework, DSB regulation appears to be evolutionarily conserved, but many details differ strikingly between organisms. **Figure 2** summarizes the known regulatory circuits, which affect DSB formation to different degrees depending on where the cell is within S phase or prophase. Each circuit is discussed in detail below.

CELL CYCLE KINASES TIE DOUBLE-STRAND BREAKS TO MEIOTIC PROGRESSION

It is clear that DSB formation is usually restricted to a specific window of time during the first meiotic prophase, based on direct detection of DSBs in yeasts and on immunostaining to detect cytological DSB markers such as Rad51 foci in other organisms (e.g., 30, 35, 65, 103, 120, 166). This constrained window is important for recombination to serve its functions in connecting homologous chromosomes. For example, recombination must be integrated with sister chromatid cohesion to form the chiasmata that hold chromosome pairs together at metaphase I (79). Proper timing is probably also important to minimize potential for genomic havoc: DSBs formed before DNA replication or after commitment to chromosome segregation may put cells at risk of mutation, aneuploidy, or meiotic arrest (59, 106).

One layer of temporal control involves developmentally regulated expression of Spo11 and other proteins required to make DSBs. Different species have evolved a range of strategies to restrict expression of these and other meiotic proteins to appropriate times, including control of transcription, splicing, mRNA stability, and translation (e.g., 23, 51, 92, 141). This type of control (gene regulation tied to differentiation itself), although important, is not considered further except for a few specific scenarios in budding and fission yeasts. Instead, we focus here and in subsequent sections on other layers of temporal control that involve more direct regulation of the activity or chromosomal association of Spo11 and its accessory factors.

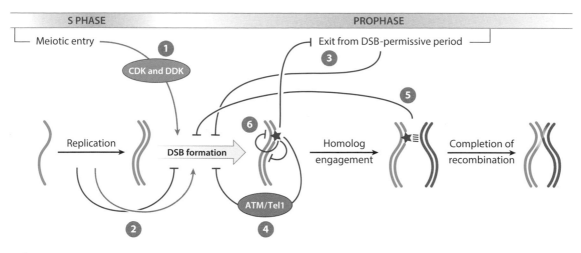

Figure 2

Overview of the network of intersecting regulatory circuits controlling timing, number, and distribution of meiotic double-strand breaks (DSBs). Circuit 1: Cell cycle regulatory kinases tie DSB formation to meiotic progression. Circuit 2: DNA replication influences the spatial and temporal patterning of DSBs (*green arrow*), and replication stress inhibits DSB formation (*red inhibitory arrow*). Circuit 3: Progression through prophase closes a window of opportunity for DSB formation. Problems in recombination and/or certain other chromosome behaviors invoke signaling pathways that extend the DSB-permissive period. Circuit 4: DSBs activate the damage-responsive kinase ATM/Tel1, which then restrains SPO11 activity via a negative feedback loop. Circuit 5: Engagement of homologous chromosomes leads to changes in chromosome structure that inhibit further DSB formation. Circuit 6: Local DSB patterning is shaped by communication between potential DSB sites, both in *cis* along the same DNA molecule and in *trans* between sister chromatids or homologous chromosomes. Abbreviations: CDK, cyclin-dependent kinase; DDK, Dbf4-dependent kinase.

The cell cycle regulatory kinases CDK (cyclin-dependent kinase) and DDK (Dbf4-dependent kinase) are key drivers of progression in meiosis, as in mitosis (93, 94). In *S. cerevisiae*, DSB formation is directly promoted by both kinases. Cdc28 (the principal cell cycle CDK), in association with the S-phase cyclins Clb5 or Clb6, phosphorylates the Spo11-accessory protein Mer2; this phosphorylation is essential for DSB formation (54, 149). Mer2 must also be phosphorylated by DDK, which comprises the kinase Cdc7 and its regulatory subunit Dbf4. DDK directly phosphorylates multiple sites on Mer2, some of which depend on prior phosphorylation of a neighboring residue by CDK (94, 139, 167, 168). Phosphorylation by both kinases apparently promotes the ability of Mer2 to interact with other proteins needed for DSB formation and thereby to recruit those proteins to chromatin (54, 124, 139).

Hsk1, the *Schizosaccharomyces pombe* ortholog of Cdc7, is essential for DSB formation and recruitment of the Spo11 ortholog Rec12 to chromatin (116, 139). Rec7 [a homolog of the Spo11-accessory protein Rec114 required for DSB formation in budding yeast (86, 102)] is phosphorylated by Hsk1, and phosphorylation-blocking *rec7* mutations reduce recombination, making Rec7 a likely target of Hsk1 relevant to DSB control (H. Masai, personal communication). Whether other targets exist is not yet known.

In mice, normal CDC7 levels are required for meiosis (78), but whether the kinase functions in DSB formation or some other process is unknown and meiotic analyses of DDK homologs have not been reported in other taxa. Furthermore, although CDKs or cyclins have been clearly implicated in recombination and/or other aspects of meiotic chromosome dynamics in many organisms (e.g., 6, 165), it has not yet been established whether DSB formation itself is controlled by CDK in species other than *S. cerevisiae*.

COORDINATING DOUBLE-STRAND BREAKS WITH DNA REPLICATION

To fulfill their functions in promoting pairing, generating connections between homologous chromosomes, and transmitting a haploid DNA content to gametes, DSBs need to form at a time when sister chromatids exist, i.e., after DNA replication has occurred locally. Cell-wide oscillation of CDK or DDK activity provides one means to control the timing of these meiotic events, but such global regulation does not by itself allow replication and DSB formation to be fully coordinated with one another. Studies in budding and fission yeast have uncovered paradigmatic mechanisms that provide this coordination and allow for error correction.

Temporospatial Coordination in Normal Meiosis

In *S. cerevisiae*, DSBs usually form approximately 90 minutes after replication (18). Pioneering studies by Lichten's group uncovered the remarkable finding that delaying replication of a chromosomal segment by deleting replication origins also delays DSB formation in that segment by the same margin (18). In strains heterozygous for the origin deletions, DSBs are delayed only on the mutated chromosome, so this temporal control of DSBs works in *cis* (105). Because DSB timing is dictated by local replication timing, replication and recombination initiation must be mechanistically coupled to one another (18, 105, 106).

A possibly related phenomenon occurs in *S. pombe* (171). If sporulation proceeds in the presence of a nitrogen source rather than in commonly used nitrogen-starvation conditions, more replication origins are utilized and relative replication times change for large swathes of the genome. How nitrogen levels effect this alteration is not known, but one consequence is clear: Regions that shift to earlier replication also display an increase in DSB formation.

One way coupling could occur is if replication is a strict prerequisite for DSBs (18, 149). However, Spo11 efficiently breaks chromosomes that remain unduplicated because the replication initiation factor Cdc6 has been depleted in *S. cerevisiae* (16, 60) or because of hydroxyurea treatment or replication factor depletion in checkpoint-defective *S. pombe* mutants (110, 116, 162). Thus, replication is dispensable for DSBs per se.

Instead, it has been proposed that temporospatial coupling of replication and DSB formation in *S. cerevisiae* operates at least in part by recruitment of DDK to the replication machinery, thereby preferentially targeting Mer2 in replicating regions for phosphorylation (106, 107) (**Figure 3a**). Mer2 binds chromatin independently of phosphorylation (54, 124) and DDK activity is limiting early in meiosis (94, 167). These features create a window of opportunity during which selective targeting of DDK to replicating chromatin could confer a head start toward DSB formation. Supporting this model, replication-DSB coordination is eliminated by overexpressing DDK (107). Coordination is also eliminated by removing the replication fork protection complex (FPC) (107), a group of proteins that travels with replisomes and helps stabilize the replication machinery during replicative stress (97). The FPC physically associates with DDK in budding and fission yeast and becomes dispensable for replication-DSB coordination if DDK is artificially tethered to replisomes (95, 107, 144). These and other findings indicate that DDK recruited by FPC to replisomes phosphorylates Mer2 in the wake of the replication fork, thus synchronizing replication with an early prerequisite for DSB formation. It remains to be seen whether *S. pombe* uses a similar mechanism.

Responding to Replication Stress

Cells also coordinate replication and DSB formation by downregulating DSB machinery in the face of replication problems. In *S. pombe* meiosis, inhibiting replication with hydroxyurea invokes

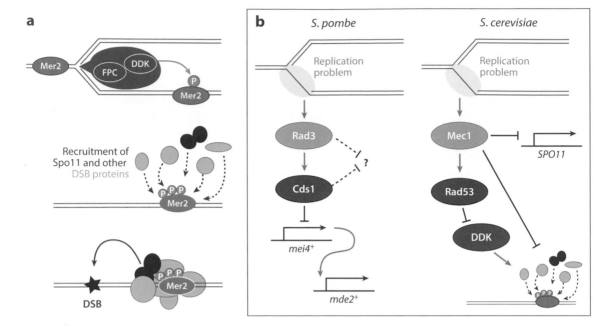

Figure 3

Coordinating replication and double-strand break (DSB) formation. (*a*) Model for temporospatial coupling of DSBs with replication in *Saccharomyces cerevisiae*. The graphic at top depicts a replication fork with a replisome (*blue*) consisting of the replication machinery and accessory proteins, such as the fork protection complex (FPC). DDK (Dbf4-dependent kinase) is recruited to the replisome via interaction with the FPC, resulting in preferential phosphorylation of the Mer2 protein bound to replicating chromatin. Mer2 phosphorylation promotes recruitment of Rec114 and other DSB proteins, including Spo11, ultimately resulting in DSB formation. (*b*) Downregulation of DSB formation in the face of replication stress. In *Schizosaccharomyces pombe*, replication defects activate the kinase Rad3 (ortholog of mammalian ATR) and its effector kinase Cds1. By an unknown mechanism, these kinases inhibit the transcription of *mei4+*, which encodes a transcription factor that is needed for expression of *mde2+*, which encodes a protein essential for DSB formation. Whether Rad3 and Cds1 also suppress DSB formation via additional pathways is unknown. In *S. cerevisiae*, replication defects activate the Rad3 ortholog Mec1, which impinges on *SPO11* transcription by an unknown mechanism. Mec1, via its effector kinase Rad53, also inhibits DDK, which likely inhibits recruitment of DSB proteins dependent on Mer2 phosphorylation (as depicted in panel *a*). Mec1 also appears to inhibit chromatin recruitment of the DSB-promoting proteins Rec114 and Mre11 by a separate, undefined pathway. Abbreviation: p, phosphorylation.

cellular responses via activation of the DNA damage response kinase Rad3 (the ortholog of mammalian ATR) and its downstream effector kinase Cds1 (108, 109). These responses include inhibition of DSB formation, attributed to Rad3- and Cds1-dependent inhibition of transcription of the *mei4+* and *mde2+* genes (102, 117) (**Figure 3*b***, left). Mei4 is a Forkhead-like transcription factor required for meiosis-specific expression of a number of meiotic genes, including *mde2+* (49, 62). Mde2 has essential functions in DSB formation: It bridges interactions between other DSB-promoting proteins and integrates DSB formation with higher-order chromosome structure (49, 102). It is not yet known how Rad3 and Cds1 activation impinge on *mei4+* transcription or whether inhibition of Mei4 expression is the sole means by which the replication checkpoint inhibits DSB formation. Interestingly, artificial expression of Mde2 is not sufficient to rescue DSB formation in the presence of hydroxyurea, suggesting there are other critical Mei4-dependent targets or that Rad3 and Cds1 have additional means of inhibiting DSB formation (K. Ohta, personal communication).

Hydroxyurea treatment also blocks DSB formation in *S. cerevisiae* via Mec1 (ortholog of *S. pombe* Rad3 and mammalian ATR) (17) (**Figure 3*b***, right). Analogous to fission yeast, replication checkpoint activation inhibits expression of a DSB-promoting protein, but in budding yeast it is *SPO11* transcription that is targeted, and only partially. As in *S. pombe*, the mechanism of transcription inhibition is unknown. Importantly, however, Mec1 and its effector kinase Rad53 (ortholog of *S. pombe* Cds1) also inhibit DDK by phosphorylating Dbf4, thereby preventing Mer2 phosphorylation and reducing or altering chromatin association of several DSB-promoting factors. Replication stress thus downregulates Spo11 activity through multiple intersecting mechanisms; this DSB inhibition may promote genome stability by preventing formation of DSBs ahead of replication forks (17). In principle, this mode of DSB regulation could contribute to spatially patterned coordination with replication. For example, ongoing replication could set up a nucleus-wide block to DSB formation that is then removed locally in conjunction with replication fork passage (59). However, it has been argued that this model does not account for properties of the FPC in the replication-DSB coordination discussed above (107). Further studies are needed to determine the extent to which Mec1- and Rad53-dependent processes contribute to DSB control when S phase is unperturbed.

CLOSING THE WINDOW OF OPPORTUNITY FOR DOUBLE-STRAND BREAK FORMATION

Restricting Spo11 activity to a specific window of prophase I requires that cells control the end of the window, not just the beginning. Insight into the regulatory systems involved has come from studies in *S. cerevisiae*, *D. melanogaster*, and *C. elegans*.

Regulated Exit from Meiotic Prophase in *Saccharomyces cerevisiae*

In budding yeast, exit from the pachytene stage of prophase is controlled by the Ndt80 transcription factor, which activates expression of more than 200 genes, including those encoding the polo-like kinase Cdc5 and M-phase cyclins Clb1 and Clb3 (34, 151, 173). Recombination products and DSBs accumulate in mutants that lack Ndt80 or that have defects in pachytene exit because of attenuated CDK activity (2, 146, 173). These observations led to the proposal that pachytene exit ends a period permissive for DSB formation, i.e., that Ndt80 acts as an indirect negative regulator of Spo11 activity (2, 54, 74) (**Figure 4a**). Recent studies confirmed this hypothesis by providing multiple lines of evidence that *ndt80* mutants make more DSBs (5, 27, 48, 132, 160).

When recombination and/or chromosome synapsis are defective, checkpoint responses mediated by Mec1 (ATR) cause Ndt80 to be hypophosphorylated and less abundant. These alterations block or attenuate Ndt80 activity, leading to meiotic arrest or delay (119, 122, 163) (**Figure 4a**). But DSB-Ndt80 cross talk is not restricted to repair-defective cells, as it clearly also occurs in normal meiosis, where it extends the length of prophase I (119). Operation of the Ndt80 circuit tends to obscure the effects of mutations that impair the DSB-forming machinery, such as partial loss-of-function *spo11* mutations, and it is thought to make DSB formation homeostatic in wild-type cells (5, 27, 48, 132). An interesting implication is that Mec1 has opposing effects depending on context: It inhibits DSB formation during S phase via control of Spo11 expression and DDK activity but promotes breakage during prophase via control of Ndt80 activity.

S. cerevisiae aborts meiosis and resumes mitotic divisions if transferred to a vegetative growth medium before pachytene exit (147). Under these conditions, DSBs disappear rapidly even if they would have continued to accumulate while still in sporulation medium (4, 142, 177; M. Lichten, personal communication). This feature implies that DSBs stop forming (74), although this has not been directly demonstrated. If so, it provides another example in which DSB potential is tied to cell cycle status (**Figure 4a**).

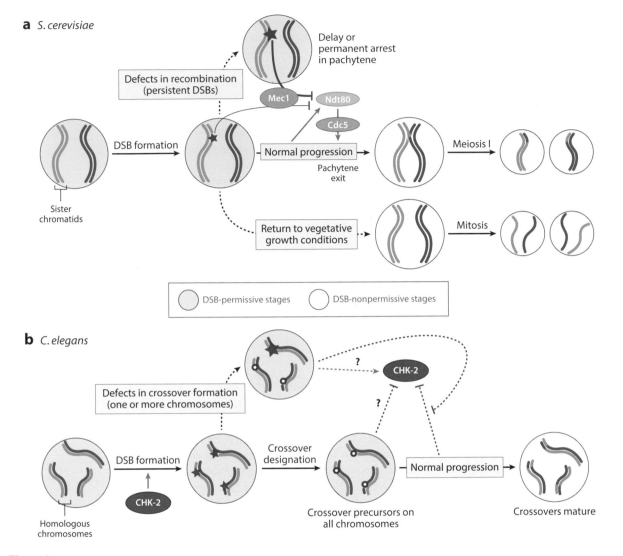

a *S. cerevisiae*

Delay or permanent arrest in pachytene

Defects in recombination (persistent DSBs)

Mec1 — Ndt80

Cdc5

Sister chromatids

DSB formation

Normal progression

Pachytene exit

Meiosis I

Return to vegetative growth conditions

Mitosis

DSB-permissive stages   DSB-nonpermissive stages

b *C. elegans*

Defects in crossover formation (one or more chromosomes)

?   CHK-2

?

Homologous chromosomes

DSB formation

CHK-2

Crossover designation

Normal progression

Crossover precursors on all chromosomes

Crossovers mature

Figure 4

Analogous but mechanistically distinct nucleus-wide systems in *Saccharomyces cerevisiae* and *Caenorhabditis elegans* link the potential to form double-strand breaks (DSBs) with meiotic progression. Nuclei in DSB-permissive stages are green and those in nonpermissive stages are white. (*a*) In *S. cerevisiae*, normal progression through meiosis witnesses activation of Ndt80, a transcription factor that governs exit from the pachytene stage of prophase. Ndt80 activation leads to synaptonemal complex (SC) disassembly and removal of DSB-promoting proteins from chromosomes, ending a DSB-permissive period as cells commit to meiotic divisions. The gene encoding the polo-like kinase Cdc5 is an Ndt80 target that may be important for downregulating DSB potential. If recombination defects result in persistent DSBs, a response mediated by the kinase Mec1 (ATR) is provoked; this attenuates Ndt80 activity and delays or blocks pachytene exit (*thick red inhibitory arrow*). Cross talk between ongoing recombination and Ndt80 activation also occurs via Mec1 to control the length of prophase in normal cells (*thin red inhibitory arrow*). If cells are exposed to appropriate nutrients before Ndt80 becomes active, they exit meiosis and resume mitotic divisions. As in normal meiotic progression, this mode of exiting prophase also appears to include nucleus-wide loss of DSB potential. (*b*) In *C. elegans*, a DSB-permissive state is marked by active CHK-2 kinase and by chromosomal association of the DSB-1 and DSB-2 proteins (not shown). A coordinated transition during the pachytene stage results in shutdown of CHK-2 activity and removal of DSB-1 and DSB-2, ending the DSB-permissive period. However, this period can be extended if one or more chromosome pairs fail to acquire an appropriate number or distribution of crossover precursors (*open red circles*). It is possible that proper acquisition of crossover precursors sends a signal that downregulates CHK-2 activity (*dashed red inhibitory arrows*) or that recombination failure sends a signal that keeps CHK-2 active (*dashed green arrow*) (adapted from 136, 152).

The Ndt80 and return-to-growth systems for shutting down DSB formation make sense because new DSBs would not be useful in either case. It is too late to induce recombination once cells move on to segregating their chromosomes in meiosis I, and recombination is not needed at all if cells are returning to vegetative growth.

These two systems also share a plausible mechanism for downregulating Spo11. When Ndt80 is activated in meiosis or when cells return to vegetative growth, meiotic chromosome structures, such as SCs, are rapidly disassembled (38, 151, 177). When caused by Ndt80 activation, this disassembly includes turnover of Spo11 and Rec114 and removal of other proteins, such as Red1, that are important for DSB formation (27, 151, 160). (Return to growth has not been as systematically evaluated.) Artificial induction of Cdc5 is sufficient to trigger this disassembly even in the absence of Ndt80 (151). Thus, DSB formation is shut down upon exit from prophase by virtue of regulated disruption of the chromosome structures and proteins needed to support Spo11 activity.

CHK-2 and Control of a Double-Strand-Break-Permissive Stage in *Caenorhabditis elegans*

Altered crossover distributions and/or persistent RAD-51 foci are seen in numerous *C. elegans* mutants that reduce or eliminate proteins needed for SC formation, or in strains heterozygous for chromosome translocations that impair synapsis (3, 28, 35, 52, 53, 56, 112). To account for these findings, Villeneuve and coworkers proposed that meiotic cells in these mutants were experiencing greater numbers of DSBs and responding to incomplete synapsis and/or lack of crossover-designated recombination intermediates by extending the time during which DSB formation can occur (53, 56, 112) (**Figure 4b**).

Recent studies provide key insights into this DSB control pathway. DSB-1 and DSB-2, proteins required for DSB formation, both localize to chromosomes during early prophase at the time DSBs normally form. However, in mutants that cannot generate crossovers, DSB-1 and DSB-2 remain on chromosomes much longer than normal, consistent with a prolonged DSB-permissive period (136, 152). Continued DSB-1 and DSB-2 localization on chromosomes occurs in mutants with very different primary defects, including those that eliminate DSB formation entirely (*spo-11*), those that disrupt recombinational repair of DSBs (e.g., *msh-5*, *rad-54*), and those that affect SC formation (e.g., *syp-1*). However, mutations eliminating HORMA-domain proteins HTP-1 and HTP-3 (homologs of yeast Hop1 and mouse HORMAD-1 and -2) cause less (DSB-1) or none (DSB-2) of this response despite causing recombination defects, suggesting that these proteins are themselves needed for the response (136, 152).

Interestingly, a crossover defect on just one chromosome is sufficient to trigger retention of DSB-1 on all chromosomes, and animals in which crossover defects happen in only a subset of nuclei show DSB-1 retention only in that defective subset. Thus, the response to crossover defects is nucleus autonomous and genome wide (28, 152). When examined, no changes in crossing over were seen on normal chromosomes in translocation heterozygotes (e.g., 98, 178), but it is important to note that excess DSBs generally do not yield additional crossovers in *C. elegans* because of extremely robust homeostatic control of crossover formation (176). Assuming extra DSBs do form on other chromosomes in translocation heterozygotes, they are presumably repaired as noncrossovers or by genetically silent recombination between sister chromatids.

The mutations that prolong DSB-1 and DSB-2 presence on chromosomes also prolong the period when the nuclear envelope protein SUN-1 is phosphorylated and rapid chromosome movements occur, suggesting that several distinct aspects of early meiotic prophase are coordinately regulated (28, 136, 169). The CHK-2 kinase (homologous to *S. pombe* Cds1 and *S. cerevisiae* Rad53) is an attractive candidate to mediate this coordinated response because CHK-2 is required for DSB formation, normal chromatin association of DSB-1 and DSB-2 proteins, and SUN-1 phosphorylation by both CHK-2 and the polo-like kinase PLK-2 (89, 125, 136, 152, 169).

These findings suggest that nuclei monitor recombination progression and, if needed, maintain CHK-2 activity to continue to initiate recombination (136, 152) (**Figure 4b**). One possibility is that acquisition of sufficient crossover-competent recombination intermediates feeds back to shut down CHK-2 activity (136). An alternative possibility is that chromosomes lacking a crossover-competent intermediate generate a signal that upregulates CHK-2 or prevents CHK-2 shutoff (152). The latter scenario would account for the ability of a single misbehaving chromosome pair to prolong DSB-1 presence. How CHK-2 activity is controlled and how it in turn affects DSB-1 and DSB-2 localization remain to be determined.

The molecular details are very different between this system and the Ndt80 circuit in *S. cerevisiae*, not least in terms of the dissimilar roles played by kinases traditionally viewed as DNA-damage responsive (CHK-2 in *C. elegans*, Mec1 in *S. cerevisiae*). Nonetheless, there are intriguing parallels: Both systems operate nucleus-wide; both have checkpoint-like properties; both require HORMA-domain proteins; both work by controlling chromosomal association of DSB-promoting factors; and both involve a coordinated transition from a state in which interhomolog interactions are favored to a state in which completion of recombination and preparation for chromosome segregation are favored. Interestingly, the duration of meiotic prophase is also extended in *A. thaliana* mutants that have recombination and/or synapsis defects, suggesting that analogous regulation occurs in plants (57, 58, 64). These similarities highlight that the imperative to integrate recombination initiation with meiotic progression is evolutionarily conserved despite extensive variation in the specific regulatory modules involved.

The Interchromosomal Effect and Precondition Mutants in *Drosophila melanogaster*

Very early studies of recombination in flies showed that suppression of crossing over on one chromosome (e.g., because of translocation heterozygosity) caused increased crossover frequencies on other chromosomes (129, 154, 155). This phenomenon, dubbed the interchromosomal effect, implies that oocytes experiencing difficulty in some aspect of crossover formation mount a nucleus-wide response. There are obvious parallels with the response to single-chromosome defects in *C. elegans* (28) but with different quantitative effects on crossing over possibly due to the less robust crossover control in flies. Global alterations in crossover distribution are also seen in flies with any of a number of mutations that interfere specifically with crossover formation. These mutations were called precondition mutants on the hypothesis that the affected genes acted before recombination to establish the normal pattern of crossing over (29). However, it was later recognized that similar changes in crossover distributions can be caused by many different defects in the core recombination machinery, including hypomorphic mutations in the *SPO11* homolog *Mei-W68* or strong alleles of the *RAD54* homolog *okra* (e.g., 13). Rather than envision that all of these recombination factors also have additional (precondition-like) roles before recombination begins, McKim and colleagues proposed that cells monitor whether chromosomes have acquired an appropriate number of crossover-designated recombination intermediates and, if not, respond by increasing the total number of DSBs (13). This idea is attractive viewed through the lens of what is known about *S. cerevisiae* and *C. elegans*. It was recognized that such a system would also explain the interchromosomal effect (13, 28), although it is also possible that these are separate phenomena (68). Direct evidence for altered DSB numbers is lacking, and indirect cytological measures show no evidence of DSB increases in translocation heterozygotes (68). Nevertheless, if changes in DSB number do occur in these pathological situations, important unanswered questions include what aspect of meiotic chromosome behavior is monitored and which signaling pathways mount the global response. Recent work suggests that the interchromosomal effect involves monitoring of chromosome axis organization by the AAA$^+$ ATPase PCH2 [homologous to *S. cerevisiae* Pch2 (pachytene checkpoint) and mouse TRIP13 proteins, discussed further below] (68).

ATM-DEPENDENT NEGATIVE FEEDBACK CONTROL OF DOUBLE-STRAND BREAK FORMATION

ATM (ataxia telangiectasia mutated) is a serine/threonine kinase defective in the cancer-prone disease ataxia telangiectasia (A-T) (140). ATM activated by DSBs triggers cell cycle checkpoints and promotes DNA repair in somatic cells (40). A-T patients display gonadal dysgenesis (143) and *Atm*$^{-/-}$ mice are sterile, with chromosome synapsis defects and arrest during meiotic prophase (7, 8, 41, 42, 175). ATM is thus essential for normal, unperturbed meiosis, but until recently it was unclear what specific roles it plays. Independent studies in mouse, flies, and yeast suggest that ATM controls a negative feedback circuit that inhibits SPO11 activity (**Figure 5a**).

In *Atm*$^{-/-}$ mice, testes have >10-fold higher steady-state levels of SPO11-oligo complexes (81) (**Figure 5b**). SPO11-oligo complexes have a long life span in wild-type spermatocytes, and the amount of the complexes is already elevated when they first appear in *Atm*$^{-/-}$ juvenile testes. Thus, the steady-state increase in SPO11-oligo complexes likely reflects increased DSB numbers rather than increased life span of the complexes. Interestingly, absence of ATM also renders DSB formation more sensitive to SPO11 expression level, suggesting that the normal robustness of DSB number control in wild-type spermatocytes depends on ATM (81).

D. melanogaster females homozygous for a temperature-sensitive allele of the fly *Atm* ortholog display 1.5- to 3-fold higher levels of γ-H2AV in oocytes and neighboring nurse cells at the restrictive temperature (69) (**Figure 5c**). γ-H2AV is the equivalent of mammalian γ-H2AX (65), a phosphorylated form of histone variant H2AX that arises in response to DSBs made in nonmeiotic contexts (133) or by SPO11 (90). In flies, meiotic γ-H2AV is generated redundantly by ATM and ATR (product of the *mei-41* gene) (69). The elevated γ-H2AV levels thus suggest that more DSBs are made in ATM-deficient flies, resulting in higher ATR activity (69).

Similarly, several independent lines of evidence point to elevated meiotic DSB formation in *S. cerevisiae* cells lacking the ATM ortholog Tel1: a higher frequency of detectable recombinants (measured in an otherwise wild-type background) and DSBs (measured in a *rad50S* background) at an artificial recombination hot spot (179), elevated DSBs at a natural hot spot and modestly higher DSB frequency on at least one whole chromosome (also measured in a *rad50S* background) (27), and a greater number of DSBs genome-wide, assayed by immunoprecipitation of covalent Spo11-oligo complexes in a *RAD50*$^+$ *SAE2*$^+$ background (N. Mohibullah & S. Keeney, unpublished results). Deletion of the *SAE2* gene or *rad50S* (for separation of function) mutations block the removal of Spo11 from DSB ends and thereby prevent DSB resection (see **Figure 1b**). Two studies reported instead that DSBs are reduced by *tel1* mutation in *rad50S* or *sae2* mutant backgrounds (5, 17). The reasons for the differences in results are unclear, but the preponderance of the data (and, in particular, data from otherwise wild-type backgrounds) indicates that DSBs are elevated in *tel1* mutant yeast, similar to ATM-deficient mice and flies.

These findings suggest that an evolutionarily conserved pathway for DSB control entails activation of ATM by DSBs, which then feeds back to inhibit further DSB formation by other SPO11 molecules (**Figure 5a**). The mechanism is not yet well understood. Is ATM kinase activity required, and if so, what is the relevant phosphorylation target(s)? One candidate target in yeast is Rec114, a meiosis-specific protein required for DSB formation: Rec114 is phosphorylated in vivo in response to DSBs. This phosphorylation does not occur in mutants lacking both Tel1 and

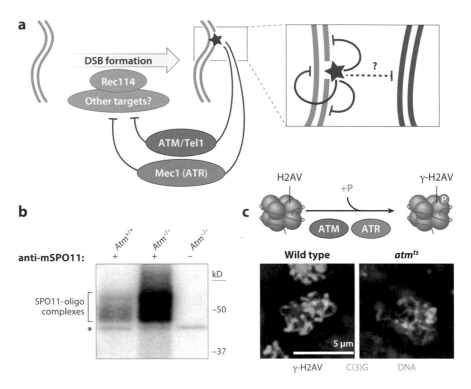

Figure 5

The inhibitory feedback circuit between the ATM/Tel1 and ATR/Mec1 kinases and SPO11 operates in *Saccharomyces cerevisiae*, mice, and *Drosophila melanogaster*. (*a*) In *S. cerevisiae*, Tel1 and Mec1 respond to double-strand breaks (DSBs) by negatively regulating local DSB formation on the same DNA molecule and on the sister chromatid, and possibly also around allelic positions on the homologous chromosome. In this pathway, phosphorylation targets of Tel1 and Mec1 may include Rec114, which is required for DSB formation. ATM acts in an analogous pathway in mice and *D. melanogaster*, but whether ATR contributes to feedback control in these organisms remains unknown. (*b*) More than 10-fold elevation in DSB levels in mice lacking ATM. The panel shows an autoradiograph of SPO11-oligo complexes isolated from wild-type and *Atm*-null mouse testes. Because two SPO11-oligo complexes are released from each DSB, they are quantitative by-products of meiotic recombination initiation that can serve as a measure of relative whole-testis DSB levels. SPO11-oligo complexes were immunopurified from testis lysates using an anti-SPO11 antibody and then radiolabeled and fractionated by electrophoresis. Asterisk indicates a SPO11-independent labeling artifact. Image adapted from Reference 81. (*c*) Increased phosphorylation of histone variant H2AV in ATM-deficient *D. melanogaster* oocytes. In response to DSBs, H2AV is phosphorylated by ATM or ATR to form γ-H2AV (equivalent of γ-H2AX in mammals). The micrographs show pachytene oocytes from a wild-type fly or an ATM-deficient fly. The cells were immunostained for γ-H2AV and C(3)G (a component of the synaptonemal complex that identifies oocytes within the ovary) and stained with DAPI to detect DNA. γ-H2AV levels are elevated in ATM-deficient oocytes and interpreted as being the result of an elevated DSB number yielding higher ATR activity. Images adapted from Reference 69 with permission of the authors. Abbreviation: p, phosphorylation.

Mec1 activity, Rec114 can be phosphorylated in vitro by Mec1 (which has similar substrate preferences as Tel1), and mutant Rec114 protein lacking putative phosphorylation sites yields signs of increased or faster DSB formation, whereas potentially phosphomimetic mutations inhibit DSB formation (27, 139). Whether Rec114 is a direct Tel1 target has not yet been established, and it remains unclear whether the absence of Rec114 phosphorylation phenocopies the absence of Tel1, so other (possibly redundant) substrates may exist. Furthermore, the fact that Mec1 can phosphorylate Rec114 suggests this kinase (like Tel1) may also inhibit DSB formation directly, either in wild-type cells or perhaps only when Mec1 is hyperactivated (i.e., in recombination mutants). This possibility adds further complexity to the mix of positive and negative roles that Mec1 plays (see above).

In flies, γ-H2AV turns over quickly in the absence of ongoing ATM and/or ATR signaling (69). Assuming similar rapid turnover for ATM targets relevant to DSB inhibition, it seems likely that ATM-dependent feedback is transient, occurring only as long as ATM itself remains active. Moreover, it is likely that this DSB inhibition works at least partly at a local level because ATM is activated in direct spatial proximity to the sites of DSBs, as judged, for example, by the location of γ-H2AX formed in mouse meiosis (90). Thus, this pathway might serve to discourage the formation of multiple DSBs near one another on the same chromatid or on sister chromatids (27, 81) (**Figure 5a**, inset). In fact, *S. cerevisiae* and *S. pombe* cells rarely cut the same chromatid at adjacent hot spots, and do so much less frequently than expected from the DSB frequencies of the individual hot spots (M. Lichten, M. Neale, G. Smith, personal communication). This behavior, i.e., interference between potential DSB sites on the same DNA molecule, requires Tel1 in both yeasts (M. Neale & G. Smith, personal communication). A further prediction, supported by DSB mapping via deep sequencing of Spo11 oligos in *S. cerevisiae* (N. Mohibullah & S. Keeney, unpublished results) and mouse (J. Lange, M. Jasin, S. Keeney, unpublished results), is that the DSB landscape in wild-type meiosis is shaped in part by the spatial patterning of ATM-dependent feedback. Additional implications of this form of DSB control are discussed in the final section.

NEGATIVE FEEDBACK TIED TO HOMOLOG ENGAGEMENT

Another pathway for restraining Spo11 activity has been illuminated by independent studies in mice, nematodes, and budding yeast. A mouse transgenic construct expressing *Spo11β* (one of several *Spo11* splicing isoforms) behaves as a hypomorphic *Spo11* mutation when the transgene insertion is in a single copy and is the only source of SPO11 protein, i.e., in mice with the genotype $Spo11^{-/-} Tg(Spo11\beta)^{+/-}$. Spermatocytes from these mice generate approximately half the normal number of DSBs as assessed by quantification of SPO11-oligo complexes or RAD51 and DMC1 foci (71). This reduction is accompanied by defects in synapsis of homologous chromosomes: Many chromosomes successfully locate their partners and form normal SCs, but in most cells several chromosomes fail to synapse properly. The unlucky chromosomes end up in topologically constrained tangles, where some axis segments remain unsynapsed and other segments show homologous or even nonhomologous synapsis (71).

Remarkably, even though the cell-wide DSB level is reduced in these mice, the unsynapsed chromosome axes in tangles continue to accumulate RAD51 foci, in some cases reaching a density comparable to wild type (**Figure 6a**). Moreover, in wild-type mice, RAD51 foci continue to accumulate on the portion of the X chromosome that is not homologous to the Y and that thus naturally remains unsynapsed. These results suggest that unsynapsed axes continue to accumulate DSBs, or put another way, that synapsis is normally accompanied by cessation of DSB formation (71). Importantly, this effect acts locally within the unsynapsed regions, which distinguishes it from the nucleus-wide responses discussed above.

C. elegans oocytes lacking X-chromosome pairing centers display X-chromosome-specific defects in homologous pairing and synapsis, plus numerous RAD-51 foci present on the X chromosomes much later than normal (88). Although it is possible that these foci are simply persisting because of repair defects, it is attractive to think that they instead represent additional DSBs above the number that would have formed in wild type. If so, these extra DSBs appear to be more numerous than those on autosomes in the same cells, which would be consistent with a local effect

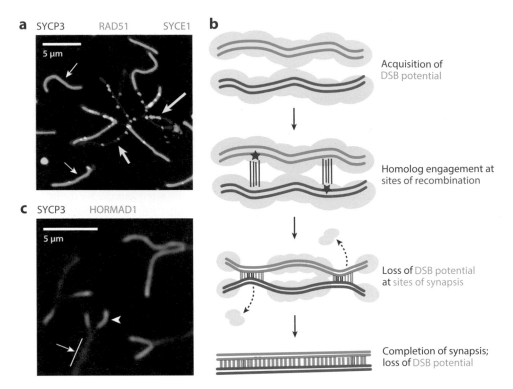

a SYCP3   RAD51   SYCE1

5 μm

c SYCP3   HORMAD1

5 μm

b

Acquisition of DSB potential

Homolog engagement at sites of recombination

Loss of DSB potential at sites of synapsis

Completion of synapsis; loss of DSB potential

Figure 6

Feedback tied to engagement of homologous chromosomes. (*a*) Continued accumulation of double-strand breaks (DSBs) on segments of mouse chromosomes that fail to synapse. The panel shows a micrograph of some of the chromosomes from a mouse spermatocyte spread on a glass slide and immunostained for SYCP3 (a component of the axial elements), SYCE1 [a component of the central element of the synaptonemal complex (SC)] and RAD51 (a strand-exchange protein). Unsynapsed axes appear red, and mature SC appears magenta from the overlap of the red SYCP3 and blue SYCE1 signals. Each green focus is a site at which a DSB has been made and is in the process of being repaired by RAD51 and other factors. This spermatocyte is from a mouse that has reduced DSB formation overall because the only source of SPO11 protein is a single copy of a transgene expressing the *Spo11β* splicing isoform. Some chromosomes synapse normally (*white arrows*), but some display synaptic failure and are trapped in tangles. The unsynapsed axes in these tangles continue to accumulate RAD51 foci (*yellow arrows*), suggesting that DSBs continue to form. Image adapted from Reference 71. (*b*) Schematic illustrating how DSB potential (i.e., the ability of chromosomes to be a substrate for SPO11; *green clouds*) is lost as homologous chromosomes engage each other during recombination and SC assembly. (*c*) Depletion of DSB-promoting protein HORMAD1 from chromosome axes after SC formation. A portion of a spread mouse spermatocyte nucleus is shown immunostained for SYCP3 and HORMAD1. Axial elements that have not yet synapsed with one another show strong staining for HORMAD1 (*arrowhead*), whereas regions that have synapsed do not (*arrow*; note that a single red signal is seen because conventional light microscopy cannot resolve the two SYCP3-staining axial elements of the synapsed homologous chromosomes). Image adapted from Reference 170 with permission of the authors.

on DSB formation specific to the misbehaving chromosomes, layered on top of the nucleus-wide prolongation of the DSB-permissive stage (28, 88, 136, 152).

These observations dovetail with independent studies of the *S. cerevisiae* ZMM proteins (Zip1–4, Msh4–5, Mer3, and others), a biochemically diverse suite of factors needed to ensure that crossover-designated recombination intermediates do indeed become crossovers (21, 87). ZMM proteins are needed for normal SC formation; in fact, one of them (Zip1) is a core structural component of the SC (156). Given these functions in recombination and synapsis, ZMM proteins were typically thought of as acting strictly downstream of DSB formation. However, surprisingly, *zmm* mutants accumulate more DSBs than wild type based on multiple lines of evidence, including elevated DSB levels by direct physical assays, increased frequency of detectable recombination products, and more Spo11-oligo complexes (160).

In principle, increased DSBs could have been an indirect consequence of the meiotic delay or arrest caused by the DSB repair defect in *zmm* mutants, which is known to trigger Mec1-dependent inhibition of Ndt80 (87, 163). However, epistasis tests show this not to be the case: *zmm ndt80* double mutants make more DSBs than either *zmm* or *ndt80* single mutants (160). These results indicate that a ZMM-dependent process that fosters engagement of homologous chromosomes is more directly responsible for inhibiting DSB formation in wild-type cells. A similar line of reasoning has been proposed to explain inferred increases in DSB formation in other yeast mutants with defects in homolog engagement (27, 82).

A plausible mechanism that accounts for results in both yeast and mouse is SC formation leading to structural changes that render chromosomes unfit substrates for Spo11 (27, 71, 160, 170) (**Figure 6b**). This model was first proposed from studies of the mouse HORMA-domain proteins HORMAD1 and HORMAD2 (170). These proteins localize to unsynapsed chromosome axes on which RAD51 foci form during the leptotene and zygotene stages but are displaced soon after synapsis (46, 145, 170) (**Figure 6c**). Likewise in *A. thaliana*, the HORMA-domain protein ASY1 becomes similarly depleted from axes after chromosomes synapse (F.C. Franklin, personal communication). Tóth and colleagues proposed that HORMAD displacement might suppress further DSB formation because homologous proteins, such as *S. cerevisiae* Hop1, are needed for normal DSB levels (61, 170) [shown later to be true for mouse HORMAD1 as well (37, 145)]. Earlier work had demonstrated that numerous DSB-promoting factors, including Hop1, Mei4 (not related to the *S. pombe* Mei4 transcription factor discussed earlier), Red1, Rec102, Rec104, and Rec114, are displaced from chromosomes after synapsis in *S. cerevisiae* (20, 73, 83, 91, 148). These yeast results were not originally interpreted in terms of DSB regulation, but they fit nicely with this model in retrospect (27, 160).

The mechanism behind the chromosome structure changes remains to be determined. In mouse and *A. thaliana*, HORMA-domain protein displacement occurs with a delay after SC formation, so synapsis cannot be an instantaneous trigger (170; F.C. Franklin, personal communication). A temporal offset for removal of DSB-promoting proteins is also seen in yeast (e.g., 73). Moreover, mouse HORMAD displacement even occurs in chromosomal segments that synapse nonhomologously, e.g., in mice lacking SPO11 (170). Thus, recombination is not essential, and although SC formation between homologous chromosomes may be the normal conduit for DSB control in wild-type cells, homolog engagement per se is also dispensable.

The AAA+ ATPase TRIP13 is required for HORMAD proteins to be displaced from synapsed axes in mice (134, 170), as is the TRIP13 ortholog PCH2 in *A. thaliana* (F.C. Franklin, personal communication). Similarly, Pch2 governs the normal nonuniform localization of Hop1 along SCs in *S. cerevisiae* (20), probably by directly remodeling Hop1 protein structure (32) (N. Kleckner, personal communication). However, available evidence does not strongly support the view that TRIP13 or Pch2 are net-negative regulators of DSB formation. In yeast, DSB numbers are not increased in *pch2* mutants (20, 44). In mouse, the fact that RAD51 foci and γH2AX are present on synapsed chromosomes during pachynema in *Trip13*-deficient spermatocytes may indicate that DSB formation continues longer than normal (71). However, global DSB levels do not appear to be increased based on SPO11-oligo quantification (S. Pacheco, M. Marcet-Ortega, J. Lange, M. Jasin, S. Keeney, I. Roig, unpublished results), so DSB persistence rather than increased DSB numbers may underlie these cytological patterns (84, 134).

It is interesting that this mode of DSB control is analogous to how Ndt80- and CHK-2-influenced circuits in *S. cerevisiae* and *C. elegans* may work, although the specific mechanisms regulating chromosomal association of DSB-promoting factors differ in each case. We also note that this feedback mechanism may provide a more permanent shutdown of DSB-forming potential compared with the proposed transience of the ATM/Tel1 circuit. Feedback tied to homolog engagement may operate over longer physical distances as well, given the potential for SC to polymerize further than ATM/Tel1 kinase signaling is likely to spread.

COMMUNICATION BETWEEN POTENTIAL DOUBLE-STRAND BREAK SITES

Four further layers of DSB regulation have been documented in *S. cerevisiae*, each involving apparent communication in *trans* between sister chromatids or homologous chromosomes, or in *cis* between regions on the same chromosome. First, DSB frequency at a hot spot can be affected by sequences at the same position on the homologous partner (26, 131, 174). In at least some cases, allelic sequences that match one another yield the most DSBs (131, 174), possibly because they also yield the most accessible (DNase I–hypersensitive) chromatin structure (77). These findings suggest that homology-dependent physical interactions between chromosomes influence chromatin structure and DSB formation, but the mechanism is not known.

Two more layers also work in *trans* between DNA molecules. Specifically, a strong hot spot can suppress DSB formation at the allelic position and at hot spots nearby (within a few kilobases) on the homologous chromosome (174, 179). Separately, numerical patterns of recombination in tetrads suggest that wild-type cells rarely break both sister chromatids at the same hot spot (179). Thus, it appears that any given cell usually experiences at most one DSB at the same place among four chromatids, even at an exceedingly strong hot spot (179). In mutants lacking Tel1 (ATM) or Mec1 (ATR), recombination patterns suggest loss of one of these control layers, i.e., DSBs can often occur at the same place on two DNA molecules in the same cell. This has been interpreted to mean that Tel1 and Mec1 convey an inhibitory signal in *trans* between homologous chromosomes (179) (**Figure 5a**, inset). However, the data are equally consistent with one or both kinases preventing breakage of both sister chromatids, as has been proposed for Tel1 in yeast (27; M. Neale, personal communication) and ATM in mouse (81). Further studies are needed to dissect the mechanisms and interplay between these pathways.

The fourth layer is revealed by the fact that creating new hot spots—whether by inserting an artificial hot spot–specifying DNA construct or by fusing Spo11 to a sequence-specific DNA binding domain to target cleavage to new positions—suppresses DSB formation in neighboring regions on the same chromosome (43, 47, 66, 118, 130, 172, 174). The magnitude of suppression decays with distance but has been detected 30–60 kilobases away and may extend further in some contexts (47, 66, 130, 172). *S. pombe* behaves similarly (153; G. Smith, personal communication). In principle, this phenomenon could involve competition between hot spots for a limiting pool of DSB-promoting factors and could occur before DSB formation. However, site-specific DSBs made by an endonuclease also suppress Spo11-generated DSBs nearby (47): Assuming that the suppression mechanism from endonuclease-directed DSBs is the same as for Spo11, this finding implies that hot-spot competition involves the spread of an inhibitory signal after DSB formation. Tel1 appears to be dispensable (N. Mohibullah & S. Keeney, unpublished results), distinguishing this form of DSB suppression from the Tel1-mediated DSB interference discussed above (i.e., less frequent double-cutting of the same chromatid than expected by chance). One possible mechanism derives from the proposal by Kleckner and colleagues that mechanical stress and stress relief drive DNA metabolic events and chromosome morphogenesis (80, 180). As a topoisomerase relative, Spo11 is predicted to be an inherently stress-sensitive enzyme (75). Thus, the same patterning forces that are proposed to shape crossover distributions (80, 180) might also shape DSB distributions (74, 75, 80).

CONCLUSIONS AND IMPLICATIONS

From the first proposals stating that DSBs might be the initiators of meiotic recombination (128, 157), it has been appreciated that this is a dangerous game for the cell to play. Work reviewed here brings into focus how cells accommodate this danger by controlling when, where, and how many DSBs are made by Spo11. The emerging view is that DSB control involves a drive toward DSB formation (promoted in part by cell cycle regulators and by development of meiosis-specific chromosome structures) that is subject to quantitative, spatial, and temporal restraints from distinct but intersecting negative influences (**Figure 2**). This view explains the basis of self-organization of recombination initiation and clarifies how DSB formation can be homeostatic and therefore robust against cell-to-cell variation and environmental perturbation.

We envision that this robustness helps cells cope with unbalanced karyotypes, such as might be encountered in outcrosses or with heterozygous de novo chromosome rearrangements. For example, the proclivity of SC to form between nonhomologous chromosome segments in maize, mice, and other organisms (e.g., 96, 104) may provide a means to eventually suppress DSB formation in regions that are unable to locate a homologous partner (160). In normal male meiosis in mammals, the X and Y chromosome share only a small segment of homology, the pseudoautosomal region (PAR) (137). The PAR must receive at least one DSB in order to support pairing and crossing over between the sex chromosomes (70), yet the cell must also cope with the presence of the large portions of the X and Y that cannot engage one another homologously. The network of DSB-regulating pathways, and in particular the ability to terminate DSB formation genome-wide at or before prophase exit, provides a means to accommodate this feature of male meiosis.

An important implication of the circuits that control Spo11 activity is that they can complicate genetic analyses, especially if a mutation impinges on multiple circuits at once. For example, *S. cerevisiae zmm* mutations simultaneously disrupt homolog engagement (removing a negative DSB regulator), inhibit Ndt80 activation via DSB repair defects (removing yet another negative DSB regulator), and hyperactivate Mec1 (ATR) [possibly adding Tel1(ATM)-like DSB inhibition] (160, 163). Thus, when homeostatic DSB control pathways respond to pathological defects in recombination or other processes, the phenotypic endpoints differ from the wild-type situation in many complex ways. By the same token, it is important to view cautiously the widespread use of recombination-defective mutant backgrounds (e.g., *rad50S* or *dmc1* in *S. cerevisiae*) to determine what DSB numbers would have been in wild type or to query the effects of other mutations on DSB numbers.

Another important implication is that Spo11-regulating processes strongly influence DSB locations by virtue of being spatially patterned, e.g., via local activity of ATM/Tel1 near DSBs or by tying feedback to homolog engagement at sites of recombination (13, 160) (N. Mohibullah & S. Keeney, unpublished results; J. Lange, M. Jasin, S. Keeney, unpublished results). We can thus divide architects of the DSB landscape into intrinsic factors (chromosomal features that govern accessibility or activity of Spo11 toward specific locations, such as loop-axis structure, nucleosome positions, or histone modifications) and more extrinsic factors layered on top (feedback and other regulatory circuits). This division fits well with the view that the factors shaping the DSB landscape work in a hierarchical fashion (123).

A related point is that modes of Spo11 regulation explain otherwise puzzling features of *set1* mutant yeast and *Prdm9*$^{-/-}$ mutant mice. In *S. cerevisiae*, trimethylation of histone H3 lysine 4 by the

Set1 methyltransferase plays a key role in directing Spo11 to cleave preferentially in nucleosome-depleted regions in promoters (1, 150). In mouse and human, DSB hot-spot locations are defined by the DNA binding specificity of PRDM9, which sports a histone methyltransferase module and a zinc-finger DNA binding domain that evolves rapidly (9, 11, 24, 111). These methyltransferases have a near-absolute role in targeting Spo11 to particular sites in the genome, yet they are almost completely dispensable for DSB formation per se: In their absence, DSBs form in relatively normal numbers but different locations (19, 24). This behavior is explained by recognizing that the default is for Spo11 to make breaks until restrained. Even if the DSB-forming machinery is crippled by absence of targeting factors such as Set1 and PRDM9, the homeostatic responses of Spo11-regulating circuits will adjust, driving break formation at high frequency but in highly abnormal positions.

Similar reasoning affects understanding of the relationship between hot-spot evolutionary dynamics and the ability to execute recombination genome-wide. Because of bias in the direction of information transfer during gene conversion (the broken chromosome copies information from the intact donor), sequence polymorphisms that inactivate a hot spot tend to be overrepresented in offspring from heterozygous individuals (e.g., 50). Thus, hot spots are expected to become colder or disappear over evolutionary timescales and new hot-spot alleles should rarely if ever go to fixation in a population. The existence of hot spots despite these headwinds has been called the hot-spot paradox (22). Many attempts to model this phenomenon implicitly or explicitly assume that recombination cannot occur if hot spots are lost (126, 164). However, the logic of DSB control makes it impossible for inactivation of even large numbers of individual hot spots to render chromosomes immune to Spo11. Thus, the need to retain recombination proficiency is unlikely by itself to be a selective constraint in favor of hot-spot retention. Moreover, when a hot spot becomes cooler or disappears, homeostatic DSB regulation is predicted to elevate DSB frequency in neighboring regions: Like a genomic Whac-a-Mole® game, new hot spots pop up whenever a hot spot decays (hat tip: M. Lichten).

The broad outlines and basic concepts of DSB regulation appear to be evolutionarily conserved. This conservation seems fitting given the universality of the risk/reward trade-off for DSB formation and recombination in meiosis. Nonetheless, detailed mechanisms differ widely between species. For example, even though involvement of DNA damage response kinases and control of the chromosomal association of DSB-promoting factors are common themes, they are often used in different contexts in different organisms and even within different regulatory pathways in the same organism. In many cases, this evolutionary plasticity clearly reflects differences between organisms' cellular and developmental constraints on sexual reproduction, such as genome size and karyotype; repetitive DNA content; recombination (in)dependence of chromosome pairing; and the organism's lifestyle, e.g., reversibility of meiotic entry in yeast versus irreversibility in metazoans. Although much remains to be learned about detailed mechanisms of the regulatory modules involved, current knowledge provides molecular frameworks to guide future experiments.

DISCLOSURE STATEMENT

The authors are not aware of any affiliations, memberships, funding, or financial holdings that might be perceived as affecting the objectivity of this review.

ACKNOWLEDGMENTS

We thank Abby Dernberg, Chris Franklin, Maria Jasin, Liisa Kauppi, Nancy Kleckner, Michael Lichten, Hong Ma, Hisao Masai, Matthew Neale, Kunihiro Ohta, Gerry Smith, Drew Thacker, and Anne Villeneuve for discussions and/or sharing unpublished data. We thank Kim McKim and Attila Tóth for permission to reproduce published images. Our research is supported by grants from the NIH (GM058673 to S.K.; GM105421 to M. Jasin and S.K.; HD053855 to S.K. and M. Jasin) and the Howard Hughes Medical Institute. J.L. was supported in part by a fellowship from the American Cancer Society (grant # PF-12-157-01-DMC). S.K. is an Investigator of the Howard Hughes Medical Institute.

LITERATURE CITED

1. Acquaviva L, Székvölgyi L, Dichtl B, Dichtl BS, de La Roche Saint André C, et al. 2013. The COMPASS subunit Spp1 links histone methylation to initiation of meiotic recombination. *Science* 339:215–18
2. Allers T, Lichten M. 2001. Differential timing and control of noncrossover and crossover recombination during meiosis. *Cell* 106:47–57
3. Alpi A, Pasierbek P, Gartner A, Loidl J. 2003. Genetic and cytological characterization of the recombination protein RAD-51 in *Caenorhabditis elegans*. *Chromosoma* 112:6–16
4. Arbel A, Zenvirth D, Simchen G. 1999. Sister chromatid–based DNA repair is mediated by *RAD54*, not by *DMC1* or *TID1*. *EMBO J.* 18:2648–58
5. Argunhan B, Farmer S, Leung WK, Terentyev Y, Humphryes N, et al. 2013. Direct and indirect control of the initiation of meiotic recombination by DNA damage checkpoint mechanisms in budding yeast. *PLOS ONE* 8:e65875
6. Azumi Y, Liu D, Zhao D, Li W, Wang G, et al. 2002. Homolog interaction during meiotic prophase I in *Arabidopsis* requires the *SOLO DANCERS* gene encoding a novel cyclin-like protein. *EMBO J.* 21:3081–95
7. Barchi M, Mahadevaiah S, Di Giacomo M, Baudat F, de Rooij DG, et al. 2005. Surveillance of different recombination defects in mouse spermatocytes yields distinct responses despite elimination at an identical developmental stage. *Mol. Cell. Biol.* 25:7203–15
8. Barlow C, Hirotsune S, Paylor R, Liyanage M, Eckhaus M, et al. 1996. *Atm*-deficient mice: a paradigm of ataxia telangiectasia. *Cell* 86:159–71
9. Baudat F, Buard J, Grey C, Fledel-Alon A, Ober C, et al. 2010. PRDM9 is a major determinant of meiotic recombination hotspots in humans and mice. *Science* 327:836–40
10. Baudat F, Imai Y, de Massy B. 2013. Meiotic recombination in mammals: localization and regulation. *Nat. Rev. Genet.* 14:794–806
11. Berg IL, Neumann R, Lam KW, Sarbajna S, Odenthal-Hesse L, et al. 2010. PRDM9 variation strongly influences recombination hot-spot activity and meiotic instability in humans. *Nat. Genet.* 42:859–63
12. Bergerat A, de Massy B, Gadelle D, Varoutas PC, Nicolas A, Forterre P. 1997. An atypical topoisomerase II from Archaea with implications for meiotic recombination. *Nature* 386:414–17
13. Bhagat R, Manheim EA, Sherizen DE, McKim KS. 2004. Studies on crossover-specific mutants and the distribution of crossing over in *Drosophila* females. *Cytogenet. Genome Res.* 107:160–71
14. Bhalla N, Dernburg AF. 2008. Prelude to a division. *Annu. Rev. Cell Dev. Biol.* 24:397–424
15. Blat Y, Protacio RU, Hunter N, Kleckner N. 2002. Physical and functional interactions among basic chromosome organizational features govern early steps of meiotic chiasma formation. *Cell* 111:791–802
16. Blitzblau HG, Chan CS, Hochwagen A, Bell SP. 2012. Separation of DNA replication from the assembly of break-competent meiotic chromosomes. *PLOS Genet.* 8:e1002643
17. Blitzblau HG, Hochwagen A. 2013. ATR/Mec1 prevents lethal meiotic recombination initiation on partially replicated chromosomes in budding yeast. *Elife* 2:e00844
18. Borde V, Goldman ASH, Lichten M. 2000. Direct coupling between meiotic DNA replication and recombination initiation. *Science* 290:806–9
19. Borde V, Robine N, Lin W, Bonfils S, Geli V, Nicolas A. 2009. Histone H3 lysine 4 trimethylation marks meiotic recombination initiation sites. *EMBO J.* 28:99–111
20. Borner GV, Barot A, Kleckner N. 2008. Yeast Pch2 promotes domainal axis organization, timely recombination progression, and arrest of defective recombinosomes during meiosis. *Proc. Natl. Acad. Sci. USA* 105:3327–32

21. Borner GV, Kleckner N, Hunter N. 2004. Crossover/noncrossover differentiation, synaptonemal complex formation, and regulatory surveillance at the leptotene/zygotene transition of meiosis. *Cell* 117:29–45

22. Boulton A, Myers RS, Redfield RJ. 1997. The hotspot conversion paradox and the evolution of meiotic recombination. *Proc. Natl. Acad. Sci. USA* 94:8058–63

23. Brar GA, Yassour M, Friedman N, Regev A, Ingolia NT, Weissman JS. 2012. High-resolution view of the yeast meiotic program revealed by ribosome profiling. *Science* 335:552–57

24. Brick K, Smagulova F, Khil P, Camerini-Otero RD, Petukhova GV. 2012. Genetic recombination is directed away from functional genomic elements in mice. *Nature* 485:642–45

25. Buhler C, Borde V, Lichten M. 2007. Mapping meiotic single-strand DNA reveals a new landscape of DNA double-strand breaks in *Saccharomyces cerevisiae*. *PLOS Biol.* 5:e324

26. Bullard SA, Kim S, Galbraith AM, Malone RE. 1996. Double strand breaks at the *HIS2* recombination hot spot in *Saccharomyces cerevisiae*. *Proc. Natl. Acad. Sci. USA* 93:13054–59

27. Carballo JA, Panizza S, Serrentino ME, Johnson AL, Geymonat M, et al. 2013. Budding yeast ATM/ATR control meiotic double-strand break (DSB) levels by down-regulating Rec114, an essential component of the DSB-machinery. *PLOS Genet.* 9:e1003545

28. Carlton PM, Farruggio AP, Dernburg AF. 2006. A link between meiotic prophase progression and crossover control. *PLOS Genet.* 2:e12

29. Carpenter AT, Sandler L. 1974. On recombination-defective meiotic mutants in *Drosophila melanogaster*. *Genetics* 76:453–75

30. Cervantes MD, Farah JA, Smith GR. 2000. Meiotic DNA breaks associated with recombination in *S. pombe*. *Mol. Cell* 5:883–88

31. Chelysheva L, Gendrot G, Vezon D, Doutriaux MP, Mercier R, Grelon M. 2007. Zip4/Spo22 is required for class I CO formation but not for synapsis completion in *Arabidopsis thaliana*. *PLOS Genet.* 3:e83

32. Chen C, Jomaa A, Ortega J, Alani EE. 2014. Pch2 is a hexameric ring ATPase that remodels the chromosome axis protein Hop1. *Proc. Natl. Acad. Sci. USA* 111:E44–53

33. Chen SY, Tsubouchi T, Rockmill B, Sandler JS, Richards DR, et al. 2008. Global analysis of the meiotic crossover landscape. *Dev. Cell* 15:401–15

34. Chu S, Herskowitz I. 1998. Gametogenesis in yeast is regulated by a transcriptional cascade dependent on Ndt80. *Mol. Cell* 1:685–96

35. Colaiacovo MP, MacQueen AJ, Martinez-Perez E, McDonald K, Adamo A, et al. 2003. Synaptonemal complex assembly in *C. elegans* is dispensable for loading strand-exchange proteins but critical for proper completion of recombination. *Dev. Cell* 5:463–74

36. Cole F, Kauppi L, Lange J, Roig I, Wang R, et al. 2012. Homeostatic control of recombination is implemented progressively in mouse meiosis. *Nat. Cell Biol.* 14:424–30

37. Daniel K, Lange J, Hached K, Fu J, Anastassiadis K, et al. 2011. Meiotic homologue alignment and its quality surveillance are controlled by mouse HORMAD1. *Nat. Cell Biol.* 13:599–610

38. Dayani Y, Simchen G, Lichten M. 2011. Meiotic recombination intermediates are resolved with minimal crossover formation during return-to-growth, an analogue of the mitotic cell cycle. *PLOS Genet.* 7:e1002083

39. de Massy B. 2013. Initiation of meiotic recombination: how and where? Conservation and specificities among eukaryotes. *Annu. Rev. Genet.* 47:563–99

40. Derheimer FA, Kastan MB. 2010. Multiple roles of ATM in monitoring and maintaining DNA integrity. *FEBS Lett.* 584:3675–81

41. Di Giacomo M, Barchi M, Baudat F, Edelmann W, Keeney S, Jasin M. 2005. Distinct DNA damage–dependent and independent responses drive the loss of oocytes in recombination-defective mouse mutants. *Proc. Natl. Acad. Sci. USA* 102:737–42

42. Elson A, Wang Y, Daugherty CJ, Morton CC, Zhou F, et al. 1996. Pleiotropic defects in ataxia-telangiectasia protein-deficient mice. *Proc. Natl. Acad. Sci. USA* 93:13084–89

43. Fan QQ, Xu F, White MA, Petes TD. 1997. Competition between adjacent meiotic recombination hotspots in the yeast *Saccharomyces cerevisiae*. *Genetics* 145:661–70

44. Farmer S, Hong EJ, Leung WK, Argunhan B, Terentyev Y, et al. 2012. Budding yeast Pch2, a widely conserved meiotic protein, is involved in the initiation of meiotic recombination. *PLOS ONE* 7:e39724

45. Froenicke L, Anderson LK, Wienberg J, Ashley T. 2002. Male mouse recombination maps for each autosome identified by chromosome painting. *Am. J. Hum. Genet.* 71:1353–68

46. Fukuda T, Daniel K, Wojtasz L, Tóth A, Hoog C. 2010. A novel mammalian HORMA domain-containing protein, HORMAD1, preferentially associates with unsynapsed meiotic chromosomes. *Exp. Cell Res.* 316:158–71

47. Fukuda T, Kugou K, Sasanuma H, Shibata T, Ohta K. 2008. Targeted induction of meiotic double-strand breaks reveals chromosomal domain-dependent regulation of Spo11 and interactions among potential sites of meiotic recombination. *Nucleic Acids Res.* 36:984–97

48. Gray S, Allison RM, Garcia V, Goldman AS, Neale MJ. 2013. Positive regulation of meiotic DNA double-strand break formation by activation of the DNA damage checkpoint kinase Mec1(ATR). *Open Biol.* 3:130019

49. Gregan J, Rabitsch PK, Sakem B, Csutak O, Latypov V, et al. 2005. Novel genes required for meiotic chromosome segregation are identified by a high-throughput knockout screen in fission yeast. *Curr. Biol.* 15:1663–69

50. Gutz H. 1971. Site specific induction of gene conversion in *Schizosaccharomyces pombe*. *Genetics* 69:317–37

51. Harigaya Y, Tanaka H, Yamanaka S, Tanaka K, Watanabe Y, et al. 2006. Selective elimination of messenger RNA prevents an incidence of untimely meiosis. *Nature* 442:45–50

52. Hayashi M, Chin GM, Villeneuve AM. 2007. *C. elegans* germ cells switch between distinct modes of double-strand break repair during meiotic prophase progression. *PLOS Genet.* 3:e191

53. Hayashi M, Mlynarczyk-Evans S, Villeneuve AM. 2010. The synaptonemal complex shapes the crossover landscape through cooperative assembly, crossover promotion and crossover inhibition during *Caenorhabditis elegans* meiosis. *Genetics* 186:45–58

54. Henderson KA, Kee K, Maleki S, Santini PA, Keeney S. 2006. Cyclin-dependent kinase directly regulates initiation of meiotic recombination. *Cell* 125:1321–32

55. Henderson KA, Keeney S. 2004. Tying synaptonemal complex initiation to the formation and programmed repair of DNA double-strand breaks. *Proc. Natl. Acad. Sci. USA* 101:4519–24

56. Henzel JV, Nabeshima K, Schvarzstein M, Turner BE, Villeneuve AM, Hillers KJ. 2011. An asymmetric chromosome pair undergoes synaptic adjustment and crossover redistribution during *Caenorhabditis elegans* meiosis: implications for sex chromosome evolution. *Genetics* 187:685–99

57. Higgins JD, Armstrong SJ, Franklin FCH, Jones GH. 2004. The *Arabidopsis* MutS homolog AtMSH4 functions at an early step in recombination: evidence for two classes of recombination in *Arabidopsis*. *Genes Dev.* 18:2557–70

58. Higgins JD, Sanchez-Morán E, Armstrong SJ, Jones GH, Franklin FCH. 2005. The *Arabidopsis* synaptonemal complex protein ZYP1 is required for normal fidelity of crossing-over and chromosome synapsis. *Genes Dev.* 19:2488–500

59. Hochwagen A, Amon A. 2006. Checking your breaks: surveillance mechanisms of meiotic recombination. *Curr. Biol.* 16:R217–28

60. Hochwagen A, Tham WH, Brar GA, Amon A. 2005. The FK506 binding protein Fpr3 counteracts protein phosphatase 1 to maintain meiotic recombination checkpoint activity. *Cell* 122:861–73

61. Hollingsworth NM, Byers B. 1989. *HOP1*: a yeast meiotic pairing gene. *Genetics* 121:445–62

62. Horie S, Watanabe Y, Tanaka K, Nishiwaki S, Fujioka H, et al. 1998. The *Schizosaccharomyces pombe* *mei4+* gene encodes a meiosis-specific transcription factor containing a Forkhead DNA-binding domain. *Mol. Cell. Biol.* 18:2118–29

63. Hunter N. 2007. Meiotic recombination. In *Molecular Genetics of Recombination*, ed. A Aguilera, R Rothstein, pp. 381–442. Berlin: Springer-Verlag

64. Jackson N, Sanchez-Morán E, Buckling E, Armstrong SJ, Jones GH, Franklin FCH. 2006. Reduced meiotic crossovers and delayed prophase I progression in AtMLH3-deficient *Arabidopsis*. *EMBO J.* 25:1315–23

65. Jang JK, Sherizen DE, Bhagat R, Manheim EA, McKim KS. 2003. Relationship of DNA double-strand breaks to synapsis in *Drosophila*. *J. Cell Sci.* 116:3069–77

66. Jessop L, Allers T, Lichten M. 2005. Infrequent co-conversion of markers flanking a meiotic recombination initiation site in *Saccharomyces cerevisiae*. *Genetics* 169:1353–67

67. Jones GH, Franklin FC. 2006. Meiotic crossing-over: obligation and interference. *Cell* 126:246–48

68. Joyce EF, McKim KS. 2010. Chromosome axis defects induce a checkpoint-mediated delay and inter-chromosomal effect on crossing over during *Drosophila* meiosis. *PLOS Genet.* 6:e1001059

69. Joyce EF, Pedersen M, Tiong S, White-Brown SK, Paul A, et al. 2011. *Drosophila* ATM and ATR have distinct activities in the regulation of meiotic DNA damage and repair. *J. Cell Biol.* 195:359–67

70. Kauppi L, Barchi M, Baudat F, Romanienko PJ, Keeney S, Jasin M. 2011. Distinct properties of the XY pseudoautosomal region crucial for male meiosis. *Science* 331:916–20

71. Kauppi L, Barchi M, Lange J, Baudat F, Jasin M, Keeney S. 2013. Numerical constraints and feedback control of double-strand breaks in mouse meiosis. *Genes Dev.* 27:873–86

72. Kauppi L, Jeffreys AJ, Keeney S. 2004. Where the crossovers are: recombination distributions in mammals. *Nat. Rev. Genet.* 5:413–24

73. Kee K, Protacio RU, Arora C, Keeney S. 2004. Spatial organization and dynamics of the association of Rec102 and Rec104 with meiotic chromosomes. *EMBO J.* 23:1815–24

74. Keeney S. 2001. Mechanism and control of meiotic recombination initiation. *Curr. Top. Dev. Biol.* 52:1–53

75. Keeney S. 2008. Spo11 and the formation of DNA double-strand breaks in meiosis. In *Recombination and Meiosis: Crossing-Over and Disjunction*, ed. R Egel, DH Lankenau, pp. 81–123. Heidelberg, Ger.: Springer-Verlag

76. Keeney S, Giroux CN, Kleckner N. 1997. Meiosis-specific DNA double-strand breaks are catalyzed by Spo11, a member of a widely conserved protein family. *Cell* 88:375–84

77. Keeney S, Kleckner N. 1996. Communication between homologous chromosomes: genetic alterations at a nuclease-hypersensitive site can alter mitotic chromatin structure at that site both in *cis* and in *trans*. *Genes Cells* 1:475–89

78. Kim JM, Takemoto N, Arai K, Masai H. 2003. Hypomorphic mutation in an essential cell-cycle kinase causes growth retardation and impaired spermatogenesis. *EMBO J.* 22:5260–72

79. Kleckner N. 2006. Chiasma formation: chromatin/axis interplay and the role(s) of the synaptonemal complex. *Chromosoma* 115:175–94

80. Kleckner N, Zickler D, Jones GH, Dekker J, Padmore R, et al. 2004. A mechanical basis for chromosome function. *Proc. Natl. Acad. Sci. USA* 101:12592–97

81. Lange J, Pan J, Cole F, Thelen MP, Jasin M, Keeney S. 2011. ATM controls meiotic double-strand-break formation. *Nature* 479:237–40

82. Lao JP, Cloud V, Huang CC, Grubb J, Thacker D, et al. 2013. Meiotic crossover control by concerted action of Rad51-Dmc1 in homolog template bias and robust homeostatic regulation. *PLOS Genet.* 9:e1003978

83. Li J, Hooker GW, Roeder GS. 2006. *Saccharomyces cerevisiae* Mer2, Mei4 and Rec114 form a complex required for meiotic double-strand break formation. *Genetics* 173:1969–81

84. Li XC, Schimenti JC. 2007. Mouse pachytene checkpoint 2 (trip13) is required for completing meiotic recombination but not synapsis. *PLOS Genet.* 3:e130

85. Lichten M. 2008. Meiotic chromatin: the substrate for recombination initiation. In *Recombination and Meiosis: Models, Means, and Evolution*, ed. R Egel, DH Lankenau, pp. 165–93. Berlin: Springer-Verlag

86. Lin Y, Larson KL, Dorer R, Smith GR. 1992. Meiotically induced *rec7* and *rec8* genes of *Schizosaccharomyces pombe*. *Genetics* 132:75–85

87. Lynn A, Soucek R, Borner GV. 2007. ZMM proteins during meiosis: crossover artists at work. *Chromosome Res.* 15:591–605

88. MacQueen AJ, Phillips CM, Bhalla N, Weiser P, Villeneuve AM, Dernburg AF. 2005. Chromosome sites play dual roles to establish homologous synapsis during meiosis in *C. elegans*. *Cell* 123:1037–50

89. MacQueen AJ, Villeneuve AM. 2001. Nuclear reorganization and homologous chromosome pairing during meiotic prophase require *C. elegans chk-2*. *Genes Dev.* 15:1674–87

90. Mahadevaiah SK, Turner JMA, Baudat F, Rogakou EP, de Boer P, et al. 2001. Recombinational DNA double strand breaks in mice precede synapsis. *Nat. Genet.* 27:271–76

91. Maleki S, Neale MJ, Arora C, Henderson KA, Keeney S. 2007. Interactions between Mei4, Rec114, and other proteins required for meiotic DNA double-strand break formation in *Saccharomyces cerevisiae*. *Chromosoma* 116:471–86

92. Margolin G, Khil PP, Kim J, Bellani MA, Camerini-Otero RD. 2014. Integrated transcriptome analysis of mouse spermatogenesis. *BMC Genomics* 15:39

93. Marston AL, Amon A. 2004. Meiosis: cell-cycle controls shuffle and deal. *Nat. Rev. Mol. Cell Biol.* 5:983–97

94. Matos J, Lipp JJ, Bogdanova A, Guillot S, Okaz E, et al. 2008. Dbf4-dependent CDC7 kinase links DNA replication to the segregation of homologous chromosomes in meiosis I. *Cell* 135:662–78

95. Matsumoto S, Ogino K, Noguchi E, Russell P, Masai H. 2005. Hsk1-Dfp1/Him1, the Cdc7-Dbf4 kinase in *Schizosaccharomyces pombe*, associates with Swi1, a component of the replication fork protection complex. *J. Biol. Chem.* 280:42536–42

96. McClintock B. 1933. The association of non-homologous parts of chromosomes in the mid-prophase of meiosis in *Zea mays*. *Z. Zellforsch. Mikrosk. Anat.* 19:191–237

97. McFarlane RJ, Mian S, Dalgaard JZ. 2010. The many facets of the Tim-Tipin protein families' roles in chromosome biology. *Cell Cycle* 9:700–5

98. McKim KS, Howell AM, Rose AM. 1988. The effects of translocations on recombination frequency in *Caenorhabditis elegans*. *Genetics* 120:987–1001

99. Mehrotra S, McKim KS. 2006. Temporal analysis of meiotic DNA double-strand break formation and repair in *Drosophila* females. *PLOS Genet.* 2:e200

100. Mets DG, Meyer BJ. 2009. Condensins regulate meiotic DNA break distribution, thus crossover frequency, by controlling chromosome structure. *Cell* 139:73–86

101. Milman N, Higuchi E, Smith GR. 2009. Meiotic DNA double-strand break repair requires two nucleases, MRN and Ctp1, to produce a single size class of Rec12 (Spo11)-oligonucleotide complexes. *Mol. Cell. Biol.* 29:5998–6005

102. Miyoshi T, Ito M, Kugou K, Yamada S, Furuichi M, et al. 2012. A central coupler for recombination initiation linking chromosome architecture to S phase checkpoint. *Mol. Cell* 47:722–33

103. Moens PB, Chen DJ, Shen Z, Kolas N, Tarsounas M, et al. 1997. Rad51 immunocytology in rat and mouse spermatocytes and oocytes. *Chromosoma* 106:207–15

104. Moses MJ, Dresser ME, Poorman PA. 1984. Composition and role of the synaptonemal complex. *Symp. Soc. Exp. Biol.* 38:245–70

105. Murakami H, Borde V, Shibata T, Lichten M, Ohta K. 2003. Correlation between premeiotic DNA replication and chromatin transition at yeast recombination initiation sites. *Nucleic Acids Res.* 31:4085–90

106. Murakami H, Keeney S. 2008. Regulating the formation of DNA double-strand breaks in meiosis. *Genes Dev.* 22:286–92

107. Murakami H, Keeney S. 2014. Temporospatial coordination of meiotic DNA replication and recombination via DDK recruitment to replisomes. *Cell* 158:861–73

108. Murakami H, Nurse P. 1999. Meiotic DNA replication checkpoint control in fission yeast. *Genes Dev.* 13:2581–93

109. Murakami H, Nurse P. 2000. DNA replication and damage checkpoints and meiotic cell cycle controls in the fission and budding yeasts. *Biochem. J.* 349:1–12

110. Murakami H, Nurse P. 2001. Regulation of premeiotic S phase and recombination-related double-strand DNA breaks during meiosis in fission yeast. *Nat. Genet.* 28:290–93

111. Myers S, Bowden R, Tumian A, Bontrop RE, Freeman C, et al. 2010. Drive against hotspot motifs in primates implicates the *PRDM9* gene in meiotic recombination. *Science* 327:876–79

112. Nabeshima K, Villeneuve AM, Hillers KJ. 2004. Chromosome-wide regulation of meiotic crossover formation in *Caenorhabditis elegans* requires properly assembled chromosome axes. *Genetics* 168:1275–92

113. Nagaoka SI, Hassold TJ, Hunt PA. 2012. Human aneuploidy: mechanisms and new insights into an age-old problem. *Nat. Rev. Genet.* 13:493–504

114. Neale MJ, Pan J, Keeney S. 2005. Endonucleolytic processing of covalent protein-linked DNA double-strand breaks. *Nature* 436:1053–57

115. Nottke AC, Beese-Sims SE, Pantalena LF, Reinke V, Shi Y, Colaiacovo MP. 2011. SPR-5 is a histone H3K4 demethylase with a role in meiotic double-strand break repair. *Proc. Natl. Acad. Sci. USA* 108:12805–10

116. Ogino K, Hirota K, Matsumoto S, Takeda T, Ohta K, et al. 2006. Hsk1 kinase is required for induction of meiotic dsDNA breaks without involving checkpoint kinases in fission yeast. *Proc. Natl. Acad. Sci. USA* 103:8131–36

117. Ogino K, Masai H. 2006. Rad3-Cds1 mediates coupling of initiation of meiotic recombination with DNA replication: Mei4-dependent transcription as a potential target of meiotic checkpoint. *J. Biol. Chem.* 281:1338–44

118. Ohta K, Wu TC, Lichten M, Shibata T. 1999. Competitive inactivation of a double-strand DNA break site involves parallel suppression of meiosis-induced changes in chromatin configuration. *Nucleic Acids Res.* 27:2175–80

119. Okaz E, Arguello-Miranda O, Bogdanova A, Vinod PK, Lipp JJ, et al. 2012. Meiotic prophase requires proteolysis of M phase regulators mediated by the meiosis-specific APC/CAma1. *Cell* 151:603–18

120. Padmore R, Cao L, Kleckner N. 1991. Temporal comparison of recombination and synaptonemal complex formation during meiosis in *S. cerevisiae*. *Cell* 66:1239–56

121. Page SL, Hawley RS. 2003. Chromosome choreography: the meiotic ballet. *Science* 301:785–89

122. Pak J, Segall J. 2002. Role of Ndt80, Sum1, and Swe1 as targets of the meiotic recombination checkpoint that control exit from pachytene and spore formation in *Saccharomyces cerevisiae*. *Mol. Cell. Biol.* 22:6430–40

123. Pan J, Sasaki M, Kniewel R, Murakami H, Blitzblau HG, et al. 2011. A hierarchical combination of factors shapes the genome-wide topography of yeast meiotic recombination initiation. *Cell* 144:719–31

124. Panizza S, Mendoza MA, Berlinger M, Huang L, Nicolas A, et al. 2011. Spo11-accessory proteins link double-strand break sites to the chromosome axis in early meiotic recombination. *Cell* 146:372–83

125. Penkner AM, Fridkin A, Gloggnitzer J, Baudrimont A, Machacek T, et al. 2009. Meiotic chromosome homology search involves modifications of the nuclear envelope protein Matefin/SUN-1. *Cell* 139:920–33

126. Pineda-Krch M, Redfield RJ. 2005. Persistence and loss of meiotic recombination hotspots. *Genetics* 169:2319–33

127. Plug AW, Xu J, Reddy G, Golub EI, Ashley T. 1996. Presynaptic association of Rad51 protein with selected sites in meiotic chromatin. *Proc. Natl. Acad. Sci. USA* 93:5920–24

128. Resnick MA. 1976. The repair of double-strand breaks in DNA: a model involving recombination. *J. Theor. Biol.* 59:97–106

129. Roberts P. 1962. Interchromosomal effects and the relation between crossing-over and nondisjunction. *Genetics* 47:1691–709

130. Robine N, Uematsu N, Amiot F, Gidrol X, Barillot E, et al. 2007. Genome-wide redistribution of meiotic double-strand breaks in *Saccharomyces cerevisiae*. *Mol. Cell. Biol.* 27:1868–80

131. Rocco V, Nicolas A. 1996. Sensing of DNA non-homology lowers the initiation of meiotic recombination in yeast. *Genes Cells* 1:645–61

132. Rockmill B, Lefrancois P, Voelkel-Meiman K, Oke A, Roeder GS, Fung JC. 2013. High throughput sequencing reveals alterations in the recombination signatures with diminishing Spo11 activity. *PLOS Genet.* 9:e1003932

133. Rogakou EP, Pilch DR, Orr AH, Ivanova VS, Bonner WM. 1998. DNA double-stranded breaks induce histone H2AX phosphorylation on serine 139. *J. Biol. Chem.* 273:5858–68

134. Roig I, Dowdle JA, Tóth A, de Rooij DG, Jasin M, Keeney S. 2010. Mouse TRIP13/PCH2 is required for recombination and normal higher-order chromosome structure during meiosis. *PLOS Genet.* 6:e1001062

135. Rosu S, Libuda DE, Villeneuve AM. 2011. Robust crossover assurance and regulated interhomolog access maintain meiotic crossover number. *Science* 334:1286–89

136. Rosu S, Zawadzki KA, Stamper EL, Libuda DE, Reese AL, et al. 2013. The *C. elegans* DSB-2 protein reveals a regulatory network that controls competence for meiotic DSB formation and promotes crossover assurance. *PLOS Genet.* 9:e1003674

137. Rouyer F, Simmler MC, Johnsson C, Vergnaud G, Cooke HJ, Weissenbach J. 1986. A gradient of sex linkage in the pseudoautosomal region of the human sex chromosomes. *Nature* 319:291–95

138. Sasaki M, Lange J, Keeney S. 2010. Genome destabilization by homologous recombination in the germ line. *Nat. Rev. Mol. Cell Biol.* 11:182–95

139. Sasanuma H, Hirota K, Fukuda T, Kakusho N, Kugou K, et al. 2008. Cdc7-dependent phosphorylation of Mer2 facilitates initiation of yeast meiotic recombination. *Genes Dev.* 22:398–410

140. Savitsky K, Bar-Shira A, Gilad S, Rotman G, Ziv Y, et al. 1995. A single ataxia telangiectasia gene with a product similar to PI-3 kinase. *Science* 268:1749–53

141. Schmid R, Grellscheid SN, Ehrmann I, Dalgliesh C, Danilenko M, et al. 2013. The splicing landscape is globally reprogrammed during male meiosis. *Nucleic Acids Res.* 41:10170–84

142. Schwacha A, Kleckner N. 1997. Interhomolog bias during meiotic recombination: meiotic functions promote a highly differentiated interhomolog-only pathway. *Cell* 90:1123–35

143. Sedgwick RP, Boder E. 1991. Ataxia-telangiectasia. In *Handbook of Clinical Neurology*, ed. JMBV de Jong, pp. 347–423. Amsterdam: Elsevier Sci. Publ.

144. Shimmoto M, Matsumoto S, Odagiri Y, Noguchi E, Russell P, Masai H. 2009. Interactions between Swi1-Swi3, Mrc1 and S phase kinase, Hsk1 may regulate cellular responses to stalled replication forks in fission yeast. *Genes Cells* 14:669–82

145. Shin YH, Choi Y, Erdin SU, Yatsenko SA, Kloc M, et al. 2010. *Hormad1* mutation disrupts synaptonemal complex formation, recombination, and chromosome segregation in mammalian meiosis. *PLOS Genet.* 6:e1001190

146. Shuster EO, Byers B. 1989. Pachytene arrest and other meiotic effects of the start mutations in *Saccharomyces cerevisiae*. *Genetics* 123:29–43

147. Simchen G. 2009. Commitment to meiosis: what determines the mode of division in budding yeast? *Bioessays* 31:169–77

148. Smith AV, Roeder GS. 1997. The yeast Red1 protein localizes to the cores of meiotic chromosomes. *J. Cell Biol.* 136:957–67

149. Smith KN, Penkner A, Ohta K, Klein F, Nicolas A. 2001. B-type cyclins *CLB5* and *CLB6* control the initiation of recombination and synaptonemal complex formation in yeast meiosis. *Curr. Biol.* 11:88–97

150. Sommermeyer V, Béneut C, Chaplais E, Serrentino ME, Borde V. 2013. Spp1, a member of the Set1 complex, promotes meiotic DSB formation in promoters by tethering histone H3K4 methylation sites to chromosome axes. *Mol. Cell* 49:43–54

151. Sourirajan A, Lichten M. 2008. Polo-like kinase Cdc5 drives exit from pachytene during budding yeast meiosis. *Genes Dev.* 22:2627–32

152. Stamper EL, Rodenbusch SE, Rosu S, Ahringer J, Villeneuve AM, Dernburg AF. 2013. Identification of DSB-1, a protein required for initiation of meiotic recombination in *Caenorhabditis elegans*, illuminates a crossover assurance checkpoint. *PLOS Genet.* 9:e1003679

153. Steiner WW, Schreckhise RW, Smith GR. 2002. Meiotic DNA breaks at the *S. pombe* recombination hot spot *M26*. *Mol. Cell* 9:847–55

154. Sturtevant AH. 1919. Inherited linkage variations in the second chromosome. In *Contributions to the Genetics of* Drosophila melanogaster, pp. 305–41. Washington, DC: Carnegie Inst. Wash.

155. Suzuki DT. 1963. Interchromosomal effects on crossing over in *Drosophila melanogaster*. II. A reexamination of X chromosome inversion effects. *Genetics* 48:1605–17

156. Sym M, Engebrecht JA, Roeder GS. 1993. ZIP1 is a synaptonemal complex protein required for meiotic chromosome synapsis. *Cell* 72:365–78

157. Szostak JW, Orr-Weaver TL, Rothstein RJ, Stahl FW. 1983. The double-strand-break repair model for recombination. *Cell* 33:25–35

158. Terasawa M, Shinohara A, Hotta Y, Ogawa H, Ogawa T. 1995. Localization of RecA-like recombination proteins on chromosomes of the lily at various meiotic stages. *Genes Dev.* 9:925–34

159. Tessé S, Storlazzi A, Kleckner N, Gargano S, Zickler D. 2003. Localization and roles of Ski8p in *Sordaria macrospora* meiosis and delineation of three mechanistically distinct steps of meiotic homolog juxtaposition. *Proc. Natl. Acad. Sci. USA* 100:12865–70

160. Thacker D, Mohibullah N, Zhu X, Keeney S. 2014. Homologue engagement controls meiotic DNA break number and distribution. *Nature* 510:241–46

161. Tischfield SE, Keeney S. 2012. Scale matters: the spatial correlation of yeast meiotic DNA breaks with histone H3 trimethylation is driven largely by independent colocalization at promoters. *Cell Cycle* 11:1496–503

162. Tonami Y, Murakami H, Shirahige K, Nakanishi M. 2005. A checkpoint control linking meiotic S phase and recombination initiation in fission yeast. *Proc. Natl. Acad. Sci. USA* 102:5797–801

163. Tung KS, Hong EJ, Roeder GS. 2000. The pachytene checkpoint prevents accumulation and phosphorylation of the meiosis-specific transcription factor Ndt80. *Proc. Natl. Acad. Sci. USA* 97:12187–92

164. Ubeda F, Wilkins JF. 2011. The Red Queen theory of recombination hotspots. *J. Evol. Biol.* 24:541–53

165. Viera A, Rufas JS, Martinez I, Barbero JL, Ortega S, Suja JA. 2009. CDK2 is required for proper homologous pairing, recombination and sex-body formation during male mouse meiosis. *J. Cell Sci.* 122:2149–59

166. Vignard J, Siwiec T, Chelysheva L, Vrielynck N, Gonord F, et al. 2007. The interplay of RecA-related proteins and the MND1-HOP2 complex during meiosis in *Arabidopsis thaliana*. *PLOS Genet.* 3:1894–906

167. Wan L, Niu H, Futcher B, Zhang C, Shokat KM, et al. 2008. Cdc28-Clb5 (CDK-S) and Cdc7-Dbf4 (DDK) collaborate to initiate meiotic recombination in yeast. *Genes Dev.* 22:386–97

168. Wan L, Zhang C, Shokat KM, Hollingsworth NM. 2006. Chemical inactivation of Cdc7 kinase in budding yeast results in a reversible arrest that allows efficient cell synchronization prior to meiotic recombination. *Genetics* 174:1767–74

169. Woglar A, Daryabeigi A, Adamo A, Habacher C, Machacek T, et al. 2013. Matefin/SUN-1 phosphorylation is part of a surveillance mechanism to coordinate chromosome synapsis and recombination with meiotic progression and chromosome movement. *PLOS Genet.* 9:e1003335

170. Wojtasz L, Daniel K, Roig I, Bolcun-Filas E, Xu H, et al. 2009. Mouse HORMAD1 and HORMAD2, two conserved meiotic chromosomal proteins, are depleted from synapsed chromosome axes with the help of TRIP13 AAA-ATPase. *PLOS Genet.* 5:e1000702

171. Wu PY, Nurse P. 2014. Replication origin selection regulates the distribution of meiotic recombination. *Mol. Cell* 53:655–62

172. Wu T-C, Lichten M. 1995. Factors that affect the location and frequency of meiosis-induced double-strand breaks in *Saccharomyces cerevisiae*. *Genetics* 140:55–66

173. Xu L, Ajimura M, Padmore R, Klein C, Kleckner N. 1995. *NDT80*, a meiosis-specific gene required for exit from pachytene in *Saccharomyces cerevisiae*. *Mol. Cell. Biol.* 15:6572–81

174. Xu L, Kleckner N. 1995. Sequence non-specific double-strand breaks and interhomolog interactions prior to double-strand break formation at a meiotic recombination hot spot in yeast. *EMBO J.* 14:5115–28

175. Xu Y, Ashley T, Brainerd EE, Bronson RT, Meyn MS, Baltimore D. 1996. Targeted disruption of *ATM* leads to growth retardation, chromosomal fragmentation during meiosis, immune defects, and thymic lymphoma. *Genes Dev.* 10:2411–22

176. Yokoo R, Zawadzki KA, Nabeshima K, Drake M, Arur S, Villeneuve AM. 2012. COSA-1 reveals robust homeostasis and separable licensing and reinforcement steps governing meiotic crossovers. *Cell* 149:75–87

177. Zenvirth D, Loidl J, Klein S, Arbel A, Shemesh R, Simchen G. 1997. Switching yeast from meiosis to mitosis: double-strand break repair, recombination and synaptonemal complex. *Genes Cells* 2:487–98

178. Zetka MC, Rose AM. 1992. The meiotic behavior of an inversion in *Caenorhabditis elegans*. *Genetics* 131:321–32

179. Zhang L, Kleckner NE, Storlazzi A, Kim KP. 2011. Meiotic double-strand breaks occur once per pair of (sister) chromatids and, via Mec1/ATR and Tel1/ATM, once per quartet of chromatids. *Proc. Natl. Acad. Sci. USA* 108:20036–41

180. Zhang L, Liang Z, Hutchinson J, Kleckner N. 2014. Crossover patterning by the beam-film model: analysis and implications. *PLOS Genet.* 10:e1004042

Cancer: Evolution Within a Lifetime

Marco Gerlinger,[1,5,*] Nicholas McGranahan,[1,2,*]
Sally M. Dewhurst,[1,*] Rebecca A. Burrell,[1]
Ian Tomlinson,[3,4] and Charles Swanton[1,6]

[1]Cancer Research UK London Research Institute, London, United Kingdom WC2A 3LY;
email: charles.swanton@cancer.org.uk

[2]Centre for Mathematics & Physics in the Life Sciences & Experimental Biology (CoMPLEX),
University College London, London, United Kingdom WC1E 6BT

[3]Molecular and Population Genetics Laboratory, Wellcome Trust Centre for Human Genetics,
University of Oxford, Oxford, United Kingdom OX3 7BN; email: iant@well.ox.ac.uk

[4]Oxford National Institute for Health Research (NIHR) Comprehensive Biomedical Research
Centre, Wellcome Trust Centre for Human Genetics, University of Oxford, Oxford,
United Kingdom OX3 7BN

[5]Present address: Translational Oncogenomics Lab, Centre for Evolution and Cancer, The
Institute of Cancer Research, London, United Kingdom SW3 6JB

[6]University College London Hospital and Cancer Institute, CRUK Lung Cancer Centre of
Excellence, London, United Kingdom WC1E 6DD

Annu. Rev. Genet. 2014. 48:215–36

First published online as a Review in Advance on
October 1, 2014

The *Annual Review of Genetics* is online at
genet.annualreviews.org

This article's doi:
10.1146/annurev-genet-120213-092314

*These authors contributed equally.

Keywords

genome instability, drug resistance, precision medicine, cancer evolution,
intratumor heterogeneity

Abstract

Subclonal cancer populations change spatially and temporally during the disease course. Studies are revealing branched evolutionary cancer growth with low-frequency driver events present in subpopulations of cells, providing escape mechanisms for targeted therapeutic approaches. Despite such complexity, evidence is emerging for parallel evolution of subclones, mediated through distinct somatic events converging on the same gene, signal transduction pathway, or protein complex in different subclones within the same tumor. Tumors may follow gradualist paths (microevolution) as well as major shifts in evolutionary trajectories (macroevolution). Although macroevolution has been subject to considerable controversy in post-Darwinian evolutionary theory, we review evidence that such nongradual, saltatory leaps, driven through chromosomal rearrangements or genome doubling, may be particularly relevant to tumor evolution. Adapting cancer care to the challenges imposed by tumor micro- and macroevolution and developing deeper insight into parallel evolutionary events may prove central to improving outcome and reducing drug development costs.

INTRODUCTION

Despite considerable progress in the mapping of the coding sequence of the human genome and rapid developments in our understanding of somatic aberrations that occur in cancers from different anatomical sites, the vast majority of metastatic solid tumors remain incurable (4). Although primary organ-confined disease in colorectal, lung, breast, and prostate carcinoma have an average five-year survival rate of 85%, metastatic disease in the same tumor types have an average five-year survival rate of merely 17% (from 2001 to 2007).

The challenge of improving survival outcomes in the metastatic setting has also contributed to the high failure rates in oncological drug development for targeted and cytotoxic therapies. This has been estimated to be three times higher than for cardiovascular medicines (74). An important component of oncology drug development is the parallel development of companion diagnostics to ensure that the correct drug is offered to a molecularly defined tumor subtype at the right time, enabling so-called stratified, precision, or personalized medicine. However, despite the promise heralded by the advent of high-throughput tumor analytical techniques, biomarker validation and clinical qualification of such biomarkers remain expensive tasks and frequently fail. Indeed, it has been estimated that less than 1% of all reported cancer biomarkers actually enter clinical practice (76).

The almost invariable development of drug resistance in the advanced disease setting combined with the difficulties in biomarker research can be explained by viewing cancer as an evolutionary process. Heritable alterations such as mutations or DNA methylation changes can be stochastically acquired by individual cancer cells. Alterations that increase cancer cell fitness can then drive the expansion of a subclone in the cancer cell population, whereas less fit subclones are likely to diminish in frequency and may go extinct. Thus, the same process of Darwinian evolution that drives the evolution of species in the wild can also drive the evolutionary adaptation of cancer. This cancer evolution has predominantly been depicted as a gradual linear process in which successive heritable alterations provide successive increases in fitness. In this review, we explore cancer within a micro- and macroevolutionary framework, highlighting the ability of cancer evolution to generate intratumor heterogeneity (ITH) through branched evolution, which may explain current shortcomings in the development of biomarkers and novel therapies.

Crucially, studies attempting to reveal specific genetic events driving key cancer phenotypes, such as metastatic ability and multi-drug resistance, have generally failed. The phenotypes associated with both metastasis and drug resistance are presumed to be complex, requiring metastatic cancer cells to migrate and to survive in novel environments and multi-drug-resistant cells to overcome the action of multiple drugs acting through distinct mechanisms. Conceivably, these phenotypes may not be achievable through single gene alterations but may require a complex pattern of cooperating genomic aberrations.

Similar difficulties in explaining the evolution of new species or radically different body plans, in the absence of intermediate forms in the fossil record, have led to the development of evolutionary concepts of large jumps, or saltations. These concepts were most notably championed by the evolutionary theorist Richard Goldschmidt, who posited that evolution could not be understood without reference to macroevolutionary events such as chromosomal rearrangements (52). Similar chromosomal rearrangements to the ones Goldschmidt postulated have long been observed in tumor cells (65, 81, 87, 141), and recent studies have highlighted the occurrence of single catastrophic events in tumor evolution (128, 154), potentially lending credence to saltationary theories in cancer.

We summarize recent progress in understanding the mechanisms driving tumor evolution, including the potential role of large-scale changes and evolutionary saltations. Finally, we discuss the clinical implications of tumor evolution and diversity within tumors.

CANCER WITHIN AN EVOLUTIONARY FRAMEWORK

The application of Charles Darwin's central hypotheses to cancer development and progression was first formally proposed by Nowell (112). Since this seminal publication, an overwhelming body of evidence has been collected demonstrating that cancers evolve during progression from a monoclonal founder clone to a clinically detectable cancer, and, in some cases, cancer evolution results in tumor recurrence and the emergence of drug-resistant disease (comprehensively reviewed in References 49, 56, and 150). In classical Darwinian theory, heritable population-level diversity provides a substrate for adaptation and selection brought about by competition for limited environmental resources. Within a tumor, if heterogeneity results in differences in fitness between cancer cells and this fitness influences survival and is heritable, such phenotypes may be subject to selection.

Types of Cancer-Causing Mutations: Drivers and Passengers

To reflect their evolutionary relevance, somatic genetic aberrations in cancer cells, from base substitutions to chromosomal-scale changes, can be categorized according to their contribution to cancer cell fitness. Driver aberrations confer a selective advantage, whereas passenger aberrations have negligible effect on cellular fitness at a particular time point in the tumor's life history.

In general terms, driver aberrations are those that play a significant role in promoting tumorigenesis or are required for tumor growth. Many driver genes have been identified in individuals with familial cancer-causing syndromes who carry germline mutations in these genes, and their functional relevance has been further defined in genetic animal models. Cancer drivers can also be identified because the affected genes or genomic regions are frequently mutated, rearranged, or subject to copy number gains or losses in a particular tumor type (81, 144). The number of driver mutations that are necessary to initiate a specific cancer and the number of drivers that are typically operative in an advanced cancer currently remain unclear. The lack of robust methods for the identification of driver aberrations as well as incomplete tumor sampling strategies contribute to this lack of clarity (47). Most approaches rely on statistical methods based on the classical McDonald-Kreitman test, which identifies genomic features that are more frequently altered than would be expected by chance, suggestive of positive evolutionary selection. However, the variability of mutation rates in different regions of the genome and the high mutational load in some cancers limit the application of this test for studies of hundreds of cancers. Novel statistical methods that adjust for local mutation loads (64) or consider the impact of replication timing and the transcriptional activity of individual genes, which both influence local mutation rates (82), have improved on some of these shortcomings. Applying such methods, an analysis of 4,742 tumor samples suggests the spectrum of driver mutations that occur in at least 2% of samples within each tumor type may be revealed by sequencing between 600 and 5,000 samples (81). Difficulties in the identification of subclonal drivers due to emerging evidence for spatial heterogeneity (48) and the tendency of apparently similar cancers to follow a number of quite different genetic pathways (144) are further reasons for the incomplete characterization of driver aberration landscapes. New sampling strategies combined with novel statistical approaches are necessary to detect subclonal mutations, some of which may be spatially separated within the same tumor.

With drivers come a much larger number of passenger mutations. Because every cell that eventually becomes a cancer has continually acquired mutations—starting from its origins as a zygote and continuing during development, adult homoeostasis, and tumorigenesis—all the pre-existing mutations in a cell with a driver mutation will hitchhike with that driver. Studies in acute myeloid leukemia (AML) (142) and mathematical modeling of mutation acquisition in cancers of self-renewing tissues (138) have revealed that the majority of passenger mutations were present before the cell of origin was transformed into a cancer cell. Passenger mutations can also be acquired during cancer progression, as illustrated by sequencing of multiple regions from each of ten clear-cell renal carcinomas. A small number of driver mutations was outnumbered by a much larger number of somatic mutations, which were likely to be passenger alterations on the trunk of the phylogenetic trees (describing mutations that had already been present in the most recent common progenitor clone of all tumor regions) and the branches (depicting mutations confined to cancer subclones) of the tumors' phylogenetic trees (48). Many of the truncal passenger mutations may have been acquired before transformation, whereas those on the branches.of the phylogenetic trees were acquired during tumor progression.

This binary classification of mutations into drivers or passengers reflects notions of tumor growth being driven by a few mutations of large effect, in turn reflecting the multistage cancer incidence models of Doll (6), Moolgavkar (103), and others. However, cancer-causing mutations of smaller effects may exist that act in addition to the effects of driver mutations. In fact, in an extreme situation, multiple mini-driver mutations might substitute for a few drivers (86). For this model to be favored, the pool of mini-drivers must be considerably larger than that of the drivers. However, it may explain why the number of known driver genes for common malignancies, such as prostate (9), breast (38), and high-grade serous ovarian cancer (19), remains surprisingly low. Alternatively, such tumors could predominantly be driven by copy number events and other structural rearrangements, as suggested for ovarian and prostate cancer (19, 55, 132).

Finally, adding an extra layer of complexity, it should be recognized that the distinction between driver and passenger mutations in a tumor may be a dynamic one, as the most advantageous (fittest) genotypes are not the same in all cancers or at all times or places in the same cancer because the selective advantage of any genotype is dependent on the environment (149). Environment must be defined broadly to include the cell's repertoire of gene and protein expression, the mutations and epigenetic changes already present in that cell, the local external microenvironment, distantly acting factors (such as hormones), and factors external to the host (such as carcinogens and exogenous therapies). For example, late-stage tumors may not always rely on an early driver event, with passenger events in the branches perhaps becoming driver events in these tumors (41, 77, 149). For some driver mutations, environmental variation may matter less because such mutations almost invariably confer a selective advantage. However, even here, there are selective constraints. For example, although p53 mutations occur in many cancer types, they are not always initiating mutations and sometimes are selected only once a degree of tumor growth and/or progression has occurred (117). In other cases, however, environmental variation may have strong influences on which genetic changes are selected. This is illustrated by the fact that many cancer genes only lead to tumors if they are mutated in the germline (113).

Cancer Is Evolution in an Expanding Population

Evolution is usually considered in a population of a fixed size in which resources are scarce. Often, this may be an accurate reflection of cancer. However, the cancer cell population overall is an expanding one. Here, some of the normal rules of mutation and selection may not apply. In expanding populations, the effects of genetic drift, which refers to the random loss or fixation of genotypes in small populations, can be magnified, and even deleterious mutants can surf on the wave front of expansion, causing high levels of genetic diversity (60). A result is that clonal replacement may not occur when new, advantageous mutations arise and are selected; hence, branching tumorigenesis may be favored. Conceivably, an expanding population of cells in a cancer may allow genotypes that are suboptimal to survive when they would not otherwise do so. Such a situation may help to maintain small, genomically unstable clones with no intrinsic selective advantage until they have acquired additional, advantageous mutations, thereby propagating diversity.

Elevated Mutation Rates in Cancer Cells

The number of mutations in any cell is determined by its age, the mutation rate, and its fate, all resulting from the effects of selection acting on the phenotypic consequences of each mutation. Although random mutations might occasionally be selectively advantageous, most mutations are likely to be either neutral or result in a reduction in fitness through direct effects on gene function or by triggering cell death pathways. Thus, a newly arisen, hypermutant tumor cell is at risk of being lost by drift or selection until it acquires its first selectively advantageous mutation(s); subsequently, genomic instability is carried along as a hitchhiking phenotype, further propagating diversity. The normal mutation rate has evolved through selection at the level of the organism to prevent transmission of deleterious mutations through the germline and to reduce the risk of cancer prior to reproductive age. However, selection in carcinogenesis principally acts at the level of the cell and thus the optimal mutation rate may be much higher than that for the whole organism. Cell-intrinsic genomic instability processes, such as mismatch repair deficiency, and exposure to carcinogens, such as cigarette smoke, increase the mutation rate and with this the probability for cancer to develop. Also concordant with this view, extrapolations from exome and kinome sequencing studies suggest that cancer cells in some tumor types may carry more than one million base-substitution mutations in the genome (20, 67). Elevated mutation rates provide greater variation across the population for different selection pressures to act upon, the potential benefits of which can be readily observed in bacteria. Co-culture experiments of bacterial strains with a mutator phenotype and wild-type strains revealed that the mutator strain has a selective advantage compared with wild-type bacteria in constant (51) and fluctuating selective environments (92). Similarly, more karyotypically diverse cell populations have been shown to have a greater tumorigenic capacity (151). However, very high mutation rates probably have deleterious effects and are likely to reduce cell viability. This has led to the concept of "just right" levels of genomic instability (17), which allow the acquisition of sufficient genetic aberrations for carcinogenesis and cancer progression but avoid the accumulation of too many deleterious alterations. Evidence from our group and others in retrospective clinical cohorts has begun to validate the presence of such "just right" thresholds of genomic instability in solid tumors (12, 119).

Branched Versus Linear Tumor Evolution

Conventionally, tumorigenesis has been thought to proceed in a linear fashion in which successive advantageous driver mutations arise within a clone, which then replaces less fit clones in the tumor in selective sweeps. Increasingly, however, the identification of genetic ITH and the reconstruction of ancestral relationships of individual subclones have revealed branched cancer evolutionary trajectories in which ubiquitous events are located on the trunk and heterogeneous alterations on the branches of a phylogenetic tree (**Figure 1**). Colon adenomas and carcinomas (135), acute leukemias (5, 111), pancreatic carcinomas (18), breast carcinomas (105, 106, 110), clear-cell renal carcinomas (48, 148), prostate cancers (59), gliomas (71), and medulloblastomas (147), among others, are increasingly seen to evolve through branched trajectories. The identification of distinct driver aberrations in several coexisting subclones has provided evidence for both functional diversification and parallel evolution in human cancers (47, 48). Branched evolution may represent adaptation to different microenvironments. However, branched evolution may also

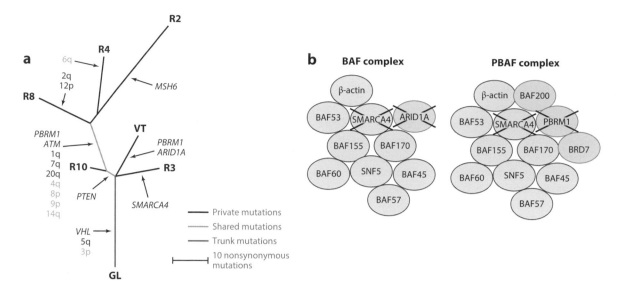

Figure 1

Branched and parallel evolution in a renal cell carcinoma of the clear-cell subtype (ccRCC). (*a*) The phylogenetic tree was inferred from the nonsynonymous somatic mutations detected by exome sequencing of six different regions from a surgically removed ccRCC. The tree is rooted at the germline (GL) DNA sequence, determined by sequencing of DNA from peripheral blood. All analyzed regions from the primary tumor (R2–R10) and from a tumor thrombus that grew from the primary tumor into the renal vein (VT) are shown. The branch lengths are proportional to the number of nonsynonymous mutations located on each branch. Genes that have acquired putative driver mutations are indicated with arrows on the branches where they occurred. ccRCC driver copy number alterations acquired by each branch are also indicated with chromosome losses in green and chromosome gains in purple. Only one truncal driver mutation in the *VHL* tumor suppressor gene and two truncal driver copy number aberrations affecting the 3p and 5q chromosome arms were detected. The remaining driver mutations and driver copy number aberrations were heterogeneous, indicating the presence of spatially separated tumor subclones and branched evolution. Parallel evolution occurred in this tumor through two distinct and spatially separated driver mutations in the *PBRM1* gene and driver mutations in the *ARID1A* and *SMARCA4* genes, which all encode members of the SWItch/Sucrose non Fermentable (SWI/SNF) chromatin remodeling complex. Adapted with permission from Reference 47. (*b*) Schematic of the BRG1-associated factor (BAF) and the Polybromo BRCG-1-associated factor (PBAF) multiprotein complexes, which are the major subclasses of the SWI/SNF complex (143). Subclass specific members are in green. Proteins affected by driver mutations are marked by a cross.

result from the possibility that there are several different genetic routes from a normal cell to late-stage cancer.

The majority of cancer types have not been subjected to detailed scrutiny and the overall prevalence of ITH and branched evolution remains unknown. Some tumor types can develop in linear or branched manners (5), showing that branched evolution is not a prerequisite for tumor development and ITH may not be a universal phenomenon. Consistent with this, estrogen receptor–positive breast cancers may be genetically relatively homogeneous and stable, based on the genetic similarity of temporally separated biopsies (38).

Subclones that evolve independently may differ in their impact on patient survival and in their sensitivity to cancer drugs. For example, the lethal clone in a metastatic prostate cancer emerged from a small, low-grade subclone that was hardly detectable at diagnosis within the predominantly high-grade primary tumor (59). Failure to accurately detect such subclones may hinder the development of better prognostic and predictive markers for personalized cancer medicine approaches. Thus, understanding spatial and temporal genetic heterogeneity is likely to be central to improving outcomes in oncology and optimizing biomarkers for clinical use.

Heterogeneity Within and Between Tumor Biopsies and Parallel Evolution

Patterns of ITH can take the form of intermingling subclones, which are detectable within tumor samples through suitable techniques such as deep sequencing and in situ molecular analyses. This pattern has been observed in ER-negative, Her2-negative breast cancer (110, 123); glioblastoma (125); and non-small-cell lung cancer (54) as well as in a minority of tumor regions sampled in clear-cell carcinoma of the kidney (47). In glioblastoma, subclones harboring an epidermal growth factor receptor (EGFR) mutation have been shown to cooperate with EGFR wild-type subclones, actively promoting heterogeneity with intermingling clones (68), and both *Drosophila melanogaster* (146) and mouse studies (25) suggest genetically distinct subpopulations can cooperate to promote tumor growth, progression, and maintenance. Such interdependencies may be exploitable for therapeutic purposes if the effective eradication of one subclone also leads to impaired viability of a dependent clone (39). However, these data also suggest that minority populations may sustain the viability of dominant clones within the tumor and imply ITH may be necessary for certain tumors to evolve.

In other tumors, the subclones may be clearly demarcated within and between primary tumors and metastatic sites (18, 32, 48, 122). The causes of the spatial separation of subclones within individual tumors have not been investigated in detail and bear striking resemblance to allopatric evolutionary events in ecology. There may be true physical barriers preventing intermixing of tumor subclones, such as an extracellular matrix, new blood vessel growth, and distinct metastatic niches. Such spatial separation of subclones may also be the result of variable microenvironmental selection pressures or due to the balance between tumor proliferation rate and the severity of ongoing genome instability. In addition, as discussed above, if cancer driver mutations occur independently in different parts of the tumor and each subclone (or evolutionary branch) is expanding and environmental resources are not limiting, it may be difficult for any one clone to displace all the others (60), particularly if these subclones have similar fitness advantages.

Data pertaining to intermetastasis heterogeneity are currently limited, but comparative lesion sequencing revealed highly similar somatic mutations of multiple metastatic sites in a patient with prostate cancer (59) and a patient with renal cancer (48) compared with clonal heterogeneity within the corresponding primary tumors. This suggests that a single clone within these primary tumors acquired the ability to metastasize efficiently. The genetic divergence that was present in both cases between the primary and metastatic sites probably results from a combination of founder effects and subsequent genetic drift in a small population (leading to the random fixation of often neutral mutations), coupled with selection for new driver mutations subsequent to spread. By contrast, Haffner and colleagues also observed that the metastatic pelvic lymph node of a prostatic carcinoma differed markedly from other sites of metastatic disease, suggesting that metastases arose from at least two separate subclonal events (59). These case studies highlight the importance of studying larger series in order to reveal whether metastases in some tumor types are similar to each other, arising from the dissemination of one subclone, or whether multiple dissemination events occur from a heterogeneous primary tumor and/or metastases.

Despite such diversity, it is becoming clear that evolution is constrained, with limitations to cancer evolutionary trajectories. Evidence for this is observed in clear-cell carcinoma of the kidney, where geographically separated subclones harbor distinct mutations in the same gene or genes within the same signal transduction pathways or protein complex (47, 48) (**Figure 1**). Recently, evidence of epistatic pair-wise gene associations was demonstrated in myelodysplastic syndrome, indicating that one mutation likely heralds the onset of the next (116). Such genetic contingencies, once understood, suggest intriguing future drug development strategies targeting the tumor's next evolutionary move.

Parallels with Microbial and Ecological Evolution

Cancer cells lack the capacity for intergenomic recombination that is characteristic of sexually reproducing organisms. Thus, cancer cell evolution may be akin to the clonal evolution of microbial populations, and as a result it has been suggested that microbial evolution experiments can foster our understanding of cancer evolution (29, 127, 133). Such experiments in which bacterial populations were grown in homogeneous selection environments have revealed important insights into the evolution of diversity (94, 95). Phenotypic diversity frequently evolved in these bacterial populations, consistent with branched evolution rather than the selection of a single highly adapted clone. Applied to cancer, this may suggest that a population of cancer cells may not converge toward a single fitness maximum but that selection pressures encountered when nutrient supplies are limited or during drug therapy, for example, could foster branched evolution. A spatially heterogeneous microenvironment can further accelerate evolutionary diversification in bacteria (78), potentially explaining the higher intratumoral diversity found in solid tumors compared with hematological cancers.

The micro-environment in which cancer cells live can be considered as complex as the environments in which organisms evolve. For instance, while organisms may be subject to predation, the cancer cell is exposed to attack by, among other things, predatory immune cells. Recent work has suggested that cancers that fail to fully evade the predation of the immune system by harboring mutations that yield neo-antigens have improved prognosis compared to cancers without immunogenic mutations (15).

Cancer Stem Cells and Tumor Evolution

Cellular hierarchies are relevant in some hematological and solid cancers, as demonstrated by the discovery of cancer stem cells (14). These are critical for cancer maintenance, as the transit-amplifying progeny are thought to have a limited replicative potential. This hierarchical stem cell–based model is often thought to conflict with the stochastic Darwinian model of tumor evolution, but recent theoretical advances have aimed to reconcile both models (for a review see 100). This cellular hierarchy not only increases nongenetic phenotypic ITH, e.g., through the intrinsic resistance of cancer stem cells to cancer therapeutics in some tumor types (22, 43), but may also affect cancer evolution, as only mutations occurring in a cancer stem cell can be propagated over a longer period of time. Thus, only the cancer stem cells can be considered to be under Darwinian selection (57). Only the size of the cancer stem cell pool and the stochastic genetic heterogeneity within this pool contribute to the evolutionary potential of cancers conforming to this cancer stem cell model. Computer modeling studies of epimutations (defined as heritable and stochastic epigenetic changes) in such hierarchically organized tumors revealed that they are likely to be more heterogeneous than tumors without a stem cell compartment because of reduced competition between intratumoral subclones, rendering them more adaptable to selection pressures, such as cancer therapy (126). Further studies are clearly needed to develop a refined model of evolution in tumors driven by cancer stem cells.

MECHANISMS DRIVING TUMOR EVOLUTION

Types of Genomic Instability in Cancer

Many cancers display evidence of genomic instability, resulting in an elevated rate of somatic mutations and structural or copy number aberrations. The most common forms of genomic instability are defective mismatch repair and chromosomal instability. Defects in mismatch repair result in highly elevated rates of frameshift and/or point mutations (40, 115). Chromosomal instability is characterized by extensive structural and numerical karyotypic heterogeneity (88) and may arise through multiple mechanisms, including DNA replication stress, DNA double-strand break repair defects, and improper chromosome attachments during mitosis (reviewed in 137). Other forms of instability, such as defects in nucleotide and base excision repair, are also observed in some tumors, although these appear to be rare in sporadic malignancies. Genomic instability may exist at the base-pair and chromosome level simultaneously, which may be attributable to one or multiple independent mechanisms (108, 110, 140). For a more detailed review on specific mechanisms contributing to genetic diversity in solid tumors, see Reference 16.

The extent and pattern of instability may vary within individual tumors and subtypes and between tumors of different tissue types over time. For example, microsatellite instability is observed more frequently in certain tissue types, such as colorectal and endometrial cancers. As different patterns of instability have different impacts upon patient prognosis and drug response (reviewed in 99), heterogeneity in genomic instability may explain some of the observed variation in outcome between different tumor types (16).

The observation that many hereditary cancer predisposition syndromes can be attributed at least in part to increased genomic instability (50, 83) has led to suggestions that genomic instability may be required for carcinogenesis (91). However, in principle a cancer could fortuitously acquire all the necessary mutations at a normal mutation rate given a relatively high rate of clonal expansion (139), and this may explain observations of an absence of overt genomic instability in some tumors (58, 85).

Alternatively, unstable clones may be selectively disadvantaged. Although the same pathways that are involved in hereditary malignancies appear to be inactivated or disrupted (albeit infrequently) in sporadic malignancies [for example, inactivation of MLH1 in colon cancer or of BRCA1/2 in ovarian cancers (61, 62)], it is not clear whether they play a role in initiating these tumors. Indeed, several studies have suggested that genomic instability may be a relatively late-onset phenotype in sporadic cancers and is involved in tumor progression rather than initiation (13, 72, 110, 145, 153). The beneficial effect of genomic instability may lie in the provision of new phenotypes that allow continual evolution in the face of new selection pressures.

Mutational Signatures of Genomic Instability Processes

Detailed analysis of the catalog of somatic mutations identified in a cancer genome can shed light on the genome instability processes that operate during the life history of a tumor. Each mutation identified in a cancer genome represents the result of an endogenous or exogenous genomic instability process that may drive tumor evolution.

Consistent with the mutagenic action of carcinogens in tobacco smoke, the tumor genomes of smokers exhibit significantly greater numbers of mutations, in particular $C > A$ transversions, compared with nonsmokers (54). Similarly, melanomas display an elevated rate of $C > T$ and $CC > TT$ mutations, which is linked to exposure to UV radiation (64). Endogenous genomic instability processes can also be associated with particular mutation types; for example, endometrial cancers with mutations in *POLE* display an elevated frequency of transversions (23), whereas upregulation of APOBEC cytidine deaminases, which normally function in innate immune responses, are thought to lead to $C > T$ and $C > G$ mutations, particularly at TpC sites. However, the underlying etiologies are unknown for many identified mutational signatures (2), and the temporal dynamics of the processes generating mutations remain unclear.

Punctuated or Catastrophic Genomic Events

It is important to distinguish between ongoing genomic instability, which results in accumulation of mutations over time, as described above, and transient catastrophic genomic rearrangement, in which mutations are acquired in sudden bursts. The development of large-scale sequencing technology has enabled hitherto unprecedented examination of patterns of mutation in cancer genomes, and it was through this technology that the recently described phenomena of chromothripsis, chromoplexy, and kataegis were revealed (3, 8, 11, 18, 90, 109, 110, 129).

Kataegis describes a catastrophic event involving localized hypermutation, preferentially involving cytosine at TpC dinucleotides (109). The APOBEC enzyme has been suggested to be responsible for this mutational phenomenon (79, 109). In some instances, kataegis has been observed to be associated with chromothripsis, a phenomenon in which one or a few chromosomes are fragmented and grossly rearranged in a manner likely to have occurred during a single catastrophic cell cycle (128). Evidence of chromothripsis was found in at least 2–3% of all cancers and up to one-quarter of bone cancers. Mechanisms responsible for chromothripsis remain unclear but may involve replication defects (90) or fragmentation of micronuclei (26). Chromoplexy is another distinct single catastrophic event inferred from detailed genomic analysis of prostate cancers, resulting in chains of translocations across multiple chromosomes, likely arising from double-stranded DNA breaks (8).

A specific advantage of such profound and presumably punctuated genomic alterations may be the ability to alter multiple cancer drivers in a single event, as shown by Stephens (128). This could precipitate major evolutionary leaps of the tumor or allow the cancer cell to access fitness maxima not achievable by sequential alterations, analogous to macroevolutionary events proposed by Goldschmidt (52).

Given that tumors are often thought to take many years to develop, it is likely that punctuated genomic alterations alone are not sufficient to cause cancer, but may act as an important rate-limiting step. As such, punctuated evolution is not inconsistent with a cancer taking years to develop. However, punctuated evolution is inconsistent with a constant mutation rate during cancer development, one of the fundamental assumptions in modeling cancer developmental timeframes.

Genome Doubling as a Mediator of Tumor Evolution

Tetraploidy may be important as an intermediate stage leading to aneuploidy in cancers (45, 120) and to promote genetic diversity and cancer progression. This is supported by studies that have shown that tetraploidy can induce tumorigenesis and affect the stability of the genome (27, 28, 34, 44, 107). In a seminal study, tetraploid cells formed tumors in nude mice containing numerical and structural chromosome aberrations, in contrast to diploid cells, which did not form tumors (44). Tetraploid cells generated through cell fusion events and oncogene overexpression also induce karyotypically diverse tumors in mice (34, 107), whereas the continual passage of tetraploid mouse ovarian cells results in aneuploid progeny (93). We have found that cancer cells that tolerate a genome doubling event also tolerate ongoing chromosome segregation errors compared with non-genome-doubled cells, resulting in increased chromosomal complexity over 18 months in culture, with genome-doubled cells evolving specific copy number variations over time (30).

Our recent analysis of multiple tumor regions from a patient with kidney cancer also revealed that the primary tumor region most similar to the chromosomally unstable metastatic sites comprised a tetraploid cell population, in contrast to the rest of the primary regions, which were diploid (48). This could support a model in which the aneuploidy that can arise following genome doubling gives rise to multiple complex phenotypes, one of which may permit metastasis.

Tetraploidy may facilitate adaptation of tumors in times of stress. The polyploid cell fraction can increase under stress in human tissues (27), such as in hepatocytes exposed to oxidative stress (53). The exact effect of polyploidy in these situations is unknown, but cycles of polyploidization and depolyploidization can generate genetically diverse progeny, which has been suggested as a mechanism to increase tolerance to the exogenous stress often encountered in the liver (37). Growth of liver nodules after induced hepatic injury in mice was due to specific aneuploidy of chromosome 16 in growing nodules (36), supporting a beneficial role for aneuploidy in resistance to environmental stress. The polyploid state could be associated with phenotypes that directly provide a survival advantage for cancer cells as well as additional benefits brought about by the resultant genetic diversity.

Whole-Genome Doubling in Evolutionary Biology: Parallels with Cancer

Work in yeast, plants, and other model systems may shed light on the role of genome doubling in cancer. Polyploidy may represent an evolutionary track distinct from the gradualist view of evolution. In plant evolution, polyploidy has been referred to as a saltation, i.e., a large-scale change from one generation to the next that is beyond the normal variation of that organism in the space of a generation (42).

Whole-genome doubling (WGD) events have been crucial in the evolution of many different complex traits in lower and higher eukaryotes (66). In *Arabidopsis thaliana*, three distinct WGD events have occurred in its evolutionary history (124) that appear to favor expansion of particular metabolic networks and the retention of duplicate enzymes for high-flux reactions (10). Genome-doubling events can facilitate novel gene functions, contributing to functional divergence of duplicated genes (1, 66). Differential evolutionary capacity of existing and newly synthesized tetraploids compared to diploids has been explored using selection for early flowering time in fireweed (*C. angustifolium*) (98). Newly synthesized tetraploids most rapidly evolved a shorter flowering time in response to an applied selection pressure, potentially due to chromosomal rearrangements and epigenetic changes after polyploidization or through altered chromosome pairing (98). Budding yeast (normally haploid) has also been used to explore the differential ability of diploid (polyploid) versus haploid yeast to adapt to novel environments, revealing that beneficial effects of an increased mutation rate are enhanced by genome doubling (136). This is likely due to the fact that recessive deleterious mutations do not affect diploid yeast to the same extent as haploids. However, haploid nonmutators were the most fit of all groups tested, suggesting that many mutations are recessive, and therefore will more rapidly benefit a haploid organism. If diploid evolution is constrained by the amount of variation, then polyploidy could enhance the capacity for genetic diversity by increasing the opportunity to mutate and change (114). The disadvantage is that recessive beneficial alleles are likely to be masked in a polyploid population.

Evidence from our own work exploring the mutational profiles of a renal tumor supports the theory that polyploidy can act as a buffer to deleterious mutations (48). The ratio of synonymous mutations to nonsynonymous mutations was lower in the polyploid-aneuploid branch of this tumor than in the diploid branch, perhaps suggesting a buffering effect for nonsynonymous mutations in polyploid cells. Genome doubling could thus provide a mechanism to mitigate Muller's ratchet (104), which refers to the gradual accumulation of deleterious mutations in asexually reproducing organisms that eventually impairs fitness. Genome-doubling events may reduce the impact of somatic genetic events with damaging phenotypes or allow the cell to lose such alterations entirely (through ensuing structural or numerical chromosomal instability), effectively compensating for disadvantageous effects of past events. Because diploids may benefit faster than polyploids from any beneficial mutations [as seen in bacteria (136)], there could be selection for a return to diploidy.

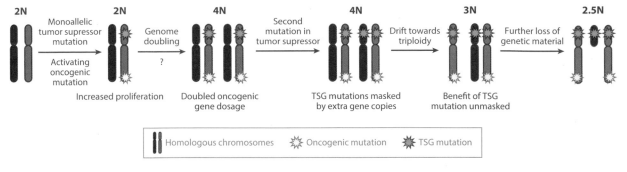

Figure 2

Representation of the effects of genome doubling on mutational load. One chromosome is represented for simplicity. Initially, both oncogenic and monoallelic tumor suppressor gene (TSG) mutations are present. After a genome-doubling event, the oncogenic gene dosage is doubled. However, the beneficial effect of further TSG mutations may be masked by the additional copies of the wild-type allele present in the cell. This could potentially select for loss of genetic material after the genome-doubling event to a triploid or near-diploid state.

This may in part explain why multiple cancer types show evidence of a previously doubled genome, yet have returned to a near-diploid genome (21, 30) (**Figure 2**). In summary, the polyploid state may allow cancer cells to evolve more complex traits and increase the genotypic space available for evolutionary adaptation (66).

Considering Micro- and Macroevolution

There is a broad spectrum of genomic mechanisms driving cancer evolution, emphasizing the need to consider microevolutionary events (such as driver genes) as well as large-scale evolutionary jumps (such as whole-genome doublings and chromosomal rearrangements) in order to understand tumor evolution.

It is conceivable that the evolution of some tumors can be reduced to mutations in specific driver genes, whereas other cancers can only be understood by considering large-scale alterations and copy number events. The existence of such a dichotomy in the evolutionary trajectories of different tumors is supported by a recent pan-cancer analysis that defined an M class of tumors, predominantly harboring somatic point mutations, and a C class, which preferentially exhibited copy number alterations (24).

Although the majority of large phenotypic leaps are deleterious, the few beneficial ones may contribute more to adaptation than mutations with small fitness effects, analogous to Goldschmidt's "hopeful monsters" (52). Thus, larger phenotypic leaps may be particularly important for evolutionary acquisition of metastatic and multi-drug-resistant phenotypes that cannot be reduced to single genetic alterations (35, 63, 84, 73) (**Figure 3**).

CLINICAL IMPLICATIONS OF CANCER EVOLUTION AND HETEROGENEITY

Intratumor Heterogeneity and Clinical Outcome

The existence of ITH in a multitude of tumor types underpins the need to consider the impact of tumor diversity on clinical outcome. ITH may be a prerequisite for cancer progression and drug-resistance evolution in some cases, rendering its measurement useful in predicting patient outcome (101).

The clinical importance of ITH has been demonstrated in the premalignant condition Barrett's esophagus, where various heterogeneity measures, including those of DNA content, loss of heterozygosity (LOH) and microsatellite shifts, mutations, and promoter hypermethylation, were each predictive of progression to cancer (97, 102). Measures of chromosomal instability (CIN), such as fluorescence in situ hybridization and flow cytometry, which assess cell population heterogeneity at the chromosomal level, hold clinical relevance across cancer types, providing additional prognostic information beyond conventional clinical parameters, such as tumor grade and stage (99). Moreover, the quantification of karyotypic heterogeneity (defined as nonclonal chromosome aberrations) of five in vitro tumor progression models, representing a range of cancer types, found a correlation of high levels of heterogeneity with high tumorigenic potential (151). A study in chronic lymphocytic leukemia also identified the presence of subclonal driver mutations as an independent risk factor of disease progression (80). Each of these studies would be consistent with a catalytic effect of ITH on cancer progression by providing selectable phenotypes for the evolution of clones with increased fitness and aggressiveness.

Genetic ITH is also likely to impinge upon the efficacy of therapies through the presence of low-frequency subclones harboring resistance mutations. In patients with non-small-cell lung cancer with activating *EGFR* mutations, the detection of heterogeneous low-frequency subclones carrying the T790M mutation, which confers resistance to EGFR tyrosine kinase inhibitors, was associated with shorter progression free survival on EGFR tyrosine kinase inhibitor therapy (96, 130). Further, the sequencing of circulating tumor DNA in peripheral blood from patients with initially *KRAS* wild-type colorectal cancer treated with anti-EGFR monoclonal antibodies found that *KRAS* mutations, which are known to confer resistance to anti-EGFR antibodies, emerged in approximately 40% of patients at the time resistance developed (31). Importantly, up to four different activating *KRAS* mutations were identified in individual patients, which shows that multiple therapy-resistant subclones can evolve in parallel, leading to polyclonal resistance. Similarly, substantial heterogeneity of kinase inhibitor resistance mutations has been documented in gastrointestinal stromal tumors (89), whereas mutations associated with resistance to imatinib mesylate, an inhibitor that targets the *BCR-ABL* kinase, were found present in minor subclones at diagnosis in chronic myeloid leukemia (118, 121). Heterogeneity in the mechanisms of targeted therapy resistance acting within one tumor has also been reported in lung cancer, for anaplastic lymphoma kinase (ALK)- and EGFR-tyrosine kinase inhibitors (7).

These studies highlight the clinical relevance of ITH, both in terms of disease progression and resistance to therapies and the need to develop clinically applicable detection methods for low-frequency subclones in order to realize precision medicine goals.

The Impact of Therapy on Tumor Evolution

Systemic cancer drug therapy rarely achieves mass extinction and cure in the metastatic setting; frequently, a period of response is followed by the development of drug resistance and progression. The initial decimation of the cancer cell population may lead to a reduction of heterogeneity akin to a population bottleneck as shown in AML, where some of the subclones, present at initial diagnosis, were no longer detectable at relapse after chemotherapy (33). Analysis of relapse specimens also revealed newly emerging subclones, undetectable before therapy, indicating that heterogeneity may increase again during population expansion. These evolutionary processes may be similar to the rapid evolution of new phenotypes observed during phases of population expansion in ecology, known as adaptive radiation (69). Studies of glioma and follicular lymphoma recurrences

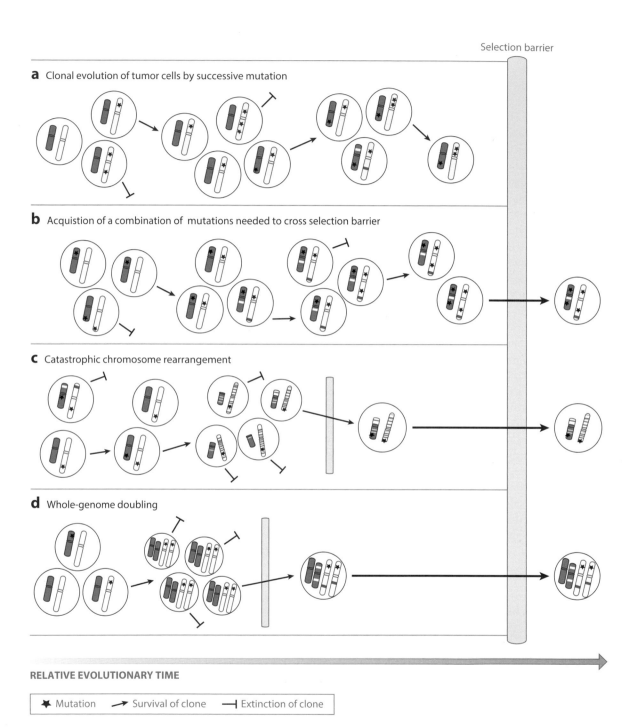

Selection barrier

a Clonal evolution of tumor cells by successive mutation

b Acquistion of a combination of mutations needed to cross selection barrier

c Catastrophic chromosome rearrangement

d Whole-genome doubling

RELATIVE EVOLUTIONARY TIME

★ Mutation   → Survival of clone   ⊣ Extinction of clone

also found that the driver mutation landscapes were often significantly different from the initially detected driver mutations, suggesting that clones initiating recurrences had branched off early in tumor evolutionary histories (71).

Thus, genomic landscapes can dramatically vary temporally during the disease course, mandating reassessments for personalized medicine approaches. AML and some gliomas sequenced at relapse after treatment with alkylating agents harbored increased numbers of mutations and mutational signatures, indicating that treatment-induced DNA damage can influence cancer evolution (33, 67, 71). The hypermutated glioma cases also acquired microsatellite instability, conferring resistance to temozolomide therapy (152). Furthermore, relapsed tumors treated with temozolomide frequently harbored mutations in key tumor suppressor pathways with mutational signatures typical of those initiated by the drug treatment itself, raising important questions regarding the use of drugs with mutagenic potential (71). A myeloma case that was serially sampled during multiple different treatment lines further demonstrated that treatment selected for the outgrowth of specific subclones (75). Yet, many of the subclones that disappeared with treatment in the myeloma case recurred after subsequent lines of treatment, which shows that effective treatment does not necessarily eradicate subclones but may only shift their clonal dominance. The ability to investigate clonal dynamics during cancer treatment will provide novel insights into the ability of cancer drugs to eradicate tumor cell subclones on the basis of specific genotypes and into whether stochastic effects allow cell survival.

The potential evolutionary cost of therapy resistance should also be considered. Melanoma xenografts harboring activating mutations in the *BRAF* driver gene developed resistance after prolonged BRAF inhibitor therapy. Surprisingly, BRAF inhibitor–resistant melanoma cells became dependent on the BRAF inhibitor, suggesting that resistance acquisition could be associated with new vulnerabilities that can exploited for therapeutic purposes (134). It has also been suggested that adaptive therapy regimens that prevent the elimination of drug-sensitive cancer clones could make it less likely that a resistant subclone will dominate the cancer cell population after therapy (46).

In conclusion, branched evolution and the resulting spatial and temporal subclonal tumor diversity are likely to have important consequences in oncology (131, 149), and novel techniques to obtain representative samples for genetic analyses, especially from metastatic disease, are urgently needed to understand the clonal dynamics of evolving tumors through the disease course.

Figure 3

Depiction of differences between micro- and macroevolutionary events. Each cell represents a major subclone within the tumor, and arrows or blocked lines indicate whether that clone survives or becomes extinct, respectively. Mixing of colors between the two chromosomes indicates a translocation event. The selection barrier represents a complex phenotype that tumor cells can acquire but that is unlikely to be accessible via single mutations. (*a*) Clonal evolution of tumor cells over time by successive mutation. Even though mutations are successively accumulated in the major clone, they are not sufficient to allow the cell to cross the selection barrier. (*b*) Evolution over time of tumor cells by successive mutation. In comparison to panel *a*, the cells acquire the combination of genetic changes needed to cross the selection barrier. (*c*) Evolution of tumor cells that undergo a catastrophic rearrangement of a few chromosomes (chromothripsis). This is likely to cause a higher attrition rate of nonviable cells with unfavorable rearrangements. However, cells that have favorable chromosome rearrangements can overcome the barrier of survival and therefore may also be able to cross the selection barrier as a result of the acquisition of multiple genomic changes in one event. (*d*) Evolution of tumor cells that undergo a whole-genome doubling event. Similar to panel *c*, there is likely a high rate of cell attrition after this event, imposing an additional selection barrier. Cells that are able to overcome this barrier, either through acquisition of a specific tolerance mechanism or through an inherent advantage in having a doubled genome, are also likely to acquire multiple genetic changes rapidly, potentially enabling them to cross the selection barrier. This is aided by the increase in chromosomal instability that occurs after whole-genome doubling.

NEXT STEPS AND SUMMARY

Genetic and epigenetic heterogeneity between cells within the same tumor provide the substrate for Darwinian evolution, allowing tumor adaptation to changing environments during progression and therapy. Recent data increasingly demonstrate that evolution often occurs in a branched manner in several tumor types, leading to intratumor diversity, with subclones differing genetically and functionally. Thus, the reductionist view of individual tumors as a population of cells that are all maintained by the same driver events can be misleading. Such heterogeneity, and the presence of subclonal driver events that may themselves be spatially separated, may explain the frequent failure to qualify prognostic and predictive biomarkers for clinical use and also the failure of many clinical trials that target specific driver events to improve patient survival.

These observations mandate the study of ITH and, most importantly, of intratumor driver event heterogeneity across all histological tumor subtypes, grades, and disease stages to reveal whether obligatory truncal drivers exist in each type and to determine the patterns of subclonal driver events that may occur with them. Focusing drug development efforts on early driver events may improve clinical benefit and decrease the failure rate of new drug trials.

New species evolve and adapt to new environments over timescales vastly greater than those witnessed in cancer evolution, which occurs within a patient's lifetime. Longitudinal cancer studies, such as the TRACERx study [TRAcking Cancer Evolution through therapy/Rx (70)], will increase our understanding of microevolutionary gene-centric and large-scale macroevolutionary genomic processes that drive cancer adaptation and identify parallel evolutionary events that might be therapeutically exploitable. Non-invasive blood-based tumor sampling methods, rebiopsies of drug-resistant cancer lesions, and the systematic collection of surgical and postmortem specimens will be necessary to further resolve cancer evolution over space and time. Such studies will provide an unprecedented opportunity to study evolution within a human lifetime, offering new insight into nature's driving force with benefits for ecology and human disease.

DISCLOSURE STATEMENT

The authors are not aware of any affiliations, memberships, funding, or financial holdings that might be perceived as affecting the objectivity of this review.

LITERATURE CITED

1. Adams KL, Wendel JF. 2005. Polyploidy and genome evolution in plants. *Curr. Opin. Plant Biol.* 8:135–41
2. Alexandrov LB, Nik-Zainal S, Wedge DC, Aparicio SA, Behjati S, et al. 2013. Signatures of mutational processes in human cancer. *Nature* 500:415–21
3. Alexandrov LB, Nik-Zainal S, Wedge DC, Campbell PJ, Stratton MR. 2013. Deciphering signatures of mutational processes operative in human cancer. *Cell Rep.* 3:246–59
4. Am. Cancer Soc. 2012. Cancer Facts & Figures 2012. Atlanta, GA: Am. Cancer Soc. **http://www.cancer.org/acs/groups/content/@epidemiologysurveilance/documents/document/acspc-031941.pdf**
5. Anderson K, Lutz C, van Delft FW, Bateman CM, Guo Y, et al. 2011. Genetic variegation of clonal architecture and propagating cells in leukaemia. *Nature* 469:356–61
6. Armitage P, Doll R. 1954. The age distribution of cancer and a multi-stage theory of carcinogenesis. *Br. J. Cancer* 8:1–12
7. Awad MM, Engelman JA, Shaw AT. 2013. Acquired resistance to crizotinib from a mutation in CD74-ROS1. *N. Engl. J. Med.* 369:1173
8. Baca SC, Prandi D, Lawrence MS, Mosquera JM, Romanel A, et al. 2013. Punctuated evolution of prostate cancer genomes. *Cell* 153:666–77

9. Barbieri CE, Baca SC, Lawrence MS, Demichelis F, Blattner M, et al. 2012. Exome sequencing identifies recurrent *SPOP*, *FOXA1* and *MED12* mutations in prostate cancer. *Nat. Genet.* 44:685–89

10. Bekaert M, Edger PP, Pires JC, Conant GC. 2011. Two-phase resolution of polyploidy in the *Arabidopsis* metabolic network gives rise to relative and absolute dosage constraints. *Plant Cell* 23:1719–28

11. Bignell GR, Greenman CD, Davies H, Butler AP, Edkins S, et al. 2010. Signatures of mutation and selection in the cancer genome. *Nature* 463:893–98

12. Birkbak NJ, Eklund AC, Li Q, McClelland SE, Endesfelder D, et al. 2011. Paradoxical relationship between chromosomal instability and survival outcome in cancer. *Cancer Res.* 71:3447–52

13. Boland CR, Sato J, Appelman HD, Bresalier RS, Feinberg AP. 1995. Microallelotyping defines the sequence and tempo of allelic losses at tumour suppressor gene loci during colorectal cancer progression. *Nat. Med.* 1:902–9

14. Bonnet D, Dick JE. 1997. Human acute myeloid leukemia is organized as a hierarchy that originates from a primitive hematopoietic cell. *Nat. Med.* 3:730–37

15. Brown SD, Warren RL, Gibb EA, Martin SD, Spinelli JJ, et al. 2014. Neo-antigens predicted by tumor genome meta-analysis correlate with increased patient survival. *Genome Res.* 24:743–50

16. Burrell RA, McGranahan N, Bartek J, Swanton C. 2013. The causes and consequences of genetic heterogeneity in cancer evolution. *Nature* 501:338–45

17. Cahill DP, Kinzler KW, Vogelstein B, Lengauer C. 1999. Genetic instability and Darwinian selection in tumours. *Trends Cell Biol.* 9:M57–60

18. Campbell PJ, Yachida S, Mudie LJ, Stephens PJ, Pleasance ED, et al. 2010. The patterns and dynamics of genomic instability in metastatic pancreatic cancer. *Nature* 467:1109–13

19. Cancer Genome Atlas Netw. 2011. Integrated genomic analyses of ovarian carcinoma. *Nature* 474:609–15

20. Cancer Genome Atlas Netw. 2012. Comprehensive molecular characterization of human colon and rectal cancer. *Nature* 487:330–37

21. Carter SL, Cibulskis K, Helman E, McKenna A, Shen H, et al. 2012. Absolute quantification of somatic DNA alterations in human cancer. *Nat. Biotechnol.* 30:413–21

22. Chen J, Li Y, Yu TS, McKay RM, Burns DK, et al. 2012. A restricted cell population propagates glioblastoma growth after chemotherapy. *Nature* 488:522–26

23. Church DN, Briggs SE, Palles C, Domingo E, Kearsey SJ, et al. 2013. DNA polymerase ε and δ exonuclease domain mutations in endometrial cancer. *Hum. Mol. Genet.* 22:2820–28

24. Ciriello G, Miller ML, Aksoy BA, Senbabaoglu Y, Schultz N, Sander C. 2013. Emerging landscape of oncogenic signatures across human cancers. *Nat. Genet.* 45:1127–33

25. Cleary AS, Leonard TL, Gestl SA, Gunther EJ. 2014. Tumour cell heterogeneity maintained by cooperating subclones in Wnt-driven mammary cancers. *Nature* 508:113–18

26. Crasta K, Ganem NJ, Dagher R, Lantermann AB, Ivanova EV, et al. 2012. DNA breaks and chromosome pulverization from errors in mitosis. *Nature* 482:53–58

27. Davoli T, de Lange T. 2011. The causes and consequences of polyploidy in normal development and cancer. *Annu. Rev. Cell Dev. Biol.* 27:585–610

28. Davoli T, Denchi EL, de Lange T. 2010. Persistent telomere damage induces bypass of mitosis and tetraploidy. *Cell* 141:81–93

29. de Bruin EC, Taylor TB, Swanton C. 2013. Intra-tumor heterogeneity: lessons from microbial evolution and clinical implications. *Genome Med.* 5:101

30. Dewhurst SM, McGranahan N, Burrell RA, Rowan AJ, Gronroos E, et al. 2014. Tolerance of whole-genome doubling propagates chromosomal instability and accelerates cancer genome evolution. *Cancer Discov.* 4:175–85

31. Diaz LA Jr, Williams RT, Wu J, Kinde I, Hecht JR, et al. 2012. The molecular evolution of acquired resistance to targeted EGFR blockade in colorectal cancers. *Nature* 486:537–40

32. Ding L, Ellis MJ, Li S, Larson DE, Chen K, et al. 2010. Genome remodelling in a basal-like breast cancer metastasis and xenograft. *Nature* 464:999–1005

33. Ding L, Ley TJ, Larson DE, Miller CA, Koboldt DC, et al. 2012. Clonal evolution in relapsed acute myeloid leukaemia revealed by whole-genome sequencing. *Nature* 481:506–10

34. Duelli DM, Padilla-Nash HM, Berman D, Murphy KM, Ried T, Lazebnik Y. 2007. A virus causes cancer by inducing massive chromosomal instability through cell fusion. *Curr. Biol.* 17:431–37

35. Duesberg P, Stindl R, Hehlmann R. 2001. Origin of multidrug resistance in cells with and without multidrug resistance genes: chromosome reassortments catalyzed by aneuploidy. *Proc. Natl. Acad. Sci. USA* 98:11283–88

36. Duncan AW, Hanlon Newell AE, Bi W, Finegold MJ, Olson SB, et al. 2012. Aneuploidy as a mechanism for stress-induced liver adaptation. *J. Clin. Investig.* 122:3307–15

37. Duncan AW, Taylor MH, Hickey RD, Hanlon Newell AE, Lenzi ML, et al. 2010. The ploidy conveyor of mature hepatocytes as a source of genetic variation. *Nature* 467:707–10

38. Ellis MJ, Ding L, Shen D, Luo J, Suman VJ, et al. 2012. Whole-genome analysis informs breast cancer response to aromatase inhibition. *Nature* 486:353–60

39. Ene CI, Fine HA. 2011. Many tumors in one: a daunting therapeutic prospect. *Cancer Cell* 20:695–97

40. Eshleman JR, Lang EZ, Bowerfind GK, Parsons R, Vogelstein B, et al. 1995. Increased mutation rate at the hprt locus accompanies microsatellite instability in colon cancer. *Oncogene* 10:33–37

41. Feldser DM, Kostova KK, Winslow MM, Taylor SE, Cashman C, et al. 2010. Stage-specific sensitivity to p53 restoration during lung cancer progression. *Nature* 468:572–75

42. France D, Hebert PDN. 1994. Hybridization and origins of polyploidy. *Proc. R. Soc. Lond. Ser. B* 258(1352):141–46

43. Frank NY, Schatton T, Frank MH. 2010. The therapeutic promise of the cancer stem cell concept. *J. Clin. Investig.* 120:41–50

44. Fujiwara T, Bandi M, Nitta M, Ivanova EV, Bronson RT, Pellman D. 2005. Cytokinesis failure generating tetraploids promotes tumorigenesis in p53-null cells. *Nature* 437:1043–47

45. Galipeau PC, Cowan DS, Sanchez CA, Barrett MT, Emond MJ, et al. 1996. 17p (p53) allelic losses, 4N (G2/tetraploid) populations, and progression to aneuploidy in Barrett's esophagus. *Proc. Natl. Acad. Sci. USA* 93:7081–84

46. Gatenby RA, Silva AS, Gillies RJ, Frieden BR. 2009. Adaptive therapy. *Cancer Res.* 69:4894–903

47. Gerlinger M, Horswell S, Larkin J, Rowan AJ, Salm MP, et al. 2014. Genomic architecture and evolution of clear cell renal cell carcinomas defined by multiregion sequencing. *Nat. Genet.* 46:225–33

48. Gerlinger M, Rowan AJ, Horswell S, Larkin J, Endesfelder D, et al. 2012. Intratumor heterogeneity and branched evolution revealed by multiregion sequencing. *N. Engl. J. Med.* 366:883–92

49. Gerlinger M, Swanton C. 2010. How Darwinian models inform therapeutic failure initiated by clonal heterogeneity in cancer medicine. *Br. J. Cancer* 103:1139–43

50. German J. 1980. Chromosome-breakage syndromes: different genes, different treatments, different cancers. *Basic Life Sci.* 15:429–39

51. Gibson TC, Scheppe ML, Cox EC. 1970. Fitness of an *Escherichia coli* mutator gene. *Science* 169:686–88

52. Goldschmidt R. 1982. *The Material Basis of Evolution: Reissued.* New Haven, CT: Yale Univ.

53. Gorla GR, Malhi H, Gupta S. 2001. Polyploidy associated with oxidative injury attenuates proliferative potential of cells. *J. Cell Sci.* 114:2943–51

54. Govindan R, Ding L, Griffith M, Subramanian J, Dees ND, et al. 2012. Genomic landscape of non-small cell lung cancer in smokers and never-smokers. *Cell* 150:1121–34

55. Grasso CS, Wu YM, Robinson DR, Cao X, Dhanasekaran SM, et al. 2012. The mutational landscape of lethal castration-resistant prostate cancer. *Nature* 487:239–43

56. Greaves M, Maley CC. 2012. Clonal evolution in cancer. *Nature* 481:306–13

57. Greaves M. 2013. Cancer stem cells as "units of selection." *Evol. Appl.* 6:102–8

58. Greenman C, Stephens P, Smith R, Dalgliesh GL, Hunter C, et al. 2007. Patterns of somatic mutation in human cancer genomes. *Nature* 446:153–58

59. Haffner MC, Mosbruger T, Esopi DM, Fedor H, Heaphy CM, et al. 2013. Tracking the clonal origin of lethal prostate cancer. *J. Clin. Investig.* 123:4918–22

60. Hallatschek O, Nelson DR. 2010. Life at the front of an expanding population. *Evolution* 64:193–206

61. Hennessy BT, Timms KM, Carey MS, Gutin A, Meyer LA, et al. 2010. Somatic mutations in *BRCA1* and *BRCA2* could expand the number of patients that benefit from poly (ADP ribose) polymerase inhibitors in ovarian cancer. *J. Clin. Oncol.* 28:3570–76

62. Herman JG, Umar A, Polyak K, Graff JR, Ahuja N, et al. 1998. Incidence and functional consequences of *hMLH1* promoter hypermethylation in colorectal carcinoma. *Proc. Natl. Acad. Sci. USA* 95:6870–75

63. Hingorani SR, Wang L, Multani AS, Combs C, Deramaudt TB, et al. 2005. $Trp53^{R172H}$ and $Kras^{G12D}$ cooperate to promote chromosomal instability and widely metastatic pancreatic ductal adenocarcinoma in mice. *Cancer Cell* 7:469–83

64. Hodis E, Watson IR, Kryukov GV, Arold ST, Imielinski M, et al. 2012. A landscape of driver mutations in melanoma. *Cell* 150:251–63

65. Holland AJ, Cleveland DW. 2009. Boveri revisited: chromosomal instability, aneuploidy and tumorigenesis. *Nat. Rev. Mol. Cell Biol.* 10:478–87

66. Huminiecki L, Conant GC. 2012. Polyploidy and the evolution of complex traits. *Int. J. Evol. Biol.* 2012:292068

67. Hunter C, Smith R, Cahill DP, Stephens P, Stevens C, et al. 2006. A hypermutation phenotype and somatic *MSH6* mutations in recurrent human malignant gliomas after alkylator chemotherapy. *Cancer Res.* 66:3987–91

68. Inda MM, Bonavia R, Mukasa A, Narita Y, Sah DW, et al. 2010. Tumor heterogeneity is an active process maintained by a mutant EGFR-induced cytokine circuit in glioblastoma. *Genes Dev.* 24:1731–45

69. Jablonski D. 2001. Lessons from the past: evolutionary impacts of mass extinctions. *Proc. Natl. Acad. Sci. USA* 98:5393–98

70. Jamal-Haniani M, Hackshaw A, Ngai Y, Shaw J, Dive C, et al. 2014. Tracking genomic cancer evolution for precision medicine: the lung TRACERx study. *PLOS Biol.* 12:e1001906

71. Johnson BE, Mazor T, Hong C, Barnes M, Aihara K, et al. 2014. Mutational analysis reveals the origin and therapy-driven evolution of recurrent glioma. *Science* 343:189–93

72. Jones AM, Thirlwell C, Howarth KM, Graham T, Chambers W, et al. 2007. Analysis of copy number changes suggests chromosomal instability in a minority of large colorectal adenomas. *J. Pathol.* 213:249–56

73. Jonkers YM, Claessen SM, Perren A, Schmid S, Komminoth P, et al. 2005. Chromosomal instability predicts metastatic disease in patients with insulinomas. *Endocr. Relat. Cancer* 12:435–47

74. Kamb A, Wee S, Lengauer C. 2007. Why is cancer drug discovery so difficult? *Nat. Rev. Drug Discov.* 6:115–20

75. Keats JJ, Chesi M, Egan JB, Garbitt VM, Palmer SE, et al. 2012. Clonal competition with alternating dominance in multiple myeloma. *Blood* 120:1067–76

76. Kern SE. 2012. Why your new cancer biomarker may never work: recurrent patterns and remarkable diversity in biomarker failures. *Cancer Res.* 72:6097–101

77. Klein A, Li N, Nicholson JM, McCormack AA, Graessmann A, Duesberg P. 2010. Transgenic oncogenes induce oncogene-independent cancers with individual karyotypes and phenotypes. *Cancer Genet. Cytogenet.* 200:79–99

78. Korona R, Nakatsu CH, Forney LJ, Lenski RE. 1994. Evidence for multiple adaptive peaks from populations of bacteria evolving in a structured habitat. *Proc. Natl. Acad. Sci. USA* 91:9037–41

79. Lada AG, Dhar A, Boissy RJ, Hirano M, Rubel AA, et al. 2012. AID/APOBEC cytosine deaminase induces genome-wide kataegis. *Biol. Direct* 7:47

80. Landau DA, Carter SL, Stojanov P, McKenna A, Stevenson K, et al. 2013. Evolution and impact of subclonal mutations in chronic lymphocytic leukemia. *Cell* 152:714–26

81. Lawrence MS, Stojanov P, Mermel CH, Robinson JT, Garraway LA, et al. 2014. Discovery and saturation analysis of cancer genes across 21 tumour types. *Nature* 505:495–501

82. Lawrence MS, Stojanov P, Polak P, Kryukov GV, Cibulskis K, et al. 2013. Mutational heterogeneity in cancer and the search for new cancer-associated genes. *Nature* 499:214–18

83. Leach FS, Nicolaides NC, Papadopoulos N, Liu B, Jen J, et al. 1993. Mutations of a *mutS* homolog in hereditary nonpolyposis colorectal cancer. *Cell* 75:1215–25

84. Lee AJ, Endesfelder D, Rowan AJ, Walther A, Birkbak NJ, et al. 2011. Chromosomal instability confers intrinsic multidrug resistance. *Cancer Res.* 71:1858–70

85. Lee RS, Stewart C, Carter SL, Ambrogio L, Cibulskis K, et al. 2012. A remarkably simple genome underlies highly malignant pediatric rhabdoid cancers. *J. Clin. Investig.* 122:2983–88

86. Leedham S, Tomlinson I. 2012. The continuum model of selection in human tumors: general paradigm or niche product? *Cancer Res.* 72:3131–34

87. Lengauer C, Kinzler KW, Vogelstein B. 1997. Genetic instability in colorectal cancers. *Nature* 386:623–27

88. Lengauer C, Kinzler KW, Vogelstein B. 1998. Genetic instabilities in human cancers. *Nature* 396:643–49

89. Liegl B, Kepten I, Le C, Zhu M, Demetri GD, et al. 2008. Heterogeneity of kinase inhibitor resistance mechanisms in GIST. *J. Pathol.* 216:64–74

90. Liu P, Erez A, Nagamani SC, Dhar SU, Kolodziejska KE, et al. 2011. Chromosome catastrophes involve replication mechanisms generating complex genomic rearrangements. *Cell* 146:889–903

91. Loeb LA. 1991. Mutator phenotype may be required for multistage carcinogenesis. *Cancer Res.* 51:3075–79

92. Loh E, Salk JJ, Loeb LA. 2010. Optimization of DNA polymerase mutation rates during bacterial evolution. *Proc. Natl. Acad. Sci. USA* 107:1154–59

93. Lv L, Zhang T, Yi Q, Huang Y, Wang Z, et al. 2012. Tetraploid cells from cytokinesis failure induce aneuploidy and spontaneous transformation of mouse ovarian surface epithelial cells. *Cell Cycle* 11:2864–75

94. Maharjan R, Seeto S, Notley-McRobb L, Ferenci T. 2006. Clonal adaptive radiation in a constant environment. *Science* 313:514–17

95. Maharjan RP, Ferenci T, Reeves PR, Li Y, Liu B, Wang L. 2012. The multiplicity of divergence mechanisms in a single evolving population. *Genome Biol.* 13:R41

96. Maheswaran S, Sequist LV, Nagrath S, Ulkus L, Brannigan B, et al. 2008. Detection of mutations in EGFR in circulating lung-cancer cells. *N. Engl. J. Med.* 359:366–77

97. Maley CC, Galipeau PC, Finley JC, Wongsurawat VJ, Li X, et al. 2006. Genetic clonal diversity predicts progression to esophageal adenocarcinoma. *Nat. Genet.* 38:468–73

98. Martin SL, Husband BC. 2012. Whole genome duplication affects evolvability of flowering time in an autotetraploid plant. *PLOS ONE* 7:e44784

99. McGranahan N, Burrell RA, Endesfelder D, Novelli MR, Swanton C. 2012. Cancer chromosomal instability: therapeutic and diagnostic challenges. "Exploring aneuploidy: the significance of chromosomal imbalance" review series. *EMBO Rep.* 13:528–38

100. Meacham CE, Morrison SJ. 2013. Tumour heterogeneity and cancer cell plasticity. *Nature* 501:328–37

101. Merlo LM, Maley CC. 2010. The role of genetic diversity in cancer. *J. Clin. Investig.* 120:401–3

102. Merlo LM, Shah NA, Li X, Blount PL, Vaughan TL, et al. 2010. A comprehensive survey of clonal diversity measures in Barrett's esophagus as biomarkers of progression to esophageal adenocarcinoma. *Cancer Prev. Res.* 3:1388–97

103. Moolgavkar SH, Knudson AG Jr. 1981. Mutation and cancer: a model for human carcinogenesis. *J. Natl. Cancer Inst.* 66:1037–52

104. Muller HJ. 1964. The relation of recombination to mutational advance. *Mutat. Res.* 106:2–9

105. Navin N, Kendall J, Troge J, Andrews P, Rodgers L, et al. 2011. Tumour evolution inferred by single-cell sequencing. *Nature* 472:90–94

106. Navin N, Krasnitz A, Rodgers L, Cook K, Meth J, et al. 2010. Inferring tumor progression from genomic heterogeneity. *Genome Res.* 20:68–80

107. Nguyen HG, Makitalo M, Yang D, Chinnappan D, St Hilaire C, Ravid K. 2009. Deregulated Aurora-B induced tetraploidy promotes tumorigenesis. *FASEB J.* 23:2741–48

108. Nikolaev SI, Sotiriou SK, Pateras IS, Santoni F, Sougioultzis S, et al. 2012. A single-nucleotide substitution mutator phenotype revealed by exome sequencing of human colon adenomas. *Cancer Res.* 72:6279–89

109. Nik-Zainal S, Alexandrov LB, Wedge DC, Van Loo P, Greenman CD, et al. 2012. Mutational processes molding the genomes of 21 breast cancers. *Cell* 149:979–93

110. Nik-Zainal S, Van Loo P, Wedge DC, Alexandrov LB, Greenman CD, et al. 2012. The life history of 21 breast cancers. *Cell* 149:994–1007

111. Notta F, Mullighan CG, Wang JC, Poeppl A, Doulatov S, et al. 2011. Evolution of human *BCR-ABL1* lymphoblastic leukaemia-initiating cells. *Nature* 469:362–67

112. Nowell PC. 1976. The clonal evolution of tumor cell populations. *Science* 194:23–28

113. OMIM. 2014. Online Mendelian inheritance in man. **http://omim.org/**

114. Otto SP, Whitton J. 2000. Polyploid incidence and evolution. *Annu. Rev. Genet.* 34:401–37

115. Palles C, Cazier JB, Howarth KM, Domingo E, Jones AM, et al. 2013. Germline mutations affecting the proofreading domains of POLE and POLD1 predispose to colorectal adenomas and carcinomas. *Nat. Genet.* 45:136–44

116. Papaemmanuil E, Gerstung M, Malcovati L, Tauro S, Gundem G, et al. 2013. Clinical and biological implications of driver mutations in myelodysplastic syndromes. *Blood* 122:3616–27

117. Rivlin N, Brosh R, Oren M, Rotter V. 2011. Mutations in the p53 tumor suppressor gene: important milestones at the various steps of tumorigenesis. *Genes Cancer* 2:466–74

118. Roche-Lestienne C, Soenen-Cornu V, Grardel-Duflos N, Lai JL, Philippe N, et al. 2002. Several types of mutations of the *Abl* gene can be found in chronic myeloid leukemia patients resistant to STI571, and they can pre-exist to the onset of treatment. *Blood* 100:1014–18

119. Roylance R, Endesfelder D, Gorman P, Burrell RA, Sander J, et al. 2011. Relationship of extreme chromosomal instability with long-term survival in a retrospective analysis of primary breast cancer. *Cancer Epidemiol. Biomark. Prev.* 20:2183–94

120. Shackney SE, Smith CA, Miller BW, Burholt DR, Murtha K, et al. 1989. Model for the genetic evolution of human solid tumors. *Cancer Res.* 49:3344–54

121. Shah NP, Nicoll JM, Nagar B, Gorre ME, Paquette RL, et al. 2002. Multiple *BCR-ABL* kinase domain mutations confer polyclonal resistance to the tyrosine kinase inhibitor imatinib (STI571) in chronic phase and blast crisis chronic myeloid leukemia. *Cancer Cell* 2:117–25

122. Shah SP, Morin RD, Khattra J, Prentice L, Pugh T, et al. 2009. Mutational evolution in a lobular breast tumour profiled at single nucleotide resolution. *Nature* 461:809–13

123. Shah SP, Roth A, Goya R, Oloumi A, Ha G, et al. 2012. The clonal and mutational evolution spectrum of primary triple-negative breast cancers. *Nature* 486:395–99

124. Simillion C, Vandepoele K, Van Montagu MC, Zabeau M, Van de Peer Y. 2002. The hidden duplication past of *Arabidopsis thaliana*. *Proc. Natl. Acad. Sci. USA* 99:13627–32

125. Snuderl M, Fazlollahi L, Le LP, Nitta M, Zhelyazkova BH, et al. 2011. Mosaic amplification of multiple receptor tyrosine kinase genes in glioblastoma. *Cancer Cell* 20:810–17

126. Sottoriva A, Vermeulen L, Tavare S. 2011. Modeling evolutionary dynamics of epigenetic mutations in hierarchically organized tumors. *PLOS Comput. Biol.* 7:e1001132

127. Sprouffske K, Merlo LM, Gerrish PJ, Maley CC, Sniegowski PD. 2012. Cancer in light of experimental evolution. *Curr. Biol.* 22:R762–71

128. Stephens PJ, Greenman CD, Fu B, Yang F, Bignell GR, et al. 2011. Massive genomic rearrangement acquired in a single catastrophic event during cancer development. *Cell* 144:27–40

129. Stephens PJ, Tarpey PS, Davies H, Van Loo P, Greenman C, et al. 2012. The landscape of cancer genes and mutational processes in breast cancer. *Nature* 486:400–4

130. Su KY, Chen HY, Li KC, Kuo ML, Yang JC, et al. 2012. Pretreatment epidermal growth factor receptor (EGFR) T790M mutation predicts shorter EGFR tyrosine kinase inhibitor response duration in patients with non-small-cell lung cancer. *J. Clin. Oncol.* 30:433–40

131. Swanton C. 2012. Intratumor heterogeneity: evolution through space and time. *Cancer Res.* 72(19):4875–82

132. Taylor BS, Schultz N, Hieronymus H, Gopalan A, Xiao Y, et al. 2010. Integrative genomic profiling of human prostate cancer. *Cancer Cell* 18:11–22

133. Taylor TB, Johnson LJ, Jackson RW, Brockhurst MA, Dash PR. 2013. First steps in experimental cancer evolution. *Evol. Appl.* 6:535–48

134. Thakur MD, Salangsang F, Landman AS, Sellers WR, Pryer NK, et al. 2013. Modelling vemurafenib resistance in melanoma reveals a strategy to forestall drug resistance. *Nature* 494(7436):251–55

135. Thirlwell C, Will OC, Domingo E, Graham TA, McDonald SA, et al. 2010. Clonality assessment and clonal ordering of individual neoplastic crypts shows polyclonality of colorectal adenomas. *Gastroenterology* 138:1441–54

136. Thompson DA, Desai MM, Murray AW. 2006. Ploidy controls the success of mutators and nature of mutations during budding yeast evolution. *Curr. Biol.* 16:1581–90

137. Thompson SL, Compton DA. 2011. Chromosomes and cancer cells. *Chromosome Res.* 19:433–44

138. Tomasetti C, Vogelstein B, Parmigiani G. 2013. Half or more of the somatic mutations in cancers of self-renewing tissues originate prior to tumor initiation. *Proc. Natl. Acad. Sci. USA* 110:1999–2004

139. Tomlinson IP, Novelli MR, Bodmer WF. 1996. The mutation rate and cancer. *Proc. Natl. Acad. Sci. USA* 93:1483–90

140. Trautmann K, Terdiman JP, French AJ, Roydasgupta R, Sein N, et al. 2006. Chromosomal instability in microsatellite-unstable and stable colon cancer. *Clin. Cancer Res.* 12:6379–85

141. von Hansemann D. 1890. Ueber asymmetriche Zelltheilung in epithel Krebsen und deren biologische Bedeutung. *Virchow's Arch. Pathol. Anat.* 119:299

142. Welch JS, Ley TJ, Link DC, Miller CA, Larson DE, et al. 2012. The origin and evolution of mutations in acute myeloid leukemia. *Cell* 150:264–78

143. Wilson BG, Roberts CW. 2011. SWI/SNF nucleosome remodellers and cancer. *Nat. Rev. Cancer* 11:481–92

144. Wood LD, Parsons DW, Jones S, Lin J, Sjoblom T, et al. 2007. The genomic landscapes of human breast and colorectal cancers. *Science* 318:1108–13

145. Woodford-Richens KL, Rowan AJ, Gorman P, Halford S, Bicknell DC, et al. 2001. SMAD4 mutations in colorectal cancer probably occur before chromosomal instability, but after divergence of the microsatellite instability pathway. *Proc. Natl. Acad. Sci. USA* 98:9719–23

146. Wu M, Pastor-Pareja JC, Xu T. 2010. Interaction between RasV12 and scribbled clones induces tumour growth and invasion. *Nature* 463:545–48

147. Wu X, Northcott PA, Dubuc A, Dupuy AJ, Shih DJ, et al. 2012. Clonal selection drives genetic divergence of metastatic medulloblastoma. *Nature* 482:529–33

148. Xu X, Hou Y, Yin X, Bao L, Tang A, et al. 2012. Single-cell exome sequencing reveals single-nucleotide mutation characteristics of a kidney tumor. *Cell* 148:886–95

149. Yap T, Gerlinger M, Futreal A, Pustzai L, Swanton C. 2012. Intratumour heterogeneity: seeing the wood for the trees. *Sci. Transl. Med.* 4:127ps10

150. Yates LR, Campbell PJ. 2012. Evolution of the cancer genome. *Nat. Rev. Genet.* 13:795–806

151. Ye CJ, Stevens JB, Liu G, Bremer SW, Jaiswal AS, et al. 2009. Genome based cell population heterogeneity promotes tumorigenicity: the evolutionary mechanism of cancer. *J. Cell. Physiol.* 219:288–300

152. Yip S, Miao J, Cahill DP, Iafrate AJ, Aldape K, et al. 2009. MSH6 mutations arise in glioblastomas during temozolomide therapy and mediate temozolomide resistance. *Clin. Cancer Res.* 15:4622–29

153. Young J, Leggett B, Gustafson C, Ward M, Searle J, et al. 1993. Genomic instability occurs in colorectal carcinomas but not in adenomas. *Hum. Mutat.* 2:351–54

154. Zack TI, Schumacher SE, Carter SL, Cherniack AD, Saksena G, et al. 2013. Pan-cancer patterns of somatic copy number alteration. *Nat. Genet.* 45:1134–40

Diverse Epigenetic Mechanisms of Human Disease

Emily Brookes[1,2] and Yang Shi[1,2]

[1]Department of Cell Biology, Harvard Medical School, Boston, Massachusetts 02115

[2]Division of Newborn Medicine, Boston Children's Hospital, Boston, Massachusetts 02115;
email: yshi@hms.harvard.edu

Annu. Rev. Genet. 2014. 48:237–68

First published online as a Review in Advance on September 5, 2014

The *Annual Review of Genetics* is online at genet.annualreviews.org

This article's doi:
10.1146/annurev-genet-120213-092518

Keywords

chromatin, transcription, cancer, neurodevelopment, fetal programming

Abstract

Epigenetic control of gene expression programs is essential for normal organismal development and cellular function. Abrogation of epigenetic regulation is seen in many human diseases, including cancer and neuropsychiatric disorders, where it can affect disease etiology and progression. Abnormal epigenetic profiles can serve as biomarkers of disease states and predictors of disease outcomes. Therefore, epigenetics is a key area of clinical investigation in diagnosis, prognosis, and treatment. In this review, we give an overarching view of epigenetic mechanisms of human disease. Genetic mutations in genes that encode chromatin regulators can cause monogenic disease or are incriminated in polygenic, multifactorial diseases. Environmental stresses can also impact directly on chromatin regulation, and these changes can increase the risk of, or directly cause, disease. Finally, emerging evidence suggests that exposure to environmental stresses in older generations may predispose subsequent generations to disease in a manner that involves the transgenerational inheritance of epigenetic information.

INTRODUCTION

5mC:
5-methylcytosine

DNMTs: DNA
methytransferases

TET: ten-eleven
translocation 2

5hmC: 5-
hydroxymethylcytosine

Epigenetics is the study of heritable changes in gene activity that are not attributable to alterations in genomic sequence. Epigenetic mechanisms are heavily implicated in human disease. The majority of cells in an individual contain an identical genome but have very different traits, enabling them to carry out their specialized functions within the body. Thus, selective utilization of the genome is required and is instigated through the establishment of appropriate gene expression patterns. It follows that pathological states may arise when the epigenetic mechanisms controlling cell-type-specific gene expression go awry.

One mediator of epigenetic control is chromatin, the complex of DNA and proteins that organizes the genome. The human genome is approximately 2 meters long and yet must fit inside cellular nuclei, the average diameter of which is a mere 6 μm. In eukaryotes, spooling DNA around histone proteins to create nucleosomes, the basic functional unit of chromatin, is the first of many stages of compaction that accomplishes this spatial feat. Nucleosomes consist of 146 bp of DNA wrapped around a core consisting of two copies each of histones H2A, H2B, H3, and H4. The linker histone H1 associates with DNA between nucleosomes. Despite the requirement for substantial genome compaction, genes and regulatory regions must be accessible for transcription, and processes such as replication and repair must be allowed to occur. Therefore, genome organization must be highly regulated, cell-type specific, and labile.

There are multiple levels of chromatin regulation to ensure correct spatial and temporal activation of genes, and those that are transmitted through cell divisions are termed epigenetic. DNA itself can undergo modification (107), as can the histone proteins around which DNA is wrapped (179). Modifications can directly affect the chromatin compaction state or can act as docking sites for reader proteins that preferentially bind specific patterns of DNA and histone modifications. In addition, nucleosome positioning (148), higher-order chromatin structure, and genome architecture within the three-dimensional nucleus (22) can be altered, which may further define active, repressed, and poised states of gene expression. Furthermore, recent evidence points to a key role for different noncoding RNA species in gene regulation and, for some classes of noncoding RNA, in defining epigenetic states (199, 212). Different epigenetic mechanisms do not function independently but are coordinately controlled through interaction and cross talk within and between different layers of regulation. Although it is clear that some chromatin modifications, such as DNA methylation and certain histone methylation marks, are heritable (242), this has not been clearly demonstrated for all the gene regulatory mechanisms we discuss in this review.

DNA Modification

DNA can be modified by the addition of methyl moieties to cytosine residues, generating 5-methylcytosine (5mC). DNA methylation is most commonly associated with silenced genomic regions, including transposable elements, repetitive sequences, imprinted genes, and the inactive X chromosome (23). The bulk of CpG dinucleotides across the genome are methylated, with the exception of those residing in CpG islands of elevated GC content, which tend to colocalize with gene promoters (46, 82). DNA methylation is catalyzed by DNA methyltransferases (DNMTs): DNMT3A and DNMT3B methylate unmethylated cytosines, whereas the maintenance methyltransferase DNMT1 recognizes hemimethylated DNA (93).

DNA methylation can be oxidized by TET (ten-eleven translocation 2) family proteins to, consecutively, 5-hydroxymethylcytosine (5hmC), formylcytosine (5fC), and carboxycytosine (5caC) (107). At present, it is unclear which reaction intermediates contain epigenetic information, although studies suggest that 5hmC profiles change with disease states (112).

Histone Modification

The histones around which DNA is wound are subject to a plethora of modifications, including methylation, acetylation, phosphorylation, and ubiquitination (202). These modifications are often located on histone tails, the N-terminal portions that extrude from the nucleosomal core. Different marks correlate with different gene activity states; this is further complicated by the effect of the site and extent of modification. For example, methylation of histone 3 lysine 4 (H3K4) residues is typically associated with active chromatin, with the trimethyl (me3) state commonly found at promoters (187), whereas the monomethyl (me1) state marks enhancers (76). Contrastingly, methylation of H3K9 or H3K27 generally marks silent chromatin (32, 155). It has been suggested that the combination of specific histone modifications constitutes a code that dictates the recruitment of factors and thereby the transcriptional state of genes (88, 179).

The writers of histone modification include lysine methyltransferases (KMTs), histone acetyltransferases (HATs), and kinases, and erasers include lysine demethylases (KDMs), histone deacetylases (HDACs), and phosphatases.

H3K4: histone 3 lysine 4

KMTs: lysine methyltransferases

HATs: histone acetyltransferases

KDMs: lysine demethylases

HDACs: histone deacetylases

Chromatin Remodeling

The spacing of nucleosomes can be regulated by chromatin remodelers, which can be critical in creating access points to the DNA sequence (148). ATP-dependent chromatin remodelers, such as the SWI/SNF family, rely on ATP hydrolysis to cause changes in nucleosome positioning or may facilitate exchange of nucleosomes or incorporation of histone variants (45, 220, 235).

In this review, we give an overview of the disparate epigenetic mechanisms that cause human disease (**Figure 1**). Genomic mutations in genes that encode regulators of the chromatin configuration are thought to exert their influences on human health through epigenetic mechanisms. These mutations can be inherited or acquired de novo, and may be directly causative of specific syndromes or may be one of several factors in polygenic disorders. Such mutations can also occur somatically in specific tissues during an individual's lifetime. The external environment can also change chromatin conformation, and the influence of nurture on disease via epigenetic mechanisms is discussed, including in the context of disease predisposition caused by early life exposures. In the interests of space, we predominantly discuss DNA methylation, histone modification, and chromatin remodeling in this review, but the principles apply to other epigenetic changes.

HERITABLE GENETIC DISORDERS CAUSED BY GERMLINE MUTATIONS IN GENES ENCODING EPIGENETIC REGULATORS

One area in which epigenetics and disease intersect is the initiation of a genetic disease by a germline mutation in an epigenetic regulator gene. In such monogenic syndromes, the causative mutation is in all cells of the body, enabling diagnosis through sequencing of easily accessible cells, such as blood cells or skin fibroblasts. Recent improvements in high-throughput sequencing have enabled unbiased identification of patient mutations, and as understanding in the field of epigenetics has progressed, mutations in more and more genes are being attributed to the epigenetic machinery. Mutations in epigenetic regulators that cause genetic syndromes can be inherited or can arise de novo in the parental germ cells or the zygote.

Disease-causing mutations affect numerous types of epigenetic regulators, including enzymes that add or remove DNA or histone modifications and those that remodel nucleosome arrays (**Table 1**). Mutations in genes encoding reader proteins, which harbor domains recognizing chromatin modifications, and regulators of higher-order chromatin structure, also have a role

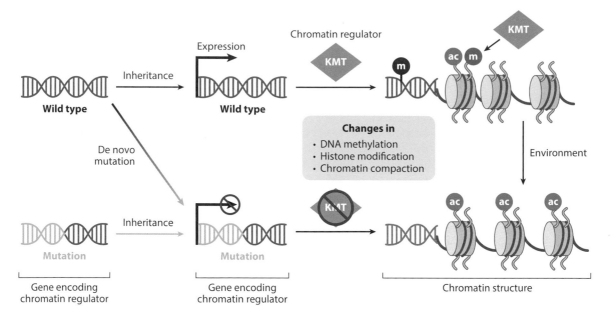

Figure 1

Diverse epigenetic mechanisms of disease. Pathological states can be caused by changes in cellular gene expression profiles, which can occur as a result of changes in chromatin configuration [for example, DNA and histone methylation (m) and histone acetylation (ac)]. This can be directly caused by environmental influences or can arise as a result of mutations in genes that encode regulators of the chromatin state [such as histone methyltransferases (KMT)]. Such mutations can be inherited or may arise de novo.

in disease. However, many enzymes are also readers, and the mechanistic function of many reader proteins and genome architecture regulators remains enigmatic and so we have not described these mutations in this review, with the exception of the well-studied Rett syndrome, which is caused by mutations in the methyl-CpG binding protein *MECP2* (5).

Common Traits of Genetic Disorders with Epigenetic Origins

Heritable genetic diseases caused by mutations in epigenetic regulator genes exhibit a variety of symptoms, pathological characteristics, and modes of inheritance, but there are unifying concepts (**Table 1**). Many of the diseases linked to mutations in epigenetic regulators show autosomal-dominant inheritance, and a substantial proportion are also X-linked. In some cases, homozygous mutations of chromatin regulators may be lethal; for implicated X-linked genes, the presence of the Y homolog of the gene (e.g., *KDM5D* for *KDM5C*) (68) may be critical for male survival given that female homozygotes are not described. These inheritance patterns demonstrate the importance of the proper balance of epigenetic modification for development.

The presence of chromatin regulator genes on the X chromosome can complicate interpretation of symptoms in heterozygous females. Random X chromosome inactivation leads to somatic mosaicism for the mutation, and disease pathology is influenced by which cells silence the mutant or the wild-type allele (4). Other X-linked mutations can skew X inactivation toward the X containing the mutant allele in carrier females (1, 66, 163, 176). Symptom variability may also be caused by the precise nature of the epigenetic regulator mutation: for example, whether the causative mutation is a missense mutation that abrogates protein function or a nonsense mutation that results in complete loss of the gene product (68, 127).

Table 1 Genetic syndromes caused by mutation in epigenetic regulator genes

Gene	Function	Syndrome	OMIM[a]	Inheritance	Symptoms			
					Cognitive	Facial	Growth	Other
DNA methylation								
DNMT1	DNMT	HSN1E (103)	614116	AD	+	−	−	Neuropathy, hearing loss
DNMT3B	DNMT	ICF1 (229)	242860	AR	+	+	−	Immune defects, centromere instability
MECP2	MBD	Rett (5)	312750	XL	+	−		Neurodevelopmental regression, stereotypic movements, seizures
MECP2	MBD	MECP2 duplication (140, 210)	300260	XL	+	+	−	Hypotonia, seizures
Histone writers								
NSD1	HMT	Sotos (110)	117550	AD	+	−	+	Heart defects
EZH2	HMT	Weaver (67)	277590	AD	+	+	+	Micrognathia, accelerated osseous maturation
MLL	HMT	Wiedemann–Steiner (90)	605130	AD	+	+	+	Hairy elbows
MLL2	HMT	Kabuki-1 (152)	147920	AD	+	+	+	Heart defects, recurrent otitis media
EHMT1	HMT	Kleefstra (100)	610253	AD	+	+	−	Heart defects, hypotonia, seizures,
CREBBP	HAT	Rubinstein–Taybi-1 (172)	180849	AD	+	+	+	Heart defects, increased tumor risk
EP300	HAT	Rubinstein–Taybi-2 (183)	613684	AD	+	+	+	
KAT6B	HAT	Genitopatellar (30)	606170	AD	+	+	+	Absent patellae, genito-urinary anomalies
KAT6B	HAT	SBBYSS (43)	603736	AD	+	+	−	Heart defects, hypotonia, skeletal problems
KANSL1	HAT	Koolen-De Vries (243)	610443	AD	+	+	−	Heart defects, hypotonia, genito-urinary and dental anomalies, seizures
ATXN7	HAT, DUB	Spinocerebellar ataxia-7 (45)	164500	AD	+	−	−	Neurodegeneration, cerebellar ataxia, vision loss
RSK2	Kinase	Coffin-Lowry (208)	303600	AD	+	−	+	Skeletal problems, hearing loss

(Continued)

Table 1 *(Continued)*

Gene	Function	Syndrome	OMIM[a]	Inheritance	Symptoms			
					Cognitive	Facial	Growth	Other
Histone erasers								
KDM6A	HDM	Kabuki-2 (116)	300867	XL	+	+	+	Heart defects
KDM5C	HDM	MRXSCJ (87)	300534	XL	+	+	+	Aggression, seizures
PHF8	HDM	MRXSSD (114)	300263	XL	+	+	−	
HDAC4	HDAC	BDMR (224)	600430	AD	+	−	+	Heart defects, brachydactyly
Chromatin remodelers								
ATR-X	Remodeler	ATR-X (65)	301040	XL	+	+	+	α-Thalassemia, genital abnormalities
ATR-X	Remodeler	MRXHF1 (64)	309580	XL	+	+	+	Hypogonadism, hearing loss, renal and skeletal abnormalities
CHD7	Remodeler	CHARGE (215)	214800	AD	+	+	−	Coloboma, choanal atresia, genito-urinary, heart, and ear abnormalities
CHD7	Remodeler	HH5 (99)	612370	AD	−	+	−	Anosmia, hypogonadotropic hypogonadism, hearing loss
ERCC6	Remodeler	Cockayne-B (134)	133540	AR	+	+	+	Cataracts, hearing loss, cellular UV sensitivity
SRCAP	Remodeler	Floating-Harbor (79)	136140	AD	+	+	+	Delayed osseous maturation
ARIDB	Remodeler	MRD12 (80)	614562	AD	+	+	+	Absent fifth fingernails and toenails, hypotonia, hirsuitism
ARIDA	Remodeler	MRD14 (209)	614607	AD	+	+	+	Absent fifth fingernails and toenails
SMARCB1	Remodeler	MRD15 (209)	614608	AD	+	+	+	Absent fifth fingernails and toenails, hirsutism, hypotonia
SMARCA4	Remodeler	MRD16 (209)	614609	AD	+	−	+	Absent fifth fingernails and toenails, hirsutism, hypotonia
SMARCA2	Remodeler	Nicolaides-Baraitser (211)	601358	AD	+	+	+	Seizures, sparse hair
SATB2	Remodeler	Glass (119)	612313	AD	+	+	+	Seizures, micrognathia

[a]Reference numbers can be found in the Online Mendelian Inheritance in Man catalog; **http://www.ncbi.nlm.nih.gov/omim**.

Abbreviations: +, present; −, absent; AD, autosomal dominant; AR, autosomal recessive; ATR-X, α-thalassemia/mental retardation syndrome, X-linked; BDMR, brachydactyly mental retardation syndrome; CHARGE, coloboma, heart anomaly, choanal atresia, retardation, genital, and ear anomalies; DNMT, DNA methyltransferase; DUB, deubiquitinase; HAT, histone acetyltransferase; HDAC, histone deacetylase; HDM, histone demethylase; HMT, histone methyltransferase; HSN1E, neuropathy, hereditary sensory, type 1E; ICF1, immunodeficiency, centromeric instability, facial anomalies syndrome 1; MBD, methyl-CpG binding domain; MRD, mental retardation, autosomal dominant; MRXHF1, mental retardation hypotonic facies syndrome, X-linked 1; MRXSCJ, mental retardation, X-linked, syndromic, Claes-Jensen type; MRXSSD, Siderius X-linked mental retardation syndrome; OMIM, Online Mendelian Inheritance in Man; SBBYSS, Ohdo syndrome, Say-Barber-Biesecker-Young-Simpson variant; XL, X-linked.

A common thread in the symptoms arising from germline mutations in epigenetic regulator genes is the presence of cognitive deficits (**Table 1**). Given that the mutation is present in all cells of the body, and given the ubiquitous expression pattern of many chromatin-modifying enzymes, these findings suggest a unique role for epigenetic regulation in brain development and function. It has been previously reported that the X chromosome is enriched for genes relating to brain development and function (238), consistent with X-linked inheritance of many mutations in epigenetic regulators (**Table 1**).

ID: intellectual disability

Intellectual disability (ID) of varying degrees of severity is seen in many of the epigenetic diseases described (**Table 1**). This can be accompanied by neurological and behavioral abnormalities (**Table 1**), such as aggression, anxiety, and epilepsy (68, 69, 101). In some instances, ID is present from birth and may reflect a neurodevelopmental role for chromatin regulation (184). In other diseases, such as neuropathy hereditary sensory 1E (226) and spinocerebellar ataxia-7 (45), cognitive problems arise with neuronal degeneration. The coincidence of epigenetic aberrations with defects in cognitive development and function is fascinating and may reflect the sensitivity of neuronal networks to signals that other, more robust, systems can withstand. However, it has made understanding the link between mutation and disease more difficult because of the inaccessibility, complexity, and cellular heterogeneity of the central nervous system.

Together with cognitive difficulties, common symptoms in epigenetic disease are craniofacial, cardiac, and growth abnormalities (including overgrowth and growth deficiency) (**Table 1**). These symptoms are commensurate with aberrant neural crest development. The neural crest is a vertebrate-specific embryonic cell group that acquires broad differentiation and migratory potential and contributes to craniofacial, peripheral nervous system, and cardiac structures (195). Haploinsufficiency of CHD7 gives rise to CHARGE (coloboma, heart defect, choanal atresia, retarded growth and development, genital hypoplasia, and ear anomalies) syndrome (215), a disorder whose symptoms fit with neural crest anomalies (**Table 1**). Interestingly, CHD7 has been shown to have a critical role in neural crest formation (11). CHD7 occupies enhancers of neural crest markers *TWIST1* and *SOX9*, and loss of CHD7 resulted in downregulation of these genes (11). Other disease-related chromatin regulators, including EZH2, mutated in Weaver syndrome, and DNMT1, mutated in ICF1 syndrome, are also implicated in neural crest regulation (136, 191). It will be informative to assess whether diseases caused by mutations in chromatin regulator genes are due to neural crest defects.

Understanding the Epigenetic Mechanisms of Heritable Genetic Disease

Germline mutations causing loss-of-function of proteins that play a role in regulating or interpreting the epigenetic code are hypothesized to induce alterations in the chromatin landscape. Changes in chromatin configuration may cause disease by altering the activity of specific genes or may precipitate other defects in genome regulation, such as an increase in the rate of DNA damage accumulation. Many studies have focused on investigating aberrant gene expression patterns in patient cells. Some examples of diseases in which the downstream targets of the mutated epigenetic regulator have been elucidated are described below.

Mutations in DNMT3B Cause ICF1 (Immunodeficiency, Centromeric instability, Facial anomalies syndrome 1). Sixty percent of ICF1 patients carry mutations in the gene encoding the de novo methyltransferase DNMT3B (162, 229). The epigenetic nature of the disorder is used in diagnosis: Patient lymphocytes exhibit hypomethylation of satellite DNA sequences in the juxtacentromeric heterochromatin of chromosomes 1, 9, and 16 (86, 229). Several studies have profiled expression and chromatin modification in lymphocytes from ICF1 patients compared with controls and identified dysregulated genes whose functions, e.g., immune function and neuro-development, were implicated in the disease phenotype (27, 51, 89). Genes with altered expression in ICF1 included DNMT3B-target genes, which showed DNA hypomethylation, and indirect targets at which no DNA methylation changes were observed (27, 51, 89). Some upregulated genes showed changes in other chromatin modifications, such as loss of repressive H3K27me3 and gain of activating H3K4me3 (89), demonstrating the interdependence of epigenetic modifications.

Mutations in EHMT1 Cause Kleefstra syndrome. Patients with Kleefstra syndrome have haploinsufficiency of the histone methyltransferase EHMT1 (100), which mediates H3K9me2 (161). Mouse models of EHMT1 loss have reduced learning and memory, hypotonia, and cranial abnormalities (13, 188). In brain and bone tissue, bone-related genes, including *Runx2*, showed an H3K9me2 decrease and transcription upregulation, consistent with being direct EHMT1 targets (13). Genome-wide analysis of brain regions lacking EHMT1 (or its binding partner EHMT2) has shown that many genes are upregulated after loss of these H3K9 methyltransferases, including neuronal and non-neuronal genes (188). Interestingly, there was aberrant reactivation of some neuronal development regulators, such as *Dach2*.

Although the downstream targets of disease-related chromatin regulators have been defined in a number of genetic disorders, we are still lacking information about tissue-specific target genes for other conditions. Gene expression aberrations in epigenetic diseases affecting the brain are often assessed in organs not directly affected by the disease, e.g., easily accessible skin or blood samples. Greater understanding of the target genes of relevant epigenetic regulators in neuronal and glial cells, and of the effect of their mutations on cellular morphology and function, is critical to understanding and treating these disorders.

Animal models, generated by mutating the disease-causing chromatin regulator gene, provide a means of investigating heritable monogenic disease because much of the epigenetic apparatus is highly conserved. Animal models have the advantage of being a whole organism that can be experimentally manipulated, and the brain can be directly investigated at different developmental stages. However, there are advantages to using clinically relevant human systems. Induced pluripotent stem (iPS) cell disease models, which take patient skin cells and reprogram them to stem cells that can be differentiated into relevant cell types, hold some promise in this regard. The genetic origin of the disease is maintained across reprogramming and differentiation, despite the erasure of epigenetic information during reprogramming (41). This strategy has been used successfully for Rett syndrome (135, 223).

Commonly Targeted Pathways: A Clue Toward Mechanism

The interplay between different types of chromatin modification, and the indirect effects of transcriptional dysregulation, can make it difficult to separate causative from consequential changes. This can be further complicated by nonhistone targets of epigenetic regulators. For example, NSD1, implicated in Sotos syndrome (110), has been shown to methylate NF-Kβ (131) as well as H3K36 (122).

Mutations in regulators that share biological function can cause similar or overlapping syndromes, and this can provide clues regarding mechanisms of disease and help to refine causative expression changes from off-target and indirect effects. Below, we have described how independent mutations can link certain syndromes to common pathways and/or gene expression changes:

1. Kleefstra syndrome is canonically caused by mutations in the H3K9me2 methyltransferase EHMT1 (100), but a recent study identified additional Kleefstra syndrome mutations in

MBD5, *MLL3*, *SMARCB1*, and *NR1I3* (102). Genetic interactions between these other epigenetic regulators and EHMT1 in *Drosophila* (102) suggest a common mechanism of disease, and it would be informative to assess the similarities in chromatin and gene expression profiles in patients or mouse models bearing these different mutations.

CHD: congenital
heart disease

2. Rubinstein-Taybi syndrome can be caused by mutations in either *CBP* [~50% patients (172)] or *EP300* [~3% (183)]. CBP and P300 are HATs with shared targets, and they also share cofactors, including neuronal factors, oncoproteins, and tumor suppressors. Note, though, that *EP300* mutants may exhibit a milder phenotype than *CBP* mutants (15).

3. Rubinstein-Taybi syndrome also shows symptom overlap with Floating-Harbor syndrome, which is caused by mutations in the chromatin remodeler *SRCAP* (79). SRCAP is a CBP coactivator (91), suggesting that changes in similar target genes may mediate the two different diseases.

4. Glass syndrome is caused by heterozygous loss-of-function of the chromatin remodeler *SATB2* (119), but the craniofacial symptoms are very similar to those seen in another ID syndrome, MRXS14 (OMIM 300676). MRXS14 is caused by mutations in *UPF3B* (204), which encodes a member of the nonsense-mediated mRNA decay complex. It was recently shown that the causative mutations in these distinct disorders operate in the same pathway: SATB2 binds to the *UPF3B* promoter and activates its expression (120).

5. Kabuki syndrome can be caused by mutations in MLL2 (152), which adds the activating modification H3K4me2/3 (84), or KDM6A (116), which removes repressive H3K27me2/3 (113). Despite targeting different amino acids, the overarching result is a more open chromatin state in both cases. H3K4 and H3K27 methylation are known to be present at the same bivalent genes in various cell types (9), suggesting that MLL2 and KDM6A could target the same gene set. This is further supported by evidence that the two proteins directly interact and share *HOX* gene targets (117).

MULTIFACTORIAL DISEASES ASSOCIATED WITH GENETIC MUTATIONS IN EPIGENETIC GENES

The next group of diseases in which epigenetic dysregulation is linked to pathology is in polygenic diseases. Mutations in specific chromatin regulators are not always sufficient to lead to disease onset, as described above, but may cause disease only when combined with environmental factors or additional mutations. Additionally, mutations may arise only in a specific cell type within an individual. As for monogenic diseases, advances in high-throughput sequencing have enabled the identification of many new mutations in multifactorial disorders. Large-scale efforts have sequenced different types of cancers and also cells from patients suffering from polygenic diseases, such as autism, schizophrenia, and congenital heart disease (CHD). Although the mutations are not, by themselves, necessarily causative in these cases, they advance the understanding of disease biology, particularly the common pathways altered across different patients. The genetic mutations associated with such complex disorders may also enable modeling of the disease using laboratory animals and may provide targets for diagnosis or treatment. Genes encoding chromatin regulators are increasingly being recognized as important contributors to multifactorial disease.

The majority of disease-related mutations in epigenetic regulators linked to multifactorial disease are de novo mutations. De novo mutations are first identified in the patient and so are not subject to evolutionary pressure, as inherited mutations are, meaning they may have more deleterious consequences. De novo mutations can occur in the germline of the parents, during embryogenesis or somatically. Somatic mutations acquired during the lifetime of an individual can arise spontaneously within a specific cell type, for example, due to errors during DNA replication, or can be induced environmentally, such as by UV irradiation.

Remarkably, paternal germline mutations seem to be a dominant cause of de novo mutations. Recent evidence points to a strong positive correlation between paternal age and number of de novo mutations in the offspring (108), as a result of ongoing spermatogenesis during the life span of males. Paternal age has also been correlated with offspring disorders including autism spectrum disorder (ASD) and schizophrenia (72, 180), in keeping with de novo mutations as a prevalent cause of such diseases.

Genetic Alterations in Epigenetic Regulators in Congenital Heart Disease

Structural cardiac defects associated with CHD are the most frequent birth defect. CHD is a multifactorial disease, with many different mutations and environmental influences being reported (6). The idea that epigenetics influences disease onset and progression comes from genetic syndromes caused by mutations in chromatin regulator genes in which congenital heart defects are a symptom, such as Rubinstein-Taybi and Kleefstra syndromes (**Table 1**). More recently, whole-exome sequencing has identified a slew of de novo CHD mutations in epigenetic regulator genes (236). Mutated genes included *MLL2*, *KDM6A*, *CHD7*, *WDR5*, *KDM5A*, and *KDM5B*, all of which regulate H3K4 methylation, and *RNF20*, *UBE2B*, and *USP44*, which are involved in the ubiquitination of H2BK120, which is itself required for H3K4 methylation.

The contribution of chromatin regulation to CHD is consistent with its critical role in heart development (164, 216), although the gene targets and precise mechanisms are as yet unknown. Interestingly, germline mutations in *MLL2* and *KDM6A* are responsible for Kabuki syndrome (116, 152), and mutations in *CHD7* are responsible for CHARGE syndrome (215), both of which are associated with heart defects (**Table 1**), demonstrating that the specific nature and context of similar mutations can give rise to different disorders with overlapping symptoms.

Genetic Alterations in Epigenetic Regulators in Autism Spectrum Disorders

ASD is a multifactorial disorder encompassing a range of neurodevelopmental anomalies that affect social interaction, behavior, and communication. One of the most frequently mutated genes in ASD is *CHD8* (109). *CHD8* encodes an ATP-dependent chromodomain helicase (206). CHD8 can recruit the linker histone H1 to repress β-catenin and p53 target genes (157, 158); WNT/β-catenin signaling is important in neuronal development. CHD8 has also been shown to form a complex with an H3K4 methyltransferase complex and may be required for this complex's recruitment to chromatin (234). CHD8 is a known binding partner of CHD7 (16) and mutations in *CHD7* are responsible for CHARGE syndrome (215), a complex syndrome in which up to two-thirds of patients have ASD (21).

De novo mutations have been described in other chromatin regulator genes in ASD and also in other neurodevelopmental disorders, such as ID, suggesting that these mutations may affect common neurodevelopmental pathways. Other CHD family members CHD1, CHD3 and CHD7; chromatin remodelers ARID1B, SMARCC1, SMARCC2, and SATB2; histone methyltransferases EHMT1 (targets H3K9), SETD2 (targets H3K36), and MLL, MLL3, MLL5, and ASH1L (targeting H3K4); and histone demethylase KDM6A (targets H3K27) have all been implicated in ASD (109). Despite the rarity of each individual mutation, their occurrence in such a diverse group of chromatin regulators suggests that there may be several different networks that can lead to deficits in cognitive development and/or function. Tying these pathways into the disparate symptoms experienced in the heterogeneous group of disorders that make up ASD

CHROMATIN PROFILING IN CANCER

Tumors display global DNA hypomethylation and site-specific DNA hypermethylation (56, 57). Hypomethylation results in genomic instability and aberrant activation of certain genes, including oncogenes (50, 182). Hypermethylation results in the silencing of tumor suppressor genes, cell cycle genes, and DNA repair genes (54). In addition, changes in DNA methylation are associated with loss of appropriate genomic imprinting (124), and hypermethylation facilitates C-to-T mutations (181). Changes in methylation status can be used as a diagnostic or prognostic marker in cancer. This is of particular use when abnormal DNA methylation patterns are early markers present in easily accessible cell types as well as the primary tumor: for example, methylation signatures in cells from the urine of prostate cancer sufferers (29). Interestingly, aberrant DNA methylation is seen in pre-malignant tissues after exposure to carcinogens such as chronic inflammation (83, 96), and methylation accumulation correlates with cancer risk (96, 189). This suggests that epigenetic changes contribute to acquired cancer predisposition.

There is also prognostic value in assessing global levels of histone modifications. Lower levels of H3K4me2 are associated with poorer outcomes in prostate, lung, and kidney cancer (14, 192, 193). Lower levels of H3K18ac and H3K9me3 predict a poorer outcome in lung and kidney cancer (192), whereas lower survival is seen for patients with lung cancer tumors expressing higher levels of H3K9ac (14). Specific patterns of H3K9me are associated with clinical outcomes in acute myeloid leukemia (146).

will be a challenge, but understanding them is crucial for progression in our understanding and treatment of such a complex collection of disorders.

Genetic Alterations in Epigenetic Regulators in Cancer

Sequencing tumor samples compared with controls has enabled identification of cancer-related mutations in many chromatin regulators. A recent study analyzed somatic mutations in nearly 5,000 human cancers and matched normal-tissue samples across 21 tumor types (115). The unprecedented depth of the study allowed identification of the majority of previously described cancer genes, plus some novel targets. Mutations in chromatin regulators that run the gamut of functions were found in many different classes of tumor. The fact that many different modification pathways seem to be targeted, plus the known interplay between different modifications, suggests that mutations in the epigenetic regulatory machinery cause global changes in chromatin composition and conformation, together with dysregulation of gene expression programs. Specific changes in chromatin configuration are known to occur during the acquisition of certain hallmarks of cancer cells (see sidebar, Chromatin Profiling in Cancer), and epigenetic regulator mutations may affect the efficacy, kinetics, or probability of these changes. Intriguingly, mutations in genes encoding core histones *HIST1H3B*, *HIST1H1E*, and *HIST1H4E* have also been identified in cancer (115). We now discuss some examples of cancer mutations in epigenetic regulator genes.

DNA modifiers in cancer. Mutations in the coding sequence of the methyltransferase *DNMT3A* are frequently found in human cancers, including in acute myeloid leukemia (233) and non-Hodgkin's lymphomas (166). In mice, *Dnmt3a* loss prevents differentiation of embryonic and hematopoietic stem cells (ESCs and HSCs, respectively), with sustained or amplified self-renewal (35, 40). This is consistent with reports that abrogation of differentiation pathways may be critical in cancer progression. Global alterations in DNA methylation and hydroxymethylation have been reported in *Dnmt3a*-null HSCs (35) and in the lymphocytic leukemia tumors that arise in these animals (171). Direct Dnmt3a targets showed decreased DNA methylation and concomitantly

HSCs: hematopoietic stem cells

increased transcription in Dnmt3a-null HSCs, including *Runx1* and *Gata3* (35), which are known to inhibit differentiation (34, 38).

Somatic mutations in *TET2*, whose gene product converts DNA 5mC to 5hmC, were first identified in myeloid cancer and are now known to be a frequent event in such malignancies (47). *TET2* mutations reduce TET2's catalytic activity, with bone marrow samples (106) and granulocytes (174) from patients with *TET2* mutations displaying lower levels of 5hmC than healthy controls. Loss of Tet2 promotes self-renewal and impairs differentiation in mice hematopoietic lineages (105, 143), ultimately leading to transformation. Similarly, *TET2* mutations abrogate differentiation of primary human myelomonocytic leukemia cells (133). These results are congruent with dramatic changes in TET2 expression, and in global levels and profiles of 5hmC, during erythroid differentiation (133).

Histone modifiers in cancer. Mutations in genes encoding histone-modifying enzymes are common in cancers, and other reviews have focused on this subject (213). Therefore here we briefly discuss H3K4 methylation enzymes that are implicated in cancer, as an example of this mechanism of disease.

The family of mixed-lineage leukemia (MLL) H3K4 histone methyltransferases is frequently mutated in human cancer (78). MLL rearrangements occurring in leukemia include fusion genes, partial tandem duplications, and amplification, and are thought to cause tumorigenesis through gain-of-function rather than loss-of-function. In some cases, the fusion partner is also an epigenetic regulator; *CBP* and *TET1* (128, 201) become fused to the N-terminal of MLL. Target genes of MLL are deregulated in MLL-linked leukemia, including the *HOX* genes (173). *HOX* genes are critical regulators of hematopoiesis and act to promote self-renewal, a hallmark of cancerous cells. Other mechanisms may also contribute to MLL gain-of-function tumorigenesis; for example, MLL fusion proteins downregulate RUNX and CBFβ (241), which are important factors in hematopoietic differentiation.

MLL family members are also implicated in cancer. Inactivating mutations in *MLL2* and *MLL3* have been found in medulloblastoma brain tumors and non-Hodgkin's lymphoma (144, 167). Furthermore, *MLL2* was one of the most frequently mutated genes identified by Lawrence et al. (115), suggesting that regulation of H3K4 methylation is a critical pathway in carcinogenesis. Interestingly, mutations in the H3K27me2/3 demethylase *KDM6A* are also seen in medulloblastoma (7); germline mutations in either *MLL2* or *KDM6A* also cause the same disease in Kabuki syndrome (116, 152). MLL2 and KDM6A directly interact and share gene targets (117), suggesting common disease pathways.

Gain-of-function mutations in the gene encoding LSD1, a demethylase for H3K4me1/2 (194), are implicated in many types of cancer, as are high levels of the LSD1 protein that are induced without genomic changes (186). Knockdown of LSD1 suppressed tumor proliferation (74), making *LSD1* a putative oncogene. Although LSD1 target genes may be important in cancer progression (74), LSD1 is also able to demethylate the tumor suppressor P53 (81).

The KDM5 demethylase family, which removes H3K4me2/3 (85), is also implicated in cancer. However, this has not been directly linked to mutations in the *KDM5A* or *KDM5B* genes, whose overexpression in tumors may result from mutations in regulatory regions, or from epigenetic effects. KDM5A (also known as RBP2) is overexpressed in gastric (239) and lung cancer (205). KDM5A directly targets homeotic genes, which control differentiation (19), and also *p27, p21,* and *p16*, which regulate cellular senescence (239). Therefore, overexpression of KDM5A is hypothesized to decrease H3K4me2/3 at the promoters of genes involved in stimulating differentiation and senescence, repressing the genes and causing cancer. KDM5A is also a binding partner of the tumor suppressor retinoblastoma (Rb) (55). Rb can activate genes repressed by RBP2 and thereby promote differentiation (19); loss of KDM5A suppresses tumorigenesis in Rb-deficient mice (125). KDM5B (also known as PLU-1) is upregulated in breast (130) and prostate cancer (228). KDM5B is linked to cellular proliferation through its repressive action on genes that negatively regulate cell growth (232). Finally, mutations in *KDM5C* were identified in kidney tumors (115).

Histone variants in cancer. One particularly striking example of chromatin regulation in cancer is in devastatingly aggressive pediatric gliomas, where the most frequent mutations are in a gene encoding the histone variant H3.3 (*H3F3A*; another gene, *H3F3B*, also encodes H3.3). Two highly specific, mutually exclusive mutations were reported, with different distribution and age characteristics; both localize to the heavily modifiable N-terminal tail of H3.3 (98, 190, 198, 227). K27M mutations were found in brainstem tumors and were more common in adolescent patients (median 11 years), whereas G34R/V mutations were found in cortical tumors of young adult patients (median 20 years). Patients bearing K27M mutations have lower survival than those with G34R/V mutations. These data suggest that K27M and G34R/V mutations arise in specific cellular subpopulations, during defined developmental windows. K27M and G34R/V tumors display different gene expression and epigenetic profiles (24, 190, 198), suggesting distinct mechanisms of tumorigenesis.

H3.3 is encoded by *H3F3B*, in addition to *H3F3A*, and in chromatin there is a mixture of H3.3 and H3. However, the presence of mutations in *H3F3A* dramatically changed global levels of histone modifications, which can occur on the canonical and variant histones. Global H3K27me3 is dramatically reduced in H3.3K27M tumors (37, 121, 214). K27M directly binds and inhibits the catalytic activity of the H3K27 methyltransferase EZH2 (18, 121), levels of which are unchanged in K27M tumors (37, 121, 214). The effect of K27M on chromatin-bound H3K27me3 and EZH2 appears to be site-specific, with local increases reported (18, 37). Additional epigenetic changes are reported in K27M tumors, including changes in DNA methylation (18, 198).

Global H3K27me3 levels are not altered in H3.3G34R/V tumors; changes in H3K36me3, located proximally to G34, are reported, but the results are inconsistent (18, 24, 37, 190). Local increases in H3K36me3 and gene expression are reported at forebrain developmental genes and at the oncogene *MYCN* in a G34V cell line (24); these changes in transcriptional program are potentially a driving mechanism for tumorigenesis.

DISORDERS LINKED TO ENVIRONMENTALLY INDUCED CHANGES IN EPIGENETIC MODIFICATIONS

In addition to genetic mutations that result in aberrations in the epigenome, there are other factors that, in the absence of immutable changes to the DNA, exert an effect on the epigenetic landscape and contribute to disease. Chromatin modifications integrate and process external signals and relay them in order to influence transcriptional regulation and the utilization of the genome. Thus, epigenetics can be viewed as a potential point of cross talk between the genome and the environment.

It has long been understood that the environment has an enormous impact on individuals. The relative contribution of a person's experiences, as opposed to their innate qualities, to their physical and behavioral traits has been widely debated. Although the discussion of nature versus nurture was first formalized thus by Francis Galton in 1895,

Nature is all that a man brings into the world; nurture is every influence from without that affects him after his birth, the juxtaposition of the two ideas goes back well before this; in Shakespeare's *The Tempest*, Prospero comments:

AD: Alzheimer's disease

PD: Parkinson's disease

A devil, a born devil on whose nature
Nurture can never stick, on whom my pains,
Humanely taken, all, all lost, quite lost.

In terms of disease, the nature versus nurture debate has been vociferous. It is known that disease susceptibility, progression, and outcome are influenced by a person's genetic predisposition and by external factors. More recently, it has become understood that the environment exerts some of its effects through epigenetics. Unlike Galton, we now consider the influence of the environment to start at conception, within the in utero milieu, and to continue after birth. For certain epigenetic programming events, there may be specific windows of susceptibility.

One of the clearest indications of the contribution of environment to disease has been through studies of monozygotic twins; genetically identical individuals can show discordancy for diseases such as Alzheimer's disease (AD, 62), autism (71), and schizophrenia (31). This disparity may be due to lifestyle differences, which can have a direct effect on health, may be mediated through alterations in the epigenome of the individual, or both (reviewed in 17). It has been shown that the epigenome of monozygotic twins is different (58).

The influence of the environment on chromatin, and hence on gene expression and phenotype, means that assessing the chromatin landscape of cells in disease may give clues to etiology. Environmental stresses and genetic mutations may cause disease through similar pathways, given that changes in the chromatin landscape may be caused either by mutations in chromatin regulators or by exposure to environmental conditions in the absence of the mutation (**Figure 1**). This suggests that epigenetic changes may be useful as biomarkers for diagnosis and prognosis or as a predictor of therapeutic efficacy (see sidebar, Chromatin Profiling in Cancer). Chromatin-modifying enzymes are also potential drug targets. We now discuss some examples in which human disease is associated with chromatin alterations in the absence of genetic mutations.

Epigenetics, Aging, and Neurodegenerative Disease

Changes in the chromatin landscape have been documented during aging and also during age-related diseases such as neurodegeneration (77, 129). Strikingly, epigenetic differences between the lymphocytes of monozygotic twins increase with age (58). The contribution of epigenetic mechanisms to the progression of diseases of aging such as AD and Parkinson's disease (PD) is linked to observations of different symptoms in monozygotic twins (62, 225). Moreover, a neuroprotective role for HDAC inhibitors in neurodegenerative disease suggests that disruptions in epigenetic regulation may be causative rather than simply correlative (142).

Alterations in gene expression, DNA methylation, histone modification, and genome organization have been associated with neurodegenerative disease in animal models and patient tissues. In postmortem brain tissue from monozygotic twins discordant for AD, and in AD patients versus controls, global levels of DNA methylation and hydroxymethylation were reduced in those with AD (42, 138). Site-specific DNA hypo- and hypermethylation in AD brains versus controls have been reported (12, 217), but the changes are relatively modest and difficult to separate from age-related alterations. Some genes subject to chromatin alterations are implicated in AD pathology, such as hypomethylation of genes implicated in amyloid-β-peptide generation, including *TMEM59* and *PSEN1* (12, 217). A recent study also demonstrated widespread loss of heterochromatin in τ-transgenic animal models and human AD, with a corresponding increase in gene expression (61). τ expression promoted oxidative stress and DNA damage, which subsequently caused the loss of heterochromatin and neuronal death (61).

In postmortem brain samples and in blood, DNA methylation was reduced at *SNCA* in sporadic PD patients versus controls (92, 203). *SNCA*, a major risk gene for PD, encodes α-synuclein, a key component of Lewy bodies. A genome-wide study of DNA methylation in blood and brain found hypo- and hypermethylated genes, including known PD-related genes, with approximately 30% agreement between the genes in the different tissues (137). In keeping with such methylation changes, nuclear DNMT1 levels are reduced in PD postmortem brains (49). In rat neuronal cells, α-synuclein overexpression relocates Dnmt1 to the cytoplasm, suggesting a mechanism for reduced DNMT1 levels in PD (49). Interestingly, availability of the methyl donor for DNA and histone methylation (measured as SAM/SAH in the blood) is related to PD symptoms; better cognitive function was related to higher methylation potential (159).

ELS: early life stress

PTSD: posttraumatic stress disorder

Early Life Stress, Epigenetics, and Disease

Exposure to early life stress (ELS) has been correlated with an enhanced risk of psychopathologies such as schizophrenia, bipolar disorder, depression, and PTSD as well as traits such as aggressive behavior and drug addiction (63). It is thought that such exposures alter gene expression programs and thereby predispose individuals to disease in later life (**Figure 2**). Chromatin modifications have the ability to be stably inherited, rapidly turned over, and influenced by external factors, and so provide a potential mechanism for such programming.

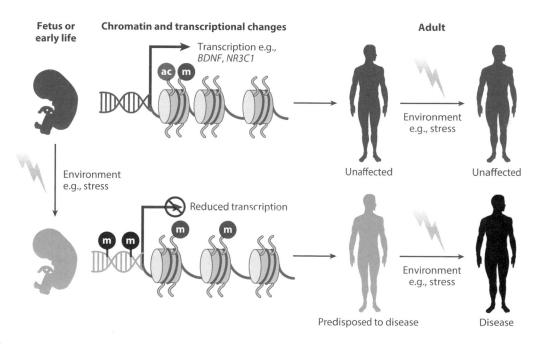

Figure 2

Epigenetic programming during developmental windows can induce predisposition to disease. Stress during fetal development or early life can cause chromatin and transcriptional changes, such as hypermethylation of *BDNF* or *NR3C1*. These changes may alter neurodevelopment or the setting of neuroendocrine axes, leading to a psychologically vulnerable phenotype and pathological responses to stressors experienced later in life. Similarly, metabolic changes in response to fetal starvation epigenetically generate a thrifty phenotype, which can cause predisposition to diabetes and hypertension. Abbreviations: ac, histone acetylation; m, methylation.

HPA: hypothalamic-pituitary-adrenal

Modeling ELS in animal models has centered on levels of maternal care during the first few weeks of life (126, 156). In rodents, as in humans, poor maternal care results in behavioral dysfunction: Adult animals that were exposed to such stresses display phenotypes indicative of depression and anxiety disorders, show inappropriate responses to stressful events, and tend to perpetuate the lack of care to their own offspring. Taking pups born to low-care mothers and fostering them with mothers who provide good maternal care rescues the phenotypes, demonstrating a nongenetic mechanism (59).

Translation of animal findings to humans has been contingent on observing similar epigenetic changes in blood samples, or in postmortem brains, which comes with its own caveats, such as whether cause of death affects the chromatin profile. However, several studies suggest that phenotypic similarities between species are mediated epigenetically. It should be noted that there are many genes whose expression and/or methylation status changes after ELS in mice, in nonhuman primates, and in humans (111, 149, 175, 219). A few gene expression changes have been looked at in detail, and these are discussed below and in **Table 2**.

Epigenetic regulation of the hypothalamic-pituitary-adrenal axis. The HPA (hypothalamic-pituitary-adrenal) axis regulates homeostatic mechanisms, including the ability to respond appropriately to adverse events (reviewed in 237). ELS animals and humans have a hyperactive neuroendocrine response to stressors later in life, making them more fearful; this is linked to altered epigenetic programming and expression of genes regulating the HPA axis (**Table 2**).

ELS animals exhibit DNA hypermethylation at the *NR3C1* gene, which encodes the glucocorticoid receptor (GR) in the hippocampus (218), and at the *ERα1b* locus in the hypothalamus (36), and they display DNA hypomethylation at the *Avp* (147) and *CRH* (39) loci in the hypothalamus. Increased expression of these genes correlates with decreased DNA methylation. DNA methylation also correlated with altered transcription factor binding: Decreased NGFI-A binding is seen at the hypermethylated *NR3C1* locus (218); decreased Stat5 binding is seen at the hypermethylated *ERα1b* (36); increased phospo-CREB binding is seen at the hypomethylated *CRH* (39); and MECP2 binding to methylated DNA is lost at the hypomethylated *Avp* locus (147). The chromatin changes were not only seen immediately after ELS in young animals but, importantly, persisted into adulthood (147, 218). For the *NR3C1* locus, it was shown that the changes are reversible not only with cross-fostering but also with HDAC inhibitor treatment (218), demonstrating the labile nature of the changes and their epigenetic origin. The genes described have critical functions in regulating the activity of the HPA axis and hence appropriate stress responses.

Human sufferers of depression, bipolar disorder, and schizophrenia show reduced *NR3C1* expression in certain brain regions, including the hippocampus (221). Development of such disorders is linked to ELS, and the effects of ELS on the *NR3C1* locus in mice (218) are seen in humans exposed to childhood abuse who later go on to commit suicide, relative to nonabused suicide victims or controls (139). DNA methylation levels at the *NR3C1* gene also correlate with childhood abuse in peripheral blood from adults with a variety of psychopathologies (169). The number of CRH and Avp expressing neurons in the hypothalamus is increased in depression (177), but epigenetic changes at their loci have not been investigated in human disease.

Epigenetic regulation of additional neuronal function genes. Chromatin changes in response to ELS have also been seen at genes associated with neuronal function outside of HPA regulation. ELS pups show DNA hypermethylation associated with decreased hippocampal *Grm1* (10) and *Gad1* expression (240). Lower promoter levels of H3K9ac and H3K4me3 are seen at *Grm1* (10), and reduced H3K9ac and NGFI-A binding are seen at *Gad1* (240). The *Grm1* gene encodes a glutamate receptor, and altered glutamate signaling after ELS results in modified synaptic strength

Table 2 Programming of disease through epigenetic changes during early life adversity

	Disease	Genes	Altered in	References
Patients				
Early life stress				
Childhood adversity	Suicide	NR3C1	Brain	(139)
Childhood adversity	Mental illness	NR3C1	Blood	(169)
Childhood adversity	Mental illness	BDNF	Blood, brain	(170)
Childhood adversity	Mental illness	FKBP5	Blood	(104)
Childhood adversity	Mental illness	SLC6A4	Blood	(95)
Childhood adversity	Mental illness	Cytokine genes	Blood	(197)
Fetal programming				
Maternal stress	Mental illness	NR3C1	Blood	(160, 178)
Prenatal famine	Metabolic disorders	IGF2	Blood	(75)
IUGR	Metabolic disorders	HNF4A	HSCs	(52)
Animal models				
Early life stress				
Poor maternal care	Anxiety/depression	NR3C1	Hippocampus	(218)
Poor maternal care	Anxiety/depression	Grm1	Hippocampus	(10)
Poor maternal care	Anxiety/depression	Gad1	Hippocampus	(240)
Poor maternal care	Anxiety/depression	ERα	Hypothalamus	(36)
Poor maternal care	Anxiety/depression	CRH	Hypothalamus	(39)
Poor maternal care	Anxiety/depression	Avp	Hypothalamus	(147)
Poor maternal care	Anxiety/depression	Bdnf	Prefrontal cortex	(185)
Poor maternal care	Anxiety/depression	MeCP2, CB1, CRFR2	Germline	(60)
Corticosteroid infusion	Anxiety/depression	FKBP5	Hippocampus, hypothalamus	(118)
Fetal programming				
Maternal stress	Anxiety/depression	CRH, NR3C1	Hypothalamus	(145)
IUGR	Metabolic disorders	Fgfr1, Vgf, Gch1, Pcsk5	Pancreas	(207)
IUGR	Metabolic disorders	NR3C1, PPARa	Liver	(123)
IUGR	Hypertension	AT_{1b}	Adrenal gland	(25)
High-fat diet	Metabolic disorders	PPARa	Liver	(231)

Abbreviations: HSCs, hematopoietic stem cells; IUGR, intrauterine growth retardation.

in the adult hippocampus (10). Deleterious *GRM1* mutations are found in schizophrenic patients (8). *GAD1* encodes glutamic acid decarboxylase, the rate-limiting enzyme in GABA synthesis; levels of this enzyme are lower in the brains of schizophrenic patients (2).

The neurotrophic factor BDNF is implicated in animal models of depression linked to chronic stress in infancy (185). Maltreated pups show decreased *Bdnf* mRNA in the adult prefrontal cortex and a corresponding increase in DNA methylation at the *Bdnf* locus (185). Decreased expression of the *BDNF* gene in the prefrontal cortex, as well as decreased expression of its receptor *TrkB*, was seen in schizophrenia patients (73) and suicide victims (53, 97). Hypermethylation of *BDNF* and *TrkB* genes was reported in the cortex of suicide subjects (53, 97). Importantly, the level of DNA methylation at the *BDNF* locus was not only higher in adults with borderline personality disorder versus controls but was shown to correlate with the severity of childhood trauma suffered (170).

Fetal Programming and Human Disease

Epidemiological evidence, including adventitiously acquired data, has suggested a link between in utero conditions and adult health for some time. For example, the Dutch Hunger Winter of 1944–1945 was severe but short lived, enabling analysis of the effects of famine during only the gestational period. Gestational famine resulted in an anticipated restriction of fetal growth but, surprisingly, also an increased incidence of type II diabetes, obesity, and cardiovascular disease in later life, particularly if the famine coincided with the early stages of pregnancy (165). The thrifty phenotype hypothesis (70) suggests that prenatal starvation preprograms the baby for a life of hardship through adaptive alterations in metabolism. However, if these children are well-nourished, such fetal adaptations result in a tendency to overaccumulate fat and energy deposits, resulting in metabolic disorders later in life. Epigenetic modifications are one mechanism of fetal programming. Nutritional and other fetal stresses during pregnancy can also have adverse effects on the mental health of the offspring and have been shown to be risk factors for ASD (132), schizophrenia (200) and other mood disorders (151).

Epigenetic changes linked to prenatal nutrition. Prenatal exposure to the Dutch Hunger Winter caused alterations in DNA methylation at the *IGF2* locus in blood, which were detected six decades after the insult occurred (75). *IGF2* is a maternally imprinted gene with a critical role in human growth and development. Interestingly, this correlation was not seen in individuals who were exposed to famine only at the end of the gestational period, suggesting that there are critical windows of development when such programming can occur. Wide-ranging changes in DNA methylation after famine-related intrauterine growth retardation (IUGR) are also seen in HSCs of human neonates, although not at the *IGF2* locus (52). *HNF4A* was hypermethylated after IUGR (52), consistent with inherited loss-of-function mutations in *HNF4A* causing an autosomal-dominant form of maturity-onset diabetes of the young (230).

IUGR can be modeled in animals through maternal diet restriction or experimental interventions such as uterine artery ligation, which result in a lower birth weight and type II diabetes in adulthood (196). Prior to the onset of adult disease, there were significant DNA methylation changes in rat pancreatic islets at genes linked to pancreatic function (207). Corresponding expression changes were seen at a panel of genes, including hypomethylation and increased expression of *Fgfr1* as well as hypermethylation and decreased expression of *Vgf*, *Gch1*, and *Pcsk5* in IUGR animals (207).

Pathways involved in lipid metabolism are also altered following fetal malnourishment. Hypomethylation and increased expression of *NR3C1* and *PPARα* genes are seen in IUGR animals (123). The GR encoded by *NR3C1* is implicated in regulating adipogenesis, and PPARs are transcription factors that regulate lipid metabolism. Hepatic *PPARα* expression was also increased in newborns but decreased at weaning in pups of mothers fed a high-fat diet (HFD) (231), a nutritional stress seemingly opposite to IUGR but with a similar phenotypic outcome of offspring metabolic syndrome. This suggests that *PPARα* may be a key target for epigenetic alteration in fetal programming of metabolism. Adult hypertension is also linked to fetal starvation, and correspondingly, the AT_{1b} angiotensin receptor gene in the adrenal gland was DNA hypomethylated and upregulated in adult rats exposed to famine prenatally (25).

Epigenetic changes linked to prenatal stress. Some of the same gene expression programs are influenced by stress in utero as in early life (**Table 2**). Hypomethylation of the *CRH* gene and hypermethylation of the *NR3C1* gene were seen in the hypothalamus of pups prenatally exposed to maternal stress (145). In humans, hypermethylation of the *NR3C1* gene is seen in cord blood cells of babies exposed to prenatal maternal depression; these babies show altered cortisol stress responses at three months (160). Moreover, *NR3C1* hypermethylation and decreased *NR3C1* expression in response to prenatal stress are maintained in the offspring for many years (178).

Transgenerational Epigenetic Inheritance of Disease Traits

An additional layer of intrigue is the concept that changes precipitated in an individual by the environment can be passed onto the next generation, originally postulated by Jean-Baptiste Lamarck. Inheritance of acquired characteristics could be mediated through epigenetics. The hypothesis is that external signals cause phenotypic adaptation through effecting changes in the epigenetic code and thereby the gene expression program. These epigenetic changes are then passed onto the offspring, who inherit the phenotype in the absence of the external signal that precipitated it.

Inheritance of chromatin-linked disease traits is easier to look at in paternal transmission, as this clearly distinguishes transgenerational inheritance from fetal programming effects. Epidemiological data in humans link the experience of famine in paternal grandfathers to obesity and cardiovascular disease two generations later (94, 168), demonstrating the relevance of this to disease. In mice, paternal diet influences gene expression and DNA methylation in the liver of the offspring, together with their metabolic profile (33). As in fetal programming, methylation at the *PPARα* gene is implicated: Males fed low-protein diets had pups exhibiting hypermethylation of a *PPARα* enhancer and a concomitant decrease in *PPARα* expression (33). Of note, the observed changes in DNA methylation were not seen in the sperm of the fathers, suggesting that the DNA methylation changes are not directly passed through the germline; changes in sperm RNA content and histone modifications were related to diet (33).

A chronic HFD in fathers led to impaired glucose tolerance in female offspring and epigenetic and transcriptional changes in pancreatic islet cells (153, 222) and adipose tissue (154). For example, DNA hypomethylation and increased expression of *Il13ra2* as well as DNA hypermethylation and decreased expression of *Pik3r1* and *Pik3ca* were seen in pancreatic cells from HFD offspring (153). Interestingly, hypermethylation of *Pik3ca* was seen in sperm from the HFD father, and hypermethylation of *Pik3ca* and *Pik3r1* was seen in E3.5 blastocysts from HFD fathers, suggesting inheritance of methylated alleles (222). Furthermore, epigenetic changes at *Pik3ca* and *Pik3r1* were seen in the next generation, whose fathers were not fed an HFD (222).

Multigenerational changes after maternal insult are also evidence for transgenerational inheritance of a phenotype. Fetal programming by prenatal famine causes glucose metabolism alterations not just in the exposed pups but in several more generations (20). DNA hypomethylation at the *NR3C1* and *PPARα* genes seen in IUGR animals (123) is also passed onto the subsequent generation (28).

CONCLUSIONS

To conclude, epigenetic changes are observed in many disease states, and chromatin modifications may act as biomarkers, disease drivers, or therapeutic targets. We have discussed what we consider to be three key areas of the epigenetic mechanisms of disease (**Figure 1**). First, genetic mutations can alter the expression of chromatin regulators, leading directly to monogenic disease via epigenetic mechanisms. Second, genetic mutations can alter the expression of chromatin regulators, leading to multifactorial disease when in combination with other events. This includes mutations that arise somatically in specific tissues, leading to cancer, and de novo mutations identified in ASD and CHD. It should be noted that we focused on enzymes that add and remove DNA and histone modifications, or that remodel chromatin directly, but there are many other relevant targets, including reader proteins that interpret the epigenetic code, noncoding RNAs that interact

Table 3 The specific nature and context of epigenetic changes determine disease[a]

Modification	Regulator	Disease caused by		
		Hereditary mutation	**Somatic mutation**	**Environmental change**
DNA methylation	DNMT1	HSNE1	Cancer	Cancer, schizophrenia
H3K27 methylation	EZH2	Weaver syndrome	Cancer	Cancer
H3K4 methylation	MLL2	Kabuki syndrome	Cancer, CHD	Cancer
Histone acetylation	CBP, P300	Rubinstein-Taybi syndrome	Cancer	Cancer, aging, neurodegeneration
Chromatin remodeling	CHD7	CHARGE syndrome	ASD, CHD, cancer	Unknown

[a]Examples of epigenetic modifications and the regulators associated with their deposition that are implicated in different diseases depending on the location and timing of changes induced in them.

Abbreviations: ASD, autism spectrum disorder; CHARGE, coloboma, heart anomaly, choanal atresia, retardation, genital, and ear anomalies; CHD, congenital heart disease.

with chromatin-modifying complexes, and enzymes that regulate the availability of precursors for chromatin modifications, such as SAM for methylation. Finally, we discussed changes in epigenetic modifications that occur in response to environmental stimuli, leading to disease predisposition, onset, or progression.

It is imperative to note that mutations in the same chromatin regulator, or changes in an epigenetic modification, can be involved in more than one disease (**Table 3**). This may be due to the timing and location of the changes or whether the changes constitute gain-of-function or loss-of-function. For example, changes in H3K27 methylation have been described in cancer, in keeping with the identification of somatic mutations in the H3K27 methyltransferase EZH2 in cancer, whereas germline *EZH2* mutations cause Weaver syndrome. Notably, some of the germline *EZH2* mutations identified in Weaver's syndrome are identical to somatic cancer mutations.

The importance of epigenetics in disease is reinforced by evidence that epigenetic drugs show efficacy in treatment of a multitude of disorders, including cancer, AD, depression, and cardiovascular disease (reviewed in 150). The promise of targeting the epigenome, even in diseases in which the chromatin-related changes may be consequence rather than cause, is that chromatin changes are able to directly affect gene expression and are, by definition, plastic and thus reversible. Challenges include the lack of specificity, and therefore the potential side effects, of drugs targeting the epigenome because the targeted enzymes are often ubiquitously expressed, critical for many biological processes, and target many different genes. In addition, there are redundancies in chromatin regulators and multiple levels of chromatin regulation, which could mean that targeting several enzymes simultaneously may be more effective.

FUTURE ISSUES

1. Much of the evidence regarding the function of chromatin modifications in human disease has been correlative. Recent genome-editing techniques, such as CRISPR/Cas and TALEs (transcription activator–like effectors), have tremendous potential in investigating the effect on chromatin conformation, gene transcription, and cellular phenotype of directing a particular epigenetic regulator to specific genomic locations (141). This will be of great advantage in understanding the molecular mechanisms by which mutations in chromatin regulator genes cause disease.

2. Patient-specific iPS cell disease models hold great promise for investigating disease mechanisms, and for drug screening, in clinically relevant human tissues. Examples include modeling of monogenic diseases caused by mutations in chromatin regulators and assessment of epigenetic changes in multifactorial diseases. Such technology is particularly pertinent for diseases with neurological symptoms, where iPS cell generation and differentiation enable molecular and cellular analysis in human neuronal cell types (26, 48, 135).

3. The understanding of the contribution of noncoding RNA species, including microRNAs and long noncoding RNAs, to gene regulation has been exponentially growing in recent years, and their contribution to disease states is an important emerging field.

4. The contribution of epigenomics to personalized medicine is a key area for ongoing research. Proof of principle has been shown in cancer, where DNA hypermethylation of the *MGMT* promoter, a gene encoding a DNA repair enzyme that protects the cell against alkylating agents, is a useful marker for predicting which patients will respond to alkylating antineoplastic treatment (3).

DISCLOSURE STATEMENT

The authors are not aware of any affiliations, memberships, funding, or financial holdings that might be perceived as affecting the objectivity of this review.

ACKNOWLEDGMENTS

We thank J.N. Anastas and A.I. Badeaux for critical reading of the manuscript. E.B. was supported in part by an EMBO Postdoctoral Fellowship and a Jérôme Lejeune Foundation Grant. Research in the Shi lab is supported by grants from the NIH (CA118478, MH096066), the Ellison Medical Foundation and the Samuel Waxman Cancer Research Foundation. Y. Shi is an American Cancer Society Research Professor.

LITERATURE CITED

1. Abidi FE, Cardoso C, Lossi AM, Lowry RB, Depetris D, et al. 2005. Mutation in the 5′ alternatively spliced region of the *XNP/ATR-X* gene causes Chudley-Lowry syndrome. *Eur. J. Hum. Genet.* 13:176–83
2. Akbarian S, Kim JJ, Potkin SG, Hagman JO, Tafazzoli A, et al. 1995. Gene expression for glutamic acid decarboxylase is reduced without loss of neurons in prefrontal cortex of schizophrenics. *Arch. Gen. Psychiatry* 52:258–66
3. Amatu A, Sartore-Bianchi A, Moutinho C, Belotti A, Bencardino K, et al. 2013. Promoter CpG island hypermethylation of the DNA repair enzyme MGMT predicts clinical response to dacarbazine in a phase II study for metastatic colorectal cancer. *Clin. Cancer Res.* 19:2265–72
4. Amir RE, Van den Veyver IB, Schultz R, Malicki DM, Tran CQ, et al. 2000. Influence of mutation type and X chromosome inactivation on Rett syndrome phenotypes. *Ann. Neurol.* 47:670–79
5. Amir RE, Van den Veyver IB, Wan M, Tran CQ, Francke U, Zoghbi HY. 1999. Rett syndrome is caused by mutations in X-linked MECP2, encoding methyl-CpG-binding protein 2. *Nat. Genet.* 23:185–88
6. Andersen TA, Troelsen KD, Larsen LA. 2013. Of mice and men: molecular genetics of congenital heart disease. *Cell. Mol. Life Sci.* 71:1327–52
7. Archer TC, Pomeroy SL. 2012. Medulloblastoma biology in the post-genomic era. *Future Oncol.* 8:1597–604

8. Ayoub MA, Angelicheva D, Vile D, Chandler D, Morar B, et al. 2012. Deleterious GRM1 mutations in schizophrenia. *PLOS ONE* 7:e32849

9. Azuara V, Perry P, Sauer S, Spivakov M, Jorgensen HF, et al. 2006. Chromatin signatures of pluripotent cell lines. *Nat. Cell Biol.* 8:532–38

10. Bagot RC, Zhang TY, Wen X, Nguyen TT, Nguyen HB, et al. 2012. Variations in postnatal maternal care and the epigenetic regulation of metabotropic glutamate receptor 1 expression and hippocampal function in the rat. *Proc. Natl. Acad. Sci. USA* 109(Suppl. 2):17200–7

11. Bajpai R, Chen DA, Rada-Iglesias A, Zhang J, Xiong Y, et al. 2010. CHD7 cooperates with PBAF to control multipotent neural crest formation. *Nature* 463:958–62

12. Bakulski KM, Dolinoy DC, Sartor MA, Paulson HL, Konen JR, et al. 2012. Genome-wide DNA methylation differences between late-onset Alzheimer's disease and cognitively normal controls in human frontal cortex. *J. Alzheimers Dis.* 29:571–88

13. Balemans MC, Ansar M, Oudakker A, van Caam AP, Bakker B, et al. 2013. Reduced euchromatin histone methyltransferase 1 causes developmental delay, hypotonia, and cranial abnormalities associated with increased bone gene expression in Kleefstra syndrome mice. *Dev. Biol.* 386:395–407

14. Barlesi F, Giaccone G, Gallegos-Ruiz MI, Loundou A, Span SW, et al. 2007. Global histone modifications predict prognosis of resected non small-cell lung cancer. *J. Clin. Oncol.* 25:4358–64

15. Bartholdi D, Roelfsema JH, Papadia F, Breuning MH, Niedrist D, et al. 2007. Genetic heterogeneity in Rubinstein-Taybi syndrome: delineation of the phenotype of the first patients carrying mutations in EP300. *J. Med. Genet.* 44:327–33

16. Batsukh T, Pieper L, Koszucka AM, von Velsen N, Hoyer-Fender S, et al. 2010. CHD8 interacts with CHD7, a protein which is mutated in CHARGE syndrome. *Hum. Mol. Genet.* 19:2858–66

17. Bell JT, Spector TD. 2011. A twin approach to unraveling epigenetics. *Trends Genet.* 27:116–25

18. Bender S, Tang Y, Lindroth AM, Hovestadt V, Jones DT, et al. 2013. Reduced H3K27me3 and DNA hypomethylation are major drivers of gene expression in K27M mutant pediatric high-grade gliomas. *Cancer Cell* 24:660–72

19. Benevolenskaya EV, Murray HL, Branton P, Young RA, Kaelin WG Jr. 2005. Binding of pRB to the PHD protein RBP2 promotes cellular differentiation. *Mol. Cell* 18:623–35

20. Benyshek DC, Johnston CS, Martin JF. 2006. Glucose metabolism is altered in the adequately-nourished grand-offspring (F3 generation) of rats malnourished during gestation and perinatal life. *Diabetologia* 49:1117–19

21. Betancur C. 2011. Etiological heterogeneity in autism spectrum disorders: more than 100 genetic and genomic disorders and still counting. *Brain Res.* 1380:42–77

22. Bickmore WA. 2013. The spatial organization of the human genome. *Annu. Rev. Genomics Hum. Genet.* 14:67–84

23. Bird A. 2002. DNA methylation patterns and epigenetic memory. *Genes Dev.* 16:6–21

24. Bjerke L, Mackay A, Nandhabalan M, Burford A, Jury A, et al. 2013. Histone H3.3 mutations drive pediatric glioblastoma through upregulation of MYCN. *Cancer Discov.* 3:512–19

25. Bogdarina I, Welham S, King PJ, Burns SP, Clark AJ. 2007. Epigenetic modification of the renin-angiotensin system in the fetal programming of hypertension. *Circ. Res.* 100:520–26

26. Brennand KJ, Simone A, Jou J, Gelboin-Burkhart C, Tran N, et al. 2011. Modelling schizophrenia using human induced pluripotent stem cells. *Nature* 473:221–25

27. Brun ME, Lana E, Rivals I, Lefranc G, Sarda P, et al. 2011. Heterochromatic genes undergo epigenetic changes and escape silencing in immunodeficiency, centromeric instability, facial anomalies (ICF) syndrome. *PLOS ONE* 6:e19464

28. Burdge GC, Slater-Jefferies J, Torrens C, Phillips ES, Hanson MA, Lillycrop KA. 2007. Dietary protein restriction of pregnant rats in the F0 generation induces altered methylation of hepatic gene promoters in the adult male offspring in the F1 and F2 generations. *Br. J. Nutr.* 97:435–39

29. Cairns P, Esteller M, Herman JG, Schoenberg M, Jeronimo C, et al. 2001. Molecular detection of prostate cancer in urine by GSTP1 hypermethylation. *Clin. Cancer Res.* 7:2727–30

30. Campeau PM, Kim JC, Lu JT, Schwartzentruber JA, Abdul-Rahman OA, et al. 2012. Mutations in KAT6B, encoding a histone acetyltransferase, cause Genitopatellar syndrome. *Am. J. Hum. Genet.* 90:282–89

31. Cannon TD, Kaprio J, Lonnqvist J, Huttunen M, Koskenvuo M. 1998. The genetic epidemiology of schizophrenia in a Finnish twin cohort. A population-based modeling study. *Arch. Gen. Psychiatry* 55:67–74

32. Cao R, Wang L, Wang H, Xia L, Erdjument-Bromage H, et al. 2002. Role of histone H3 lysine 27 methylation in Polycomb-group silencing. *Science* 298:1039–43

33. Carone BR, Fauquier L, Habib N, Shea JM, Hart CE, et al. 2010. Paternally induced transgenerational environmental reprogramming of metabolic gene expression in mammals. *Cell* 143:1084–96

34. Challen GA, Goodell MA. 2010. Runx1 isoforms show differential expression patterns during hematopoietic development but have similar functional effects in adult hematopoietic stem cells. *Exp. Hematol.* 38:403–16

35. Challen GA, Sun D, Jeong M, Luo M, Jelinek J, et al. 2012. Dnmt3a is essential for hematopoietic stem cell differentiation. *Nat. Genet.* 44:23–31

36. Champagne FA, Weaver IC, Diorio J, Dymov S, Szyf M, Meaney MJ. 2006. Maternal care associated with methylation of the estrogen receptor-α1b promoter and estrogen receptor-α expression in the medial preoptic area of female offspring. *Endocrinology* 147:2909–15

37. Chan KM, Fang D, Gan H, Hashizume R, Yu C, et al. 2013. The histone H3.3K27M mutation in pediatric glioma reprograms H3K27 methylation and gene expression. *Genes Dev.* 27:985–90

38. Chen D, Zhang G. 2001. Enforced expression of the GATA-3 transcription factor affects cell fate decisions in hematopoiesis. *Exp. Hematol.* 29:971–80

39. Chen J, Evans AN, Liu Y, Honda M, Saavedra JM, Aguilera G. 2012. Maternal deprivation in rats is associated with corticotrophin-releasing hormone (CRH) promoter hypomethylation and enhances CRH transcriptional responses to stress in adulthood. *J. Neuroendocrinol.* 24:1055–64

40. Chen T, Ueda Y, Dodge JE, Wang Z, Li E. 2003. Establishment and maintenance of genomic methylation patterns in mouse embryonic stem cells by Dnmt3a and Dnmt3b. *Mol. Cell. Biol.* 23:5594–605

41. Cherry AB, Daley GQ. 2012. Reprogramming cellular identity for regenerative medicine. *Cell* 148:1110–22

42. Chouliaras L, Mastroeni D, Delvaux E, Grover A, Kenis G, et al. 2013. Consistent decrease in global DNA methylation and hydroxymethylation in the hippocampus of Alzheimer's disease patients. *Neurobiol. Aging* 34:2091–99

43. Clayton-Smith J, O'Sullivan J, Daly S, Bhaskar S, Day R, et al. 2011. Whole-exome-sequencing identifies mutations in histone acetyltransferase gene *KAT6B* in individuals with the Say-Barber-Biesecker variant of Ohdo syndrome. *Am. J. Hum. Genet.* 89:675–81

44. Cote J, Quinn J, Workman JL, Peterson CL. 1994. Stimulation of GAL4 derivative binding to nucleosomal DNA by the yeast SWI/SNF complex. *Science* 265:53–60

45. David G, Abbas N, Stevanin G, Durr A, Yvert G, et al. 1997. Cloning of the *SCA7* gene reveals a highly unstable CAG repeat expansion. *Nat. Genet.* 17:65–70

46. Deaton AM, Bird A. 2011. CpG islands and the regulation of transcription. *Genes Dev.* 25:1010–22

47. Delhommeau F, Dupont S, Della Valle V, James C, Trannoy S, et al. 2009. Mutation in TET2 in myeloid cancers. *N. Engl. J. Med.* 360:2289–301

48. Derosa BA, Van Baaren JM, Dubey GK, Vance JM, Pericak-Vance MA, Dykxhoorn DM. 2012. Derivation of autism spectrum disorder–specific induced pluripotent stem cells from peripheral blood mononuclear cells. *Neurosci. Lett.* 516:9–14

49. Desplats P, Spencer B, Coffee E, Patel P, Michael S, et al. 2011. α-Synuclein sequesters Dnmt1 from the nucleus: a novel mechanism for epigenetic alterations in Lewy body diseases. *J. Biol. Chem.* 286:9031–37

50. Eden A, Gaudet F, Waghmare A, Jaenisch R. 2003. Chromosomal instability and tumors promoted by DNA hypomethylation. *Science* 300:455

51. Ehrlich M, Buchanan KL, Tsien F, Jiang G, Sun B, et al. 2001. DNA methyltransferase 3B mutations linked to the ICF syndrome cause dysregulation of lymphogenesis genes. *Hum. Mol. Genet.* 10:2917–31

52. Einstein F, Thompson RF, Bhagat TD, Fazzari MJ, Verma A, et al. 2010. Cytosine methylation dysregulation in neonates following intrauterine growth restriction. *PLOS ONE* 5:e8887

53. Ernst C, Deleva V, Deng X, Sequeira A, Pomarenski A, et al. 2009. Alternative splicing, methylation state, and expression profile of tropomyosin-related kinase B in the frontal cortex of suicide completers. *Arch. Gen. Psychiatry* 66:22–32

54. Esteller M, Corn PG, Baylin SB, Herman JG. 2001. A gene hypermethylation profile of human cancer. *Cancer Res.* 61:3225–29

55. Fattaey AR, Helin K, Dembski MS, Dyson N, Harlow E, et al. 1993. Characterization of the retinoblastoma binding proteins RBP1 and RBP2. *Oncogene* 8:3149–56

56. Feinberg AP, Vogelstein B. 1983. Hypomethylation distinguishes genes of some human cancers from their normal counterparts. *Nature* 301:89–92

57. Fernandez AF, Assenov Y, Martin-Subero JI, Balint B, Siebert R, et al. 2012. A DNA methylation fingerprint of 1628 human samples. *Genome Res.* 22:407–19

58. Fraga MF, Ballestar E, Paz MF, Ropero S, Setien F, et al. 2005. Epigenetic differences arise during the lifetime of monozygotic twins. *Proc. Natl. Acad. Sci. USA* 102:10604–9

59. Francis D, Diorio J, Liu D, Meaney MJ. 1999. Nongenomic transmission across generations of maternal behavior and stress responses in the rat. *Science* 286:1155–58

60. Franklin TB, Russig H, Weiss IC, Graff J, Linder N, et al. 2010. Epigenetic transmission of the impact of early stress across generations. *Biol. Psychiatry* 68:408–15

61. Frost B, Hemberg M, Lewis J, Feany MB. 2014. Tau promotes neurodegeneration through global chromatin relaxation. *Nat. Neurosci.* 17:357–66

62. Gatz M, Pedersen NL, Berg S, Johansson B, Johansson K, et al. 1997. Heritability for Alzheimer's disease: the study of dementia in Swedish twins. *J. Gerontol. A Biol. Sci. Med. Sci.* 52:M117–25

63. Gershon A, Sudheimer K, Tirouvanziam R, Williams LM, O'Hara R. 2013. The long-term impact of early adversity on late-life psychiatric disorders. *Curr. Psychiatry Rep.* 15:352

64. Gibbons RJ, Higgs DR. 2000. Molecular-clinical spectrum of the ATR-X syndrome. *Am. J. Med. Genet.* 97:204–12

65. Gibbons RJ, Picketts DJ, Villard L, Higgs DR. 1995. Mutations in a putative global transcriptional regulator cause X-linked mental retardation with α-thalassemia (ATR-X syndrome). *Cell* 80:837–45

66. Gibbons RJ, Suthers GK, Wilkie AO, Buckle VJ, Higgs DR. 1992. X-linked α-thalassemia/mental retardation (ATR-X) syndrome: localization to Xq12-q21.31 by X inactivation and linkage analysis. *Am. J. Hum. Genet.* 51:1136–49

67. Gibson WT, Hood RL, Zhan SH, Bulman DE, Fejes AP, et al. 2012. Mutations in EZH2 cause Weaver syndrome. *Am. J. Hum. Genet.* 90:110–18

68. Goncalves TF, Goncalves AP, Fintelman Rodrigues N, dos Santos JM, Pimentel MM, Santos-Reboucas CB. 2014. KDM5C mutational screening among males with intellectual disability suggestive of X-linked inheritance and review of the literature. *Eur. J. Med. Genet.* 57:138–44

69. Hagberg B, Aicardi J, Dias K, Ramos O. 1983. A progressive syndrome of autism, dementia, ataxia, and loss of purposeful hand use in girls: Rett's syndrome: report of 35 cases. *Ann. Neurol.* 14:471–79

70. Hales CN, Barker DJ. 1992. Type 2 (non-insulin-dependent) diabetes mellitus: the thrifty phenotype hypothesis. *Diabetologia* 35:595–601

71. Hallmayer J, Cleveland S, Torres A, Phillips J, Cohen B, et al. 2011. Genetic heritability and shared environmental factors among twin pairs with autism. *Arch. Gen. Psychiatry* 68:1095–102

72. Hare EH, Moran PA. 1979. Raised parental age in psychiatric patients: evidence for the constitutional hypothesis. *Br. J. Psychiatry* 134:169–77

73. Hashimoto T, Bergen SE, Nguyen QL, Xu B, Monteggia LM, et al. 2005. Relationship of brain-derived neurotrophic factor and its receptor TrkB to altered inhibitory prefrontal circuitry in schizophrenia. *J. Neurosci.* 25:372–83

74. Hayami S, Kelly JD, Cho HS, Yoshimatsu M, Unoki M, et al. 2011. Overexpression of LSD1 contributes to human carcinogenesis through chromatin regulation in various cancers. *Int. J. Cancer* 128:574–86

75. Heijmans BT, Tobi EW, Stein AD, Putter H, Blauw GJ, et al. 2008. Persistent epigenetic differences associated with prenatal exposure to famine in humans. *Proc. Natl. Acad. Sci. USA* 105:17046–49

76. Heintzman ND, Stuart RK, Hon G, Fu Y, Ching CW, et al. 2007. Distinct and predictive chromatin signatures of transcriptional promoters and enhancers in the human genome. *Nat. Genet.* 39:311–18

77. Hernandez DG, Nalls MA, Gibbs JR, Arepalli S, van der Brug M, et al. 2011. Distinct DNA methylation changes highly correlated with chronological age in the human brain. *Hum. Mol. Genet.* 20:1164–72

78. Hess JL. 2004. MLL: a histone methyltransferase disrupted in leukemia. *Trends Mol. Med.* 10:500–7

79. Hood RL, Lines MA, Nikkel SM, Schwartzentruber J, Beaulieu C, et al. 2012. Mutations in SRCAP, encoding SNF2-related CREBBP activator protein, cause Floating-Harbor syndrome. *Am. J. Hum. Genet.* 90:308–13

80. Hoyer J, Ekici AB, Endele S, Popp B, Zweier C, et al. 2012. Haploinsufficiency of ARID1B, a member of the SWI/SNF-A chromatin-remodeling complex, is a frequent cause of intellectual disability. *Am. J. Hum. Genet.* 90:565–72

81. Huang J, Sengupta R, Espejo AB, Lee MG, Dorsey JA, et al. 2007. p53 is regulated by the lysine demethylase LSD1. *Nature* 449:105–8

82. Illingworth RS, Bird AP. 2009. CpG islands: a rough guide. *FEBS Lett.* 583:1713–20

83. Issa JP, Ahuja N, Toyota M, Bronner MP, Brentnall TA. 2001. Accelerated age-related CpG island methylation in ulcerative colitis. *Cancer Res.* 61:3573–77

84. Issaeva I, Zonis Y, Rozovskaia T, Orlovsky K, Croce CM, et al. 2007. Knockdown of ALR (MLL2) reveals ALR target genes and leads to alterations in cell adhesion and growth. *Mol. Cell. Biol.* 27:1889–903

85. Iwase S, Lan F, Bayliss P, de la Torre-Ubieta L, Huarte M, et al. 2007. The X-linked mental retardation gene *SMCX/JARID1C* defines a family of histone H3 lysine 4 demethylases. *Cell* 128:1077–88

86. Jeanpierre M, Turleau C, Aurias A, Prieur M, Ledeist F, et al. 1993. An embryonic-like methylation pattern of classical satellite DNA is observed in ICF syndrome. *Hum. Mol. Genet.* 2:731–35

87. Jensen LR, Amende M, Gurok U, Moser B, Gimmel V, et al. 2005. Mutations in the *JARID1C* gene, which is involved in transcriptional regulation and chromatin remodeling, cause X-linked mental retardation. *Am. J. Hum. Genet.* 76:227–36

88. Jenuwein T, Allis CD. 2001. Translating the histone code. *Science* 293:1074–80

89. Jin B, Tao Q, Peng J, Soo HM, Wu W, et al. 2008. DNA methyltransferase 3B (DNMT3B) mutations in ICF syndrome lead to altered epigenetic modifications and aberrant expression of genes regulating development, neurogenesis and immune function. *Hum. Mol. Genet.* 17:690–709

90. Jones WD, Dafou D, McEntagart M, Woollard WJ, Elmslie FV, et al. 2012. De novo mutations in MLL cause Wiedemann-Steiner syndrome. *Am. J. Hum. Genet.* 91:358–64

91. Johnston H, Kneer J, Chackalaparampil I, Yaciuk P, Chrivia J. 1999. Identification of a novel SNF2/SWI2 protein family member, SRCAP, which interacts with CREB-binding protein. *J. Biol. Chem.* 274:16370–76

92. Jowaed A, Schmitt I, Kaut O, Wullner U. 2010. Methylation regulates α-synuclein expression and is decreased in Parkinson's disease patients' brains. *J. Neurosci.* 30:6355–59

93. Jurkowska RZ, Jurkowski TP, Jeltsch A. 2011. Structure and function of mammalian DNA methyltransferases. *ChemBioChem* 12:206–22

94. Kaati G, Bygren LO, Edvinsson S. 2002. Cardiovascular and diabetes mortality determined by nutrition during parents' and grandparents' slow growth period. *Eur. J. Hum. Genet.* 10:682–88

95. Kang HJ, Kim JM, Stewart R, Kim SY, Bae KY, et al. 2013. Association of SLC6A4 methylation with early adversity, characteristics and outcomes in depression. *Prog. Neuropsychopharmacol. Biol. Psychiatry* 44:23–28

96. Katsurano M, Niwa T, Yasui Y, Shigematsu Y, Yamashita S, et al. 2012. Early-stage formation of an epigenetic field defect in a mouse colitis model, and non-essential roles of T- and B-cells in DNA methylation induction. *Oncogene* 31:342–51

97. Keller S, Sarchiapone M, Zarrilli F, Videtic A, Ferraro A, et al. 2010. Increased BDNF promoter methylation in the Wernicke area of suicide subjects. *Arch. Gen. Psychiatry* 67:258–67

98. Khuong-Quang DA, Buczkowicz P, Rakopoulos P, Liu XY, Fontebasso AM, et al. 2012. K27M mutation in histone H3.3 defines clinically and biologically distinct subgroups of pediatric diffuse intrinsic pontine gliomas. *Acta Neuropathol.* 124:439–47

99. Kim HG, Kurth I, Lan F, Meliciani I, Wenzel W, et al. 2008. Mutations in CHD7, encoding a chromatin-remodeling protein, cause idiopathic hypogonadotropic hypogonadism and Kallmann syndrome. *Am. J. Hum. Genet.* 83:511–19

100. Kleefstra T, Brunner HG, Amiel J, Oudakker AR, Nillesen WM, et al. 2006. Loss-of-function mutations in euchromatin histone methyl transferase 1 (EHMT1) cause the 9q34 subtelomeric deletion syndrome. *Am. J. Hum. Genet.* 79:370–77

101. Kleefstra T, van Zelst-Stams WA, Nillesen WM, Cormier-Daire V, Houge G, et al. 2009. Further clinical and molecular delineation of the 9q subtelomeric deletion syndrome supports a major contribution of EHMT1 haploinsufficiency to the core phenotype. *J. Med. Genet.* 46:598–606

102. Kleefstra T, Kramer JM, Neveling K, Willemsen MH, Koemans TS, et al. 2012. Disruption of an EHMT1-associated chromatin-modification module causes intellectual disability. *Am. J. Hum. Genet.* 91:73–82

103. Klein CJ, Botuyan MV, Wu Y, Ward CJ, Nicholson GA, et al. 2011. Mutations in DNMT1 cause hereditary sensory neuropathy with dementia and hearing loss. *Nat. Genet.* 43:595–600

104. Klengel T, Mehta D, Anacker C, Rex-Haffner M, Pruessner JC, et al. 2013. Allele-specific FKBP5 DNA demethylation mediates gene-childhood trauma interactions. *Nat. Neurosci.* 16:33–41

105. Ko M, Bandukwala HS, An J, Lamperti ED, Thompson EC, et al. 2011. Ten-eleven-translocation 2 (TET2) negatively regulates homeostasis and differentiation of hematopoietic stem cells in mice. *Proc. Natl. Acad. Sci. USA* 108:14566–71

106. Ko M, Huang Y, Jankowska AM, Pape UJ, Tahiliani M, et al. 2010. Impaired hydroxylation of 5-methylcytosine in myeloid cancers with mutant TET2. *Nature* 468:839–43

107. Kohli RM, Zhang Y. 2013. TET enzymes, TDG and the dynamics of DNA demethylation. *Nature* 502:472–79

108. Kong A, Frigge ML, Masson G, Besenbacher S, Sulem P, et al. 2012. Rate of de novo mutations and the importance of father's age to disease risk. *Nature* 488:471–75

109. Krumm N, O'Roak BJ, Shendure J, Eichler EE. 2013. A de novo convergence of autism genetics and molecular neuroscience. *Trends Neurosci.* 37:95–105

110. Kurotaki N, Imaizumi K, Harada N, Masuno M, Kondoh T, et al. 2002. Haploinsufficiency of NSD1 causes Sotos syndrome. *Nat. Genet.* 30:365–66

111. Labonte B, Suderman M, Maussion G, Navaro L, Yerko V, et al. 2012. Genome-wide epigenetic regulation by early-life trauma. *Arch. Gen. Psychiatry* 69:722–31

112. Laird A, Thomson JP, Harrison DJ, Meehan RR. 2013. 5-hydroxymethylcytosine profiling as an indicator of cellular state. *Epigenomics* 5:655–69

113. Lan F, Bayliss PE, Rinn JL, Whetstine JR, Wang JK, et al. 2007. A histone H3 lysine 27 demethylase regulates animal posterior development. *Nature* 449:689–94

114. Laumonnier F, Holbert S, Ronce N, Faravelli F, Lenzner S, et al. 2005. Mutations in PHF8 are associated with X linked mental retardation and cleft lip/cleft palate. *J. Med. Genet.* 42:780–86

115. Lawrence MS, Stojanov P, Mermel CH, Robinson JT, Garraway LA, et al. 2014. Discovery and saturation analysis of cancer genes across 21 tumour types. *Nature* 505:495–501

116. Lederer D, Grisart B, Digilio MC, Benoit V, Crespin M, et al. 2012. Deletion of KDM6A, a histone demethylase interacting with MLL2, in three patients with Kabuki syndrome. *Am. J. Hum. Genet.* 90:119–24

117. Lee MG, Villa R, Trojer P, Norman J, Yan KP, et al. 2007. Demethylation of H3K27 regulates polycomb recruitment and H2A ubiquitination. *Science* 318:447–50

118. Lee RS, Tamashiro KL, Yang X, Purcell RH, Harvey A, et al. 2010. Chronic corticosterone exposure increases expression and decreases deoxyribonucleic acid methylation of Fkbp5 in mice. *Endocrinology* 151:4332–43

119. Leoyklang P, Suphapeetiporn K, Siriwan P, Desudchit T, Chaowanapanja P, et al. 2007. Heterozygous nonsense mutation SATB2 associated with cleft palate, osteoporosis, and cognitive defects. *Hum. Mutat.* 28:732–38

120. Leoyklang P, Suphapeetiporn K, Srichomthong C, Tongkobpetch S, Fietze S, et al. 2013. Disorders with similar clinical phenotypes reveal underlying genetic interaction: SATB2 acts as an activator of the *UPF3B* gene. *Hum. Genet.* 132:1383–93

121. Lewis PW, Muller MM, Koletsky MS, Cordero F, Lin S, et al. 2013. Inhibition of PRC2 activity by a gain-of-function H3 mutation found in pediatric glioblastoma. *Science* 340:857–61

122. Li Y, Trojer P, Xu CF, Cheung P, Kuo A, et al. 2009. The target of the NSD family of histone lysine methyltransferases depends on the nature of the substrate. *J. Biol. Chem.* 284:34283–95

123. Lillycrop KA, Phillips ES, Jackson AA, Hanson MA, Burdge GC. 2005. Dietary protein restriction of pregnant rats induces and folic acid supplementation prevents epigenetic modification of hepatic gene expression in the offspring. *J. Nutr.* 135:1382–86

124. Lim DH, Maher ER. 2010. Genomic imprinting syndromes and cancer. *Adv. Genet.* 70:145–75

125. Lin W, Cao J, Liu J, Beshiri ML, Fujiwara Y, et al. 2011. Loss of the retinoblastoma binding protein 2 (RBP2) histone demethylase suppresses tumorigenesis in mice lacking Rb1 or Men1. *Proc. Natl. Acad. Sci. USA* 108:13379–86

126. Liu D, Diorio J, Tannenbaum B, Caldji C, Francis D, et al. 1997. Maternal care, hippocampal glucocorticoid receptors, and hypothalamic-pituitary-adrenal responses to stress. *Science* 277:1659–62

127. Lopez-Atalaya JP, Gervasini C, Mottadelli F, Spena S, Piccione M, et al. 2012. Histone acetylation deficits in lymphoblastoid cell lines from patients with Rubinstein-Taybi syndrome. *J. Med. Genet.* 49:66–74

128. Lorsbach RB, Moore J, Mathew S, Raimondi SC, Mukatira ST, Downing JR. 2003. TET1, a member of a novel protein family, is fused to MLL in acute myeloid leukemia containing the t(10;11)(q22;q23). *Leukemia* 17:637–41

129. Lu H, Liu X, Deng Y, Qing H. 2013. DNA methylation, a hand behind neurodegenerative diseases. *Front. Aging Neurosci.* 5:85

130. Lu PJ, Sundquist K, Baeckstrom D, Poulsom R, Hanby A, et al. 1999. A novel gene (*PLU-1*) containing highly conserved putative DNA/chromatin binding motifs is specifically up-regulated in breast cancer. *J. Biol. Chem.* 274:15633–45

131. Lu T, Jackson MW, Wang B, Yang M, Chance MR, et al. 2010. Regulation of NF-κB by NSD1/FBXL11-dependent reversible lysine methylation of p65. *Proc. Natl. Acad. Sci. USA* 107:46–51

132. Lyall K, Schmidt RJ, Hertz-Picciotto I. 2014. Maternal lifestyle and environmental risk factors for autism spectrum disorders. *Int. J. Epidemiol.* 43:443–64

133. Madzo J, Liu H, Rodriguez A, Vasanthakumar A, Sundaravel S, et al. 2014. Hydroxymethylation at gene regulatory regions directs stem/early progenitor cell commitment during erythropoiesis. *Cell Rep.* 6:231–44

134. Mallery DL, Tanganelli B, Colella S, Steingrimsdottir H, van Gool AJ, et al. 1998. Molecular analysis of mutations in the *CSB* (*ERCC6*) gene in patients with Cockayne syndrome. *Am. J. Hum. Genet.* 62:77–85

135. Marchetto MC, Carromeu C, Acab A, Yu D, Yeo GW, et al. 2010. A model for neural development and treatment of Rett syndrome using human induced pluripotent stem cells. *Cell* 143:527–39

136. Martins-Taylor K, Schroeder DI, LaSalle JM, Lalande M, Xu RH. 2012. Role of DNMT3B in the regulation of early neural and neural crest specifiers. *Epigenetics* 7:71–82

137. Masliah E, Dumaop W, Galasko D, Desplats P. 2013. Distinctive patterns of DNA methylation associated with Parkinson disease: identification of concordant epigenetic changes in brain and peripheral blood leukocytes. *Epigenetics* 8:1030–38

138. Mastroeni D, McKee A, Grover A, Rogers J, Coleman PD. 2009. Epigenetic differences in cortical neurons from a pair of monozygotic twins discordant for Alzheimer's disease. *PLOS ONE* 4:e6617

139. McGowan PO, Sasaki A, D'Alessio AC, Dymov S, Labonte B, et al. 2009. Epigenetic regulation of the glucocorticoid receptor in human brain associates with childhood abuse. *Nat. Neurosci.* 12:342–48

140. Meins M, Lehmann J, Gerresheim F, Herchenbach J, Hagedorn M, et al. 2005. Submicroscopic duplication in Xq28 causes increased expression of the *MECP2* gene in a boy with severe mental retardation and features of Rett syndrome. *J. Med. Genet.* 42:e12

141. Mendenhall EM, Williamson KE, Reyon D, Zou JY, Ram O, et al. 2013. Locus-specific editing of histone modifications at endogenous enhancers. *Nat. Biotechnol.* 31:1133–36

142. Meng J, Li Y, Camarillo C, Yao Y, Zhang Y, et al. 2014. The anti-tumor histone deacetylase inhibitor SAHA and the natural flavonoid curcumin exhibit synergistic neuroprotection against amyloid-β toxicity. *PLOS ONE* 9:e85570

143. Moran-Crusio K, Reavie L, Shih A, Abdel-Wahab O, Ndiaye-Lobry D, et al. 2011. Tet2 loss leads to increased hematopoietic stem cell self-renewal and myeloid transformation. *Cancer Cell* 20:11–24

144. Morin RD, Mendez-Lago M, Mungall AJ, Goya R, Mungall KL, et al. 2011. Frequent mutation of histone-modifying genes in non-Hodgkin lymphoma. *Nature* 476:298–303

145. Mueller BR, Bale TL. 2008. Sex-specific programming of offspring emotionality after stress early in pregnancy. *J. Neurosci.* 28:9055–65

146. Muller-Tidow C, Klein HU, Hascher A, Isken F, Tickenbrock L, et al. 2010. Profiling of histone H3 lysine 9 trimethylation levels predicts transcription factor activity and survival in acute myeloid leukemia. *Blood* 116:3564–71

147. Murgatroyd C, Patchev AV, Wu Y, Micale V, Bockmuhl Y, et al. 2009. Dynamic DNA methylation programs persistent adverse effects of early-life stress. *Nat. Neurosci.* 12:1559–66

148. Narlikar GJ, Sundaramoorthy R, Owen-Hughes T. 2013. Mechanisms and functions of ATP-dependent chromatin-remodeling enzymes. *Cell* 154:490–503

149. Naumova OY, Lee M, Koposov R, Szyf M, Dozier M, Grigorenko EL. 2012. Differential patterns of whole-genome DNA methylation in institutionalized children and children raised by their biological parents. *Dev. Psychopathol.* 24:143–55

150. Nebbioso A, Carafa V, Benedetti R, Altucci L. 2012. Trials with "epigenetic" drugs: an update. *Mol. Oncol.* 6:657–82

151. Neugebauer R, Hoek HW, Susser E. 1999. Prenatal exposure to wartime famine and development of antisocial personality disorder in early adulthood. *J. Am. Med. Assoc.* 282:455–62

152. Ng SB, Bigham AW, Buckingham KJ, Hannibal MC, McMillin MJ, et al. 2010. Exome sequencing identifies MLL2 mutations as a cause of Kabuki syndrome. *Nat. Genet.* 42:790–93

153. Ng SF, Lin RC, Laybutt DR, Barres R, Owens JA, Morris MJ. 2010. Chronic high-fat diet in fathers programs β-cell dysfunction in female rat offspring. *Nature* 467:963–66

154. Ng SF, Lin RC, Maloney CA, Youngson NA, Owens JA, Morris MJ. 2014. Paternal high-fat diet consumption induces common changes in the transcriptomes of retroperitoneal adipose and pancreatic islet tissues in female rat offspring. *FASEB J.* 28:1830–41

155. Nielsen SJ, Schneider R, Bauer UM, Bannister AJ, Morrison A, et al. 2001. Rb targets histone H3 methylation and HP1 to promoters. *Nature* 412:561–65

156. Nishi M, Horii-Hayashi N, Sasagawa T. 2014. Effects of early life adverse experiences on the brain: implications from maternal separation models in rodents. *Front. Neurosci.* 8:166

157. Nishiyama M, Oshikawa K, Tsukada Y, Nakagawa T, Iemura S, et al. 2009. CHD8 suppresses p53-mediated apoptosis through histone H1 recruitment during early embryogenesis. *Nat. Cell Biol.* 11:172–82

158. Nishiyama M, Skoultchi AI, Nakayama KI. 2012. Histone H1 recruitment by CHD8 is essential for suppression of the Wnt-β-catenin signaling pathway. *Mol. Cell. Biol.* 32:501–12

159. Obeid R, Schadt A, Dillmann U, Kostopoulos P, Fassbender K, Herrmann W. 2009. Methylation status and neurodegenerative markers in Parkinson disease. *Clin. Chem.* 55:1852–60

160. Oberlander TF, Weinberg J, Papsdorf M, Grunau R, Misri S, Devlin AM. 2008. Prenatal exposure to maternal depression, neonatal methylation of human glucocorticoid receptor gene (*NR3C1*) and infant cortisol stress responses. *Epigenetics* 3:97–106

161. Ogawa H, Ishiguro K, Gaubatz S, Livingston DM, Nakatani Y. 2002. A complex with chromatin modifiers that occupies E2F- and Myc-responsive genes in G0 cells. *Science* 296:1132–36

162. Okano M, Bell DW, Haber DA, Li E. 1999. DNA methyltransferases Dnmt3a and Dnmt3b are essential for de novo methylation and mammalian development. *Cell* 99:247–57

163. Ounap K, Puusepp-Benazzouz H, Peters M, Vaher U, Rein R, et al. 2012. A novel c.2T > C mutation of the *KDM5C/JARID1C* gene in one large family with X-linked intellectual disability. *Eur. J. Med. Genet.* 55:178–84

164. Paige SL, Thomas S, Stoick-Cooper CL, Wang H, Maves L, et al. 2012. A temporal chromatin signature in human embryonic stem cells identifies regulators of cardiac development. *Cell* 151:221–32

165. Painter RC, Roseboom TJ, Bleker OP. 2005. Prenatal exposure to the Dutch famine and disease in later life: an overview. *Reprod. Toxicol.* 20:345–52

166. Palomero T, Couronne L, Khiabanian H, Kim MY, Ambesi-Impiombato A, et al. 2014. Recurrent mutations in epigenetic regulators, RHOA and FYN kinase in peripheral T cell lymphomas. *Nat. Genet.* 46:166–70

167. Parsons DW, Li M, Zhang X, Jones S, Leary RJ, et al. 2011. The genetic landscape of the childhood cancer medulloblastoma. *Science* 331:435–39

168. Pembrey ME, Bygren LO, Kaati G, Edvinsson S, Northstone K, et al. 2006. Sex-specific, male-line transgenerational responses in humans. *Eur. J. Hum. Genet.* 14:159–66

169. Perroud N, Paoloni-Giacobino A, Prada P, Olie E, Salzmann A, et al. 2011. Increased methylation of glucocorticoid receptor gene (*NR3C1*) in adults with a history of childhood maltreatment: a link with the severity and type of trauma. *Transl. Psychiatry* 1:e59

170. Perroud N, Salzmann A, Prada P, Nicastro R, Hoeppli ME, et al. 2013. Response to psychotherapy in borderline personality disorder and methylation status of the BDNF gene. *Transl. Psychiatry* 3:e207

171. Peters SL, Hlady RA, Opavska J, Klinkebiel D, Pirruccello SJ, et al. 2013. Tumor suppressor functions of Dnmt3a and Dnmt3b in the prevention of malignant mouse lymphopoiesis. *Leukemia* 28:1138–45

172. Petrij F, Giles RH, Dauwerse HG, Saris JJ, Hennekam RC, et al. 1995. Rubinstein-Taybi syndrome caused by mutations in the transcriptional co-activator CBP. *Nature* 376:348–51

173. Poppe B, Vandesompele J, Schoch C, Lindvall C, Mrozek K, et al. 2004. Expression analyses identify MLL as a prominent target of 11q23 amplification and support an etiologic role for MLL gain of function in myeloid malignancies. *Blood* 103:229–35

174. Pronier E, Almire C, Mokrani H, Vasanthakumar A, Simon A, et al. 2011. Inhibition of TET2-mediated conversion of 5-methylcytosine to 5-hydroxymethylcytosine disturbs erythroid and granulomonocytic differentiation of human hematopoietic progenitors. *Blood* 118:2551–55

175. Provencal N, Suderman MJ, Guillemin C, Massart R, Ruggiero A, et al. 2012. The signature of maternal rearing in the methylome in rhesus macaque prefrontal cortex and T cells. *J. Neurosci.* 32:15626–42

176. Qiao Y, Liu X, Harvard C, Hildebrand MJ, Rajcan-Separovic E, et al. 2008. Autism-associated familial microdeletion of Xp11.22. *Clin. Genet.* 74:134–44

177. Raadsheer FC, Hoogendijk WJ, Stam FC, Tilders FJ, Swaab DF. 1994. Increased numbers of corticotropin-releasing hormone expressing neurons in the hypothalamic paraventricular nucleus of depressed patients. *Neuroendocrinology* 60:436–44

178. Radtke KM, Ruf M, Gunter HM, Dohrmann K, Schauer M, et al. 2011. Transgenerational impact of intimate partner violence on methylation in the promoter of the glucocorticoid receptor. *Transl. Psychiatry* 1:e21

179. Rando OJ. 2012. Combinatorial complexity in chromatin structure and function: revisiting the histone code. *Curr. Opin. Genet. Dev.* 22:148–55

180. Reichenberg A, Gross R, Weiser M, Bresnahan M, Silverman J, et al. 2006. Advancing paternal age and autism. *Arch. Gen. Psychiatry* 63:1026–32

181. Rideout WM 3rd, Coetzee GA, Olumi AF, Jones PA. 1990. 5-Methylcytosine as an endogenous mutagen in the human LDL receptor and p53 genes. *Science* 249:1288–90

182. Rodriguez J, Frigola J, Vendrell E, Risques RA, Fraga MF, et al. 2006. Chromosomal instability correlates with genome-wide DNA demethylation in human primary colorectal cancers. *Cancer Res.* 66:8462–9468

183. Roelfsema JH, White SJ, Ariyurek Y, Bartholdi D, Niedrist D, et al. 2005. Genetic heterogeneity in Rubinstein-Taybi syndrome: mutations in both the CBP and EP300 genes cause disease. *Am. J. Hum. Genet.* 76:572–80

184. Ronan JL, Wu W, Crabtree GR. 2013. From neural development to cognition: unexpected roles for chromatin. *Nat. Rev. Genet.* 14:347–59

185. Roth TL, Lubin FD, Funk AJ, Sweatt JD. 2009. Lasting epigenetic influence of early-life adversity on the *BDNF* gene. *Biol. Psychiatry* 65:760–69

186. Rotili D, Mai A. 2011. Targeting histone demethylases: a new avenue for the fight against cancer. *Genes Cancer* 2:663–79

187. Santos-Rosa H, Schneider R, Bannister AJ, Sherriff J, Bernstein BE, et al. 2002. Active genes are tri-methylated at K4 of histone H3. *Nature* 419:407–11

188. Schaefer A, Sampath SC, Intrator A, Min A, Gertler TS, et al. 2009. Control of cognition and adaptive behavior by the GLP/G9a epigenetic suppressor complex. *Neuron* 64:678–91

189. Schulmann K, Sterian A, Berki A, Yin J, Sato F, et al. 2005. Inactivation of p16, RUNX3, and HPP1 occurs early in Barrett's-associated neoplastic progression and predicts progression risk. *Oncogene* 24:4138–48

190. Schwartzentruber J, Korshunov A, Liu XY, Jones DT, Pfaff E, et al. 2012. Driver mutations in histone H3.3 and chromatin remodelling genes in paediatric glioblastoma. *Nature* 482:226–31

191. Schwarz D, Varum S, Zemke M, Scholer A, Baggiolini A, et al. 2014. Ezh2 is required for neural crest-derived cartilage and bone formation. *Development* 141:867–77

192. Seligson DB, Horvath S, McBrian MA, Mah V, Yu H, et al. 2009. Global levels of histone modifications predict prognosis in different cancers. *Am. J. Pathol.* 174:1619–28

193. Seligson DB, Horvath S, Shi T, Yu H, Tze S, et al. 2005. Global histone modification patterns predict risk of prostate cancer recurrence. *Nature* 435:1262–66

194. Shi Y, Lan F, Matson C, Mulligan P, Whetstine JR, et al. 2004. Histone demethylation mediated by the nuclear amine oxidase homolog LSD1. *Cell* 119:941–53

195. Siebert JR, Graham JM Jr, MacDonald C. 1985. Pathologic features of the CHARGE association: support for involvement of the neural crest. *Teratology* 31:331–36

196. Simmons RA, Templeton LJ, Gertz SJ. 2001. Intrauterine growth retardation leads to the development of type 2 diabetes in the rat. *Diabetes* 50:2279–86

197. Smith AK, Conneely KN, Kilaru V, Mercer KB, Weiss TE, et al. 2011. Differential immune system DNA methylation and cytokine regulation in post-traumatic stress disorder. *Am. J. Med. Genet. B Neuropsychiatr. Genet.* 156B:700–8

198. Sturm D, Witt H, Hovestadt V, Khuong-Quang DA, Jones DT, et al. 2012. Hotspot mutations in H3F3A and IDH1 define distinct epigenetic and biological subgroups of glioblastoma. *Cancer Cell* 22:425–37

199. Stuwe E, Toth KF, Aravin AA. 2014. Small but sturdy: small RNAs in cellular memory and epigenetics. *Genes Dev.* 28:423–31

200. Susser ES, Lin SP. 1992. Schizophrenia after prenatal exposure to the Dutch Hunger Winter of 1944–1945. *Arch. Gen. Psychiatry* 49:983–88

201. Taki T, Sako M, Tsuchida M, Hayashi Y. 1997. The t(11;16)(q23;p13) translocation in myelodysplastic syndrome fuses the MLL gene to the CBP gene. *Blood* 89:3945–50

202. Tan M, Luo H, Lee S, Jin F, Yang JS, et al. 2011. X Identification of 67 histone marks and histone lysine crotonylation as a new type of histone modification. *Cell* 146:1016–28

203. Tan YY, Wu L, Zhao ZB, Wang Y, Xiao Q, et al. 2013. Methylation of α-synuclein and leucine-rich repeat kinase 2 in leukocyte DNA of Parkinson's disease patients. *Parkinsonism Relat. Disord.* 20:308–13

204. Tarpey PS, Raymond FL, Nguyen LS, Rodriguez J, Hackett A, et al. 2007. Mutations in UPF3B, a member of the nonsense-mediated mRNA decay complex, cause syndromic and nonsyndromic mental retardation. *Nat. Genet.* 39:1127–33

205. Teng YC, Lee CF, Li YS, Chen YR, Hsiao PW, et al. 2013. Histone demethylase RBP2 promotes lung tumorigenesis and cancer metastasis. *Cancer Res.* 73:4711–21

206. Thompson BA, Tremblay V, Lin G, Bochar DA. 2008. CHD8 is an ATP-dependent chromatin remodeling factor that regulates β-catenin target genes. *Mol. Cell. Biol.* 28:3894–904

207. Thompson RF, Fazzari MJ, Niu H, Barzilai N, Simmons RA, Greally JM. 2010. Experimental intrauterine growth restriction induces alterations in DNA methylation and gene expression in pancreatic islets of rats. *J. Biol. Chem.* 285:15111–18

208. Trivier E, De Cesare D, Jacquot S, Pannetier S, Zackai E, et al. 1996. Mutations in the kinase Rsk-2 associated with Coffin-Lowry syndrome. *Nature* 384:567–70

209. Tsurusaki Y, Okamoto N, Ohashi H, Kosho T, Imai Y, et al. 2012. Mutations affecting components of the SWI/SNF complex cause Coffin-Siris syndrome. *Nat. Genet.* 44:376–78

210. Van Esch H, Bauters M, Ignatius J, Jansen M, Raynaud M, et al. 2005. Duplication of the MECP2 region is a frequent cause of severe mental retardation and progressive neurological symptoms in males. *Am. J. Hum. Genet.* 77:442–53

211. Van Houdt JK, Nowakowska BA, Sousa SB, van Schaik BD, Seuntjens E, et al. 2012. Heterozygous missense mutations in SMARCA2 cause Nicolaides-Baraitser syndrome. *Nat. Genet.* 44:445–49; S1

212. Vance KW, Ponting CP. 2014. Transcriptional regulatory functions of nuclear long noncoding RNAs. *Trends Genet.* 30:348–55

213. Varier RA, Timmers HT. 2011. Histone lysine methylation and demethylation pathways in cancer. *Biochim. Biophys. Acta* 1815:75–89

214. Venneti S, Garimella MT, Sullivan LM, Martinez D, Huse JT, et al. 2013. Evaluation of histone 3 lysine 27 trimethylation (H3K27me3) and enhancer of Zest 2 (EZH2) in pediatric glial and glioneuronal tumors shows decreased H3K27me3 in H3F3A K27M mutant glioblastomas. *Brain Pathol.* 23:558–64

215. Vissers LE, van Ravenswaaij CM, Admiraal R, Hurst JA, de Vries BB, et al. 2004. Mutations in a new member of the chromodomain gene family cause CHARGE syndrome. *Nat. Genet.* 36:955–57

216. Wamstad JA, Alexander JM, Truty RM, Shrikumar A, Li F, et al. 2012. Dynamic and coordinated epigenetic regulation of developmental transitions in the cardiac lineage. *Cell* 151:206–20

217. Wang SC, Oelze B, Schumacher A. 2008. Age-specific epigenetic drift in late-onset Alzheimer's disease. *PLOS ONE* 3:e2698

218. Weaver IC, Cervoni N, Champagne FA, D'Alessio AC, Sharma S, et al. 2004. Epigenetic programming by maternal behavior. *Nat. Neurosci.* 7:847–54

219. Weaver IC, Meaney MJ, Szyf M. 2006. Maternal care effects on the hippocampal transcriptome and anxiety-mediated behaviors in the offspring that are reversible in adulthood. *Proc. Natl. Acad. Sci. USA* 103:3480–85

220. Weber CM, Henikoff S. 2014. Histone variants: dynamic punctuation in transcription. *Genes Dev.* 28:672–82

221. Webster MJ, Knable MB, O'Grady J, Orthmann J, Weickert CS. 2002. Regional specificity of brain glucocorticoid receptor mRNA alterations in subjects with schizophrenia and mood disorders. *Mol. Psychiatry* 7:985–94; 24

222. Wei Y, Yang CR, Wei YP, Zhao ZA, Hou Y, et al. 2014. Paternally induced transgenerational inheritance of susceptibility to diabetes in mammals. *Proc. Natl. Acad. Sci. USA* 111:1873–78

223. Williams EC, Zhong X, Mohamed A, Li R, Liu Y, et al. 2014. Mutant astrocytes differentiated from Rett syndrome patients: Specific iPSCs have adverse effects on wild type neurons. *Hum. Mol. Genet.* 23:2968–80

224. Williams SR, Aldred MA, Der Kaloustian VM, Halal F, Gowans G, et al. 2010. Haploinsufficiency of HDAC4 causes brachydactyly mental retardation syndrome, with brachydactyly type E, developmental delays, and behavioral problems. *Am. J. Hum. Genet.* 87:219–28

225. Wirdefeldt K, Gatz M, Reynolds CA, Prescott CA, Pedersen NL. 2011. Heritability of Parkinson disease in Swedish twins: a longitudinal study. *Neurobiol. Aging* 32:1923.e1–.e8

226. Wright A, Dyck PJ. 1995. Hereditary sensory neuropathy with sensorineural deafness and early-onset dementia. *Neurology* 45:560–62

227. Wu G, Broniscer A, McEachron TA, Lu C, Paugh BS, et al. 2012. Somatic histone H3 alterations in pediatric diffuse intrinsic pontine gliomas and non-brainstem glioblastomas. *Nat. Genet.* 44:251–53

228. Xiang Y, Zhu Z, Han G, Ye X, Xu B, et al. 2007. JARID1B is a histone H3 lysine 4 demethylase up-regulated in prostate cancer. *Proc. Natl. Acad. Sci. USA* 104:19226–31

229. Xu GL, Bestor TH, Bourc'his D, Hsieh CL, Tommerup N, et al. 1999. Chromosome instability and immunodeficiency syndrome caused by mutations in a DNA methyltransferase gene. *Nature* 402:187–91

230. Yamagata K, Furuta H, Oda N, Kaisaki PJ, Menzel S, et al. 1996. Mutations in the hepatocyte nuclear factor-4α gene in maturity-onset diabetes of the young (MODY1). *Nature* 384:458–60

231. Yamaguchi R, Nakagawa Y, Liu YJ, Fujisawa Y, Sai S, et al. 2010. Effects of maternal high-fat diet on serum lipid concentration and expression of peroxisomal proliferator-activated receptors in the early life of rat offspring. *Horm. Metab. Res.* 42:821–25

232. Yamane K, Tateishi K, Klose RJ, Fang J, Fabrizio LA, et al. 2007. PLU-1 is an H3K4 demethylase involved in transcriptional repression and breast cancer cell proliferation. *Mol. Cell* 25:801–12

233. Yamashita Y, Yuan J, Suetake I, Suzuki H, Ishikawa Y, et al. 2010. Array-based genomic resequencing of human leukemia. *Oncogene* 29:3723–31

234. Yates JA, Menon T, Thompson BA, Bochar DA. 2010. Regulation of HOXA2 gene expression by the ATP-dependent chromatin remodeling enzyme CHD8. *FEBS Lett.* 584:689–93

235. Yen K, Vinayachandran V, Batta K, Koerber RT, Pugh BF. 2012. Genome-wide nucleosome specificity and directionality of chromatin remodelers. *Cell* 149:1461–73

236. Zaidi S, Choi M, Wakimoto H, Ma L, Jiang J, et al. 2013. De novo mutations in histone-modifying genes in congenital heart disease. *Nature* 498:220–23

237. Zannas AS, West AE. 2013. Epigenetics and the regulation of stress vulnerability and resilience. *Neuroscience* 264:157–70

238. Zechner U, Wilda M, Kehrer-Sawatzki H, Vogel W, Fundele R, Hameister H. 2001. A high density of X-linked genes for general cognitive ability: a run-away process shaping human evolution? *Trends Genet.* 17:697–701

239. Zeng J, Ge Z, Wang L, Li Q, Wang N, et al. 2010. The histone demethylase RBP2 is overexpressed in gastric cancer and its inhibition triggers senescence of cancer cells. *Gastroenterology* 138:981–92

240. Zhang TY, Hellstrom IC, Bagot RC, Wen X, Diorio J, Meaney MJ. 2010. Maternal care and DNA methylation of a glutamic acid decarboxylase 1 promoter in rat hippocampus. *J. Neurosci.* 30:13130–37

241. Zhao X, Chen A, Yan X, Zhang Y, He F, et al. 2014. Down-regulation of RUNX1/CBFβ by MLL fusion proteins enhances HSC self-renewal. *Blood* 123:1729–38

242. Zhu B, Reinberg D. 2011. Epigenetic inheritance: uncontested? *Cell Res.* 21:435–41

243. Zollino M, Orteschi D, Murdolo M, Lattante S, Battaglia D, et al. 2012. Mutations in KANSL1 cause the 17q21.31 microdeletion syndrome phenotype. *Nat. Genet.* 44:636–38

From Egg to Gastrula: How the Cell Cycle Is Remodeled During the *Drosophila* Mid-Blastula Transition

Jeffrey A. Farrell[1] and Patrick H. O'Farrell[2]

[1]Department of Molecular and Cellular Biology, Harvard University, Cambridge, Massachusetts 02138; email: jfarrell@g.harvard.edu

[2]Department of Biophysics and Biochemistry, University of California, San Francisco 94158; email: ofarrell@cgl.ucsf.edu

Annu. Rev. Genet. 2014. 48:269–94

First published online as a Review in Advance on September 5, 2014

The *Annual Review of Genetics* is online at genet.annualreviews.org

This article's doi: 10.1146/annurev-genet-111212-133531

Keywords

Drosophila, MBT, maternal-zygotic transition, replication, G2, cell cycle

Abstract

Many, if not most, embryos begin development with extremely short cell cycles that exhibit unusually rapid DNA replication and no gap phases. The commitment to the cell cycle in the early embryo appears to preclude many other cellular processes that only emerge as the cell cycle slows just prior to gastrulation at a major embryonic transition known as the mid-blastula transition (MBT). As reviewed here, genetic and molecular studies in *Drosophila* have identified changes that extend S phase and introduce a postreplicative gap phase, G2, to slow the cell cycle. Although many mysteries remain about the upstream regulators of these changes, we review the core mechanisms of the change in cell cycle regulation and discuss advances in our understanding of how these might be timed and triggered. Finally, we consider how the elements of this program may be conserved or changed in other organisms.

INTRODUCTION

The early embryonic cell cycles of most organisms are extensively modified so that they are unusually fast. These unusual cycles slow just as morphogenesis begins at gastrulation. Although there are exceptions and differences in the details for each organism (83, 121), this pattern of development is widespread in organisms that lay eggs that develop externally, with many prominent examples, including insects (35, 73), amphibians (78, 79), and fish (55). In studying this process in *Xenopus laevis*, John Gerhart dubbed the dramatic slowing of the cell cycle and associated onset of various cellular activities the mid-blastula transition (MBT), after the developmental stage at which the transition occurs in *Xenopus laevis* embryos (42).

Long before the transition had been named, Boveri, in 1892, recognized a distinctive feature of these early events. He crossed different species of sea urchin, examined the early divisions of resulting embryos, and concluded that the form and rate of cleavage are determined wholly by the mother, whereas later aspects of development show influences of both parents (118). This early insight exemplifies a widespread feature of early embryogenesis: The mother preloads the egg with material that directs the early rapid cell cycles that subdivide the large eggs of nearly all species. Somehow, the preloaded package of gene products executes a precisely timed dynamic program in which the egg goes through a species-specific number of rapid divisions before remodeling the cell cycle.

In this review, we set out to first describe the phenomenon of the MBT as it occurs in *Drosophila melanogaster*, an organism where it has been studied extensively with the aid of excellent genetic and cell biological tools. After describing the cell cycle and development of the early *Drosophila* embryo, we describe changes in the cell cycle machinery that occur during early development and present a mechanistic model for the slowing of the cell cycle. We then discuss how this transition could potentially be timed in the *Drosophila* embryo, and then finally, we compare the proposed mechanisms to what is known in other organisms.

EARLY DEVELOPMENT IN *DROSOPHILA*

Like other externally developing eggs, the *Drosophila* egg is large—approximately 100,000 times the size of an average somatic cell in the adult fly. After fertilization, its nuclei multiply and divide in synchrony in a massive cytoplasm (referred to as a syncytium) (88). The nuclei share the cytoplasm and are not separated by plasma membranes; consequently, mitosis occurs without cytokinesis. Although this mode of development may seem unfamiliar, it is a paradigm commonly used in insects. The early morphology and cell cycle of the embryo are diagrammed in **Figure 1**.

The earliest nuclear cycles are exceptionally rapid—the nuclei divide every 8.6 minutes (35, 88). Thus, whereas an average tissue culture or somatic cell takes 8–24 hours to go through the cell cycle, a *Drosophila* embryo manages 14 cell cycles in only 1.5 hours, producing an embryo with thousands of cells. How does the early embryo achieve this notable speed? In short, by modifying the cell cycle such that the process of DNA replication occurs much more quickly than in most cell cycles (6, 72, 94) and by omitting the gap phases between replication and mitosis (26, 72). So, instead of pausing in gap phases, the nuclei alternate between mitosis and extraordinarily short S phases, during which the entire genome replicates simultaneously in as little as 3.4 minutes (**Figure 1**) (6). These cell cycles divide the large preexisting cytoplasm without growth (82). Moreover, the cell cycle is essentially synchronous throughout the entire embryo during these cycles.

During interphases 8 and 9, most of the nuclei migrate outward, forming a shell of nuclei near the plasma membrane by interphase 10, known as the blastoderm (**Figure 1**) (35, 88). The blastoderm cycles resemble the preblastoderm cycles in that they are quick and synchronous, but

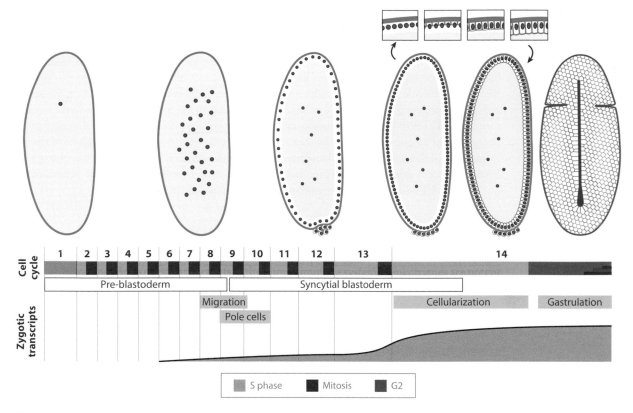

Figure 1

Early development in *Drosophila*. A diagram of the first 14 cycles of *Drosophila* development, with notable morphological stages illustrated at the top (anterior up and posterior down). Note that although most embryos are displayed as sections through the middle of the embryo with the ventral side to the left, the final illustration is a surface view, with ventral in the midline, facing the reader. The process of cellularization is diagrammed in more detail in the insets. A horizontal bar representing the first 200 min of embryogenesis shows the duration of each phase of the cell cycle: S phase (*green*), mitosis (*red*), and G2 (*blue*). Mitosis 14 is represented as a series of small bars because the embryo is no longer synchronous at this time and individual groups of cells enter mitosis at different times according to a developmentally programmed schedule. The preblastoderm and blastoderm stages are indicated with white boxes. The timing of notable morphological events is demarcated in gray boxes: the migration of the nuclei to the blastoderm, the isolation of the germline by cellularization of the pole cells, the cellularization of the blastoderm nuclei, and gastrulation movements, whose onset is marked by formation of the ventral furrow. The approximate number of genes for which zygotic transcripts have been detected over time is represented by the purple curve at the bottom.

once the nuclei reach the surface and form the blastoderm, the cell cycle begins to slow gradually, with interphase progressively lengthening slightly each cycle, from approximately 9 minutes in cycle 10 to 14 minutes in cycle 13 (35, 88). These cycles still lack gap phases.

However, between cycles 13 and 14, a much more dramatic change in the regulation of the cell cycle slows it considerably and desynchronizes it—although interphase 13 was the same length for every nucleus (20 minutes) (35), interphase 14 is a minimum of 70 minutes, and its length differs for distinct groups of cells in a spatially stereotyped pattern (34). This slowing occurs via two mechanisms: First, S phase becomes much longer but remains synchronous, increasing from 14 minutes in cycle 13 to 50 minutes in cycle 14 (94), and second, interphase 14 exhibits the first G2 gap phase (25, 26), where cells pause for a self-determined, developmentally programmed length of time between the completion of replication and entry into mitosis (**Figure 1**) (26, 34).

Blastoderm: the region of the egg that gives rise to embryonic tissues. In *Drosophila*, a shell of nuclei surrounding the yolk

There is still no G1 gap phase between mitosis and replication (26)—it will not be introduced for several more cell cycles in a separate regulatory change (14, 58, 59).

During interphase 14, not only does the cell cycle change, morphogenesis begins. During the long S-phase 14, the process of cellularization envelops each nucleus in a plasma membrane by moving existing membrane down between the nuclei as well as depositing new membrane, thereby dividing the syncytial blastoderm into more than 6,000 cells (**Figure 1**). Then, following cellularization, the first gastrulation movements begin during G2 of interphase 14. These include formation of a ventral furrow, where the mesodermal precursors internalize, and a cephalic furrow that transiently demarcates the cells that will make up the cephalon (or head region) of the fly (**Figure 1**). Throughout these gastrulation movements, stereotyped groups of cells known as mitotic domains exit their G2 phases and enter mitosis 14 in a developmentally programmed order, in their first departure from synchronous mitotic cycles (26, 34).

TRANSITIONS IN EMBRYONIC REGULATION

Early development is extremely fast, and consequently, many events often occur nearly simultaneously and processes with progressive steps can appear abrupt. Often, groups of changes in early embryonic development have been lumped together as abrupt transitions that, after more detailed analysis, can be seen to result from several distinct regulatory processes. Here, we try and put the major transitions of the early embryo into this context.

At fertilization, the embryonic genome is nearly quiescent, and maternally loaded gene products direct development. As the embryo progresses, however, zygotic gene products become required for developmental events and the progress of the cell cycle; thus, control of development and the cell cycle is handed off from the maternal genome to the zygotic genome, often called the maternal-zygotic transition (MZT). But is it really a single transition? Early definitions of the MZT were made based on the first obvious derangement of an embryo (usually a failure to continue cell division) when the input of the zygotic genome was blocked by inhibiting transcription (45, 67). By focusing on the first obvious deviation from normal, this approach defines a single transition point but could erroneously collapse a progressive sequence of events into a single moment. Although neither the loss of any single zygotic gene nor any large region of the genome perturbs the earliest steps of development, an ever increasing number of mutants display phenotypes as development progresses. This shows that dependency on zygotic gene activity first becomes obvious in cycle 14 and that there is a continuous and progressive emergence of zygotic control of more and more processes (38, 110, 116).

Because transcription must underlie the function of the zygotic genome, molecular assays for the first zygotic gene expression, often called zygotic genome activation (ZGA), have also been used as a measure of MZT. A major increase in transcription accompanies the progression of early embryonic divisions, and experimental measures in numerous systems have defined a point of first activation. The first zygotic phenotype does not necessarily become apparent at the time of the changes in zygotic expression that caused it, and this has resulted in multiple conflicting times defined as the MZT in *Drosophila*. For example, even though zygotic phenotypes are first described in cycle 14, cycle 10 has been designated as the first point of zygotic transcription in *Drosophila* based on the incorporation of radioactive precursors into mature transcripts (27, 124). Additionally, an assay's ability to define a time of ZGA is limited by its sensitivity, and several considerations suggest that cycle 10 is simply the point in a progressive process at which an increase in transcriptional activity first makes it detectable by this methodology. This is probably further compounded by the increase in nuclear content over the early cycles—when looking at transcriptional output, a similar activation of the genome is twice as difficult to detect each

Cellularization: the process of dividing the syncytial cytoplasm into many cells and enveloping each nucleus in its own cell membrane

Maternal-zygotic transition (MZT): the switch of developmental control from maternal gene products to zygotic gene products, defined either by functional requirements or zygotic transcription

Zygotic genome activation (ZGA): the stage when significant levels of transcription are observed from the initially relatively quiescent zygotic genome

cycle prior because there is half as much template and consequently half as many transcripts produced.

Several findings suggest that transcription can and does occur before cycle 10 in *Drosophila*. For instance, the transcriptional cascade leading to sex determination has begun by cycle 8 with zygotic expression of *sisA* (29, 30) or potentially even earlier (16, 29, 84). The promoters of a number of patterning genes, including *ftz*, are also active before cycle 10 (9, 86). Additionally, the transcription of other genes has been detected before cycle 10 (86, 113), and as more sensitive assays are applied, the list continues to grow (1, 15, 64). It seems as if there is a progressive increase in transcription over the development of the embryo, where the transcription of new genes can be detected in each cycle. Such early transcriptional activity is not unique to *Drosophila*: When biochemical and autoradiographic methods were pushed to high sensitivity, transcription was detected in *Xenopus* eggs before the MBT (56), which was previously considered the time of ZGA (78, 79).

The summarized analyses of specific genes, as well as genome-wide studies of gene expression (64), suggest that different genes initiate expression at different times throughout much of early development. The complexity of transcriptional activation argues for multiple events rather than a discrete genome-wide transition in transcription (86, 120, 121). Furthermore, genome-wide analyses of mutants that affect transcriptional onset in the early embryo have found that some, but not all, genes are affected in each mutant, and the affected genes are in partially overlapping groups (3, 62, 65, 80), supporting the idea that each individual gene's transcriptional onset could be determined by a promoter-specific combination of multiple inputs, leading to several different programs determining transcriptional onset.

Thus, at the genome-wide level, the switch from maternal to zygotic control seems to be a progressive process rather than a sharp transition, regardless of whether one considers the requirement for zygotic gene function or the onset of zygotic expression. Although there is a general and steep increase in embryonic transcription that culminates in high transcriptional activity in cycle 14, this increase is achieved in multiple increments distributed over early development, combined with an exponential expansion of coding capacity and increased opportunity for transcription as the cell cycle slows. Not surprisingly, use of the term MZT has varied depending on the emphasis of the investigators and the experimental criteria that they use to assess it. Although the concept of a global MZT is perhaps outdated, the idea of a maternal-to-zygotic transition can remain useful, so long as it is focused on a very specific process in which one can define a time of onset of zygotic transcription of one or more particular genes required for the execution of the specific process.

Like the MZT, the MBT itself was once defined as a transition in which seemingly everything that was important in early development changed in a single concerted event. The changes used to define the MBT in the early *Xenopus* work were the onset of transcription, the initial slowing of the cell cycle, and the onset of cellular movements (78, 79). However, subsequent observations showed that these events could be uncoupled in *Xenopus* (56) and in *Drosophila* as well (23). Moreover, the initial departure from precisely repeating cell cycles to gradually slowing cycles, which was used to mark the MBT in the original *Xenopus* literature, was followed by another dramatic change in the cell cycle just prior to gastrulation called the early gastrula transition (EGT) (50). In *Drosophila*, the most dramatic and abrupt changes occur at cycle 14, and this has been conventionally called the MBT in this organism, a convention to which we adhere. However, it is important to note that the changes at cycle 10 in *Drosophila* embryos, which we call pre-MBT slowing, might be more analogous to the MBT changes of *Xenopus*, and the changes in *Drosophila* cycle 14 embryos might be more properly compared with the EGT changes in *Xenopus*.

The concept of the MBT as a single, concerted transition in the embryo has been eroded by the recognition of numerous steps in early development. However, it remains true that the fourteenth cell cycle of *Drosophila* marks an important transition in embryogenesis, accompanied by a major change in cell cycle regulation as well as a consequent slowing of the cell cycle, a dramatic upregulation of zygotic transcription, increased turnover of maternal messages, and the onset of gastrulation movements. Given that these concurrent events are not necessarily co-regulated, we suggest that it is important to study the mechanistic underpinnings of each event; here, we focus on the mechanistic basis of the slowing of the cell cycle, the defining event of the MBT.

WHY HAVE A SPECIALIZED, EXCEPTIONALLY RAPID EARLY CELL CYCLE PROGRAM?

It's impossible to be sure of the evolutionary pressures that led to modern biological phenomena. However, a major problem encountered by eggs laid in the external environment could potentially explain the need for rapid cell cycles: Eggs cannot eat. Instead, they must subsist entirely on the limited supply of maternal nutrients that are contained in the egg. Thus, in order to support the developmental program until it generates an organism that can feed itself, eggs have evolved to be very large and nutrient-rich. But this creates a new problem—before hatching, the egg has no means of escape or active defense, and it presents a highly nutritious target for predators. The rapid cell cycles of early development are possibly a response to this, as they serve to minimize the pre-hatching vulnerable stage and proceed to hatching as quickly as possible.

ACHIEVING SPEED

As mentioned briefly in the introduction, the rapidity of the pre-MBT cell cycles is achieved through a potent combination of specializations of the cell cycle: (*a*) their dependence entirely on maternal contributions, (*b*) their lack of gap phases, and (*c*) their unusually speedy replication of the genome.

Dependence Only on Maternal Contributions

During the more leisurely paced cell cycles of later development, stage-specific transcription of genes advances the cycle to the next phase. For example, transcription of S-phase genes, promoted by E2F, contributes strongly to the G1-to-S transition (20, 21). However, because the rate of production of completed transcripts is limited by the elongation rate and maximal packing of RNA polymerase onto the gene, in an egg with much more cytoplasm per copy of the genome and a very short interphase, it is considerably more difficult to make sufficient numbers of transcripts to control the cytoplasm during each cell cycle. Large somatic cells often circumvent this problem by amplifying their entire genome (37, 49, 101), amplifying selected genes in the genome (10, 103), or carrying repeated arrays of genes that are particularly in demand (8, 53), but these strategies have not been widely adopted in eggs. Instead, the early embryo takes a shortcut and relies primarily on translation of maternally provided transcripts and post-translational modification to direct the early cycles.

Lack of Gap Phases

In the canonical cell cycle, two gap phases—G1 and G2—serve as pauses before entry into DNA replication and mitosis, respectively. Cells often use them as a time to grow, or to exit the cell cycle if they do not need to undergo further divisions (76). However, the cells of the early embryo, with no external source of nutrition for growth and a clear imperative to proliferate, neither

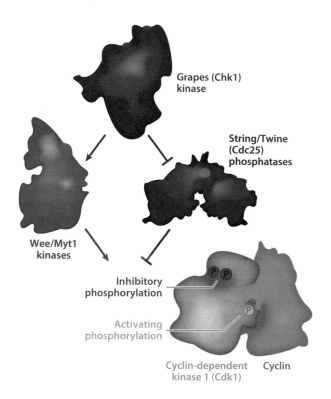

Figure 2

Regulation of cyclin-dependent kinase 1 (Cdk1). Cdk1 has diverse inputs that can regulate its kinase activity, which are diagrammed here. First, it requires a cyclin partner. Second, it requires activating phosphorylation in order to be functional. Third, it must be free of inhibitory phosphorylation, which blocks the ATP-binding pocket. This inhibitory phosphorylation is added by Wee and Myt1 kinases (which are thus inhibitors of Cdk1) and removed by String and Twine (Cdc25) phosphatases (which are thus activators of Cdk1). Grapes (Chk1) kinase can inhibit Cdk1 by inhibiting Cdc25 and activating Wee, thereby promoting the inhibitory phosphorylation of Cdk1.

grow nor pause as they subdivide their generous cytoplasms into smaller cells. Thus, the gap phases are dispensed with, thereby hastening the cell cycle, until cycle 14, when a G2 phase—a pause between completion of replication and entry into mitosis—is introduced (26, 72). Pausing in G2 requires that cyclin-dependent kinase 1 (Cdk1) be inactive after the completion of DNA replication; otherwise, it performs its characteristic role and triggers entry into mitosis (25, 26, 75, 76). The appearance of G2 in cell cycle 14 coincides with inhibitory phosphorylation of essentially all Cdk1 and inactivation of this mitotic activator (**Figure 2**) (28). Such widespread inhibitory phosphorylation does not occur in the early cycles, but Wee kinase, which is localized to the nucleus, still contributes, perhaps by locally constraining Cdk1 activity to prevent nuclear progress to mitosis (106, 107).

Rapid S Phase

The speed of DNA replication in early *Drosophila* embryos is truly astounding—in 3.4 minutes, the embryo replicates its entire genome, a process that takes 50 minutes immediately following the MBT and in differentiated somatic cells can take 8 hours or more (6, 94). Two differences contribute to the extension of the embryonic S phase from 3.4 to 50 minutes. First, replication

Replication origin: genomic location at which DNA replication begins by building a replication fork; licensed by assembling pre-replication complexes on the DNA prior to replication

origins are slightly more tightly packed—on average, they are found every 7.9 kb in the preblastoderm embryo (6), compared with 10.6 kb in a cycle 14 embryo (72). Thus, with a slightly longer distance for each replication fork to travel, it would take slightly longer to replicate the DNA between origins in cycle 14. However, this ∼30% increase in origin spacing does little to explain the ∼1,500% change in S-phase length between the initial rapid cycles and cell cycle 14.

A larger contribution seems to come from a change in replication timing of different sequences. In most S phases, such as those after the MBT, not all sequences in the DNA replicate at the same time. Although the euchromatin begins to replicate immediately upon entering S phase, sequences in heterochromatin typically wait until later in S phase to replicate and are called late-replicating sequences (**Figure 3a**) (94). The prime example of late-replicating sequences in *Drosophila* is the satellite sequences—stretches of megabases of highly repetitive DNA that account for nearly 30% of the genome (63, 94). In the preblastoderm cycles, all sequences—both the euchromatin and the satellite sequences—replicate at essentially the same time; given that all regions of the DNA are copied simultaneously, the job is completed quite quickly. As development progresses, the satellite sequences shift from being early replicating (and replicating simultaneously with the euchromatin) to being late replicating (**Figure 3a**) (70, 94, 123). The shift is subtle and gradual in the pre-MBT cycles, where the replication of different sequences still largely overlaps, but with slight offsets, and then is dramatic after the MBT, when different clusters of satellite sequence replicate in a prolonged sequence (70, 94, 123). In cycle 14, some of the satellite sequences do not begin to replicate until 15 or 30 minutes after S phase begins—a delay that is longer than the entirety of S-phase 13 (94). These delays by the late-replicating sequences account for most of the lengthening of S phase.

Surprisingly, the shift in the satellite sequences from early replicating to late replicating has also been tied to the activity of Cdk1 (32). This was initially unexpected because it is generally thought that Cdk2 is the regulator of S phase, whereas Cdk1 regulates mitosis (75, 76). However, manipulations of Cdk1 levels in early embryos changed the timing of the replication of the satellite sequences (32). In cycle 14, the satellite sequences usually replicate late and Cdk1 activity is usually low (28, 94); increased Cdk1 activity, however, was able to accelerate S phase and promote earlier replication of the satellite sequences (32). Conversely, before the MBT, the satellite sequences usually replicate early, and Cdk1 activity is present throughout the cycle (28, 94); decreasing Cdk1 activity during the early cycles lengthened S phase significantly through, it seemed, delaying the replication of satellite sequences (32). Despite the thought that Cdk2 usually regulates S phase, injection of activated Cdk2 did not affect the timing of DNA replication (32). This has led to the idea that the satellite sequences are intrinsically late replicating but that modest Cdk1 activity during S phase can override that preference and cause them to replicate early.

THE CONSEQUENCES OF SPEED

The devotion of the early embryo to cell proliferation quickly generates the cells needed to begin specifying and building the tissues of the embryo, but it also seems to preclude and defer other

---

Figure 3

Model for mid-blastula transition cell cycle slowing. (*a*) Approximations across developmental time (*marked at the top*) show replication timing of early and late-replicating sequences (94), protein levels of cyclin (28) and Cdc25 (31), and Cdk1 activity (28). Also listed are the approximate number of genomes and the transcriptional onset of genes discussed extensively in the text. (*b*) A graphic of the model espoused in this review for how the cell cycle changes mechanistically during *Drosophila* early development. Note that cycles 10–12 are omitted and cycle 14 is separated into 3 sections. Abbreviation: CKI, cyclin-dependent kinase inhibitor.

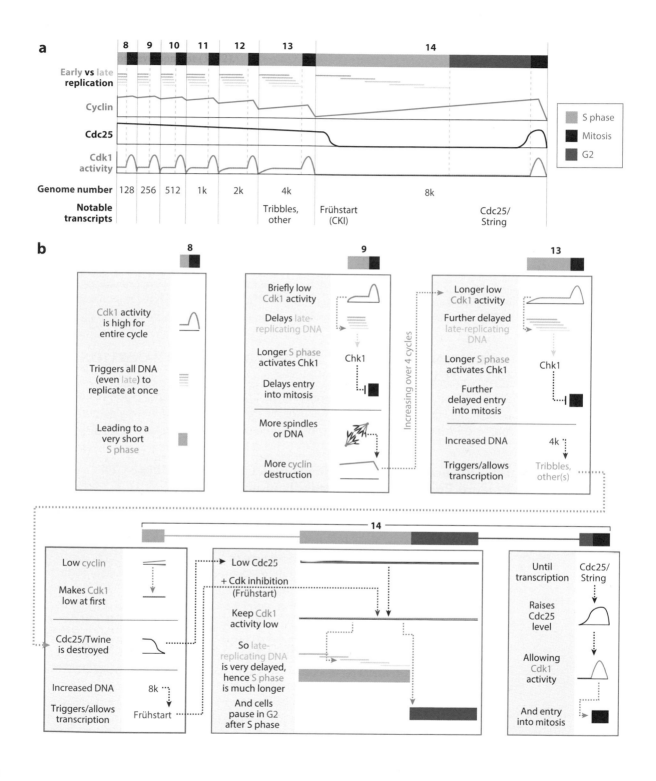

critical activities. Mitosis is extremely disruptive: It appropriates the cytoskeleton to build a mitotic spindle and cleavage furrow, and during the event, the Golgi is disassembled and translation is suppressed (82). Given that many morphogenetic events require specialized cytoskeletal rearrangements, they must wait for the cell cycle to slow. For example, the movement of the plasma membrane during cellularization is directed by massive cytoskeletal structures (35, 89, 99), and cellularization is disrupted and forced to restart if mitosis is induced during the process (23). Additionally, the shape changes that drive internalization of cells into the ventral furrow are directed by changes in the cytoskeleton as well (110), and mitosis during the process of formation of the furrow often forces cells back out, preventing their proper internalization (47, 68, 92). Thus, the cell cycle must be slowed to create conditions conducive to morphogenesis.

Furthermore, many morphogenetic events, including cellularization and gastrulation, require localized zygotic transcription (61). However, transcription is suppressed during mitosis, and transcripts in the process of being extended—nascent transcripts—are aborted in mitosis and must be restarted in the following interphase. As a result, only short transcripts can be completed within the brief interphases of the early cycles (95). Additionally, intense replication activity may interfere with transcription in interphase—it is notable that in both *Xenopus* and *Drosophila*, the earliest observed transcription was seen only in nuclei with early prophase characteristics, suggesting that in these early cycles transcription may be confined to the very end of interphase, as replication is completed and nuclei are preparing for mitosis (56, 86). Thus, when the cell cycle slows at the MBT, it provides an increased opportunity for transcription (23, 95).

A MECHANISTIC VIEW OF PRE-MID-BLASTULA TRANSITION CELL CYCLE SLOWING

So, given that it is developmentally critical to end this program of rapid divisions, how does the embryo slow them? There are two phases of cell cycle slowing: first, the gradual slowing of the cycles before the MBT and then the dramatic slowing at the MBT. We discuss each of these in turn.

DNA Replication Acts via the Replication Checkpoint to Time Cell Cycle Progress Prior to the Mid-Blastula Transition

Each cell cycle prior to the MBT slows by a slightly bigger factor than the one before, which gives the impression of a progressive process that leads up to the dramatic, switch-like change in cycle 14. This progressive slowing appears to be part of the transition in a number of species, suggesting that it may be an integral component of the early cycles.

Excellent insight into how this slowing occurs comes from studies of mutations in *grapes*, which encodes the *Drosophila* homolog of Chk1—a checkpoint kinase that is activated in response to incomplete DNA replication (36, 98). Once activated, Grapes suppresses cyclin:Cdk1 activity by promoting its inhibitory phosphorylation (36, 98). It does this by enhancing the activity of Wee kinase, thereby promoting the addition of the inhibitory phosphates, and suppressing the activity of Cdc25 phosphatase, which would normally remove those inhibitory phosphates (**Figure 2**). In most situations, Chk1 kinases are dispensable and perform essential roles only during DNA damage or replication stress; in accord with this, *grapes* mutants are viable unless so stressed (36, 97, 98).

However, female *grapes* mutant flies lay eggs that never hatch. Without their maternal supply of Grapes, these embryos (hereafter called *grapes* embryos) suffer a mitotic catastrophe in mitosis 13 when they attempt to divide prior to completing the replication of their DNA (36, 98, 108, 122). This phenotype seems to result from an inability of *grapes* embryos to induce inhibitory phosphorylation of Cdk1 (36, 98) and thereby delay mitosis until completion of replication, as it is similar to the phenotype observed in embryos from *wee* mutant mothers (85, 105). More importantly for the timing of the pre-MBT cycles, *grapes* embryos fail to extend cycles 11–13 to the degree seen in normal embryos (98). By cycle 13, the mismatch between the increasing duration of S phase and incompletely lengthened interphase leads *grapes* embryos to enter mitosis with incompletely replicated chromosomes, an event that is visualized as extensive DNA bridging between chromosomes trying to separate in anaphase. This suggests that inhibitory phosphorylation driven by Grapes mediates pre-MBT interphase lengthening. Moreover, when S phase is artificially lengthened by injection of aphidicolin, the early cycles are extended in a *grapes*-dependent manner (98). This suggests that an extended S phase can lengthen interphase by activating Grapes. Indeed, mutations in other genes in the Grapes-dependent checkpoint pathway give the same early defects in cell cycle lengthening, arguing for a model in which the checkpoint pathway extends interphase (85, 97, 105). Given that DNA replication becomes gradually slower during these cycles (94), the model suggests that continuing S phase prevents mitotic entry by activating the checkpoint pathway that inhibits Cdk1, thereby coupling interphase length to S-phase length. Accordingly, it is S-phase length that sets the pace of these early cycles.

Further confirmation that DNA replication determines the length of the early cycles comes from experiments in which S phase was deleted by injection of the inhibitor Geminin (70). Geminin prevents the assembly of pre-replication complexes in the following cycle (71, 87). Thus, upon entry into the next S phase, there are no licensed origins, so no replication begins (70). Note that deletion of S phase is different from inhibition of replication (such as with aphidicolin); when replication is inhibited, it begins, but progress of DNA polymerase is slowed, which induces replication stress, activates the checkpoint, and prevents cell cycle progress. In contrast, when S phase is deleted, even though no replication forks are built and no replication occurs, the cell appears to lack signals that replication failed and the cell cycle goes on, although of course abnormally because mitosis occurs with unreplicated chromosomes (70). Deleting S phase during the early cycles shortens the early interphase to the same degree as mutation of *grapes* but has no consequence on the duration of interphase in embryos lacking *grapes* function (70). This further supports the idea that continued DNA replication causes the signal that activates the DNA replication checkpoint in each cycle and thereby lengthens the pre-MBT cycles slowly.

Why Does DNA Replication Slow During the Early Cycles?

Initial mechanistic proposals for the gradual slowing of the cell cycle involved titration of a maternal factor; the thought was that as the DNA doubled each cycle, the amount of a maternally loaded component would eventually become insufficient to support continued rapid cell cycles (42). Early models suggested that titration of replication factors could slow S phase by decreasing origin utilization and increasing the distance traversed by each replication fork (79). However, as discussed above, the change in origin frequency is not substantial (6, 72), and it is now apparent that S phase lengthens as a result of an increasing delay in the initiation of replication of satellite sequences (94, 123). So then, why is there a progressive delay in the timing of initiation of satellite DNA replication in cycles 10–13? The finding that cyclin:Cdk1 drives early replication of satellite sequences prior to the MBT and that downregulation of cyclin:Cdk1 triggers S-phase extension at the MBT has led us to view some old findings in a new light (32).

The levels of cyclins and the activation of Cdk1 in the early embryo were examined in detail some years ago (28). A role for cyclin in timing the blastoderm cycles was revealed by genetic reductions in cyclin dose, which modestly increased the lengths of these cycles (28). At the time, the data were interpreted in light of cyclin's requirement to enter mitosis; it was thought that in these cycles, entry into mitosis required the accumulation of a particular amount of cyclin,

Aphidicolin:
a reversible inhibitor of eukaryotic nuclear DNA replication that slows or stalls the progress of DNA polymerase

Pre-replication complex: a protein complex formed on origins that, after maturation through additional protein recruitment, subsequently directs assembly of replication forks and initiation of replication

and reduction in the amount of cyclin mRNA loaded by the mother would increase the time required to translate that amount. However, further study has revealed that cyclin is not limiting for the entry into mitosis; when RNAi knockdown is used to reduce the level of cyclin mRNA by approximately two-thirds in *grapes* mutant embryos, the timing of the preblastoderm cycles is not changed (122). Thus, cyclins are in excess, and even with a significant reduction in the level of their message, they accumulate rapidly enough to trigger mitosis in *grapes* embryos, which occurs even earlier than in wild-type embryos. The finding that cyclin levels do not directly govern mitotic entry is in concert with the more recent findings indicating that S-phase duration is the key determinant of interphase length. Thus, we revisit the early studies to consider how cyclin levels and Cdk1 activity might influence S-phase duration.

Early observations showed that the cyclins are abundant in the early embryo, and neither their bulk levels nor the bulk activity of Cdk1 kinase oscillate in the preblastoderm cycles (**Figure 3a**) (28). However, cyclin ubiquitination is required to exit mitosis even in these early cycles without obvious oscillations in cyclin level, which has led to the unconfirmed but widely accepted assumption that localized cyclin destruction centered on the mitotic spindle allows mitotic exit (107). During the progressively slowing blastoderm cycles, the increased destruction of cyclins becomes apparent and progressively more intense with each cycle (28). Thus, if cyclin destruction is coupled to the mitotic apparatus, perhaps the increasing DNA leads to greater destruction of the cyclins with each mitosis, and this results in the lower cyclin levels observed at the beginning of each blastoderm interphase. Even though mitotic cyclin destruction is far from complete, once mitotic decline in cyclins becomes apparent, all Cdk1 in the embryo loses its activating phosphorylation and loses activity for a brief period at the beginning of each blastoderm cycle (**Figure 3a**) (28). The early replication of the satellite sequences depends on Cdk1 activity (32), and in the blastoderm cycles, brief delays in the onset of satellite sequence replication become apparent (94, 123). Thus, we suggest that the increasingly long period of Cdk1 inactivity at the beginning of each blastoderm cycle—seemingly caused by the progressively more thorough cyclin destruction at each mitosis and more prolonged loss of the activating phosphorylation—results in an increasing offset in the time of initiation of satellite replication in each successive cycle (**Figure 3b**). This gradually makes S phase longer, resulting in more prolonged activation of the replication checkpoint in successive cycles, thereby further delaying the entry into mitosis in each cycle and lengthening it.

THE DRAMATIC LENGTHENING OF THE CELL CYCLE AT THE MID-BLASTULA TRANSITION

Although the gradual lengthening of the pre-MBT cycles seems to be a part of a progressive program in which each successive cycle lengthens slightly, the transition in cycle 14 is much more abrupt. The duration of S phase lengthens from 15 minutes to 50 minutes—a much starker increase than in the preceding cycles—and a G2 phase is introduced (26, 94) (**Figure 1**). Additionally, for the first time, rather than entering mitosis synchronously, different domains of cells exhibit different schedules of mitosis (34). Both of these changes—the longer S phase and G2—result from the downregulation of Cdk1 (26, 28, 32). However, in cycle 14, Cdk1 activity is regulated differently than in the pre-MBT cycles, during which progressively increasing cyclin destruction and slight and progressive delays in satellite replication appear to be responsible for the pre-MBT slowing.

Cdc25 Phosphatase Activity Determines the Length of Interphase 14

Given the apparent ubiquity of the inhibitory kinases, Wee and Myt1, Cdc25 phosphatase, which removes inhibitory phosphates from Cdk1, is necessary for cyclin:Cdk1 activity (**Figure 2**)

(19, 25, 26, 28, 43, 90). Gel shift assays first detect inhibited phospho-isoforms of Cdk1 in cycle 14 and show full conversion of Cdk1 as S phase of cycle 14 progresses (28). In contrast, these inhibited isoforms are rare in the preceding cycles (28, 105). This suggests a change in the regulation of Cdk1. Indeed, isolation of mutants in *string*, which encodes one of the two *Drosophila* homologs of Cdc25 phosphatase (**Figure 2**), revealed that entry into mitosis 14 required zygotic transcription of *cdc25/string* (25, 26), and premature *string* expression from an inducible transgene was sufficient to shorten G2 and trigger early entry into mitosis (26). This showed that progression to mitosis 14 is limited by a requirement for Cdc25 activity. Furthermore, RNA in situ hybridization and immunohistochemistry showed that *string* transcription and subsequent accumulation of String protein foreshadowed the exit from the G2 of cycle 14 and the entry into mitosis for each cell (22, 24, 25). This further supports the idea that developmentally regulated expression of *cdc25/string* determines the length of G2 for each cell and triggers its entry into mitosis.

Injection of *cdc25* mRNA in early cycle 14 showed that Cdc25 activity could have a second effect on shortening the cell cycle, aside from triggering exit from G2. Cdc25 activity in cycle 14 triggered early replication of the satellite sequences and thereby vastly shortened S phase (32). This effect was phenocopied when mRNA encoding a variant of Cdk1 that is resistant to inhibitory phosphorylation was injected (32). Conversely, lowering Cdk1 activity in cycle 13 lengthened the normally short pre-MBT S phase and seemed to prematurely trigger the onset of late replication (32). This suggested that a reduction in Cdc25 and the consequent reduction in Cdk1 activity between interphase 13 and 14 lengthened cycle 14 both through slowing DNA replication and promoting G2.

Maternal Cdc25 Is Depleted via Destruction of Cdc25/Twine Protein

Why does Cdc25 suddenly become limiting in cycle 14, so as to permit inhibitory phosphorylation of Cdk1? Early work had shown that the egg has significant maternal RNA stores (22) encoding two different Cdc25 proteins, String (25) and Twine (2, 12). Additionally, immunoblots showed that protein levels of String (28) and Twine (17, 31) were also high during the pre-MBT cycles. Although dramatic destruction of both maternal transcripts occurs in cycle 14 (22), RNAi knockdown of *string* and *twine* mRNA in the early embryo did not cause cell cycle arrest (31), suggesting that downregulation of these messages is not a sufficient explanation for the cell cycle change, although it likely acts to reinforce the change (22, 77). String protein declines slowly during the blastoderm cycles and becomes essentially undetectable in cycle 13 (28), but Twine protein is relatively stable before the MBT, and its level remains high until early in cycle 14, when it is abruptly destabilized and destroyed (17, 31). Thus, it seems that Twine protein destruction likely allows accumulation of the inhibitory phosphorylation on Cdk1—catalyzed by inhibitory kinases, Wee and Myt1 (**Figure 2**) (52, 85, 105)—to inhibit its activity, lengthen S phase, and add G2 (**Figure 3b**).

Inhibitory Phosphorylation Collaborates with an Inhibitor of Cdk1

Genetic studies further support involvement of Twine and other regulators of Cdk1 in the trigger that prolongs the cell cycle at the MBT. Embryos laid by mothers that were germline deficient for *string* and heterozygous for *twine* (i.e., mothers that had one copy of a *cdc25* gene instead of four) often cellularized one cycle early (22). Reciprocally, embryos from mothers with two additional copies of *twine* (and so approximately 1.5 times the normal *cdc25* gene dose) occasionally executed the MBT one cycle late (22). The frequency of this extra pre-MBT division increased dramatically when embryos were mutant for *frühstart*, a cyclin-dependent kinase inhibitor (CKI)

α-**Amanitin:**
an inhibitor of RNA
synthesis with
particularly potent
action on RNA
polymerase II

that is transcribed in early cycle 14 (41, 46, 47). These results further confirm that the amount of maternal Cdc25 and embryonic Cdk1 activity influence the timing of the MBT and that Cdc25 and Cdk1 downregulation are most likely the key timers of cell cycle slowing (**Figure 3***b*).

THE TRANSITION IS ACTIVE AND DEPENDENT ON TRANSCRIPTION

Although the changes to cell cycle regulators that result in the extension of interphase at the MBT are described above, it does not explain why the transition happens in cycle 14. Major historical interest has been focused on the dependence of the cell cycle on transcription, which also seems to be a major component involved in triggering the MBT. Requirement for transcription has been probed in many systems by testing the consequence of its inhibition by α-amanitin, an inhibitor of RNA polymerase II (13, 22, 78, 94). In *Drosophila*, injection of α-amanitin before MBT prevents the prolongation of S phase and the MBT, resulting in an additional rapid synchronous cell cycle (22, 94). As described above, Cdc25 elimination is critical for these processes, and α-amanitin also prevented the programmed destruction of Twine protein in cycle 14 (31). The exact transcriptional cascade that triggers destruction of Twine has not been fully elucidated, although candidate factors have been identified, including *tribbles*, which acts through an unknown mechanism (31, 109). Regardless, these results argue that Twine destruction and the MBT-associated prolongation of the cell cycle are active processes that require transcription in *Drosophila*.

TIMING THE MID-BLASTULA TRANSITION

The Nuclear to Cytoplasmic Ratio

Early and influential experiments that physically manipulated the number of nuclei in frog egg fragments provided evidence that changing the ratio of nuclei to cytoplasm (hereafter, the N:C ratio) alters the time of MBT events (78, 79), suggesting that some events require accumulation of particular amounts of DNA to occur. Studies of *Drosophila* strains that produce haploid embryos support the idea that the N:C ratio is important for determining the time of slowing the cell cycle in *Drosophila* as well (23). These embryos, which have half as much DNA, require one more cell cycle (presumably for one more round of DNA replication) to trigger slowing of the cell cycle (23). An even more detailed analysis that used aneuploidy to generate embryos with varying amounts of DNA suggests that there is a threshold of approximately 70% of the DNA that is normally present in a cycle 14 embryo—embryos with less than 75% the normal amount of DNA regularly went through an extra cycle before the MBT, and embryos with more than 134% often went through one fewer cycle before the MBT (65). However, it does not seem likely that the N:C ratio is read out at a single time—the N:C ratio influences the activation of transcription of the patterning gene *ftz* in cycle 9 (86), but detailed analyses of aneuploid embryos indicate that it also is monitored in cycle 14 (65).

Multiple Timers for Multiple Events

Summaries of the early work exploring the influence of N:C on the MBT tend to ignore an important aspect of the studies of haploid *Drosophila* embryos—not all MBT-associated events were delayed a cycle. Cellularization actually began in cycle 14 as normal, but the abnormally early mitosis aborted it and forced it to restart and complete in the now long cycle 15 (23). Similarly, an analysis of transcriptional activation suggested that a dramatic upregulation of zygotic transcription still occurred in cycle 14 (23). Thus, some MBT-associated events do not depend on reaching a threshold N:C ratio. Instead, it appeared that the lower N:C ratio, by allowing a continuation of rapid cell cycles, might indirectly interfere with many other processes normally scheduled to begin as the cell cycle slowed.

A recent experiment emphasizes that N:C does not affect all MBT events. Embryos whose cell cycles were prematurely arrested in interphase 12 or 13 using RNAi against the mitotic cyclins initiated cellularization, an MBT event, at the normal time—neither advanced nor delayed (69). This finding that the experimental downregulation of Cdk1 activity can bypass the need for cycle 14 N:C ratio for some MBT events suggests that the only requirement for N:C is to trigger the downregulation of Cdk1 or to trigger some consequence of the downregulation of Cdk1, such as slowing of the cell cycle. Furthermore, the normal timing of MBT events that occurred in this bypass argues that there are other timers that are independent of N:C and cell cycle progress. Recognition that there is such a timer is somewhat disquieting because we have no idea how it works. However, numerous events occur with amazing temporal precision in the early embryo—even before the approach of the MBT—that are not significantly affected by inhibition of transcription or deletion of large segments of the genome (22, 74, 117). These events, such as the migration of the nuclei to the syncytial blastoderm (35, 104), provide an independent and compelling argument for timers that are distinct from the MBT and use maternally provided products to both time and execute developmental events.

Could Nuclear-to-Cytoplasmic-Ratio-Dependent Transcription Time the Cell Cycle Transition at the Mid-Blastula Transition?

As argued above, the onset of zygotic transcription seems to be a progressive process with different genes initiating expression at different times. A study that used high-throughput methods to compare expression of several hundred genes in wild-type embryos and haploid mutants has shown that the expression of some genes is deferred for one cycle in haploid embryos, whereas others initiate transcription in cycle 14 regardless of the accumulated DNA (65). These were classified as N:C-dependent and time-dependent genes, respectively. Thus, at least some transcripts expressed at cycle 14 are regulated by the N:C ratio. Given that transcription is required for cell cycle slowing, one attractive model is that N:C-dependent transcription governs onset of Twine destruction, and delayed Twine destruction in haploid embryos supports this idea (31). Furthermore, the *tribbles* transcript, although provided maternally, is upregulated by N:C-dependent zygotic expression in cycle 13 (65), and it functions by unknown mechanisms to promote destruction of String and Twine (47, 68, 92). RNAi experiments suggest that it contributes to, but is not essential for, the onset of destruction of Twine at the MBT (31). Thus, N:C-dependent *tribbles* transcription could contribute to timing Twine destruction, although there appear to be other transcriptional inputs into its destruction given that α-amanitin is more effective than *tribbles* RNAi at suppressing Twine destruction (31). Additionally, transcription of the cyclin:Cdk inhibitor *frühstart* begins in cycle 14 (41, 46, 47) and seems to be dependent on the N:C ratio (65). Its action as an inhibitor of Cdk1 seems to further contribute to and reinforce slowing of the cell cycle. Thus, the onset of a small collection of N:C-dependent transcripts could be the trigger for the slowing of the cell cycle (**Figure 3b**).

N:C-dependent genes that are expressed before the MBT are excellent candidates for the regulation of the cell cycle. However, many N:C-dependent genes are expressed after the MBT, and they may only be indirectly dependent on the N:C ratio if their expression depends on an MBT event, such as cell cycle slowing to enable sufficient time for elongation of their transcripts. Thus, a few directly N:C-dependent genes could create a positive feedback loop by slowing the cell cycle and enabling the transcription of other indirectly N:C-dependent genes.

Although a model based on N:C-dependent transcription as a timer of cell cycle slowing is attractive, several cautions are in order. First, a demonstration that transcription is required for cell cycle slowing does not require that transcription be the timing signal. Second, thorough experiments that created aneuploid embryos missing different parts of the genome have failed to identify any single locus that is required for cell cycle slowing at the MBT (74, 117). Third, at present it has not been determined whether N:C-dependent transcripts respond directly or indirectly to the N:C ratio. Thus, it remains formally possible that the cell cycle fails to slow in cycle 14 in haploids because of the failure to express N:C-dependent transcripts or that the cell cycle might fail to slow in cycle 14 in haploids for an independent reason, and the appearance of the transcripts is then indirectly delayed because of the altered developmental program. Nonetheless, the most attractive model at present is that increasing transcription of several different loci that are dependent on the N:C ratio, including *tribbles* and *frühstart*, is involved in the change in regulation of the cell cycle and its slowing in cycle 14 (**Figure 3***b*).

THE MYSTERIOUS IDENTITIES OF THE TIMERS

The question of how the N:C is detected and could result in transcription that affects or triggers the MBT remodeling of the cell cycle is still a major question in the field. Several prominent proposals have been made, but they are all unsatisfying in some way. We present them below, as well as their caveats. However, we consider this question to be one of the foremost remaining puzzles in understanding the timing of the *Drosophila* MBT.

Titration Models

The most influential models for how the N:C ratio could be detected and trigger a transition have been titration models, in which some maternally loaded component in the embryo is limiting (42). In their initial forms, they revolved around limiting cell cycle components and the idea that the increasing DNA in the embryo would eventually become insufficient for the maternal cell cycle component to continue to support rapid cell cycles, thus slowing the cell cycle (79). Recent results have shown that overexpression of some DNA replication components can delay the onset of the MBT in *Xenopus*, which could potentially be consistent with such a mechanism (11). However, results in *Drosophila* argue that the titration of a cell cycle component is not directly responsible for lengthening cycle 14. Reintroduction of Cdc25 at the MBT is sufficient to drive a short cycle, which shows that it is the only component limiting for a short cell cycle (32). It, however, does not seem to be passively titrated by increased DNA but rather actively destroyed in cycle 14 (31). Thus, if a component is titrated in the *Drosophila* embryo, it is probably not a cell cycle component that mechanistically slows the cell cycle. However, titration of a maternal component could still act as a sensor for N:C, only now the new findings argue that the signal generated by this sensor must trigger an active process to bring on the destruction of Twine and the MBT. One potential candidate could be an inhibitor of transcription of some kind whose titration triggers transcription of N:C-dependent transcripts. The inhibitor *tramtrack* is thought to act in this fashion to allow transcription of a small collection of genes in cycle 11 (86); perhaps different inhibitors that remain to be identified act in cycles 13 and 14.

Inhibition by the Short Length of the Rapid Cycles

Another popular model has been that the sheer shortness of the early cycles made transcription impossible because entry into mitosis aborts nascent transcription (95). Accordingly, N:C might activate transcription indirectly by slowing the cell cycle and making interphase long enough that there is sufficient time to complete the transcription of the genes important for triggering the MBT. This certainly contributes to the reinforcement of the onset of transcription after the cell cycle is actively slowed; some genes transcribed in cycle 14 are too long to be completed during a short cell cycle, and so require the slowing of the cell cycle as a permissive condition to allow their transcription (95). However, experiments using RNAi against cyclins to prematurely lengthen the cell cycle suggest that an inability to complete elongation does not limit the onset of MBT events—it did not advance the time of Twine destruction, which suggests that the transcripts involved in Twine destruction were not advanced (31). Moreover, it did not advance the timing of cellularization or gastrulation, suggesting that perhaps no MBT events are timed by such a model (69).

Zelda or Vielfaltig: A DNA-Binding Protein

A mutant of great interest for the timing of the MBT is the zinc-finger DNA-binding protein Zelda (also known as Vielfaltig), which is found highly enriched at genes that are expressed in cycle 14 and also those genes expressed during the pre-MBT cycles (which typically have more Zelda binding sites) (48, 80). Indeed, increasing or decreasing the number of Zelda binding sites near a gene can advance or delay (respectively) the time of onset of transcription of that gene (113), and adding Zelda binding motifs to a GFP (green fluorescent protein) reporter can confer pre-MBT expression on the reporter (15). A percentage of *zelda* embryos go through an extra rapid division, suggesting that Zelda regulates the transcription of one or more genes involved in timing the MBT (62, 109). However, Zelda binds to most of its target genes in cycle 8, long before they become activated (48), and the current model for its function is that it assists the binding of other transcription factors (40, 91, 119). Thus, it seems less likely that Zelda is the switch that changes at the time of transcription to turn on expression of particular genes, but more likely it instead creates a permissive condition that allows those genes to be activated by another trigger—perhaps the binding of other transcription factors.

Smaug: An RNA-Binding Protein

One mutant whose role at the MBT is rather confusing is the *smaug* mutant. Embryos from mothers mutant for *smaug* (hereafter, *smaug* embryos) do not slow their cell cycle or execute the MBT. Additionally, expression of *smaug* in a gradient results in a gradient of cellularization, so Smaug has been proposed as a timer of the MBT (3). *smaug* embryos exhibit defects in activation of the DNA replication checkpoint—they do not exhibit pre-MBT slowing or respond to aphidicolin, similar to *grapes* embryos (3). As discussed previously, the DNA replication checkpoint has an important role in regulating the cell cycle of the early embryo, so it is possible that Smaug's role in regulating the cell cycle acts through Grapes/Chk1. However, *smaug* embryos also have penetrant defects in the onset of zygotic transcription of a number of genes (3). Indeed, *tribbles* and *frühstart* expression are thought to be decreased in *smaug* embryos, as are some regulators of cellularization (3). However, the basis of Smaug's connection to the DNA replication checkpoint and transcription phenotypes is not well understood. Smaug is an RNA-binding protein, thought to repress translation of its targets through its binding and promote their destruction by recruiting the CCR4/POP2/NOT deadenylase complex and thereby shortening their poly(A) tail (93, 100, 111). There is no understood connection, however, between Smaug's known biochemical function in repressing translation of or destabilizing transcripts and the onset of zygotic transcription or the activity of the DNA replication checkpoint. Moreover, there is no explanation for how Smaug would detect or be regulated by the N:C ratio. Clearly, more investigation into Smaug's potential role in the transition is needed.

AN EVOLUTIONARY PERSPECTIVE ON THE MID-BLASTULA TRANSITION

The genetics and detailed developmental studies of *Drosophila* have offered us a fairly detailed view of the MBT, but how does it relate to events in other organisms? As presented at the outset of this article, rapid cleavage divisions are a widespread feature of early embryos (83), as if there were a consistent benefit to the basic strategy of rapid early cell proliferation throughout the evolution of animals with externally deposited eggs. Indeed, whereas *Drosophila* goes through 13 rapid cell cycles of lengths ranging from 8–20 minutes prior to dramatic slowing and desynchronization, *Xenopus laevis* goes through 12 rapid cell cycles of approximately 30 minutes each, and zebrafish undergo 10 rapid cell cycles that are 15 minutes each (55, 57). Thus, for all of these organisms, within three hours of fertilization, they have subdivided the massive maternally provided cytoplasm into thousands of cells.

However, convergent evolution might have created this behavior independently in different lineages, and even if the programs are related by descent, it is unlikely that the regulatory program would be conserved unchanged. Organisms have dramatically different eggs with associated distinctions in early embryogenesis. Thus, we might expect the control of embryonic cell cycle slowing to have undergone considerable modification in parallel with the changes in organization of early embryogenesis (60, 126). In fact, sequence homology comparison suggests that Twine and String diverged quite recently, and Twine may be unique to *Drosophila*, as only other *Drosophila* species have Cdc25 homologs whose protein sequences are more homologous to Twine than String (J.A. Farrell & P.H. O'Farrell, personal observation). This suggests that at least some details of control of the MBT must be different in other Diptera, and one might expect more dramatic differences over larger evolutionary time spans.

Xenopus

The *Xenopus* early cycles show superficial similarity to those in *Drosophila*. The *Xenopus* embryo undergoes 11 rapid cycles, each approximately 30 minutes long, and a twelfth cycle that is slightly longer and is defined as the MBT. Cycles 13–15 get progressively longer (50, 99, and 253 min, respectively) (50). The EGT occurs early in the greatly extended cycle 15 and is associated with onset of gastrulation. The *Xenopus* MBT is dependent on the N:C ratio, as determined by ligation of eggs, and is not dependent on zygotic transcription, based on injection of α-amanitin (78). The detailed cell cycle changes associated with cycle prolongation immediately after the *Xenopus* MBT are somewhat unclear. An early report that both G1 and G2 were introduced at the MBT (44) was contradicted by two later reports (39, 51); however, it is not clear that any of these studies have the temporal resolution and sensitivity to define the specific cycle at which either a G1 or G2 is introduced. Nonetheless, all of the studies agree that extension of S phase is an early change, and that by the time of gastrulation, embryos acquire both a G1 and a G2. At the time of the *Xenopus* MBT, Chk1 becomes activated and is required for cell cycle lengthening (96), similar to *Drosophila*. Moreover, Chk1-dependent inhibitory phosphorylation of Cdk1 becomes evident at the MBT (33), and accumulates between the time of the MBT and EGT (96). This involvement of the checkpoint pathway in the initial slowing of the cell cycle is similar to Grapes/Chk1-dependent pre-MBT slowing of the cell cycle in *Drosophila*. Perhaps, as in *Drosophila*, Chk1 activation in *Xenopus* acts to defer mitosis as S phase prolongs. However, in *Xenopus*, Chk1 phosphorylation both inhibits Cdc25 activity and promotes its destruction (96, 115), whereas in *Drosophila* the abrupt destruction of Twine/Cdc25 does not require Chk1 activity (31).

Recent work in *Xenopus* has found that overexpression of four DNA replication factors— Dbp11/Cut5, Treslin/SLD3, Drf1, and RecQ4—can trigger additional short, pre-MBT like cycles in *Xenopus*, but does not accelerate the pre-MBT cycles (11). The new finding indicates that the maturation step influenced by increasing the abundance of these factors limits progress of the post-MBT cell cycle, but not the pre-MBT cell cycle. These factors promote a step in the maturation of the pre-replication complex necessary to build functional replication forks. This step is also influenced by Cdk activity (7). It is possible, as the recent report suggests, that increasing amounts of DNA titrate the manipulated replication factors such that they become limiting at the MBT, and this titration causes the slowing of the cycle (11). An alternate interpretation of that data that is more consistent with the observations in *Drosophila* would be that another change in the control of the cell cycle—such as a decrease in Cdk activity—slows a step in the maturation of pre-replication complexes and causes the prolongation of S-phase at the MBT; this regulatory change could also cause the cell cycle to become sensitive to an experimental increase in proteins that promote this maturation step because such an increase would bypass the newly introduced restriction on cell cycle progression.

Zebrafish

The zebrafish embryo goes through 9 rapid cycles, slows very slightly in interphases 10–11, and exhibits an extreme slowing in interphase 12. It first begins to exhibit cell cycle asynchrony in cycle 11 (55). Partial enucleation experiments suggest that the MBT slowing of the cell cycle is also timed by the N:C ratio in zebrafish, as in *Drosophila* and *Xenopus* (55). The dramatic slowing of the cell cycle does not seem to depend on transcription, as embryos injected with α-amanitin did not exhibit signs of continued cell division (13, 54). The embryos acquire a G1 phase at the MBT, but in a transcription-dependent fashion, suggesting that it is not the entry into G1 that regulates the dramatic slowing of the cell cycle at the MBT in zebrafish (125). Instead, as in *Drosophila*, slowing of the cell cycle seems to result from a longer S phase at the MBT (125) and acquisition of a G2 phase at the MBT, although this G2 phase does not require zygotic transcription (13). Cdc25 and Cdk1 activity become limiting at the MBT, like in *Drosophila*, as overexpression of either Cdc25a or inhibitory phosphorylation-resistant Cdk1 causes continued rapid divisions, whereas phosphorylation-resistant Cdk2 does not (13). However, upregulation of Cdk1 activity is not capable of preventing the lengthening of S phase in zebrafish, and so inhibitory-resistant Cdk1 causes cells to enter mitosis with unreplicated DNA and undergo mitotic catastrophe (13). This suggests that although downregulation of Cdc25 and Cdk1 may be responsible for adding a G2 phase and for the observed change in cell cycle timing at the MBT, similar to *Drosophila*, the mechanism that causes S phase to get longer is different and not dependent on Cdk1. Neither the connection between the N:C ratio and the lengthening of S phase nor the connection between the N:C ratio and the downregulation of Cdc25 and Cdk1 is known in zebrafish.

Other Connections

There are additional hints in the literature that some of the mechanisms uncovered in *Drosophila* may be important in the early development of other organisms. For instance, there is a suggestion that, in urochordate embryos, transcription may change the length of S phase and therefore introduce the first asynchrony into the cell cycle (18). Furthermore, there are suggestions that Cdk1 modulation of replication timing may occur in mammalian systems as well, as tested in cell culture (114). Additionally, the specific requirement for Chk1 kinase in early embryogenesis extends all the way to mammals (112), suggesting that this may be a critically conserved component of slowing embryonic cell cycles.

Eutheria: the clade of mammals that includes the placental mammals (versus the Metatheria, which is predominantly marsupials)

The Mid-Blastula Transition Might Not Look the Same in Every Organism

There are numerous organisms that do not seem to exhibit a classic MBT. However, perhaps some of these organisms may retain elements of the program but in somewhat differing contexts. For instance, in segmented annelids and short germband insects, cells do not all slow rapid cell cycles synchronously. Instead, the posterior cells continue proliferation and begin morphogenesis in a protracted sequence. For instance, studies in the leech (*Helobdella triserialis*) have detailed cell cycle slowing in a very different context (5). The major slowing of the cell cycle that can be observed in leeches occurs during the asymmetric divisions of the teloblast cells, which produce the segmented ectodermal structures. The teloblast cells are large and divide rapidly and asymmetrically approximately once an hour, with an 11-min S phase and 21-min G2. The large daughter of the division, in a stem cell pattern, continues the rapid teloblast cell cycle program. The much smaller daughter, however, is called a primary blast cell and has a very different cell cycle—a 4.7-h S phase and a 16–28-h G2 (5). Thus, these smaller cells exhibit a dramatic extension of S phase and acquisition of a long G2, in conjunction with a massive upregulation in zygotic transcription, like the cycle 14 cells of a *Drosophila* embryo. This cell cycle slowing and transcriptional activation are viewed as representing the parallel with the MBT in leeches (4, 5). Because the change occurs repeatedly in a continuing lineage, the regulation—especially its connection with timing—differs markedly from that in *Drosophila*. Nonetheless, these changes have a striking association with the abrupt change in N:C that occurs with the birth of the small blast cell. Thus, even if the MBT is not always a synchronous, embryo-wide event, it may nonetheless retain a fundamental input from N:C and exhibit similar changes in cell cycle phases to prolong the cycle.

Mammalian embryos are also often viewed as not having an MBT because of their initially slow cycles upon fertilization. However, a broader consideration of chordates suggests that the rapid early cycles and abrupt slowing of the cycle in association with gastrulation are widespread. It is seen in the protochordates such as *Ciona*, as well as in fish, amphibians, and birds. Did it suddenly disappear in parallel with the late evolution of eutherian development, or was it changed in ways that make it less recognizable? The early stages of mammalian development cannot be compared with the development of even relatively close non-mammalian vertebrates because new steps were added at the beginning of development to generate the extraembryonic tissues that are a central feature of mammalian development. This insertion into early development displaced gastrulation and its associated processes to a later time; at gastrulation and afterward, many features of gene expression and mammalian morphogenesis can be compared with other vertebrates (83). Exceedingly rapid cell cycles occur just prior to gastrulation in mammals (66, 102), and several features argue that these rapid cycles represent the evolutionary vestige of the pre-MBT cycles and that slowing of the cycle at gastrulation may have parallels to the MBT (83). But just as we have seen in other organisms, the regulation of MBT-associated events is likely to have changed in conjunction with the evolution of the mammalian program of early development. One type of change stands out. In externally developing eggs, the transition from maternal control to zygotic control is more or less coincident with the slowing of the cell cycle, but the mammalian embryo transitions to dependence on zygotic gene expression during the early cleavage cycles before it begins to generate the extraembryonic tissues and long before gastrulation (81). Nonetheless, after completion of the uniquely eutherian early program, onset of expression of particular zygotic genes, independent of ZGA, could generate conditions that recapitulate the pre-gastrulation programs that are part of the legacy imprinted on metazoan development over hundreds of millions of years of evolution.

SUMMARY POINTS

1. In early development, many embryos exhibit a period of rapid cell division enabled by a modified cell cycle with an unusually rapid S phase and a lack of gap phases.

2. In *Drosophila*, the cell cycle first slows gradually, as late-replicating sequences begin to delay replication slightly, lengthening S phase and triggering the DNA replication checkpoint, which delays entry into mitosis.

3. Upon achieving a particular N:C ratio, the *Drosophila* embryo triggers an active and transcription-dependent process that inhibits Cdk1 via destruction of the Cdc25 isoform Twine, elimination of *cdc25* mRNAs, and transcription of a CDK inhibitor, *frühstart*.

4. This inhibition of Cdk1 triggers the cell cycle slowing at the MBT: Late-replicating sequences begin to delay replication significantly and a G2 gap phase is introduced during which cells now depend on the transcription of *cdc25* to enter mitosis.

5. It is not fully clear how the embryo determines when to initiate this program to inhibit Cdk1, but we propose that the best current model is that transcription of specific zygotic transcripts times it and thus links the process to the major activation of the *Drosophila* genome in cycles 13 and 14.

6. A number of other changes in the embryo, including the onset of morphogenesis, are timed by an unknown mechanism other than the N:C ratio.

FUTURE ISSUES

1. How does the *Drosophila* embryo detect the N:C ratio and transduce it into a change in transcriptional activity?

2. What are the many mechanisms that trigger the onset of zygotic transcription?

3. Does transcription truly time the changes to the cell cycle at the MBT, or is it merely permissive?

4. How does Cdk1 activity affect the timing of late-replicating sequences?

DISCLOSURE STATEMENT

The authors are not aware of any affiliations, memberships, funding, or financial holdings that might be perceived as affecting the objectivity of this review.

ACKNOWLEDGMENTS

We thank members of the O'Farrell lab for helpful discussions and Alexander Schier, Katherine Rogers, and James Gagnon for comments on the manuscript. This research was supported by National Institutes of Health grant GM037193 to P.H.O'F. and a National Science Foundation Graduate Research Fellowship and Jane Coffin Childs Memorial Fund Postdoctoral Fellowship to J.A.F.

LITERATURE CITED

1. Ali-Murthy Z, Lott SE, Eisen MB, Kornberg TB. 2013. An essential role for zygotic expression in the pre-cellular *Drosophila* embryo. *PLOS Genet.* 9(4):e1003428

2. Alphey L, Jimenez J, White-Cooper H, Dawson I, Nurse P, Glover DM. 1992. Twine, a *cdc25* homolog that functions in the male and female germline of *Drosophila*. *Cell* 69(6):977–88

3. Benoit B, He CH, Zhang F, Votruba SM, Tadros W, et al. 2009. An essential role for the RNA-binding protein Smaug during the *Drosophila* maternal-to-zygotic transition. *Development* 136(6):923–32

4. Bissen ST, Weisblat DA. 1987. Early differences between alternate N blast cells in leech embryo. *J. Neurobiol.* 18(3):251–69

5. Bissen ST, Weisblat DA. 1989. The durations and compositions of cell cycles in embryos of the leech, *Helobdella triserialis*. *Development* 106(1):105–18

6. Blumenthal AB, Kriegstein HJ, Hogness DS. 1974. The units of DNA replication in *Drosophila melanogaster* chromosomes. *Cold Spring Harb. Symp. Quant. Biol.* 38:205–23

7. Boos D, Sanchez-Pulido L, Rappas M, Pearl LH, Oliver AW, et al. 2011. Regulation of DNA replication through Sld3-Dpb11 interaction is conserved from yeast to humans. *Curr. Biol.* 21(13):1152–57

8. Brown DD, Wensink PC, Jordan E. 1971. Purification and some characteristics of 5S DNA from *Xenopus laevis*. *Proc. Natl. Acad. Sci. USA* 68(12):3175–79

9. Brown JL, Sonoda S, Ueda H, Scott MP, Wu C. 1991. Repression of the *Drosophila fushi tarazu* (*ftz*) segmentation gene. *EMBO J.* 10(3):665–74

10. Calvi BR, Lilly MA, Spradling AC. 1998. Cell cycle control of chorion gene amplification. *Genes Dev.* 12(5):734–44

11. Collart C, Allen GE, Bradshaw CR, Smith JC, Zegerman P. 2013. Titration of four replication factors is essential for the *Xenopus laevis* midblastula transition. *Science* 341(6148):893–96

12. Courtot C, Fankhauser C, Simanis V, Lehner CF. 1992. The *Drosophila cdc25* homolog *twine* is required for meiosis. *Development* 116(2):405–16

13. Dalle Nogare DE, Pauerstein PT, Lane ME. 2009. G2 acquisition by transcription-independent mechanism at the zebrafish midblastula transition. *Dev. Biol.* 326(1):131–42

14. de Nooij JC, Letendre MA, Hariharan IK. 1996. A cyclin-dependent kinase inhibitor, Dacapo, is necessary for timely exit from the cell cycle during *Drosophila* embryogenesis. *Cell* 87(7):1237–47

15. De Renzis S, Elemento O, Tavazoie S, Wieschaus EF. 2007. Unmasking activation of the zygotic genome using chromosomal deletions in the *Drosophila* embryo. *PLOS Biol.* 5(5):e117

16. Deshpande G, Stukey J, Schedl P. 1995. Scute (*sis-b*) function in *Drosophila* sex determination. *Mol. Cell. Biol.* 15(8):4430–40

17. Di Talia S, She R, Blythe SA, Lu X, Zhang QF, Wieschaus EF. 2013. Posttranslational control of Cdc25 degradation terminates *Drosophila's* early cell-cycle program. *Curr. Biol.* 23(2):127–32

18. Dumollard R, Hebras C, Besnardeau L, McDougall A. 2013. β-Catenin patterns the cell cycle during maternal-to-zygotic transition in urochordate embryos. *Dev. Biol.* 384(2):331–42

19. Dunphy WG, Kumagai A. 1991. The cdc25 protein contains an intrinsic phosphatase activity. *Cell* 67(1):189–96

20. Duronio RJ, O'Farrell PH. 1995. Developmental control of the G1 to S transition in *Drosophila*: Cyclin E is a limiting downstream target of E2F. *Genes Dev.* 9(12):1456–68

21. Dyson N. 1994. PRB, p107 and the regulation of the E2F transcription factor. *J. Cell Sci. Suppl.* 18:81–87

22. Edgar BA, Datar SA. 1996. Zygotic degradation of two maternal Cdc25 mRNAs terminates *Drosophila's* early cell cycle program. *Genes Dev.* 10(15):1966–77

23. Edgar BA, Kiehle CP, Schubiger G. 1986. Cell cycle control by the nucleo-cytoplasmic ratio in early *Drosophila* development. *Cell* 44(2):365–72

24. Edgar BA, Lehman DA, O'Farrell PH. 1994. Transcriptional regulation of *string* (*cdc25*): a link between developmental programming and the cell cycle. *Development* 120(11):3131–43

25. Edgar BA, O'Farrell PH. 1989. Genetic control of cell division patterns in the *Drosophila* embryo. *Cell* 57(1):177–87

26. Edgar BA, O'Farrell PH. 1990. The three postblastoderm cell cycles of *Drosophila* embryogenesis are regulated in G2 by *string*. *Cell* 62(3):469–80

27. Edgar BA, Schubiger G. 1986. Parameters controlling transcriptional activation during early *Drosophila* development. *Cell* 44(6):871–77

28. Edgar BA, Sprenger F, Duronio RJ, Leopold P, O'Farrell PH. 1994. Distinct molecular mechanisms regulate cell cycle timing at successive stages of *Drosophila* embryogenesis. *Genes Dev.* 8(4):440–52

29. Erickson JW, Cline TW. 1993. A bZIP protein, sisterless-a, collaborates with bHLH transcription factors early in *Drosophila* development to determine sex. *Genes Dev.* 7(9):1688–702

30. Erickson JW, Cline TW. 1998. Key aspects of the primary sex determination mechanism are conserved across the genus *Drosophila*. *Development* 125(16):3259–68

31. Farrell JA, O'Farrell PH. 2013. Mechanism and regulation of Cdc25/Twine protein destruction in embryonic cell-cycle remodeling. *Curr. Biol.* 23(2):118–26

32. Farrell JA, Shermoen AW, Yuan K, O'Farrell PH. 2012. Embryonic onset of late replication requires Cdc25 down-regulation. *Genes Dev.* 26(7):714–25

33. Ferrell JE, Wu M, Gerhart JC, Martin GS. 1991. Cell cycle tyrosine phosphorylation of p34cdc2 and a microtubule-associated protein kinase homolog in *Xenopus* oocytes and eggs. *Mol. Cell. Biol.* 11(4):1965–71

34. Foe VE. 1989. Mitotic domains reveal early commitment of cells in *Drosophila* embryos. *Development* 107(1):1–22

35. Foe VE, Alberts BM. 1983. Studies of nuclear and cytoplasmic behaviour during the five mitotic cycles that precede gastrulation in *Drosophila* embryogenesis. *J. Cell Sci.* 61:31–70

36. Fogarty P, Campbell SD, Abu-Shumays R, Phalle BS, Yu KR, et al. 1997. The *Drosophila* grapes gene is related to checkpoint gene *chk1/rad27* and is required for late syncytial division fidelity. *Curr. Biol.* 7(6):418–26

37. Follette PJ, Duronio RJ, O'Farrell PH. 1998. Fluctuations in cyclin E levels are required for multiple rounds of endocycle S phase in *Drosophila*. *Curr. Biol.* 8(4):235–38

38. Follette PJ, O'Farrell PH. 1997. Cdks and the *Drosophila* cell cycle. *Curr. Opin. Genet. Dev.* 7(1):17–22

39. Frederick DL, Andrews MT. 1994. Cell cycle remodeling requires cell-cell interactions in developing *Xenopus* embryos. *J. Exp. Zool.* 270(4):410–16

40. Fu S, Nien CY, Liang HL, Rushlow C. 2014. Co-activation of microRNAs by Zelda is essential for early *Drosophila* development. *Development* 141(10):2108–18

41. Gawliński P, Nikolay R, Goursot C, Lawo S, Chaurasia B, et al. 2007. The *Drosophila* mitotic inhibitor Frühstart specifically binds to the hydrophobic patch of cyclins. *EMBO Rep.* 8(5):490–96

42. Gerhart JC. 1980. Mechanisms regulating pattern formation in the amphibian egg and early embryo. In *Biological Regulation and Development*, Vol. 2. *Molecular Organization and Cell Function*, ed. RF Goldberger, pp. 133–316. New York: Plenum Press. 2nd ed.

43. Gould KL, Nurse P. 1989. Tyrosine phosphorylation of the fission yeast cdc2$^+$ protein kinase regulates entry into mitosis. *Nature* 342(6245):39–45

44. Graham CF, Morgan RW. 1966. Changes in the cell cycle during early amphibian development. *Dev. Biol.* 14(3):439–60

45. Gross PR, Cousineau GH. 1964. Macromolecule synthesis and the influence of actinomycin on early development. *Exp. Cell Res.* 33:368–95

46. Grosshans J, Müller HAJ, Wieschaus E. 2003. Control of cleavage cycles in *Drosophila* embryos by *frühstart*. *Dev. Cell* 5(2):285–94

47. Grosshans J, Wieschaus E. 2000. A genetic link between morphogenesis and cell division during formation of the ventral furrow in *Drosophila*. *Cell* 101(5):523–31

48. Harrison MM, Li X-Y, Kaplan T, Botchan MR, Eisen MB. 2011. Zelda binding in the early *Drosophila melanogaster* embryo marks regions subsequently activated at the maternal-to-zygotic transition. *PLOS Genet.* 7(10):e1002266

49. Hayashi S. 1996. A Cdc2 dependent checkpoint maintains diploidy in *Drosophila*. *Development* 122(4):1051–58

50. Howe JA, Howell M, Hunt T, Newport JW. 1995. Identification of a developmental timer regulating the stability of embryonic cyclin A and a new somatic A-type cyclin at gastrulation. *Genes Dev.* 9(10):1164–76

51. Iwao Y, Uchida Y, Ueno S, Yoshizaki N, Masui Y. 2005. Midblastula transition (MBT) of the cell cycles in the yolk and pigment granule-free translucent blastomeres obtained from centrifuged *Xenopus* embryos. *Dev. Growth Differ.* 47(5):283–94

52. Jin Z, Homola EM, Goldbach P, Choi Y, Brill JA, Campbell SD. 2005. *Drosophila* Myt1 is a Cdk1 inhibitory kinase that regulates multiple aspects of cell cycle behavior during gametogenesis. *Development* 132(18):4075–85

53. Kafatos FC, Mitsialis SA, Spoerel N, Mariani B, Lingappa JR, Delidakis C. 1985. Studies on the developmentally regulated expression and amplification of insect chorion genes. *Cold Spring Harb. Symp. Quant. Biol.* 50:537–47

54. Kane DA, Hammerschmidt M, Mullins MC, Maischein HM, Brand M, et al. 1996. The zebrafish epiboly mutants. *Development* 123:47–55

55. Kane DA, Kimmel CB. 1993. The zebrafish midblastula transition. *Development* 119(2):447–56

56. Kimelman D, Kirschner M, Scherson T. 1987. The events of the midblastula transition in *Xenopus* are regulated by changes in the cell cycle. *Cell* 48(3):399–407

57. Kimmel CB, Ballard WW, Kimmel SR, Ullmann B, Schilling TF. 1995. Stages of embryonic development of the zebrafish. *Dev. Dyn.* 203(3):253–310

58. Knoblich JA, Sauer K, Jones L, Richardson H, Saint R, Lehner CF. 1994. Cyclin E controls S phase progression and its down-regulation during *Drosophila* embryogenesis is required for the arrest of cell proliferation. *Cell* 77(1):107–20

59. Lane ME, Sauer K, Wallace K, Jan YN, Lehner CF, Vaessin H. 1996. Dacapo, a cyclin-dependent kinase inhibitor, stops cell proliferation during *Drosophila* development. *Cell* 87(7):1225–35

60. Laugsch M, Schierenberg E. 2004. Differences in maternal supply and early development of closely related nematode species. *Int. J. Dev. Biol.* 48(7):655–62

61. Leptin M. 2005. Gastrulation movements: the logic and the nuts and bolts. *Curr. Opin. Cell Biol.* 8(3):305–20

62. Liang H-L, Nien C-Y, Liu H-Y, Metzstein MM, Kirov N, Rushlow C. 2008. The zinc-finger protein Zelda is a key activator of the early zygotic genome in *Drosophila*. *Nature* 456(7220):400–3

63. Lohe AR, Hilliker AJ, Roberts PA. 1993. Mapping simple repeated DNA sequences in heterochromatin of *Drosophila melanogaster*. *Genetics* 134(4):1149–74

64. Lott SE, Villalta JE, Schroth GP, Luo S, Tonkin LA, Eisen MB. 2011. Noncanonical compensation of zygotic X transcription in early *Drosophila melanogaster* development revealed through single-embryo RNA-seq. *PLOS Biol.* 9(2):e1000590

65. Lu X, Li JM, Elemento O, Tavazoie S, Wieschaus EF. 2009. Coupling of zygotic transcription to mitotic control at the *Drosophila* mid-blastula transition. *Development* 136(12):2101–10

66. Mac Auley A, Werb Z, Mirkes PE. 1993. Characterization of the unusually rapid cell cycles during rat gastrulation. *J. Embryol. Exp. Morphol.* 117(3):873–83

67. Manes C. 1973. The participation of the embryonic genome during early cleavage in the rabbit. *Dev. Biol.* 32(2):453–59

68. Mata J, Curado S, Ephrussi A, Rørth P. 2000. Tribbles coordinates mitosis and morphogenesis in *Drosophila* by regulating String/CDC25 proteolysis. *Cell* 101(5):511–22

69. McCleland ML, O'Farrell PH. 2008. RNAi of mitotic cyclins in *Drosophila* uncouples the nuclear and centrosome cycle. *Curr. Biol.* 18(4):245–54

70. McCleland ML, Shermoen AW, O'Farrell PH. 2009. DNA replication times the cell cycle and contributes to the mid-blastula transition in *Drosophila* embryos. *J. Cell Biol.* 187(1):7–14

71. McGarry TJ, Kirschner MW. 1998. Geminin, an inhibitor of DNA replication, is degraded during mitosis. *Cell* 93(6):1043–53

72. McKnight SL, Miller OL. 1977. Electron microscopic analysis of chromatin replication in the cellular blastoderm *Drosophila melanogaster* embryo. *Cell* 12(3):795–804

73. McKnight SL, Miller OL. 1979. Post-replicative nonribosomal transcription units in *D. melanogaster* embryos. *Cell* 17(3):551–63

74. Merrill PT, Sweeton D, Wieschaus E. 1988. Requirements for autosomal gene activity during precellular stages of *Drosophila melanogaster*. *Development* 104(3):495–509

75. Morgan D. 1997. Cyclin-dependent kinases: engines, clocks, and microprocessors. *Annu. Rev. Cell Dev. Biol.* 13(1):261–91

76. Morgan DO. 2007. *The Cell Cycle*. London: New Science Ltd. 1st ed.

77. Nabel-Rosen H, Toledano-Katchalski H, Volohonsky G, Volk T. 2005. Cell divisions in the *Drosophila* embryonic mesoderm are repressed via posttranscriptional regulation of *string/cdc25* by HOW. *Curr. Biol.* 15(4):295–302

78. Newport J, Kirschner M. 1982. A major developmental transition in early *Xenopus* embryos: I. Characterization and timing of cellular changes at the midblastula stage. *Cell* 30(3):675–86

79. Newport J, Kirschner M. 1982. A major developmental transition in early *Xenopus* embryos: II. Control of the onset of transcription. *Cell* 30(3):687–96

80. Nien C-Y, Liang H-L, Butcher S, Sun Y, Fu S, et al. 2011. Temporal coordination of gene networks by Zelda in the early *Drosophila* embryo. *PLOS Genet.* 7(10):e1002339

81. Nothias JY, Majumder S, Kaneko KJ. 1995. Regulation of gene expression at the beginning of mammalian development. *J. Biol. Chem.* 270(38):22077–80

82. O'Farrell PH. 2004. How metazoans reach their full size: the natural history of bigness. In *Cell Growth: Control of Cell Size*, ed. MN Hall, M Raff, G Thomas, pp. 1–22. New York: Cold Spring Harbor Press

83. O'Farrell PH, Stumpff J, Su TT. 2004. Embryonic cleavage cycles: How is a mouse like a fly? *Curr. Biol.* 14(1):R35–R45

84. Parkhurst SM, Lipshitz HD, Ish-Horowicz D. 1993. *achaete-scute* feminizing activities and *Drosophila* sex determination. *Development* 117(2):737–49

85. Price D, Rabinovitch S, O'Farrell PH, Campbell SD. 2000. *Drosophila wee1* has an essential role in the nuclear divisions of early embryogenesis. *Genetics* 155(1):159–66

86. Pritchard DK, Schubiger G. 1996. Activation of transcription in *Drosophila* embryos is a gradual process mediated by the nucleocytoplasmic ratio. *Genes Dev.* 10(9):1131–42

87. Quinn LM, Herr A, McGarry TJ, Richardson H. 2001. The *Drosophila* Geminin homolog: roles for Geminin in limiting DNA replication, in anaphase and in neurogenesis. *Genes Dev.* 15(20):2741–54

88. Rabinowitz M. 1941. Studies on the cytology and early embryology of the egg of *Drosophila melanogaster*. *J. Morphol.* 69(1):1–49

89. Rose LS, Wieschaus E. 1992. The *Drosophila* cellularization gene *nullo* produces a blastoderm-specific transcript whose levels respond to the nucleocytoplasmic ratio. *Genes Dev.* 6(7):1255–68

90. Russell P, Nurse P. 1986. Cdc25⁺ functions as an inducer in the mitotic control of fission yeast. *Cell* 45(1):145–53

91. Satija R, Bradley RK. 2012. The TAGteam motif facilitates binding of 21 sequence-specific transcription factors in the *Drosophila* embryo. *Genome Res.* 22(4):656–65

92. Seher TC, Leptin M. 2000. Tribbles, a cell-cycle brake that coordinates proliferation and morphogenesis during *Drosophila* gastrulation. *Curr. Biol.* 10(11):623–29

93. Semotok JL, Cooperstock RL, Pinder BD, Vari HK, Lipshitz HD, Smibert CA. 2005. Smaug recruits the CCR4/POP2/NOT deadenylase complex to trigger maternal transcript localization in the early *Drosophila* embryo. *Curr. Biol.* 15(4):284–94

94. Shermoen AW, McCleland ML, O'Farrell PH. 2010. Developmental control of late replication and S phase length. *Curr. Biol.* 20(23):2067–77

95. Shermoen AW, O'Farrell PH. 1991. Progression of the cell cycle through mitosis leads to abortion of nascent transcripts. *Cell* 67(2):303–10

96. Shimuta K, Nakajo N, Uto K, Hayano Y, Okazaki K, Sagata N. 2002. Chk1 is activated transiently and targets Cdc25A for degradation at the *Xenopus* midblastula transition. *EMBO J.* 21(14):3694–703

97. Sibon OC, Laurençon A, Hawley R, Theurkauf WE. 1999. The *Drosophila* ATM homologue Mei-41 has an essential checkpoint function at the midblastula transition. *Curr. Biol.* 9(6):302–12

98. Sibon OC, Stevenson VA, Theurkauf WE. 1997. DNA-replication checkpoint control at the *Drosophila* midblastula transition. *Nature* 388(6637):93–97

99. Simpson L, Wieschaus E. 1990. Zygotic activity of the *nullo* locus is required to stabilize the actin-myosin network during cellularization in *Drosophila*. *Development* 110(3):851–63

100. Smibert CA, Wilson JE, Kerr K, Macdonald PM. 1996. Smaug protein represses translation of unlocalized *nanos* mRNA in the *Drosophila* embryo. *Genes Dev.* 10(20):2600–9

101. Smith AV, Orr-Weaver TL. 1991. The regulation of the cell cycle during *Drosophila* embryogenesis: the transition to polyteny. *Development* 112(4):997–1008

102. Snow MHL. 1977. Gastrulation in the mouse: growth and regionalization of the epiblast. *J. Embryol. Exp. Morphol.* 42(1):293–303

103. Spradling AC. 1981. The organization and amplification of two chromosomal domains containing *Drosophila* chorion genes. *Cell* 27:193–201

104. Stiffler LA, Ji JY, Trautmann S, Trusty C, Schubiger G. 1999. Cyclin A and B functions in the early *Drosophila* embryo. *Development* 126(23):5505–13

105. Stumpff J, Duncan T, Homola E, Campbell SD, Su TT. 2004. *Drosophila* Wee1 kinase regulates Cdk1 and mitotic entry during embryogenesis. *Curr. Biol.* 14(23):2143–48

106. Su TT, Campbell SD, O'Farrell PH. 1999. *Drosophila grapes*/CHK1 mutants are defective in cyclin proteolysis and coordination of mitotic events. *Curr. Biol.* 9(16):919–22

107. Su TT, Sprenger F, DiGregorio PJ, Campbell SD, O'Farrell PH. 1998. Exit from mitosis in *Drosophila* syncytial embryos requires proteolysis and cyclin degradation, and is associated with localized dephosphorylation. *Genes Dev.* 12(10):1495–503

108. Sullivan W, Fogarty P, Theurkauf W. 1993. Mutations affecting the cytoskeletal organization of syncytial *Drosophila* embryos. *Development* 118(4):1245–54

109. Sung H-W, Spangenberg S, Vogt N, Grosshans J. 2013. Number of nuclear divisions in the *Drosophila* blastoderm controlled by onset of zygotic transcription. *Curr. Biol.* 23(2):133–38

110. Sweeton D, Parks S, Costa M, Wieschaus E. 1991. Gastrulation in *Drosophila*: the formation of the ventral furrow and posterior midgut invaginations. *Development* 112(3):775–89

111. Tadros W, Goldman AL, Babak T, Menzies F, Vardy L, et al. 2007. SMAUG is a major regulator of maternal mRNA destabilization in *Drosophila* and its translation is activated by the PAN GU kinase. *Dev. Cell* 12(1):143–55

112. Takai H, Tominaga K, Motoyama N, Minamishima YA, Nagahama H, et al. 2000. Aberrant cell cycle checkpoint function and early embryonic death in $Chk1^{-/-}$ mice. *Genes Dev.* 14(12):1439–47

113. ten Bosch JR, Benavides JA, Cline TW. 2006. The TAGteam DNA motif controls the timing of *Drosophila* pre-blastoderm transcription. *Development* 133(10):1967–77

114. Thomson AM, Gillespie PJ, Blow JJ. 2010. Replication factory activation can be decoupled from the replication timing program by modulating Cdk levels. *J. Cell Biol.* 188(2):209–21

115. Uto K, Inoue D, Shimuta K, Nakajo N, Sagata N. 2004. Chk1, but not Chk2, inhibits Cdc25 phosphatases by a novel common mechanism. *EMBO J.* 23(16):3386–96

116. Wieschaus E. 1996. Embryonic transcription and the control of developmental pathways. *Genetics* 142(1):5–10

117. Wieschaus E, Sweeton D. 1988. Requirements for X-linked zygotic gene activity during cellularization of early *Drosophila* embryos. *Development* 104(3):483–93

118. Wilson EB. 1925. *The Cell in Development and Heredity*. New York: Macmillan. 3rd ed.

119. Xu Z, Chen H, Ling J, Yu D, Struffi P, et al. 2014. Impacts of the ubiquitous factor Zelda on Bicoid-dependent DNA binding and transcription in *Drosophila*. *Genes Dev.* 28(6):608–21

120. Yasuda GK, Baker J, Schubiger G. 1991. Temporal regulation of gene expression in the blastoderm *Drosophila* embryo. *Genes Dev.* 5(10):1800–12

121. Yasuda GK, Schubiger G. 1992. Temporal regulation in the early embryo: Is MBT too good to be true? *Trends Genet.* 8(4):124–27

122. Yuan K, Farrell JA, O'Farrell PH. 2012. Different cyclin types collaborate to reverse the S-phase checkpoint and permit prompt mitosis. *J. Cell Biol.* 198(6):973–80

123. Yuan K, Shermoen AW, O'Farrell PH. 2014. Illuminating DNA replication during *Drosophila* development using TALE-lights. *Curr. Biol.* 24(4):R144–45

124. Zalokar M. 1976. Autoradiographic study of protein and RNA formation during early development of *Drosophila* eggs. *Dev. Biol.* 49(2):425–37

125. Zamir E, Kam Z, Yarden A. 1997. Transcription-dependent induction of G1 phase during the zebrafish midblastula transition. *Mol. Cell. Biol.* 17(2):529–36

126. Zhurov V, Terzin T, Grbić M. 2007. (In)discrete charm of the polyembryony: evolution of embryo cloning. *Cell. Mol. Life Sci.* 64(21):2790–98

Cellular and Molecular Mechanisms of Single and Collective Cell Migrations in *Drosophila*: Themes and Variations

Shirin M. Pocha and Denise J. Montell

Molecular, Cellular and Developmental Biology, University of California, Santa Barbara, California; 93106-9625; email: shirin.pocha@lifesci.ucsb.edu, denise.montell@lifesci.ucsb.edu

Annu. Rev. Genet. 2014. 48:295–318

The *Annual Review of Genetics* is online at genet.annualreviews.org

This article's doi:
10.1146/annurev-genet-120213-092218

Keywords

chemotaxis, EMT, development, tubulogenesis

Abstract

The process of cell migration is essential throughout life, driving embryonic morphogenesis and ensuring homeostasis in adults. Defects in cell migration are a major cause of human disease, with excessive migration causing autoimmune diseases and cancer metastasis, whereas reduced capacity for migration leads to birth defects and immunodeficiencies. Myriad studies in vitro have established a consensus view that cell migrations require cell polarization, Rho GTPase–mediated cytoskeletal rearrangements, and myosin-mediated contractility. However, in vivo studies later revealed a more complex picture, including the discovery that cells migrate not only as single units but also as clusters, strands, and sheets. In particular, the role of E-Cadherin in cell motility appears to be more complex than previously appreciated. Here, we discuss recent advances achieved by combining the plethora of genetic tools available to the *Drosophila* geneticist with live imaging and biophysical techniques. Finally, we discuss the emerging themes such studies have revealed and ponder the puzzles that remain to be solved.

INTRODUCTION

For decades, the process of cell migration has captivated scientists working in different branches of biology. Developmental biologists discovered cell migrations that shape embryos, immunologists revealed the migratory nature of immune cells, and cell biologists strove to understand the underlying mechanisms. Their combined efforts have resulted in a vast body of literature describing the global effects of migration at the organismal level and the molecular mechanisms governing migration at the biochemical level. Although many initial studies dissecting migration mechanisms were performed using cultured cells, the advent of fluorescent tagging and live imaging enabled the extension of this work to cells and tissues in living organisms. Some of these studies corroborated findings from cell culture experiments, whereas others were contradictory, highlighting the diversity of cell migrations and their underlying mechanisms in vivo.

A fascinating outcome of these studies has been the identification of diverse types of migrating entities: the single cell migrating alone and the cluster, strand, or sheet of cells migrating collectively. These cells migrate on a variety of different substrates and through diverse terrains. In this review, we discuss the work in *Drosophila* that has combined genetic approaches and live imaging to reveal cellular and molecular mechanisms for single and collective cell migrations in different tissues and stages of development. Comparing and contrasting the mechanisms leads us to ask whether there are unifying principles underlying most or all migrations or whether each migration mechanism is unique. The answer is that there are clearly common themes upon which evolution has built a multitude of variations. Common themes include specification of leader cells and followers in collective cell movements; variations occur in whether the leader fate is permanent or transient, in how tightly the leaders and followers are connected, and in the organization of and means of communication between the followers. Central roles for Rho family GTPases represent a key molecular theme; however, the details of their functions vary from one cell type to the other. Cadherins and integrins are critically important adhesion molecules in virtually every cell movement, but the contributions of classical cadherins are more varied and interesting than previously imagined. Here, we explore the themes and variations of individual and collective cell movements revealed by genetic approaches in *Drosophila*.

THE BASICS OF CELL MIGRATION

Studies on the migration of cells in culture have led to the development of a generic view: Migrating cells polarize toward a chemoattractive signal and/or away from repulsive signals. Part of this polarization is the protrusion of an actin-rich leading edge, which then adheres to the substrate. Retraction of the rear of the cell is then driven by the disassembly of old adhesions while actomyosin contractility of the rear and cell body allows forward motion. These processes occur continuously during migration so that the cell can respond to changing environmental stimuli. Some of the molecules involved in migration regulate more than one of these steps and so spatial control of their activation is essential. For example, Rho family GTPases are small GTP-hydrolases that are active when GTP-bound and inactive upon GTP hydrolysis. In the most basic model, the Rho GTPase Rac is active at the front of the cell, where it stimulates actin polymerization. In fibroblasts migrating on glass, Rac activation results in the formation of broad ruffling protrusions called lamellae (43). As the founding member of the Rho GTPase family, Rho is related to Rac in amino acid sequence, but Rho is thought to act predominantly in the cell body and rear to activate actomyosin contractility. A third Rho family member, Cdc42, stimulates production of long, finger-like projections called filopodia, which sense extracellular cues and polarize cells in response (43). This view is enticingly simple and elegant, and is supported by observations in many cells; however, increasingly sophisticated techniques, such as the live imaging of cells in their natural environment, have revealed many exceptions to this dogma. As we discuss, a far more complex picture is emerging in vivo, where variations on this theme occur in virtually every cell type studied in detail.

Cell Migration in Human Diseases

Cell migration not only is an essential biological process, it is often also a driving cause of human disease. Increases in migratory capacity drive metastasizing tumor cells (135) and autoimmune diseases, such as multiple sclerosis (30, 90), whereas reduced migratory capacity can cause birth defects (56, 126), immunodeficiencies (2), and impaired wound healing (138). The ability to specifically impede aberrant cell migrations and promote desirable movements would thus benefit many branches of modern medicine. Elucidating the detailed mechanisms for a variety of migrating cells and understanding their similarities and differences is a crucial step in this direction. As touched upon above, in vitro studies may have provided a limited understanding of the scale of complexity found in vivo, and use of clinical samples, although diagnostically informative, provides only a snapshot of the disease state, with limited scope for manipulation. Therefore, the establishment of disease paradigms in model organisms is invaluable.

Drosophila has proved an excellent animal in which to dissect complex biological phenomena at a molecular level because of its rapid generation time and minimal genetic redundancy as well as the sheer number of experimental tools developed over the century of its use as a model organism by thousands of investigators. Here, we review how genetic studies in *Drosophila* have advanced our understanding of cellular and molecular mechanisms controlling both single and collective cell migrations during development.

THE GENETICS OF SINGLE-CELL MIGRATIONS

Gastrulation

The early *Drosophila* embryo consists of a single epithelial sheet, which undergoes extensive morphogenetic movements to form the tissues and organs of a fully developed embryo. The first of these movements is gastrulation, during which invagination of the ventral epithelium is followed by a form of epithelial-to-mesenchymal transition (EMT), the migration of individual mesenchymal cells along the ectoderm (**Figures 1** and **2**), and the cessation of migration as secondary epithelial structures are formed by mesenchymal-to-epithelial transitions (6).

The mechanism of ventral furrow formation and epithelial-to-mesenchymal transition. A maternally provided gradient of the transcription factor Dorsal initiates a network of signaling pathways and transcription factors that establish the dorso-ventral axis of the fly embryo (122). Two such transcription factors, Twist and Snail, are expressed in the ventral-most ectoderm and initiate changes in cell shape, polarity, and adhesion that drive invagination and formation of a ventral furrow (67). First found in *Drosophila* genetic screens, Snail and Twist are highly conserved regulators of EMT in many tissues and organisms as well as in tumors (140). Invagination of cells during fly gastrulation is driven predominantly by Twist-mediated apical flattening and constriction. Twist promotes cell shape changes, mesodermal cell fate, and mesenchymal cell behavior during furrow ingression, whereas Snail represses transcription of ectodermal determinants (68, 110). After ventral furrow formation, two Twist targets, the FGF (fibroblast growth factor) receptor Heartless and Downstream of FGF (Dof), drive EMT (113, 131). FGF signaling induces the mesenchymal cells to spread out onto the extracellular matrix underlying the ectoderm. The Heartless ligands, FGF8-like1 and 2, are expressed in a gradient along the dorsal and lateral ectoderm, leading Heartless-expressing mesenchymal cells to spread radially along the ectoderm (121, 129). Close proximity to the ectoderm is essential, as it also provides differentiation cues that induce specific cell fates at specific locations along the dorsal/ventral axis. For example, the signal from Dpp (a *Drosophila* BMP-like protein produced by dorsal ectoderm cells) initiates the differentiation of dorsal mesoderm cells into cardiac muscle cells (32).

Hemocytes

Primordial germ cells

Border cells

Gastrulation

Dorsal closure

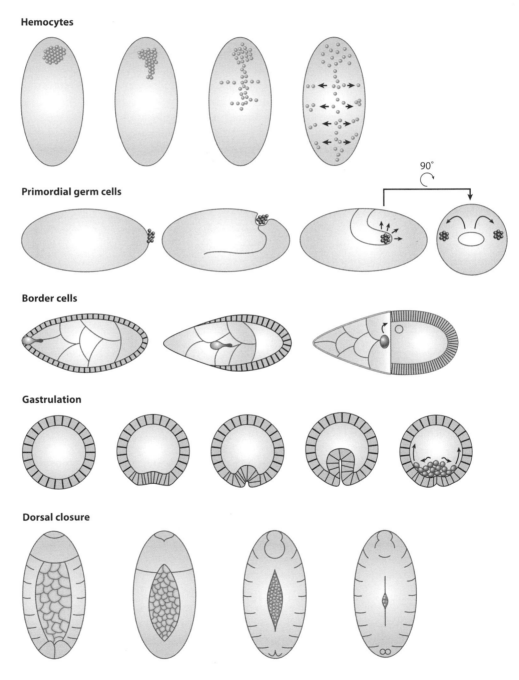

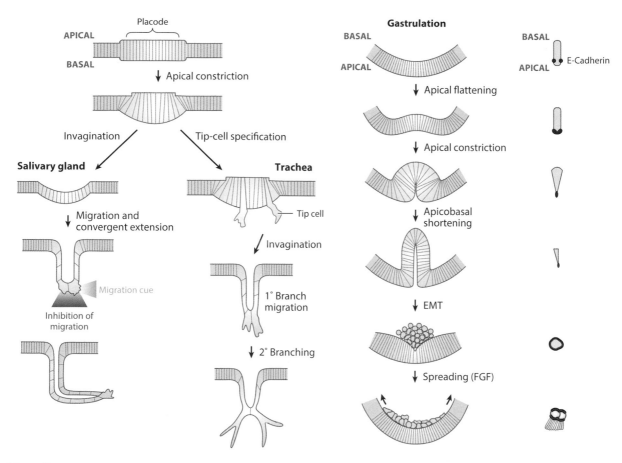

Figure 2

Migrations initiated by epithelial invagination. Comparison of the three modes of tissue formation following epithelial invagination to form the salivary glands, trachea, and mesoderm. Cell shape and E-Cadherin changes during gastrulation are shown on the right. Abbreviations: EMT, epithelial-to-mesenchymal transition; FGF, fibroblast growth factor.

Figure 1

Single and collective migrations in *Drosophila* development. Hemocytes originate in the head and migrate in a highly stereotypic manner along the ventral side of the embryo. Primordial germ cells move at first passively with the midgut primordium, with the global movements caused by germband extension. They then polarize and transmigrate as single cells through the midgut epithelium. Next, they sort bilaterally and migrate to the two sets of somatic gonad precursors on either side of the embryo. Border cells migrate in between nurse cells of the developing follicle, toward the oocyte, turning dorsally at the end of migration. Gastrulation starts with apical flattening and constriction, followed by an epithelial-to-mesenchymal transition and the dorsal spreading of mesodermal cells. Dorsal closure: The ectoderm migrates over the constricting amnioserosa to seal a hole by fusing of the two sides and forming a complete ectodermal epithelium. Arrows show paths of migration.

Regulation of cell adhesion during gastrulation. During gastrulation, mesoderm cells lose epithelial characteristics, such as apical adherens junctions and apicobasal polarity (91). During this process, both Twist and Snail modify Cadherin levels. Snail represses zygotic E-Cadherin expression, leaving only low levels of maternally provided protein during ventral furrow formation (91, 93). In contrast, Twist upregulates N-Cadherin (49). However, the exact role of the E- and N-Cadherin switch during gastrulation is still unclear. The loss of adherens junctions is necessary for cells of an epithelium to singularize; however, it is not clear whether this indeed happens to all invaginating cells during fly gastrulation (87, 109). Using live imaging and photoactivatable GFP (green fluorescent protein) to label single cells, Murray & Saint (87) showed that following invagination, the cells in contact with the ectoderm migrate as a group to spread over the ectoderm. In contrast, cells at the tip of the ventral furrow round up and migrate randomly over the already spreading mesoderm cells to gain access to the ectoderm layer. The authors conclude that both of these steps are dependent on FGF signaling through Heartless and that a combination of FGF chemotactic signals and differential adhesion of migrating mesoderm cells to mesoderm versus ectoderm drive both processes. McMahon et al. (79) employed two-photon imaging combined with detailed quantification to track 100,000 cells per embryo via Histone2A-GFP. The authors describe individual trajectories for each cell, allowing comparison between embryos. This confirmed the two cell behaviors described by Murray & Saint (87): the collective migration of cells closest to the ectoderm and the random migration of cells at the top of the furrow. However, using Heartless mutants, the individual cell trajectories of many cells allowed McMahon et al. (79) to reveal two distinct cell behaviors in the absence of FGF signaling. The cells in contact with the ectoderm (those that invaginate last during furrow formation) are not affected by the lack of Heartless, eventually achieving the same positions as seen in wild-type embryos. In contrast, cells that invaginated earlier and are therefore higher up in the furrow frequently failed to collapse and thus fail to contact the ectoderm. As a result, they then remain in a clump and fail to migrate or intercalate to form a monolayer in contact with the ectoderm. When mutant cells do contact the ectoderm, they tend to migrate and intercalate normally, suggesting that a key function of FGF signaling is promoting the collapse and spreading of the mesoderm cells to ensure contact with the ectoderm. Exactly which cell biological changes characterize this collapse remain to be clarified. Comparing these two experimental approaches, the authors arrived at similar overall conclusions and models; however, the meticulous tracking of each cell revealed a more specific defect in *heartless* mutant animals, leading to a more precise understanding of the role of FGF signaling.

The finding that some mesodermal cells migrate collectively raises the question as to whether they really undergo EMT or whether their behavior would best be described by another term. However, owing to the involvement of the well-characterized EMT regulators Twist and Snail in *Drosophila* gastrulation, describing the various cell behaviors as an EMT spectrum seems appropriate. A potential indicator of where a process lies on the EMT spectrum is its dependence on different Cadherin molecules. The upregulation of N-Cadherin and downregulation of E-Cadherin is a well-established step in classical, full EMT (37). A recent study into the function of the Cadherin switch during *Drosophila* gastrulation found no role for zygotic E-Cadherin or N-Cadherin during both mesoderm invagination and spreading (109). Schäfer et al. (109) overexpressed E-Cadherin in the ventral ectoderm and found that spreading of the ectoderm was not compromised. This led the authors to propose that *Drosophila* gastrulation occurs by a partial EMT, during which cell-cell contacts need not be entirely dissolved for completion of gastrulation. Mesoderm cell fate was, however, defective upon overexpression of E- but not N-Cadherin, and Schäfer et al. (109) conclude that Snail-mediated downregulation of E-Cadherin may normally have an indirect effect on Wingless (*Drosophila* Wnt)-mediated mesoderm differentiation.

Adherens junctions: intercellular adhesions between neighboring epithelial cells containing E-cadherin and its binding partners, β-catenin and α-catenin

This is because Armadillo (Arm; *Drosophila* β-Catenin) serves both as a component of adherens junctions and as a limiting component of Wnt signaling. Reduction of E-Cadherin releases Arm from junctions, potentially increasing the proportion of Arm available to participate in Wnt signaling. It is interesting that N-Cadherin overexpression does not perturb mesoderm differentiation. Schäfer et al. (109) propose that this may be due to the previously demonstrated lower affinity of N-Cadherin for Arm, the different subcellular locations at which the Cadherins sequester Arm (the junction by E-Cadherin and intracellular membranes by N-Cadherin), or to the fact that N-Cadherin binds to an alternative subpopulation of Arm that is not used in the Wnt pathway. However, conclusions based purely on overexpression experiments should be drawn with caution.

Rho GTPases in gastrulation. Both Rho and the Rho-specific guanine nucleotide exchange factor (GEF) RhoGEF2 are required for activation of myosin contractility, which drives the apical constriction that in turn drives ventral furrow formation (4, 42). Cell division is actively inhibited during invagination of the furrow (40) but restarts upon initiation of EMT and depends upon another RhoGEF, Pebble. In addition to a role in cytokinesis, Pebble is required separately for EMT and the spreading of mesodermal cells along the ectoderm (115), with mesodermal cells lacking Pebble remaining more tightly adhered to one another and failing to extend dorsal protrusions. The Rho-family GTPase Rap1 and its GEF Dizzy/PDZ-GEF are also required for ventral furrow formation (115, 117). Although Rho-dependent actomyosin contractility occurs in the absence of Rap1, the failure to form a coherent adherens junction belt without Rap1 (22) renders actomyosin contractions futile, as F-actin is no longer linked to the cell periphery. Following furrow formation, both Rap1 and Dizzy are downregulated to allow mesodermal spreading (117), suggesting that cell-cell adhesions must remain plastic at some level for mesoderm development, despite the controversy surrounding the exact role of the cadherin switch and EMT.

Notable EMTs in vertebrate development include mesendoderm migration during gastrulation, delamination of trunk neural crest cells, and muscle precursor cell migration into limb buds. Homologs of Twist and Snail are clearly involved in these movements and when studied in as much molecular detail as *Drosophila* gastrulation, complexities such as the EMT spectrum may also prove to be conserved. Indeed, a recent study revealed that cells disseminating from cultured murine mammary epithelial cysts induced by Twist retain E-Cadherin expression (112). Moreover, reduction of E-Cadherin levels in these cells inhibits their migration as single cells, suggesting a full EMT does not drive their dissemination and that retention of junctional proteins is in fact necessary. Although it is not clear precisely what purpose E-Cadherin serves in a single cell, one possibility is that CIL (contact inhibition of locomotion) interactions between individual cells require E-Cadherin, α-Catenin, and β-Catenin, and promote their dispersion. Thus, an E-Cadherin-dependent program can promote motility of normal and cancer cells, suggesting an alternative mechanism to EMT for tumor metastasis.

Germ Cells

The primordial germ cells (PGCs) of *Drosophila* are specified very early in development, at the posterior end of blastoderm-stage embryos. The PGCs form prior to and at some distance from the somatic gonadal precursors (SGPs) and the process by which the two are united in the embryonic gonad has been the focus of intense study for many years (104). Interactions between germ cells and somatic gonad tissues promote germline stem cell specification, self-renewal, and differentiation into egg and sperm, thereby ensuring successful transmission of totipotent genetic material to the next generation.

Regulation of primordial germ cell adhesion. Shortly after specification, PGCs adopt a migratory morphology, remodel their actin cytoskeleton, and extend polarized pseudopods (51). However, the initial stage of their movement into the embryo is largely passive, and during gastrulation, as PGCs are carried with the invaginating posterior midgut primordium (**Figure 1**), they round up, adhere to one another in an E-Cadherin-dependent manner (24), and display increasingly uniform cortical actin and decreased protrusive activity (62). At this stage of development, E-Cadherin levels are sensitive to H_2O_2. Peroxiredoxin degrades H_2O_2, and, consequently, *peroxiredoxin* mutant flies phenocopy maternal E-Cadherin mutants. Once inside the forming midgut, PGCs again change morphology, clustering tightly together in a radial arrangement, with their tails together in the middle of the cluster and extending protrusions toward the midgut from their outward-facing leading edges (62). This arrangement requires GPCR (G-protein coupled receptor) signaling from Tre1 to alter the localization of Rho1, actin, and E-Cadherin, which initially localize to the entire membrane of PGCs but are excluded from the leading edges of PGCs during radial rearrangement (62, 63). Thereafter, PGCs disperse, transmigrate as single cells in between cells of the midgut epithelium into the mesoderm, and then sort bilaterally to the sides of the embryo to join the two groups of SGPs that are specified at dorsolateral segments of the mesoderm.

Primordial germ cell transmigration. The transmigration step has been extensively studied due to similarities with invasive cancer cells, which transmigrate across endothelial cell layers into and out of blood vessels during metastasis. Two genes that play a key role in PGC transmigration are *Wunen1* and *Wunen2* (118, 143, 144). Both Wunen genes encode fly homologs of mammalian lipid phosphate phosphatases (LPPs) that hydrolyze phospholipids within the extracellular leaflet of the plasma membrane but show different substrate specificities (14). The exquisite variety of genetic tools available to the *Drosophila* geneticist has allowed a precise dissection of the contributions that somatically expressed Wunens make to PGC migration and survival versus those of Wunens provided by the germline (101, 143, 144). These studies highlight the value of studying migration in vivo, as well as the need for independent genetic manipulation of the migrating cells and their environs.

Chemotaxis of primordial germ cells: Wunens and lipid modification. Wunens are expressed in restricted locations that PGCs normally avoid and were thus initially thought to encode repellents. However, it now appears more likely that they degrade a phospholipid attractant, creating a gradient such that PGCs migrate toward high phospholipid concentrations and away from regions of high Wunen expression (118). After transmigration through the midgut, which requires the active loosening of junctions between midgut epithelial cells (17, 111), PGCs encounter the central nervous system (CNS) at the embryonic midline. High levels of *Wunen 1* and *Wunen 2* in the CNS drive their bilateral sorting away from the midline (107). Germ cells themselves also express both Wunens, which causes them to repel each other (101) and leads to their dispersal. Following their transendothelial migration Wunen-mediated PGC-PGC repulsion ensures equal bilateral sorting to the SGPs (101). In addition to their roles in directing PGC migration, autonomous expression of Wun1 and 2 is required for PGC survival during migration (44, 101, 102). PGCs express Wun1 and 2 from maternally provided mRNA, which allows them to compete with somatically expressed Wunens for phospholipids (101). Wun2 allows the PGCs to take up the phospholipid and convert it into a survival signal (44, 102). The purpose of this connection between migration and survival could be to ensure that cells die if they end up in the wrong place. Alternatively or in addition, this mode of migration may have evolved from chemotaxis mechanisms present in primitive single-celled organisms that allow them to sense and move toward nutrients.

In addition to the repulsive action of somatically expressed Wunens, the lateral mesoderm and SGPs generate signals to attract and capture the PGCs in the latter half of their migration. Although the exact nature of the chemoattractant(s) remains elusive, a screen for PGC migration defects identified one gene hypothesized to process it. 3-hydroxy-3-methylglutaryl coenzyme A reductase (HMGCR), best known for its role in mammalian cholesterol biosynthesis, is expressed dynamically during development in target tissues of the migrating PGCs (130). Flies are sterol auxotrophs (46); the enzymes needed for cholesterol synthesis are missing from the fly genome (108). Thus, *Drosophila* HMGCR is only involved in lipid-modifying pathways that do not result in cholesterol synthesis and the role of HMGCR in isoprenoid synthesis has been implicated in PGC migration (108). The discovery that Shifted, a secreted protein that facilitates the long-distance transmission of Hedgehog, acts in concert with the HMGCR pathway to ensure PGC migration (25) added to the body of evidence that a lipid/lipidated molecule(s) attracts PGCs to the embryonic gonad. Processing and secretion of this attractant via the Ste24 prenylprotease and Mdr49, an ABC transporter, show striking functional homology to the production of a-type mating factor in yeast and lipidated peptides in mammals (103). Thus, lipids and lipidated proteins play particularly important roles in *Drosophila* PGC migration.

Lipids also function as chemoattractants in the human immune system. Leukotrienes are potent proinflamatory lipids that act as chemokines for leukocytes and increase their endothelial adherence and activation state (26). Indeed, when produced in excess, leukotrienes are a major cause of both acute and chronic inflammatory diseases, such as asthma, inflammatory bowel disease, and rheumatoid arthritis (26). Lipids are also produced by pathogens, and the immune system has evolved to use these as chemoattractants. The major outer membrane lipid component of gram-negative bacteria, lipopolysaccharide (LPS), elicits an inflammatory response by initiating leukocyte chemotaxis as well as activating leukocytes to produce proinflamatory cytokines (18). Thus, lipids are used in a variety of settings during development and adult homeostasis to direct cell migration.

> **Sterol auxotroph:** an organism incapable of synthesizing sterols and that must obtain them through their diet

Hemocyte Migration

Hemocytes, the *Drosophila* embryonic immune cells akin to mammalian macrophages, have provided researchers with an excellent system for dissecting the genetics of single-cell migrations in vivo. Originating in the procephalic mesoderm, hemocytes migrate along highly reproducible routes during development to take up their final sentinel positions (**Figure 1**). Once evenly distributed throughout the embryo, hemocytes remain highly motile, patrolling for infection or dying cells (127). Early studies suggested that PDGF (platelet-derived growth factor)- and VEGF (vascular endothelial growth factor)-related factors (PVFs) serve as chemoattractants for the stereotypical embryonic migration pattern. Loss of the PVF receptor (PVR) from hemocytes appeared to inhibit migration, loss of the PVF ligands phenocopies loss of the receptor, and ectopic expression of PVFs is capable of misdirecting migration (21). However, subsequent detailed analysis of the PVR mutant phenotype showed that PVR was important for hemocyte viability and that dying hemocytes attract and redirect additional hemocytes midway through their developmental migration (13). Mutant and rescue analyses demonstrated that the primary role for PVR signaling is to promote hemocyte survival rather than chemotaxis (13, 92), with the single exception that activation of PVR by PVFs expressed by the hindgut epithelium of the extended germband is required for hemocytes to cross that epithelial barrier (92).

Rho GTPases in hemocyte migration. In addition to developmental signals, sites of injury or infection attract hemocytes. Using laser ablation to create epithelial wounds in conjunction

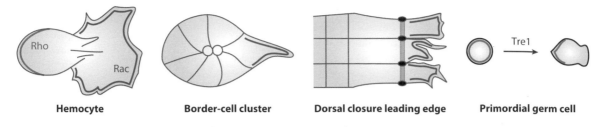

Figure 3

Morphological similarities between single and collectively migrating cells.

with live imaging, Stramer and colleagues developed an elegant model to study the inflammatory response in *Drosophila*. They used this system to dissect the roles of the Rho GTPases and revealed some exciting confirmations of and contradictions to previous models. The small round cells described in the initial characterization of hemocyte origins (21) turned out to be merely the cell bodies of large, highly protrusive and dynamic cells (124). As predicted from cell culture studies, formation of broad, ruffling lamellae that polarize toward chemoattractants requires all three *Drosophila* Rac proteins, and Rho is required for the detachment and retraction of the cell rear (**Figure 3**). However, contrary to in vitro data that suggest a requirement for Cdc42 in chemotaxis, in vivo, in embryos mutant for the single *cdc42* gene, hemocytes disperse normally during development and accumulate at wounds despite reduced persistence of migration (124). Therefore, the requirement for Cdc42 may be context dependent; subtle and/or maternal Cdc42 protein, which the mother loads into the egg during oogenesis, may persist long enough to support hemocyte movements.

Chemotaxis during hemocyte development and wound responses. Further investigation into the genes required for hemocyte migration identified a changing hierarchy of competing guidance cues during development (84). A major attractant to wound sites is H_2O_2 released by dying cells (84), whereas developmental migration can occur in the absence of cell death (21). So what happens when wounding occurs during developmental migration? Are some migration cues more attractive than others? The answer depends on timing. There is a refractory period early in hemocyte migration, during which developmental signals dominate and even wounds near the normal migration path are incapable of recruiting hemocytes (84). Following this refractory period, a rapid and robust redirection of hemocytes away from their normal path to wounds is observed. Strikingly, the refractory period does not apply to dying cells, because apoptotic cells at the margins of the developmental path are able to divert neighboring hemocytes from their stereotypical routes. This finding fits with the data described above on the recruitment of hemocytes to dying hemocytes in *pvr* mutant embryos.

Directional movement of hemocytes is not only achieved by chemotaxis. More recent findings explain hemocyte dispersal during development on the basis of a mechanism called CIL (23, 123). CIL was initially described in the 1950s by Abercrombie (1) in his studies into the social lives of cells. Two different types of behavior occur when migrating fibroblasts contact one another. In Type I CIL, cells stop following contact, whereas in Type II CIL, cells repolarize and migrate away from one another (1). In the ensuing half-century, little progress was made to characterize the molecular mechanisms underlying Abercrombie's initial observations. However, in recent years there has been a renewed interest in these concepts. Using genetic manipulations and live imaging, Stramer et al. (123) identified a specific type of cellular protrusion, which mediates CIL and is dependent on microtubules and the microtubule bundling protein Clasp, between migrating hemocytes. Furthermore, computer simulations of this phenomenon reveal that this CIL alone could in principle account for the developmental dispersal of hemocytes observed many years earlier (23). CIL also determines the final positioning of collectively migrating cranial neural crest cells in *Xenopus* (128). Neural crest cells chemotax toward sensory organ placodal cells, which express the chemokine Sdf1. However, the neural crest cells actively repulse placodal cells, a type II CIL event in which the repolarization after collision is regulated by Wnt/PCP (planar cell polarity) and N-Cadherin activity. Thus, CIL appears to determine the final positions of both hemocytes and neural crest and placode cells. In *Drosophila*, CIL seems to function to distribute hemocytes evenly throughout the embryo, whereas in *Xenopus*, it appears to maintain placodal tissue boundaries and ensure correct neural crest cell migration.

THE GENETICS OF COLLECTIVE MIGRATIONS

Salivary Glands

The *Drosophila* embryonic salivary glands (SGs) form from a group of cells that, once specified as the SG placode, cease dividing (88) and attain their final tubular morphology, organization, and location through a combination of convergent extension and collective migration (2). Following invagination from the SG placode, the salivary cells move together in a highly stereotypical pattern: at first migrating dorsally and then turning and traveling toward the posterior so that the distal end of the SG stops at the level of the third thoracic segment (11). Although the size and shape of the SG are established concurrently with its migration, in mutant animals lacking the function of genes involved in migration, its length and diameter are not dramatically compromised (12). Therefore, SG migration is required only for final positioning of the gland and is not a driving force for convergent extension that generates its shape. The first gene found to regulate SG migration was *ribbon*, which, owing to a genetic interaction with *zipper* (nonmuscle myosin heavy chain), was suggested to act upon the actomyosin cytoskeleton (10). *Ribbon* encodes a member of the BTB-domain family of transcription factors and upregulates expression of the apical determinant Crumbs and downregulates activity, but not expression, of Moesin, a cross-linker that attaches F-actin to the plasma membrane, thereby increasing cortical stiffness (57). Using a combination of live imaging and computer modeling, *ribbon* mutants were indeed shown to display increased apical stiffness (19). This was attributed to increased Moesin activity and reduced Crumbs levels and hypothesized to cause the defective SG tube elongation and lumen formation in *ribbon* mutants (19).

Chemotaxis I: fibroblast growth factor signaling and salivary gland directional migration. Insight into the posterior migration of the SG was obtained from a screen that identified *heartless* as essential for SG shape and positioning (12). Interestingly, neither *heartless* nor its downstream effector, *heartbroken*, is expressed in the SG, instead showing high expression in the visceral mesoderm. Further analysis of the *heartless* and *heartbroken* phenotypes, in which the visceral mesoderm fails to form correctly, revealed that the visceral mesoderm itself provides the migration cue for the SG to begin the posterior portion of its migration. Furthermore, SG migration requires the integrin PS1 expressed in the SG and the integrin PS2 expressed in both mesoderm and SG cells (12). This suggests that, as with many single-cell migrations, SG cells require interactions with the ECM for their collective migration. In the mesoderm, by contrast, integrin has been proposed to organize the ECM upon which the SG migrate (12).

Chemotaxis II: *slit* and *Netrin* signaling and salivary gland directional migration. Two genes, *slit* and *Netrin*, classically known for their roles in axon guidance (59) are also involved in

Planar cell polarity (PCP): polarization within the plane of an epithelium

Placode: an area within an epithelium, thicker than neighboring cells, that gives rise to a tissue/organ during embryogenesis

Convergent extension: the simultaneous lengthening of a tissue along one axis and narrowing of the perpendicular axis by cell intercalation

guiding the collective posterior migration of SG (60). Netrin, expressed in the visceral mesoderm and CNS midline, acts as a chemo-attractant for the migrating SG, causing it to migrate along the visceral mesoderm and parallel to the CNS. Slit is expressed in the CNS midline and acts as a chemorepellant. A model is proposed in which a balance between the two ensures the precise positioning of the SG during development (60). In addition to Slit, Wnt5 expressed in the CNS also repels the migrating SG through its receptor Derailed, which is expressed in the migrating SG tip (45). The report on Slit and Netrin was the first to attribute a non-neuronal guidance function to Netrin in *Drosophila*. More recently, however, Slit, Netrin, and the Netrin receptor Deleted in Colorectal Cancer (DCC) have been found downregulated in numerous human cancers, where they correlate with tumor progression, revealing potential roles as tumor suppressors (80). Although the mechanisms underlying Slit and Netrin function in tumor progression are unknown, both have been linked to tumor cell migration. Slit overexpression inhibits medulloblastoma cell migration (136) and Netrin1 stimulates the migration of pancreatic adenocarcinoma cells (29). Any connection between *Drosophila* SG development or axon pathfinding and tumor progression would have been hard to predict a priori, but the conservation of these key molecular pathways highlights the value of studying simple model organisms such as flies and worms (9). These systems offer tremendous technical advantages for dissecting the molecular mechanisms governing fundamental and evolutionarily conserved processes like cell migration in an in vivo setting.

Rho GTPases in migration and convergent extension of the salivary glands. Rac functions in the distal (leading) cells of the SGs, where it promotes the extension of actin-rich basal protrusions, as it does for many migratory cell types (96). However, the role for Rac in the SG as a whole is more complex. Rac is required throughout the SG to promote collective migration by reducing E-Cadherin protein levels downstream of integrin signaling (96). Slightly lowering E-Cadherin levels is thought to facilitate the junctional rearrangements that are essential for the cells to be able to converge and extend, elongating the gland, as they migrate (95, 96). Rho1 also plays a role in maintaining epithelial identity and integrity during migration by ensuring the correct localization of apical determinants Crumbs and aPKC (139). These data suggest that the entire tissue contributes to the collective migration of the SG, as opposed to an active leading edge dragging a passive tissue behind it. The front senses direction, whereas the back actively rearranges by convergent extension.

Trachea

Similar to the SG, the tracheal network forms from placodes of the ectodermal epithelium. However, whereas only two salivary placodes develop, multiple tracheal placodes form, one at each side of each of the body segments. The tracheal placodes constrict their apical surfaces and invaginate to form tracheal sacs, after which no further cell division occurs. Similar to the SG, the entire tracheal system elaborates by cell elongation, intercalation, and migration. From each tracheal sac, six primary branches sprout and undergo further rounds of branching as the tracheal system develops. Interestingly, although each branching step requires a similar set of genes, they use different cellular mechanisms to produce distinct types of tubes, and therefore tracheal development is not an iterative process (106). The first migratory steps occur at the beginning of tubulogenesis. For each one of the six primary branches, two tracheal sac cells extend basal protrusions and begin migration, towing 4–20 placodal cells to produce the primary branch. Tubulation occurs concomitant with migration through a combination of cell elongation and intercalation so that eventually most primary branches form tubes that are one cell thick (2) (35).

Chemotaxis I: fibroblast growth factor signaling and tracheal migration and branching.
Migration away from the tracheal sac and branching are both regulated by the FGF receptor
Breathless, which is expressed by tracheal cells that migrate toward sites of Branchless (FGF)
expression (106, 125). Upon reaching the source of Branchless, the tubules branch and the two
smaller branches migrate further (125). The two migrating cells are termed tip cells, and in contrast
to the cells they tow (which display only epithelial characteristics) tip cells maintain a polarized
migratory morphology throughout tracheal development (66). As the primary branch is forming,
the tip cells retain the constricted apical surface that initially drove placode invagination. The tip-
cell apicobasal axis lies parallel to the direction of migration: The apical surface maintains contact
with the trailing cells, whereas the basal surface accumulates F-actin as it extends forward protrusions (66). In contrast, cells further back in the branch rapidly lose apical constriction, orient their
apicobasal axis perpendicular to the direction of migration and localize F-actin predominantly to
the apical junctional region. Therefore, only a small proportion of cells acquire classical migratory
characteristics.

Lateral inhibition:
the prevention of one
cell from adopting a
specific fate or
behavior through
specific contacts with
its neighbor

Breathless/Branchless signaling makes multiple contributions to tracheal migration and
branching. There is initially a competition between tracheal pit cells, the winners of which exhibit
greater Breathless expression and consequently receive higher levels of Breathless/Branchless signaling and so acquire tip-cell identity (36). Moreover, in an entirely *breathless* mutant embryo,
expression of Breathless in a single cell is sufficient for a branch to form (66). High levels of
Breathless signaling in the tip cells alter their pattern of gene expression to drive tip-cell behavior (48). Notch-mediated lateral inhibition limits the number of tip cells (36), as does the
competition for Breathless signaling. The *Drosophila* homolog of the human tumor suppressor
gene *von Hippel-Lindau* regulates cell surface levels of Breathless by regulating its endocytosis in
a hypoxia-independent fashion early in tracheal development (47).

Chemotaxis II: slit and robo guidance of tracheal branch migration. In addition to Breath-
less/Branchless, developing tracheal cells respond to Slit expressed in the CNS, which repels some
tracheal branches and attracts others, depending on the expression levels of distinct receptors. The
tip cells of different branches express different levels of the two Slit receptors Robo and Robo2.
Activation of Robo repels migrating tip cells, whereas Robo2 activation attracts tip cells to the
source of Slit (31). Therefore, the combination of these attractive and repulsive signals ensures an
even distribution of branches throughout the embryo.

Rho GTPases and tracheal branch migration. Rac is activated in the tip cells, where it destabilizes adherens junctions and promotes polarization and migration (20, 36). Indeed, the exclusion
of cells that express dominant-negative Rac from the tips of migrating tracheal branches (66)
supports previous findings that Rac is an essential downstream effector of Breathless signaling in
these cells (20). Both Rho and Cdc42 maintain the localization of key epithelial polarity proteins
(E-Cadherin, Crumbs, and α-Catenin) in branch cells, and expression of a dominant-negative
form of either GTPase results in loss of epithelial characteristics, increased migration speed, and
breaks in individual branches (66). In contrast, constitutively active forms of Rho and Cdc42 result
in extremely compact branches that migrate slower than wild-type branches and these cells rarely
make it to tip-cell positions (66).

Together, data gathered over many years have revealed a picture in which one or two tip cells
gain a highly migratory phenotype while remaining attached to neighboring cells that organize
themselves into a multicellular tube to form a tracheal branch. At a molecular and morphological
level, the migratory tip cells polarize in the direction of migration (with high Rac activity and motile
front actin-based protrusions), whereas the branch cells maintain their epithelial characteristics

via Rho and Cdc42-mediated stabilization of adherens junctions. Using a combination of genetic manipulation and laser dissection, Caussinus et al. (16) assessed the physical properties of the migrating branches and determined that the migrating tip cells exert a tensile stress throughout the branch, which acts as a driving force for the cell intercalation events that are necessary for tubulogenesis and branch elongation.

During zebrafish development, the posterior lateral line primordium (pllp) also undergoes a collective migration along the body of the fish to reach the tail (38). Similar to the developing *Drosophila* SG and trachea, the pllp cells are connected by tight junctions (61), which ensures collective movement. However, dynamic protrusions are observed not only at the tip of the pllp but also from cells at the side, suggesting that all of the cells in the pllp provide motive force and may even sense direction to some extent, and that steering and maintaining pllp integrity are not mutually exclusive (41). However, mosaic experiments similar to those used to study *Drosophila* tracheal development have shown that remarkably few leading cells are required to ensure directional migration of the pllp (41). This highlights a conserved mechanism used in different organisms, whereby collective migration requires only a few leading cells to be competent in sensing direction as long as the following cells remain attached strongly enough for the group to follow.

Border Cells

Border cells of the *Drosophila* ovary migrate as a cluster of four to eight motile cells surrounding two nonmotile cells called polar cells (see 83 for a recent review). They originate within the epithelial layer of somatic follicle cells, which surround and support developing germ cells in a structure called an egg chamber (**Figure 1**). The sixteen germ cells per egg chamber are siblings, one of which becomes the oocyte and the others differentiate into enormous polyploid nurse cells. All these cells remain connected by cytoplasmic bridges through which nurse cells donate their cytoplasm to the oocyte so that each egg chamber produces a single egg. The follicle cells generally provide yolk proteins, patterning signals, and the eggshell. The polar cells and border cells, in particular, produce the eggshell structure called the micropyle through which the sperm enters. Therefore, if border cells or polar cells fail to reach the oocyte, the female is sterile.

Border-cell migration seems deceptively simple because the cells migrate in a straight line. However, to accomplish this simple goal, the cells must extract themselves from the follicular epithelium and detach from the basal lamina that surrounds the entire egg chamber. They must extend protrusions in between the nurse cells, stay on the straight path, choose a leader, coordinate the activities of the individual cells, and carry the polar cells; finally, when they reach the oocyte, they take a turn toward the dorsal side. Much has been learned about how the border cells are specified and how they sense direction collectively, yet many intriguing questions remain. Specifically, the mechanisms by which they initially exit the epithelium remain mysterious, as are those that cause them to stop. The similarities between border-cell migration and the dissemination of tumor nests make these particularly interesting questions to address using the powerful approaches this system affords.

JAK/STAT signaling and border-cell specification. The molecular mechanisms by which 4 to 8 migratory border cells are specified out of the 600 to 900 immobile epithelial follicle cells have been clarified. Early in the development of each egg chamber, a special pair of cells forms at each pole. These polar cells secrete the cytokine Unpaired, which activates the JAK/STAT pathway in neighboring cells (78, 114). The follicle cells closest to the polar cells receive the highest levels of JAK/STAT signaling, and at the anterior pole this activates a gene expression program that allows them to begin their invasive migration in between the nurse cells toward the oocyte (8, 133).

Feedback mechanisms convert this initially graded JAK/STAT signal into sharp on and off states (82, 119, 120, 141). The spatially localized JAK/STAT signal is integrated with a developmental timing cue, the steroid hormone ecdysone, to produce motility of the correct cells and at the appropriate stage of development (3, 52).

As the border cells initiate their migration, one or two cells extend a long protrusion in between the anterior nurse cells, perhaps competing to take the lead (33). The border cells must dramatically reorganize their cell-cell and cell-matrix adhesions at this point, detach from the laminin- and collagen-rich basal lamina and break their adhesions with neighboring epithelial cells while remaining attached to the polar cells and to one another. The border cells retain apicobasal polarity and apical adherens junctions even as their lateral edges in contact with nurse cells become protrusive and contractile (89, 94, 98). Detachment requires Notch activity (98, 132) and Par-1 kinase-mediated activation of myosin (73, 75).

Receptor tyrosine kinases and chemotaxis of border cells. Guidance of the collective border cell migration has been extensively studied. Multiple secreted ligands are produced in the germline, at highest concentration in the oocyte, and activate the two receptor tyrosine kinases (RTKs) PVR and EGFR, which are expressed on the surfaces of all follicle cells, including the border cells (27, 28). Mutation of either receptor alone causes distinct but mild migration defects. However, mutation of both receptors causes a dramatic defect in which the cluster loses its sense of direction. Moreover, ectopic sources of ligands for either receptor can redirect border-cell migration (76, 77). However, the difference in receptor activation between the front and back of the cluster is so slight that it can only be detected if the receptor is overexpressed (54). Rac activity is higher at the front than the back (99, 134), and cell morphology and behavior is dramatically different between the front and the back (97, 98). Therefore, there should be an amplification mechanism downstream of the receptors to magnify the difference in Rac activity and protrusion stability between the front and the back.

Rho GTPases in border-cell migration. The key role for Rac for cell migration in vivo, and specifically its role in generating protrusions, was first demonstrated in the border cells (86). It is now clear that the cell with the highest Rac activity sets the direction of migration for the whole cluster because photoactivation of Rac in one cell is sufficient to redirect the cluster, and local inhibition of Rac in the lead cell confuses all the cells (134). The lead cell normally silences protrusive behavior in the rest of the cells, and this communication requires the activity of the JNK pathway (134), Rab11, and Moesin (99), although a detailed mechanism remains to be elucidated. Although at any given moment one cell takes the lead, unlike the tracheal system this is not a permanent cell fate and different cells can occupy the lead position at different times during the migration. Rho and Cdc42 both appear to promote cluster cohesion by organizing the distributions of epithelial polarity proteins, F-actin, integrin, and cadherin (5, 71).

Regulation of border-cell adhesion. The presence of cell-cell adhesions is a defining feature of collectively migrating cells, which, unlike single cells, must adhere together as a group while interacting dynamically with a substrate. The regulation of E-Cadherin is particularly complex and interesting for migrating border cells because they use it not only to connect the individual migrating cells to one another but also as a cell-substrate adhesion molecule as they migrate directly upon the surfaces of the nurse cells. Although it has been clear for some time that genetic ablation of E-Cadherin from either migrating border cells or their nurse cell substrate impairs their migration (89), a recent study reveals multiple intricate ways in which E-Cadherin promotes the collective guidance of border-cell migration. Cai et al. (15) developed an in vivo optical sensor of mechanical tension across E-Cadherin molecules, which they combined with cell-type-specific RNAi, photoactivatable Rac, and live imaging to tease apart multiple distinct roles for E-Cadherin, each in a different subcellular location. They discovered that adhesion between border cells and the nurse cells works in a positive feedback loop with Rac to amplify guidance receptor signaling at the leading edge, stabilizing forward-directed protrusion. In contrast, E-Cadherin-mediated adhesion between individual migratory border cells mechanically couples them and is essential for the lead cell to communicate direction to the followers. Adhesion between motile cells and nonmigratory polar cells prevents the cluster from splitting apart and polarizes each individual cell such that it protrudes outward. These activities work in concert to coordinate collective direction sensing. Thus, E-Cadherin can perform different functions in different locations within the same cell, depending on its relative concentration and the local biochemical environment (i.e., the presence or absence of guidance receptor signaling). An intriguing notion is that these distinct roles for E-Cadherin need not always occur together in the same migration event. Single cells migrating over the surfaces of other cells might only use the feedback amplification between Rac and E-Cadherin, without adhesion to other migrating cells. Collectively, migrating cells moving through ECM might employ cell-cell adhesion for coordinating their movement while using integrin in a feedback loop with Rac to stabilize front protrusions. Thus, the distinct mechanisms could function as modules and might be combined in various ways with cell-matrix adhesion and chemotaxis to diversify morphogenetic cell movements in vivo.

Dorsal closure. The process of dorsal closure in *Drosophila* embryonic development is an excellent example of the migration of a sheet of cells. Rather than a cluster of epithelial cells leaving the epithelium, as we have seen for border cells, the entire epidermal epithelium migrates over the underlying amnioserosa to seal a dorsal hole in the epidermis (**Figure 1**). Although many cell migrations occur over substrates that passively provide traction, in this case the extraembryonic amnioserosa cells actively participate in the process by constricting apically, which shrinks the surface over which epithelial cells must migrate and to which they are attached by adherens-type junctions (39).

Changes in cell shape and polarization at the onset of dorsal closure. As the amnioserosa cells constrict, shrink, and pull, the migrating epithelial cells also undergo shape changes at the onset of dorsal closure. The dorsal-most (leading edge) epidermal cells stretch along the dorso-ventral axis (105, 142). Rearrangements of the actin cytoskeleton precede cell shape changes in both the amnioserosa and epidermis. For the migrating epithelial cells, those at the leading edge undergo the most dramatic changes, as they form the protrusive, exploratory morphology characteristic of single-cell migrations (50). Like border cells, leading-edge cells retain epithelial characteristics during migration, remaining attached to their neighbors and maintaining apicobasal polarity and PCP. The latter depends on noncanonical Wnt signaling, which drives the aforementioned cell shape changes of the leading edge and initiates accumulation of actin nucleating foci at the dorsal-most junction between the epithelium and amnioserosa (55, 85). Interestingly, in addition to the polarization of proteins commonly associated with planar polarity (such as Flamingo and Disheveled), some apicobasal polarity proteins also become planar polarized, with Canoe, Discs large, and FasIII being lost from the dorsal edge (55, 64). This suggests that septate junctions are actively remodeled as leading-edge cells elongate. Thereafter, Ena (34), Cdc42 (50), Rac, and Rac-mediated Jnk signaling (137) act upon the actin cytoskeleton to form filopodial and lamellipodial protrusions that extended from the leading edge cells over the amnioserosa surface (3).

In the final step of dorsal closure, leading-edge cell microtubule-dependent (53) filopodia facilitate the zippering of the opposing epithelial sheets as the dorsal hole is closed and ensure the correct matching of segmental and parasegmental stripes (50, 81). The unconventional myosin Myosin XV delivers E-Cadherin and microtubule regulators to the leading-edge filopodia, thus ensuring adhesion between the opposing epithelial sheets when they are in close proximity (70).

Actomyosin cable formation. Concurrent with the initial shape changes of the leading-edge cells, a supracellular actomyosin cable is formed at the interface between amnioserosa and epithelium (142). The actin cable has been proposed to act as a purse string, which, together with myosin II–mediated contraction of the cable, pulls together the two sheets of the epidermis (7). However, this is not the only force acting during dorsal closure, as interference with the cable integrity by laser ablation (58) or reduction of Rho activity (72) fails to completely halt dorsal closure. An additional function of the actin cable is to maintain the integrity of the epithelium leading edge and, perhaps counterintuitively, to restrain the movements of the leading-edge cells so that they move in a coordinated fashion (50). More recently, live imaging revealed the pulsatile nature of the actin cable contractions. This led to the proposal of a ratchet model, in which the actin cable does not just act as a purse string continually pulling on the edges of the epithelium but rather contracts briefly to restrain the contractile pulses that are occurring apically in the amnioserosa and thereby prevent apical relaxation (116). This joins an increasing body of work that has attributed pulsatile contractions of actomyosin networks to a wide variety of epithelial movements (69, 74, 100).

CONCLUSIONS

As we have seen, there are some striking similarities between the mechanisms different cell types use to migrate, whether they migrate alone or as part of a group. Indeed, some aspects of the cell migration dogma that arose from studies of cultured cells (65) hold true for many cell types in vivo. However, the great diversity of migratory cell types and arrangements found in vivo, as well as the terrains through which they move, demands variations in the cellular and molecular mechanisms.

Within living organisms, cells can migrate as individuals, like PGCs, or in groups of various sizes. A sheet of epithelial cells can march in a straight line and in unison—like the epidermis during dorsal closure—or an epithelial sheet can invaginate—like mesodermal cells during gastrulation and the ectodermal cells of the SG and trachea. Migratory groups of cells can vary in size from six border cells to hundreds of cells moving in concert. Some cells detach from their original neighbors, and others do not.

The environments through which cells move are equally diverse. Cells can migrate on a basement membrane secreted by another cell layer or through a loose matrix of ECM proteins, or they can squeeze in between other cells that are in direct contact. Some cells, such as PGCs, interact with each of these environments at different points in their trajectories. The composition of the environment can affect migratory mechanisms. Integrin-mediated adhesion and signaling dominate cell-matrix interactions, whereas cadherin-mediated adhesion and signaling dominate cell-cell interactions. And if this complexity were not enough, it is important to keep in mind that at any particular stage of embryonic development, multiple migrations could be happening concurrently, so the guidance mechanisms need to be diverse enough to prevent unwanted cross talk and confusion.

It is then not surprising that guidance factors can be protein or lipid or small molecules like H_2O_2, and they can be attractive or repulsive. We propose that this variety helps to ensure that multiple complex migratory events can occur accurately within the same embryo at the same time. In addition, many chemotactic cues promote differentiation or affect survival when cells arrive at their final destination. The specific chemoattractant or repellent selected for each migratory cell type may depend on multiple properties and functions.

Given the diversity and complexity of migratory cell types, arrangements, and microenvironments, one might then predict that each cell type would require its own unique set of molecular mechanisms. Instead, we find a set of themes and variations. Cells follow paths defined by combinations of a relatively limited number of chemoattractants and repellents. The receptors for these extracellular signals converge onto Rho family GTPases, which then organize and regulate cytoskeletal dynamics within cells, where there is arguably less need for diversity. A nearly universal observation, first documented in vivo in the border cells, is that Rac stimulates protrusion. In general, leading cells in different tissues exhibit high levels of Rac activity, which leads to similar exploratory and protrusive characteristics as well as the ability to exert force on trailing cells and affect their behavior. Rho-mediated activation of myosin contractility is also important in many migrating cells. The role of Cdc42 is a bit more complicated, generating sensory filopodia in some cells and regulating apicobasal polarity in others.

The regulation of cell-cell and cell-matrix adhesion in different migratory cell types varies according to whether cells are migrating individually or in groups and whether they are migrating through ECM, on a basement membrane, or directly on the surfaces of other cells. In particular, recent findings suggest that E-Cadherin function is far more nuanced than previously appreciated because it can promote or inhibit both single and collective cell migration in a variety of contexts. Moreover, as shown for the border cells, E-Cadherin can play multiple distinct positive roles in collective chemotaxis, serving different functions in different locations within the same cell. Perhaps then we should revisit the roles of other epithelial polarity proteins in cell migration as well. Initially thought to maintain a static apicobasal polarity to ensure epithelial integrity and barrier function, polarity proteins of the Par, Crumbs, and Scribble complexes might also prove much more dynamic and play multiple roles in migrating cells. Live imaging will likely reveal important changes in polarity protein localization and dynamics that may well be cell-type specific. Deciphering the various molecular mechanisms by which E-Cadherin promotes motility in some contexts while inhibiting it in others will also be an interesting topic for future research.

Together genetic studies of diverse cell migrations in *Drosophila* and in other organisms demonstrate that without in-depth exploration of each type of migration we would not be able to discern the common themes and unique variations. It is likely that healing cells and metastasizing tumor cells alike employ these same themes and variations and thus that a thorough understanding is necessary for us to gain molecular control of both beneficial and harmful cell movements for the improvement of human health.

DISCLOSURE STATEMENT

The authors are not aware of any affiliations, memberships, funding, or financial holdings that might be perceived as affecting the objectivity of this review.

ACKNOWLEDGMENTS

We would like to thank Brian Stramer, Deborah Andrew, Angela Stathopoulos, and Ruth Lehmann for critical comments on the manuscript. Work in the lab of D.J.M. is supported by NIH grants R01RM46425 and R01GM73164. S.M.P. is supported by an award from the Errett Fisher Foundation.

LITERATURE CITED

1. Abercrombie M, Heaysman JEM. 1953. Observations on the social behaviour of cells in tissue culture. I. Speed of movement of chick heart fibroblasts in relation to their mutual contacts. *Exp. Cell Res.* 5:111–31

2. Badolato R. 2013. Defects of leukocyte migration in primary immunodeficiencies. *Eur. J. Immunol.* 43:1436–40

3. Bai J, Uehara Y, Montell DJ. 2000. Regulation of invasive cell behavior by taiman, a *Drosophila* protein related to AIB1, a steroid receptor coactivator amplified in breast cancer. *Cell* 103:1047–58

4. Barrett K, Leptin M, Settleman J. 1997. The Rho GTPase and a putative RhoGEF mediate a signaling pathway for the cell shape changes in *Drosophila* gastrulation. *Cell* 91:905–15

5. Bastock R, Strutt D. 2007. The planar polarity pathway promotes coordinated cell migration during *Drosophila* oogenesis. *Development* 134:3055–64

6. Baum B, Settleman J, Quinlan MP. 2008. Transitions between epithelial and mesenchymal states in development and disease. *Semin. Cell Dev. Biol.* 19:294–308

7. Bement WM, Forscher P, Mooseker MS. 1993. A novel cytoskeletal structure involved in purse string wound closure and cell polarity maintenance. *J. Cell Biol.* 121:565–78

8. Borghese L, Fletcher G, Mathieu J, Atzberger A, Eades WC, et al. 2006. Systematic analysis of the transcriptional switch inducing migration of border cells. *Dev. Cell* 10:497–508

9. Bradford D, Cole SJ, Cooper HM. 2009. Netrin-1: diversity in development. *Int. J. Biochem. Cell Biol.* 41:487–93

10. Bradley PL, Andrew DJ. 2001. ribbon encodes a novel BTB/POZ protein required for directed cell migration in *Drosophila melanogaster*. *Development* 128:3001–15

11. Bradley PL, Haberman AS, Andrew DJ. 2001. Organ formation in *Drosophila*: specification and morphogenesis of the salivary gland. *BioEssays* 23:901–11

12. Bradley PL, Myat MM, Comeaux CA, Andrew DJ. 2003. Posterior migration of the salivary gland requires an intact visceral mesoderm and integrin function. *Dev. Biol.* 257:249–62

13. Bruckner K, Kockel L, Duchek P, Luque CM, Rorth P, Perrimon N. 2004. The PDGF/VEGF receptor controls blood cell survival in *Drosophila*. *Dev. Cell* 7:73–84

14. Burnett C, Howard K. 2003. Fly and mammalian lipid phosphate phosphatase isoforms differ in activity both in vitro and in vivo. *EMBO Rep.* 4:793–99

15. Cai D, Chen SC, Prasad M, He L, Wang X, et al. 2014. Mechanical feedback through E-Cadherin promotes direction sensing during collective cell migration. *Cell* 157:1143–59

16. Caussinus E, Colombelli J, Affolter M. 2008. Tip-cell migration controls stalk-cell intercalation during *Drosophila* tracheal tube elongation. *Curr. Biol.* 18:1727–34

17. Chanet S, Schweisguth F. 2012. Regulation of epithelial polarity by the E3 ubiquitin ligase Neuralized and the Bearded inhibitors in *Drosophila*. *Nat. Cell Biol.* 14:467–76

18. Chen LY, Pan ZK. 2009. Synergistic activation of leukocytes by bacterial chemoattractants: potential drug targets. *Endocr. Metab. Immune Disord. Drug Targets* 9:361–70

19. Cheshire AM, Kerman BE, Zipfel WR, Spector AA, Andrew DJ. 2008. Kinetic and mechanical analysis of live tube morphogenesis. *Dev. Dyn.* 237:2874–88

20. Chihara T, Kato K, Taniguchi M, Ng J, Hayashi S. 2003. Rac promotes epithelial cell rearrangement during tracheal tubulogenesis in *Drosophila*. *Development* 130:1419–28

21. Cho NK, Keyes L, Johnson E, Heller J, Ryner L, et al. 2002. Developmental control of blood cell migration by the *Drosophila* VEGF Pathway. *Cell* 108:865–76

22. Choi W, Harris NJ, Sumigray KD, Peifer M. 2013. Rap1 and Canoe/afadin are essential for establishment of apical-basal polarity in the *Drosophila* embryo. *Mol. Biol. Cell* 24:945–63

23. Davis JR, Huang C-Y, Zanet J, Harrison S, Rosten E, et al. 2012. Emergence of embryonic pattern through contact inhibition of locomotion. *Development* 139:4555–60

24. DeGennaro M, Hurd TR, Siekhaus DE, Biteau B, Jasper H, Lehmann R. 2011. Peroxiredoxin stabilization of DE-Cadherin promotes primordial germ cell adhesion. *Dev. Cell* 20:233–43

25. Deshpande G, Zhou K, Wan JY, Friedrich J, Jourjine N, et al. 2013. The *hedgehog* pathway gene *shifted* functions together with the *hmgcr*-dependent isoprenoid biosynthetic pathway to orchestrate germ cell migration. *PLOS Genet.* 9:e1003720

26. Di Gennaro A, Haeggstrom JZ. 2012. The leukotrienes: immune-modulating lipid mediators of disease. *Adv. Immunol.* 116:51–92

27. Duchek P, Rorth P. 2001. Guidance of cell migration by EGF receptor signaling during *Drosophila* oogenesis. *Science* 291:131–33

28. Duchek P, Somogyi K, Jekely G, Beccari S, Rorth P. 2001. Guidance of cell migration by the *Drosophila* PDGF/VEGF receptor. *Cell* 107:17–26

29. Dumartin L, Quemener C, Laklai H, Herbert J, Bicknell R, et al. 2010. Netrin-1 mediates early events in pancreatic adenocarcinoma progression, acting on tumor and endothelial cells. *Gastroenterology* 138:1595–606

30. Engelhardt B, Ransohoff RM. 2012. Capture, crawl, cross: the T cell code to breach the blood-brain barriers. *Trends Immunol.* 33:579–89

31. Englund C, Steneberg P, Falileeva L, Xylourgidis N, Samakovlis C. 2002. Attractive and repulsive functions of Slit are mediated by different receptors in the *Drosophila* trachea. *Development* 129:4941–51

32. Frasch M. 1995. Induction of visceral and cardiac mesoderm by ectodermal Dpp in the early *Drosophila* embryo. *Nature* 374:464–67

33. Fulga TA, Rorth P. 2002. Invasive cell migration is initiated by guided growth of long cellular extensions. *Nat. Cell Biol.* 4:715–19

34. Gates J, Mahaffey JP, Rogers SL, Emerson M, Rogers EM, et al. 2007. Enabled plays key roles in embryonic epithelial morphogenesis in *Drosophila*. *Development* 134:2027–39

35. Ghabrial A, Luschnig S, Metzstein MM, Krasnow MA. 2003. Branching morphogenesis of the *Drosophila* tracheal system. *Annu. Rev. Cell Dev. Biol.* 19:623–47

36. Ghabrial AS, Krasnow MA. 2006. Social interactions among epithelial cells during tracheal branching morphogenesis. *Nature* 441:746–49

37. Gheldof A, Berx G. 2013. Cadherins and epithelial-to-mesenchymal transition. *Prog. Mol. Biol. Transl. Sci.* 116:317–36

38. Ghysen A, Dambly-Chaudiere C. 2004. Development of the zebrafish lateral line. *Curr. Opin. Neurobiol.* 14:67–73

39. Gorfinkiel N, Arias AM. 2007. Requirements for adherens junction components in the interaction between epithelial tissues during dorsal closure in *Drosophila*. *J. Cell Sci.* 120:3289–98

40. Grosshans J, Wieschaus E. 2000. A genetic link between morphogenesis and cell division during formation of the ventral furrow in *Drosophila*. *Cell* 101:523–31

41. Haas P, Gilmour D. 2006. Chemokine signaling mediates self-organizing tissue migration in the zebrafish lateral line. *Dev. Cell* 10:673–80

42. Hacker U, Perrimon N. 1998. DRhoGEF2 encodes a member of the Dbl family of oncogenes and controls cell shape changes during gastrulation in *Drosophila*. *Genes Dev.* 12:274–84

43. Hall A. 2012. Rho family GTPases. *Biochem. Soc. Trans.* 40:1378–82

44. Hanyu-Nakamura K, Kobayashi S, Nakamura A. 2004. Germ cell–autonomous Wunen2 is required for germline development in *Drosophila* embryos. *Development* 131:4545–53

45. Harris KE, Beckendorf SK. 2007. Different Wnt signals act through the Frizzled and RYK receptors during *Drosophila* salivary gland migration. *Development* 134:2017–25

46. Hobson RP. 1935. On a fat-soluble growth factor required by blow-fly larvae: identity of the growth factor with cholesterol. *Biochem. J.* 29:2023–26

47. Hsouna A, Nallamothu G, Kose N, Guinea M, Dammai V, Hsu T. 2010. *Drosophila* von Hippel-Lindau tumor suppressor gene function in epithelial tubule morphogenesis. *Mol. Cell. Biol.* 30:3779–94

48. Ikeya T, Hayashi S. 1999. Interplay of Notch and FGF signaling restricts cell fate and MAPK activation in the *Drosophila* trachea. *Development* 126:4455–63

49. Iwai Y, Usui T, Hirano S, Steward R, Takeichi M, Uemura T. 1997. Axon patterning requires DN-cadherin, a novel neuronal adhesion receptor, in the *Drosophila* embryonic CNS. *Neuron* 19:77–89

50. Jacinto A, Wood W, Balayo T, Turmaine M, Martinez-Arias A, Martin P. 2000 . Dynamic actin-based epithelial adhesion and cell matching during *Drosophila* dorsal closure. *Curr. Biol.* 10:1420–26

51. Jaglarz MK, Howard KR. 1995. The active migration of *Drosophila* primordial germ cells. *Development* 121:3495–503

52. Jang AC, Chang YC, Bai J, Montell D. 2009. Border-cell migration requires integration of spatial and temporal signals by the BTB protein Abrupt. *Nat. Cell Biol.* 11:569–79

53. Jankovics F, Brunner D. 2006. Transiently reorganized microtubules are essential for zippering during dorsal closure in *Drosophila melanogaster*. *Dev. Cell* 11:375–85

54. Janssens K, Sung HH, Rorth P. 2010. Direct detection of guidance receptor activity during border cell migration. *Proc. Natl. Acad. Sci. USA* 107:7323–28

55. Kaltschmidt JA, Lawrence N, Morel V, Balayo T, Fernandez BG, et al. 2002. Planar polarity and actin dynamics in the epidermis of *Drosophila*. *Nat. Cell Biol.* 4:937–44

56. Kawauchi T, Shikanai M, Kosodo Y. 2013. Extra–cell cycle regulatory functions of cyclin-dependent kinases (CDK) and CDK inhibitor proteins contribute to brain development and neurological disorders. *Genes Cells* 18:176–94

57. Kerman BE, Cheshire AM, Myat MM, Andrew DJ. 2008. Ribbon modulates apical membrane during tube elongation through Crumbs and Moesin. *Dev. Biol.* 320:278–88

58. Kiehart DP, Galbraith CG, Edwards KA, Rickoll WL, Montague RA. 2000. Multiple forces contribute to cell sheet morphogenesis for dorsal closure in *Drosophila*. *J. Cell Biol.* 149:471–90

59. Killeen MT, Sybingco SS. 2008. Netrin, Slit, and Wnt receptors allow axons to choose the axis of migration. *Dev. Biol.* 323:143–51

60. Kolesnikov T, Beckendorf SK. 2005. NETRIN and SLIT guide salivary gland migration. *Dev. Biol.* 284:102–11

61. Kollmar R, Nakamura SK, Kappler JA, Hudspeth AJ. 2001. Expression and phylogeny of claudins in vertebrate primordia. *Proc. Natl. Acad. Sci. USA* 98:10196–201

62. Kunwar PS, Sano H, Renault AD, Barbosa V, Fuse N, Lehmann R. 2008. Tre1 GPCR initiates germ cell transepithelial migration by regulating *Drosophila melanogaster* E-Cadherin. *J. Cell Biol.* 183:157–68

63. Kunwar PS, Starz-Gaiano M, Bainton RJ, Heberlein U, Lehmann R. 2003. Tre1, a G protein–coupled receptor, directs transepithelial migration of *Drosophila* germ cells. *PLOS Biol.* 1:E80

64. Laplante C, Nilson LA. 2011. Asymmetric distribution of Echinoid defines the epidermal leading edge during *Drosophila* dorsal closure. *J. Cell Biol.* 192:335–48

65. Lauffenburger DA, Horwitz AF. 1996. Cell migration: a physically integrated molecular process. *Cell* 84:359–69

66. Lebreton G, Casanova J. 2014. Specification of leading and trailing cell features during collective migration in the *Drosophila* trachea. *J. Cell Sci.* 127:465–74

67. Leptin M, Grunewald B. 1990. Cell shape changes during gastrulation in *Drosophila*. *Development* 110:73–84

68. Leptin M, Roth S. 1994. Autonomy and non-autonomy in *Drosophila* mesoderm determination and morphogenesis. *Development* 120:853–59

69. Levayer R, Lecuit T. 2013. Oscillation and polarity of E-Cadherin asymmetries control actomyosin flow patterns during morphogenesis. *Dev. Cell* 26:162–75

70. Liu R, Woolner S, Johndrow JE, Metzger D, Flores A, Parkhurst SM. 2008. Sisyphus, the *Drosophila* myosin XV homolog, traffics within filopodia transporting key sensory and adhesion cargos. *Development* 135:53–63

71. Llense F, Martin-Blanco E. 2008. JNK signaling controls border cell cluster integrity and collective cell migration. *Curr. Biol.* 18:538–44

72. Magie CR, Meyer MR, Gorsuch MS, Parkhurst SM. 1999. Mutations in the Rho1 small GTPase disrupt morphogenesis and segmentation during early *Drosophila* development. *Development* 126:5353–64

73. Majumder P, Aranjuez G, Amick J, McDonald JA. 2012. Par-1 controls myosin-II activity through myosin phosphatase to regulate border cell migration. *Curr. Biol.* 22:363–72

74. Martin AC, Kaschube M, Wieschaus EF. 2009. Pulsed contractions of an actin-myosin network drive apical constriction. *Nature* 457:495–99

75. McDonald JA, Khodyakova A, Aranjuez G, Dudley C, Montell DJ. 2008. PAR-1 kinase regulates epithelial detachment and directional protrusion of migrating border cells. *Curr. Biol.* 18:1659–67

76. McDonald JA, Pinheiro EM, Kadlec L, Schupbach T, Montell DJ. 2006. Multiple EGFR ligands participate in guiding migrating border cells. *Dev. Biol.* 296:94–103

77. McDonald JA, Pinheiro EM, Montell DJ. 2003. PVF1, a PDGF/VEGF homolog, is sufficient to guide border cells and interacts genetically with Taiman. *Development* 130:3469–78

78. McGregor JR, Xi R, Harrison DA. 2002. JAK signaling is somatically required for follicle cell differentiation in *Drosophila*. *Development* 129:705–17

79. McMahon A, Supatto W, Fraser SE, Stathopoulos A. 2008. Dynamic analyses of *Drosophila* gastrulation provide insights into collective cell migration. *Science* 322:1546–50

80. Mehlen P, Delloye-Bourgeois C, Chedotal A. 2011. Novel roles for Slits and netrins: axon guidance cues as anticancer targets? *Nat. Rev. Cancer* 11:188–97

81. Millard TH, Martin P. 2008. Dynamic analysis of filopodial interactions during the zippering phase of *Drosophila* dorsal closure. *Development* 135:621–26

82. Monahan AJ, Starz-Gaiano M. 2013. Socs36E attenuates STAT signaling to optimize motile cell specification in the *Drosophila* ovary. *Dev. Biol.* 379:152–66

83. Montell DJ, Yoon WH, Starz-Gaiano M. 2012. Group choreography: mechanisms orchestrating the collective movement of border cells. *Nat. Rev. Mol. Cell Biol.* 13:631–45

84. Moreira S, Stramer B, Evans I, Wood W, Martin P. 2010. Prioritization of competing damage and developmental signals by migrating macrophages in the *Drosophila* embryo. *Curr. Biol.* 20:464–70

85. Morel V, Arias AM. 2004. Armadillo/β-Catenin-dependent Wnt signalling is required for the polarisation of epidermal cells during dorsal closure in *Drosophila*. *Development* 131:3273–83

86. Murphy AM, Montell DJ. 1996. Cell type–specific roles for Cdc42, Rac, and RhoL in *Drosophila* oogenesis. *J. Cell Biol.* 133:617–30

87. Murray MJ, Saint R. 2007. Photoactivatable GFP resolves *Drosophila* mesoderm migration behaviour. *Development* 134:3975–83

88. Myat MM, Andrew DJ. 2000. Organ shape in the *Drosophila* salivary gland is controlled by regulated, sequential internalization of the primordia. *Development* 127:679–91

89. Niewiadomska P, Godt D, Tepass U. 1999. DE-Cadherin is required for intercellular motility during *Drosophila* oogenesis. *J. Cell Biol.* 144:533–47

90. Norman MU, Hickey MJ. 2005. Mechanisms of lymphocyte migration in autoimmune disease. *Tissue Antigens* 66:163–72

91. Oda H, Tsukita S, Takeichi M. 1998. Dynamic behavior of the cadherin-based cell-cell adhesion system during *Drosophila* gastrulation. *Dev. Biol.* 203:435–50

92. Parsons B, Foley E. 2013. The *Drosophila* platelet-derived growth factor and vascular endothelial growth factor–receptor related (Pvr) protein ligands Pvf2 and Pvf3 control hemocyte viability and invasive migration. *J. Biol. Chem.* 288:20173–83

93. Peinado H, Ballestar E, Esteller M, Cano A. 2004. Snail mediates E-Cadherin repression by the recruitment of the Sin3A/histone deacetylase 1 (HDAC1)/HDAC2 complex. *Mol. Cell. Biol.* 24:306–19

94. Pinheiro EM, Montell DJ. 2004. Requirement for Par-6 and Bazooka in *Drosophila* border cell migration. *Development* 131:5243–51

95. Pirraglia C, Jattani R, Myat MM. 2006. Rac function in epithelial tube morphogenesis. *Dev. Biol.* 290:435–46

96. Pirraglia C, Walters J, Ahn N, Myat MM. 2013. Rac1 GTPase acts downstream of αPS1βPS integrin to control collective migration and lumen size in the *Drosophila* salivary gland. *Dev. Biol.* 377:21–32

97. Poukkula M, Cliffe A, Changede R, Rorth P. 2011. Cell behaviors regulated by guidance cues in collective migration of border cells. *J. Cell Biol.* 192:513–24

98. Prasad M, Montell DJ. 2007. Cellular and molecular mechanisms of border cell migration analyzed using time-lapse live-cell imaging. *Dev. Cell* 12:997–1005

99. Ramel D, Wang X, Laflamme C, Montell DJ, Emery G. 2013. Rab11 regulates cell-cell communication during collective cell movements. *Nat. Cell Biol.* 15:317–24

100. Rauzi M, Lenne PF, Lecuit T. 2010. Planar polarized actomyosin contractile flows control epithelial junction remodelling. *Nature* 468:1110–14

101. Renault AD, Kunwar PS, Lehmann R. 2010. Lipid phosphate phosphatase activity regulates dispersal and bilateral sorting of embryonic germ cells in *Drosophila*. *Development* 137:1815–23

102. Renault AD, Sigal YJ, Morris AJ, Lehmann R. 2004. Soma-germ line competition for lipid phosphate uptake regulates germ cell migration and survival. *Science* 305:1963–66

103. Ricardo S, Lehmann R. 2009. An ABC transporter controls export of a *Drosophila* germ cell attractant. *Science* 323:943–46

104. Richardson BE, Lehmann R. 2010. Mechanisms guiding primordial germ cell migration: strategies from different organisms. *Nat. Rev. Mol. Cell Biol.* 11:37–49

105. Ring JM, Martinez Arias A. 1993. *puckered*, a gene involved in position-specific cell differentiation in the dorsal epidermis of the *Drosophila* larva. *Development* 19(Suppl.):251–59

106. Samakovlis C, Hacohen N, Manning G, Sutherland DC, Guillemin K, Krasnow MA. 1996. Development of the *Drosophila* tracheal system occurs by a series of morphologically distinct but genetically coupled branching events. *Development* 122:1395–407

107. Sano H, Renault AD, Lehmann R. 2005. Control of lateral migration and germ cell elimination by the *Drosophila melanogaster* lipid phosphate phosphatases Wunen and Wunen 2. *J. Cell Biol.* 171:675–83

108. Santos AC, Lehmann R. 2004. Isoprenoids control germ cell migration downstream of HMGCoA reductase. *Dev. Cell* 6:283–93

109. Schäfer G, Narasimha M, Vogelsang E, Leptin M. 2014. Cadherin switching during the formation and differentiation of the *Drosophila* mesoderm: implications for epithelial mesenchymal transitions. *J. Cell Sci.* 127:1511–22

110. Seher TC, Narasimha M, Vogelsang E, Leptin M. 2007. Analysis and reconstitution of the genetic cascade controlling early mesoderm morphogenesis in the *Drosophila* embryo. *Mech. Dev.* 124:167–79

111. Seifert JR, Lehmann R. 2012. *Drosophila* primordial germ cell migration requires epithelial remodeling of the endoderm. *Development* 139:2101–6

112. Shamir ER, Pappalardo E, Jorgens DM, Coutinho K, Tsai WT, et al. 2014. Twist1-induced dissemination preserves epithelial identity and requires E-Cadherin. *J. Cell Biol.* 204:839–56

113. Shishido E, Higashijima S, Emori Y, Saigo K. 1993. Two FGF-receptor homologues of *Drosophila*: one is expressed in mesodermal primordium in early embryos. *Development* 117:751–61

114. Silver DL, Montell DJ. 2001. Paracrine signaling through the JAK/STAT pathway activates invasive behavior of ovarian epithelial cells in *Drosophila*. *Cell* 107:831–41

115. Smallhorn M, Murray MJ, Saint R. 2004. The epithelial-mesenchymal transition of the *Drosophila* mesoderm requires the Rho GTP exchange factor Pebble. *Development* 131:2641–51

116. Solon J, Kaya-Copur A, Colombelli J, Brunner D. 2009. Pulsed forces timed by a ratchet-like mechanism drive directed tissue movement during dorsal closure. *Cell* 137:1331–42

117. Spahn P, Ott A, Reuter R. 2012. The PDZ-GEF protein Dizzy regulates the establishment of adherens junctions required for ventral furrow formation in *Drosophila*. *J. Cell Sci.* 125:3801–12

118. Starz-Gaiano M, Cho NK, Forbes A, Lehmann R. 2001. Spatially restricted activity of a *Drosophila* lipid phosphatase guides migrating germ cells. *Development* 128:983–91

119. Starz-Gaiano M, Melani M, Meinhardt H, Montell D. 2009. Interpretation of the UPD/JAK/STAT morphogen gradient in *Drosophila* follicle cells. *Cell Cycle* 8:2917–25

120. Starz-Gaiano M, Melani M, Wang X, Meinhardt H, Montell DJ. 2008. Feedback inhibition of Jak/STAT signaling by apontic is required to limit an invasive cell population. *Dev. Cell* 14:726–38

121. Stathopoulos A, Tam B, Ronshaugen M, Frasch M, Levine M. 2004. *pyramus* and *thisbe*: FGF genes that pattern the mesoderm of *Drosophila* embryos. *Genes Dev.* 18:687–99

122. Stathopoulos A, Van Drenth M, Erives A, Markstein M, Levine M. 2002. Whole-genome analysis of dorsal-ventral patterning in the *Drosophila* embryo. *Cell* 111:687–701

123. Stramer B, Moreira S, Millard T, Evans I, Huang C-Y, et al. 2010. Clasp-mediated microtubule bundling regulates persistent motility and contact repulsion in *Drosophila* macrophages in vivo. *J. Cell Biol.* 189:681–89

124. Stramer B, Wood W, Galko MJ, Redd MJ, Jacinto A, et al. 2005. Live imaging of wound inflammation in *Drosophila* embryos reveals key roles for small GTPases during in vivo cell migration. *J. Cell Biol.* 168:567–73

125. Sutherland D, Samakovlis C, Krasnow MA. 1996. *branchless* encodes a *Drosophila* FGF homolog that controls tracheal cell migration and the pattern of branching. *Cell* 87:1091–101

126. Takahashi Y, Sipp D, Enomoto H. 2013. Tissue interactions in neural crest cell development and disease. *Science* 341:860–63

127. Tepass U, Fessler LI, Aziz A, Hartenstein V. 1994. Embryonic origin of hemocytes and their relationship to cell death in *Drosophila*. *Development* 120:1829–37

128. Theveneau E, Steventon B, Scarpa E, Garcia S, Trepat X, et al. 2013. Chase-and-run between adjacent cell populations promotes directional collective migration. *Nat. Cell Biol.* 15:763–72

129. Tulin S, Stathopoulos A. 2010. Analysis of Thisbe and Pyramus functional domains reveals evidence for cleavage of *Drosophila* FGFs. *BMC Dev. Biol.* 10:83

130. Van Doren M, Broihier HT, Moore LA, Lehmann R. 1998. HMG-CoA reductase guides migrating primordial germ cells. *Nature* 396:466–69

131. Vincent Sp, Wilson R, Coelho C, Affolter M, Leptin M. 1998. The *Drosophila* protein Dof is specifically required for FGF signaling. *Mol. Cell* 2:515–25

132. Wang X, Adam JC, Montell D. 2007. Spatially localized Kuzbanian required for specific activation of Notch during border cell migration. *Dev. Biol.* 301:532–40

133. Wang X, Bo J, Bridges T, Dugan KD, Pan TC, et al. 2006. Analysis of cell migration using whole-genome expression profiling of migratory cells in the *Drosophila* ovary. *Dev. Cell* 10:483–95

134. Wang X, He L, Wu YI, Hahn KM, Montell DJ. 2010. Light-mediated activation reveals a key role for Rac in collective guidance of cell movement in vivo. *Nat. Cell Biol.* 12:591–97

135. Wells A, Grahovac J, Wheeler S, Ma B, Lauffenburger D. 2013. Targeting tumor cell motility as a strategy against invasion and metastasis. *Trends Pharmacol. Sci.* 34:283–89

136. Werbowetski-Ogilvie TE, Seyed Sadr M, Jabado N, Angers-Loustau A, Agar NY, et al. 2006. Inhibition of medulloblastoma cell invasion by Slit. *Oncogene* 25:5103–12

137. Woolner S, Jacinto A, Martin P. 2005. The small GTPase Rac plays multiple roles in epithelial sheet fusion: dynamic studies of *Drosophila* dorsal closure. *Dev. Biol.* 282:163–73

138. Xu F, Zhang C, Graves DT. 2013. Abnormal cell responses and role of TNF-α in impaired diabetic wound healing. *BioMed Res. Int.* 2013:754802

139. Xu N, Keung B, Myat MM. 2008. Rho GTPase controls invagination and cohesive migration of the *Drosophila* salivary gland through Crumbs and Rho-kinase. *Dev. Biol.* 321:88–100

140. Yang J, Weinberg RA. 2008. Epithelial-mesenchymal transition: at the crossroads of development and tumor metastasis. *Dev. Cell* 14:818–29

141. Yoon WH, Meinhardt H, Montell DJ. 2011. miRNA-mediated feedback inhibition of JAK/STAT morphogen signalling establishes a cell fate threshold. *Nat. Cell Biol.* 13:1062–69

142. Young PE, Richman AM, Ketchum AS, Kiehart DP. 1993. Morphogenesis in *Drosophila* requires nonmuscle myosin heavy chain function. *Genes Dev.* 7:29–41

143. Zhang N, Zhang J, Cheng Y, Howard K. 1996. Identification and genetic analysis of *wunen*, a gene guiding *Drosophila melanogaster* germ cell migration. *Genetics* 143:1231–41

144. Zhang N, Zhang J, Purcell KJ, Cheng Y, Howard K. 1997. The *Drosophila* protein Wunen repels migrating germ cells. *Nature* 385:64–67

The Structure and Regulation of Flagella in *Bacillus subtilis*

Sampriti Mukherjee and Daniel B. Kearns

Department of Biology, Indiana University, Bloomington, Indiana 47405;
email: dbkearns@indiana.edu

Annu. Rev. Genet. 2014. 48:319–40

First published online as a Review in Advance on
September 10, 2014

The *Annual Review of Genetics* is online at
genet.annualreviews.org

This article's doi:
10.1146/annurev-genet-120213-092406

Keywords

motility, flagella, *Bacillus*, bistability, homeostasis

Abstract

Bacterial flagellar motility is among the most extensively studied physiological systems in biology, but most research has been restricted to using the highly similar Gram-negative species *Escherichia coli* and *Salmonella enterica*. Here, we review the recent advances in the study of flagellar structure and regulation of the distantly related and genetically tractable Gram-positive bacterium *Bacillus subtilis*. *B. subtilis* has a thicker layer of peptidoglycan and lacks the outer membrane of the Gram-negative bacteria; thus, not only phylogenetic separation but also differences in fundamental cell architecture contribute to deviations in flagellar structure and regulation. We speculate that a large number of flagella and the absence of a periplasm make *B. subtilis* a premier organism for the study of the earliest events in flagellar morphogenesis and the type III secretion system. Furthermore, *B. subtilis* has been instrumental in the study of heterogeneous gene transcription in subpopulations and of flagellar regulation at the translational and functional level.

INTRODUCTION

Bacillus subtilis has two forms of active movement, swimming and swarming motility, that are powered by rotating flagella (73, 113). Swimming motility takes place via individual cells moving in three dimensions of a liquid volume. Swarming motility, by contrast, takes place via groups of cells moving in two dimensions atop solid surfaces (75). Commonly used domesticated laboratory strains have lost the ability to swarm as a result of mutations in two genetic loci: One mutation disrupts Sfp, which is involved in the synthesis of a lipopeptide surfactant that reduces surface tension to facilitate spreading, and the other mutation disrupts SwrA, the putative master regulator of flagellar synthesis (118). Simultaneous repair of both mutations restores various laboratory strains to swarming motility that is indistinguishable from undomesticated ancestral strains (72). The ancestral strains not only swim and swarm but also spread over surfaces via a process called sliding (77). Sliding is the product of colony growth, hydration of an extracellular polysaccharide (EPS) capsule, and reduced surface tension conferred by the surfactant (77, 126). In this review, we focus on the structure and regulation of *B. subtilis* flagella and motility to highlight the depth, breadth, and complexity of hierarchical structural assembly in bacteria.

The flagellar structure is complex and is considered to have three architectural domains: the basal body, the hook, and the filament (**Figure 1a**). The basal body is embedded in the cell envelope, houses the secretion apparatus that exports the more distal flagellar components, and provides the power for flagellar rotation. The hook is a flexible universal joint connected to the basal body that changes the angle of rotation of the long helical filament that acts as a propeller. Thirty-two genes required for basal body synthesis are concentrated in the large 27-kb *fla/che* operon expressed via the action of RNA polymerase and the vegetative sigma factor sigma A (σ^A). Once hook assembly is complete, the alternative sigma factor sigma D (σ^D) is activated and enables expression of another set of genes dedicated to filament assembly and rotation (**Figure 1b**). Much of the work on flagellar structure has been conducted in Gram-negative model organisms and was recently reviewed (28). We discuss structural domains prior to discussing regulation, starting from the most distal components and working toward the cell interior by comparing the Gram-negative paradigm with studies in *B. subtilis*.

FLAGELLAR STRUCTURE

Filament

The filament is a helical propeller composed of a repeating protein monomer called flagellin (151). Flagellin is polymerized by interaction between protomer N-terminal and C-terminal domains to make 11 protofilaments that form a hollow cylinder (20 nm in diameter with a 2-nm channel). Although the filament is considered rigid, it can transition between various polymorphic forms in response to environmental conditions and torsional load and, importantly, during direction reversals, which govern motile behavior (33, 145). *B. subtilis* encodes two homologs of flagellin, *yvzB* and *hag*. The *yvzB* gene, if expressed, encodes a partial C-terminal domain of flagellin that is insufficient for polymerization. The *yvzB* gene, however, is not required for motility, has no known function, and may be a pseudogene, although its expression is occasionally reported in transcriptional profiling experiments, perhaps because of the near identity of sequence it shares with *hag*, the other flagellin gene (74, 118). The *hag* gene, by contrast, is essential for flagellar assembly and encodes the flagellin monomer protein Hag (short for H-antigen) (86).

Hag monomers are secreted in an unfolded state through the hollow flagellar rod and hook by the flagellar type III secretion (T3S) apparatus within the basal body (42, 137). Flagellin readily oligomerizes into filaments in vitro; but in vivo, polymerization is restricted by intracellular and

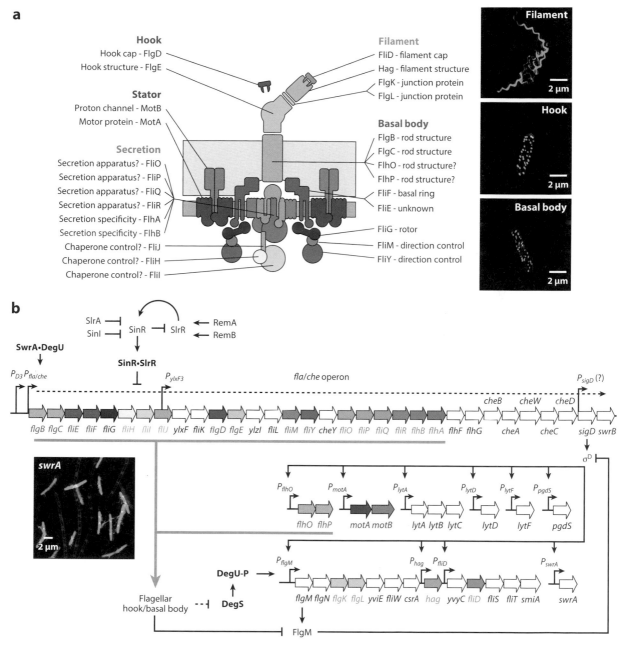

Figure 1

Flagellar structure and genetic hierarchy. (*a*) Graphic depicting the putative structure of the *Bacillus subtilis* flagellum based on empirical data and similarity to *Salmonella enterica*. Peptidoglycan is indicated in light gray. Membrane is indicated in dark gray. Flagellar components are colored and labeled. Micrographs are super-resolution fluorescence images of the indicated structures in wild-type cells: filament (maleimide-stained Hag^{T209C}), hook (maleimide-strained FlgE^{T123C}), and basal body (FliM-GFP fusion protein). (*b*) Graphic depicting genetic hierarchy of flagellar genes in *B. subtilis*. Open arrows are genes. Bent arrows are promoters. Closed arrows indicate activation; T bars indicate inhibition. Genes are color coded to match structures in panel *a*. Micrograph indicates bistable gene expression found in a *swrA* mutant of *B. subtilis*. Membranes are red (stained with FM 4–64 dye). P_{bag} expression is colored green cytoplasmically (P_{bag}-YFP). Filaments are colored green extracellularly (maleimide-stained Hag^{T209C}). Only a subpopulation of cells expresses the filament gene *hag* and assembles filaments.

extracellular chaperones. Secretion is enhanced by the intracellular chaperone FliS that binds and delivers Hag to the T3S apparatus (4, 107). Once a Hag monomer emerges from the secretion conduit, it encounters the extracellular chaperone FliD that catalyzes flagellin folding and ushers flagellin assembly at the tip of the nascent filament (39, 64). FliD serves as the filament cap, which when absent, results in secretion and accumulation of flagellin in the extracellular medium (63, 107). FliD polymerizes atop two structural transition proteins called FlgK and FlgL that bridge the flagellar hook and filament.

Hook

The hook is a curved hollow cylinder situated between the basal body and the filament that acts as a universal joint (11, 36, 81). The FlgE protein constitutes the primary structural subunit of the hook consistent with its high abundance in biochemical preparations of purified *B. subtilis* hook–basal body complexes (30, 81). Like flagellin, the hook subunits are secreted by the flagellar basal body through the rod and polymerized underneath a specific capping protein, FlgD (30, 116). The three-dimensional structure of FlgE suggests that the polymerized hook tolerates expansion on one side and compression on the other, enabling the structure to transmit and reorient motor torque on the helical filament (125).

The length of the polymerized hook is regulated by FliK. In the absence of FliK, hook secretion and polymerization proceed unchecked, resulting in extended structures of FlgE called polyhooks (30). FliK has been likened to a molecular ruler, as the length of the FliK primary sequence is correlated to the length of the flagellar hook (40, 98, 105). The lengths of *B. subtilis* FliK and flagellar hook are each roughly 25% longer than that of *Salmonella enterica* (30). Furthermore, FliK is secreted by the flagellum and interacts with both the flagellar hook protein and the T3S component FlhB. The mechanism may involve intermittent secretion of FliK during hook assembly to periodically monitor the status of hook length. Once the hook is complete, FliK activates autoproteolysis of FlhB, which in turn triggers a substrate-specificity switch that allows recognition of filament-class substrates for filament secretion and assembly (46, 69, 130). Thus, FliK controls the secretion duration of rod-hook class structural subunits and governs the transition to the assembly of the filament at individual basal bodies.

Basal Body

The basal body is the part of the flagellum that is anchored in the cell membrane and transits the peptidoglycan (**Figure 1a**). Each basal body serves as the structural anchor, polymerization platform, secretion conduit for the hook and filament structural proteins, and rotor for torque generation. Relatively little research has been done on the Gram-positive basal body save that it was purified from *B. subtilis* and was found both to be similar in structure and to contain proteins related to those found in flagella of *S. enterica* (81, 82). As such, most of what is discussed here summarizes research in Gram-negative model organisms, and differences in the *B. subtilis* basal body structure or function are specifically emphasized.

Flagellar basal body assembly likely begins with the T3S apparatus, which is composed of FliO, FliP, FliQ, FliR, FlhA, and FlhB (89, 97). Besides FliO, the other five components are conserved in the T3S found in the needle complex/injectisome of pathogens but are less understood than the cargo proteins that they secrete. Flagellar secretion requires the proton motive force, so presumably one or more of the proteins forms a proton channel (99, 119). FliO seems to be specific to flagellar secretion systems and may stabilize FliP (10). FlhA is the best understood protein in the complex, and it interacts with the FliHIJ chaperone-stripping complex to control the secretion of filament-class flagellar proteins (6, 76, 141). FlhA may also form the nucleation center around which the basal body protein FliF assembles (89).

The base of the basal body is a membrane-anchored ring polymerized from multiple subunits of the protein FliF (68, 146). FliF has two transmembrane segments and a large central extracellular domain that, along with a protein of unknown function (FliE), likely forms a fitting for the rod (109, 138, 146). Beneath FliF sits FliG, which polymerizes into a cytoplasmic gear-like rotor (48, 88, 140). Beneath FliG are two rings of protein made of FliM and FliY that interact with the chemotaxis system and control the direction of flagellar rotation (120, 131). Residues throughout the FliG primary sequence are required for flagellar biosynthesis for unknown reasons, but charged residues particularly in the C terminus of FliG interact with MotA to generate the torque for flagellar rotation (65, 90, 91). Torque is then imparted through FliF to the flagellar rod.

The rod serves as an axle to transmit basal body rotation to the hook and filament. The rod structure is assembled from four related proteins with the inferred assembly order from cell proximal to distal: FlgB, FlgC, FlgF, and FlgG (59, 82). FlgB is thought to be near FliF, as FlgB was found to interact with FliE (82, 100, 109). FlgG is thought to be adjacent to the hook, as FlgG was the only rod protein to be copurified with the hook when rods were sheared. Further, FlgG was separately implicated in outer membrane transit (29, 117). Rod assembly order was difficult to observe directly in *Escherichia coli*, however, as mutations in any of the rod genes abolished the assembly of all of the rod proteins, leading to the assumption that the entire structure was metastable depending on completion (82). Alternatively, the absence of a subunit may render partial rod structures susceptible to proteolytic degradation in the Gram-negative periplasm (58). Recent cryo-EM (electron microscopy) tomography studies of partial rod structures enabled the determination of rod assembly order in *Borrelia burgdorferi* suggesting that rod instability/proteolysis may be particular to *E. coli* (154).

B. subtilis encodes four putative rod structural proteins annotated as FlgB, FlgC, FlhO, and FlhP because of their similarity to each other and their homology to the rod-hook structural class. Furthermore, all four putative rod proteins copurified with the *B. subtilis* hook–basal body (81). FlgB and FlgC have not been directly investigated, but FlhO and FlhP are likely part of the rod, upstream of hook assembly, as mutation of either protein abolishes hook synthesis and results in secretion of hook subunits into the extracellular environment (30). Whereas the rod genes are often coexpressed with the other basal genes, in *B. subtilis* only *flgB* and *flgC* are encoded in the *fla/che* operon, whereas the *flhO* and *flhP* genes, by contrast, reside as a dicistron elsewhere in the genome that is expressed at a very low level and amplified by a σ^D-dependent promoter (30). The unusual genetic architecture of the *B. subtilis* rod genes may be pertinent to complex regulation of flagellar assembly particular to Gram-positive bacteria.

The *B. subtilis* basal body differs from that of *S. enterica* and *E. coli* in two critical ways. First, *B. subtilis* appears to lack bushing proteins. The bushings in Gram-negative bacteria are two different proteins that form separate rings in the peptidoglycan layer and outer membrane that allow the rod to transit and spin freely in the context of the cell envelope (135). The bushings may also function as a torque stabilizer by direct interaction with the proton channel stators (57). *B. subtilis* does not encode homologs of the bushing proteins, and electron microscopy of purified basal bodies does not seem to indicate ring-like densities that could potentially be attributed to bushing-like structures (38, 81). Perhaps *B. subtilis* encodes as-yet-undiscovered bushing proteins that do not resemble those of Gram-negative bacteria. Alternatively, the Gram-positive envelope structure alone may serve as a sufficient bearing and stabilizer for flagellar rotation.

The second structure that the *B. subtilis* basal body appears to lack, when compared with Gram-negative bacteria, is the rod cap. In *S. enterica*, the rod cap protein FlgJ has two functions conferred by separate domains. It is thought that FlgJ is loaded first as a cap on the nascent rod such that the N-terminal domain acts as a chaperone to usher rod protein polymerization (56). *B. subtilis* lacks a homolog of FlgJ. Perhaps *B. subtilis* encodes an as-yet-undiscovered rod cap, but it is noteworthy that *B. burgdorferi*, a relative of *B. subtilis* phylogenetically closer than *S. enterica*, encodes a homolog of FlgJ that is not essential for rod assembly (153). Thus, a rod cap may be unnecessary; however, why this would be is unclear, as polymerization of two other flagellar structures, the hook and the filament, require specific cap chaperones.

The C-terminal domain of FlgJ in Gram-negative bacteria is a peptidoglycan hydrolase oriented such that as the rod extends, FlgJ contacts and cleaves the peptidoglycan to permit rod penetration (111). Peptidoglycan remodeling is thought to be essential for flagellar assembly, as the diameter of the rod (8–14 nm) exceeds the pore size of peptidoglycan (4 nm) (35, 138, 154). The absence of FlgJ in *B. subtilis*, therefore, creates complications for cell envelope transit of the flagellum. Whereas FlgJ mounted on the end of the rod provides an ideal solution for spatially restricting peptidoglycan hydrolysis to the precise place and time of flagellar assembly in *S. enterica*, *B. subtilis* could encode a related hydrolase with a different mechanism of spatiotemporal control. *B. subtilis* is thought to encode at least 35 peptidoglycan hydrolases that are commonly called autolysins because when native regulation is disrupted their uncontrolled activity results in cell lysis (133).

Autolysins have been implicated in promoting *B. subtilis* motility, as mutants defective in synthesis and secretion of multiple autolysins are nonmotile and grow in long, unseparated chains of cells (44, 45). Although chaining mutants defective in multiple autolysins were later found to be disrupted for the transcriptional regulator SinR (see below), three individual autolysins have been specifically implicated in the control of motility: LytC (an *N*-acetylmuramoyl-L-alanine amidase), LytD (an endo-β-*N*-acetylglucosaminidase), and LytF (a γ-D-glutamate *meso*-diaminopimelate muropeptidase) (13, 84, 87, 93, 94, 124). Expression of the genes encoding LytC, LytD, and LytF in whole or in part requires the flagellar sigma factor σ^D, consistent with the reasonable supposition that these enzymes facilitate flagellar peptidoglycan penetration (44, 122). Recently, however, a *lytC lytD lytF* triple mutant was shown to be proficient in flagellar biosynthesis, indicating that if peptidoglycan remodeling is required, the activity is conducted by as-yet-undiscovered enzymes (25).

LytC, LytD, and LytF are nonetheless required for motility. One model suggested that the autolysins might promote motility simply by separating cells in chains that would otherwise be unable to coordinate their motility and inadvertently generate conflicting forces (13). LytF appears to be the only autolysin of the three that is necessary and sufficient for cell separation, however, and cells simultaneously lacking LytC and LytD, although poorly motile, do not appear to have a substantial cell separation defect (25). Rather, motility of the LytC LytD double mutant could be improved by disruption of the intracellular protease LonA (25). The target of LonA in motility regulation is unknown, but the involvement of LonA suggests that LytC and LytD may play a regulatory role in addition to, or instead of, a purely structural role in peptidoglycan remodeling. Ultimately, the role of autolysins in motility is unclear, but thus far no peptidoglycan hydrolase has been found to be essential for flagellar synthesis. How the flagellum transits the 30–40-nm-thick Gram-positive peptidoglycan remains unknown (133).

Cell Biology

Cytological tools have been generated for the study of flagellar assembly in *B. subtilis* (**Figure 1a**). Introduction of a unique cysteine residue into exposed surfaces of Hag (Hag^{T209C}) and FlgE (FlgE^{T123C}) enables fluorescent labeling of the flagellar filament and hook, respectively, with a cysteine-reactive dye (15, 30, 144). Basal bodies are fluorescently labeled by fusing green fluorescent protein (GFP) to FliM (54). Labeling each structural domain of the *B. subtilis* flagellum enables determination of whether and where any particular motility mutant is defective in the process of flagellar assembly. In addition, each stage has been observed sequentially, and it was determined that basal bodies form within 5 minutes, hooks form within 10 minutes, and filaments begin within 15 minutes of flagellar gene induction (54). Filament elongation is slow, however, such that both long filaments and motile cells are first observed only after 40 minutes of assembly. Thus, when *B. subtilis* is growing rapidly and dividing roughly every 20 minutes, the flagella initiated in a mother cell will be functional in the granddaughter cells.

B. subtilis synthesizes approximately 20 flagellar basal bodies along the length of the 4-μm cell (54). Although the peritrichous arrangement is often considered to be synonymous with random positioning, measurements indicate that the *B. subtilis* basal bodies have a nonrandom distribution. Basal bodies are separated from one another by a minimum distance greater than predicted by chance, conferring a grid-like organization to the overall pattern (54). Furthermore, although the basal bodies are more symmetrically distributed with respect to the midcell than predicted by chance, they also appear to be anomalously underrepresented at the cell poles (54). The overall pattern of flagellar position in *B. subtilis* is yet to be fully characterized, but basal body distribution is genetically controlled and may be disrupted by mutation of either FlhF or FlhG encoded within the *fla/che* operon (54).

FlhF and FlhG are known regulators of flagellar positioning in Gram-negative bacteria with a single polar flagellum (70). FlhF is a SIMIBI family signal recognition particle-like GTPase that localizes to the site of nascent flagellar synthesis and recruits FliF in *Vibrio cholerae* (51). In *B. subtilis*, FlhF GTPase activity is stimulated upon contact with FlhG, a SIMIBI family putative ATPase related to the MinD protein that governs cell division site selection in some bacteria (7, 8). Mutation of FlhF abolishes symmetrical placement of flagella in *B. subtilis* and causes flagella to accumulate at, rather than be excluded from, the cell poles (54). By contrast, mutation of FlhG abolishes flagellar spacing, causing flagella to aggregate and originate from a single locus (54). FlhG seems to antagonize FlhF activity, as mutation of FlhF is epistatic to mutation of FlhG in double mutants (54). The mechanism by which the interaction of FlhF and FlhG controls flagellar patterning needs further investigation.

Flagellate bacteria have a genetically programmed number and pattern of flagella that are sufficiently robust such that these characteristics were considered definitive traits for numerical taxonomy (54). Presumably, each particular flagellar pattern confers a selective advantage to each particular species. The peritrichous flagellar arrangement is correlated with the ability of bacteria to swarm over surfaces, but severe perturbations in flagellar patterning do not reduce swarming in *B. subtilis* (54). Rather, flagellar patterning may maintain productive motility of individual cells by ensuring the inheritance of a sufficient number of flagella after binary fission (54). Flagellar structure is an important reservoir of epigenetic information, and unequal flagellar inheritance has been recently demonstrated to have pleiotropic regulatory effects (83). Thus, the function of specific flagellar patterns is poorly understood but may have consequences beyond simply the regulation of motility.

FLAGELLAR REGULATION

The regulation of motility is complex in *B. subtilis*. As with many flagellar motility systems, elaborate feedback mechanisms couple the flagellar structural assembly state to gene expression. *B. subtilis*, however, has added complexity wherein late flagellar genes are expressed only in a subpopulation (74). During exponential growth, the population bifurcates such that cells are either joined end-to-end in sessile cell chains or grow as single motile individuals (**Figure 1b**). The mixed population of physiologically distinct sessile chains and motile cells is perhaps advantageous for instantaneous environmental sampling: Chains can colonize the current location, form biofilms, and enjoy enhanced resistance to protozoan grazing, and motile cells can disperse to new and potentially favorable niches. Because motile and sessile cells freely mix in liquid culture, population heterogeneity presents technical complications, as the physiology of each subpopulation is difficult to study in isolation.

TRANSCRIPTIONAL REGULATION

Population heterogeneity is under the control of the alternate sigma factor σ^D in *B. subtilis* (55, 74, 95). Cells mutated for σ^D grow exclusively as long nonmotile chains and thereby lack heterogeneity (55). Furthermore, single cell analyses using transcriptional fusions of fluorescent proteins to the σ^D-dependent promoters P_{hag} and P_{lytF} indicate that these genes were expressed only in the motile subpopulation (25, 94, 101). Cells that are "ON" for σ^D activity express Hag to complete flagellum assembly and express LytF to separate individuals from chains. Cells that are "OFF" for σ^D activity do not express either Hag or LytF and grow as aflagellate chains (25). Thus, the activity of σ^D governs the fate of individuals in each subpopulation.

The relative frequencies of motile and nonmotile cells in a *B. subtilis* population differ depending on the strain because of their intrinsic bias in σ^D-dependent gene expression (50, 74). Commonly used laboratory strains are biased toward OFF cells and grow predominantly as long chains, whereas the ancestral strain is biased toward ON cells and grows predominantly as motile individuals (74). The frequency of the σ^D ON state is increased by the biasing protein SwrA, and laboratory strains favor the OFF state because of the inheritance of a high-frequency frameshift mutation that abolishes SwrA function (20, 72, 74, 152). Indeed, the *swrA* gene was first discovered as a spontaneous mutation in laboratory strains with an improved motility allele called *ifm* that was later determined to be a revertant that restored the *swrA* open reading frame (20, 50). The allele *ifm* that repaired SwrA function was so named for increased frequency of motility, further supporting its role in controlling population bias (50).

SwrA controls population bias by transcriptionally activating the *fla/che* operon. Cells mutated for *swrA* reduce expression from the $P_{fla/che}$ promoter and basal body number roughly twofold, whereas cells that artificially overexpress *swrA* increase expression from the $P_{fla/che}$ promoter and basal body number roughly twofold (54, 74, 127). Furthermore, spontaneous mutations that bypass the absence of SwrA increase expression of the *fla/che* operon either by mutation of the promoter closer to consensus or by deletion of an upstream rho-independent terminator that allows transcriptional read-through (3, 74). Finally, transcriptomic analysis shows that SwrA elevates the expression of the *fla/che* operon (including the *sigD* gene), which in turn increases expression of the σ^D regulon (74). SwrA may activate the *fla/che* operon indirectly by binding to and modulating the response regulator DegU (3, 104, 114, 143).

SwrA biases σ^D-dependent gene expression by raising σ^D levels above a threshold such that motile cells contain high levels and chaining cells contain low levels (31). σ^D protein levels are low in chains because of a failure to express the *sigD* gene, the penultimate gene in the long 27-kb *fla/che* transcript (which when fully extended measures 9 μm in length, more than twice the length of the cell) (1, 96, 147). Transcript abundance decreases along the *fla/che* operon such that the position of the *sigD* gene is critical for bias, and the number of ON cells in the population may be increased simply by moving the *sigD* gene forward in the operon, closer to the promoter (**Figure 1b**) (31). Further, the two subpopulations are differentiated by a fourfold increase in expression magnitude throughout the operon in motile cells relative to chaining cells (31). Thus, there appears to be a threshold of σ^D protein that is exceeded only in the motile subpopulation, and heterogeneity could arise by a combination of noise in the promoter activation and the positioning of a key regulator at the end of an operon.

The mechanism that causes the gradual decrease in *fla/che* operon transcript abundance is controlled, at least in part, by SlrA (**Figure 1b**). SlrA is a small peptide, which when overexpressed, results in an enhanced distance-dependent decrease in *fla/che* operon transcript abundance, undetectable levels of σ^D protein, an OFF state for the entire σ^D regulon, and aflagellate cell chains (32, 78). Furthermore, mutation of *slrA* increases the frequency of σ^D ON cells in a *swrA* mutant background (32). Thus, when SwrA is absent there is a reduced rate of transcript initiation from the $P_{fla/che}$ promoter such that the gradual decrease in *fla/che* transcript abundance causes σ^D protein to fall below a threshold level. When SlrA is also absent, the gradual decrease in the *fla/che* transcript does not occur and σ^D protein level remains high, thereby compensating for the absence of SwrA. SlrA inhibits *fla/che* operon transcript through a paralogous pair of DNA binding proteins: SinR and SlrR (23, 32, 112).

SinR is the master transcriptional repressor of biofilm gene expression in *B. subtilis* and also represses its paralog SlrR (71, 78). When repression by SinR is antagonized, SlrR is expressed and a SinR · SlrR heteromer is formed (23). The SinR · SlrR heteromer appears to adopt a new, perhaps heteromeric, binding sequence and thereby retargets the regulators. For example, the heteromeric complex, but neither protein alone, was shown to bind to and repress the promoters of the σ^D-dependent genes expressing Hag and LytF as well as LytC (which is under partial σ^D control) (23, 87). The benefit of directly repressing individual genes under σ^D control, however, is unclear given that SlrA-SinR-SlrR also inhibits the accumulation of σ^D protein and, as a consequence, deactivates the entire σ^D regulon (29, 74, 128). Further, low-level artificial expression of σ^D was able to override Hag and autolysin repression, suggesting that SlrA-SinR-SlrR primarily acts upstream of σ^D and that repression at individual downstream promoters, if relevant, is either weak or easily reversed (32). The complete set of targets and the binding site for the SinR · SlrR heteromeric complex remain to be determined.

SlrA manipulation was used to create a homogenously OFF population to demonstrate two properties of bistability: hysteresis and hypersensitivity. Hysteresis is a property of bistable systems in which each state, once acquired, resists switching to the other in the absence of a history-dependent stimulus. Hysteresis was confirmed when artificial ectopic induction of σ^D overrode the inhibitory effects of SlrA for 20 cell generations after induction was removed (32). Once a system exceeds a threshold level of stimulus that differs depending on the state history, hypersensitivity ensures a rapid transition and prevents the system from dwelling in intermediate states. Hypersensitivity was demonstrated when a linear increase in the induction of the *fla/che* operon gave a nonlinear response in σ^D-dependent gene expression (32). Hysteresis and hypersensitivity are often mechanistically related.

To account for hysteresis, biological bistable regulatory systems experience feed-forward regulation. Feed-forward regulation was indicated when transient expression from an artificially induced gene encoding σ^D maintained an ON state but only when the native copy of the σ^D gene was also present in the chromosome (32). Multiple possibilities for feed-forward regulation on the native *sigD* gene have been reported (**Figure 1b**). First, a σ^D-dependent promoter enhances expression of the activator SwrA, which in turn activates the *fla/che* operon (21, 104). Second, a σ^D-dependent promoter (P_{D3}) is located upstream of the $P_{fla/che}$ promoter, which when deleted simultaneously with SwrA, causes a motility defect (41, 104, 148). Third, σ^D-dependent promoter activity, P_{ylxF3}, was found upstream of genes encoding flagellar hook components within the *fla/che* operon, but the precise location of the promoter and its relevance to bistability have not been determined (32). Finally, a σ^D-dependent promoter was reported to be immediately upstream of the *sigD* gene itself, but subsequent work has been unable to detect the putative promoter's transcriptional activity (2, 32, 148). Either all or a combination of these σ^D-dependent promoters may contribute to increasing *sigD* gene expression and σ^D protein levels above a threshold.

Hypersensitivity in a system is often mediated by cooperative protein interactions. Cooperativity is difficult to explain for sigma factors, as most function as monomers bound to RNA polymerase. Unlike other sigma factors, however, σ^D has been shown to bind to promoters in the absence of the RNA polymerase core enzyme (12, 27, 129). Binding of σ^D to promoter DNA shows super-shifted banding patterns, suggesting multiple proteins may bind in a cooperative complex (27). In addition, σ^D activity is enhanced in response to cooperative assembly of the flagellar hook–basal body structure, and certain proteins required for hook assembly are transcribed primarily from a σ^D-dependent promoter (9, 30). Thus, when σ^D becomes active, more hook–basal body complexes are completed, and σ^D activity becomes hysteretically self-amplifying.

The activity of σ^D is enhanced by hook–basal body completion through antagonism of the antisigma factor FlgM (22). FlgM acts as an antisigma factor by binding to σ^D and inhibiting association with RNA polymerase (12, 134). One way FlgM is antagonized is by the completion of the hook–basal body complex (**Figure 2a**). In *S. enterica*, the hook–basal body antagonizes FlgM when completion of the flagellar hook changes the specificity of the flagellar secretion apparatus to secrete FlgM from the cytoplasm and liberate σ^D to direct transcription (61, 85). In contrast, FlgM secretion has not been reported in *B. subtilis*. Furthermore, the N terminus of FlgM from *S. enterica* is disordered to promote secretion, whereas the N terminus of FlgM from *B. subtilis*

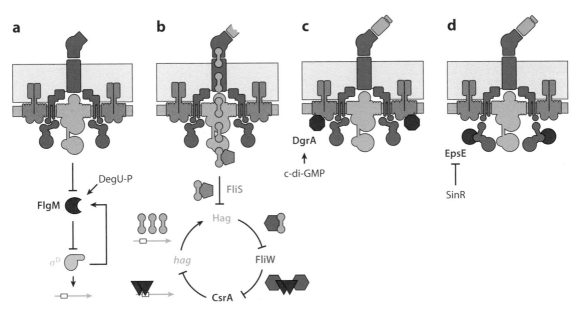

Figure 2

Flagellar regulation. (*a*) FlgM (*red*) antagonizes σ^D (*cyan*), but σ^D activates FlgM expression prioritized by DegU-P. The autoinhibitory loop on σ^D is broken by completed hook–basal body structure by an unknown mechanism, and the *hag* transcript (*green arrow*) is synthesized. (*b*) CsrA (*red triangle*) binds to the Shine-Dalgarno sequence (*open box*) of the *hag* transcript and inhibits translation. When Hag (*green barbell*), aided by the secretion chaperone FliS (*orange pentagon*), is secreted, FliW (*blue hexagon*) switches partners and antagonizes CsrA, which allows high-level Hag translation for the duration of filament assembly. (*c*) DgrA (*red octagon*) is activated by binding c-di-GMP and interacts with MotA to inhibit flagellar rotation. (*d*) EpsE (*red circle*) is relieved from SinR repression during biofilm formation and acts as a clutch by binding to FliG and disengaging the rotor (*purple*) from MotA (*brown*).

is structured, perhaps suggesting an alternative mechanism of control that remains unknown (12, 34).

A second way in which FlgM is regulated in *B. subtilis* is via its expression by σ^D, the same sigma factor that it inhibits to seemingly create a regulatory conflict (60, 102). To resolve the conflict, FlgM expression is activated by the two-component system response regulator DegU (**Figure 2a**) (60). When phosphorylated by its cognate histidine kinase, DegS, DegU-P lowers the σ^D threshold at the P_{flgM} promoter such that *flgM* expression is prioritized over other members of the σ^D regulon (60, 106). Although the signal input for the soluble DegS kinase is unknown, DegS may be activated either by incomplete flagella or impedance of flagellar rotation, establishing a quality control system for assembly analogous to that proposed for *Campylobacter jejuni* (17, 19, 24, 62). Phosphorylation of DegU further inhibits motility at the $P_{fla/che}$ promoter, and perhaps other DegU phenotypes, such as biofilm formation, hydrophobin synthesis, protease secretion, and polyglutamate production, are activated in response to the flagellar defect cue (3, 19, 110, 143).

Regulation of flagellar gene expression involves at least two other poorly understood proteins, SwrB and YmdB. SwrB is a single-pass transmembrane protein expressed immediately downstream of σ^D in the *fla/che* operon (**Figure 1b**) (147). Mutation of SwrB specifically abolishes swarming motility and appears to reduce late-class flagellar gene expression (72, 147). Although the mechanism of SwrB function is unknown, it synergizes with SwrA as a *swrA swrB* double mutant results in a uniform population of nonmotile chains (74). YmdB resembles a phosphodiesterase that inhibits motility gene expression such that when it is mutated, *hag* (flagellin) expression increases dramatically (37). How and where YmdB inhibits *hag* is unknown, but YmdB mutants are also severely defective in biofilm formation, perhaps suggesting that YmdB could mediate its effects through SlrA-SinR-SlrR or other biofilm regulators.

POST-TRANSCRIPTIONAL REGULATION

Flagellar assembly is an energy-expensive process primarily because of the estimated 20,000 subunits of flagellin that are required to assemble a single filament (92). To support the high structural demand, flagellin genes are expressed from strong promoters and flagellin proteins are translated from near-consensus ribosome binding sites. The energetic cost of flagellar synthesis is particularly high in *B. subtilis*, as it synthesizes roughly 20 flagella per cell (54). To offset the high metabolic cost, the flagellin primary sequence has evolved to contain a higher proportion of energy-efficient amino acids, and flagellin expression has been shown to be highly regulated in every organism in which it has been studied in detail (132). Thus, flagellin must be expressed at a high level during filament assembly and yet flagellin levels must be held in check to prevent not only wasteful consumption of nutrients but also consumption of molecular space in the cytoplasm.

In *B. subtilis*, cytoplasmic levels of the Hag flagellin protein are homeostatically auto-inhibited by a negative feedback loop that includes FliW and CsrA (108) (**Figure 2b**). Hag binds to the FliW protein, causing FliW to release its alternative interacting partner: CsrA (108, 142). CsrA in turn binds to two stem loops in the 5′ UTR of the *hag* transcript, occludes the Shine-Dalgarno sequence, and inhibits Hag translation (150). As Hag translation decreases, the reduced levels of Hag result in a stoichiometric excess of FliW. Free FliW in turn binds to CsrA and inhibits CsrA's ability to bind to *hag* mRNA, causing relief of Hag translational inhibition. The cycle is believed to rapidly repeat, causing levels of the Hag protein to oscillate over a low and narrow threshold concentration in the cytoplasm (108).

Hag homeostasis is disrupted during periods of filament assembly. The flagellar filament chaperone FliS binds to Hag and ushers it for export through the flagellar secretion apparatus (6, 107).

Hag secretion results in a transient decrease in the cytoplasmic pool of Hag, allowing FliW to antagonize CsrA activity and permit high-level translation of *hag* transcript for the duration of filament assembly. Conversely, it has been speculated that when filament assembly is completed, a transient increase in the Hag cytoplasmic pool sequesters FliW to release CsrA activity and repress Hag translation such that homeostasis autocorrects. As it is important to the model that secretion regulates homeostasis, it was shown that both FliW and FliS bind Hag monomer units simultaneously, as competition between the two proteins triggers regulatory consequences that limit filament assembly (107).

CsrA is a highly conserved RNA binding protein, and regulation of flagellar assembly may be its ancestral function. The *csrA* gene shows a remarkable phylogenetic distribution being restricted primarily to those bacterial genomes that also encode flagellin (108). FliW may be the ancestral form of CsrA antagonism, as the genes encoding the two proteins are often translationally coupled and found tightly linked to filament structural genes in the deepest branches of the bacterial tree (108). The γ-proteobacteria in which CsrA is most commonly studied regulate CsrA activity, not by FliW, but by small RNAs that bind and sequester multiple CsrA dimers simultaneously to pleiotropically regulate biofilm formation, carbon storage, virulence gene expression, and motility (5). Perhaps some organisms, like the γ-proteobacteria, acquired CsrA in the absence of its native antagonist FliW and evolved the most expedient regulators in the form of competitive inhibitors that are merely permutations of CsrA binding sites (108).

The role of CsrA in *B. subtilis* is thought to coordinate flagellin expression with the assembly state of the flagellar basal body, a task assigned to FlgM in *S. enterica* (61, 85). In *B. subtilis*, both FlgM and CsrA synergize to control flagellin expression (108). We note however that translational regulation is more immediate than transcriptional regulation for the assembly of structures, and whereas CsrA is flagellin specific, FlgM controls the entire σ^D regulon of genes. Thus, the role of FlgM may be more directly related to coordinating other aspects of cell physiology with flagellar assembly than to coordinating filament assembly per se. Further, it can be speculated that post-transcriptional regulation is perhaps more prevalent than commonly appreciated in limiting structural assemblies such as flagellar rod and hook, secretion apparatuses, pili, ribosomes, and cell division machinery.

FUNCTIONAL REGULATION

Functional regulators control flagellar rotation. The most fundamental members of the functional regulatory class are the stators that transduce the power of chemiosmotic ion motive force into the mechanical power of flagellar rotation. Eight to eleven membrane-bound stator complexes composed of four subunits of the protein MotA and two subunits of the protein MotB surround each flagellum (14). MotB binds to peptidoglycan and accepts a proton to cause a conformational change in MotA and impart torque on the rotor FliG (79, 80). *B. subtilis* encodes two pairs of paralogous proteins MotP/MotS and MotA/MotB. MotP and MotS consume the sodium motive force and conditionally support flagellar mediated motility (24, 67, 139). MotA and MotB, by contrast, consume the proton motive force, are required to power both swimming and swarming motility, and functionally require the same residues as their *E. coli* counterparts (19, 24, 103). Thus, the mechanism of flagellar force generation seems to be conserved from the proteobacteria to the firmicutes.

New functional regulators of the flagellum are being discovered in the context of biofilm formation (53). Biofilms are multicellular aggregates of bacteria held together by an extracellular matrix often composed of EPS (47). Flagellate cells lose motility during the transition to the sessile biofilm state. The motility-to-biofilm transition seems rapid and thus turning off transcription, or even translation, may be too slow to inhibit motility in the short term. To complicate matters further, flagella are often numerous and highly stable structures such that the inhibition of de novo synthesis should not be expected to stop rotation of the pre-existent flagella. By contrast, functional regulators are ideal to rapidly inhibit motility because as little as a single protein is required to interact with the basal body and inhibit rotation. In *B. subtilis*, two functional flagellar inhibitors have been discovered: DgrA and EpsE.

DgrA (aka YpfA) is a di-quanylate receptor protein that binds to the small signaling molecule c-di-GMP through a conserved PilZ domain (26, 49). When c-di-GMP is bound, DgrA inhibits flagellar rotation seemingly by direct interaction with MotA (**Figure 2c**) (26). Similar activity has been observed by YcgR, the DgrA homolog in *E. coli*, which slows flagellar rotation using a brake-like mechanism (16, 43, 121). DgrA may be involved in the motility-to-biofilm transition, as c-di-GMP accumulation often mediates the physiological changes that accompany biofilm formation. In *B. subtilis*, however, mutations that abolish or artificially elevate the cytoplasmic pool of c-di-GMP have no obvious effect on biofilms (26, 49). Thus, the physiological roles of both c-di-GMP and motility inhibition by DgrA in *B. subtilis* are currently unknown.

A functional flagellar regulator more explicitly related to biofilm formation is EpsE (**Figure 2d**). EpsE is a bifunctional glycosyltransferase that requires the conserved active site residues for synthesis of the *B. subtilis* biofilm EPS matrix and separately requires a set of residues predicted to occupy an exposed α helix to inhibit motility (52). EpsE inhibits motility by binding to the rotor protein FliG, and biophysical studies indicate that EpsE functions as a clutch by disengaging the rotor and stator components (15). EpsE is encoded within the 15 gene *eps* operon for biofilm EPS synthesis, the transcription of which is repressed and activated by the master biofilm regulators, SinR and RemA, respectively (71, 149). In addition, an RNA antiterminator is required for expression of *eps* genes encoded downstream of *epsE*, perhaps to ensure that motility is inhibited before proceeding to matrix synthesis (66). Thus, the motility-to-biofilm transition is conceptually unified in the regulation of *epsE*.

The inhibition of flagellar rotation also results in increased secretion of poly-γ-glutamate (PGA), a polymer that forms a capsule in *Bacillus anthracis* but manifests in *B. subtilis* as a slime layer (18). Mutation of MotA/MotB or inhibition of flagellar rotation by expression of *epsE* results in highly mucoid colonies because of PGA overproduction (19, 24). PGA overproduction is due to activation of the operon responsible for PGA synthesis, which is dependent on the two-component system DegS/DegU (19, 24). How the DegS kinase might sense flagellar rotation is unclear, and why PGA production appears to be linked to flagellar rotation is unknown. PGA, however, has been linked to surface adhesion, and DegU has been shown to interfere with SinR and SlrR (115, 136). Thus, the inverse regulation between flagellar function and PGA synthesis may be another facet of the motility-to-biofilm transition.

Chemotaxis is another form of functional regulation that controls the direction of flagellar rotation. Cells move up chemical gradients by biasing the time spent either running in a relatively straight direction or erratically tumbling to acquire a new trajectory. Running and tumbling are mediated by rotating the flagella counterclockwise and clockwise, respectively, and a signal transduction system controls the frequency at which flagella change direction. In *E. coli*, the CheA/CheY two-component system generates clockwise rotation for a tumble, but in *B. subtilis*, the system is reversed and CheA/CheY generates counterclockwise rotation for a run. Further, *E. coli* gradually methylates its chemoreceptors in response to stimuli, but *B. subtilis* rapidly demethylates its receptors before a slower remethylation step. Finally, to aid adaptation, *B. subtilis* has two proteins, a soluble CheC and the flagellar bound FliY, that deactivate CheY to restore tumbling and reset the system. The differences between the chemotactic responses of *B. subtilis* and *E. coli* have been recently reviewed in detail (123).

SUMMARY POINTS

1. Electron microscopy of purified Gram-positive and Gram-negative bacterial flagella indicates that the two structures are different likely because of the demands of cell envelope architecture. *B. subtilis* nonetheless encodes many flagellar structural proteins that have homologs in the Gram-negative bacteria but presumably with different regulation and with different mechanisms for accommodating the thick peptidoglycan and the lack of an outer membrane.

2. Cell biological tools have been developed for observing *B. subtilis* flagellar basal bodies, hooks, and filaments. Wild-type cells of *B. subtilis* produce roughly 20 flagella per cell that are distributed nonrandomly over the cell surface. Flagellar number is increased by SwrA. FlhF and FlhG control flagellar positioning.

3. Late-class flagellar filament genes and genes for cell-separating autolysins under the control of the sigma factor σ^D are transcribed in a subpopulation of cells. Heterogeneity maintains a subpopulation of motile cells that permit rapid response to environmental change, as flagella take up to three generations to synthesize de novo. Bistability of σ^D-dependent gene expression is controlled by the *fla/che* operon architecture, SwrA, SwrB, DegS/DegU, YmdB, and the SlrA/SinR/SlrR system.

4. Flagellin is energetically costly and homeostatically restricted in the cytoplasm by a partner-switching mechanism involving the Hag flagellin, FliW, and the RNA binding protein CsrA. CsrA specifically inhibits Hag translation, and translation inhibition is relieved during periods of Hag secretion and active filament synthesis. The regulation of flagellar assembly may be the ancestral role for CsrA, a highly pleiotropic regulator of virulence in γ-proteobacteria.

5. *B. subtilis* has been an important organism for the study of functional regulators that control flagellar rotation. During biofilm formation, the bifunctional glycosyltransferase/clutch EpsE disengages the rotor from the stators to depower rotation. In response to c-di-GMP, DgrA (YpfA, YcgR) may act as a brake by binding to the flagellar stator. The inhibition of flagellar function may have cascading effects, including an increase in the extracellular polymer poly-γ-glutamate.

FUTURE ISSUES

1. Rod assembly order, mechanism of polymerization, and length control are poorly understood. *B. subtilis* appears to lack the rod cap, peptidoglycan bushings, and flagellar-specific peptidoglycan hydrolase. How the *B. subtilis* rod penetrates and is accommodated by peptidoglycan is unknown. The mechanistic contribution of the six proteins at the core of the T3S apparatus is unknown. How T3S activity is coordinated during flagellar assembly is unknown. Gram-positive bacteria may be ideal for studying the earliest events in T3S, as they lack a periplasm and outer membrane.

2. *B. subtilis* undergoes epigenetic development in mid-log phase as it bifurcates into motile and nonmotile subpopulations, but the feedback mechanisms that govern bistable hysteresis and hypersensitivity are poorly understood. The mechanisms and advantage of regulators that bias the proportion of the two subpopulations are poorly understood. The pool of flagellin subunits is homeostatically restricted in the cytoplasm, and homeostatic control of other structural subunits, such as the rod and hook, is possible but unexplored. Post-transcriptional regulation may be common in the assembly of multicomponent *trans*-envelope machines but is understudied.

3. Flagellar number and patterning are species specific, but the mechanisms of number and spatial control are poorly understood. Flagellar positioning appears to be determined cytoplasmically but must be coordinated with peptidoglycan penetration and anchoring. The physiological or ecological relevance of particular numbers and patterns of flagella is poorly understood. Whereas flagellar structural genes are well conserved in the bacterial domain, the distribution of regulators of flagellar expression and assembly appears to be narrow. For *B. subtilis*, the functions of SwrA, SwrB, DegS/DegU, YmdB, and the Sin/Slr system require further study. Genetic architecture and operon length may be general factors that contribute to the regulation of motility.

4. The motility-to-biofilm transition provides a venue for the study of motility inhibition and has led to the discovery of an important class of flagellar functional regulators. Understanding when, where, and how motility is inhibited will be important for developing anti-biofilm therapeutics. Importantly, flagellar function at the level of the stator proteins seems to be coupled to other aspects of physiology, but how stator information is sensed and transduced is unknown.

DISCLOSURE STATEMENT

The authors are not aware of any affiliations, memberships, funding, or financial holdings that might be perceived as affecting the objectivity of this review.

ACKNOWLEDGMENTS

We thank Dr. Liz Sockett for comments and discussions during the preparation of this review. Work in Dan Kearns's lab is supported by NIH Grant GM093030 to D.B.K.

LITERATURE CITED

1. Albertini AM, Caramori T, Crabb WD, Scoffone F, Galizzi A. 1991. The *flaA* locus of *Bacillus subtilis* is part of a large operon coding for flagellar structures, motility functions, and an ATPase-like polypeptide. *J. Bacteriol.* 173(11):3573–79

2. Allmansberger R. 1997. Temporal regulation of *sigD* from *Bacillus subtilis* depends on a minor promoter in front of the gene. *J. Bacteriol.* 179(20):6531–35

3. Amati G, Bisicchia P, Galizzi A. 2004. DegU-P represses expression of the motility *fla-che* operon in *Bacillus subtilis*. *J. Bacteriol.* 186(18):6003–14

4. Auvray F, Thomas J, Fraser GM, Hughes C. 2001. Flagellin polymerization control by a cytosolic export chaperone. *J. Mol. Biol.* 308:221–29

5. Babitzke P, Romeo T. 2007. CsrB sRNA family: sequestration of RNA-binding regulatory proteins. *Curr. Opin. Microbiol.* 10:156–63

6. Bange G, Kummerer N, Engel C, Bozkurt G, Wild K, et al. 2010. FlhA provides the adaptor for coordinated delivery of late flagella building blocks to the type III secretion system. *Proc. Natl. Acad. Sci. USA* 107:11295–300

7. Bange G, Kummerer N, Grudnik P, Lindner R, Petzold G, et al. 2011. Structural basis for the molecular evolution of SRP-GTPase activation by protein. *Nat. Struct. Mol. Biol.* 18(12):1376–80

8. Bange G, Petzold G, Wild K, Parlitz RO, Sinning I. 2007. The crystal structure of the third signal-recognition GTPase FlhF reveals a homodimer with bound GTP. *Proc. Natl. Acad. Sci. USA* 104(34):13621–25

9. Barilla D, Caramori T, Galizzi A. 1994. Coupling of flagellin gene transcription to flagellar assembly in *Bacillus subtilis*. *J. Bacteriol.* 176(15):4558–64

10. Barker CS, Meshcheryakova IV, Kostyukova AS, Samatey FA. 2010. FliO regulation of FliP in the formation of the *Salmonella enterica* flagellum. *PLOS Genet.* 6(9):e1001143

11. Berg HC, Anderson RA. 1973. Bacteria swim by rotating their flagellar filaments. *Nature* 245:380–82

12. Bertero MG, Gonzales B, Tarricone C, Ceciliani F, Galizzi A. 1999. Overproduction and characterization of the *Bacillus subtilis* anti-sigma factor FlgM. *J. Biol. Chem.* 274:12103–7

13. Blackman SA, Smith TJ, Foster SJ. 1998. The role of autolysins during vegetative growth of *Bacillus subtilis* 168. *Microbiology* 144:73–82

14. Blair DF. 2003. Flagellar movement driven by proton translocation. *FEBS Lett.* 545:86–95

15. Blair KM, Turner L, Winkelman JT, Berg HC, Kearns DB. 2008. A molecular clutch disables flagella in the *Bacillus subtilis* biofilm. *Science* 320:1636–38

16. Boehm A, Kaiser M, Li H, Spangler C, Kasper CA, et al. 2010. Second messenger-mediated adjustment of bacterial swimming velocity. *Cell* 141:107–16

17. Boll JM, Hendrixson DR. 2013. A regulatory checkpoint during flagellar biogenesis in *Campylobacter jejuni* initiates signal transduction to activate transcription of flagellar genes. *Mbio* 4(5):e00432–13

18. Buescher JM, Margaritis A. 2007. Microbial biosynthesis of polyglutamic acid biopolymer and applications in the biopharmaceutical, biomedical and food industries. *Crit. Rev. Biotechnol.* 27:1–19

19. Cairns LS, Marlow VL, Bissett E, Ostrowski A, Stanley-Wall NR. 2013. A mechanical signal transmitted by the flagellum controls signaling in *Bacillus subtilis*. *Mol. Microbiol.* 90(1):6–21

20. Calvio C, Celandroni F, Ghelardi E, Amati G, Salvetti S, et al. 2005. Swarming differentiation and swimming motility in *Bacillus subtilis* are controlled by *swrA*, a newly identified dicistronic operon. *J. Bacteriol.* 187(15):5356–66

21. Calvio C, Osera C, Amati G, Galizzi A. 2008. Autoregulation of *swrAA* and motility in *Bacillus subtilis*. *J. Bacteriol.* 190(16):5720–28

22. Caramori T, Barilla D, Nessi C, Sacchi L, Galizzi A. 1996. Role of FlgM in σ^D-dependent gene expression in *Bacillus subtilis*. *J. Bacteriol.* 178(11):3113–18

23. Chai Y, Norman T, Kolter R, Losick R. 2010. An epigenetic switch governing daughter cell separation in *Bacillus subtilis*. *Genes Dev.* 24:754–65

24. Chan J, Guttenplan SB, Kearns DB. 2014. Defects in the flagellar motor increase synthesis of poly-γ-glutamate in *Bacillus subtilis*. *J. Bacteriol.* 196(4):740–53

25. Chen R, Guttenplan SB, Blair KM, Kearns DB. 2008. Role of the σ^D-dependent autolysins in *Bacillus subtilis* population heterogeneity. *J. Bacteriol.* 191:5775–84

26. Chen Y, Chai Y, Guo J, Losick R. 2012. Evidence for cyclic-di-GMP-mediated signaling in *Bacillus subtilis*. *J. Bacteriol.* 194(18):5080–90

27. Chen Y, Helmann JD. 1995. The *Bacillus subtilis* flagellar regulatory protein σ^D: overproduction, domain analysis and DNA-binding properties. *J. Mol. Biol.* 249:743–53

28. Chevance FFV, Hughes KT. 2008. Coordinating assembly of a bacterial macromolecular machine. *Nat. Rev. Microbiol.* 6(6):455–65

29. Chevance FFV, Takahashi N, Karlinsey JE, Gnerer J, Hirano T, et al. 2007. The mechanism of outer membrane penetration by the eubacterial flagellum and implications for spirochete evolution. *Genes Dev.* 21(18):2326–35

30. Courtney CR, Cozy LM, Kearns DB. 2012. Molecular characterization of the flagellar hook in *Bacillus subtilis*. *J. Bacteriol.* 194:4619–29

31. Cozy LM, Kearns DB. 2010. Gene position in a long operon governs motility development in *Bacillus subtilis*. *Mol. Microbiol.* 76(2):273–85

32. Cozy LM, Phillips AM, Calvo RA, Bate AR, Hsueh Y, et al. 2012. SlrA/SinR/SlrR inhibits motility gene expression upstream of a hypersensitive and hysteretic switch at the level of σ^D in *Bacillus subtilis*. *Mol. Microbiol.* 83(6):1210–28

33. Darnton NC, Berg HC. 2007. Force-extension measurements on bacterial flagella: triggering polymorphic transformations. *Biophys. J.* 92:2230–36

34. Daughdrill GW, Chadsey MS, Karlinsey JE, Hughes KT, Dahlquist FW. 1997. The C-terminal half of the anti-sigma factor, FlgM, becomes structured when bound to its target, σ^{28}. *Nat. Struct. Biol.* 4:285–91

35. Demchick P, Koch AL. 1996. The permeability of the wall fabric of *Escherichia coli* and *Bacillus subtilis*. *J. Bacteriol.* 178(3):768–73

36. Depamphilis ML, Adler J. 1971. Purification of intact flagella from *Escherichia coli* and *Bacillus subtilis*. *J. Bacteriol.* 105(1):376–83

37. Diethmaier C, Pietack N, Gunka K, Wrede C, Lehnik-Habrink M, et al. 2011. A novel factor controlling bistability in *Bacillus subtilis*: the YmdB protein affects flagellin expression and biofilm formation. *J. Bacteriol.* 193(21):5997–6007

38. Dimmit K, Simon M. 1971. Purification and thermal stability of intact *Bacillus subtilis* flagella. *J. Bacteriol.* 105(1):369–75

39. Emerson SU, Tokuyasu K, Simon MI. 1970. Bacterial flagella: polarity of elongation. *Science* 169:190–92

40. Erhardt M, Singer HM, Wee DH, Keener JP, Hughes KT. 2011. An infrequent molecular ruler controls flagellar hook length in *Salmonella enterica*. *EMBO J.* 30(14):2948–61

41. Estacio W, Anna-Arriola S, Adedipe M, Marquez-Magaña LM. 1998. Dual promoters are responsible for transcription initiation of the *fla/che* operon in *Bacillus subtilis*. *J. Bacteriol.* 180(14):3548–55

42. Evans LDB, Poulter S, Terentjev EM, Hughes C, Fraser GM. 2013. A chain mechanism for flagellum growth. *Nature* 504:287–90

43. Fang X, Gomelsky M. 2010. A post-translational c-di-GMP-dependent mechanism regulating flagellar motility. *Mol. Microbiol.* 76(5):1295–305

44. Fein JE. 1979. Possible involvement of bacterial autolytic enzymes in flagellar morphogenesis. *J. Bacteriol.* 137(2):933–46

45. Fein JE, Rogers HJ. 1976. Autolytic enzyme-deficient mutants of *Bacillus subtilis* 168. *J. Bacteriol.* 127(13):1427–42

46. Ferris HU, Furukawa Y, Minamino T, Kroetz MB, Kihara M, et al. 2005. FlhB regulates ordered export of flagellar components via autocleavage mechanism. *J. Biol. Chem.* 280(50):41236–42

47. Flemming H, Wingender J. 2010. The biofilm matrix. *Nat. Rev. Microbiol.* 8:623–33

48. Francis NR, Irikura VM, Yamaguchi S, DeRosier DJ, Macnab RM. 1992. Localization of the *Salmonella typhimurium* flagellar switch protein FliG to the cytoplasmic M-ring face of the basal body. *Proc. Natl. Acad. Sci. USA* 89:6304–8

49. Gao X, Mukherjee S, Matthews PM, Hammad LA, Kearns DB, et al. 2013. Functional characterization of core components of the *Bacillus subtilis* cyclic-di-GMP signaling pathway. *J. Bacteriol.* 195(21):4782–92

50. Grant GF, Simon MI. 1969. Synthesis of bacterial flagella II. PBSI transduction of flagella-specific markers in *Bacillus subtilis*. *J. Bacteriol.* 99(1):116–24

51. Green JCD, Kahramanoglou C, Rahman A, Pender AMC, Charbonnel N, et al. 2009. Recruitment of the earliest component of the bacterial flagellum to the old cell division pole by a membrane-associated signal recognition particle family GTP-binding protein. *J. Mol. Biol.* 391:679–90

52. Guttenplan SB, Blair KM, Kearns DB. 2010. The EpsE flagellar clutch is bifunctional and synergizes with EPS biosynthesis to promote *Bacillus subtilis* biofilm formation. *PLOS Genet.* 6(12):e1001243

53. Guttenplan SB, Kearns DB. 2013. Regulation of flagellar motility during biofilm formation. *FEMS Microbiol. Rev.* 37:849–71

54. Guttenplan SB, Shaw S, Kearns DB. 2013. The cell biology of peritrichous flagella in *Bacillus subtilis*. *Mol. Microbiol.* 87(1):211–29

55. Helmann JD, Marquez LM, Chamberlin MJ. 1988. Cloning, sequencing and disruption of the *Bacillus subtilis* σ^{28} gene. *J. Bacteriol.* 170(4):1568–74

56. Hirano T, Minamino T, Macnab RM. 2001. The role in flagellar rod assembly of the N-terminal domain of *Salmonella* FlgJ, a flagellum-specific muramidase. *J. Mol. Biol.* 312:359–69

57. Hizukuri Y, Kojima S, Homma M. 2010. Disulphide cross-linking between the stator and the bearing components in the bacterial flagellar motor. *J. Biochem.* 148(3):309–18

58. Hizukuri Y, Yakushi T, Kawagishi I, Homma M. 2006. Role of the intramolecular disulfide bond in FlgI, the flagellar P-ring component of *Escherichia coli*. *J. Bacteriol.* 188(12):4190–97

59. Homma M, Kutsukake K, Hasebe M, Iino T, Macnab RM. 1990. FlgB, FlgC, FlgF and FlgG: a family of structurally related proteins in the flagellar basal body of *Salmonella typhimurium*. *J. Mol. Biol.* 211:465–77

60. Hsueh Y, Cozy LM, Sham L, Calvo RA, Gutu AD, et al. 2011. DegU-phosphate activates expression of the anti-sigma factor FlgM in *Bacillus subtilis*. *Mol. Microbiol.* 81(4):1092–108

61. Hughes KT, Gillen KL, Semon MJ, Karlinsey JE. 1993. Sensing structural intermediates in bacterial flagellar assembly by export of a negative regulator. *Science* 262:1277–80

62. Iino T. 1969. Polarity of flagellar growth in *Salmonella*. *J. Gen. Microbiol.* 56:227–39

63. Ikeda T, Oosawa K, Hotani H. 1996. Self-assembly of the filament capping protein, FliD, of bacterial flagella into an annular structure. *J. Mol. Biol.* 259:679–86

64. Ikeda T, Yamaguchi S, Hotani H. 1993. Flagellar growth in a filament-less *Salmonella fliD* mutant supplemented with hook-associated protein 2. *J. Biochem.* 114:39–44

65. Irikura VM, Kihara M, Yamaguchi S, Sockett H, Macnab RM. 1993. *Salmonella typhimurium fliG* and *fliN* mutations causing defects in assembly, rotation and switching of the flagellar motor. *J. Bacteriol.* 175(3):802–10

66. Irnov I, Winkler WC. 2010. A regulatory RNA required for antitermination of biofilm and capsular polysaccharide operons in Bacillales. *Mol. Microbiol.* 76(3):559–75

67. Ito M, Hicks DB, Henkin TM, Guffanti AA, Powers BD, et al. 2004. MotPS is the stator-force generator for motility of alkaliphilic *Bacillus*, and its homologue is a second functional Mot in *Bacillus subtilis*. *Mol. Microbiol.* 53(4):1035–49

68. Jones CJ, Macnab RM, Okino H, Aizawa S. 1990. Stoichiometric analysis of the flagellar hook-(basal body) complex of *Salmonella typhimurium*. *J. Mol. Biol.* 212:377–87

69. Journet L, Agrain C, Broz P, Cornelis GR. 2003. The needle length of bacterial injectisomes is determined by a molecular ruler. *Science* 302:1757–60

70. Kazmierczak BI, Hendrixson DR. 2013. Spatial and numerical regulation of flagellar biosynthesis in polarly flagellated bacteria. *Mol. Microbiol.* 88(4):655–63

71. Kearns DB, Chu F, Branda SS, Kolter R, Losick R. 2005. A master regulator for biofilm formation by *Bacillus subtilis*. *Mol. Microbiol.* 55(3):739–49

72. Kearns DB, Chu F, Rudner R, Losick R. 2004. Genes governing swarming in *Bacillus subtilis* and evidence for a phase variation mechanism controlling surface motility. *Mol. Microbiol.* 52:357–69

73. Kearns DB, Losick R. 2003. Swarming motility in undomesticated *Bacillus subtilis*. *Mol. Microbiol.* 49:581–90

74. Kearns DB, Losick R. 2005. Cell population heterogeneity during growth of *Bacillus subtilis*. *Genes Dev.* 19:3083–94

75. Kearns DB. 2010. A field guide to bacterial swarming motility. *Nat. Rev. Microbiol.* 8:634–44

76. Kinoshita M, Hara N, Imada K, Namba K, Minamino T. 2013. Interactions of bacterial flagellar chaperone-substrate complexes with FlhA contribute to co-ordinating assembly of the flagellar filament. *Mol. Microbiol.* 90:1249–61

77. Kinsinger RF, Shirk MC, Fall R. 2003. Rapid surface motility in *Bacillus subtilis* is dependent on extracellular surfactin and potassium ion. *J. Bacteriol.* 185(18):5627–31

78. Kobayashi K. 2008. SlrR/SlrA controls the initiation of biofilm formation in *Bacillus subtilis*. *Mol. Microbiol.* 69(6):1399–410

79. Kojima S, Blair DF. 2001. Conformational change in the stator of the bacterial flagellar motor. *Biochemistry* 40:13041–50

80. Kojima S, Imada K, Sakuma M, Sudo Y, Kojima C, et al. 2009. Stator assembly and activation mechanism of the flagellar motor by the periplasmic region of MotB. *Mol. Microbiol.* 73(4):710–18

81. Kubori T, Okumura M, Kobayashi N, Nakamura D, Iwakura M, Aizawa SI. 1997. Purification and characterization of the flagellar hook–basal body complex of *Bacillus subtilis*. *Mol. Microbiol.* 24:399–410

82. Kubori T, Shinamoto N, Yamaguchi S, Namba K, Aizawa S. 1992. Morphological pathway of flagellar assembly in *Salmonella typhimurium*. *J. Mol. Biol.* 226:433–46

83. Kulasekara BR, Kamischke C, Kulasekara HD, Christen M, Wiggins PA, et al. 2013. C-di-GMP heterogeneity is generated by the chemotaxis machinery to regulate flagellar motility. *Elife* 2:e01402

84. Kuroda A, Sekiguchi J. 1991. Molecular cloning and sequencing of a major *Bacillus subtilis* autolysin gene. *J. Bacteriol.* 173:7304–12

85. Kutsukake K. 1994. Excretion of the anti-sigma factor through a flagellar substructure couples flagellar gene expression with flagellar assembly in *Salmonella typhimurium*. *Mol. Gen. Genet.* 243:605–12

86. LaVallie ER, Stahl ML. 1989. Cloning of the flagellin gene from *Bacillus subtilis* and complementation studies of an in vitro–derived deletion mutation. *J. Bacteriol.* 171(6):3085–94

87. Lazarevic V, Margot P, Soldo B, Karamata D. 1992. Sequencing and analysis of the *Bacillus subtilis lytRABC* divergon: a regulatory unit encompassing the structural genes of the N-acetylmuramoyl-L-alanine amidase and its modifier. *J. Gen. Microbiol.* 138:1949–61

88. Lee LK, Ginsburg MA, Crovace C, Donohoe M, Stock D. 2010. Structure of the torque ring of the flagellar motor and the molecular basis for rotational switching. *Nature* 466:996–1000

89. Li H, Sourjik V. 2011. Assembly and stability of flagellar motor in *Escherichia coli*. *Mol. Microbiol.* 80(4):886–99

90. Lloyd SA, Blair DF. 1997. Charged residues of the rotor protein FliG essential for torque generation in the flagellar motor of *Escherichia coli*. *J. Mol. Biol.* 266:733–44

91. Lloyd SA, Tang H, Wang X, Billings S, Blair DF. 1996. Torque generation in the flagellar motor of *Escherichia coli*: evidence of a direct role for FliG but not for FliM or FliN. *J. Bacteriol.* 178(1):223–31

92. Macnab RM. 1992. Genetics and biogenesis of bacterial flagella. *Annu. Rev. Genet.* 26:131–58

93. Margot P, Mauel C, Karamata D. 1994. The gene of the N-acteylglucosaminidase, a *Bacillus subtilis* 168 cell wall hydrolase not involved in vegetative cell autolysis. *Mol. Microbiol.* 12:535–45

94. Margot P, Pagni M, Karamata D. 1999. *Bacillus subtilis* 168 gene *lytF* encodes a γ–D-glutamate-meso-diaminopimelate muropeptidase expressed by the alternative vegetative sigma factor σ^D. *Microbiology* 145:57–65

95. Marquez LM, Helmann JD, Ferrari E, Parker HM, Ordal GW, et al. 1990. Studies of σ^D-dependent functions in *Bacillus subtilis*. *J. Bacteriol.* 172(6):3435–43

96. Marquez-Magaña LM, Chamberlin MJ. 1994. Characterization of the *sigD* transcription unit of *Bacillus subtilis*. *J. Bacteriol.* 176(8):2427–34

97. Minamino T, Macnab RM. 1999. Components of the *Salmonella* flagellar export apparatus and classification of export substrates. *J. Bacteriol.* 181(5):1388–94

98. Minamino T, Moria N, Hirano T, Hughes KT, Namba K. 2009. Interaction of FliK with the bacterial flagellar hook is required for efficient export specificity switching. *Mol. Microbiol.* 74(1):239–51

99. Minamino T, Namba K. 2008. Distinct roles of the FliI ATPase and proton motive force in bacterial flagellar protein export. *Nature* 451:485–88

100. Minamino T, Yamaguchi S, Macnab RM. 2000. Interaction between FliE and FlgB, a proximal rod component of the flagellar basal body of *Salmonella*. *J. Bacteriol.* 182(11):3029–36

101. Mirel DB, Chamberlin MJ. 1989. The *Bacillus subtilis* flagellin gene (*hag*) is transcribed by the σ^{28} form of RNA polymerase. *J. Bacteriol.* 171(6):3095–101

102. Mirel DB, Lauer P, Chamberlin MJ. 1994. Identification of flagellar synthesis regulatory and structural genes in a σ^D-dependent operon of *Bacillus subtilis*. *J. Bacteriol.* 176:4492–500

103. Mirel DB, Lustre VM, Chamberlin MJ. 1992. An operon of *Bacillus subtilis* motility genes transcribed by the σ^D form of RNA polymerase. *J. Bacteriol.* 174:4197–204

104. Mordini S, Osera C, Marini S, Scavone F, Bellazzi R, et al. 2013. The role of SwrA, DegU and P_{D3} in *fla/che* expression in *B. subtilis*. *PLOS ONE* 8:e85065

105. Moriya N, Minamino T, Hughes KT, Macnab RM, Namba K. 2006. The type III flagellar export specificity switch is dependent on FliK ruler and a molecular clock. *J. Mol. Biol.* 359:466–77

106. Mukai K, Kawata M, Tanak T. 1990. Isolation and phosphorylation of the *Bacillus subtilis degS* and *degU* gene products. *J. Bio. Chem.* 265(32):20000–6

107. Mukherjee S, Babitzke P, Kearns DB. 2013. FliW and FliS function independently to control cytoplasmic flagellin levels in *Bacillus subtilis*. *J. Bacteriol.* 195:297–306

108. Mukherjee S, Yakhnin H, Kysela D, Sokoloski J, Babitzke P, et al. 2011. CsrA-FliW interaction governs flagellin homeostasis and a checkpoint on flagellar morphogenesis in *Bacillus subtilis*. *Mol. Microbiol.* 82:447–61

109. Muller V, Jones CJ, Kawagishi I, Aizawa S, Macnab RM. 1992. Characterization of the *fliE* genes of *Escherichia coli* and *Salmonella typhimurium* and identification of the FliE protein as a component of the flagellar hook-basal body complex. *J. Bacteriol.* 174(7):2298–304

110. Murray EJ, Kiley TB, Stanley-Wall NR. 2009. A pivotal role for the response regulator DegU in controlling multicellular behavior. *Microbiology* 155:1–8

111. Nambu T, Minamino T, Macnab RM, Kutsukake K. 1999. Peptidoglycan-hydrolyzing activity of the FlgJ protein, essential for flagellar rod formation in *Salmonella typhimurium*. *J. Bacteriol.* 181(5):1555–61

112. Newman JA, Rodrigues C, Lewis RJ. 2013. Molecular basis of the activity of SinR protein, the master regulator of biofilm formation in *Bacillus subtilis*. *J. Bio. Chem.* 288(15):10766–78

113. Nishihara T, Freese E. 1975. Motility of *Bacillus subtilis* during growth and sporulation. *J. Bacteriol.* 123:366–71

114. Ogura M, Tsukahara K. 2012. SwrA regulates assembly of *Bacillus subtilis* DegU via its interaction with N-terminal domain of DegU. *J. Biochem.* 151(6):643–55

115. Ogura M, Yoshikawa H, Chibazakura T. 2014. Regulation of the response regulator gene *degU* through the binding of SinR/SlrR and exclusion of SinR/SlrR by DegU in *Bacillus subtilis*. *J. Bacteriol.* 196(4):873–81

116. Ohnishi K, Ohto Y, Aizawa S, Macnab RM, Iino T. 1994. FlgD is a scaffolding protein needed for flagellar hook assembly in *Salmonella typhimurium*. *J. Bacteriol.* 176(8):2272–81

117. Okino H, Isomura M, Yamaguchi S, Magariyama Y, Kudo S, et al. 1989. Release of flagellar filament-hook-rod complex by a *Salmonella typhimurium* mutant defective in the M ring of the basal body. *J. Bacteriol.* 171(4):2075–82

118. Patrick JE, Kearns DB. 2009. Laboratory strains of *Bacillus subtilis* do not exhibit swarming motility. *J. Bacteriol.* 191(22):7129–33

119. Paul K, Erhardt M, Hirano T, Blair DF, Hughes KT. 2008. Energy source of flagellar type III secretion. *Nature* 451:489–92

120. Paul K, Gonzalez-Bonet G, Bilwes AM, Crane BR, Blair D. 2011. Architecture of the flagellar motor. *EMBO J.* 30:2962–71

121. Paul K, Nieto V, Carlquist WC, Blair DF, Harshey RM. 2010. The c-di-GMP binding protein YcgR controls flagellar motor direction and speed to affect chemotaxis by a "backstop brake" mechanism. *Cell* 38:128–39

122. Pooley HM, Karamata D. 1984. Genetic analysis of autolysin-deficient and flagellaless mutants of *Bacillus subtilis*. *J. Bacteriol.* 160(3):1123–29

123. Rao CV, Glekas GD, Ordal GW. 2008. The three adaptation systems of *Bacillus subtilis* chemotaxis. *Trends Microbiol.* 16(10):480–87

124. Rashid MH, Sekiguchi J. 1996. *flaD* (SinR) mutations affect SigD-dependent functions at multiple points in *Bacillus subtilis*. *J. Bacteriol.* 178(22):6640–43

125. Samatey FA, Matusnami H, Imada K, Nagashima S, Shaikh TR, et al. 2004. Structure of the bacterial flagellar hook and implications for the molecular universal joint mechanism. *Nature* 431:1062–68

126. Seminara A, Angelini TE, Wilking JN, Vlamakis H, Ebrahim S, et al. 2012. Osmotic spreading of *Bacillus subtilis* biofilms driven by an extracellular matrix. *Proc. Natl. Acad. Sci. USA* 109(4):1116–21

127. Senesi S, Ghelardi E, Celandroni F, Salvetti S, Parisio E, et al. 2004. Surface-associated flagellum formation and swarming differentiation in *Bacillus subtilis* are controlled by the *ifm* locus. *J. Bacteriol.* 186(4):1158–64

128. Serizawa M, Yamamoto H, Yamaguchi H, Fujita Y, Kobayashi K, et al. 2004. Systematic analysis of SigD-regulated genes in *Bacillus subtilis* by DNA microarray and Northern blotting analyses. *Gene* 329:125–36

129. Sevim E, Gaballa A, Belduz O, Helmann JD. 2011. DNA-binding properties of the *Bacillus subtilis* and *Aeribacillus pallidus* AC6 σ^D proteins. *J. Bacteriol.* 193(2):575–79

130. Shibata S, Takahashi N, Chevance FFV, Karlinsey JE, Hughes KT, et al. 2007. FliK regulates flagellar hook length as an internal ruler. *Mol. Microbiol.* 64(5):1404–15

131. Sircar R, Greenswag AR, Bilwes AM, Gonzalez-Bonet G, Crane BR. 2013. Structure and activity of the flagellar rotor protein FliY: a member of the CheC phosphatase family. *J. Biol. Chem.* 288(19):13493–502

132. Smith DR, Chapman MR. 2010. Economical evolution: microbes reduce the synthetic cost of extracellular proteins. *Mbio* 1(3):e00131–10

133. Smith TJ, Blackman SA, Foster SJ. 2000. Autolysins of *Bacillus subtilis*: multiple enzymes with multiple functions. *Microbiology* 146:249–62

134. Sorenson MK, Ray SS, Darst SA. 2004. Crystal structure of the flagellar σ/anti-σ complex σ^{28}/FlgM reveals an intact σ factor in an inactive conformation. *Mol. Cell* 14:127–38

135. Stallmeyer MJB, Aizawa S, Macnab RM, DeRosier DJ. 1989. Image reconstruction of the flagellar basal body of *Salmonella typhimurium*. *J. Mol. Biol.* 205:519–28

136. Stanley NR, Lazazzera BA. 2005. Defining the genetic differences between wild and domestic strains of *Bacillus subtilis* that affect poly-γ-DL-glutamic acid production and biofilm formation. *Mol. Microbiol.* 57(4):1143–58

137. Stern AS, Berg HC. 2013. Single-file diffusion of flagellin in flagellar filaments. *Biophys. J.* 105:182–84

138. Suzuki H, Yonekura K, Murata K, Hirai T, Oosawa K, et al. 1998. A structural feature in the central channel of the bacterial flagellar FliF ring complex is implicated in type III protein export. *J. Struct. Biol.* 124:104–14

139. Terahara N, Fujisawa M, Powers B, Henkin TM, Krulwich TA, et al. 2006. An intergenic stem-loop mutation in the *Bacillus subtilis* ccpA-motPS operon increases motPS transcription and the MotPS contribution to motility. *J. Bacteriol.* 188(7):2701–5

140. Thomas DR, Francis NR, Xu C, DeRosier DJ. 2006. The three-dimensional structure of the flagellar rotor from a clockwise-locked mutant of *Salmonella enterica* serovar *typhimurium*. *J. Bacteriol.* 188(20):7039–48

141. Thomas J, Stafford GP, Hughes C. 2004. Docking of cytosolic chaperone-substrate complexes at the membrane ATPase during flagellar type III protein export. *Proc. Natl. Acad. Sci. USA* 101(11):3945–50

142. Titz B, Rajagopala SV, Ester C, Häuser R, Uetz P. 2006. Novel conserved assembly factor of the bacterial flagellum. *J. Bacteriol.* 188:7700–6

143. Tsukahara K, Ogura M. 2008. Promoter selectivity of the *Bacillus subtilis* response regulator DegU, a positive regulator of the *fla/che* operon and *sacB*. *BMC Microbiol.* 8:8

144. Turner L, Ryu WS, Berg HC. 2000. Real-time imaging of fluorescent flagellar filaments. *J. Bacteriol.* 182(10):2793–801

145. Turner L, Stern AS, Berg HC. 2012. Growth of flagellar filaments is independent of flagellar length. *J. Bacteriol.* 194:2437–42

146. Ueno T, Oosawa K, Aizawa S. 1992. M ring, S ring and proximal rod of the flagellar basal body of *Salmonella typhimurium* are composed of subunits of a single protein, FliF. *J. Mol. Biol.* 227:672–77

147. Werhane H, Lopez P, Mendel M, Zimmer M, Ordal GW, Marquez-Magaña LM. 2004. The last gene of the *fla/che* operon in *Bacillus subtilis*, *ylxL*, is required for maximal σ^D function. *J. Bacteriol.* 186(12):4025–29

148. West JT, Estacio W, Marquez-Magaña L. 2000. Relative roles of the *fla/che* P_A, P_{D-3}, and P_{sigD} promoters in regulating motility and *sigD* expression in *Bacillus subtilis*. *J. Bacteriol.* 182(17):4841–48

149. Winkelman JT, Bree AC, Bate AR, Eichenberger P, Gourse RL, et al. 2013. RemA is a DNA-binding protein that activates biofilm matrix gene expression in *Bacillus subtilis*. *Mol. Microbiol.* 88(5):984–97

150. Yakhnin H, Pandit P, Petty TJ, Baker CS, Romeo T, et al. 2007. CsrA of *Bacillus subtilis* regulates translation initiation of the gene encoding the flagellin protein (*hag*) by blocking ribosome binding. *Mol. Microbiol.* 64:1605–20

151. Yonekura K, Maki-Yonekura S, Namba K. 2003. Complete atomic model of the bacterial flagellar filament by electron cryomicroscopy. *Nature* 424:643–50

152. Zeigler DR, Pragai Z, Rodriguez S, Chevreux B, Muffler A, et al. 2008. The origins of 168, W23, and other *Bacillus subtilis* legacy strains. *J. Bacteriol.* 190(21):6983–95

153. Zhang K, Tong BA, Liu J, Li C. 2012. A single-domain FlgJ contributes to flagellar hook and filament formation in the Lyme disease spirochete *Borrelia burgdorferi*. *J. Bacteriol.* 194(4):866–74

154. Zhao X, Zhang K, Boquoi T, Hu B, Motaleb MA, et al. 2013. Cryoelectron tomography reveals the sequential assembly of bacterial flagella in *Borrelia burgdorferi*. *Proc. Natl. Acad. Sci. USA* 110(35):14390–95

Transcription-Associated Mutagenesis

Sue Jinks-Robertson[1] and Ashok S. Bhagwat[2]

[1]Department of Molecular Genetics and Microbiology, Duke University Medical Center, Durham, North Carolina 27710; email: sue.robertson@duke.edu

[2]Department of Chemistry and Department of Microbiology and Immunology, Wayne State University, Detroit, Michigan 48202; email: axb@chem.wayne.edu

Annu. Rev. Genet. 2014. 48:341–59

First published online as a Review in Advance on September 10, 2014

The *Annual Review of Genetics* is online at genet.annualreviews.org

This article's doi:
10.1146/annurev-genct-120213-092015

Keywords

DNA damage, topoisomerase, somatic hypermutation, class-switch recombination, cytosine deamination

Abstract

Transcription requires unwinding complementary DNA strands, generating torsional stress, and sensitizing the exposed single strands to chemical reactions and endogenous damaging agents. In addition, transcription can occur concomitantly with the other major DNA metabolic processes (replication, repair, and recombination), creating opportunities for either cooperation or conflict. Genetic modifications associated with transcription are a global issue in the small genomes of microorganisms in which noncoding sequences are rare. Transcription likewise becomes significant when one considers that most of the human genome is transcriptionally active. In this review, we focus specifically on the mutagenic consequences of transcription. Mechanisms of transcription-associated mutagenesis in microorganisms are discussed, as is the role of transcription in somatic instability of the vertebrate immune system.

INTRODUCTION

Transcription-associated mutagenesis (TAM): localized changes in DNA that are associated with transcription of the target sequence

RNA polymerase (RNAP): the enzyme/complex that makes an RNA copy of a DNA template; RNAP II specifically synthesizes mRNA

R-loop: three-strand structure in which RNA is base-paired with one strand of duplex DNA, leaving the other DNA strand unpaired

Nontranscribed strand (NTS): DNA strand that has the same sequence as the RNA transcript; also referred to as the coding strand

The maintenance of genome integrity is usually considered in relation to the three Rs: replication, repair, and recombination. This review focuses on how the other major DNA metabolic process—transcription—affects stability of the underlying DNA template. This is of particular significance when one considers that functional genes make up only 1% of the human genome, and yet recent estimates indicate that up to 80% of the human genome may be transcriptionally active (24). An effect of transcription on mutagenesis was first recognized in microorganisms almost a half-century ago, but the diverse causes and potential evolutionary implications of transcription-associated mutagenesis (TAM) have only recently been appreciated.

Transcription typically copies only one DNA strand, leaving the other (nontranscribed) strand in a transiently single-stranded state (32) that renders it chemically reactive and vulnerable to endogenous DNA damage. Within the transcription bubble transiently created by RNA polymerase (RNAP), only a short tract of single-stranded DNA (ssDNA) is exposed (**Figure 1**). Following RNA polymerization, the transcript and its complementary DNA strand exit RNAP through separate channels (100), thereby disrupting short RNA:DNA hybrids and promoting the reannealing of DNA strands. However, very long tracts of ssDNA can form if the transcript threads back and stably base pairs with its template. The resulting three-strand structure is referred to as an R-loop (**Figure 1**). Within R-loops the nontranscribed strand (NTS) is not only vulnerable to damage, it can also assume secondary structures that perturb or trigger other DNA metabolic processes. In prokaryotes, transcription and translation are coupled, with immediate transcript engagement by ribosomes preventing stable R-loop formation (34). In eukaryotes, where transcription and translation occur in separate cellular compartments, cotranscriptional processing of transcripts (e.g., splicing and nuclear transport) similarly discourages R-loop formation (58). Finally, transcription produces twin domains of positive and negative supercoiling (63). Positive supercoils are generated ahead of the transcription machinery and reflect overwinding of the helix as DNA strands are separated (**Figure 1**). Behind the machinery, the corresponding underwound state of DNA leads to the accumulation of negative supercoils. Underwinding exposes both DNA strands to endogenous damage and promotes R-loop formation, whereas overwinding can impede further strand separation. Supercoils are relaxed by topoisomerases, which nick and reseal one or both strands of DNA (98).

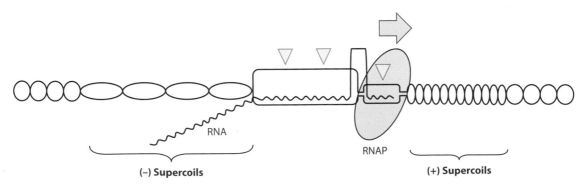

Figure 1

Effects of transcription on the DNA template. The transcription bubble and a trailing R-loop are indicated as small and large rectangles, respectively. Circles indicate normal intertwining of DNA strands; compressed or extended ovals correspond to over- or underwound strands, respectively, and the regions of associated positive (+) or negative (−) supercoils are indicated. RNAP (RNA polymerase) is depicted as a blue oval, and the blue arrow indicates its direction of movement on the DNA template. DNA and RNA strands are black and red, respectively; yellow triangles indicate damage to ssDNA.

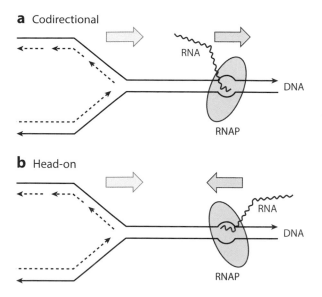

a Codirectional

RNA

DNA

RNAP

b Head-on

RNA

DNA

RNAP

Figure 2

Conflicts between the replication and transcription machineries. Movement of the replisome and RNAP (RNA polymerase) in the same or opposite direction can cause (*a*) codirectional or (*b*) head-on conflicts, respectively. Red and black lines represent RNA and DNA, respectively; dashed lines depict newly synthesized DNA; and blue ovals represent RNAP. Yellow and blue arrows indicate the direction of replication-fork and RNAP movement, respectively.

The potential effects of transcription on DNA stability become more complex if one considers that transcription occurs concurrently with and is influenced by other major DNA metabolic processes. Superimposed on top of transcription-associated DNA damage, for example, is the occurrence of transcription-coupled repair (TCR). TCR is triggered by damage in the transcribed strand (TS) that blocks RNAP and leads to recruitment of the nucleotide-excision repair (NER) machinery. This results in the preferential repair of lesions in the transcribed relative to the NTS of DNA (36). Potential conflicts between transcription and DNA replication have attracted particular attention; these conflicts are exacerbated by R-loop formation and are the major source of transcription-associated recombination (reviewed in 1). Transcription-replication conflicts are defined as codirectional if the replication fork moves in the same direction as the transcription machinery and as head-on if the two converge (**Figure 2**). In the codirectional orientation, the TS is the leading strand of replication; in the head-on orientation, the TS is the lagging strand of replication. Head-on conflicts are generally considered more detrimental than codirectional conflicts and have been invoked to explain the co-orientation of most bacterial genes with replication-fork movement, especially the highly transcribed ribosomal RNA operons (66). However, because the rate of bacterial replication is approximately ten times faster than that of transcription, codirectional conflicts may also occur when the replication apparatus overtakes RNAP. In eukaryotes, the transcription and replication machineries move at similar rates, and transcription and replication are usually temporally separated within the cell cycle. Even so, transcription-replication conflicts within very long genes are inevitable and have been linked to common fragile sites in mammalian genomes (41). Here, only contributions of replication-transcription conflicts to localized mutagenesis are considered, and we refer the reader to several excellent reviews that deal with transcription-associated recombination and gross chromosome alterations (1, 40).

Transcription-coupled repair (TCR): NER subpathway that specifically removes damage from the transcribed strand of active genes

Transcribed strand (TS): strand of DNA copied by RNAP

Nucleotide-excision repair (NER): removes a lesion-containing oligonucleotide, leaving a 20–25-nt gap that is filled in using the undamaged strand as template

Although TAM clearly has pathological effects on genome integrity, it has been harnessed during evolution to drive localized and very rapid genetic change. This is particularly evident in the vertebrate immune system, where transcription is required for somatic hypermutation (SHM) and class-switch recombination (CSR) within immunoglobulin (Ig) genes. TAM also provides a potential source of replication-independent genetic change in nongrowing cells, and this has been implicated in stress-induced mutation in bacteria (104) and in trinucleotide-repeat instability in eukaryotes (59). Below, we summarize the characterized sources of TAM in bacteria, discuss specific mechanisms of TAM that have been uncovered in budding yeast, and consider the specific example of transcription-associated instability in the vertebrate immune system.

TRANSCRIPTION-ASSOCIATED MUTAGENESIS IN BACTERIA

The first suggestions of TAM date to the early 1970s, when it was reported that induction of the *lac* operon increased reversion caused by the frameshift mutagen ICR-191 in *Escherichia coli* (43). It similarly was reported that derepression of *his* genes increased UV-induced reversion in *Salmonella typhimurium* (87). Another 20 years passed, however, before a link between mutagenesis and transcription was definitively established.

The detection of TAM in bacteria (as well as in yeast; see below) has relied primarily on selective systems in which transcription can be varied at will. Most studies have used reversion assays, which are inherently limited because they detect only a subset of all possible mutations. A potential complication in reversion assays is that a transcription-driven increase in the corresponding gene product may shorten the time needed to express the selected phenotype and thereby artificially inflate the measured rate. In contrast to the functional restoration required by reversion assays, forward mutation assays select against the encoded protein. Although the spectrum of mutation types detected is much broader, elevated transcription can exacerbate an associated phenotypic lag (i.e., the wild-type gene product must be diluted out before the mutant phenotype is expressed), and this has the potential to underestimate or completely mask TAM. Finally, in addition to inherent biases associated with a given assay, the magnitude and/or mechanism of TAM may be affected by the location of the reporter on a plasmid versus the chromosome, the orientation of the reporter relative to replication-fork movement, and the specific growth conditions used. These issues should be borne in mind in the TAM descriptions that follow.

DNA Damage and Strand-Related Asymmetries in Mutation Accumulation

An early observation of a strand-related bias in mutation accumulation was made in *E. coli*, where the sequence change diagnostic of hydrolytic cytosine deamination to uracil was strongly biased to the NTS of *lacI* (26). In this and other experiments that have focused on a particular type of damage, strain backgrounds that are defective in its repair are often used. By convention, the sequence of the NTS, which is identical to that of the mRNA, is the sequence reported. As illustrated in **Figure 3**, deamination of cytosine on the NTS results in C > T mutations, whereas deamination of cytosine on the TS generates G > A sequence changes. Thus, by comparing the accumulation of C > T versus G > A changes, one can infer relative deamination of the NTS versus TS strand. Comparative analysis of genes in *E. coli* and *S. enterica* indicates that the cytosine deamination bias primarily reflects an asymmetry associated with transcription rather than replication (28). Although it was not possible to infer whether the strand-associated asymmetry reflected preferential damage of the NTS and/or biased repair of the TS via TCR, a subsequent comparison of mutation patterns in expressed versus nonexpressed DNA was more consistent with the former (29). The enhanced

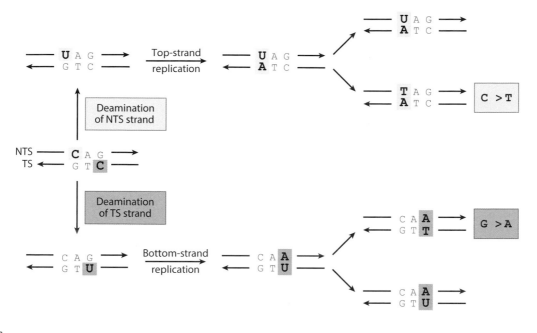

Figure 3

Inferring strand specificity from mutation patterns associated with cytosine deamination. Yellow and purple boxes indicate consequences of cytosine deamination on the nontranscribed strand (NTS) and transcribed strand (TS), respectively.

deamination of the NTS inferred in vivo is consistent with much faster cytosine deamination in ssDNA than in double-stranded DNA in vitro (30).

The first direct demonstration of a correlation between transcription level and preferential deamination of cytosine on the NTS came through analyzing reversion of a missense allele under control of the highly inducible *tac* promoter (6). It was found that the bias for cytosine deamination on the NTS was evident only if transcription was highly activated. Importantly, the NTS bias was maintained when the direction of transcription through the reporter was reversed and was thus independent of the sequence of the NTS. That the NTS has single-strand characteristics relative to the TS is additionally supported by its enhanced sensitivity to enzymatic deamination following expression of a mammalian cytosine deaminase (see below).

In addition to the specific case of cytosine deamination, an NTS bias is also evident for spontaneous oxidative lesions (53) and for damage generated by the alkylating agent methylmethane sulfonate (MMS) (25). Oxidative damage to guanine generates 7,8-dihydro-8-oxo-guanine (8-oxoG), which mispairs frequently with adenine and gives rise to GC > TA mutations. Reversion of a TGA stop codon via 8-oxoG-associated G > T transversions was examined at a reporter inserted in both orientations relative to the strong *tac* promoter. Significantly, transcription from P_{tac} elevated G > T transversions only if the stop codon was on the NTS, consistent with enhanced transcription-associated damage to this strand (53). Interestingly, high levels of transcription reduced reversion when the TGA was on the TS, suggesting that preferential repair of lesions on the TS via TCR may also contribute to some of the strand specificity. A strand bias of MMS-induced mutations was similarly assayed by scoring reversion of a CCA missense allele inserted in either orientation relative to P_{tac} (25). Inducing transcription caused mutations only at cytosines in the NTS, consistent with methylation targeted to ssDNA. Whether TAM in these systems primarily reflects ssDNA within the transcription bubble or more extensive ssDNA exposed within R-loops has not been specifically addressed.

The observation that many different types of base substitutions accumulate in a transcription-dependent fashion underscores the generality of TAM (46, 52). Mutagenesis is initiated more frequently on the NTS than on the TS strand of active genes, but whether all nucleotides on the NTS are equally mutable is unclear. It has been suggested, for example, that the folding of ssDNA into stem-loop structures exposes bases in single-strand loops to endogenous damage and renders them hypermutable. The *mfd* program developed by Wright and colleagues uses the free energy of all possible stem-loop structures to derive a mutability index for each base within a short stretch of ssDNA (106). Correlations have been observed between the calculated mutability index and reversion rates at specific sites in *E. coli* reporters (11, 88), and a similar correlation has been noted for highly mutable sites in the p53 tumor suppressor gene (108) and in Ig genes (107, 109).

Effects of Starvation/Stress on Mutagenesis

TAM is readily observed when transcription is induced to high levels in a reporter fused to a heterologous promoter, but elevated transcription is also a natural response of bacteria to amino acid starvation. Starvation generally induces/derepresses only those genes relevant to biosynthesis of the corresponding amino acid and is additionally modulated as part of the ppGpp-mediated stringent response. An effect of the stringent response was found when examining reversion of *leuB* and *argH* alleles in *E. coli*, and it was speculated that this could reflect an associated increase in transcription (103). A correlation between reversion and starvation-induced transcription was subsequently established (105), and the relationship between the two is linear (82). One important consequence of starvation-stimulated transcription is that mutagenesis is higher in those genes in which changes can potentially be beneficial (reviewed in 104), and it is possible that a similar phenomenon may underlie some examples of adaptive mutation (17).

A relationship between stress-associated transcription and reversion has also been reported in *Bacillus subtilis*. In this case, the correlation was made under prolonged starvation conditions in which mutations accumulated in a replication-independent, but time-dependent, manner (81). It has been argued that such stress-induced mutations represent an adaptive response that fosters rapid evolutionary change (reviewed in 31). Interestingly, recent work in *E. coli* has demonstrated the importance of R-loops in mutagenesis that occurs in stressed cells (101). In this case, a novel mechanism of transcription-initiated genetic instability was proposed in which an exposed 3′ end of the RNA within an R-loop is used to initiate origin-independent replication. A subsequent encounter of DNA polymerase with a nick on the template strand is hypothesized to generate a double-strand end that then initiates recombination-associated mutagenesis.

Replication-Transcription Conflicts

Head-on encounters between highly transcribed ribosomal RNA genes and replication forks slow DNA synthesis and affect overall fitness (93). A strong codirectional orientation bias also has been reported for a set of core genes common to diverged *B. subtilis* strains (76). Comparative analyses of these strains suggested that nonsynonymous changes accumulate faster in core genes with the head-on orientation and that there is a positive correlation between these changes and transcript abundance. In the few cases in which the direction of replication-fork movement on mutagenesis within a defined reporter has been directly examined, mutation rates were higher in the head-on orientation than in the codirectional orientation. In *B. subtilis*, for example, reversion of a *hisC* nonsense allele was affected by the direction of replication but only under conditions of high transcription (76). In *E. coli*, forward mutations in *rpoB* were similarly higher in the head-on orientation than in the codirectional orientation, but the specific contribution of transcription was not examined (93). Although the reason why head-on conflicts are more mutagenic than codirectional conflicts is not known, studies in yeast suggest there may be a recombination connection. In particular, recombination-associated DNA synthesis is more error prone than replicative DNA synthesis (44, 96), and head-on transcription-replication conflicts stimulate recombination more than do codirectional conflicts (80).

TRANSCRIPTION-ASSOCIATED MUTAGENESIS IN YEAST

TAM has been well documented in *Saccharomyces cerevisiae*, but similar reports have not emerged from the other major yeast model, *Schizosaccharomyces pombe*. Although all the data described below were obtained using budding yeast, there is no a priori reason to suspect that results will not be widely applicable to other eukaryotes. An early indication that TAM occurs in yeast came 20 years after initial reports in bacteria. As in bacteria, it was found that limiting a specific amino acid was associated with elevated reversion of a gene in the corresponding biosynthetic pathway, and it similarly was speculated that this might be related to starvation-associated induction of transcription (55). Definitive evidence of TAM was obtained following fusion of a forward- or reverse-mutation reporter to the highly inducible, galactose-regulated *pGAL* promoter (16). Subsequent studies of TAM have used *pGAL* or the heterologous tetracycline/doxycycline-regulated *pTET* promoter. As in bacterial cells, there is a direct proportionality between the transcript level and mutagenesis (48). Although head-on encounters between transcription and replication forks in yeast also slow DNA replication more than do codirectional encounters (19, 80), reversing the direction of replication through a *pTET*-driven frameshift reporter had no significant effect on the reversion rate. There were, however, orientation-specific effects evident in the corresponding spectra (48, 49). On a genome-wide scale, the accumulation of DNA polymerase correlates with high transcription, indicating that both head-on and codirectional conflicts slow replication in yeast (4).

A central question has been whether results obtained with a small number of reporter genes are relevant on an evolutionary timescale. This has recently been addressed through comparative analysis of *S. cerevisiae* and *Saccharomyces paradoxus* genomes, as well as by sequencing spontaneous mutations that accumulate over hundreds of generations in budding yeast. Both types of analysis revealed a positive correlation between transcription and mutagenesis (74). Importantly, analyses were confined to intronic sequences, thereby removing confounding selective constraints on the analyzed sequences. Below, we focus on the diverse mechanisms that contribute to TAM in budding yeast.

DNA Damage as a Source of Transcription-Associated Mutagenesis

The most extensive TAM studies have been done using *LYS2*-based frameshift reversion assays that detect either net +1 or −1 events. Genetic studies with these systems (16, 50, 67), as well as recent experiments with nonsense reversion assays (2, 51), have implicated DNA damage as a major source of TAM. Key observations have been that TAM increases when an error-free mechanism of lesion bypass (i.e., template switch or homologous recombination) is impaired and when either NER or base-excision repair (BER) is inactivated, and it decreases in the absence of the error-prone translesion synthesis (TLS) DNA polymerase Pol ζ (for a review of repair/bypass pathways in yeast, see 9). In nonsense reversion assays, all detectable base substitutions were elevated, but a strong proportional increase in transversions at GC base pairs was noted (2, 51). Although no preferential accumulation of spontaneous damage on the NTS was evident, it should be noted that nonsense

Base-excision repair (BER): incises the DNA backbone adjacent to an abasic site and initiates its replacement with a nucleotide specified by the complementary strand

Translesion synthesis (TLS): polymerization of DNA opposite lesions by specialized, low-fidelity DNA polymerases

reversion assays are incapable of detecting the CG > TA mutations characteristic of cytosine deamination. In relation to possible strand specificity, enzymatic deamination of cytosine by human activation-induced deaminase (AID) was reported to target both DNA strands, suggesting that negative supercoiling behind the transcription machinery may be relevant. However, deamination occurred preferentially on the NTS when conditions favoring R-loop formation were used (33).

Topoisomerase 1 as a Mutagen in Transcriptionally Active DNA

Sequence analysis in a *pGAL-LYS2* forward-mutation assay revealed that most, if not all, mutation types were elevated by transcription, but established small deletions of 2–5 bp as a specific signature of TAM. These events made up ~25% of mutations if a reporter was highly transcribed but were absent if transcription occurred at very low levels (61). Subsequent analyses of mutagenesis in *pGAL-CAN1* and *pTET-CAN1* reporters confirmed the short-deletion TAM signature and demonstrated that events accumulate at discrete tandem-repeat hot spots (62, 97). The size of the deletion corresponded to the size of the repeat unit, and the repeat was present in only two to four copies prior to the deletion event. Furthermore, the primary sequence of the repeat unit was highly variable, indicating that any repeat can potentially harbor a transcription-associated deletion. Significantly, short deletions were completely eliminated upon loss of topoisomerase 1 (Top1), an enzyme that resolves transcription-associated supercoils by nicking and resealing one strand of DNA. Subsequent work demonstrated that the Top1-dependent hot spots are of two distinct types: those that reflect processing of a covalently trapped Top1 cleavage complex and those that reflect incision at a ribonucleoside monophosphate (rNMP) embedded in duplex DNA (**Figure 4**) (15). It should be noted that this particular TAM signature is expected to be associated only with a eukaryotic-specific type 1B enzyme, which forms a $3'$-phosphotyrosyl link to the nicked DNA. It is possible, however, that other types of topoisomerase-mediated damage may have mutagenic consequences that have yet to be defined.

RNA: DNA Hybrids Initiate Complex Mutations in Highly Transcribed DNA

The mutagenic consequences of Top1 incision at an rNMP are most evident in the absence of RNase H2, an enzyme that initiates error-free removal of 1–3 rNMPs from DNA as well as the degradation of the RNA component of R-loops (13, 92). Studies in RNase H2-deficient strains have revealed the occurrence of complex mutations, which are characterized by multiple, simultaneous sequence changes that extend identity between the arms of an imperfect inverted repeat.

Activation-induced deaminase (AID): enzyme that deaminates cytosine in DNA to uracil and is required for postinfection genetic alterations in Ig genes

Topoisomerase 1 (Top1): eukaryotic Type 1B enzyme that relaxes supercoils and forms a $3'$-phosphotyrosyl linkage when it nicks DNA

Figure 4

Mechanisms of Top1 mutagenesis in transcriptionally active DNA. Two distinct mechanisms of Top1-dependent mutagenesis are shown, with a hypothetical dinucleotide repeat highlighted in gray. When Top1 incision occurs, the active-site tyrosine forms a covalent linkage to the $3'$-PO_4 on one side of the DNA nick, leaving a $5'$-OH on the other side. (*a*) (Step I) Top1 becomes trapped as a stabilized cleavage complex, and (Step II) its removal by unknown proteins generates a 2-nt gap within the 2-bp tandem repeat. (Step III) Realignment of the DNA strands converts the gap to a nick, which (Step IV) facilitates ligation and produces the mutation intermediate. (Step V) Replication of the newly ligated strand results in a permanent, 2-bp deletion; replication of the other strand is of no genetic consequence. (*b*) Top1 incises at the position of an rNMP (*red R*). (Step I) The $2'$-OH of ribose attacks the phosphotyrosyl bond, releasing Top1 and generating a $2',3'$-cyclic phosphate (*red triangle*). (Step II) A second incision by Top1 upstream of the nick releases the intervening oligonucleotide and (Step III) transiently traps the covalent enzyme-DNA intermediate. (Step IV) Realignment of the two DNA strands by the repeat sequence correctly orients the Top1-DNA complex and the $5'$-OH, enabling efficient Top1-mediated rejoining of the ends. (Step V) Replication of the top strand fixes the 2-bp deletion.

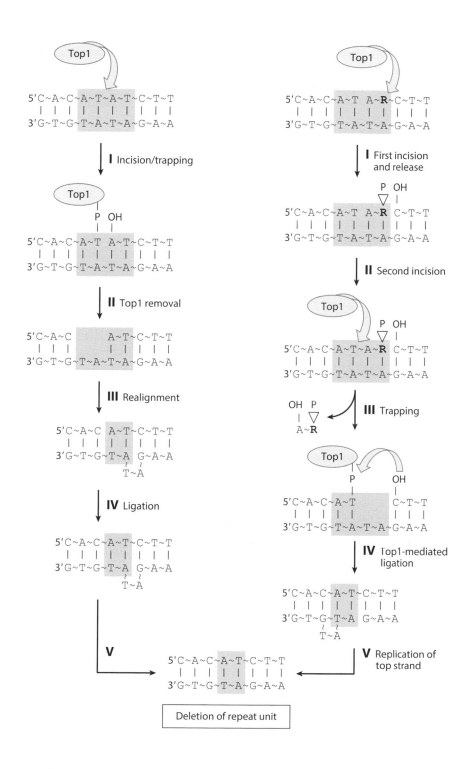

Apurinic/
apyrimidinic (AP)
site: site in DNA that
is missing a base; also
referred to as an abasic
site

Uracil N-glycosylase
(UNG): enzyme that
removes uracil from
DNA, creating an AP
site

Although these mutations are evident only under conditions of highly activated transcription, they are mechanistically distinct from rNMP-initiated small deletions, in that they do not require Top1 activity and are strongly affected by the direction of replication-fork movement (49). The effect of replication direction coupled with multiple sequence changes suggests a template-switch mechanism, with persistent rNMPs in the DNA template being the likely trigger. An additional requirement for RNase H1, which only processes more extensive RNA:DNA hybrids, for the generation of complex mutations suggests that either the RNA primers of Okazaki fragments or cotranscriptional R-loops may be the source of the relevant rNMPs (49).

Replacement of Thymine with Uracil in Transcriptionally Active DNA

Hydrolytic or enzymatic release of a base from the phosphodiester backbone generates an apurinic/ apyrimidinic (AP) site that is a potent block to both DNA and RNAPs. Genetic studies with a frameshift reversion assay revealed that TAM increased when AP-site repair was disrupted, indicating that AP sites are one type of damage that initiates TAM (67). Yeast has five DNA N-glycosylases that remove abnormal or damaged bases from DNA (reviewed in 9), and each was eliminated to determine its contribution to AP-site formation. The only glycosylase relevant to TAM was uracil N-glycosylase (UNG), which specifically excises uracil from DNA (50). Uracil in DNA can result from cytosine deamination, as noted previously, or can arise through use of dUTP in place of dTTP during DNA synthesis. The relevance of the latter was demonstrated by showing a reduction in TAM upon overproduction of Dut1, an enzyme that hydrolyzes dUTP to prevent its use during DNA synthesis (50). It should be noted that the assays used did not exclude introduction of uracil into transcriptionally active DNA via spontaneous cytosine deamination as well. The reason why elevated levels of uracil are incorporated into DNA under high-transcription conditions remains a subject of investigation.

Does Transcription Contribute to Mutagenesis in Nongrowing or Stressed Cells?

Budding yeast does not have a stress response analogous to the SOS system of bacterial cells, which promotes global mutagenesis through the activation of TLS polymerases (31). Nevertheless, in yeast, as in bacterial cells, starvation for a specific nutrient or provision of a specific carbon source can induce the expression of genes encoding the corresponding biosynthetic or catabolic activities (reviewed in 45). It seems likely that TAM will be relevant to at least some examples of so-called adaptive mutation in yeast (39), but this issue has not been specifically addressed. Indeed, genome-wide positions of DNA turnover in stationary-phase cells have been correlated with transcription in microarray-based analyses (18). Such replication-independent DNA synthesis likely reflects repair reactions, which in turn may reflect transcription-associated damage to the DNA template. A relevant observation may be the recent report that abnormal transcription in non-dividing yeast cells contributes to trinucleotide repeat instability (114). The specific relationship between transcription and the instability of sequences that can adopt non-B secondary structures (e.g., stem-loops, triplexes, or G-quadruplexes) has been the subject of a recent review (7) and is not further considered here.

TRANSCRIPTION-ASSOCIATED MUTAGENESIS IN HIGHER EUKARYOTES

Given the universality of DNA structure and the high conservation in basic DNA metabolic processes, it seems likely that many of the TAM mechanisms documented in microorganisms will extend to higher eukaryotes. Attempts to link elevated transcription to increased forward mutation in a specific target gene have been unsuccessful, however (60, 72). This could reflect either a correspondingly lengthened phenotypic lag or, given the proportionality between transcription and mutagenesis observed in microbial systems, an insufficient level of transcription to detect an effect. More recently, a very specific case of TAM emerged when examining the strand specificity of UV-induced mutations in the *hprt* gene (42). Furthermore, comparative genome analyses suggest a global link between transcription and mutagenesis (35, 79). Although a focus on TAM in individual genes has generally been unsuccessful, evolution has co-opted this process to drive maturation of the vertebrate immune system. The remainder of this review focuses on this specialized case of TAM, in which some of the lessons learned may be more generally applicable.

Switch (S) regions: GC-rich, repetitive regions upstream of the constant segment of Ig heavy-chain exons where class-switch recombination occurs; several kb in length

Features of Somatic Hypermutation and Class-Switch Recombination

Vertebrate antibody genes undergo three genetic alterations that result in antibody maturation (**Figure 5**): SHM, gene conversion (GC), and CSR. We summarize here recent work regarding the transcription dependence of SHM and CSR, focusing mainly on mammalian systems. More extensive discussion of relevant literature and detailed models for the role of transcription in AID-generated hypermutations may be found in other reviews (86, 94, 95).

The variable segment of an Ig molecule interacts with an antigen, and SHM fine-tunes antibody-antigen interactions through point mutations that are confined to an $\sim$1,500-bp variable segment of Ig genes that begins $\sim$150 bp downstream of Ig promoters. CSR is a region-specific recombination event that replaces the default constant segment of the Ig heavy chain (μ) with one of the other constant segments, thereby changing the functional consequence of an antibody-antigen interaction.

AID is a B-cell-specific deaminase that converts cytosines to uracils in ssDNA (10, 14, 20, 91) and is required to initiate SHM and CSR (3, 37, 68). During SHM, AID-generated uracils are either not repaired, leading to CG > TA transitions, or are repaired by error-prone pathways to create other types of mutations. During CSR, processing of AID-generated uracils within noncoding switch (S) regions creates double-strand breaks that initiate genetic rearrangement; collateral base substitutions are also acquired within the S regions. Both SHM and S-region mutations occur at frequencies that are several orders of magnitude higher than the normal somatic mutation frequency and both require transcription of the target sequences.

The Role of Transcription in Somatic Hypermutation and Class-Switch Recombination

AID immunoprecipitates with RNAP II (70), and ChIP-seq analysis demonstrates that AID associates with nearly 6,000 genes in stimulated murine B cells (112). Genes associated with AID have a corresponding mRNA abundance 40 times greater than that of genes that did not recruit the protein. In transcriptionally active genes, AID and RNAP II peaked at the transcription start site, and AID occupancy mirrored RNAP II density along individual genes. Although most of the non-Ig genes that recruited AID were not hypermutated in stimulated wild-type B cells, mutations did accumulate in a UNG$^{-/-}$ background (112). These and earlier results (8, 64, 75, 89) demonstrate that AID targets many RNAP II–transcribed genes but that the level of deamination-associated, off-target mutagenesis is much lower in other genes than in Ig genes.

A key question regarding the transcription dependence of SHM is how AID is recruited to transcribed DNA. At least three possibilities have been considered, and these are not mutually exclusive: (*a*) transcription promotes formation of non-B DNA structures to which AID preferentially

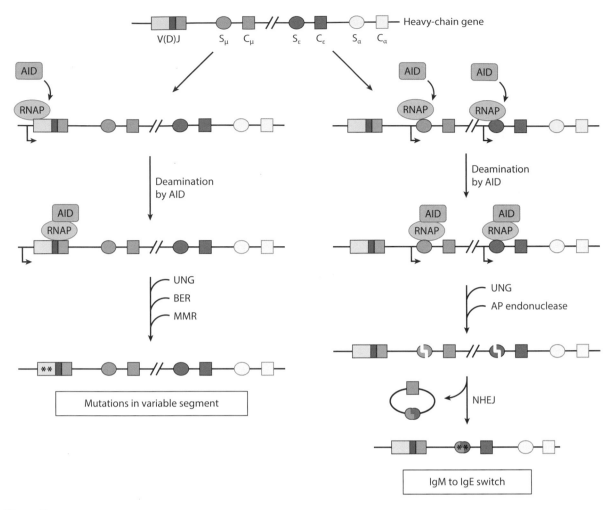

Figure 5

A model for somatic hypermutation (SHM) and class-switch recombination (CSR) during antibody maturation. Different sequence elements are shown as rectangles or ovals of different colors. The direction of transcription is indicated by a rightward arrow. AID travels with RNAP (RNA polymerase) II as it transcribes DNA and converts cytosine to uracil on each DNA strand. Uracil is excised by uracil N-glycosylase (UNG) and processed by error-prone base-excision repair (BER) or by mismatch repair (MMR) to introduce point mutations, which are indicated by asterisks. To initiate CSR, double-strand breaks are mainly generated by AP endonuclease (65) incision at UNG-generated apurinic/apyrimidinic (AP) sites. Broken ends are ligated by the nonhomologous end-joining (NHEJ) pathway, and the intervening DNA is released as a switch circle. Abbreviations: AID, activation-induced deaminase; C, constant segment (only the μ, ε, α regions are shown); S, switch region preceding each C segment; V(D)J, variable segment.

binds, (*b*) transcription-associated chromatin modifications recruit AID, and (*c*) specific transcription-associated protein factors recruit AID. As mentioned previously, transcription can generate a variety of non-B DNA structures. It has been suggested, for example, that formation of stem-loop structures in the variable segment of Ig genes sensitizes bases in single-strand loops to deamination by AID (107, 109). Within S regions, the asymmetric distribution of guanines on the NTS promotes R-loop formation (113), with the displaced strand furthermore having the ability to form G-quadruplex DNA (21, 22). AID may interact specifically with the ssDNA within these structures, or these structures may cause the elongating RNAP II to pause or stall. AID may also interact with DNA in stalled transcription bubbles (23). One difficulty in explaining SHM based on transcription-associated, non-B DNA structures alone is that these structures are inherently asymmetric and primarily affect only one DNA strand. By contrast, AID-associated cytosine deamination lacks a strand bias (89, 110), and closely spaced nicks on both DNA strands of S regions are likely required to create the double-strand breaks required for CSR. One possibility is that negative supercoiling behind the transcription machinery renders both DNA strands accessible to AID, and this might explain the inverse correlation between Top1 level and SHM frequency (54).

There are complex changes in chromatin regions of Ig genes during antibody maturation, and these are more fully discussed in other reviews (for example, 57). It is useful to note, however, that in heavy-chain genes, different S regions are transcribed in response to different cytokine stimuli, and only the transcribed regions undergo switching and hypermutation (110). Although activating histone marks and germline transcripts are found in the S_μ region even in unstimulated B cells, active hypermutation of other S regions requires cytokine activation of B cells (69). Interestingly, treatment of hypermutating cells with the histone deacetylase inhibitor, trichostatin A, increased histone H4 acetylation and resulted in hypermutation of a normally unmutated constant-segment exon (102).

Numerous proteins or protein complexes have reported interactions with AID (56). Many of the AID-associated nuclear proteins are involved in transcription and RNA processing, and it is possible that AID is part of transcription "factories" that contain many transcription-related protein factors (73). The DRB sensitivity-inducing factor (DSIF) complex, for example, causes RNAP II to pause shortly after promoter clearance and, following release from this pause, DSIF travels with the elongating polymerase (38, 78, 111). AID interacts with the Spt5 component of DSIF in vitro and in vivo and the co-occupancy of genes with AID and RNAP II depends on the presence of Spt5 (77). Although Spt5 occupancy of genes correlates with hypermutation frequencies associated with CSR (77), recent work demonstrated that Spt5 knockdown slightly increased SHM (99). With regard to SHM, it was suggested that knockdown of Spt5 reduced RNAP II processivity, promoting transcription termination and RNA degradation by the exosome.

Components of the RNA exosome complex interact with AID (5), and their knockdown reduces both CSR (5) and SHM (99). Furthermore, addition of RNA exosome-enriched extracts to an in vitro transcription system enhances the ability of AID to target both DNA strands (5). An attractive model is that degradation of pre-mRNA by the RNA exosome upon stalling of RNAP II makes both DNA strands accessible to AID. Several other factors may also play a role in AID targeting. ChIP-Seq analysis has shown that replication protein A (RPA), an AID cofactor in CSR, is localized to Ig switch regions but not to most non-Ig AID-targeted genes (112). In addition, depletion of a specific isoform of the pre-mRNA splicing factor SRSF1 (serine/arginine rich splicing factor-1) has been reported to suppress SHM in chicken DT40 cells without affecting off-target mutagenesis. It was suggested that an isoform-associated reduction of splicing specifically at variable segments creates cotranscriptional R-loops, thereby generating the requisite ssDNA substrates for AID (47).

It is likely that local changes in DNA structure and chromatin remodeling, along with the help of protein chaperones and enzymes, together conspire to hypermutate a transcribed gene in an AID-dependent manner. Although the focus here has been on the specific relationship between AID and transcription during targeted mutagenesis in the vertebrate immune system, it is important to note that AID is only one member of the larger APOBEC (apolipoprotein B mRNA editing enzyme, catalytic polypeptide-like) family of cytosine deaminases (90). APOBECs have been implicated in the generation of clustered mutations (kataegis) in tumor cells (83), with persistent ssDNA generated during double-strand break repair providing the substrate for a similar phenomenon

Replication protein A (RPA): heterotrimeric complex that binds ssDNA and prevents pairing between complementary strands

Apolipoprotein B mRNA editing enzyme, catalytic polypeptide-like (APOBEC): family of vertebrate enzymes that deaminate cytosines in RNA or DNA, and includes AID

in budding yeast (12, 71, 84, 85). It seems likely that the ssDNA component of cotranscriptional DNA structures will also be a target of APOBECs as well as endogenous DNA damage.

SUMMARY POINTS

1. Transcription creates transient regions of ssDNA, which is more chemically reactive and damage accessible than duplex DNA. ssDNA exists within the transcription bubble created by RNAP, is enhanced by negative supercoiling, and is associated with R-loop structures.

2. In bacterial cells, TAM preferentially targets the NTS of active genes and is influenced by the direction of replication-fork movement. Importantly, transcription may be relevant to stress responses and adaptation to adverse/novel environments.

3. In budding yeast, transcription elevates all mutation types. Documented causes of TAM include an increase in associated DNA damage, an elevation in direct dUMP incorporation into the underlying DNA template, and recruitment of Top1 to relieve associated supercoiling.

4. Genetic alterations associated with SHM and CSR in the vertebrate immune system provide an example of how transcription regulates genetic instability.

5. Comparative genome analyses suggest that transcription modifies the mutation landscape in both prokaryotes and eukaryotes on an evolutionary timescale.

DISCLOSURE STATEMENT

The authors are not aware of any affiliations, memberships, funding, or financial holdings that might be perceived as affecting the objectivity of this review.

ACKNOWLEDGMENTS

A.S.B. would like to thank Sophia Shalhout (Wayne State University) for comments on the manuscript. S.J.R. gratefully acknowledges the intellectual and experimental contributions of past and present lab members. A.S.B. has been supported by National Institutes of Health grant GM057200 and by a Grants-Plus award from Wayne State University. Transcription-related work in the S.J.R. lab has been funded by National Institutes of Health grants GM038464, GM093197, and GM101690.

LITERATURE CITED

1. Aguilera A, Garcia-Muse T. 2012. R loops: from transcription byproducts to threats to genome stability. *Mol. Cell* 46:115–24

2. Alexander MP, Begins KJ, Crall WC, Holmes MP, Lippert MJ. 2013. High levels of transcription stimulate transversions at GC base pairs in yeast. *Environ. Mol. Mutagen.* 54:44–53

3. Arakawa H, Hauschild J, Buerstedde JM. 2002. Requirement of the activation-induced deaminase (AID) gene for immunoglobulin gene conversion. *Science* 295:1301–6

4. Azvolinsky A, Giresi PG, Lieb JD, Zakian VA. 2009. Highly transcribed RNA polymerase II genes are impediments to replication fork progression in *Saccharomyces cerevisiae*. *Mol. Cell* 34:722–34

5. Basu U, Meng FL, Keim C, Grinstein V, Pefanis E, et al. 2011. The RNA exosome targets the AID cytidine deaminase to both strands of transcribed duplex DNA substrates. *Cell* 144:353–63

6. Beletskii A, Bhagwat AS. 1996. Transcription-induced mutations: increase in C to T mutations in the nontranscribed strand during transcription in *Escherichia coli*. *Proc. Natl. Acad. Sci. USA* 93:13919–24

7. Belotserkovskii BP, Mirkin SM, Hanawalt PC. 2013. DNA sequences that interfere with transcription: implications for genome function and stability. *Chem. Rev.* 113:8620–37

8. Bemark M, Neuberger MS. 2000. The c-MYC allele that is translocated into the IgH locus undergoes constitutive hypermutation in a Burkitt's lymphoma line. *Oncogene* 19:3404–10

9. Boiteux S, Jinks-Robertson S. 2013. DNA repair mechanisms and the bypass of DNA damage in *Saccharomyces cerevisiae*. *Genetics* 193:1025–64

10. Bransteitter R, Pham P, Scharff MD, Goodman MF. 2003. Activation-induced cytidine deaminase deaminates deoxycytidine on single-stranded DNA but requires the action of RNase. *Proc. Natl. Acad. Sci. USA* 100:4102–7

11. Burkala E, Reimers JM, Schmidt KH, Davis N, Wei P, Wright BE. 2007. Secondary structures as predictors of mutation potential in the *lacZ* gene of *Escherichia coli*. *Microbiology* 153:2180–89

12. Burns MB, Lackey L, Carpenter MA, Rathore A, Land AM, et al. 2013. APOBEC3B is an enzymatic source of mutation in breast cancer. *Nature* 494:366–70

13. Cerritelli SM, Crouch RJ. 2009. Ribonuclease H: the enzymes in eukaryotes. *FEBS J.* 276:1494–505

14. Chaudhuri J, Tian M, Khuong C, Chua K, Pinaud E, Alt FW. 2003. Transcription-targeted DNA deamination by the AID antibody diversification enzyme. *Nature* 422:726–30

15. Cho JE, Kim N, Li YC, Jinks-Robertson S. 2013. Two distinct mechanisms of topoisomerase 1–dependent mutagenesis in yeast. *DNA Repair* 12:205–11

16. Datta A, Jinks-Robertson S. 1995. Association of increased spontaneous mutation rates with high levels of transcription in yeast. *Science* 268:1616–19

17. Davis BD. 1989. Transcriptional bias: a non-Lamarckian mechanism for substrate-induced mutations. *Proc. Natl. Acad. Sci. USA* 86:5005–9

18. de Morgan A, Brodsky L, Ronin Y, Nevo E, Korol A, Kashi Y. 2010. Genome-wide analysis of DNA turnover and gene expression in stationary-phase *Saccharomyces cerevisiae*. *Microbiology* 156:1758–71

19. Deshpande AM, Newlon CS. 1996. DNA replication fork pause sites dependent on transcription. *Science* 272:1030–33

20. Dickerson SK, Market E, Besmer E, Papavasiliou FN. 2003. AID mediates hypermutation by deaminating single stranded DNA. *J. Exp. Med.* 197:1291–96

21. Duquette ML, Handa P, Vincent JA, Taylor AF, Maizels N. 2004. Intracellular transcription of G-rich DNAs induces formation of G-loops, novel structures containing G4 DNA. *Genes Dev.* 18:1618–29

22. Duquette ML, Pham P, Goodman MF, Maizels N. 2005. AID binds to transcription-induced structures in c-MYC that map to regions associated with translocation and hypermutation. *Oncogene* 24:5791–98

23. Eddy J, Vallur AC, Varma S, Liu H, Reinhold WC, et al. 2011. G4 motifs correlate with promoter-proximal transcriptional pausing in human genes. *Nucleic Acids Res.* 39:4975–83

24. ENCODE Proj. Consort., Bernstein BE, Birney E, Dunham I, Green ED, et al. 2012. An integrated encyclopedia of DNA elements in the human genome. *Nature* 489:57–74

25. Fix D, Canugovi C, Bhagwat AS. 2008. Transcription increases methylmethane sulfonate-induced mutations in *alkB* strains of *Escherichia coli*. *DNA Repair* 7:1289–97

26. Fix DF, Glickman BW. 1987. Asymmetric cytosine deamination revealed by spontaneous mutational specificity in an Ung- strain of *Escherichia coli*. *Mol. Gen. Genet.* 209:78–82

27. Fraenkel S, Mostoslavsky R, Novobrantseva TI, Pelanda R, Chaudhuri J, et al. 2007. Allelic "choice" governs somatic hypermutation in vivo at the immunoglobulin kappa-chain locus. *Nat. Immunol.* 8:715–22

28. Francino MP, Chao L, Riley MA, Ochman H. 1996. Asymmetries generated by transcription-coupled repair in enterobacterial genes. *Science* 272:107–9

29. Francino MP, Ochman H. 2001. Deamination as the basis of strand-asymmetric evolution in transcribed *Escherichia coli* sequences. *Mol. Biol. Evol.* 18:1147–50

30. Frederico LA, Kunkel TA, Shaw BR. 1990. A sensitive genetic assay for the detection of cytosine deamination: determination of rate constants and the activation energy. *Biochemistry* 29:2532–37

6. Demonstrated that cytosine deamination is more frequent on the NTS in *E. coli*.

31. Galhardo RS, Hastings PJ, Rosenberg SM. 2007. Mutation as a stress response and the regulation of evolvability. *Crit. Rev. Biochem. Mol. Biol.* 42:399–435

32. Gnatt AL, Cramer P, Fu J, Bushnell DA, Kornberg RD. 2001. Structural basis of transcription: an RNA polymerase II elongation complex at 3.3 Å resolution. *Science* 292:1876–82

33. Gomez-Gonzalez B, Aguilera A. 2007. Activation-induced cytidine deaminase action is strongly stimulated by mutations of the THO complex. *Proc. Natl. Acad. Sci. USA* 104:8409–14

34. Gowrishankar J, Harinarayanan R. 2004. Why is transcription coupled to translation in bacteria? *Mol. Microbiol.* 54:598–603

35. Green P, Ewing B, Miller W, Thomas PJ, NISC Comp. Seq., Green ED. 2003. Transcription-associated mutational asymmetry in mammalian evolution. *Nat. Genet.* 33:514–17

36. Hanawalt PC, Spivak G. 2008. Transcription-coupled DNA repair: two decades of progress and surprises. *Nat. Rev. Mol. Cell Biol.* 9:958–70

37. Harris RS, Sale JE, Petersen-Mahrt SK, Neuberger MS. 2002. AID is essential for immunoglobulin V gene conversion in a cultured B cell line. *Curr. Biol.* 12:435–38

38. Hartzog GA, Fu J. 2013. The Spt4-Spt5 complex: a multi-faceted regulator of transcription elongation. *Biochim. Biophys. Acta* 1829:105–15

39. Heidenreich E. 2007. Adaptive mutation in *Saccharomyces cerevisiae*. *Crit. Rev. Biochem. Mol. Biol.* 42:285–311

40. Helmrich A, Ballarino M, Nudler E, Tora L. 2013. Transcription-replication encounters, consequences and genomic instability. *Nat. Struct. Mol. Biol.* 20:412–18

41. Helmrich A, Ballarino M, Tora L. 2011. Collisions between replication and transcription complexes cause common fragile site instability at the longest human genes. *Mol. Cell* 44:966–77

42. Hendriks G, Calleja F, Besaratinia A, Vrieling H, Pfeifer GP, et al. 2010. Transcription-dependent cytosine deamination is a novel mechanism in ultraviolet light-induced mutagenesis. *Curr. Biol.* 20:170–75

43. Herman RK, Dworkin NB. 1971. Effect of gene induction on the rate of mutagenesis by ICR-191 in *Escherichia coli*. *J. Bacteriol.* 106:543–50

44. Hicks WM, Kim M, Haber JE. 2010. Increased mutagenesis and unique mutation signature associated with mitotic gene conversion. *Science* 329:82–85

45. Hinnebusch AG. 1992. General and pathway-specific regulatory mechanisms controlling the synthesis of amino acid biosynthesis enzymes in *Saccharomyces cerevisiae*. In *The Molecular and Cellular Biology of the Yeast* Saccharomyces: *Gene Expression*, ed. EW Jones, JR Pringle, JR Broach, pp. 319–414. Cold Spring Harbor, NY: Cold Spring Harbor Lab. Press

46. Hudson RE, Bergthorsson U, Ochman H. 2003. Transcription increases multiple spontaneous point mutations in *Salmonella enterica*. *Nucleic Acids Res.* 31:4517–22

47. Kanehiro Y, Todo K, Negishi M, Fukuoka J, Gan W, et al. 2012. Activation-induced cytidine deaminase (AID)-dependent somatic hypermutation requires a splice isoform of the serine/arginine-rich (SR) protein SRSF1. *Proc. Natl. Acad. Sci. USA* 109:1216–21

48. Kim N, Abdulovic AL, Gealy R, Lippert MJ, Jinks-Robertson S. 2007. Transcription-associated mutagenesis in yeast is directly proportional to the level of gene expression and influenced by the direction of DNA replication. *DNA Repair* 6:1285–96

49. Kim N, Cho JE, Li YC, Jinks-Robertson S. 2013. RNA:DNA hybrids initiate quasi-palindrome-associated mutations in highly transcribed yeast DNA. *PLoS Genet.* 9:e1003924

50. Kim N, Jinks-Robertson S. 2009. dUTP incorporation into genomic DNA is linked to transcription in yeast. *Nature* 459:1150–53

51. Kim N, Jinks-Robertson S. 2010. Abasic sites in the transcribed strand of yeast DNA are removed by transcription-coupled nucleotide excision repair. *Mol. Cell. Biol.* 30:3206–15

52. Klapacz J, Bhagwat AS. 2002. Transcription-dependent increase in multiple classes of base substitution mutations in *Escherichia coli*. *J. Bacteriol.* 184:6866–72

53. Klapacz J, Bhagwat AS. 2005. Transcription promotes guanine to thymine mutations in the non-transcribed strand of an *Escherichia coli* gene. *DNA Repair* 4:806–13

33. Examined the interplay between transcription, R-loop formation, and AID expression in yeast.

50. Demonstrated increased incorporation of uracil into transcriptionally active yeast DNA.

54. Kobayashi M, Sabouri Z, Sabouri S, Kitawaki Y, Pommier Y, et al. 2011. Decrease in topoisomerase I is responsible for activation-induced cytidine deaminase (AID)-dependent somatic hypermutation. *Proc. Natl. Acad. Sci. USA* 108:19305–10

55. Korogodin VI, Korogodin VL, Fajszi C, Chepurnoy AI, Mikhova-Tsenova N, Simonyan NV. 1991. On the dependence of spontaneous mutation rates on the functional state of genes. *Yeast* 7:105–17

56. Larijani M, Martin A. 2012. The biochemistry of activation-induced deaminase and its physiological functions. *Semin. Immunol.* 24:255–63

57. Li G, Zan H, Xu Z, Casali P. 2013. Epigenetics of the antibody response. *Trends Immunol.* 34:460–70

58. Li X, Manley JL. 2006. Cotranscriptional processes and their influence on genome stability. *Genes Dev.* 20:1838–47

59. Lin Y, Hubert L Jr, Wilson JH. 2009. Transcription destabilizes triplet repeats. *Mol. Carcinog.* 48:350–61

60. Lippert MJ, Chen Q, Liber HL. 1998. Increased transcription decreases the spontaneous mutation rate at the thymidine kinase locus in human cells. *Mutat. Res.* 401:1–10

61. Lippert MJ, Freedman JA, Barber MA, Jinks-Robertson S. 2004. Identification of a distinctive mutation spectrum associated with high levels of transcription in yeast. *Mol. Cell. Biol.* 24:4801–9

62. Lippert MJ, Kim N, Cho JE, Larson RP, Schoenly NE, et al. 2011. Role for topoisomerase 1 in transcription-associated mutagenesis in yeast. *Proc. Natl. Acad. Sci. USA* 108:698–703

63. Liu LF, Wang JC. 1987. Supercoiling of the DNA template during transcription. *Proc. Natl. Acad. Sci. USA* 84:7024–27

64. Liu M, Duke JL, Richter DJ, Vinuesa CG, Goodnow CC, et al. 2008. Two levels of protection for the B cell genome during somatic hypermutation. *Nature* 451:841–45

65. Masani S, Han L, Yu K. 2013. Apurinic/apyrimidinic endonuclease 1 is the essential nuclease during immunoglobulin class switch recombination. *Mol. Cell Biol.* 33:1468–73

66. Merrikh H, Zhang Y, Grossman AD, Wang JD. 2012. Replication-transcription conflicts in bacteria. *Nat. Rev. Microbiol.* 10:449–58

67. Morey NJ, Greene CN, Jinks-Robertson S. 2000. Genetic analysis of transcription-associated mutation in *Saccharomyces cerevisiae*. *Genetics* 154:109–20

68. Muramatsu M, Kinoshita K, Fagarasan S, Yamada S, Shinkai Y, Honjo T. 2000. Class switch recombination and hypermutation require activation-induced cytidine deaminase (AID), a potential RNA editing enzyme. *Cell* 102:553–63

69. Nagaoka H, Muramatsu M, Yamamura N, Kinoshita K, Honjo T. 2002. Activation-induced deaminase (AID)-directed hypermutation in the immunoglobulin Sμ region: implication of AID involvement in a common step of class switch recombination and somatic hypermutation. *J. Exp. Med.* 195:529–34

70. Nambu Y, Sugai M, Gonda H, Lee CG, Katakai T, et al. 2003. Transcription-coupled events associating with immunoglobulin switch region chromatin. *Science* 302:2137–40

71. Nik-Zainal S, Alexandrov LB, Wedge DC, Van Loo P, Greenman CD, et al. 2012. Mutational processes molding the genomes of 21 breast cancers. *Cell* 149:979–93

72. Palombo F, Kohfeldt E, Calcagnile A, Nehls P, Dogliotti E. 1992. N-methyl-N-nitrosourea-induced mutation in human cells: effects of the transcriptional activity of the target gene. *J. Mol. Biol.* 223:587–94

73. Papantonis A, Cook PR. 2013. Transcription factories: genome organization and gene regulation. *Chem. Rev.* 113:8683–705

74. Park C, Qian W, Zhang J. 2012. Genomic evidence for elevated mutation rates in highly expressed genes. *EMBO Rep.* 13:1123–29

75. Pasqualucci L, Migliazza A, Fracchiolla N, William C, Neri A, et al. 1998. BCL-6 mutations in normal germinal center B cells: evidence of somatic hypermutation acting outside Ig loci. *Proc. Natl. Acad. Sci. USA* 95:11816–21

76. Paul S, Million-Weaver S, Chattopadhyay S, Sokurenko E, Merrikh H. 2013. Accelerated gene evolution through replication-transcription conflicts. *Nature* 495:512–15

77. Pavri R, Gazumyan A, Jankovic M, Di Virgilio M, Klein I, et al. 2010. Activation-induced cytidine deaminase targets DNA at sites of RNA polymerase II stalling by interaction with Spt5. *Cell* 143:122–33

78. Peterlin BM, Price DH. 2006. Controlling the elongation phase of transcription with P-TEFb. *Mol. Cell* 23:297–305

62. Along with Reference 97, reported that Top1 is the major source of TAM in yeast.

70. Demonstrated that AID coimmunoprecipitates with RNAP II.

76. Documented the evolutionary consequences of transcription-replication conflicts in bacteria.

79. Polak P, Querfurth R, Arndt PF. 2010. The evolution of transcription-associated biases of mutations across vertebrates. *BMC Evol. Biol.* 10:187

80. Prado F, Aguilera A. 2005. Impairment of replication fork progression mediates RNA polII transcription-associated recombination. *EMBO J.* 24:1267–76

81. Pybus C, Pedraza-Reyes M, Ross CA, Martin H, Ona K, et al. 2010. Transcription-associated mutation in *Bacillus subtilis* cells under stress. *J. Bacteriol.* 192:3321–28

82. Reimers JM, Schmidt KH, Longacre A, Reschke DK, Wright BE. 2004. Increased transcription rates correlate with increased reversion rates in *leuB* and *argH Escherichia coli* auxotrophs. *Microbiology* 150:1457–66

83. Roberts SA, Gordenin DA. 2014. Clustered and genome-wide transient mutagenesis in human cancers: hypermutation without permanent mutators or loss of fitness. *BioEssays* 36:382–93

84. Roberts SA, Sterling J, Thompson C, Harris S, Mav D, et al. 2012. Clustered mutations in yeast and in human cancers can arise from damaged long single-strand DNA regions. *Mol. Cell* 46:424–35

85. Sakofsky CJ, Roberts SA, Malc E, Mieczkowski PA, Resnick MA, et al. 2014. Break-induced replication is a source of mutation clusters underlying kataegis. *Cell Rep.* 7:1640–48

86. Samaranayake M, Bujnicki JM, Carpenter M, Bhagwat AS. 2006. Evaluation of molecular models for the affinity maturation of antibodies: roles of cytosine deamination by AID and DNA repair. *Chem. Rev.* 106:700–19

87. Savic DJ, Kanazir DT. 1972. The effect of a histidine operator-constitutive mutation on UV-induced mutability within the histidine operon of *Salmonella typhimurium*. *Mol. Gen. Genet.* 118:45–50

88. Schmidt KH, Reimers JM, Wright BE. 2006. The effect of promoter strength, supercoiling and secondary structure on mutation rates in *Escherichia coli*. *Mol. Microbiol.* 60:1251–61

89. Shen HM, Tanaka A, Bozek G, Nicolae D, Storb U. 2006. Somatic hypermutation and class switch recombination in Msh6$^{-/-}$ Ung$^{-/-}$ double-knockout mice. *J. Immunol.* 177:5386–92

90. Smith HC, Bennett RP, Kizilyer A, McDougall WM, Prohaska KM. 2012. Functions and regulation of the APOBEC family of proteins. *Semin. Cell Dev. Biol.* 23:258–68

91. Sohail A, Klapacz J, Samaranayake M, Ullah A, Bhagwat AS. 2003. Human activation-induced cytidine deaminase causes transcription-dependent, strand-biased C to U deaminations. *Nucleic Acids Res.* 31:2990–94

92. Sparks JL, Chon H, Cerritelli SM, Kunkel TA, Johansson E, et al. 2012. RNase H2-initiated ribonucleotide excision repair. *Mol. Cell* 47:980–86

93. Srivatsan A, Tehranchi A, MacAlpine DM, Wang JD. 2010. Co-orientation of replication and transcription preserves genome integrity. *PLoS Genet.* 6:e1000810

94. Storb U. 2014. Why does somatic hypermutation by AID require transcription of its target genes? *Adv. Immunol.* 122:253–77

95. Storb U, Shen HM, Michael N, Kim N. 2001. Somatic hypermutation of immunoglobulin and non-immunoglobulin genes. *Philos. Trans. R. Soc. Lond. B* 356:13–19

96. Strathern JN, Shafer B, McGill CB. 1995. DNA synthesis errors associated with double-strand-break repair. *Genetics* 140:965–72

97. Takahashi T, Burguiere-Slezak G, Van der Kemp PA, Boiteux S. 2011. Topoisomerase 1 provokes the formation of short deletions in repeated sequences upon high transcription in *Saccharomyces cerevisiae*. *Proc. Natl. Acad. Sci. USA* 108:692–97

98. Wang JC. 2002. Cellular roles of DNA topoisomerases: a molecular perspective. *Nat. Rev. Mol. Cell Biol.* 3:430–40

99. Wang X, Fan M, Kalis S, Wei L, Scharff MD. 2014. A source of the single-stranded DNA substrate for activation-induced deaminase during somatic hypermutation. *Nat. Commun.* 5:4137

100. Westover KD, Bushnell DA, Kornberg RD. 2004. Structural basis of transcription: separation of RNA from DNA by RNA polymerase II. *Science* 303:1014–16

101. Wimberly H, Shee C, Thornton PC, Sivaramakrishnan P, Rosenberg SM, Hastings PJ. 2013. R-loops and nicks initiate DNA breakage and genome instability in non-growing *Escherichia coli*. *Nat. Commun.* 4:2115

102. Woo CJ, Martin A, Scharff MD. 2003. Induction of somatic hypermutation is associated with modifications in immunoglobulin variable region chromatin. *Immunity* 19:479–89

97. Along with Reference 62, reported that Top1 is the major source of TAM in yeast.

103. Wright BE. 1996. The effect of the stringent response on mutation rates in *Escherichia coli* K-12. *Mol. Microbiol.* 19:213–19

104. Wright BE. 2004. Stress-directed adaptive mutations and evolution. *Mol. Microbiol.* 52:643–50

105. Wright BE, Longacre A, Reimers JM. 1999. Hypermutation in derepressed operons of *Escherichia coli* K12. *Proc. Natl. Acad. Sci. USA* 96:5089–94

106. Wright BE, Reschke DK, Schmidt KH, Reimers JM, Knight W. 2003. Predicting mutation frequencies in stem-loop structures of derepressed genes: implications for evolution. *Mol. Microbiol.* 48:429–41

107. Wright BE, Schmidt KH, Davis N, Hunt AT, Minnick MF. 2008. II. Correlations between secondary structure stability and mutation frequency during somatic hypermutation. *Mol. Immunol.* 45:3600–8

108. Wright BE, Schmidt KH, Hunt AT, Lodmell JS, Minnick MF, Reschke DK. 2011. The roles of transcription and genotoxins underlying p53 mutagenesis in vivo. *Carcinogenesis* 32:1559–67

109. Wright BE, Schmidt KH, Minnick MF, Davis N. 2008. I. VH gene transcription creates stabilized secondary structures for coordinated mutagenesis during somatic hypermutation. *Mol. Immunol.* 45:3589–99

110. Xue K, Rada C, Neuberger MS. 2006. The in vivo pattern of AID targeting to immunoglobulin switch regions deduced from mutation spectra in msh2$^{-/-}$ ung$^{-/-}$ mice. *J. Exp. Med.* 203:2085–94

111. Yamaguchi Y, Shibata H, Handa H. 2013. Transcription elongation factors DSIF and NELF: promoter-proximal pausing and beyond. *Biochim. Biophys. Acta* 1829:98–104

112. Yamane A, Resch W, Kuo N, Kuchen S, Li Z, et al. 2011. Deep-sequencing identification of the genomic targets of the cytidine deaminase AID and its cofactor RPA in B lymphocytes. *Nat. Immunol.* 12:62–69

113. Yu K, Roy D, Bayramyan M, Haworth IS, Lieber MR. 2005. Fine-structure analysis of activation-induced deaminase accessibility to class switch region R-loops. *Mol. Cell. Biol.* 25:1730–36

114. Zhang Y, Shishkin AA, Nishida Y, Marcinkowski-Desmond D, Saini N, et al. 2012. Genome-wide screen identifies pathways that govern GAA/TTC repeat fragility and expansions in dividing and nondividing yeast cells. *Mol. Cell* 48:254–65

Gastrointestinal Microbiota–Mediated Control of Enteric Pathogens

Sophie Yurist-Doutsch,[1] Marie-Claire Arrieta,[1] Stefanie L. Vogt,[1] and B. Brett Finlay[1,2,3]

[1]Michael Smith Laboratories, The University of British Columbia, Vancouver, British Columbia, Canada, V6T 1Z4; email: yurists@msl.ubc.ca, marrieta@msl.ubc.ca, svogt@msl.ubc.ca, bfinlay@msl.ubc.ca

[2]Department of Microbiology and Immunology, The University of British Columbia, Vancouver, British Columbia, Canada, V6T 1Z4

[3]Department of Biochemistry and Molecular Biology, The University of British Columbia, Vancouver, British Columbia, Canada, V6T 1Z4

Annu. Rev. Genet. 2014. 48:361–82

First published online as a Review in Advance on September 10, 2014

The *Annual Review of Genetics* is online at genet.annualreviews.org

This article's doi: 10.1146/annurev-genet-120213-092421

Keywords

intestinal microbiota, colonization resistance, *Salmonella enterica*, *Escherichia coli*, *Clostridium difficile*

Abstract

The gastrointestinal (GI) microbiota is a complex community of microorganisms residing within the mammalian gastrointestinal tract. The GI microbiota is vital to the development of the host immune system and plays a crucial role in human health and disease. The composition of the GI microbiota differs immensely among individuals yet specific shifts in composition and diversity have been linked to inflammatory bowel disease, obesity, atopy, and susceptibility to infection. In this review, we describe the GI microbiota and its role in enteric diseases caused by pathogenic *Escherichia coli*, *Salmonella enterica*, and *Clostridium difficile*. We discuss the central role of the GI microbiota in protective immunity, resistance to enteric pathogens, and resolution of enteric colitis.

THE GASTROINTESTINAL MICROBIOTA: DEFINITION AND COMPOSITION

The human gastrointestinal tract (GIT) is home to a complex microbial community with as many as 100 trillion prokaryotic members (119). Although mammals are colonized by many single-cell organisms at various anatomical locations, the GIT harbors the highest number and density of bacteria (67). This complex ecosystem is initially acquired during and for some time after birth. Although the GIT composition varies during the first year of life, at around one year of age the GIT microbiota of an individual becomes stable (87) and remains largely unchanged unless perturbed by antimicrobial treatment or disease (18).

The composition of the gastrointestinal (GI) microbiota is highly variable among individuals. Analysis of stool samples collected from healthy subjects showed that the vast majority of bacteria in the GIT belong to only four phyla: Bacteroidetes, Firmicutes, Proteobacteria, and Actinobacteria (27, 30, 38, 65). At the genus level, the diversity is so high that no two stool samples (of more than 250 tested samples in the Human Microbiome Project) have the same composition (38). Within the GIT, microbial communities differ in the stomach, small intestine, cecum, and colon, with higher levels of bacterial diversity in the lower segments of the GIT and higher levels of aerobic and facultative anaerobic bacteria in the stomach and upper small intestine (30, 35). Additionally, the bacteria present in stool differ from those that are adherent to the GI mucosa (27).

Despite the high variability in GIT microbiota composition among individuals, metagenomic analysis reveals high levels of similarity between subjects when comparing various metabolic pathways within microbiota genes (38, 57), suggesting a high level of metabolic redundancy among different bacterial taxa.

IMPACT OF THE GASTROINTESTINAL TRACT MICROBIOTA ON THE IMMUNE SYSTEM

The Mammalian Immune System

The immune system is defined as the host's defense against destructive forces from both outside (e.g., bacteria, viruses, parasites) and within (e.g., malignant and autoreactive cells) the body. Immune responses are generally classified as either innate or acquired. The components and cells that make up these two arms of the immune system are presented in **Table 1**.

The innate immune system provides immunity to invading organisms without the need for prior exposure to these antigens and includes physical barriers, such as the skin and mucous membranes; cell-mediated barriers, including phagocytic cells, inflammatory cells, dendritic cells, and natural killer cells; and soluble mediators, such as cytokines, complement proteins, and acute-phase proteins (24). This arm of the immune system provides the early phases of host defense that protect the organism during the four to five days it takes for lymphocytes to become activated.

The acquired, or adaptive, immune system develops over an individual's lifetime. Lymphocytes are an important cellular component of this arm of the immune system that modulate the function of other immune cells or directly destroy cells infected with intracellular pathogens (**Table 1**). Each developing T or B cell generates a unique receptor, or recognition molecule, such that a set of cells expressing a vast array of diverse receptors is produced, allowing immune cells to selectively eliminate virtually any foreign antigen that enters the body (24). B cells, abundant in lymph nodes, recognize foreign antigen through membrane-bound antibodies, or immunoglobulins, and upon activation become antibody-secreting plasma cells to effectively remove soluble bacteria/antigens (24).

Table 1 The immune system

Arm of immune system	Defenses	Components	Functions
Innate immune system	Physical barriers	Skin Mucous membranes	Prevent the entry of antigens into systemic circulation

	Cell-mediated barriers	Phagocytic cells,
		  e.g., neutrophils,
		  macrophages
			Engulf foreign antigens

		Inflammatory cells,
		  e.g., basophils, mast cells
		Natural killer cells
		Dendritic cells
			Release inflammatory mediators,
			  e.g., histamine, prostaglandins

	Soluble factors	Cytokines Complement proteins Acute-phase proteins	Destroy infected or malignant cells Present antigens to lymphocytes
Acquired immune system	B lymphocytes T lymphocytes	Plasma cells	Activate/recruit other cells Enhance phagocytosis Promote repair of damaged tissue
		CD4$^+$ T cells Th1 cells Th2 cells Th17 cells Tregs	Secrete antibodies Induce activation of lymphocytes Promote cell-mediated responses Promote humoral (antibody) responses
		CD8$^+$ T cells Cytotoxic T cells Suppressor T cells	Peripheric tolerance Destroy infected or malignant cells Suppress activity of lymphocytes

T cells express a T-cell receptor (TCR) that recognizes a foreign antigen presented in complex with a major histocompatibility complex (MHC) molecule on the surface of an antigen-presenting cell (APC) (24). Subpopulations of T cells include the helper T (Th) cells, which are identified by the presence of the membrane glycoprotein CD4, and cytotoxic T cells that express the CD8 glycoprotein (24). CD4$^+$ cells are classified into a number of Th types: Th1, Th2, Th17, and Treg (16, 24, 68). Viral infections or microbes that infect macrophages or natural killer (NK) cells elicit a Th1 response. Th1 lymphocytes secrete interferon gamma (IFN-γ) and tumor necrosis factor beta (TNF-β). A Th2 response elicits Th2 lymphocytes in response to helminths, allergens, and extracellular microbes. Th2 lymphocytes produce cytokines interleukin (IL)-4, 5, and 13, among others. Th17 cells are lymphocytes that produce IL-17 to recruit neutrophils and macrophages and to trigger an inflammatory response in different organs in order to remove extracellular pathogens from the body. Additionally, there is a regulatory arm characterized by regulatory T cells (Tregs) that produce IL-10 to downregulate all other immune arms with the objective of preventing an exacerbated and injury-provoking immune response. Achieving a balance between all these arms is what enables an adequate immune response to fight pathogens and prevents immune-mediated diseases, such as diabetes type 1, rheumatoid arthritis, inflammatory bowel disease (IBD), allergies, asthma, etc.

The Gut-Associated Lymphoid Tissues

The mucosal immune system is strategically located in areas where external pathogens and antigens may gain access to the body. This includes the mucosal-associated lymphoid tissues, which

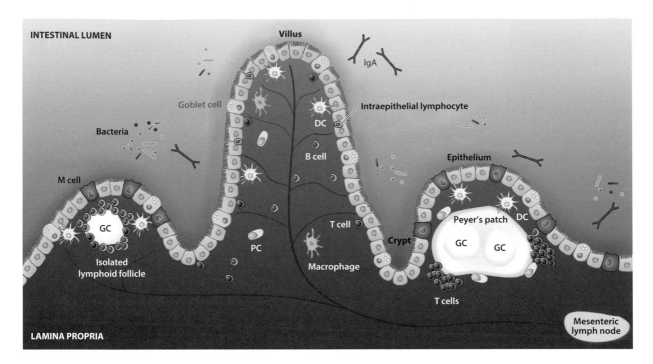

Figure 1

Diagram of the gut-associated lymphoid tissue (GALT). The GALT consists of aggregated lymphoid follicles (Peyer's patches and isolated lymphoid follicles) and diffuse or nonaggregated tissue (lamina propria). The epithelium of aggregated follicles contains M cells, which are specialized in antigen sampling. Underneath is the subepithelial dome, which consists of germinal centers (GCs) of B cells in different stages of maturation, as well as dendritic cells (DCs) and T cells. Aggregated follicles are the inductive sites of immunity, where naïve T and B cells are activated to then migrate to the mesenteric lymph node and into the effector sites of the GALT, the lamina propria. Within the lamina propria, mature B cells or plasma cells (PCs) produce IgA, which gets actively transported into the intestinal lumen. DCs in the lamina propria also sample antigens and present them directly to T and B cells. Macrophages are important innate immune cells capable of engulfing and killing microorganisms by phagocytosis. Figure adapted from Reference 2.

protect sites such as the respiratory, urinary, and reproductive tracts, and the gut-associated lymphoid tissues (GALTs), which protect the intestine. As the intestine is the first line of defense from the environment and must integrate complex interactions among diet, external pathogens, and local immunological processes, it is critical that protective immune responses are mounted against potential pathogens, yet it is equally important that hypersensitivity reactions to dietary and commensal microbial antigens are minimized. The GALT is composed of aggregated tissue in the form of Peyer's patches and solitary lymphoid follicles, nonaggregated cells in the lamina propria and intraepithelial regions of the intestine, and mesenteric lymph nodes (58) (**Figure 1**).

Peyer's patches are aggregates of lymphoid follicles found throughout the mucosa and submucosa of the small intestine. These patches contain both CD4$^+$ and CD8$^+$ T cells, as well as naïve B cells, plasma cells, macrophages, and dendritic cells (58). Overlying the Peyer's patches are specialized epithelial cells known as M cells, which endocytose, transport, and release antigens from the gut into the Peyer's patches, where these antigens are presented on APCs to T and B cells (50, 58).

The lamina propria harbors a diffuse population of T and B cells, plasma cells, dendritic cells, mast cells, and macrophages, all covered by a single layer of epithelial cells (58). Intestinal epithelial cells (IECs) or enterocytes are intimately involved in digestion, absorption, and transport of nutrients but also play a major role in immune regulation of the underlying cells of the lamina propria.

IECs are known to produce different chemokines and cytokines depending on the type of microbial molecules that come in contact with the epithelium (40). IECs have been shown to condition the immune response of dendritic cells by inducing, through an unknown mechanism, the secretion of tolerogenic signals (IL-10) or active immunity signals (IL-12) by the dendritic cells (91). IECs are also involved in antigen transport to underlying antigen-presenting cells, or they can present the antigen themselves to lymphocytes.

The Interactions Between the Microbiota and the Gut-Associated Lymphoid Tissues

Life without symbiotic microorganisms has a profound effect on the host immune system. Experiments in germ-free mice have consistently shown that many components of the immune system remain underdeveloped until bacterial colonization occurs. Structurally, the aggregated lymphoid structures of the GALT, mesenteric lymph nodes (MLNs), and isolated lymphoid follicles and Peyer's patches are reduced in size and cellularity in germ-free mice (11, 72). Functionally, these mice are depleted in the production of many cytokines, CD4[+] T helper cells, Tregs, B cells, Th17 cells, and antimicrobial peptides, and in expression of the MHC class II, etc. (26, 31, 43, 77, 83, 92, 107). Many of these defects can be reverted upon inoculation with a single species of bacteria (72, 112), indicating that it is the interactions with the microbiota that kick-start the postnatal phase of immune development. Gut bacteria also have critical effects on the development of IECs. IECs express microbial pattern recognition receptors (PRRs) that bind to microbial signals and upregulate the expression of mucus, cytokines, and other immune components (90). In axenic mice, IECs have blunted microvilli, and they display reduced expression of PRRs (1, 122).

Colonization of the host GIT by the microbiota produces inflammatory and tolerogenic signals that stimulate and regulate the different arms of the immune response in order to achieve homeostasis, a state that is favorable to both the microbiota and the host. The Th17 immune response provides a good example of these opposing functions. Despite the fact that Th17 cells are critical in controlling extracellular bacterial and fungal infections (100, 126), overproduction of Th17 cytokines IL-17 and IL-23 is associated with colitis (3, 13) and autoimmunity (64, 118, 126). Thus, the appropriate level of the Th17 response must be achieved to prevent microbial attack and avoid uncontrolled inflammation. An example of a member of the microbiota involved in regulating the Th17 response has been described in segmented filamentous bacteria (SFB), recently named *Candidatus* Savagella. SFB are a clostridial species that resides in the murine ileum and are a potent stimulator of the Th17 response (31, 126). Colonization with SFB induced the upregulation of cytokines, antimicrobial peptides, and serum amyloid A (SAA), an acute-phase protein secreted during inflammation. SAA is believed to promote the Th17 differentiation of CD4[+] T cells (5, 42). Not surprisingly, microbial signals have also been found to limit the Th17 immune response. Unspecified commensal bacteria induce IECs to produce IL-25, which inhibits IL-23 production by dendritic cells, thus preventing Th17 differentiation (128). In addition, polysaccharide A (PSA) from *Bacteroides fragilis* downregulated the Th17 response and induced propagation of Tregs in mice that are genetically predisposed to develop autoimmune encephalomyelitis. This treatment was sufficient to prevent disease in this animal model (85). These examples show how microbial signals can act as the tipping forces that increase or decrease Th17 immune cell activation.

Although certain microbes induce a proinflammatory response and others a tolerogenic one, how the immune response differentiates between them is unknown. One way that microbiota may induce a tolerogenic immune response is via their interactions with Tregs. Tregs have a central role in regulating inflammatory responses, and they have been involved in the modulation of disease in animal models of asthma (94) and colitis (4) upon changes in the intestinal microbiome. Although a large proportion of Tregs are developed in the thymus with the purpose of generating tolerance toward self-antigens and preventing autoimmune responses, a subset of Tregs are trained by the GIT microbiota. Analysis of several repertoires of the TCR α chain revealed that the TCRs from Treg cells of the colonic lamina propria are highly heterogeneous compared with Treg cells from other lymphoid organs. Moreover, the study showed that the mouse microbiota was essential for the induction of this particular population of colonic Treg cells from naïve T cells, implying that there may be post-thymic mechanisms of immune cell education that occur via interactions with the commensal microbiota (60). Thus, besides protecting the host from microbial attack, postnatal bacterial colonization may serve the immune system with another important function: to be used as a training ground for a specific population of immune cells.

THE ROLE OF THE MICROBIOTA IN ENTERIC INFECTION

Although the GIT microbiota is often referred to as commensal, with the implication that these organisms are neither harmful nor beneficial to their host, numerous studies clearly demonstrate that the composition and diversity of the GIT microbiota are crucial to host health and disease. Increasing numbers of studies provide evidence for the importance of the GIT microbiota in many human disorders, including immune-mediated and metabolic diseases such as inflammatory bowel disease (30, 54, 70, 120), obesity (65, 110, 129), asthma, atopy (10, 51, 93), etc. In the following sections, we focus on the key role of the GIT microbiota in the susceptibility to and recovery from enteric infection.

Colonization Resistance

The microbiota can determine host susceptibility to bacterial infections. Germ-free and antibiotic-treated mice are significantly more susceptible to infectious disease caused by intestinal pathogens, including *Shigella flexneri* (103), *Citrobacter rodentium* (125), *Listeria monocytogenes* (127), and *Salmonella enterica* (29, 81, 99). The microbiota can prevent or ameliorate infection by direct microbial antagonism or by indirectly promoting appropriate host immune defenses. Direct microbial effects, also defined as colonization resistance (114), include competition for nutrients and host receptors and secretion of antimicrobial substances (97). For example, the probiotic strain *Escherichia coli* Nissle 1917 reduces *Salmonella* Typhimurium colonization of the mouse intestine by competing for iron, a limiting nutrient in vivo (25). As an example of indirect effects via immune modulation, administration of the Toll-like receptor (TLR) 5 ligand flagellin to antibiotic-treated mice inhibited colonization with vancomycin-resistant *Enterococcus* (VRE) by upregulating the secretion of the bactericidal lectin REGIIIγ in IECs and Paneth cells (56). Flagellin stimulation has also been shown to induce differentiation of nonspecific IgA$^+$ plasma cells via activation of TLR5 in intestinal dendritic cells (111). Colonization with *B. fragilis* is sufficient to prevent murine experimental colitis induced by *Helicobacter hepaticus*. *B. fragilis* is coated with PSA, and this capsular component was shown to suppress IL-17, increase IL-10 production, and ameliorate disease in this model (73). Collectively, these findings suggest that the GI microbiota alter the course of bacterial infections both directly and indirectly via the host's immune response. In the following section, we discuss in detail interactions between the GIT microbiota and three major enteric pathogens: *E. coli*, *S.* Typhimurium, and *Clostridium difficile*.

Attaching/Effacing Pathogens: Enterohemorrhagic *Escherichia coli*–Enteropathogenic *Escherichia coli*–*Citrobacter rodentium*

E. coli is a common commensal inhabitant of the mammalian intestinal tract that can become a GI or extraintestinal pathogen by genetically acquiring pathogenicity islands and plasmids (21). Among the several *E. coli* pathotypes that cause enteric infections, two of the best studied are enteropathogenic *E. coli* (EPEC) and enterohemorrhagic *E. coli* (EHEC). Both EPEC and EHEC are attaching and effacing (A/E) pathogens, which are characterized by their ability to adhere tightly to the host intestinal epithelium and cause localized destruction (effacement) of microvilli (20, 53). These phenotypes are conferred by the chromosomal pathogenicity island known as the locus of enterocyte effacement (LEE), which encodes a type III secretion system (T3SS) and a variety of effector proteins that are secreted into host cells, leading to rearrangement of the actin cytoskeleton (95). Additionally, EHEC but not EPEC carries the *stx* genes, encoding the Shiga toxin that is responsible for the more severe outcomes of EHEC infection, such as hemolytic uremic syndrome (21). Because EPEC and EHEC are human-specific pathogens that do not cause representative infections in model hosts such as mice (78), *C. rodentium*, which carries the LEE and is a natural pathogen of mice, is frequently used as a surrogate to study these host-pathogen and microbiota-pathogen interactions and mechanisms (79).

C. rodentium infections of mice have provided convincing evidence that the microbiota plays a critical role in determining the outcome of infection with A/E pathogens. As one extreme example, germ-free mice have an impaired ability to clear *C. rodentium* infections. Specific pathogen-free C57BL/6 mice, possessing an intact microbiota, clear *C. rodentium* infections within 22 days, whereas germ-free mice of the same strain remain colonized with *C. rodentium* even 42 days post infection (dpi) (52). Less drastic perturbations of the microbiota, such as treatment with antibiotics, can also affect the outcome of infection. Treating mice with metronidazole prior to *C. rodentium* infection leads to more severe cecal inflammation 6 dpi (125). Notably, the overall abundance of gut microbiota was not altered by the metronidazole treatment in this study, suggesting that specific microbes whose abundance was altered by metronidazole play an important role in protecting or predisposing the host to *C. rodentium*–induced colitis (125).

Even differences in microbiota composition between strains of mice can have drastic effects on *C. rodentium* infection. Mouse strains have long been known to vary in their susceptibility to *C. rodentium*: Resistant strains, such as NIH Swiss and C57BL/6, survive and clear the infection, whereas *C. rodentium* infection is lethal to susceptible strains, such as C3H/HeJ (113). In order to assess whether differences in gut microbiota composition in these mouse strains are responsible for the different infection outcomes, microbiota (fecal) transplantation was performed. In these experiments, recipient mice were first treated with antibiotics to deplete their natural microbiota, then repeatedly gavaged with donor fecal extracts from either the same or a different strain of mice to establish a new gut microbial community. Remarkably, transplantation of resistant microbiota into a susceptible recipient significantly increased survival of *C. rodentium* infection compared with mock-treated susceptible mice (33, 121).

There are several mechanisms by which the microbiota can affect the ability of A/E pathogens to infect their hosts (**Figure 2a**). One of these mechanisms is competition for nutrients. Other enterobacteria, particularly commensal *E. coli*, are especially effective at competing for nutrients with A/E pathogens. Indeed, precolonization of mice with three commensal strains of *E. coli* prevents subsequent colonization by EHEC (63). There are likely several limiting nutrients for which

Figure 2

The gastrointestinal (GI) microbiota during *Citrobacter rodentium* infection in mice. (*a*) Early stage of infection: *C. rodentium* enters the GI tract (GIT), where short-chain fatty acids (SCFAs) (as well as other potential mechanisms) lead to an increase in T3SS (type III secretion system) expression and attachment to the epithelium. Variation in GI microbiota composition between mice strains results in varying susceptibility to the pathogen. A GI microbiota poor in commensal *Escherichia coli* allows superior pathogen proliferation in the GIT. (*b*) Late stage of infection: Two populations of *C. rodentium* are present in the GIT. One is attached to the epithelium and the other replicates in the lumen. Microbiota-mediated increase in mucin secretion by goblet cells helps detach the pathogen and push it out of the GIT. (*c*) End stage of infection: The regeneration of the GIT microbiota leads to effective pathogen exclusion. Abbreviation: A/E, attaching and effacing.

a Early

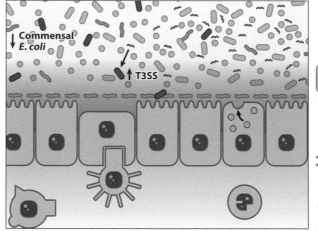

Commensal *E. coli*

T3SS

b Late

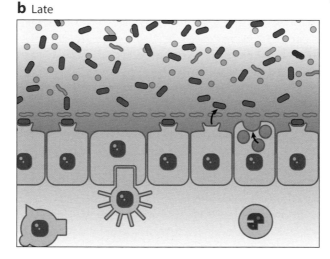

c End

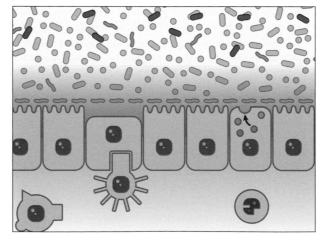

 A/E pathogen

 Enterocyte

 M cell

 Goblet cell

 Dendritic cell

 Macrophage

 Neutrophil

 Segmented filamentous bacteria

 Microbiota

Short-chain fatty acid

EHEC and commensal enterobacteria compete, including proline and mono- and disaccharides (47, 76). The importance of competition for sugars is supported by studies of germ-free mice monocolonized with *C. rodentium* (52). When these mice are subsequently colonized with commensal *E. coli*, which like *C. rodentium* preferentially metabolizes monosaccharides, the intestinal load of *C. rodentium* is reduced by several orders of magnitude. However, colonization with *Bacteroides thetaiotaomicron*, which can metabolize polysaccharides that are not used by *C. rodentium*, fails to decrease *C. rodentium* colonization unless mice are fed on a simple sugar diet that forces the bacteria to compete for monosaccharides (52). Colonization of the epithelial surface, rather than the lumen of the intestine or the outer mucus layer where the microbiota reside (45), may be one strategy that A/E pathogens employ to reduce competition for nutrients. *C. rodentium* T3S mutants that cannot adhere to the host epithelium can colonize germ-free mice but are unable to establish an infection in mice with an intact microbiota (52). Association with the epithelium may allow A/E pathogens to access nutrients that are not available to microbiota in the lumen of the intestine, such as the products of plasma membrane damage (75).

The GIT microbiota also exerts indirect effects on A/E pathogens via the host immune system (**Figure 2b**). The barrier posed by the mucus layer is a central part of the defense against A/E pathogens (and most other enteric pathogens). Mucus secretion from goblet cells in the colon is enhanced during the clearance phase of *C. rodentium* infections, and *Muc2*$^{-/-}$ mice that lack the major intestinal mucin protein have a higher rate of mortality during *C. rodentium* infection than wild-type mice (8, 36). The microbiota appear to play an important role in the maintenance of the mucus layer, as mice treated with the antibiotic metronidazole have a thinner inner mucus layer and are more susceptible to *C. rodentium*–induced colitis (125). The microbiota from resistant mouse strains also promotes the production of proinflammatory cytokines, such as TNF-α and MIP-2α (33, 121). These cytokines enhance host defenses against *C. rodentium* by increasing the production of antimicrobial peptides Reg3β and Reg3γ and by promoting infiltration of neutrophils into the colonic mucosa and submucosa (33, 121). IL-22 appears to be particularly important for protection against *C. rodentium* because treatment of susceptible mice with anti-IL-22 antibodies abrogates the protection normally provided by transplantation with resistant microbiota (121).

In addition to changing the ability of A/E pathogens to survive in the GIT by competition and modulation of host immunity, the microbiota also releases metabolites that alter A/E pathogen gene expression. Short-chain fatty acids (SCFAs) are major microbiota-derived metabolites produced from the anaerobic fermentation of dietary fiber in the cecum and the colon (23). When added to culture media at a concentration similar to that found in the human colon, the SCFAs acetate, propionate, and butyrate activate expression of several EHEC virulence genes, including the LEE-encoded T3S proteins EspB and Tir and the chromosomally encoded adhesin Iha (37, 80). SCFAs, particularly butyrate, produced by intestinal microbiota could therefore be a cue for EHEC to trigger the expression of genes needed for adherence and colonization of the colon. Mucus-derived sugars that are liberated by the microbiota could be another signal that EHEC uses to sense its location. Fucose, which is cleaved from intestinal mucin by species such as *B. thetaiotaomicron*, represses expression of the LEE via the two-component system FusKR (86). Because expression of the LEE represents a significant metabolic burden, this fucose-mediated repression could help EHEC to conserve energy while it is present in the mucus layer rather than in close proximity to the epithelium where the T3SS is needed (86). In addition to these examples, there remain additional uncharacterized metabolites produced by the microbiota that affect EHEC virulence gene expression. For example, growth of EHEC either in a human microbiota–conditioned medium or in the presence of probiotic *Lactobacillus* or *Bifidobacterium* species represses the expression of the Stx2 Shiga toxin (14, 23). Further studies are needed to reveal the full scope of how the microbiota affects infections by A/E pathogens.

Salmonella enterica

Unlike the A/E pathogens that generally remain in the intestinal lumen, *S. enterica* are a group of facultative intracellular pathogens that can invade the host tissue as well as proliferate in the gut and as such can generate illnesses ranging from mild gastroenteritis to acute systemic colonization, also known as typhoid fever. Once the pathogen reaches the host GIT, it uses a T3SS encoded by *Salmonella* pathogenicity island 1 (SPI-1) to invade M cells in Peyer's patches along the small intestine (46). Once internalized, a second T3SS encoded by *Salmonella* pathogenicity island 2 (SPI-2) is activated by the pathogen, allowing it to cross the epithelial barrier. In the lamina propria, *S. enterica* is internalized by macrophages that transfer the pathogen to the internal organs (22). Moreover, *S. enterica* can be sampled from the intestine by dendritic cells. These phagocytic cells internalize the pathogen directly from the gut lumen and are hijacked by *S. enterica* for systemic transfer and colonization (115) (**Figure 3***a*).

Different serovars of *S. enterica* are linked to different manifestations of disease. In typhoidal salmonellosis, caused in humans by serovars Typhi and Paratyphi, the pathogen invades the internal organs without generating intestinal inflammation or significant colonization of the GIT. Non-typhoidal salmonellosis results in bacterial GI colonization accompanied by inflammation and pathology, but there is little to no systemic infiltration of the pathogen. *S. enterica* serovar Typhimurium (*S.* Typhimurium), an agent of gastroenteritis in humans, can be used to model both typhoidal and non-typhoidal salmonellosis in mice; the shift between the two diseases is moderated by the GIT microbiota.

Oral inoculation of C57B/6 mice with *S.* Typhimurium results in high levels of colonization of the liver, spleen, and mesenteric lymph nodes but not intestinal colonization or pathology. However, if mice are treated 24 hours before infection with a high dose of the antibiotic streptomycin (20 mg), the animals develop GI colonization and inflammation (6). Low doses of antibiotic treatment (streptomycin, vancomycin, or metronidazole at doses of 150–600 mg/L of drinking water) also result in increased susceptibility to *S.* Typhimurium gastroenteritis, accompanied by GI pathology and proinflammatory cytokine secretion (TNF-α, MCP-1, KC, IL-6) (29, 99). The low-dose streptomycin treatment alters the GIT microbiota composition and reduces microbial diversity but does not significantly reduce the total bacterial count in the gut (29, 32, 99). Similar to the case with *C. rodentium*, these data suggest that it is not the overall abundance of GIT microbiota but rather their diversity or the presence or absence of certain organisms that determines susceptibility to *S.* Typhimurium gastroenteritis. Antibiotic treatment also affects *S.* Typhimurium colonization of other mouse strains, such as 129 SvJ. Unlike the C57B/6 mice that succumb to *S.* Typhimurium infection within several days, the 129 SvJ mice are more resistant because of host genetic differences and can recover and clear the pathogen. Following *S.* Typhimurium

---

Figure 3

The gastrointestinal (GI) microbiota during *Salmonella enterica* serovar Typhimurium gastroenteritis in mice. (*a*) Early stage of infection: *S.* Typhimurium invades M cells in Peyer's patches or is taken up by dendritic cells. The initial composition of the GI microbiota is critical for *S.* Typhimurium proliferation in the GI tract (GIT). Only mice pretreated with antibiotics or colonized by a low-complexity microbiota develop gastroenteritis following *S.* Typhimurium infection. Microbiota-mediated release of fucose provides the pathogen with a unique carbon source in the post-antibiotic-treated GIT. (*b*) Late stage of infection: *S.* Typhimurium–induced inflammation leads to the release of thiosulfate and ethanolamine, which are used selectively by the pathogen for growth. Furthermore, neutrophil recruitment to the gut lumen and RegIIIβ further reduce the GI microbiota and thus aid *S.* Typhimurium colonization of the GIT. (*c*) End stage of infection: The GI microbiota regenerates and pushes the pathogen out of the host. Abbreviation: T3SS, type III secretion system.

a Early

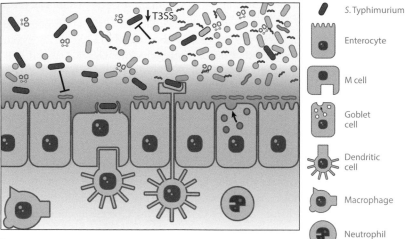

↓T3SS

b Late

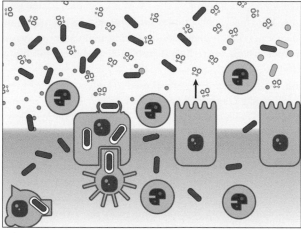

c End

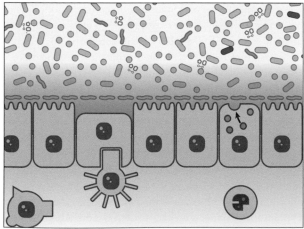

	S. Typhimurium
	Enterocyte
	M cell
	Goblet cell
	Dendritic cell
	Macrophage
	Neutrophil
	Segmented filamentous bacteria
	Microbiota
	Short-chain fatty acid
	Thiosulfate Fucose Ethanolamine
	RegIIIβ

infection, up to 25% of infected 129 SvJ mice develop a supershedder phenotype characterized by long-term GI colonization and inflammation as well as continuous fecal shedding of the pathogen. Treatment of the mice with streptomycin or neomycin (5 mg) before infection resulted in all mice having the supershedder phenotype. Even mice that were allowed to recover for seven days after neomycin treatment, a time frame in which the microbiota generally reverts to the pretreated state, remained susceptible to GI colonization, demonstrating the importance of the microbiota for the ability of 129 SvJ mice to clear *S.* Typhimurium infection (62). Antibiotic treatment was also shown to promote *S.* Typhimurium GIT colonization in the FvB murine strain (19). In this model, increased susceptibility was observed even after three weeks of post-antibiotic recovery.

Perhaps the clearest evidence for the importance of a complex GIT microbiota in preventing/reducing *S.* Typhimurium colonization was shown using low-complexity microbiota (LCM) mice. These animals are bred germ free and are colonized at birth with the Altered Schaedler Flora, a combination of eight bacterial strains that are representative of the major phylogenetic groups found in conventional murine microbiota. The LCM mice have significant reduction in GI microbial diversity (8 versus >500 strains in conventionally bred mice) but similar bacterial counts in the gut. These animals also develop gastroenteritis when infected with *S.* Typhimurium without the need for antibiotic pretreatment (104). When housed with conventionally bred animals (which facilitates microbiota transfer between animals), the LCM mice gained partial resistance to *S.* Typhimurium–induced colitis. The transfer of complex GIT microbiota from the conventionally bred mice to the LCM mice was not complete, and lingering population changes, specifically increased levels of Enterobacteriaceae, were linked with increased susceptibility to *S.* Typhimurium colitis (104).

A simplified model of the post-antibiotic-treated gut was used to understand why it was permissive to *S.* Typhimurium colonization (82). Germ-free mice were colonized with *B. thetaiotaomicron* and infected with *S.* Typhimurium. *S.* Typhimurium isolated from the gut of these mice was shown to upregulate genes involved in sialic acid and fucose catabolism compared with *S.* Typhimurium isolated from uncolonized germ-free mice. Furthermore, mutants in these metabolic pathways have reduced virulence in both the *B. thetaiotaomicron* monocolonized mice and in antibiotic-treated animals. Consistently, the GI content of both the *B. thetaiotaomicron* monocolonized mice and the antibiotic-treated mice contained elevated levels of sialic acid. *B. thetaiotaomicron* is able to free N-acetylneuraminic acid (Neu5Ac) from intestinal mucin by the action of a specific sialidase. Under normal conditions Neu5Ac is used as a carbon source by various members of the GIT microbiota, however in monocolonized or antibiotic-treated mice the absence of Neu5A metabolizing strains leaves the sugar available for *S.* Typhimurium consumption.

Hydrogen is another metabolite that has been shown to promote *S.* Typhimurium proliferation in the gut of mice with a simplified microbiota. LCM mice were infected with an *S.* Typhimurium transposon mutant library, and mutants of the *hyb* hydrogen utilization operon were found to have attenuated virulence in the model (69). In competition with infections with the wild-type strain, the *S.* Typhimurium *hyb* operon mutant displayed attenuated virulence in both LCM and conventional mice. However, in animals treated with a high dose of streptomycin, the *hyb* mutant had similar fitness to the wild-type strain. Hydrogen is produced by the GIT microbiota during carbon fermentation and was shown in this study to be important in the early stages of *S.* Typhimurium GIT colonization. Other microbiota-induced metabolic changes in the gut may also affect *S.* Typhimurium colitis. A directed metabolomics survey of low-dose-streptomycin-treated mice found a reduction in SCFA concentration in the ceca (32). The SCFAs propionate and butyrate have been shown to suppress SPI-1 expression in vitro (39, 61), and propionate was also demonstrated to kill *S.* Typhimurium under gut-like in vitro conditions (17). Collectively, these studies strongly support the notion that antibiotic treatment promotes metabolic changes in the GIT that can be conducive to *S.* Typhimurium colonization (**Figure 3a**).

GI inflammation induced by *S.* Typhimurium has been generally viewed as a mechanism for pathogen control and expulsion. Recent studies show instead that the pathogen actually manipulates inflammation as a source of nutrients and to out-compete the GIT microbiota in the gut lumen (**Figure 2b**). In antibiotic-treated animals, wild-type *S.* Typhimurium induces colitis and inflammation and out-competes the GIT microbiota (98, 105). An *S.* Typhimurium strain lacking the SPI-1 and SPI-2 secretion systems (the avir strain) can establish initial colonization in the gut of antibiotic-treated mice, but the infection is not sustained and the pathogen is out-competed and expelled by the GIT microbiota. However, in mice that develop spontaneous colitis or if gut inflammation is induced by T-cell transfer, *S.* Typhimurium[avir] can colonize the GIT to the same extent as the wild-type strain (105). This indicates that gut inflammation is actually supporting *S.* Typhimurium colonization. In the gastroenteritis model, *S.* Typhimurium infection strongly induces the expression of RegIIIβ, a lectin secreted by the GI mucosa that was also shown to have bacteriocin activity against gram-positive and gram-negative bacteria, including *E. coli*, but is harmless to *S.* Typhimurium itself (106). The elevation of RegIIIβ in the intestine was shown to be inflammation dependent, as it does not occur in immune-deficient mice and may be one mechanism by which the pathogen uses the immune response to clear the GIT microbiota. An alternative or parallel mechanism involves *S.* Typhimurium–induced neutrophil infiltration into the gut lumen. Neutrophil recruitment during *S.* Typhimurium infection is virulence factor and colitis dependent (98), and elimination of neutrophils by antibody treatment results in a diminished ability of *S.* Typhimurium to expel the GIT microbiota (34). Furthermore, neutrophil elastase, a serine protease produced by neutrophils during inflammation, was demonstrated to shift the GIT microbiota of mice in a manner that supported *S.* Typhimurium GIT colonization (34). Neutrophil efflux to the GIT lumen is also accompanied by increased production of reactive oxygen species (ROS) and nitric oxide (NO). Previously believed to be important in controlling pathogens, ROS and NO in fact assist *S.* Typhimurium in its competition with the GIT microbiota. The GIT microbiota produces H_2S, which is subsequently metabolized into thiosulfate by the gut lumen. Inflammation-induced release of ROS and NO oxidizes thiosulfate into tetrathionate, an electron acceptor used selectively by *S.* Typhimurium that also limits the ability of other bacteria to replicate (124). Indeed, the *ttrA* gene responsible for tetrathionate utilization is crucial to the ability of *S.* Typhimurium to out-compete the GIT microbiota. Tetrathionate respiration in vivo is coupled with the consumption of ethanolamine, a nutrient that is not fermented by the GIT microbiota and as such serves as an abundant carbon source for *S.* Typhimurium in the GIT (109). The ability of *S.* Typhimurium to consume ethanolamine was demonstrated to promote colonization of the inflamed GIT but was dependent on the pathogen's ability to reduce tetrathionate. The picture emerging from all of the above is that the mucosal defense benefits *S.* Typhimurium by helping the pathogen to out-compete the GIT microbiota.

Mice infected with an SPI-2 *S.* Typhimurium mutant present a self-limiting infection similar to *S.* Typhimurium gastroenteritis in humans. In this model of the disease, the microbiota begins to regenerate and out-competes the pathogen (**Figure 3c**). The starting point for microbiota regrowth remains unknown. It may result from a reduction in the availability of the nutrients required by *S.* Typhimurium for GIT proliferation, such as thiosulfate or sialic acid, but may also be the result of other, yet undiscovered signals. What is clear is that it is a GIT microbiota–, rather than immune–, dependent process. T- and B-cell-depleted mice can clear *S.* Typhimurium GIT infection but only if they are colonized with a complex microbiota prior to pathogen inoculation (28). In the model described in this study, all of the mice were treated with high-dose streptomycin prior to *S.* Typhimurium infection. How exactly the GIT microbiota exerts immune-independent protection on the host even though it is cleared from the GI before infection is just one of the issues that require further study in the field of host-*Salmonella*-microbiota interactions.

Clostridium difficile

Unlike *S. enterica* or pathogenic *E. coli*, *C. difficile* is often present in the GIT microbiota of healthy individuals, ranging from 1% to 50% of the population (7, 41, 74). As a minor member of the GIT microbiota, *C. difficile* has no adverse effects on human health. High colonization by the bacteria, however, leads to a condition known as *C. difficile*–associated diarrhea (CDAD) and is the main cause of pseudomembranous colitis, a severe and sometimes deadly form of antibiotic-associated diarrhea (55). *C. difficile* is a spore-forming gram-positive bacteria that induces severe colitis through the action of two unique toxins: Toxin A and Toxin B (117). Both toxins are glucosyltransferases that are delivered into host cells, where they modify Rho and Ras family GTPase and induce epithelial cell necrosis (44, 48, 49). Furthermore, these toxins have been shown to disrupt epithelial barrier function by modifying tight junction protein interaction (84).

It remains unclear whether patients who develop CDAD acquire *C. difficile* from the environment or from overgrowth of indigenous bacteria (123). It is, however, remarkably clear that the GIT microbiota provides the host with a high level of resistance to this disease. CDAD in adults is almost exclusively found in antibiotic-treated or otherwise susceptible hospitalized patients (88). CDAD is more common in infants under one year of age (59), even without additional risk factors, but children of that age have been reported to harbor an unstable GIT microbiota (87). Furthermore, the vast majority of animal models of *C. difficile* colonization involve either germ-free mice or various animals that require aggressive antimicrobial therapy prior to inoculation with the pathogen (9) (**Figure 4a**).

Similar to *S.* Typhimurium– and *C. rodentium*–induced colitis, increased susceptibility to CDAD can be linked with population shifts of the GIT microbiota rather than complete microbial depletion. Stool samples of hospitalized CDAD patients that were collected prior to the onset of disease harbored a microbial population of reduced diversity with diminished proportions of the phylum Bacteroidetes and families Bacteroidaceae and Clostridiales as well as increased levels of the family Enterococcaceae (116). Additionally, a study that sampled the GIT microbiota of patients suffering from CDAD found that those who experience repeated infection have a higher level of *C. difficile* colonization of the gut and an overall reduction in microbiota diversity when compared with healthy subjects as well as with patients suffering an initial episode of the disease (15). In an animal study, mice treated with a single dose of clindamycin became susceptible to *C. difficile* colitis (12). Similar to the effect of low-dose streptomycin treatment, clindamycin administration did not reduce overall bacterial counts in the gut but did significantly lower population diversity. The post-treatment microbial population was dominated by Enterobacteriaceae species and largely reduced in Lachnospiraceae and *Barnesiella* populations.

A key question in the study of CDAD is how antimicrobial treatment and the subsequent change in the GIT microbiota break the natural resistance to *C. difficile* over-colonization. Ng et al. (82) reported that sialic acid metabolism by *C. difficile* is critical for colonization of antibiotic-treated mice as well as germ-free mice monocolonized with *B. thetaiotaomicron* (82). In both murine populations, sialic acid is more available than in conventionally colonized untreated animals. A more recent study employed metabolomics to survey the *C. difficile* susceptible GIT (108). Significant alterations in many metabolites were observed in mice treated with cefoperazonein in comparison to untreated animals or those that were allowed to recover from the treatment to the point at which they were once again resistant to *C. difficile*. The bile acid taurocholate was among

a Early

C. difficile

Enterocyte

M cell

Goblet cell

Dendritic cell

Macrophage

Neutrophil

Segmented filamentous bacteria

Microbiota

Antibiotic

Sialic acid

Toxin A

Toxin B

b Late

c End

Fecal transfer

Figure 4

The gastrointestinal (GI) microbiota during *Clostridium difficile* infection. (*a*) Early stage of infection: Antibiotic treatment increases susceptibility to *C. difficile* colonization, possibly by reducing concentrations of microbiota-produced factors that repress *C. difficile* spore germination. Sialic acid, released by the GI microbiota in the antibiotic-treated GI tract (GIT), is used by the pathogen as an energy source. (*b*) Late stage of infection: *C. difficile* release of Toxin A and Toxin B leads to epithelial necrosis and tight junction disruption. There is an increase in concentrations of *C. difficile* spore germination–promoting molecules. (*c*) End stage of infection: Transfer of healthy microbiota (fecal transplant) and restoration of gut homeostasis are currently the best treatments for *C. difficile* infection.

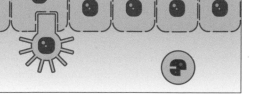

the metabolites significantly increased in the cecum of susceptible mice. In vitro assays confirmed that taurocholate is an inducer of *C. difficile* spore germination. This is consistent with previous in vitro studies that reported primary and secondary bile acids as activators and inhibitors of *C. difficile* spore germination, respectively (101, 102). The metabolomic screen of Theriot et al. (108) also found increases in various sugars in the antibiotic-treated ceca, and the ability of *C. difficile* to grow on these carbon sources was confirmed. As such, a possible mechanism of *C. difficile* expansion in the post-antibiotic-treated gut is that increased availability of various nutrients and changes in spore germination signals together lead to increased pathogen proliferation (**Figure 4***b*). The GIT microbiota may also have other means to control *C. difficile* population levels. *Bacillus thuringiensis* isolated from stool of a healthy human was shown to produce thuricin CD, a bacteriocin that can kill *C. difficile* but not other gram-positive bacteria common to the human GIT (89). Although this study does not link the presence or absence of *B. thuringiensis* to CDAD, it does demonstrate the potential role of direct bacteria-bacteria interactions in controlling pathogens in the gut.

Although several antibiotics can be used to treat CDAD, recurrent infections are frequent because of *C. difficile*'s ability to form resistant spores. Approximately 15–30% of CDAD patients treated with broad-spectrum antibiotics, such as vancomycin and metronidazole, experience recurrence of infection after the first episode and repeated recurrence occurs in as many as 60% of patients (96). New narrow-spectrum antibiotics, such as fidaxomicin, are emerging as improved treatments for CDAD infection (66). Nevertheless, fecal transplant, the transfer of an entire gut population directly to the patient GIT, is currently the most efficient therapy for the disease (71) (**Figure 4***c*).

CONCLUDING REMARKS

The GIT microbiota plays a crucial role during the various stages of bacterial enteric colitis. Although pathogenic *E. coli*, *S.* Typhimurium, and *C. difficile* have very different virulence pathways and, appropriately, unique interactions with the host and its microbiota, commonalities do exist. The susceptibility to all three pathogens can be increased by a single dose of an antibiotic that does not clear the GIT microbiota but rather shifts it to a composition more permissive to pathogen colonization. Furthermore, all three pathogens are able to utilize metabolites made available through alterations to the GIT bacterial population. Finally, recovery from gastric bacterial infection depends on the restoration of a complex microbial flora in the GIT. In recent years, much data have accumulated about the host-mediated interaction between the GIT microbiota and bacterial pathogens. Future efforts will likely turn toward deciphering the direct molecular interaction between the microbes harbored by the human body and those that try to attack it.

DISCLOSURE STATEMENT

The authors are not aware of any affiliations, memberships, funding, or financial holdings that might be perceived as affecting the objectivity of this review.

ACKNOWLEDGMENTS

We would like to thank Nat F. Brown for figure preparation and Eric M. Brown for critical reading of the review. The Finlay lab is funded by grants from the Canadian Institutes of Health Research (CIHR).

LITERATURE CITED

1. Abrams GD, Bauer H, Sprinz H. 1963. Influence of the normal flora on mucosal morphology and cellular renewal in the ileum. A comparison of germ-free and conventional mice. *Lab. Invest.* 12:355–64

2. Arrieta Mendez M-C. 2011. *The role of small intestinal permeability in the pathogenesis of colitis in the interleukin-10 gene deficient mouse.* PhD Diss. Univ. Alberta, Edmonton AB, Can.

3. Atarashi K, Nishimura J, Shima T, Umesaki Y, Yamamoto M, et al. 2008. ATP drives lamina propria T(H)17 cell differentiation. *Nature* 455:808–12

4. Atarashi K, Tanoue T, Shima T, Imaoka A, Kuwahara T, et al. 2011. Induction of colonic regulatory T cells by indigenous *Clostridium* species. *Science* 331:337–41

5. Ather JL, Ckless K, Martin R, Foley KL, Suratt BT, et al. 2011. Serum amyloid A activates the NLRP3 inflammasome and promotes Th17 allergic asthma in mice. *J. Immunol.* 187:64–73

6. Barthel M, Hapfelmeier S, Quintanilla-Martinez L, Kremer M, Rohde M, et al. 2003. Pretreatment of mice with streptomycin provides a *Salmonella enterica* serovar Typhimurium colitis model that allows analysis of both pathogen and host. *Infect. Immun.* 71:2839–58

7. Beaugerie L, Flahault A, Barbut F, Atlan P, Lalande V, et al. 2003. Antibiotic-associated diarrhoea and *Clostridium difficile* in the community. *Aliment. Pharmacol. Thera.* 17:905–12

8. Bergstrom KSB, Kissoon-Singh V, Gibson DL, Ma C, Montero M, et al. 2010. Muc2 protects against lethal infectious colitis by disassociating pathogenic and commensal bacteria from the colonic mucosa. *PLOS Pathog.* 6:e1000902

9. Best EL, Freeman J, Wilcox MH. 2012. Models for the study of *Clostridium difficile* infection. *Gut Microbes* 3:145–67

10. Bisgaard H, Li N, Bonnelykke K, Chawes BL, Skov T, et al. 2011. Reduced diversity of the intestinal microbiota during infancy is associated with increased risk of allergic disease at school age. *J. Allergy Clinic. Immunol.* 128:646–52; e1–5

11. Bouskra D, Brezillon C, Berard M, Werts C, Varona R, et al. 2008. Lymphoid tissue genesis induced by commensals through NOD1 regulates intestinal homeostasis. *Nature* 456:507–10

12. Buffie CG, Jarchum I, Equinda M, Lipuma L, Gobourne A, et al. 2012. Profound alterations of intestinal microbiota following a single dose of clindamycin results in sustained susceptibility to *Clostridium difficile*–induced colitis. *Infect. Immun.* 80:62–73

13. Buonocore S, Ahern PP, Uhlig HH, Ivanov II, Littman DR, et al. 2010. Innate lymphoid cells drive interleukin-23-dependent innate intestinal pathology. *Nature* 464:1371–75

14. Carey CM, Kostrzynska M, Ojha S, Thompson S. 2008. The effect of probiotics and organic acids on Shiga-toxin 2 gene expression in enterohemorrhagic *Escherichia coli* O157:H7. *J. Microbiol. Methods* 73:125–32

15. Chang JY, Antonopoulos DA, Kalra A, Tonelli A, Khalife WT, et al. 2008. Decreased diversity of the fecal microbiome in recurrent *Clostridium difficile*–associated diarrhea. *J. Infect. Dis.* 197:435–38

16. Chen Z, Cobbold SP, Waldmann H, Metcalfe SM. 1994. Tolerance induction in concordant heart-xenografted mice by CD4 and CD8 monoclonal antibodies. *Transplant. Proc.* 26:1199–200

17. Cherrington CA, Hinton M, Pearson GR, Chopra I. 1991. Short-chain organic acids at ph 5.0 kill *Escherichia coli* and *Salmonella* spp. without causing membrane perturbation. *J. Appl. Bacteriol.* 70:161–65

18. Costello EK, Lauber CL, Hamady M, Fierer N, Gordon JI, Knight R. 2009. Bacterial community variation in human body habitats across space and time. *Science* 326:1694–97

19. Croswell A, Amir E, Teggatz P, Barman M, Salzman NH. 2009. Prolonged impact of antibiotics on intestinal microbial ecology and susceptibility to enteric *Salmonella* infection. *Infect. Immun.* 77:2741–53

20. Croxen MA, Finlay BB. 2010. Molecular mechanisms of *Escherichia coli* pathogenicity. *Nat. Rev. Microbiol.* 8:26–38

21. Croxen MA, Law RJ, Scholz R, Keeney KM, Wlodarska M, Finlay BB. 2013. Recent advances in understanding enteric pathogenic *Escherichia coli*. *Clin. Microbiol. Rev.* 26:822–80

22. de Jong HK, Parry CM, van der Poll T, Wiersinga WJ. 2012. Host-pathogen interaction in invasive salmonellosis. *PLOS Pathog.* 8:e1002933

23. de Sablet T, Chassard C, Bernalier-Donadille A, Vareille M, Gobert AP, Martin C. 2009. Human microbiota-secreted factors inhibit Shiga toxin synthesis by enterohemorrhagic *Escherichia coli* O157:H7. *Infect. Immun.* 77:783–90

24. Delves PJ, Roitt IM. 2000. The immune system. First of two parts. *New Engl. J. Med.* 343:37–49

25. Deriu E, Liu JZ, Pezeshki M, Edwards RA, Ochoa RJ, et al. 2013. Probiotic bacteria reduce *Salmonella typhimurium* intestinal colonization by competing for iron. *Cell Host Microbe* 14:26–37

26. Dobber R, Hertogh-Huijbregts A, Rozing J, Bottomly K, Nagelkerken L. 1992. The involvement of the intestinal microflora in the expansion of CD4+ T cells with a naive phenotype in the periphery. *Dev. Immunol.* 2:141–50

27. Eckburg PB, Bik EM, Bernstein CN, Purdom E, Dethlefsen L, et al. 2005. Diversity of the human intestinal microbial flora. *Science* 308:1635–38

28. Endt K, Stecher B, Chaffron S, Slack E, Tchitchek N, et al. 2010. The microbiota mediates pathogen clearance from the gut lumen after non-typhoidal *Salmonella* diarrhea. *PLOS Pathog.* 6:e1001097

29. Ferreira RB, Gill N, Willing BP, Antunes LC, Russell SL, et al. 2011. The intestinal microbiota plays a role in *Salmonella*-induced colitis independent of pathogen colonization. *PLOS ONE* 6:e20338

30. Frank DN, St. Amand AL, Feldman RA, Boedeker EC, Harpaz N, Pace NR. 2007. Molecular-phylogenetic characterization of microbial community imbalances in human inflammatory bowel diseases. *Proc. Natl. Acad. Sci. USA* 104:13780–85

31. Gaboriau-Routhiau V, Rakotobe S, Lecuyer E, Mulder I, Lan A, et al. 2009. The key role of segmented filamentous bacteria in the coordinated maturation of gut helper T cell responses. *Immunity* 31:677–89

32. Garner CD, Antonopoulos DA, Wagner B, Duhamel GE, Keresztes I, et al. 2009. Perturbation of the small intestine microbial ecology by streptomycin alters pathology in a *Salmonella enterica* serovar *typhimurium* murine model of infection. *Infect. Immun.* 77:2691–702

33. Ghosh S, Dai C, Brown K, Rajendiran E, Makarenko S, et al. 2011. Colonic microbiota alters host susceptibility to infectious colitis by modulating inflammation, redox status, and ion transporter gene expression. *Am. J. Physiol. Gastrointest. Liver Physiol.* 301:G39–49

34. Gill N, Ferreira RB, Antunes LC, Willing BP, Sekirov I, et al. 2012. Neutrophil elastase alters the murine gut microbiota resulting in enhanced *Salmonella* colonization. *PLOS ONE* 7:e49646

35. Gu S, Chen D, Zhang JN, Lv X, Wang K, et al. 2013. Bacterial community mapping of the mouse gastrointestinal tract. *PLOS ONE* 8:e74957

36. Gustafsson JK, Navabi N, Rodriguez-Piñeiro AM, Alomran AHA, Premaratne P, et al. 2013. Dynamic changes in mucus thickness and ion secretion during *Citrobacter rodentium* infection and clearance. *PLOS ONE* 8:e84430

37. Herold S, Paton JC, Srimanote P, Paton AW. 2009. Differential effects of short-chain fatty acids and iron on expression of *iha* in Shiga-toxigenic *Escherichia coli*. *Microbiology* 155:3554–63

38. Hum. Microbiome Proj. C. 2012. Structure, function and diversity of the healthy human microbiome. *Nature* 486:207–14

39. Hung CC, Garner CD, Slauch JM, Dwyer ZW, Lawhon SD, et al. 2013. The intestinal fatty acid propionate inhibits *Salmonella* invasion through the post-translational control of HilD. *Mol. Microbiol.* 87:1045–60

40. Hurley BP, McCormick BA. 2004. Intestinal epithelial defense systems protect against bacterial threats. *Curr. Gastroenterol. Rep.* 6:355–61

41. Iizuka M, Konno S, Itou H, Chihara J, Toyoshima I, et al. 2004. Novel evidence suggesting *Clostridium difficile* is present in human gut microbiota more frequently than previously suspected. *Microbiol. Immunol.* 48:889–92

42. Ivanov II, Atarashi K, Manel N, Brodie EL, Shima T, et al. 2009. Induction of intestinal Th17 cells by segmented filamentous bacteria. *Cell* 139:485–98

43. Ivanov II, Frutos Rde L, Manel N, Yoshinaga K, Rifkin DB, et al. 2008. Specific microbiota direct the differentiation of IL-17-producing T-helper cells in the mucosa of the small intestine. *Cell Host Microbe* 4:337–49

44. Jank T, Aktories K. 2008. Structure and mode of action of clostridial glucosylating toxins: the ABCD model. *Trends Microbiol.* 16:222–29

45. Johansson MEV, Phillipson M, Petersson J, Velcich A, Holm L, Hansson GC. 2008. The inner of the two Muc2 mucin-dependent mucus layers in colon is devoid of bacteria. *Proc. Natl. Acad. Sci. USA* 105:15064–69

46. Jones BD, Ghori N, Falkow S. 1994. *Salmonella typhimurium* initiates murine infection by penetrating and destroying the specialized epithelial M cells of the Peyer's patches. *J. Exp. Med.* 180:15–23

47. Jones SA, Jorgensen M, Chowdhury FZ, Rodgers R, Hartline J, et al. 2008. Glycogen and maltose utilization by *Escherichia coli* O157:H7 in the mouse intestine. *Infect. Immun.* 76:2531–40

48. Just I, Richter HP, Prepens U, von Eichel-Streiber C, Aktories K. 1994. Probing the action of *Clostridium difficile* toxin B in *Xenopus laevis* oocytes. *J. Cell Sci.* 107(Pt. 6):1653–59

49. Just I, Selzer J, Wilm M, von Eichel-Streiber C, Mann M, Aktories K. 1995. Glucosylation of Rho proteins by *Clostridium difficile* toxin B. *Nature* 375:500–3

50. Kagnoff MF. 1993. Immunology of the intestinal tract. *Gastroenterology* 105:1275–80

51. Kalliomaki M, Kirjavainen P, Eerola E, Kero P, Salminen S, Isolauri E. 2001. Distinct patterns of neonatal gut microflora in infants in whom atopy was and was not developing. *J. Allergy Clin. Immunol.* 107:129–34

52. Kamada N, Kim Y-G, Sham HP, Vallance BA, Puente JL, et al. 2012. Regulated virulence controls the ability of a pathogen to compete with the gut microbiota. *Science* 336:1325–29

53. Kaper JB, Nataro JP, Mobley HL. 2004. Pathogenic *Escherichia coli*. *Nat. Rev. Microbiol.* 2:123–40

54. Kassinen A, Krogius-Kurikka L, Makivuokko H, Rinttila T, Paulin L, et al. 2007. The fecal microbiota of irritable bowel syndrome patients differs significantly from that of healthy subjects. *Gastroenterology* 133:24–33

55. Kelly CP, Pothoulakis C, LaMont JT. 1994. *Clostridium difficile* colitis. *New Engl. J. Med.* 330:257–62

56. Kinnebrew MA, Ubeda C, Zenewicz LA, Smith N, Flavell RA, Pamer EG. 2010. Bacterial flagellin stimulates toll-like receptor 5-dependent defense against vancomycin-resistant enterococcus infection. *J. Infect. Dis.* 201:534–43

57. Kurokawa K, Itoh T, Kuwahara T, Oshima K, Toh H, et al. 2007. Comparative metagenomics revealed commonly enriched gene sets in human gut microbiomes. *DNA Res.* 14:169–81

58. Langkamp-Henken B, Glezer JA, Kudsk KA. 1992. Immunologic structure and function of the gastrointestinal tract. *Nutr. Clin. Pract.* 7:100–8

59. Larson HE, Barclay FE, Honour P, Hill ID. 1982. Epidemiology of *Clostridium difficile* in infants. *J. Infect. Dis.* 146:727–33

60. Lathrop SK, Bloom SM, Rao SM, Nutsch K, Lio CW, et al. 2011. Peripheral education of the immune system by colonic commensal microbiota. *Nature* 478:250–54

61. Lawhon SD, Maurer R, Suyemoto M, Altier C. 2002. Intestinal short-chain fatty acids alter *Salmonella typhimurium* invasion gene expression and virulence through BarA/SirA. *Mol. Microbiol.* 46:1451–64

62. Lawley TD, Bouley DM, Hoy YE, Gerke C, Relman DA, Monack DM. 2008. Host transmission of *Salmonella enterica* serovar Typhimurium is controlled by virulence factors and indigenous intestinal microbiota. *Infect. Immun.* 76:403–16

63. Leatham MP, Banerjee S, Autieri SM, Mercado-Lubo R, Conway T, Cohen PS. 2009. Precolonized human commensal *Escherichia coli* strains serve as a barrier to *E. coli* O157:H7 growth in the streptomycin-treated mouse intestine. *Infect. Immun.* 77:2876–86

64. Lee YK, Menczes JS, Umesaki Y, Mazmanian SK. 2010. Microbes and Health Sackler Colloquium: proinflammatory T-cell responses to gut microbiota promote experimental autoimmune encephalomyelitis. *Proc. Natl. Acad. Sci. USA* 108(Suppl. 1):4615–22

65. Ley RE, Turnbaugh PJ, Klein S, Gordon JI. 2006. Microbial ecology: human gut microbes associated with obesity. *Nature* 444:1022–23

66. Louie TJ, Cannon K, Byrne B, Emery J, Ward L, et al. 2012. Fidaxomicin preserves the intestinal microbiome during and after treatment of *Clostridium difficile* infection (CDI) and reduces both toxin reexpression and recurrence of CDI. *Clin. Infect. Dis.* 55(Suppl. 2):S132–42

67. Luckey TD. 1972. Introduction to intestinal microecology. *Am. J. Clin. Nutr.* 25:1292–94

68. Macdonald JC, Torriani FJ, Morse LS, Karavellas MP, Reed JB, Freeman WR. 1998. Lack of reactivation of cytomegalovirus (CMV) retinitis after stopping CMV maintenance therapy in AIDS patients with sustained elevations in CD4 T cells in response to highly active antiretroviral therapy. *J. Infect. Dis.* 177:1182–87

69. Maier L, Vyas R, Cordova CD, Lindsay H, Schmidt TS, et al. 2013. Microbiota-derived hydrogen fuels *Salmonella typhimurium* invasion of the gut ecosystem. *Cell Host Microbe* 14:641–51

70. Malinen E, Rinttila T, Kajander K, Matto J, Kassinen A, et al. 2005. Analysis of the fecal microbiota of irritable bowel syndrome patients and healthy controls with real-time PCR. *Am. J. Gastroenterol.* 100:373–82

71. Mattila E, Uusitalo-Seppala R, Wuorela M, Lehtola L, Nurmi H, et al. 2012. Fecal transplantation, through colonoscopy, is effective therapy for recurrent *Clostridium difficile* infection. *Gastroenterology* 142:490–96

72. Mazmanian SK, Liu CH, Tzianabos AO, Kasper DL. 2005. An immunomodulatory molecule of symbiotic bacteria directs maturation of the host immune system. *Cell* 122:107–18

73. Mazmanian SK, Round JL, Kasper DL. 2008. A microbial symbiosis factor prevents intestinal inflammatory disease. *Nature* 453:620–25

74. McFarland LV, Mulligan ME, Kwok RY, Stamm WE. 1989. Nosocomial acquisition of *Clostridium difficile* infection. *New Engl. J. Med.* 320:204–10

75. Miranda RL, Conway T, Leatham MP, Chang DE, Norris WE, et al. 2004. Glycolytic and gluconeogenic growth of *Escherichia coli* O157:H7 (EDL933) and *E. coli* K-12 (MG1655) in the mouse intestine. *Infect. Immun.* 72:1666–76

76. Momose Y, Hirayama K, Itoh K. 2008. Competition for proline between indigenous *Escherichia coli* and *E. coli* O157:H7 in gnotobiotic mice associated with infant intestinal microbiota and its contribution to the colonization resistance against *E. coli* O157:H7. *Antonie van Leeuwenhoek* 94:165–71

77. Moreau MC, Ducluzeau R, Guy-Grand D, Muller MC. 1978. Increase in the population of duodenal immunoglobulin A plasmocytes in axenic mice associated with different living or dead bacterial strains of intestinal origin. *Infect. Immun.* 21:532–39

78. Mundy R, Girard F, FitzGerald AJ, Frankel G. 2006. Comparison of colonization dynamics and pathology of mice infected with enteropathogenic *Escherichia coli*, enterohaemorrhagic *E. coli* and *Citrobacter rodentium*. *FEMS Microbiol. Lett.* 265:126–32

79. Mundy R, MacDonald TT, Dougan G, Frankel G, Wiles S. 2005. *Citrobacter rodentium* of mice and man. *Cell. Microbiol.* 7:1697–706

80. Nakanishi N, Tashiro K, Kuhara S, Hayashi T, Sugimoto N, Tobe T. 2009. Regulation of virulence by butyrate sensing in enterohaemorrhagic *Escherichia coli*. *Microbiology* 155:521–30

81. Nardi RM, Silva ME, Vieira EC, Bambirra EA, Nicoli JR. 1989. Intragastric infection of germfree and conventional mice with *Salmonella typhimurium*. *Braz. J. Med. Biol. Res.* 22:1389–92

82. Ng KM, Ferreyra JA, Higginbottom SK, Lynch JB, Kashyap PC, et al. 2013. Microbiota-liberated host sugars facilitate post-antibiotic expansion of enteric pathogens. *Nature* 502:96–99

83. Niess JH, Leithauser F, Adler G, Reimann J. 2008. Commensal gut flora drives the expansion of proinflammatory CD4 T cells in the colonic lamina propria under normal and inflammatory conditions. *J. Immunol.* 180:559–68

84. Nusrat A, von Eichel-Streiber C, Turner JR, Verkade P, Madara JL, Parkos CA. 2001. *Clostridium difficile* toxins disrupt epithelial barrier function by altering membrane microdomain localization of tight junction proteins. *Infect. Immun.* 69:1329–36

85. Ochoa-Reparaz J, Mielcarz DW, Wang Y, Begum-Haque S, Dasgupta S, et al. 2010. A polysaccharide from the human commensal *Bacteroides fragilis* protects against CNS demyelinating disease. *Mucosal Immunol.* 3:487–95

86. Pacheco AR, Curtis MM, Ritchie JM, Munera D, Waldor MK, et al. 2012. Fucose sensing regulates bacterial intestinal colonization. *Nature* 492:113–17

87. Palmer C, Bik EM, DiGiulio DB, Relman DA, Brown PO. 2007. Development of the human infant intestinal microbiota. *PLOS Biol.* 5:e177

88. Pothoulakis C, LaMont JT. 1993. *Clostridium difficile* colitis and diarrhea. *Gastroenterol. Clin. N. Am.* 22:623–37

89. Rea MC, Sit CS, Clayton E, O'Connor PM, Whittal RM, et al. 2010. Thuricin CD, a posttranslationally modified bacteriocin with a narrow spectrum of activity against *Clostridium difficile*. *Proc. Natl. Acad. Sci. USA* 107:9352–57

90. Rescigno M. 2011. The intestinal epithelial barrier in the control of homeostasis and immunity. *Trends Immunol.* 32:256–64

91. Rimoldi M, Chieppa M, Larghi P, Vulcano M, Allavena P, Rescigno M. 2005. Monocyte-derived dendritic cells activated by bacteria or by bacteria-stimulated epithelial cells are functionally different. *Blood* 106:2818–26

92. Round JL, Mazmanian SK. 2009. The gut microbiota shapes intestinal immune responses during health and disease. *Nat. Rev. Immunol.* 9:313–23

93. Russell SL, Finlay BB. 2012. The impact of gut microbes in allergic diseases. *Curr. Opin. Gastroenterol.* 28:563–69

94. Russell SL, Gold MJ, Hartmann M, Willing BP, Thorson L, et al. 2012. Early life antibiotic-driven changes in microbiota enhance susceptibility to allergic asthma. *EMBO Rep.* 13:440–47

95. Schmidt MA. 2010. LEEways: tales of EPEC, ATEC and EHEC. *Cell. Microbiol.* 12:1544–52

96. Sears P, Ichikawa Y, Ruiz N, Gorbach S. 2013. Advances in the treatment of *Clostridium difficile* with fidaxomicin: a narrow spectrum antibiotic. *Ann. N. Y. Acad. Sci.* 1291:33–41

97. Sekirov I, Finlay BB. 2009. The role of the intestinal microbiota in enteric infection. *J. Physiol.* 587:4159–67

98. Sekirov I, Gill N, Jogova M, Tam N, Robertson M, et al. 2010. *Salmonella* SPI-1-mediated neutrophil recruitment during enteric colitis is associated with reduction and alteration in intestinal microbiota. *Gut Microbes* 1:30–41

99. Sekirov I, Tam NM, Jogova M, Robertson ML, Li Y, et al. 2008. Antibiotic-induced perturbations of the intestinal microbiota alter host susceptibility to enteric infection. *Infect. Immun.* 76:4726–36

100. Smith K, McCoy KD, Macpherson AJ. 2007. Use of axenic animals in studying the adaptation of mammals to their commensal intestinal microbiota. *Semin. Immunol.* 19:59–69

101. Sorg JA, Sonenshein AL. 2008. Bile salts and glycine as cogerminants for *Clostridium difficile* spores. *J. Bacteriol.* 190:2505–12

102. Sorg JA, Sonenshein AL. 2009. Chenodeoxycholate is an inhibitor of *Clostridium difficile* spore germination. *J. Bacteriol.* 191:1115–17

103. Sprinz H, Kundel DW, Dammin GJ, Horowitz RE, Schneider H, Formal SB. 1961. The response of the germfree guinea pig to oral bacterial challenge with *Escherichia coli* and *Shigella flexneri*. *Am. J. Pathol.* 39:681–95

104. Stecher B, Chaffron S, Kappeli R, Hapfelmeier S, Freedrich S, et al. 2010. Like will to like: abundances of closely related species can predict susceptibility to intestinal colonization by pathogenic and commensal bacteria. *PLOS Pathog.* 6:e1000711

105. Stecher B, Robbiani R, Walker AW, Westendorf AM, Barthel M, et al. 2007. *Salmonella enterica* serovar Typhimurium exploits inflammation to compete with the intestinal microbiota. *PLOS Biol.* 5:2177–89

106. Stelter C, Kappeli R, Konig C, Krah A, Hardt WD, et al. 2011. *Salmonella*-induced mucosal lectin RegIIIβ kills competing gut microbiota. *PLOS ONE* 6:e20749

107. Strauch UG, Obermeier F, Grunwald N, Gurster S, Dunger N, et al. 2005. Influence of intestinal bacteria on induction of regulatory T cells: lessons from a transfer model of colitis. *Gut* 54:1546–52

108. Theriot CM, Koenigsknecht MJ, Carlson PE Jr, Hatton GE, Nelson AM, et al. 2014. Antibiotic-induced shifts in the mouse gut microbiome and metabolome increase susceptibility to *Clostridium difficile* infection. *Nat. Commun.* 5:3114

109. Thiennimitr P, Winter SE, Winter MG, Xavier MN, Tolstikov V, et al. 2011. Intestinal inflammation allows *Salmonella* to use ethanolamine to compete with the microbiota. *Proc. Natl. Acad. Sci. USA* 108:17480–85

110. Turnbaugh PJ, Hamady M, Yatsunenko T, Cantarel BL, Duncan A, et al. 2009. A core gut microbiome in obese and lean twins. *Nature* 457:480–84

111. Uematsu S, Fujimoto K, Jang MH, Yang BG, Jung YJ, et al. 2008. Regulation of humoral and cellular gut immunity by lamina propria dendritic cells expressing Toll-like receptor 5. *Nat. Immunol.* 9:769–76

112. Umesaki Y, Okada Y, Matsumoto S, Imaoka A, Setoyama H. 1995. Segmented filamentous bacteria are indigenous intestinal bacteria that activate intraepithelial lymphocytes and induce Mhc class II molecules and fucosyl asialo Gm1 glycolipids on the small intestinal epithelial cells in the ex-germ-free mouse. *Microbiol. Immunol.* 39:555–62

113. Vallance BA, Deng W, Jacobson K, Finlay BB. 2003. Host susceptibility to the attaching and effacing bacterial pathogen *Citrobacter rodentium*. *Infect. Immun.* 71:3443–53

114. van der Waaij D, Berghuis-de Vries JM, Lekkerkerk-van der Wees JEC. 1971. Colonization resistance of the digestive tract in conventional and antibiotic-treated mice. *J. Hyg.* 69:405–11

115. Vazquez-Torres A, Jones-Carson J, Baumler AJ, Falkow S, Valdivia R, et al. 1999. Extraintestinal dissemination of *Salmonella* by CD18-expressing phagocytes. *Nature* 401:804–8

116. Vincent C, Stephens DA, Loo VG, Edens TJ, Behr MA, et al. 2013. Reductions in intestinal Clostridiales precede the development of nosocomial *Clostridium difficile* infection. *Microbiome* 1:18

117. Voth DE, Ballard JD. 2005. *Clostridium difficile* toxins: mechanism of action and role in disease. *Clin. Microbiol. Rev.* 18:247–63

118. Wen L, Ley RE, Volchkov PY, Stranges PB, Avanesyan L, et al. 2008. Innate immunity and intestinal microbiota in the development of type 1 diabetes. *Nature* 455:U1109–10

119. Whitman WB, Coleman DC, Wiebe WJ. 1998. Prokaryotes: the unseen majority. *Proc. Natl. Acad. Sci. USA* 95:6578–83

120. Willing BP, Dicksved J, Halfvarson J, Andersson AF, Lucio M, et al. 2010. A pyrosequencing study in twins shows that gastrointestinal microbial profiles vary with inflammatory bowel disease phenotypes. *Gastroenterology* 139:1844–54; e1

121. Willing BP, Vacharaksa A, Croxen M, Thanachayanont T, Finlay BB. 2011. Altering host resistance to infections through microbial transplantation. *PLOS ONE* 6:e26988

122. Willing BP, Van Kessel AG. 2007. Enterocyte proliferation and apoptosis in the caudal small intestine is influenced by the composition of colonizing commensal bacteria in the neonatal gnotobiotic pig. *J. Anim. Sci.* 85:3256–66

123. Wilson KH. 1993. The microecology of *Clostridium difficile*. *Clin. Infect. Dis.* 16(Suppl. 4):S214–18

124. Winter SE, Thiennimitr P, Winter MG, Butler BP, Huseby DL, et al. 2010. Gut inflammation provides a respiratory electron acceptor for *Salmonella*. *Nature* 467:426–29

125. Wlodarska M, Willing B, Keeney KM, Menendez A, Bergstrom KS, et al. 2011. Antibiotic treatment alters the colonic mucus layer and predisposes the host to exacerbated *Citrobacter rodentium*–induced colitis. *Infect. Immun.* 79:1536–45

126. Wu HJ, Ivanov II, Darce J, Hattori K, Shima T, et al. 2010. Gut-residing segmented filamentous bacteria drive autoimmune arthritis via T helper 17 cells. *Immunity* 32:815–27

127. Zachar Z, Savage DC. 1979. Microbial interference and colonization of the murine gastrointestinal tract by *Listeria monocytogenes*. *Infect. Immun.* 23:168–74

128. Zaph C, Du Y, Saenz SA, Nair MG, Perrigoue JG, et al. 2008. Commensal-dependent expression of IL-25 regulates the IL-23-IL-17 axis in the intestine. *J. Exp. Med.* 205:2191–98

129. Zhang H, DiBaise JK, Zuccolo A, Kudrna D, Braidotti M, et al. 2009. Human gut microbiota in obesity and after gastric bypass. *Proc. Natl. Acad. Sci. USA* 106:2365–70

The Relations Between Recombination Rate and Patterns of Molecular Variation and Evolution in *Drosophila*

Brian Charlesworth and José L. Campos

Institute of Evolutionary Biology, School of Biological Sciences, University of Edinburgh, Edinburgh EH9 3JT, United Kingdom; email: brian.charlesworth@ed.ac.uk, jcampos@staffmail.ed.ac.uk

Annu. Rev. Genet. 2014. 48:383–403

First published online as a Review in Advance on September 10, 2014

The *Annual Review of Genetics* is online at genet.annualreviews.org

This article's doi: 10.1146/annurev-genet-120213-092525

Keywords

codon usage, crossing over, efficacy of selection, Hill-Robertson interference, nucleotide site diversity, repetitive DNA

Abstract

Genetic recombination affects levels of variability and the efficacy of selection because natural selection acting at one site affects evolutionary processes at linked sites. The variation in local recombination rates across the *Drosophila* genome provides excellent material for testing hypotheses concerning the evolutionary consequences of recombination. The current state of knowledge from studies of *Drosophila* genomics and population genetics is reviewed here. Selection at linked sites has influenced the relations between recombination rates and patterns of molecular variation and evolution, such that higher rates of recombination are associated with both higher levels of variability and a greater efficacy of selection. It seems likely that background selection against deleterious mutations is a major factor contributing to these patterns in genome regions in which crossing over is rare or absent, whereas selective sweeps of positively selected mutations probably play an important role in regions with crossing over.

INTRODUCTION

The evolutionary significance of sexual reproduction primarily reflects the fact that it allows progeny to transmit a mixture of genetic material from their two parents through the process of genetic recombination. Recombination can enhance the effectiveness of natural selection through its effects on a variety of population genetics mechanisms (18, 40, 61, 113). The purpose of this paper is to describe the lessons learned from studies of *Drosophila* about the evolutionary consequences of recombination, with a special emphasis on the relations between rates of recombination and patterns of DNA sequence variation and evolution in different regions of the genome.

Research on *Drosophila* species has probably contributed more to our understanding of genetics and the genetic basis of evolutionary change than work on any other single type of organism. This, of course, can be traced to the pioneering studies of *Drosophila melanogaster* carried out in Thomas Hunt Morgan's Fly Lab. Of critical importance for the topic of this paper was the discovery of crossing over in females in most regions of the genome (104), as well as the absence of crossing over in males (105) and from the small (dot) fourth chromosome in both sexes (26) (reviewed in 10). In addition, the elucidation of the chromosomal basis of sex-linked inheritance led to the realization that the *Drosophila* Y chromosome lacks most of the genes present on the X chromosome, despite the fact that the X and Y chromosomes pair and disjoin from each other during the first division of meiosis in males (25). The Y chromosome is thus genetically degenerate compared with the X, a property shared with many other taxa (12, 27). H.J. Muller (106) recognized early on that this degeneracy is associated with the fact that the Y is normally transmitted strictly from father to son, so that there is no exchange of genetic material between the X and Y chromosomes in *Drosophila*.

Later work led to the discovery of heterochromatin as a cytologically distinctive component of chromosomes, which is mainly found in the regions around the centromeres (the pericentric heterochromatin) (68). Genes capable of mutations that lead to recognizable phenotypic differences from wild type were found to be largely lacking from the heterochromatin, as opposed to the gene-rich euchromatin, and crossing over was found to be either nonexistent or very infrequent in the heterochromatin (10, 109, 112). The heterochromatin proximal to the centromeres of the X chromosome and autosome is classified as α-heterochromatin and contains large quantities of tandemly repeated satellite sequences (10, 128) as well as transposable element (TE) sequences (53, 128). The β-heterochromatin borders on the euchromatin and is highly enriched in TE sequences (101, 147). Recent sequencing efforts have, however, shown that the autosomal and X chromosomal heterochromatin of *D. melanogaster* also contains approximately 400 functional genes (128). The Y contains only a few functional protein-coding genes, none of which have X homologs, and is largely made up of α-heterochromatin (10, 12, 83).

The use of polytene chromosomes to identify the physical locations of genes through mutations associated with deficiencies and chromosome rearrangements showed that tightly linked genes can be quite distant from each other on the chromosome itself (94, 109, 112). This reflects the fact that crossing over in the euchromatin close to the centromere is strongly suppressed in *D. melanogaster* (94). The rate of crossing over per unit physical distance [measured as centiMorgans per megabase (cM/Mb), using the reference genome sequence to establish physical map distances] tends to increase away from the centromere to a maximum in the middle of the chromosome arm, although there is considerable local variation around this general pattern (49). There is a rather sudden decline to a low value close to the tips of the chromosomes (the telomeres), especially on the X chromosome (**Figure 1**). The suppression of crossing over in the euchromatin proximal to the centromeres is caused by the centromere itself, rather than by the heterochromatin (10, 19, 112, 145). Regions of the euchromatic genome with very low rates of crossing over have higher densities of TEs than regions with high rates of crossing over (17, 84, 115, 118).

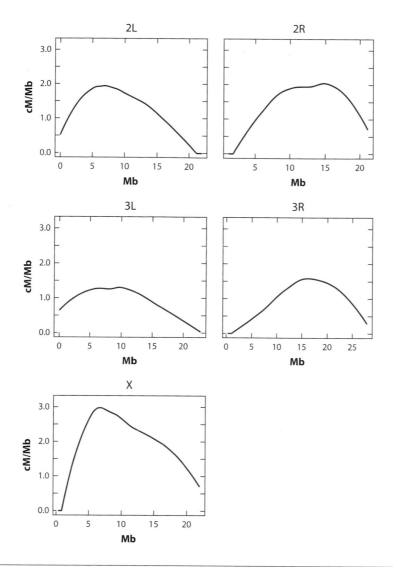

Figure 1

Plots of the rate of crossing over per unit physical distance (cM/Mb) versus location along the *Drosophila melanogaster* reference genome sequence (release 5.36) for each of the major chromosome arms (L and R indicate left and right arm, respectively, for the two major autosomes). For X, 2L, and 3L, position zero denotes the telomere; for 2R and 3R, it denotes the most centromere-proximal part of the sequenced genome. The rates of crossing over displayed were obtained using the high-density single-nucleotide polymorphism map described in Reference 49; rates of crossing over for individual loci were smoothed using Loess regressions to eliminate noise and local fluctuations in recombination rates (31). The crossover rates were adjusted by a factor of two-thirds for the X and one-half for the autosomes in order to obtain effective rates that reflect the absence of crossing over in *Drosophila* males and the fractions of their respective times in the population that an X chromosome and an autosome spend in females (89).

Interestingly, rates of gene conversion seem to be much less variable across the genome than rates of crossing over because recombination events in regions where crossing over is heavily suppressed have been detected by both conventional genetic mapping (49) and analyses of patterns of DNA sequence variability in a number of different species (9, 22, 23, 31, 34, 72, 88,

125, 140). However, the short physical distances affected by gene conversion events [the mean conversion tract length in *D. melanogaster* is estimated to be in the region of 350 to 500 bp (49, 71, 102)] mean that these events are likely to have much less influence on evolutionary processes than crossing over.

These general features of the genome are by no means unique to *Drosophila*; for example, many species of flowering plants show even stronger pericentric suppression of crossing over than *Drosophila* (6). The richness of the genetic and genomic resources available in *Drosophila* (especially *D. melanogaster*) means, however, that far more detailed information on the population genetics and evolutionary consequences of differences in rates of crossing over is available than in other systems.

Several other properties of *Drosophila* have also proved extremely useful for this purpose. In particular, comparative genetic, cytological, and genome sequencing studies of *Drosophila* species, initiated by Sturtevant (137, 138) and Muller (108), have shown that the basic chromosome complement consists of five acrocentric major chromosome arms (Muller's elements A–E) plus the small dot chromosome (element F), whose gene content is highly conserved across the genus (10, 55). Derived states have arisen from fusions between elements. In particular, there is an ancestral X and Y chromosome pair (the X representing element A), but several species have neo-X and neo-Y chromosomes resulting from fusions between an X and an autosome (e.g., *Drosophila pseudoobscura*), a Y and an autosome (*Drosophila miranda*), or both (*Drosophila albomicans*) (10, 12). Because the autosomal arm that becomes a neo-Y chromosome is immediately shut off from both gene conversion and crossing over with its partner, these cases provide illuminating natural experiments on the evolutionary effects of a complete lack of recombination over a large eukaryotic chromosome with more than 3,000 genes, given that the neo-Y chromosome is transmitted exclusively through males.

THEORETICAL BACKGROUND

In order to help understand how these features of *Drosophila* genetics and genomics can be used to test ideas about the evolutionary significance of recombination, we give a brief outline of the most important population genetics processes that are likely to be involved.

Hill-Robertson Interference

As proposed by Felsenstein (61), the process known as Hill-Robertson interference (HRI) is probably a major player in the interaction between recombination and selection. This term refers to the fact that selection acting on one site in the genome interferes with evolution at other genetically linked sites. This was first modeled in detail by Hill & Robertson (70), although the basic idea was proposed earlier by Fisher (62) and Muller (107). HRI is a consequence of the fact that only a limited number of combinations of variants at different sites can exist at a given time in a finite population (61), which creates nonrandom associations between variants at different sites [linkage disequilibrium (LD)]. This means, for example, that a favorable variant at one site may be present in a genotype with a deleterious variant at another site, causing negative LD between the selectively favorable variants at each site; this reduces the additive genetic variance in fitness, on which the response to selection depends (28). Recombination breaks down such LD and thereby enhances the population's ability to respond to selection (18, 28, 40, 50, 61). **Table 1** lists the main types of evolutionary processes that can cause HRI and is based on the classification described in Reference 40. Other evolutionary processes that can contribute to the evolutionary effects of recombination are discussed in References 18 and 113.

A useful, but somewhat crude, way of looking at HRI is to think of it in terms of the effective population size (N_e). This measures the number of individuals that successfully transmit genes

Table 1 The main types of Hill-Robertson interference effects

1. Selective sweeps of positively selected mutations
The spread of a favorable mutation drags to fixation any closely linked variants that are initially associated with it on the same haplotype, providing that any deleterious fitness effects that these variants exert do not overcome its selective advantage. Variability is reduced, the spread of other favorable mutations is impaired, and the fixation by drift of deleterious mutations is accelerated.
2. Background selection caused by deleterious mutations
Deleterious mutations enter the population at sites distributed over the genomic region in question and are removed by purifying selection with high probability. Neutral or weakly selected variants that are closely linked to these mutations are eliminated from the population at the same time so that the effective population size is reduced relative to the case when no selection is acting.
3. Hill-Robertson interference among deleterious mutations
In a sufficiently large genomic region with little or no recombination, sites causing background selection experience mutual interference, allowing drift to affect the frequencies of selectively deleterious variants. With reversible mutations between favored and disfavored alternatives, the population eventually approaches a statistical equilibrium under drift, selection, and mutation, with far more sites fixed for deleterious mutations than with high levels of recombination, and with reduced variability at neutral or nearly neutral sites.
4. Muller's ratchet
This is similar to 2, except that haplotypes lacking deleterious mutations are assumed to be sufficiently rare that they can be lost from the population by drift. It differs from 3 in the assumption that mutation is assumed to be unidirectional, from favored to disfavored alleles. In the absence of recombination, this implies that the zero-mutation class cannot be restored. The next best class then replaces it and is in turn lost in a process of successive, irreversible clicks of the ratchet. Each such loss is quickly followed by fixation of a deleterious mutation on the chromosome. Unlike 3, no equilibrium is ever reached because there are no reverse mutations. The ratchet is thus likely to apply to irreversible types of mutations, such as insertions or deletions, rather than single-nucleotide mutations.

between generations and is often much smaller than the number of breeding individuals (36, 144). Selection on heritable traits reduces N_e because genetic variants are preferentially transmitted in association with the fittest genotypes in the population down successive generations, even with free recombination (119). Close linkage between a given segregating nucleotide site and other sites that are under selection causes especially large effects on the focal site, because the effect of genetic variation in fitness is maintained longer when such linkage is tight (120).

This reduction in N_e influences both the level of neutral variability and the efficacy of selection. The expected equilibrium neutral diversity per nucleotide site is equal to the product of $4N_e$ and the mutation rate u per site, provided that $4N_e u \ll 1$ (80). Furthermore, the probability that genetic drift fixes a deleterious mutation that reduces the fitness of its homozygous carriers by s is close to the value for a neutral mutation when $N_e s \ll 1$ but is negligible when $N_e s \gg 1$ (62, 78). Similarly, the chance that a selectively favorable mutation with selective advantage s becomes established in a population is close to the neutral value when $N_e s \ll 1$ but approaches the value for an infinitely large population when $N_e s > 1$. Although not all features of HRI effects on patterns of variability and the efficacy of selection can be interpreted in terms of a reduction in N_e (36, 46, 63, 73, 74, 110, 124, 141), it nevertheless provides a useful heuristic.

If HRI is operating, and if other things (especially gene density) are equal, these theoretical results mean that regions of the genome with low rates of recombination should show both reduced levels of DNA sequence variability (especially at sites where mutations are neutral or nearly neutral) and signatures of a reduced efficacy of selection, involving positive selection (when new, favorable mutations spread to high frequencies or fixation) and purifying selection (when selection tends to eliminate deleterious mutations). The effects on the efficacy of selection are likely to be strongest when a mutation at a focal site is weakly selected relative to variants at linked sites so that it cannot break through the constraints imposed by selection at these sites (73, 77, 141, 149). We concentrate below on examining the extent to which HRI effects explain the effects of recombination rates on patterns of DNA sequence variation and evolution in *Drosophila* genomes.

Other Effects of Recombination on Genome Evolution

It is, however, important to recognize that not all aspects of the effects of recombination on properties of the genome reflect HRI. First, tandemly repeated satellite DNA sequences are likely to be selectively nearly neutral and hence are relatively insensitive to HRI as far as their probabilities of fixation are concerned (45), but they nevertheless accumulate in low recombination regions, as described above. This is probably a consequence of the fact that unequal crossing over between homologous arrays of tandem repeats generates daughter arrays with different numbers of repeats, some larger and some smaller than the parental values. In a finite population, there is therefore a finite probability that the population will become fixed for a haplotype with a single copy of the sequence; this is a terminal state in the absence of any amplification process that creates multiple copies de novo (42, 45). Models of the interaction between unequal crossing over, genetic drift, weak purifying selection against high copy numbers, and amplification processes show that large arrays of repeats are most likely to be maintained in regions that lack crossing over (131, 132). It seems likely that this explains why large arrays of satellite repeats are regularly found in very low recombination regions of the genome (45). It is, however, very hard to test this hypothesis directly, and so the population genetics of tandem arrays are not considered further.

Second, TEs that belong to the same family, and thus have a high degree of sequence homology, can pair and undergo exchange with each other, even if they are located in different places in the genome; such events occur at very low frequencies but can be detected experimentally (103). Ectopic exchanges cause chromosome rearrangements, such as duplications, deletions, inversions, and translocations, depending on the orientation of the TEs with respect to each other and whether or not they are located on the same chromosome (45, 89, 103). Many such rearrangements have deleterious effects on the fitness of the resulting progeny, creating a selective disadvantage to the TEs in question (89, 103). Other things being equal, it is likely that TEs are subject to less ectopic exchange in genome regions with low rates of recombination, and hence tend to accumulate there under the pressure of transpositional increase in their copy number and genetic drift (45, 89). Of course, the weakening through HRI of the efficacy of purifying selection against any negative fitness effects of TE insertions would also tend to cause their accumulation in low recombination regions (118); it is a challenge to distinguish between this effect and that of reduced frequencies of ectopic exchange (see below).

POPULATION GENOMIC CORRELATES OF RECOMBINATION RATES IN *DROSOPHILA*

We now describe some of the major patterns of relationship between recombination rates and measures of DNA sequence variation and evolution in *Drosophila*.

Levels of Genetic Variability

Early studies of natural variation at the DNA sequence level, mostly using restriction mapping of cloned regions of the genome of *D. melanogaster* and *Drosophila ananassae*, found that variability was severely reduced in telomeric and pericentric euchromatic regions of the major chromosomes, where the rate of crossing over is low (2, 3, 136), and on the dot chromosome of *D. melanogaster*

(21). Begun & Aquadro (20) then showed that there was a positive correlation across the *D. melanogaster* genome between the local rate of crossing over experienced by a gene and its level of silent variability.

These results have been confirmed by later studies (for example, see 117), including recent results from whole-genome resequencing projects on *D. melanogaster* populations (31, 90, 97), as can be seen in **Figure 2**; similar patterns have been found in *D. pseudoobscura* (99). The most extreme reduction in variability is found in regions such as the dot chromosome, heterochromatin, and Y or neo-Y chromosomes, which completely lack crossing over under normal conditions [noncrossover (NC) regions]; an approximately tenfold reduction in silent site variability has been found on the dot chromosome in several different species (9, 22, 31, 58, 72, 125, 140), with the illuminating exception of *Drosophila willistoni*, where the dot is fused to element E (3R in *D. melanogaster*) and experiences crossing over (116).

Comparable reductions are seen for genes in the other NC regions in *D. melanogaster*, such as the pericentric heterochromatin of the two major autosomes and the telomere of the X chromosome (1–3, 23, 31, 88, 90, 97), with the exception of the X-chromosome pericentric heterochromatin, where the reduction is approximately half this value (31). Interestingly, there is an 80-kb region close to the telomere of the X chromosome where recombination rates increase considerably relative to the rest of the distal portion of the X, which is associated with an increase in variability (5). A nearly 100-fold reduction in diversity relative to the rest of the genome is seen on the *D. miranda* neo-Y chromosome (15). This is an entire Muller element C that has become fused to the ancestral Y chromosome following the split of *D. miranda* from *D. pseudoobscura* (10, 130). The sex-ratio distorter X chromosome of *Drosophila recens*, which is associated with a chromosome-wide suppression of crossing over caused by a complex of inversions, similarly exhibits drastically reduced levels of DNA sequence variability (57). There is also evidence for greatly reduced variability on the Y chromosomes of several *Drosophila* species (14, 91).

Importantly, the level of silent site divergence of *D. melanogaster* from closely related *Drosophila* species has little relation to recombination rate (20, 92) and even shows a slight increase in the NC regions (30, 66) (see **Figure 2**). Because the rate of neutral or nearly neutral evolution is strongly determined by the mutation rate (79), this implies that the relation between diversity and crossover rate cannot be caused by any mutagenic effect of recombination (20). It follows that some form of selection at linked sites must be responsible for the relation between recombination rates and variability, such that N_e is higher in genomic regions with higher recombination rates (see the above section Theoretical Background).

Codon Usage

When the sites that cause HRI are much more strongly selected than the sites that experience it, the reduction in the effectiveness of selection at the latter sites should be closely related to the reduction in N_e for neutral sites (73, 77, 135, 149). Codon usage bias (CUB) is an example of a trait for which the intensity of selection in *Drosophila* is very small; even for regions of the genome with normal crossing over rates, the typical intensity of selection on synonymous variants causing a change in codon usage (measured by $N_e s$) has recently been estimated to be approximately 0.25 (32, 51, 67, 148). According to the standard Li-Bulmer model of selection on CUB (29, 93), the equilibrium level of CUB of a gene under the effects of mutation, selection, and drift is determined jointly by $N_e s$ and the level of mutational bias (the ratio of the mutation rate from preferred to unpreferred codons to the reverse mutation rate). Given that codons preferred by natural selection tend to end in G or C in *Drosophila* (4), the GC content (defined as the sum of the frequencies of G and C nucleotides) of third coding positions is similarly affected.

A tenfold reduction in N_e, such as that seen in most of the NC regions of the *D. melanogaster* genome (see above), should therefore reduce $N_e s$ for synonymous mutations to close to zero. If CUB is close to equilibrium, NC regions should show little evidence for CUB, and have a GC content in the third coding position that is much smaller than that in regions with normal levels of crossing over, given the strong mutational bias from GC to AT base pairs in *Drosophila* (76, 123, 148). This is indeed observed in *D. melanogaster* (30, 66, 82). Gene expression levels do not seem to be reduced in the NC regions (31), implying that this effect is not a reflection of the well-known relation between CUB and gene expression (56).

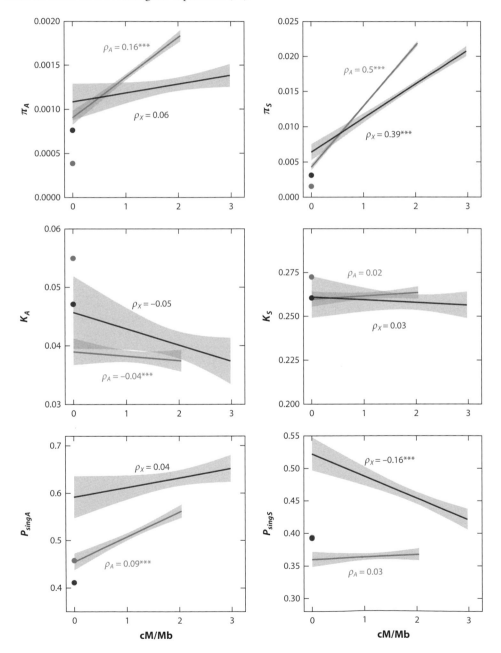

The Li-Bulmer model also predicts a positive correlation between CUB and crossing-over rate in genomic regions that do experience crossing over (C regions). This is not seen in the data on *D. melanogaster*, where there is little or no such correlation in the autosomal C regions, and a slightly negative correlation for the X chromosome C region (32, 127). The reason for these discrepancies is currently unclear. One possibility is that the recombinational landscape in *D. melanogaster* has evolved over a timescale that is much shorter than the one required for the equilibration of codon usage (127); this is qualitatively consistent with the substantial differences in the amount and chromosomal distribution of crossovers between *D. melanogaster* and its close relatives (47, 137, 139). Another possibility is that codon usage is subject to stabilizing selection rather than directional selection, as proposed by Kimura (81); weak stabilizing selection with mutational bias and free recombination has recently been shown to lead to insensitivity to N_e of both the trait mean and $N_e s$ for individual nucleotide variants (39). This would lead to a lack of correlation between CUB and recombination rate, except when recombination rates are drastically reduced and HRI effects are important.

Purifying Selection on Nonsynonymous Mutations

The positive correlation between recombination rate and N_e discussed above suggests that low recombination regions should show evidence of a reduced efficacy of purifying selection, allowing deleterious mutations to reach higher frequencies within the population than would otherwise be possible. In agreement with this expectation, K_A (the fraction of nonsynonymous sites that differ between a pair of related species) is elevated in the NC regions of the *D. melanogaster* genome relative to its value in C regions, whereas there is only a small increase in K_S (the fraction of synonymous sites that differ) (30, 66), as shown in **Figure 2**.

Similarly, the neo-Y chromosome of *D. miranda* shows an accelerated rate of fixation of nonsynonymous mutations since its divergence from its common ancestor with the neo-X (146, 150). One difficulty in interpreting this observation is that many genes on the neo-Y are either inactivated by loss-of-function mutations or are expressed at a low level relative to their neo-X counterparts (150). Both of these events would allow the accumulation of otherwise deleterious nonsynonymous mutations, and indeed the fastest rates of protein sequence evolution are seen in such genes; nevertheless, functional and expressed genes on the neo-Y chromosome also show a higher rate of nonsynonymous substitutions than their homologs on the neo-X chromosome (150).

These observations could also be explained by a higher rate of fixation of advantageous nonsynonymous mutations in NC regions, especially for the *D. miranda* neo-Y chromosome,

Figure 2

Plots of various population genetics and evolutionary statistics against the effective rates of crossing over per megabase shown in **Figure 1** for regions of the autosomes (*green*) and X chromosome (*red*) of the *Drosophila melanogaster* genome that have nonzero rates of crossing over (the estimated rates of crossing over in females have been multiplied by 2/3 for the X and by 1/2 for the autosomes). The green and red dots are the mean values for the noncrossover regions of the autosomes and X chromosome, respectively. The lines are linear regression lines, and the shaded areas represent 95% confidence intervals for the predicted means of the dependent variables. Subscripts A and S for the dependent variables refer to differences at nonsynonymous and synonymous sites, respectively; K is the divergence per nucleotide site from *Drosophila yakuba*; π is the level of diversity per nucleotide site in the Rwandan population analyzed in Reference 31; P_{sing} is the proportion of nucleotide variants present only once at a site in the Rwandan sample of 17 sequenced haploid genomes; ρ is the Spearman rank correlation coefficient between the pairs of variables in each plot, excluding the noncrossover regions. The three asterisks (***) indicate statistical significance at the $p = 0.001$ level for the Spearman rank correlations (ρ) shown in the figure, excluding the noncrossover regions.

where some previously autosomal genes may be becoming adapted to transmission exclusively through males (150). Several lines of evidence suggest, however, that a relaxed efficacy of purifying selection in a low recombination environment is primarily responsible. First, as shown in **Figure 2**, the diversity per nucleotide site for nonsynonymous variants (π_A) is reduced in the NC regions of *D. melanogaster* but to a smaller extent than that for synonymous variants (π_S) (31, 117). This suggests that, in the absence of crossing over, selection is less effective at eliminating slightly deleterious amino acid mutations (whose frequencies are influenced by drift as well as selection); if segregating nonsynonymous mutations were so strongly disadvantageous that their frequencies are determined solely by mutation-selection balance, their nucleotide site diversity would be independent of N_e (96). Second, the ratio of the number of derived nonsynonymous variants to the number of derived synonymous variants is much higher in the *D. melanogaster* NC regions compared with the C regions, even for variants that are present at intermediate frequencies and are thus sufficiently weakly selected that their frequencies are influenced by genetic drift (31). Third, for the C regions of the *D. melanogaster* autosomes, measures of the abundance of rare nonsynonymous variants relative to common ones, such as the proportion of variants at a site that are singletons (i.e., present only once at a given site in the sample), show a positive correlation with crossover rates (31), indicating that selection is more effective at keeping nonsynonymous variants at low frequencies when recombination rates are high (see **Figure 2**). Finally, as shown in **Figure 3**, the use of statistical methods for estimating the distribution of the effects on fitness of new, deleterious nonsynonymous mutations from polymorphism data (59, 85) reveals that the proportion of these that fall in the nearly neutral range of $N_e s$ values (between 0 and 1) is strongly negatively correlated with the recombination rate in the C regions of the *D. melanogaster* genome (31).

Positive Selection on Nonsynonymous Mutations

There is also evidence that the efficacy of positive selection in favor of new nonsynonymous mutations is enhanced by recombination. This comes from the application of methods that use data on within-species polymorphism and between-species divergence, for both nonsynonymous and synonymous variants, to infer the extent to which nonsynonymous mutations are subject to positive selection (59, 60, 85, 122, 129, 142). Two measures are commonly used: α, the proportion of nonsynonymous differences between a pair of related species that have been fixed by positive selection (129), and ω_α, the ratio of the rate of substitution of positively selected nonsynonymous

---

Figure 3

Plots of the values of P_{neut}, α, and ω_α for bins of genes of *Drosophila melanogaster* against the effective rates of crossing over per megabase shown in **Figure 1** for regions of the autosomes (*green*) and X chromosome (*red*) of the genome that have nonzero rates of crossing over (the estimated rates of crossing over in females have been multiplied by 2/3 for the X and by 1/2 for the autosomes). The bins were chosen to contain approximately equal numbers of genes (31). The green and red dots are the mean values for the noncrossover regions of the autosomes and X chromosome, respectively. The lines are linear regression lines, and the shaded areas represent 95% confidence intervals for the predicted means of the dependent variables; P_{neut} is the proportion of new, nonsynonymous mutations that have $N_e s$ values between 0 and 1, and hence behave as nearly neutral (estimated as described in Reference 59); α is the estimated proportion of nonsynonymous differences between *Drosophila melanogaster* and *Drosophila yakuba* that have been fixed by positive selection; ω_α is the ratio of the rate of substitution of positively selected nonsynonymous mutations to the rate for synonymous substitutions (which are presumed to be nearly neutral); ρ is the Spearman rank correlation coefficient between the pairs of variables in each plot, excluding the noncrossover regions. The single (*), double (**), and triple asterisks (***) indicate statistical significance at the $p = 0.05, 0.01$, and 0.001 levels, respectively, for the Spearman rank correlations (ρ) shown in the figure, excluding the noncrossover regions.

mutations to the rate of synonymous substitutions (which are presumed to be nearly neutral) (65). These measures are not independent of each other because $\omega_\alpha = \alpha K_A / K_S$.

For the *D. melanogaster* NC regions analyzed from the Malawi (90) and Rwandan (31) genome-wide polymorphism data sets, and for the dot chromosome in several species (9, 22), there is no evidence for significantly positive values of either of these statistics, as can be seen in **Figure 3** (note that these estimates are noisy because there is only a limited number of truly independent

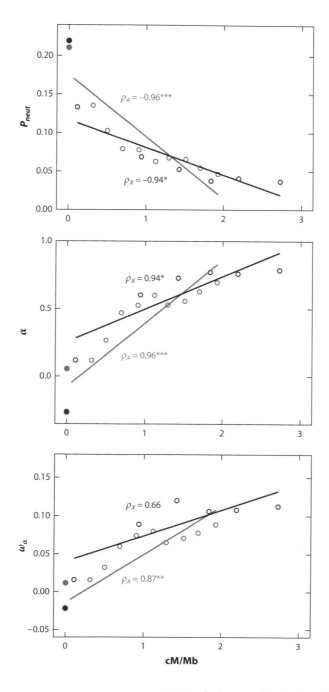

data points, owing to the strong correlations among different sites within an NC region; negative values can thus arise purely by chance). This is in contrast to the significantly positive values found across most of the C regions, as described next.

These studies used a method that compares the ratio of the numbers of nonsynonymous and synonymous variants to the corresponding ratio for fixed differences between species (60, 129, 142); this is downwardly biased in the absence/reduction of purifying selection against deleterious mutations (100). This bias can be avoided by using the DFE-α method, which compares the observed number of fixations of nonsynonymous variants to the number of fixations of deleterious nonsynonymous mutations predicted from the estimated distribution of mutational effects on fitness (59, 85). Although this method may itself be biased in the absence of crossing over because of the high degree of LD among sites in NC regions, it should be valid when there is some crossing over, given that LD is relatively small in magnitude in the C regions (122). For the C regions in the Rwandan population of *D. melanogaster*, both α and ω_α are positively correlated with the rate of recombination (31), as shown in **Figure 3**. Similar patterns are found using the alternative method for both the Malawi (90) and Rwandan (31) populations.

Transposable Element Abundances in Low Recombination Regions

As mentioned in the Introduction, there is an excess of TEs or TE-derived sequences in the centromere-proximal and β-heterochromatin of *D. melanogaster*, where crossing over is rare or absent. Population genetics surveys have shown that lower rates of genetic recombination are associated with both higher frequencies of TEs at segregating sites and more sites fixed for TEs than elsewhere in the genome (16, 43, 84, 98, 115). The TEs involved tend to be eroded in length, suggesting a long duration at their sites of insertion without transpositional activity; some heterochromatic elements have also experienced tandem duplication events subsequent to their insertion (41, 52, 147).

Furthermore, it seems likely that the negative association between intron length and rate of recombination exhibited by the genome of *D. melanogaster* (33, 48) is at least partly caused by the accumulation of TE insertions in regions with low recombination, because the density of TE-derived sequences is enriched within introns in these regions, especially on the fourth chromosome, which is a major contributor to this association (17). This effect is especially large for genes located in the heterochromatin, whose introns are large and mainly composed of TE-derived sequences (54, 128). A spectacular example of this is provided by the giant (megabase long) introns of the fertility factors of the Y chromosome, notably in *Drosophila hydei*, whose megabase introns are made up largely of TE-derived sequences (87).

The newly evolved neo-Y chromosome of *D. miranda* provides an opportunity to study the early stages of TE accumulation in a nonrecombining part of the genome. Sequences of 2.5 Mb of BAC clones from the neo-sex chromosomes of *D. miranda* showed that approximately 20% of the DNA sequence of the neo-Y chromosome is TE-derived, compared with only approximately 1% of the homologous portion of the neo-X chromosome (13); the TEs on the neo-Y are longer than on the neo-X, consistent with their being recent insertions of active elements. Population surveys of a subset of these elements show that they are fixed on the neo-Y but mostly polymorphic elsewhere in the genome (11).

WHAT CAUSES THE RELATIONS BETWEEN RECOMBINATION RATE, LEVELS OF VARIABILITY, AND EFFICACY OF SELECTION?

The results described above show that the rate of genetic recombination in a region of the genome affects the amount of variability and extent of adaptation at the level of DNA and protein sequences.

This raises the question of which of the processes listed in **Table 1** are likely to be causally involved. Several of the patterns described above have been used in attempts to answer this question.

Causes of the Relations Between Levels of Variability and the Recombination Rate

For more than twenty years, researchers on *Drosophila* population genetics have debated the relative importance of selective sweeps versus background selection (processes 1 and 2 in **Table 1**) in causing the positive correlation between silent site diversity and the local rate of crossing over; evidence for effects of both of these processes is reviewed in Reference 134. Several different approaches have been used. One is to fit the predictions of either type of model to the observed relation between the level of silent nucleotide site diversity and the local rate of crossing over, mainly in C regions. Using the relatively sparse data on *D. melanogaster* available in the mid-1990s, it proved possible to fit both types of model to the data, using plausible assumptions (35, 133, 143). This approach thus proved inconclusive; however, a recent study suggests that background selection explains a large fraction of the variation in nucleotide site diversity across chromosome arms in *D. melanogaster* (47).

Causes of Distorted Site Frequency Spectra in Noncrossover Regions

An alternative approach has been to examine whether or not the site frequency spectrum (SFS) for supposedly neutral or nearly neutral silent sites in NC regions fits the predictions of the different models (the SFS describes the proportions of nucleotide sites with variants at different frequencies). An almost complete lack of recombination means that a recent hard selective sweep (involving an advantageous mutation that arose in a unique haplotype) completely wipes out variability. [Note that sweeps can be caused by selection at the level of gametes as well as individuals; distorted segregation associated with centromeric and telomeric drive has been proposed as an explanation for reduced variability in NC regions (5, 151); direct evidence for this is currently lacking.] Any variability in such a region therefore represents the input of new mutations, which will not have had time to drift to high frequencies, meaning there will be many more polymorphic sites with variants at low frequencies than expected under equilibrium between mutation and genetic drift (24, 126). Evidence for such a pattern in low recombination regions has been obtained in some conventional resequencing studies (8, 23) as well as in the recent genome-wide resequencing studies (31, 90, 97).

Unfortunately, background selection can also produce a considerable excess of rare neutral variants when recombination is infrequent and selection is weak (46, 64, 74, 111, 124), so this excess does not distinguish between the two processes. However, studies of patterns of variation on the dot chromosome in *D. melanogaster*, *Drosophila simulans* (72), and *Drosophila americana* (22), as well as in the five *D. melanogaster* NC regions (31), show that the skew in the SFS toward rare variants is not sufficiently large to be consistent with a hard selective sweep.

This conclusion must be interpreted with care. In particular, as mentioned above, there is evidence for gene conversion in the NC regions, meaning that the assumption of no recombination is too strong; recombination during a sweep is likely to reduce the skew in the SFS (31). However, calculations of the effect of a hard sweep in an NC region using current estimates of gene conversion rates show that an implausibly large selective advantage of the sweeping mutation is needed to produce the observed reduction in diversity in NC regions of *D. melanogaster* (31). Of course, the sweep could have been soft rather than hard, with the advantageous mutations concerned arising on several different haplotypes, leading to a much lower skew than with a hard sweep (69, 114).

However, this would also lead to a smaller reduction in diversity than under a hard sweep so that even stronger selection would be required, suggesting that this possibility is unlikely. The lack of evidence for adaptive evolution of protein sequences in the NC regions of the X chromosome and autosomes of *D. melanogaster* and on the dot chromosomes of some other chromosomes, discussed above, also suggests that sweeps due to positive selection at the individual level are rare in the NC regions, although gametic selection (5, 151) is not ruled out.

It is thus hard to reconcile the data on the NC regions with a selective sweep model. Can background selection explain the observations? Calculations using estimates of the distribution of fitness effects of deleterious nonsynonymous mutations and the mutation rate for *D. melanogaster* show that the standard background selection model predicts an approximately 1,000-fold reduction in N_e for the dot chromosome, given the number of genes (around 80) that it contains (95); this is far greater than the estimate of an approximately tenfold reduction from surveys of silent site diversity. Similar results apply to the other NC regions (31).

This dilemma can be resolved as follows. The standard background selection model assumes that the deleterious mutations involved are at equilibrium under the deterministic balance between mutation and selection (37, 44). In a genomic region that lacks crossing over, HRI occurs among these mutations. This is equivalent to a reduction in the effectiveness of selection, leading to an increased rate of fixation of deleterious mutations, higher frequencies at segregating sites, and a reduction in their ability to reduce the level of variability at linked neutral or nearly neutral sites (74).

Use of a mutation rate and distribution of selection coefficients for nonsynonymous mutations that are close to those estimated for *Drosophila* shows that the expected reductions in neutral variability for regions that correspond in size to the five independent NC regions of *Drosophila melanogaster* are much smaller than the standard background selection predictions (74) and fit the observed values for the Rwandan population of *Drosophila melanogaster* reasonably well (31). Importantly, as the number of selected sites in an NC region increases, the expected reduction in neutral diversity approaches an asymptotic value, presumably because of the increasing intensity of HRI among the mutations involved. With the *Drosophila* selection and mutation parameters, the asymptotic level of neutral diversity approaches 1% of its value with free recombination; this prediction agrees with the findings for the neo-Y of *D. miranda* (15, 74), which still contains more than 1,000 active genes (150).

It thus seems that typical levels of diversity in NC regions can be explained without appealing to selective sweeps. However, African populations of *D. melanogaster* seem to have experienced a recent selective sweep on their Y chromosome because they exhibit greatly reduced diversity compared with non-African populations, even though these also show reduced variability on the Y compared with the rest of the genome (91).

Causes of Distorted Site Frequency Spectra in Regions with Crossing Over

The interpretation of the relation between the site frequency spectra and recombination rates in the C regions of the genome is not straightforward. The autosomes show little effect of recombination rate on the degree of distortion in favor of rare synonymous variants, whereas the X chromosome shows a significant decline with increasing recombination, consistent with a greater intensity of hitchhiking effects when recombination is infrequent (31), as shown in **Figure 2**. The level of distortion for the X chromosome is also larger than for autosomal genes, even for similar effective rates of crossing over (see the caption for **Figure 1** for an explanation of this measure).

The only obvious explanation for this larger distortion of the SFS on the X chromosome is a higher intensity of hitchhiking effects due to positive selection, consistent with the substantially higher values of α and ω_α in the C regions of the X compared with the autosomes (31) (see

Figure 3). Other evidence also suggests that the SFSs of genes in the C regions of both the X chromosome and the autosomes have been distorted by sweeps (90), especially the negative relation between K_A and the synonymous diversity of a gene in *D. melanogaster*, at least for genes with high K_A values (7, 67). The overall patterns of levels of variability and degrees of distortion of the SFSs along the X and autosomes in this population are, however, quite complex and difficult to interpret without appealing to the joint effects of background selection, selective sweeps, and past changes in population size (31). Detailed modeling of the chromosome-wide effects of all of these processes is needed before we can reach definitive conclusions.

Accumulation of Transposable Elements in Nonrecombining Genomic Regions

As already mentioned, there is long-standing evidence for the accumulation of TEs in regions of the *Drosophila* genome that largely or completely lack crossing over. However, early studies using in situ hybridization to study TE frequencies in populations found it hard to discriminate between the hypothesis that HRI causes less effective selection against the deleterious fitness effects of insertions in these regions (118) and the alternative possibility that a lack of ectopic exchange in such regions removes any selection against insertions into sequences where they do not directly affect fitness, and can therefore be fixed by drift (89). More recent evidence from population surveys using genomic methods seems to favor the ectopic exchange hypothesis, especially as the overwhelming majority of TE insertions occur outside coding sequences, where their direct fitness effects are likely to be small (16, 84, 115). In particular, families of TEs that are greater in length, and families with higher mean copy numbers in the genome, tend to have insertions that segregate at lower frequencies within the population than other families; this is what is expected under the ectopic exchange model, given that members of such families are more likely to encounter each other and undergo an exchange (84, 115).

DISCUSSION

The evidence described above shows that selection at linked sites has strongly influenced the relations between recombination rates within different genome regions and their patterns of molecular variation and adaptation across the *Drosophila* genome. It seems clear that higher rates of recombination facilitate both higher levels of nucleotide site variability and more effective positive and purifying selection. This is in agreement with theoretical predictions and underscores the evolutionary significance of sexual reproduction. Both background selection and selective sweeps appear to play a role, with the former probably predominating in genome regions where crossing over is negligible.

Because of space limitations, little has been said about other phenomena in which the evolutionary consequences of recombination play an important role in shaping the *Drosophila* genome, such as the forces that maintain inversion polymorphisms (86, 121) and the causes of the loss of gene function on evolving Y chromosomes, where Muller's ratchet (**Table 1**, Section 4) has probably played an important role (75). In addition, we have not described the contribution of selection at linked sites to the distinctive patterns of differences in levels of variability between X chromosomes and autosomes seen in different *Drosophila* species; this was discussed in Reference 38 in the context of background selection.

DISCLOSURE STATEMENT

The authors are not aware of any affiliations, memberships, funding, or financial holdings that might be perceived as affecting the objectivity of this review.

ACKNOWLEDGMENTS

We thank Deborah Charlesworth for her comments on the manuscript and the Biotechnology and Biological Sciences Research Council of the United Kingdom for financial support.

LITERATURE CITED

1. Aguadé M, Meyers W, Long AD, Langley CH. 1994. Reduced DNA sequence polymorphism in the *su(s)* and *su(wa)* regions of *Drosophila melanogaster* as revealed by SSCP and stratified DNA sequencing. *Proc. Natl. Acad. Sci. USA* 91:4658–62
2. Aguadé M, Miyashita N, Langley CH. 1989. Reduced variation in the *yellow-achaete-scute* region in natural populations of *Drosophila melanogaster*. *Genetics* 122:607–15
3. Aguadé M, Miyashita N, Langley CH. 1989. Restriction-map variation at the *zeste-tko* region in natural populations of *Drosophila melanogaster*. *Mol. Biol. Evol.* 6:123–30
4. Akashi H. 1994. Synonymous codon usage in *Drosophila melanogaster*: natural selection and translational accuracy. *Genetics* 136:927–35
5. Anderson JA, Song Y-S, Langley CH. 2008. Molecular population genetics of *Drosophila* subtelomeric DNA. *Genetics* 178:477–87
6. Anderson LK, Doyle GG, Brigham B, Carter J, Hooker KD, et al. 2003. High resolution crossover maps for each bivalent of *Zea mays* using recombination nodules. *Genetics* 165:849–65
7. Andolfatto P. 2007. Hitchhiking effects of recurrent beneficial amino acid substitutions in the *Drosophila melanogaster* genome. *Genome Res.* 17:1755–62
8. Andolfatto P, Przeworski M. 2001. Regions of lower crossing over harbor more rare variants in African populations of *Drosophila melanogaster*. *Genetics* 158:657–65
9. Arguello JR, Zhang Y, Kado T, Fan CZ, Zhao RP, et al. 2010. Recombination yet inefficient selection along the *Drosophila melanogaster* subgroup's fourth chromosome. *Mol. Biol. Evol.* 27:848–61
10. Ashburner M, Golic KG, Hawley RS. 2005. *Drosophila. A Laboratory Handbook*. Cold Spring Harbor, NY: Cold Spring Harbor Press
11. Bachtrog D. 2003. Accumulation of Spock and Worf, two novel non-LTR retrotransposons, on the neo-Y chromosome of *Drosophila miranda*. *Mol. Biol. Evol.* 20:173–81
12. Bachtrog D. 2013. Y-chromosome evolution: emerging insights into processes of Y-chromosome degeneration. *Nat. Rev. Genet.* 14:113–24
13. Bachtrog D, Hom E, Wong KM, Maside X, De Jong P. 2008. Genomic degradation of a young Y chromosome in *Drosophila miranda*. *Genome Biol.* 9:R30
14. Bachtrog D, Thornton K, Clark A, Andolfatto P. 2006. Extensive introgression of mitochondrial DNA relative to nuclear genes in the *Drosophila yakuba* species group. *Evolution* 60:292–302
15. Bartolomé C, Charlesworth B. 2006. Evolution of amino-acid sequences and codon usage on the *Drosophila miranda* neo-sex chromosomes. *Genetics* 174:2033–44
16. Bartolomé C, Maside X. 2004. The lack of recombination drives the fixation of transposable elements on the fourth chromosome of *Drosophila melanogaster*. *Genet. Res.* 83:91–100
17. Bartolomé C, Maside X, Charlesworth B. 2002. On the abundance and distribution of transposable elements in the genome of *Drosophila melanogaster*. *Mol. Biol. Evol.* 19:926–37
18. Barton NH. 2010. Genetic linkage and natural selection. *Philos. Trans. R. Soc. B* 365:2559–69
19. Beadle GW. 1932. A possible influence of the spindle fibre on crossing-over in *Drosophila*. *Proc. Natl. Acad. Sci. USA* 18:160–65
20. Begun DJ, Aquadro CF. 1992. Levels of naturally occurring DNA polymorphism correlate with recombination rate in *Drosophila melanogaster*. *Nature* 356:519–20
21. Berry AJ, Ajioka JW, Kreitman M. 1991. Lack of polymorphism on the *Drosophila* fourth chromosome resulting from selection. *Genetics* 129:1111–17
22. Betancourt AJ, Welch JJ, Charlesworth B. 2009. Reduced effectiveness of selection caused by lack of recombination. *Curr. Biol.* 19:655–60
23. Braverman J, Lazzaro BP, Aguadé M, Langley CH. 2005. DNA sequence polymorphism and divergence at the *erect wing* and *suppressor of sable* loci of *Drosophila melanogaster* and *D. simulans*. *Genetics* 170:1153–65

24. Braverman JM, Hudson RR, Kaplan NL, Langley CH, Stephan W. 1995. The hitchhiking effect on the site frequency spectrum of DNA polymorphism. *Genetics* 140:783–96

25. Bridges CB. 1918. Non-disjunction as proof of the chromosome theory of heredity. *Genetics* 1:1–52,107–63

26. Bridges CB. 1935. The mutants and linkage data of chromosome four of *Drosophila melanogaster. Biol. Zh.* 4:401–20

27. Bull JJ. 1983. *Evolution of Sex Determining Mechanisms*. Menlo Park, CA: Benjamin Cummings

28. Bulmer MG. 1974. Linkage disequilibrium and genetic variability. *Genet. Res.* 23:281–89

29. Bulmer MG. 1991. The selection-mutation-drift theory of synonymous codon usage. *Genetics* 129:897–907

30. Campos JL, Charlesworth B, Haddrill PR. 2012. Molecular evolution in nonrecombining regions of the *Drosophila* genome. *Genome Biol. Evol.* 4:278–88

31. Campos JL, Halligan DL, Haddrill PR, Charlesworth B. 2014. The relationship between recombination rate and patterns of molecular evolution and variation in *Drosophila melanogaster. Mol. Biol. Evol.* 31:1010–28

32. Campos JL, Zeng K, Parker DJ, Charlesworth B, Haddrill PR. 2013. Codon usage bias and effective population sizes on the X chromosome versus the autosomes in *Drosophila melanogaster. Mol. Biol. Evol.* 30:811–23

33. Carvalho AB, Clark AG. 1999. Intron size and natural selection. *Nature* 401:344

34. Chan AH, Jenkins PA, Song YS. 2012. Genome-wide fine-scale recombination rate variation in *Drosophila melanogaster. PLOS Genet.* 8:e1003090

35. Charlesworth B. 1996. Background selection and patterns of genetic diversity in *Drosophila melanogaster. Genet. Res.* 68:131–50

36. Charlesworth B. 2009. Effective population size and patterns of molecular evolution and variation. *Nat. Rev. Genet.* 10:195–205

37. Charlesworth B. 2012. The effects of deleterious mutations on evolution at linked sites. *Genetics* 190:1–18

38. Charlesworth B. 2012. The role of background selection in shaping patterns of molecular evolution and variation: evidence from the *Drosophila* X chromosome. *Genetics* 191:233–46

39. Charlesworth B. 2013. Stabilizing selection, purifying selection and mutational bias in finite populations. *Genetics* 194:955–71

40. Charlesworth B, Betancourt AJ, Kaiser VB, Gordo I. 2010. Genetic recombination and molecular evolution. *Cold Spring Harb. Symp. Quant. Biol.* 74:177–86

41. Charlesworth B, Jarne P, Assimacopoulos S. 1994. The distribution of transposable elements within and between chromosomes in a population of *Drosophila melanogaster*. III. Element abundances in heterochromatin. *Genet. Res.* 64:183–97

42. Charlesworth B, Langley CH, Stephan W. 1986. The evolution of restricted recombination and the accumulation of repeated DNA sequences. *Genetics* 112:947–62

43. Charlesworth B, Lapid A, Canada D. 1992. The distribution of transposable elements within and between chromosomes in a population of *Drosophila melanogaster*. II. Inferences on the nature of selection against elements. *Genet. Res.* 60:115–30

44. Charlesworth B, Morgan MT, Charlesworth D. 1993. The effect of deleterious mutations on neutral molecular variation. *Genetics* 134:1289–303

45. Charlesworth B, Sniegowski P, Stephan W. 1994. The evolutionary dynamics of repetitive DNA in eukaryotes. *Nature* 371:215–20

46. Charlesworth D, Charlesworth B, Morgan MT. 1995. The pattern of neutral molecular variation under the background selection model. *Genetics* 141:1619–32

47. Comeron J. 2014. Background selection as baseline for nucleotide variation across the *Drosophila* genome. *PLOS Genet.* 10:e1004434

48. Comeron JM, Kreitman M. 2000. The correlation between intron length and recombination in *Drosophila*: equilibrium between mutational and selective forces. *Genetics* 156:1175–90

49. Comeron JM, Ratnappan R, Bailin S. 2012. The many landscapes of recombination in *Drosophila melanogaster. PLOS Genet.* 8:e1002905

50. Comeron JM, Williford A, Kliman RM. 2008. The Hill-Robertson effect: evolutionary consequences of weak selection in finite populations. *Heredity* 100:19–31

51. de Procé SM, Zeng K, Betancourt AJ, Charlesworth B. 2012. Selection on codon usage and base composition in *Drosophila americana*. *Biol. Lett.* 8:82–85

52. Dimitri P. 1997. Constitutive heterochromatin and transposable elements in *Drosophila melanogaster*. *Genetica* 100:85–93

53. Dimitri P, Corradini N, Rossi F, Mei E, Zhimulev IF, Verni F. 2005. Transposable elements as architects of the heterochromatic genome in *Drosophila melanogaster*. *Cytogenet. Genome Res.* 110:165–72

54. Dimitri P, Junakovic N, Arca B. 2003. Colonization of heterochromatic genes by transposable elements in *Drosophila*. *Mol. Biol. Evol.* 20:503–12

55. *Drosophila* 12 Genomes Proj. Consort. 2007. Evolution of genes and genomes on the *Drosophila* phylogeny. *Nature* 450:203–18

56. Drummond DA, Wilke CO. 2008. Mistranslation-induced protein misfolding as a dominant constraint on coding-sequence evolution. *Cell* 13:341–52

57. Dyer KA, Charlesworth B, Jaenike J. 2007. Chromosome-wide linkage disequilibrium as a consequence of meiotic drive. *Proc. Natl. Acad. Sci. USA* 104:1587–92

58. Dyer KA, White BE, Bray MJ, Piqué DG, Betancourt AJ. 2011. Molecular evolution of a Y chromosome to autosome gene duplication in *Drosophila*. *Mol. Biol. Evol.* 28:1293–306

59. Eyre-Walker A, Keightley PD. 2009. Estimating the rate of adaptive mutations in the presence of slightly deleterious mutations and population size change. *Mol. Biol. Evol.* 26:2097–108

60. Fay J, Wykhoff GJ, Wu C-I. 2002. Testing the neutral theory of molecular evolution with genomic data from *Drosophila*. *Nature* 415:1024–26

61. Felsenstein J. 1974. The evolutionary advantage of recombination. *Genetics* 78:737–56

62. Fisher RA. 1930. *The Genetical Theory of Natural Selection*. Oxford: Oxford Univ. Press

63. Gordo I, Charlesworth B. 2001. The speed of Muller's ratchet with background selection, and the degeneration of Y chromosomes. *Genet. Res.* 78:149–62

64. Gordo I, Navarro A, Charlesworth B. 2002. Muller's ratchet and the pattern of variation at a neutral locus. *Genetics* 161:835–48

65. Gossmann TI, Song B-H, Windsor AJ, Mitchell-Olds T, Dixon CJ, et al. 2010. Genome wide analyses reveal little evidence for adaptive evolution in many plant species. *Mol. Biol. Evol.* 27:1822–32

66. Haddrill PR, Halligan DL, Tomaras D, Charlesworth B. 2007. Reduced efficacy of selection in regions of the *Drosophila* genome that lack crossing over. *Genome Biol.* 8:R18.1–7

67. Haddrill PR, Zeng K, Charlesworth B. 2011. Determinants of synonymous and nonsynonymous variability in three species of *Drosophila*. *Mol. Biol. Evol.* 28:1731–43

68. Heitz E. 1933. Über α- and β-heterochromatin sowie Konstanz und Bau der Chromomeren bei *Drosophila*. *Biol. Zentralblatt* 54:588–609

69. Hermisson J, Pennings PS. 2005. Soft sweeps: molecular population genetics of adaptation from standing genetic variation. *Genetics* 169:2335–52

70. Hill WG, Robertson A. 1966. The effect of linkage on limits to artificial selection. *Genet. Res.* 8:269–94

71. Hilliker AJ, Harauz G, Reaume AG, Gray M, Clark SH, Chovnick A. 1994. Meiotic gene conversion tract length distribution within the *rosy* locus of *Drosophila melanogaster*. *Genetics* 137:1019–26

72. Jensen MA, Charlesworth B, Kreitman M. 2002. Patterns of genetic variation at a chromosome 4 locus of *Drosophila melanogaster* and *D. simulans*. *Genetics* 160:493–507

73. Johnson T, Barton NH. 2002. The effect of deleterious alleles on adaptation in asexual populations. *Genetics* 162:395–411

74. Kaiser VB, Charlesworth B. 2009. The effects of deleterious mutations on evolution in non-recombining genomes. *Trends Genet.* 25:9–12

75. Kaiser VB, Charlesworth B. 2010. Muller's ratchet and the degeneration of the *Drosophila miranda* neo-Y chromosome. *Genetics* 185:339–48

76. Keightley PD, Trivedi M, Thomson M, Oliver F, Kumar S, Blaxter ML. 2009. Analysis of the genome sequences of three *Drosophila melanogaster* spontaneous mutation accumulation lines. *Genome Res.* 19:1195–201

77. Kim Y. 2004. Effect of strong directional selection on weakly selected mutations at linked sites: implication for synonymous codon usage. *Mol. Biol. Evol.* 21:286–94

78. Kimura M. 1962. On the probability of fixation of a mutant gene in a population. *Genetics* 47:713–19

79. Kimura M. 1968. Evolutionary rate at the molecular level. *Nature* 217:624–26

80. Kimura M. 1971. Theoretical foundations of population genetics at the molecular level. *Theor. Popul. Biol.* 2:174–208

81. Kimura M. 1981. Possibility of extensive neutral evolution under stabilizing selection with special reference to non-random usage of synonymous codons. *Proc. Natl. Acad. Sci. USA* 78:454–58

82. Kliman RM, Hey J. 1993. Reduced natural selection associated with low recombination in *Drosophila melanogaster*. *Mol. Biol. Evol.* 10:1239–58

83. Koerich L, Wang X, Clark AG, Carvalho AB. 2008. Low conservation of gene content in the *Drosophila* Y chromosome. *Nature* 456:949–51

84. Kofler R, Betancourt AJ, Schlötterer C. 2012. Sequencing of pooled DNA samples (Pool-Seq) uncovers complex dynamics of transposable element insertions in *Drosophila melanogaster*. *PLOS Genet.* 8:e1002487

85. Kousathanas A, Keightley PD. 2013. A comparison of models to infer the distribution of fitness effects of new mutations. *Genetics* 193:1197–208

86. Krimbas CB, Powell JR, eds. 1992. *Drosophila Inversion Polymorphism*. Boca Raton, FL: CRC Press

87. Kurek R, Reugels A, Lammermann U, Bünemann H. 2000. Molecular aspects of intron evolution in dynein encoding mega-genes on the heterochromatic Y chromosome of *Drosophila* spp. *Genetica* 109:113–23

88. Langley CH, Lazzaro BP, Phillips W, Heikkinen E, Braverman JM. 2000. Linkage disequilibria and the site frequency spectra in the *su(s)* and *su(wa)* regions of the *Drosophila melanogaster* X chromosome. *Genetics* 156:1837–52

89. Langley CH, Montgomery EA, Hudson RR, Kaplan NL, Charlesworth B. 1988. On the role of unequal exchange in the containment of transposable element copy number. *Genet. Res.* 52:223–35

90. Langley CH, Stevens K, Cardeno C, Lee YCG, Schrider DR, et al. 2012. Genomic variation in natural populations of *Drosophila melanogaster*. *Genetics* 192:533–98

91. Larracuente AM, Clark AG. 2013. Surprising differences in the variability of Y chromosomes in African and cosmopolitan populations of *Drosophila melanogaster*. *Genetics* 193:201–14

92. Larracuente AM, Sackton TB, Greenberg AJ, Wong A, Singh ND, et al. 2008. Evolution of protein-coding genes in *Drosophila*. *Trends Genet.* 24:114–23

93. Li W-H. 1987. Models of nearly neutral mutations with particular implications for non-random usage of synonymous codons. *J. Mol. Evol.* 24:337–45

94. Lindsley DL, Sandler L. 1977. The genetic analysis of meiosis in female *Drosophila*. *Philos. Trans. R. Soc. B* 277:295–312

95. Loewe L, Charlesworth B. 2007. Background selection in single genes may explain patterns of codon bias. *Genetics* 175:1381–93

96. Loewe L, Charlesworth B, Bartolomé C, Nöel V. 2006. Estimating selection on nonsynonymous mutations. *Genetics* 172:1079–92

97. Mackay TFC, Richards S, Stone EA, Barbadilla A, Ayroles JF, et al. 2012. The *Drosophila melanogaster* Genetic Reference Panel. *Nature* 482:173–78

98. Maside X, Assimacopoulos S, Charlesworth B. 2005. Fixations of transposable elements in the *D. melanogaster* genome. *Genet. Res.* 85:195–203

99. McGaugh SE, Heil CSS, Manzano-Winkler B, Loewe L, Goldstein S, et al. 2012. Recombination modulates how selection affects linked sites in *Drosophila*. *PLOS Biol.* 10:e1001422

100. Messer PW, Petrov DA. 2013. Frequent adaptation and the McDonald-Kreitman test. *Proc. Natl. Acad. Sci. USA* 110:8615–20

101. Miklos GLG, Cotsell JN. 1990. Chromosome structure at interfaces between major chromatin types: alpha- and beta-heterochromatin. *BioEssays* 12:1–6

102. Miller EL, Takeo S, Nandana K, Paulson A, Gogol MA, et al. 2012. A whole-chromosome analysis of meiotic recombination in *Drosophila melanogaster*. *G3* 2:249–60

103. Montgomery EA, Huang S-M, Langley CH, Judd BH. 1991. Chromosome rearrangement by ectopic recombination in *Drosophila melanogaster*: genome structure and evolution. *Genetics* 129:1085–98

104. Morgan TH. 1911. Random segregation versus coupling in Mendelian inheritance. *Science* 33:384

105. Morgan TH. 1912. Complete linkage in the second chromosome of the male of *Drosophila*. *Science* 36:719–20

106. Muller HJ. 1914. A gene for the fourth chromosome of *Drosophila*. *J. Exp. Zool.* 17:325–26

107. Muller HJ. 1932. Some genetic aspects of sex. *Am. Nat.* 66:118–38

108. Muller HJ. 1940. Bearing of the *Drosophila* work on systematics. In *The New Systematics*, ed. JS Huxley, pp. 185–268. Oxford: Oxford Univ. Press

109. Muller HJ, Painter TS. 1932. The differentiation of the sex chromosomes of *Drosophila* into genetically active and inert regions. *Z. Indukt. Abstamm. Vererb.* 62:316–65

110. Neher RA. 2013. Genetic draft, selective interference and population genetics of rapid adaptation. *Annu. Rev. Ecol. Evol. Syst.* 44:195–215

111. Nicolaisen LE, Desai M. 2012. Distortions in genealogies due to purifying selection. *Mol. Biol. Evol.* 29:3589–600

112. Offerman CA, Muller HJ. 1932. Regional differences in crossing over as a function of the chromosome structure. *Proc. 6th Int. Cong. Genet.*, Vol. 2, ed. DF Jones, pp. 143–45. New York: Brooklyn Bot. Garden

113. Otto SP, Lenormand T. 2002. Resolving the paradox of sex and recombination. *Nat. Rev. Genet.* 3:256–61

114. Pennings PS, Hermisson J. 2006. Soft sweeps III: the signature of positive selection from recurrent mutation. *PLOS Genet.* 2:1998–2012

115. Petrov DA, Fiston-Lavier A, Lipatov M, Lenkov K, González J. 2011. Population genomics of transposable elements in *Drosophila melanogaster*. *Mol. Biol. Evol.* 28:1633–44

116. Powell JR, Dion K, Papaceit M, Vicario S, Garrick RC. 2011. Nonrecombining genes in a recombination environment: the *Drosophila* "dot" chromosome. *Mol. Biol. Evol.* 28:825–33

117. Presgraves D. 2005. Recombination enhances protein adaptation in *Drosophila melanogaster*. *Curr. Biol.* 15:1651–56

118. Rizzon C, Marais G, Gouy M, Biémont C. 2002. Recombination rate and the distribution of transposable elements in the *Drosophila melanogaster* genome. *Genome Res.* 12:400–7

119. Robertson A. 1961. Inbreeding in artificial selection programmes. *Genet. Res.* 2:189–94

120. Santiago E, Caballero A. 1998. Effective size and polymorphism of linked neutral loci in populations under selection. *Genetics* 149:2105–17

121. Schaeffer SW. 2008. Selection in heterogeneous environments maintains the gene arrangement polymorphism of *Drosophila pseudoobscura*. *Evolution* 62:3082–99

122. Schneider A, Charlesworth B, Eyre-Walker A, Keightley PD. 2011. A method for inferring the rate of occurrence and fitness effects of advantageous mutations. *Genetics* 189:1427–37

123. Schrider DR, Houle D, Lynch M, Hahn MW. 2013. Rates and genomic consequences of spontaneous mutational events in *Drosophila melanogaster*. *Genetics* 194:937–54

124. Seger J, Smith WA, Perry JJ, Hunn K, Kaliszewska ZA, et al. 2010. Gene genealogies distorted by weakly interfering mutations in constant environments. *Genetics* 184:529–45

125. Sheldahl LE, Weinreich DM, Rand DM. 2003. Recombination, dominance and selection on amino-acid polymorphisms in the *Drosophila* genome: contrasting patterns on the X and fourth chromosomes. *Genetics* 165:1195–208

126. Simonsen KL, Churchill GA, Aquadro CF. 1995. Properties of statistical tests of neutrality for DNA polymorphism data. *Genetics* 141:413–29

127. Singh ND, Davis JC, Petrov DA. 2005. Codon bias and noncoding GC content correlate negatively with recombination rate on the *Drosophila* X chromosome. *J. Mol. Evol.* 61:315–24

128. Smith CD, Shu S, Mungall CJ, Karpen GH. 2007. The Release 5.1 annotation of *Drosophila melanogaster* heterochromatin. *Science* 316:1587–91

129. Smith NGC, Eyre-Walker A. 2002. Adaptive protein evolution in *Drosophila*. *Nature* 415:1022–24

130. Steinemann M, Steinemann S. 1998. Enigma of Y chromosome degeneration: neo-Y and neo-X chromosomes of *Drosophila miranda* a model for sex chromosome evolution. *Genetica* 102/103:409–20

131. Stephan W. 1986. Recombination and the evolution of satellite DNA. *Genet. Res.* 47:167–74

132. Stephan W. 1987. Quantitative variation and chromosomal location of satellite DNAs. *Genet. Res.* 50:41–52

133. Stephan W. 1995. An improved method for estimating the rate of fixation of favorable mutations based on DNA polymorphism data. *Mol. Biol. Evol.* 12:959–62

134. Stephan W. 2010. Genetic hitchhiking versus background selection: the controversy and its implications. *Philos. Trans. R. Soc. B* 365:1245–53

135. Stephan W, Charlesworth B, McVean GAT. 1999. The effect of background selection at a single locus on weakly selected, partially linked variants. *Genet. Res.* 73:133–46

136. Stephan W, Langley CH. 1989. Molecular genetic variation in the centromeric region of the X chromosome in three *Drosophila ananassae* populations. I. Contrasts between the *vermilion* and *forked* loci. *Genetics* 121:89–99

137. Sturtevant AH. 1929. The genetics of *Drosophila simulans*. *Carnegie Inst. Wash. Publ.* 399:1–62

138. Sturtevant AH, Tan CC. 1937. The comparative genetics of *Drosophila pseudoobscura* and *Drosophila melanogaster*. *J. Genet.* 34:415–31

139. True JR, Mercer JM, Laurie CC. 1996. Differences in crossover frequency and distribution among three sibling species of *Drosophila*. *Genetics* 142:507–23

140. Wang W, Thornton K, Berry A, Long M. 2002. Nucleotide variation along the *Drosophila melanogaster* fourth chromosome. *Science* 295:134–37

141. Weissman DB, Barton NH. 2012. Limits to the rate of adaptive substitution in sexual populations. *PLOS Genet.* 8:e1002740

142. Welch JJ. 2006. Estimating the genomewide rate of adaptive protein evolution in *Drosophila*. *Genetics* 173:821–27

143. Wiehe THE, Stephan W. 1993. Analysis of a genetic hitchhiking model and its application to DNA polymorphism data. *Mol. Biol. Evol.* 10:842–54

144. Wright S. 1931. Evolution in Mendelian populations. *Genetics* 16:97–159

145. Yamamoto M-T, Miklos GLG. 1978. Genetic studies on heterochromatin and their implications for the functions of satellite DNA. *Chromosoma* 66:71–98

146. Yi S, Charlesworth B. 2000. Contrasting patterns of molecular evolution of the genes on the new and old sex chromosomes of *Drosophila miranda*. *Mol. Biol. Evol.* 17:703–17

147. Zann V, Emery A, Coiffet M, Zytnicki M, Luyten I, et al. 2013. Distribution, evolution and diversity of retrotransposons at the *flamenco* locus reflect the regulatory properties of piRNA clusters. *Proc. Natl. Acad. Sci. USA* 110:19842–47

148. Zeng K. 2010. A simple multiallele model and its application to identifying preferred-unpreferred codons using polymorphism data. *Mol. Biol. Evol.* 27:1327–37

149. Zeng K, Charlesworth B. 2010. The effects of demography and linkage on the estimation of selection and mutation parameters. *Genetics* 186:1411–24

150. Zhou Q, Bachtrog D. 2012. Sex-specific adaptation drives early sex chromosome evolution in *Drosophila*. *Science* 337:341–45

151. Zwick ME, Salstrom JL, Langley CH. 1999. Genetic variation in rates of nondisjunction: association of two naturally occurring polymorphisms in the chromokinesin *nod* with increased rates of nondisjunction in *Drosophila melanogaster*. *Genetics* 152:1605–14

The Genetics of *Neisseria* Species

Ella Rotman and H. Steven Seifert*

¹Department of Microbiology-Immunology, Northwestern University Feinberg School of Medicine, Chicago, Illinois 60611; email: e-rotman@northwestern.edu, h-seifert@northwestern.edu

Annu. Rev. Genet. 2014. 48:405–31

First published online as a Review in Advance on September 10, 2014

The *Annual Review of Genetics* is online at genet.annualreviews.org

This article's doi: 10.1146/annurev-genet-120213-092007

*Corresponding author

Keywords

antigenic variation, phase variation, DNA transformation, gonorrhea, meningitis, horizontal gene transfer

Abstract

Neisseria gonorrhoeae and *Neisseria meningitidis* are closely related organisms that cause the sexually transmitted infection gonorrhea and serious bacterial meningitis and septicemia, respectively. Both species possess multiple mechanisms to alter the expression of surface-exposed proteins through the processes of phase and antigenic variation. This potential for wide variability in surface-exposed structures allows the organisms to always have subpopulations of divergent antigenic types to avoid immune surveillance and to contribute to functional variation. Additionally, the *Neisseria* are naturally competent for DNA transformation, which is their main means of genetic exchange. Although bacteriophages and plasmids are present in this genus, they are not as effective as DNA transformation for horizontal genetic exchange. There are barriers to genetic transfer, such as restriction-modification systems and CRISPR loci, that limit particular types of exchange. These host-restricted pathogens illustrate the rich complexity of genetics that can help define the similarities and differences of closely related organisms.

INTRODUCTION

Members of the genus *Neisseria* are Gram-negative β-proteobacteria, commonly found as commensals on animal and human mucosal surfaces, such as the nasopharynx or the genital epithelium. Most members of the *Neisseria* exist as small diplococci of approximately 0.6–1 microns that can also occur as monococci or, occasionally, tetrads. The two clinically relevant and most studied species are *Neisseria gonorrhoeae* (the gonococcus), the sole causative agent of the sexually transmitted infection gonorrhea, and *Neisseria meningitidis* (the meningococcus), which can cause cerebrospinal meningitis and septicemia. Other human commensal species include *Neisseria lactamica*, *Neisseria polysaccharea*, *Neisseria cinerea*, *Neisseria subflava*, *Neisseria flavescens*, *Neisseria perflava*, *Neisseria mucosa*, *Neisseria elongata*, and *Neisseria sicca*.

Gonorrhea is the second most prevalent sexually acquired infection in the United States, with more than 300,000 reported cases per year (27) and an estimated 106 million cases worldwide (198). *N. gonorrhoeae* is an exclusively human pathogen that primarily colonizes the urogenital epithelia. Symptomatic infection is characterized by a robust inflammatory response and a purulent discharge composed almost entirely of gonococci and recruited polymorphonuclear leukocytes (PMNs; neutrophils) (36, 149). The purulent exudate from the urogenital tract is usually observed in infected males, but is less often noticed in infected females. Gonococci can ascend to the upper genital tract, leading to serious diseases, such as epididymitis in men and cervicitis, endometriosis, and pelvic inflammatory disease in women (a major cause of infertility). In rare cases, gonococci can disseminate to the blood stream from the initial site of infection, causing disseminated gonococcal infection (DGI) and the associated complications of arthritis and endocarditis (27, 198). *N. gonorrhoeae* can also infect the eyes of newborns as they pass through the birth canal, resulting in ocular gonorrhea, which is a leading cause of infectious blindness in the developing world (202).

The meningococci, first recognized in the nineteenth century, are also restricted to humans. Meningococcal disease affects fewer than 1,000 people per year in the United States but up to tens of thousands of people a year in epidemics around the world, with the highest rates occurring in sub-Saharan Africa (67, 197). Ten to twenty percent of children and young adults are carriers for *N. meningitidis* worldwide, but the majority of those carriers do not experience invasive disease (31, 170). Invasive disease can present as septicemia or meningitis and is a serious infection with 10% mortality (50% if untreated) and neurological problems in up to 20% of survivors (133). Determining why some strains are hyperinvasive and identifying vaccine targets are active areas of meningococcal research.

The human commensal *Neisseria* are also restricted to the human nasopharynx and are closely related to *N. meningitidis* and *N. gonorrhoeae*, although they rarely cause disease in immunocompetent people. *N. lactamica*, the closest relative to the pathogenic species, is often used in studies comparing pathogenic with nonpathogenic *Neisseria*. *N. lactamica* typically colonizes children (55) and may protect against *N. meningitidis* colonization (50). Many of the virulence traits associated with the pathogenic *Neisseria* are also present in the commensal species, such as iron scavenging and immunoreactive surface proteins (101, 159), presumably because they are required for colonization, growth, and survival of the organisms within the human host. Interestingly, some known virulence factors are not absolutely required for virulence. For example, although the capsule is considered a major virulence factor of meningococci, the unencapsulated *N. meningitidis* strain α14 can still cause invasive disease (143). The determination of which genes (or combinations of genes) are responsible for virulence is an active field of investigation.

Like many bacteria, the *Neisseria* contain a repertoire of genetic elements that enable increased genomic plasticity. There are conserved genetic systems in the pathogenic *Neisseria* that likely contribute to host colonization and survival, such as DNA transformation and variation of displayed antigens. The pathogenic *Neisseria* use their basic recombination and replication machinery to create a system of complex intrastrain diversity that enhances spread and colonization. The pathogenic *Neisseria* have very low linkage disequilibrium (154), a measurement of how closely related chromosomal loci are relative to chance (68), which means there must be extensive recombination between strains in mixed infections. Thus, *N. gonorrhoeae* is essentially nonclonal, whereas *N. meningitidis* shows localized clonality during outbreaks but also does not exist in long-term, clonal lineages.

Although most gonococcal strains still respond to some regime of antibiotic treatment, resistance to currently available antimicrobials is common enough to complicate treatment and multidrug-resistant strains continue to emerge (5, 94, 203). Antibiotic resistance is presently less of an issue in meningococci than in gonococci, but this is likely to change considering the over-prescription of antibiotics that occurs worldwide. There are 13 different meningococcal serogroups, classified by capsule, but disease-causing isolates are primarily types A, B, and C (67, 170). There are highly effective capsular-based vaccines available against meningococcal serotypes A and C (67), and a new multicomponent serotype B vaccine has been approved in many countries but not in the United States (115). There are no known effective vaccine strategies for gonococci. Therefore, studying how the mechanisms of antigenic variation and horizontal gene transfer relate to immune system evasion is of prime importance for the development of novel vaccine strategies.

This review discusses the genome organization of the *Neisseria* and mechanisms of horizontal genetic exchange. The extensive phase and pilin antigenic variation (pilin Av) processes are the second focus of this review, and these systems illustrate a series of genetic processes that have allowed the *Neisseria* to become successful pathogens.

GENOME ORGANIZATION

The *Neisseria* typically contain an ~2.2-Mbp circular chromosome. However, both *N. meningitidis* and *N. gonorrhoeae* are polyploid, containing two to five genome equivalents per growing monococcal cell, with an estimated two chromosomal copies per coccal cell unit (185). However, these organisms are still genetically haploid and cannot be made to carry two different alleles in the same genomic locus (186). Interestingly, the nonpathogenic *N. lactamica* is monoploid (186), leading to the possibility that polyploidy is a virulence trait. In the bacteria *Deinococcus radiodurans*, polyploidy is used to repair radiation-induced chromosomal damage via homologous recombination (37). The presence of extra template DNA could facilitate the processes of antigenic variation and other recombination-mediated gene exchanges in *Neisseria*, but this remains to be directly demonstrated.

Elements and Islands

One of the more salient features of the *Neisseria* is the amount and variety of repetitive DNA sequences present in their genomes (101). Hundreds of copies of repeat elements are present, which play diverse roles, such as serving as phage integration sites or alternative terminators, and these DNA elements can be used as a means to trace strain lineages and divergence.

Repeat elements. The most abundant repeated sequence in neisserial genomes is a nonpalindromic 10-bp sequence known as the DNA uptake sequence (DUS) (56). There are approximately 2,000 10-mer DUSs found in neisserial chromosomes, amounting to approximately one every kilobase of DNA (101), although they are not evenly spaced. The DUSs function in mediating species-specific DNA transformation (discussed below).

A 20-bp sequence called dRS3 (duplicated repeat sequence 3) is the second most abundant repeat, appearing 200 times in *N. gonorrhoeae* (101) and up to 700 times throughout the *N. meningitidis* genome (12, 119, 101). Inverted dRS3 sequences flank different 50–150-bp RS (repeat sequence) elements, together known as a neisserial intergenic mosaic element (NIME) (119). The dRS3 can act as a site for phage integration (12, 85) and has been mapped to points of genome rearrangements (139, 163).

The Correia repeat (CR) is a 26-bp sequence found hundreds of times in the *N. gonorrhoeae* and *N. meningitidis* genomes (21, 33, 34, 99). The CRs often flank a 100–150-bp DNA cassette as terminal inverted repeats, collectively known as a Correia repeat enclosed element (CREE) (99) or a *Neisseria* miniature insertion sequence (NEMIS) (106) (described below).

Insertion sequences, minimal mobile elements, and Correia repeat enclosed elements. Like many bacteria, the *Neisseria* contain insertion sequences (ISs), which are small transposable elements capable of movement within the chromosome. ISs only encode genes for transposition and are usually flanked by inverted repeats. The *Neisseria* species contain on the order of 10–50 ISs in total (101), comparable to other bacterial species. IS elements are at sites of major inversion events in *N. meningitidis* (12, 20, 139) and define differences between *N. gonorrhoeae* strains (163). Different IS element profiles of *N. meningitidis* have been used to track the separation of the meningococci from *N. gonorrhoeae* and *N. lactamica* (139).

Minimal mobile elements (MMEs) demonstrate the genomic flexibility of the *Neisseria*. An MME is a cassette flanked by protein-encoding genes whose contents differ between strains or species. The cassettes can be exchanged and incorporated by homologous recombination in the conserved flanking regions (138, 158). In this manner, a rare insertion event originally independent of homology can spread throughout a population, especially in naturally competent bacteria.

The CREEs are abundant in the meningococci, constituting approximately 2% of the genome (119, 183), but are not as numerous in *N. gonorrhoeae* or *N. lactamica* (42, 101), indicating that changes in these elements followed the evolutionary divergence of the three species. Although not thought to encode a protein product, the CREE contains an IHF (integration host factor) binding site and has features commonly associated with transposons (such as inverted terminal repeats) (21, 42). Inverted Correia elements can potentially act as transcription terminators (53), and, interestingly, the CREEs also contain promoter elements (150, 156, 160). The possibility exists that CREEs could influence the transcription of adjacent genes, but this was not found to be the case for a CREE located adjacent to genes encoding minor pilin proteins (98). The CREEs have also been located at sites of chromosomal inversions between different *N. gonorrhoeae* strains (163), suggesting they provide either regions of homology or an active function for genomic translocations.

The gonococcal genetic island. The gonococcal genetic island (GGI) is an approximately 57-kb DNA element present in 80% of *N. gonorrhoeae* isolates and 17% of *N. meningitidis* strains but is not normally found in commensal *Neisseria* (45, 200). The GGI has a G+C content of 43% compared to the average 52% for *N. gonorrhoeae* and fewer DUSs than average for the *Neisseria* genome, suggesting horizontal acquisition (45, 63, 157). The GGI is integrated at the *dif* site near the terminus of replication, causing a duplication of the *dif* gene into a functional *difA* and a divergent *difB* (46, 157). The action of the XerCD site-specific recombinase can excise the island at low frequencies (63).

The GGI encodes a type IV secretion system (T4SS), containing transfer and secretion genes related to the F factor (45). The T4SS actively secretes single-stranded DNA (ssDNA) into the extracellular environment, (45, 64), providing an efficient substrate for DNA transformation.

There is considerable variability in the GGI, with some T4SS versions of the GGI encoding the AtlA peptidoglycan hydrolase, whereas others contain the peptidoglycan endopeptidase *eppA* gene, and some contain a serum-resistant allele of *traG* (called *sac-4*) (45). The presence of both *atlA* and *sac-4* occurs more often in DGI isolates than in other clinical isolates (45). In *N. gonorrhoeae*, the island also contains genes encoding two DNA methylases, a topoisomerase, a helicase, orthologs to plasmid partitioning proteins, and an ssDNA binding protein (63, 77). The GGIs in *N. meningitidis* are more varied and contain deletions, which may indicate this element is being lost from this species (200). Although structural proteins of the T4SS play a role in TonB-independent iron acquisition of *N. gonorrhoeae*, aiding survival in human epithelial cells (207), the presence of the T4SS does not increase the adhesion and invasiveness of meningococci to human epithelial cells (200). Whether the GGI has other role(s) in neisserial virulence in addition to its role in genetic transfer remains to be determined.

Bacteriophages

Only a handful of double-stranded and M13-like filamentous phages and prophages have been identified in the *Neisseria*, and the lack of lytic or transducing phages precludes the use of bacteriophages for genetic manipulation. However, the ease of transformation has made it unnecessary to use phages to generate knockouts and insertions or to cleanly backcross mutations. Although research on neisserial phages has been slow to yield genetic tools, recently the *N. gonorrhoeae* Ngoφ6 has been cloned as a phagemid and can function productively in a wide range of Gram-negative species, such as *Escherichia coli*, *Haemophilus influenzae*, *N. sicca*, and *Pseudomonas* species (122).

The first reports of neisserial bacteriophages were in nonpathogenic species, such as *N. flavescens* and *N. perflava* (120, 169, 175). Phage heads and plaques were later detected in culture supernatants of *N. meningitidis* (26), and although plaques were observed in lawns of *N. gonorrhoeae*, no phage particles were identified (24). Genome sequence analyses showed that many phage-related genes are present in neisserial genomes arranged as apparent prophages. Some phage gene clusters in *N. gonorrhoeae* and *N. meningitidis* are bordered by sites for a recombinase called Piv or Irg (85, 152) and others are integrated into the sites of the repeat sequence dRS3 (16). Prophage sequences present at the rearrangement break points suggest a possible active or passive role in recombination (48, 140, 163). However, there still remains little evidence that these prophages are competent to produce bacteriophage particles.

Of the identified phages, there are several examples of Mu-like prophages in the *N. meningitidis* genome. Phage Mu is a transposable element encoding its own transposase as well as more traditional phage genes like those for head, tail, and lytic proteins. Analysis of genomes revealed the Mu-like PMN1 and PMN2 phages from Z2491 (serogroup A) (119), the characterized MuMenB, which encodes membrane-associated antigens, from MC58 (serogroup B) (104), and a phage found in FAM18 (serogroup C) (48). Although Mu-like phages were not found in *N. gonorrhoeae* genomic sequences (104), five different double-stranded DNA (dsDNA) lysogenic phages were identified in strain FA1090 (Ngoφ1-5), and their repressor genes were shown to be functional in *E. coli* (121). A repressor from Ngoφ4 has been suggested to affect *N. gonorrhoeae* pathogenesis by regulating the ability of gonococci to adhere and invade human cervical cells (38). Although the dsDNA phages were not able to form plaques in commensal *Neisseria*, lambda-like particles from *N. gonorrhoeae* culture supernatants, presumed to be from the integrated prophages, were visualized by electron microscopy (121).

Both *N. meningitidis* and *N. gonorrhoeae* contain integrated filamentous prophages in their genomes. An 8-kb filamentous phage, known as Nf1 (85), is present in multiple *N. meningitidis* serotypes, some with several copies per strain. This phage contains an MDA (meningococcal disease-associated) island whose presence is correlated with hyperinvasive phylogenetic groups in adults (15, 16), but a mechanistic basis for this correlation has not been elucidated. A circular double-stranded replicative form and a single-stranded mature form of the DNA were detected at low levels by polymerase chain reaction (PCR), but phage particles have not been demonstrated. Several plasmids isolated from *N. lactamica* and *N. meningitidis* had genes and organization similar to the MDA phage (193), suggesting horizontal gene transfer of MDA between the commensals and pathogenic species. Filamentous phages with similar organization have been reported in *N. gonorrhoeae* (123), but their roles in pathogenicity remain to be determined.

MECHANISMS OF HORIZONTAL GENE TRANSFER

DNA Transformation

DNA transformation is the dedicated process of taking up DNA from the environment and incorporating it, either by recombining it with homologous sequences or, in the case of plasmids, by sometimes establishing a new episome. Bacteria that are capable of transformation are said to be naturally competent, whereas methods to introduce DNA into bacteria like *E. coli* are called artificial transformation. Whereas most bacterial competence is transient and regulated, the *Neisseria* are constitutively able to uptake and incorporate DNA during all growth phases (17, 162), although not all cells in a population readily take up DNA (54a). Natural transformation has many possible benefits for the bacterial cell, such as a means to spread antibiotic-resistance determinants and other beneficial alleles or as a source of nutrients. It has also been suggested that uptake of homologous DNA can provide a template for use in DNA repair (161), although it is surprising that this has never been directly demonstrated.

Because the *Neisseria* are human-specific organisms, homologous DNA would come from other bacteria that colonize humans, with the highest probability being uptake of DNA from the commensal *Neisseria* that inhabit the nasopharynx or from other *N. gonorrhoeae* in the genital tract. *N. gonorrhoeae* may donate DNA to the environment by direct secretion or through cell lysis (62), and both ssDNA and dsDNA can be taken up from the environment (29, 165). The amount of homology between the transforming DNA and the chromosome influences the efficiency of transformation. A minimum of 73 nucleotides of flanking homology are required for inefficient transformation of *N. meningitidis* (5a, 134), but at least 500 nucleotides of flanking homology are required for efficient transformation frequencies (44).

DNA uptake sequence–stimulated transformation. As mentioned above, *Neisseria* genomes contain a nonpalindromic 10-bp DUS (56), which occurs at a frequency of approximately one per kilobase pair. Islands of DNA thought to have been acquired via horizontal gene transfer have fewer DUSs, and DUSs found within coding sequences of genome maintenance genes are over-represented relative to other coding regions (39). In *N. gonorrhoeae*, 76% of the 10-mer DUSs contain an additional two nucleotides, which provide slightly increased transformation efficiency (6). The DUS may also have a regulatory role in transcription, as half of 12-mer DUSs are arranged as inverted repeats at the ends of genes. It is likely that these inverted repeat DUSs function as Rho-independent transcriptional terminators (6, 56, 153), although this remains to be experimentally confirmed.

Although the DUS itself is not the target of a known restriction system, the *Neisseria* show preferential selectivity for the incorporation of DUS-containing DNA. *N. gonorrhoeae* can be inefficiently transformed with non-DUS-containing DNA (19); however, the efficiency of transformation is increased by orders of magnitude when the incoming DNA contains a DUS

(47, 165). Different *Neisseria* species show DUS variations (dialects) that affect the degree to which the DNA is transformed (28, 54); there is also variation in the transformation efficiency of the same piece of DNA between strains (47), and the location and number of DUS, but not orientation, also have mild effects on the transformation efficiency (5a).

Plasmids

Naturally occurring plasmids are found in *N. gonorrhoeae*, *N. meningitidis*, and *N. lactamica* (193). There is a small 4.2-kb stably maintained plasmid called pJD1 found in 96% of all gonococci (128). Because the plasmid does not contain antibiotic resistance or identifiable virulence genes, it is also known as the cryptic plasmid. Other neisserial species contain cryptic plasmids of different sizes that hybridize to part of pJD1 from *N. gonorrhoeae* (76). The pJD1-encoded *cppB* (cryptic plasmid protein B) gene was used as a PCR target for *N. gonorrhoeae* diagnostics, although other nucleic acid amplification targets proved more specific for diagnosis (181, 188). pJD1 also carries a virulence-associated *vapDX* toxin-antitoxin system (91), of which the VapX antitoxin neutralizes the VapD ribonuclease. Some other antibiotic-resistant conjugative plasmids contain the *vapD* gene without an associated *vapX* gene (116), limiting them to exist only within pJD1-containing *N. gonorrhoeae*. The presence of the *vapDX* toxin-antitoxin system likely provides the selective pressure to maintain pJD1 within *N. gonorrhoeae* populations.

Plasmid-mediated antibiotic resistance. The commensal *Neisseria* are thought to be a reservoir of genetic material for the pathogenic strains, and plasmid transfer has been shown to occur between the pathogenic and commensal species (130, 192, 193). Plasmids are one way to exchange antibiotic-resistance elements and have contributed to β-lactamase (49) and tetracycline resistance in *N. gonorrhoeae* (112), as well as sulfonamide resistance in *N. meningitidis* (51, 129). Tetracycline resistance is mediated by the ubiquitous *tetM* gene and is usually carried on a conjugative plasmid (116) whose backbone can also mobilize the transfer of other plasmids containing other antibiotic-resistance markers (116, 129). Although much of the current resistance of *N. gonorrhoeae* to fluoroquinolones and cephalosporins is carried on the chromosome and not on plasmids, it is likely that resistance to future antibiotics may be acquired and rapidly spread via plasmids. Interestingly, there are fewer plasmid-encoded determinants of antibiotic resistance in the *Neisseria* as compared to enteric bacteria. The reasons for this are unclear, but it is likely that the selection pressures for gut microbes and their accessibility to other bacteria with antibiotic-resistance plasmids differ from those of bacteria that reside in the nasopharynx or genital tract.

Plasmids as genetic tools. Despite the presence of natural plasmids and the ease of transformation, plasmids have not been commonly used as tools for genetic complementation or protein expression in *Neisseria*. Unless there is homology in the *Neisseria* genome, the plasmid is lost following transformation (57). Instead, plasmids are frequently used as tools to construct gene knockouts and to place complementing genes on the *Neisseria* chromosome. The gene of interest is cloned into a vector flanked by regions of chromosomal homology, manipulated in vitro, and then transformed into *Neisseria*. Through double crossover events, the region between the flanking homologous DNA is replaced on the chromosome and the remaining vector is degraded (44). The result is a stable copy of the locus integrated into the chromosome. Thus, this strategy makes it so that constant selective pressure is not required for maintenance. There are some plasmids that have been constructed for laboratory use with neisserial plasmids as backbones: These include shuttle vectors between *N. gonorrhoeae* and either *Escherichia coli* (168) or *Haemophilus* species (117) and the Hermes system for complementation via a conjugative plasmid (92), as well as a vector for the constitutive expression of green fluorescence protein (GFP) (164). Although not used as often as in *E. coli*, plasmids do have a role in the genetic manipulation of *Neisseria* in the laboratory.

Barriers to Genetic Transfer

Despite the *Neisseria* being naturally competent and having a marked preference for DUS-containing DNA, several barriers exist that reduce the transformation efficiency of DNA from other neisserial species and strains. The DUS dialect from one species corresponds to the ComP receptor of that species, and slight variations in sequence can reduce transformation efficiency (13). Additionally, each neisserial strain has multiple restriction-modification (R-M) systems to prevent entry of incorrectly modified DNA, and it was recently found that *N. meningitidis* has a CRISPR locus that functions against specific spacers.

Restriction-modification.

R-M systems cleave foreign DNA that is not protected at a sequence-specific site. There are four main types of R-M systems in bacteria based on the types of subunits, sequence specificity, cleavage position, and cofactor requirements. Type I R-M consists of a three-subunit enzyme, generally encoded by the *hsdM*, *hsdS*, and *hsdR* genes, that recognizes asymmetric bipartite sequences. Type II R-M systems are the simplest and most common. They consist of separate methylation and restriction enzymes that both recognize the same 4–8-bp sequence, usually palindromic. Frequently, these are the enzymes used in laboratory cloning. The Type III R-M systems have a two-subunit enzyme consisting of modification and restriction subunits, and Type IV is a diverse family consisting only of methyl-dependent restriction enzymes (100). Types I–III have all been found in the *Neisseria* (167). Because chromosomal DNA is primarily converted to ssDNA as it enters the cell, it is usually not restricted; however, plasmids enter neisserial cells predominantly as dsDNA (18), making them a target for restriction (5a, 166). Consequently, chromosomal DNA has higher transformation efficiencies than plasmid DNA (62).

Some *N. meningitidis* isolates carry a *dam* gene whose product methylates the adenine of GATC sequences, such as in *E. coli*; however, the *N. meningitidis* Dam activity plays no role in mismatch correction (MMC), origin sequestration, or other regulatory functions attributed to the *E. coli* Dam. *N. meningitidis* strains without the *dam* gene carry an alternative gene called *drg* (for *dam*-replacing gene) at the same genomic locus. *drg* was originally reported to be associated with invasive strains, but a correlation was not observed when a larger sample size was subsequently examined (81). Drg is notable in that it cleaves GATC sequences with methyl groups, an activity opposite to that of Dam. The function(s) of Drg and Dam is not understood nor is the reason why different strains have these opposing allelic gene replacements.

Whereas in most bacterial strains, one or two R-M systems are sufficient, the pathogenic *Neisseria* can have at least fourteen different systems per strain (131, 167). For a genus that is generally receptive to its own DNA, it is unclear why there are so many redundant, strain-specific barriers to genetic transfer. The acquisition and spread of R-M systems have been used to determine phylogenetic clades of *N. meningitidis* (20), suggesting that the *Neisseria* exploit these systems to increase the stability of different strains. Like insertion elements and transposons, R-M systems have been proposed to be selfish genetic elements replicating in genomes solely for their own propagation (89). This does not seem like a sufficient reason for the *Neisseria* to have so many relative to many other bacterial species. Another line of thought is that the R-M systems confer a benefit to the bacterial host, whether as a defense against foreign DNA or as a means to preserve clonal lineages. It may be that there is a middle ground where *Neisseria* easily acquire selfish DNA that is then adapted to benefit the host cell. It is notable that other naturally competent genera, such as *Bacillus*, *Helicobacter*, and *Haemophilus*, also have an abundance of R-M systems (131).

CRISPR. Clustered regularly interspaced short palindromic repeat (CRISPR) loci in bacteria and archaea offer adaptive, sequence-based immunity against incoming DNA such as that found in plasmids and phages (78). These consist of 24–48-bp repeat spacer sequences separated by unique spacers from foreign elements. The locus is flanked by CRISPR-associated proteins (Cas) that assist in the processing and maturation of CRISPR-RNA. Although no CRISPR systems have been identified in *N. gonorrhoeae*, they have been found in *N. lactamica* and six strains of *N. meningitidis* (58, 206). The type II-C CRISPR system of *N. meningitidis* is unique in that each CRISPR repeat carries its own promoter, streamlining the CRISPR RNA maturation pathway by bypassing the requirement for pre-crRNA processing. The NmCas9 protein from the *N. meningitidis* system has recently been adapted for use in RNA-directed genome engineering of human pluripotent stem cells (73), showing promise as a genetic tool.

Given that natural transformation is the major means of horizontal gene transfer in the *Neisseria*, it is interesting to note that the CRISPR system does interfere with transformation of chromosomal DNA (206), in effect limiting the amount of genomic exchange that occurs in this usually receptive organism. It is likely that this CRISPR-mediated restriction of transformation occurs after ssDNA is integrated into a homologous DNA duplex in the cell. Self-targeting spacers, i.e., a spacer sequence that matches chromosomal DNA, are usually found with degenerate CRISPR elements (172); however, sequences that are identical to closely related organisms act to limit horizontal gene transfer. Spacer sequences in *N. meningitidis* match chromosomal sequences found in other *Neisseria* species (206). Perhaps this is a mechanism to maintain the clonality of *N. meningitidis* and other commensal inhabitants of the nasopharynx. *N. gonorrhoeae*, generally isolated in the genitourinary tract, does not often come in contact with other neisserial DNA and is not known to form clonal lineages. Therefore, it is likely that *N. gonorrhoeae* has no need for a CRISPR system to maintain clonality.

MECHANISMS OF DIVERSITY GENERATION

Antigenic variation is one of the more effective strategies used by pathogens to evade immune surveillance. The variability in surface-exposed structures allows the organisms to always have a subpopulation with divergent antigenic types. By changing the outer surface components, the *Neisseria* can avoid recognition by the adaptive immune system, prolonging a current infection and enabling reinfection to occur. Moreover, these antigenic variation systems can also provide functional variants that alter specific interactions with virulence factors and host molecules. The two major types of antigenic variation in the *Neisseria* are the reversible phase variation of hundreds of genes and the nonreciprocal gene conversion at the *pilE* locus. The variability of the Type IV pili, the lipooligosaccharide (LOS) antigen, and the outer membrane Opa proteins are major advantages to *N. meningitidis* and *N. gonorrhoeae*, facilitating their success in pathogenesis.

Phase Variation

Phase variation is the reversible change between defined expression states of genes, often a switch between an expressed (ON) and unexpressed (OFF) state or a switch between two forms of a gene product (113). Phase variation in bacteria can be mediated by invertible DNA segments, differential methylation, and changes in stretches of homopolymeric nucleotides or short tandem repeats (microsatellites) that alter the repeat number. If the repeat is in the coding region, the change causes a shift in reading frame, leading to the switch between ON and OFF states (**Figure 1a,b**). Most phase variation in the *Neisseria* is mediated by polynucleotide repeat expansion or contraction during replication in a process called slipped-strand mispairing, in which one repeat is paired with an adjacent repeat on the opposite strand (97), independent of homologous recombination functions (114). Depending on the gene measured, phase variation occurs at a rate between 10^{-2}–10^{-6}/cell/generation (105, 126), and there is a correlation between increased frequency of phase variation and a higher number of repeats (41, 114, 127). Neisserial genomes contain a large number of phase-variable genes; it is estimated that there are more than 100 phase-regulated genes altogether in the pathogenic *Neisseria*, with approximately 80 per strain (84, 137, 155). Many genes identified as virulence factors are phase variable, containing homopolymeric tracts or short tandem repeats. The combination of products from phase-variable genes contributes to the continual change in the expressed repertoire of surface antigens, facilitating immune evasion. Evidence for immune avoidance by phase variation includes the in vitro resistance of *N. meningitidis*

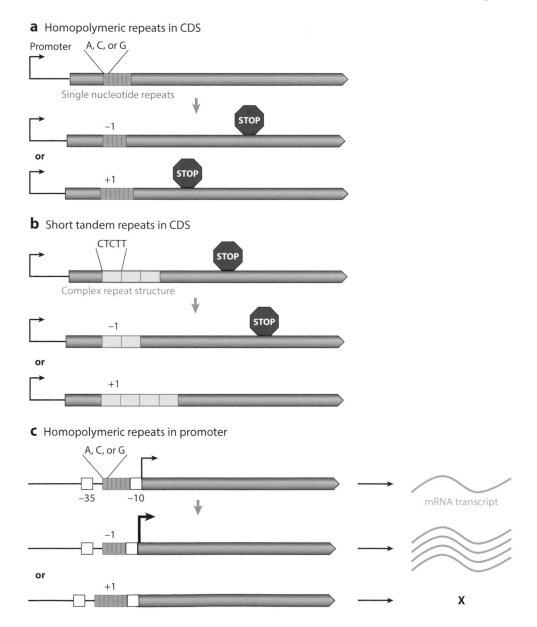

a Homopolymeric repeats in CDS

b Short tandem repeats in CDS

c Homopolymeric repeats in promoter

to bactericidal antibodies against the LgtG LOS modifier (10) and against the outer membrane porin PorA (182). However, it should be noted that not all phase-variable genes express surface antigens and they are not all directly involved in pathogenesis.

Phase variation through homopolymeric repeats. Genes with homopolymeric repeats typically contain runs of A, C, or G on the coding strand (**Figure 1a**; **Table 1**), and there can be up to 17 repeats (127). Examples of genes containing homopolymeric repeats within the reading frame are those that encode for the glycosylation of pilin and LOS (*pgl* and *lgt* genes), the pilus accessory protein *pilC*, and the capsule polysialyltransferase *siaA* (**Table 1**). Polynucleotide repeats can also occur in promoter regions (**Figure 1c**), where changes in repeat number alter the required spacing of the −10 and −35 sequences to affect transcription efficiency. Examples of this sort of phase variation occur in the *porA* and *opc* genes encoding outer membrane proteins and in the siderophore receptor encoding gene *fetA* (**Table 1**). Some variable genes, such as *pilC*, have been modified in the laboratory to replace every third nucleotide of the homopolymeric tract so as not to be variable (30, 111).

Phase variation of lipooligosaccharides. The LOS moiety is an endotoxin composed of a lipid A anchor in the outer membrane and an oligosaccharide core that is differentially glycosylated. The LOS is immunologically reactive, and there are 12 recognized LOS immunotypes in *Neisseria* (142), which have different sensitivities to human serum. The *lgt* genes of *N. meningitidis* and *N. gonorrhoeae* encode different LOS biosynthesis genes that transfer sugars to the LOS core, and a subset of the *lgt* genes contain homopolymeric repeats in their 5′ ends (**Table 1**). Phase variation allows combinations of possible glycosyltransferases, which produce different LOS structures (79). LOS is one of the main variable antigens of the *Neisseria*, along with Opa proteins and the Type IV pili, which are described below.

Phase variation through polynucleotide repeats. In addition to homopolymeric repeats, there are several types of polynucleotide repeats that contribute to phase variation in the *Neisseria* by using more complex repeat structures of 3–5-bp elements (**Figure 1b**). These tandem repeats of short nucleotide sequences are less common than homopolymeric tracts (84, 155). Changes in polynucleotide repeat number in a promoter region can alter the binding site for a regulatory protein, such as the binding of IHF to the meningococcal *nadA* gene altering adhesin expression

Figure 1

Types of phase variation. The blue arrow represents the coding region of the gene in the direction of translation. The small black arrow is the promoter. Orange rectangles represent a single nucleotide repeat, and yellow rectangles represent a more complex repeat structure. Wavy blue lines represent mRNA transcript. (*a*) Homopolymeric repeats. The addition or deletion of a single nucleotide changes the reading frame of the gene. In this example, a normally translated gene is shifted out of frame and truncates early. (*b*) Short tandem repeats. The addition or deletion of a short DNA tract changes the reading frame of the gene if the number of nucleotides is not a multiple of three. In this example, the gene is out of frame. Deletion of one pentanucleotide repeat shifts the reading frame but not to the correct one, whereas addition of the repeat restores the proper frame. (*c*) Homopolymeric repeats in the promoter region alter the distance between the −10 and −35 polymerase recognition sites, affecting the degree of binding. In this example, changes to the promoter length alter transcription levels. Addition of one nucleotide causes more efficient transcription, whereas removal of one nucleotide leads to loss of transcription, denoted by X. Abbreviation: CDS, coding sequence.

Table 1 Examples of phase-variable regulated genes in *Neisseria*

Gene	Function	Repeat	Location of repeat	Organism	Reference
pglB	Pilin glycosylation	poly-A	5′ coding region	Nm	(124)
pglG,H	Pilin glycosylation	poly-C	5′ coding region	Nm	(124)
pglA/pgtA	Pilin glycosylation	poly-G	5′ coding region	Nm/Ng	(9, 80)
lgtA,C,D,G	Lipooligosaccharide biosynthesis	poly-G	5′ coding region	Ng, Nm	(79, 204)
pilC	Pilin associated	poly-G	5′ coding region	Ng, Nm	(82)
siaA	Capsule polysialylase	poly-C	5′ coding region	Nm	(65)
porA	Porin	poly-G	Promoter	Nm	(190, 191)
opc	Outer membrane protein	poly-C	Promoter	Nm	(136)
fetA	Siderophore receptor	poly-C	Promoter	Ng	(25)
nadA	Adhesin/invasin	TAAA	Upstream promoter binding	Nm	(102, 108)
opa	Opacity protein	CTCTT	5′ coding region	Ng, Nm, Nl	(173)
lip	Lipoprotein	15 bp	5′ coding region	Ng, Nm	(7, 201)
pilQ	Secretin	24 bp	5′ coding region	Ng, Nm	(187)
dcaC	Cell wall division	108 bp	Coding region	Ng, Nm, Nl	(160)
frpAC	Iron-regulated RTX family protein	27 bp	3′ coding region	Nm	(184)
modA	Type III DNA methylase	AGCC	5′ coding region	Ng, Nm	(164)
modB	Type III DNA methylase	CCCAA	5′ coding region	Ng, Nm	(164)

Abbreviations: Ng, *Neisseria gonorrhoeae*; Nl, *Neisseria lactamica*; Nm, *Neisseria meningitidis*.

(102, 108). One of the well-characterized tandem nucleotide repeats occurs in the *opa* genes that encode the opacity proteins on the surface of the cell (see below).

   Neisseria also encode tandem repeats that have multiples of three nucleotides, such as the *lip*, *pilQ*, *dcaC*, and *frpAC* genes (**Table 1**). Although the mechanism of slipped-strand mispairing alters the repeat number, these genes are not subjected to reading-frame shifts. The structural effect on the resulting proteins can be considered another means of diversity generation (83). This mechanism is more similar to the well-known triplet repeat expansion genes common in neurological disorders (110).

Phase and antigenic variation of *opa* genes. The *opa* genes encode 25–30-kDa outer membrane proteins [some of which provide an opaque colony morphology on agar plates (177)], which bind to and signal through a variety of host cell receptors (reviewed in 135). In *N. meningitidis*, there are four *opa* genes (119, 183), whereas *N. gonorrhoeae* can encode up to 11 different *opa*s (14, 43). The *opa* genes each contain adjacent CTCTT repeats in the DNA encoding the signal sequence (**Table 1**) (171). There are from 2 to 20 CTCTT repeats in each *opa* (8, 171), and each gene can independently phase vary ON and OFF using slipped-strand mispairing. Thus, each individual Opa ON/OFF variation is mediated by a phase variation event (which results in the expression of the 11 different variants) and therefore represents an antigenic variation process mediated through phase variation. Additionally, there is a low frequency of recombination between *opa* gene copies that can contribute to strain diversification (3). The promoter strength has been mapped for three *opa*s in *N. gonorrhoeae* and correlated with the frequency of repeat changes (11). However, there is no correlation between orientation or distance of these three *opa* genes from the origin of replication and their expression levels. Strains containing deletions in all *opa* genes have been created in both *N. meningitidis* (134) and *N. gonorrhoeae* (8, 95), and strains constitutively expressing specific *opa* genes have been constructed (32, 93) to facilitate studies examining the contribution of specific proteins to pathogenesis without the issue of phase variation confounding the results.

Phase-variable regulatory proteins. Phase variation can also have a regulatory function, as seen in the phasevarion. First described in *H. influenzae*, the phasevarion uses a phase-variable DNA methyltransferase to control the expression of other unlinked genes through differential methylation of promoter sequences. In the pathogenic *Neisseria*, the *mod* genes encode a Type III R-M DNA methyltransferase and are phase variable by containing polynucleotide repeat tracts (**Table 1**). Differential methylation by ModA and ModB of promoter regions directly affects expression of dozens of genes, including those that encode virulence factors, such as lactoferrin binding proteins in *N. meningitidis* and genes that mediate oxidative stress resistance in *N. gonorrhoeae* (164).

Impact of mismatch correction on phase variation. Although the genes of the MMC system are not phase regulated themselves, loss-of-function mutations to *mutL* or *mutS* can result in a 10–1,000-fold increase in phase variation (61, 96). Because MutS only recognizes stem-loop structures smaller than 4 bp (118), MMC mutations have an effect on homopolymeric tracts and short tandem repeats but not on the pentamer repeats of the *opa* gene family (103). Disruptions of the MMC pathway also result in a general mutator phenotype because mutations that arise from replication errors are not corrected (141). Although an increase in mutation rate can have a fitness cost for an organism, in the case of the pathogenic *Neisseria*, the benefit of an increase in phase variation may provide other selective advantages in the host (113). In support of this idea, an analysis of 95 serogroup A meningococci demonstrated that 54 had elevated mutation rates, of which 21 were due to defective *mutS* or *mutL* genes (127), a high percentage compared with other pathogenic bacteria (61). It is unclear whether these mutator strains form stable lineages that can transmit from host-to-host or exist only to provide a short term advantage within a host.

Pilin Antigenic Variation

Pilin Av is a representative type of diversity generation that some pathogens use to avoid immune surveillance (194). Although antigenic variation can be mediated by independent phase variation of multiple related genes, such as the LOS biosynthesis genes and the *opa* loci, pilin Av differs from phase variation in that the system promotes changes to a gene that alter the peptide sequence rather than the expression level, resulting in multiple forms of the antigen. The genetic information for producing the different variants is present within the cell, but only one variant is usually expressed at a time. Pilin Av is mediated by a gene conversion process, which is a nonreciprocal transfer of DNA from a donor homolog to a recipient locus without the donor locus being changed in the process. Gene conversion does not define a mechanism but instead refers to nonreciprocal events that occur during a recombination process. Some examples of directed gene conversion systems include the mating-type switch between MATa and MATα cells of the haploid yeast *Saccharomyces cerevisiae*, which incorporates one of two types of MAT genes into an expressed locus (60); the Ig gene diversity in chicken B cells, in which the single light-chain-combined V_λ and J_λ segments generate diversity with donor DNA from 25 nearby pseudo-V_λ genes (125); and the *vls* locus in the Lyme disease bacterium *Borrelia burgdorferi*, which incorporates 1 of 15 silent cassettes into the *vlsE* gene expressing the VlsE surface-exposed lipoprotein (205).

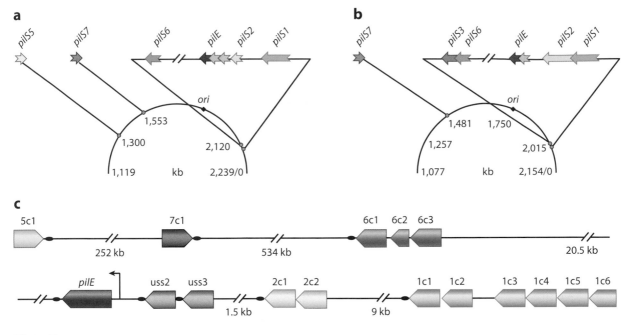

Figure 2

Chromosomal organization of pilin genes in *Neisseria gonorrhoeae*. A representative map of *pilE* and *pilS* loci arranged according to chromosomal position. Colored shapes indicate *pilE* and *pilS* gene copies, with the pointed end showing direction of translation. The coding sequence of *pilS* copies ranges from 300 to 400 bp in length, and the *pilE* coding region is 500 bp in length. The small black arrow in front of *pilE* indicates a functional promoter. The black oval represents the conserved *Sma/Cla* repeat. *ori* is the predicted origin of replication, experimentally determined for FA1090 (185). Numbers indicate chromosomal position in kilobase pairs. (*a*) Position and orientation of pilin loci with respect to the replication origin in strain MS11. (*b*) Position of pilin loci in strain FA1090. The presence and orientation are similar between the strains, except FA1090 does not have *pilS5* but does have a *pilS3* locus, more silent copies within *pilS2*, and only one upstream silent copy (uss). (*c*) Graphic representation of silent copies within each locus and the relative distances between them for strain MS11. The diagrams for MS11 and FA1090 are based on sequence data from GenBank (Accession numbers CP003909.1 and AE004969.1, respectively).

Pilin Av in *Neisseria* is one of the best-studied systems of gene diversification. The type IV pili of *Neisseria* are long filamentous surface structures involved in DNA transformation, twitching motility, and adherence to epithelial cells (162, 176, 199). Pilin (or PilE), the major component of the type IV pilus, is encoded by the chromosomal locus *pilE*. In the gonococci and other *Neisseria*, there are clusters of promoterless, truncated pilin genes scattered around the chromosome in different loci (**Figure 2**). These silent *pilS* copies share significant sequence similarity to the *pilE* gene and act as reservoirs of variant genetic information. *N. gonorrhoeae* has as many as 19 silent copies in 4 or 5 silent loci (66), whereas most meningococci have 4–6 silent copies in 1 locus. Given the number of silent gene copies in *N. gonorrhoeae*, it is surprising that direct recombination between them does not occur more frequently. Although some evidence does point to gross chromosomal rearrangements at *pilS* loci between gonococcal strains (163), the silent copies remain fairly stable within a strain.

Transfer of DNA sequences from a *pilS* copy into *pilE* results in changes to variable regions of the *pilE* gene, without reciprocal changes occurring in the *pilS* copy (59). The variable regions of *pilE* encode exposed regions on the pilus surface that are recognized by antibodies (52). The type of pilin expressed in *N. gonorrhoeae* is called class I, whereas some meningococci and all commensal *Neisseria* express a related class II pilin (4). The class II pilins lack the hypervariable region and they do not undergo pilin Av (2, 69), although *N. meningitidis* expressing class II pili can still colonize and cause invasive disease. Differences in the exposed part of meningococcal pilin can affect cell signaling in endothelial and epithelial cell types (109).

Measuring pilin antigenic variation. The piliation state of *N. gonorrhoeae* was first described in the middle of the twentieth century as different colony phenotypes that correlated to the ability to infect people (86). Piliated cells form small domed colonies due to pilus-pilus interactions between cells, whereas nonpiliated cells grow faster and spread out to form large flat colonies (180). As colonies are allowed to grow over extended periods of time, a subset of bacteria convert to a nonpiliated phenotype, leading to irregular borders due to differences in growth rate and appearance (145). Serial passage in the laboratory can lead to the conversion of piliated colonies to nonpiliated and back to piliated again (179). In the course of human infection, gonococci continuously change pilin sequences (66, 146).

Factors that affect the frequency of pilin Av can be determined by measuring the appearance of nonpiliated colonies, a subset of the total possible variants, over time. These can be observed as blebs on the edge of a colony or as the percent of nonpiliated cells in a population. A reduction in the number of nonpiliated cells usually corresponds to reduced frequencies of pilin Av. However, nonpiliated variants can also be produced by phase variation of the *pilC* genes (82) or by the nonreversible deletion of the *pilE* gene (178), leading to situations in which the appearance of nonpiliated variants is not an exact measure of pilin Av. Moreover, nonpiliated variants are only produced from a subset of *pilS* copies, which can also result in misleading results. Pilin Av has also been assayed by a variety of Southern blot, PCR, and qRT-PCR approaches (71, 132, 147, 196), but the best way to measure the frequency is through assessing changes in *pilE* sequences via traditional or next-generation sequencing (35, 40). *pilE* sequencing provides an exact frequency (approximately 10% of variation) and the rate of antigenic variation can be calculated if the number of generations is known ($\sim 6.8 \times 10^{-3}$ recombination events per CFU per generation for *N. gonorrhoeae* strain FA1090, variant 1-81-S2) (35). This is one of the highest rates of change reported for a prokaryotic diversity generation system. The median rate reported for *N. gonorrhoeae* strain MS11 (1.7×10^{-3} events/CFU/generation), and *N. meningitidis* MC58 (1.6×10^{-3} events/CFU/generation), is lower but still results in a significant subpopulation of variants ($\sim 5\%$) (69). These rates are in the same range as the variation systems expressed by eukaryotic pathogens *Trypanosoma brucei* (189) and *Plasmodium falciparum* (72).

Factors required for pilin Av. There are general and specific factors known to be involved with recombination at the *pilE* locus that change the *pilE* sequence while keeping the silent copies intact. General factors include members of the classic set of proteins used in bacterial homologous recombination, including, but not limited to, RecA, RecX, RecO, RecR, RecJ, RecQ, RecG, and RuvABC (71, 90, 107, 144, 145, 151, 174). RecA is the master protein for recombination, catalyzing strand invasion, and annealing, which is enhanced by RecX in *N. gonorrhoeae*. RecO, RecR, RecJ, and RecQ are part of the RecF-like pathway in *Neisseria* (there is no RecF ortholog). The RecF pathway, characterized in *E. coli*, utilizes gapped DNA homologous recombination, and also assists RecA, whereas RecG and RuvABC catalyze resolution of the Holliday junctions created by the recombination events (91a). Inactivation of any of these proteins reduces the frequency of pilin Av, but not all mutants have the same magnitude of pilin Av reduction.

Although most changes to laboratory conditions, such as temperature, oxygen, or carbon source, do not significantly affect Av, the loss of iron in the medium boosted all recombination, including Av (148), suggesting that the iron-starved conditions of the human body activate pilin Av. The molecular basis of this stimulation by iron starvation is unknown.

Sequences required for pilin Av. At the end of each locus is a conserved 65-bp sequence known as the *Sma*/*Cla* repeat (flanked by the *Sma*I and *Cla*I restriction sites in *N. gonorrhoeae*) (**Figure 2c**), which may be involved in pilin Av (196). Deletions of this repeat lower the frequency of pilin Av measured by a PCR-based assay, and the repeat has been shown to bind an unknown protein (195). It is presently unclear whether this repeat has any role in pilin Av in addition to providing extended regions of homology.

A genetic screen suggested that a *cis*-acting site was present in the region immediately upstream of *pilE* (87, 145), and subsequent directed mutational analysis identified a 16-nucleotide G-rich sequence in the region that forms a G4 (guanine tetraplex or quadruplex) structure in vitro (22). Individual point mutations at any of the G-C base pairs in this sequence disrupt pilin Av, whereas mutation of the T-A base pairs has no effect. These mutations altered the formation of the G4 structure in vitro, and treatment with *N*-methyl mesoporphyrin IX (NMM), which specifically interacts with G4 structures and not other forms of nucleic acids, inhibited pilin Av. Mutations that disrupted the G4 structure or NMM treatment also inhibited the appearance of nicks at the G4 sequence, whereas mutations in *recJ* and *recQ* enhanced the detection of nicks in a wild-type cell but not in a G4 point mutant (22). The G4 structure is the first genetic element shown to be specific for pilin Av and has led to the suggestion that alternative DNA structures may be involved in other recombination-based diversity generation systems.

Interestingly, there is a small RNA (sRNA) that overlaps with the *pilE* G4 motif that is required for the G4 structure to function during pilin Av (23). Mutations that block transcription of the G4 sRNA also block pilin Av, but expression of the G4 sRNA in an ectopic locus does not complement these promoter mutations. Altering the direction of transcription of the G4 sRNA or the strand it is expressed from still allows expression of the *pilE* gene, but prevents pilin Av from occurring, suggesting that this sRNA, and possibly the G4 structure, can act in only one orientation. It will be interesting to learn how the G4 and sRNA affect pilin genes in other neisserial species compared to the system found in *N. gonorrhoeae*.

Models for pilin Av. Several models have been proposed to explain the unidirectional recombination reactions that lead to pilin Av (reviewed in 70, 88). All models start with a break in the *pilE* sequence and alignment with a *pilS* copy (**Figure 3**). Unwinding at the break allows a 3′ ssDNA end of *pilE* to invade *pilS* at a region of microhomology. In the unequal crossing-over model (**Figure 3a**), the displaced *pilS* strand becomes a template for the repair at the broken *pilE*. In the successive half-crossing-over model (**Figure 3c**), one of the single-stranded *pilE* ends forms a direct link (a half-crossover) to the *pilS* in a RecF-like-dependent manner, which the second *pilE* end invades a short distance away for a second half-crossover, destroying the donor DNA.

The hybrid intermediate model (**Figure 3b**) invokes a several-step recombination process occurring between two sister chromosomes and may be the reason for the diploid nature of the pathogenic *Neisseria*. This model was formulated to explain the detection of a putative recombination intermediate that was detected when pilin Av was blocked by a promoterless *cat* gene inserted into a silent copy (75). Pilin Av is initiated when a single crossover event occurs between a *pilS* copy and *pilE* at a region of microhomology to form a hybrid intermediate (**Figure 3b**). Because the *pilS* and *pilE* loci are 5–450 kb apart, and deletions between these loci are not normally observed, this event must lead to destruction of the remainder of the donor chromosome, but it has not been directly tested. The creation of the hybrid intermediate is independent of (or requires little) RecA protein (74). The hybrid intermediate was proposed to be targeted for recombination with *pilE* on a different chromosome, as it recombined with a higher efficiency at *pilE* than at the originating *pilS* copy, and this targeting was RecA dependent. The recombination of the hybrid intermediate with *pilE* requires a crossover exchange at an extended region of homology and a

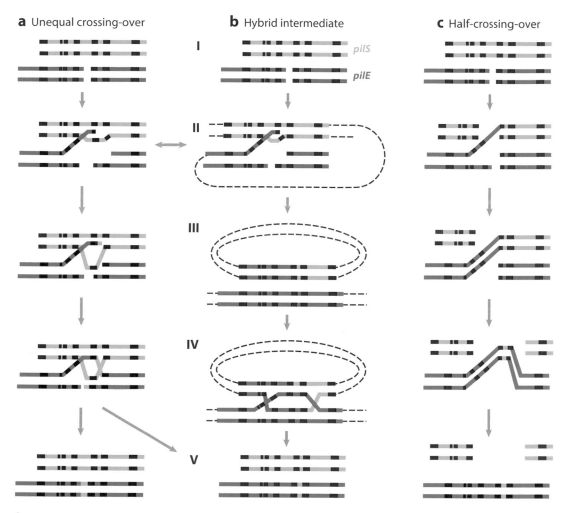

a Unequal crossing-over

b Hybrid intermediate

c Half-crossing-over

pilS

pilE

I

II

III

IV

V

Figure 3

Models of pilin antigenic variation. Each pair of lines represents duplex DNA. Black lines represent regions of homology between *pilE* and *pilS*, and blue and red lines represent sequences unique to *pilS* and *pilE*, respectively. (*a*) The unequal crossing-over model depicts the displaced *pilS* strand acting as a template for repair at the broken *pilE*. After branch migration, the Holliday junctions are resolved into two molecules, with an unchanged pilS and a variant pilE. (*b*) The hybrid intermediate model involves first a recombination event to create a hybrid and two to resolve it. I. The first recombination event takes place intrachromosomally between *pilE* and any of the *pilS* copies within a locus. This could be initiated by a double-stranded break (*shown*) or a single-strand nick that produces a gap (*not shown*). II. A single strand of *pilE* invades the *pilS* locus at one of the regions of microhomology. III. A hybrid intermediate is formed with a crossover at the region of microhomology, with the remainder of the chromosome lost. The size of the intermediate depends on the distance between *pilE* and the *pilS* locus. IV. Two recombination events occur between the hybrid intermediate and the chromosomal copy of *pilE*: one at an extended homology that is RecA dependent and the second at a region of microhomology, which results in complete exchange between the two. Although depicted as two RecA-dependent events, the second recombination event at the microhomology is likely similar to the crossover formed in step III. V. The hybrid is lost, and the chromosome contains an intact *pilS* locus and a variant *pilE* sequence. (*c*) The half-crossing-over model consists of two successive half crossovers with little or no DNA synthesis. The second crossover occurs a short distance away from the first and results in replacement of the *pilS* donor sequence into *pilE* and destruction of the chromosome containing *pilS*. Modified from Reference 88.

second recombination at a region of microhomology (**Figure 3***b*), similar to the initiating event, and produces a variant *pilE* sequence with the original *pilS* locus apparently unchanged (**Figure 3***b*). How recombination at regions of microhomology is achieved is presently unknown but could be mediated by a resection and annealing process. It will be interesting to see how pilin Av compares to mechanisms of antigenic variation in other pathogenic organisms.

SUMMARY POINTS

1. *N. gonorrhoeae* and *N. meningitidis* are the two pathogenic members of an otherwise commensal genus of bacteria. *N. gonorrhoeae* is the sole agent of the sexually transmitted infection gonorrhea, and *N. meningitidis* can cause bacterial meningitis and septicemia.

2. The *Neisseria* genomes contain repeat elements, insertion sequences, and bacteriophages that contribute to genomic rearrangements and differences between species. The genomes also have multiple R-M systems and some species contain a CRISPR system.

3. The *Neisseria* are naturally competent during all phases of growth. DNA containing a 10-bp neisserial DUS is preferentially transformed by *Neisseria*. The DUS is common in neisserial genomes, occurring approximately once every kilobase.

4. Phase variation, or the random switching between the ON and OFF state of genes, occurs by slipped-strand synthesis at polynucleotide tracts and short tandem repeats. More than 100 genes are said to be phase regulated in the *Neisseria*, and many virulence factors are phase variable. There are two major surface antigens, LOS and Opa, that undergo antigenic variation through a multigene phase variation process.

5. Pilin Av is the unidirectional recombination of silent gene copies into the expressed *pilE* locus encoding the surface pili. The high rate of pilin antigenic variation is a means the *Neisseria* use to avoid immune surveillance.

DISCLOSURE STATEMENT

The authors are not aware of any affiliations, memberships, funding, or financial holdings that might be perceived as affecting the objectivity of this review.

ACKNOWLEDGMENTS

We would like to thank Adrienne Chen and Elizabeth Stohl for helpful discussions and comments on the manuscript. Our work has been funded by grants from the National Institutes of Health: R37 AI033493 and R01 AI044239 to H.S.S. and F32 AI0I94945 to E.R.

LITERATURE CITED

1. Deleted in proof

2. Aho EL, Botten JW, Hall RJ, Larson MK, Ness JK. 1997. Characterization of a class II pilin expression locus from *Neisseria meningitidis*: evidence for increased diversity among pilin genes in pathogenic *Neisseria* species. *Infect. Immunity* 65:2613–20

3. Aho EL, Dempsey JA, Hobbs MM, Klapper DG, Cannon JG. 1991. Characterization of the *opa* (class 5) gene family of *Neisseria meningitidis*. *Mol. Microbiol.* 5:1429–37

4. Aho EL, Keating AM, McGillivray SM. 2000. A comparative analysis of pilin genes from pathogenic and nonpathogenic *Neisseria* species. *Microb. Pathog.* 28:81–88

5. Allen VG, Mitterni L, Seah C, Rebbapragada A, Martin IE, et al. 2013. *Neisseria gonorrhoeae* treatment failure and susceptibility to cefixime in Toronto, Canada. *J. Am. Med. Assoc.* 309:163–70

5a. Ambur OH, Frye SA, Nilsen M, Hovland E, Tonjum T. 2012. Restriction and sequence alterations affect DNA uptake sequence-dependent transformation in *Neisseria meningitidis*. *PLOS ONE* 7:e39742

6. Ambur OH, Frye SA, Tonjum T. 2007. New functional identity for the DNA uptake sequence in transformation and its presence in transcriptional terminators. *J. Bacteriol.* 189:2077–85

7. Baehr W, Gotschlich EC, Hitchcock PJ. 1989. The virulence-associated gonococcal H.8 gene encodes 14 tandemly repeated pentapeptides. *Mol. Microbiol.* 3:49–55

8. Ball LM, Criss AK. 2013. Constitutively Opa-expressing and Opa-deficient *Neisseria gonorrhoeae* strains differentially stimulate and survive exposure to human neutrophils. *J. Bacteriol.* 195:2982–90

9. Banerjee A, Wang R, Supernavage SL, Ghosh SK, Parker J, et al. 2002. Implications of phase variation of a gene (*pgtA*) encoding a pilin galactosyl transferase in gonococcal pathogenesis. *J. Exp. Med.* 196:147–62

10. Bayliss CD, Hoe JC, Makepeace K, Martin P, Hood DW, Moxon ER. 2008. *Neisseria meningitidis* escape from the bactericidal activity of a monoclonal antibody is mediated by phase variation of *lgtG* and enhanced by a mutator phenotype. *Infect. Immunity* 76:5038–48

11. Belland RJ, Morrison SG, Carlson JH, Hogan DM. 1997. Promoter strength influences phase variation of neisserial *opa* genes. *Mol. Microbiol.* 23:123–35

12. Bentley SD, Vernikos GS, Snyder LA, Churcher C, Arrowsmith C, et al. 2007. Meningococcal genetic variation mechanisms viewed through comparative analysis of serogroup C strain FAM18. *PLOS Genet.* 3:e23

13. Berry JL, Cehovin A, McDowell MA, Lea SM, Pelicic V. 2013. Functional analysis of the interdependence between DNA uptake sequence and its cognate ComP receptor during natural transformation in *Neisseria* species. *PLOS Genet.* 9:e1004014

14. Bhat KS, Gibbs CP, Barrera O, Morrison SG, Jahnig F, et al. 1991. The opacity proteins of *Neisseria gonorrhoeae* strain MS11 are encoded by a family of 11 complete genes. *Mol. Microbiol.* 5:1889–901

15. Bille E, Ure R, Gray SJ, Kaczmarski EB, McCarthy ND, et al. 2008. Association of a bacteriophage with meningococcal disease in young adults. *PLOS ONE* 3:e3885

16. Bille E, Zahar JR, Perrin A, Morelle S, Kriz P, et al. 2005. A chromosomally integrated bacteriophage in invasive meningococci. *J. Exp. Med.* 201:1905–13

17. Biswas GD, Sox T, Blackman E, Sparling PF. 1977. Factors affecting genetic transformation of *Neisseria gonorrhoeae*. *J. Bacteriol.* 129:983–92

18. Biswas GD, Sparling PF. 1981. Entry of double-stranded deoxyribonucleic acid during transformation of *Neisseria gonorrhoeae*. *J. Bacteriol.* 145:638–40

19. Boyle-Vavra S, Seifert HS. 1996. Uptake-sequence-independent DNA transformation exists in *Neisseria gonorrhoeae*. *Microbiology* 142(Pt. 10):2839–45

20. Budroni S, Siena E, Dunning Hotopp JC, Seib KL, Serruto D, et al. 2011. *Neisseria meningitidis* is structured in clades associated with restriction modification systems that modulate homologous recombination. *Proc. Natl. Acad. Sci. USA* 108:4494–99

21. Buisine N, Tang CM, Chalmers R. 2002. Transposon-like Correia elements: structure, distribution and genetic exchange between pathogenic *Neisseria* sp. *FEBS Lett.* 522:52–58

22. Cahoon LA, Seifert HS. 2009. An alternative DNA structure is necessary for pilin antigenic variation in *Neisseria gonorrhoeae*. *Science* 325:764–67

23. Cahoon LA, Seifert HS. 2013. Transcription of a *cis*-acting, noncoding, small RNA is required for pilin antigenic variation in *Neisseria gonorrhoeae*. *PLOS Pathogens* 9:e1003074

24. Campbell LA, Short HB, Young FE, Clark VL. 1985. Autoplaquing in *Neisseria gonorrhoeae*. *J. Bacteriol.* 164:461–65

25. Carson SD, Stone B, Beucher M, Fu J, Sparling PF. 2000. Phase variation of the gonococcal siderophore receptor FetA. *Mol. Microbiol.* 36:585–93

26. Cary SG, Hunter DH. 1967. Isolation of bacteriophages active against *Neisseria meningitidis*. *J. Virol.* 1:538–42

27. CDC. 2013. *Sexually Transmitted Disease Surveillance 2012.* Atlanta, GA: CDC

28. Cehovin A, Simpson PJ, McDowell MA, Brown DR, Noschese R, et al. 2013. Specific DNA recognition mediated by a type IV pilin. *Proc. Natl. Acad. Sci. USA* 110:3065–70

29. Chaussee MS, Hill SA. 1998. Formation of single-stranded DNA during DNA transformation of *Neisseria gonorrhoeae*. *J. Bacteriol.* 180:5117–22

30. Cheng Y, Johnson MD, Burillo-Kirch C, Mocny JC, Anderson JE, et al. 2013. Mutation of the conserved calcium-binding motif in *Neisseria gonorrhoeae* PilC1 impacts adhesion but not piliation. *Infect. Immunity* 81:4280–89

31. Claus H, Maiden MC, Wilson DJ, McCarthy ND, Jolley KA, et al. 2005. Genetic analysis of meningococci carried by children and young adults. *J. Infect. Dis.* 191:1263–71

32. Cole JG, Fulcher NB, Jerse AE. 2010. Opacity proteins increase *Neisseria gonorrhoeae* fitness in the female genital tract due to a factor under ovarian control. *Infect. Immunity* 78:1629–41

33. Correia FF, Inouye S, Inouye M. 1986. A 26-base-pair repetitive sequence specific for *Neisseria gonorrhoeae* and *Neisseria meningitidis* genomic DNA. *J. Bacteriol.* 167:1009–15

34. Correia FF, Inouye S, Inouye M. 1988. A family of small repeated elements with some transposon-like properties in the genome of *Neisseria gonorrhoeae*. *J. Biol. Chem.* 263:12194–98

35. Criss AK, Kline KA, Seifert HS. 2005. The frequency and rate of pilin antigenic variation in *Neisseria gonorrhoeae*. *Mol. Microbiol.* 58:510–19

36. Criss AK, Seifert HS. 2012. A bacterial siren song: intimate interactions between *Neisseria* and neutrophils. *Nat. Rev. Microbiol.* 10:178–90

37. Daly MJ, Minton KW. 1995. Interchromosomal recombination in the extremely radioresistant bacterium *Deinococcus radiodurans*. *J. Bacteriol.* 177:5495–505

38. Daou N, Yu C, McClure R, Gudino C, Reed GW, Genco CA. 2013. *Neisseria* prophage repressor implicated in gonococcal pathogenesis. *Infect. Immunity* 81:3652–61

39. Davidsen T, Rodland EA, Lagesen K, Seeberg E, Rognes T, Tonjum T. 2004. Biased distribution of DNA uptake sequences towards genome maintenance genes. *Nucleic Acids Res.* 32:1050–58

40. Davies JK, Harrison PF, Lin YH, Bartley S, Khoo CA, et al. 2014. The use of high-throughput DNA sequencing in the investigation of antigenic variation: application to *Neisseria* species. *PLOS ONE* 9:e86704

41. De Bolle X, Bayliss CD, Field D, van de Ven T, Saunders NJ, et al. 2000. The length of a tetranucleotide repeat tract in *Haemophilus influenzae* determines the phase variation rate of a gene with homology to type III DNA methyltransferases. *Mol. Microbiol.* 35:211–22

42. De Gregorio E, Abrescia C, Carlomagno MS, Di Nocera PP. 2003. Asymmetrical distribution of *Neisseria* miniature insertion sequence DNA repeats among pathogenic and nonpathogenic *Neisseria* strains. *Infect. Immunity* 71:4217–21

43. Dempsey JA, Litaker W, Madhure A, Snodgrass TL, Cannon JG. 1991. Physical map of the chromosome of *Neisseria gonorrhoeae* FA1090 with locations of genetic markers, including *opa* and *pil* genes. *J. Bacteriol.* 173:5476–86

44. Dillard JP. 2011. Genetic manipulation of *Neisseria gonorrhoeae*. *Curr. Protoc. Microbiol.* doi: 10.1002/9780471729259.mc04a02s00

45. Dillard JP, Seifert HS. 2001. A variable genetic island specific for *Neisseria gonorrhoeae* is involved in providing DNA for natural transformation and is found more often in disseminated infection isolates. *Mol. Microbiol.* 41:263–77

46. Dominguez NM, Hackett KT, Dillard JP. 2011. XerCD-mediated site-specific recombination leads to loss of the 57-kilobase gonococcal genetic island. *J. Bacteriol.* 193:377–88

47. Duffin PM, Seifert HS. 2010. DNA uptake sequence-mediated enhancement of transformation in *Neisseria gonorrhoeae* is strain dependent. *J. Bacteriol.* 192:4436–44

48. Dunning Hotopp JC, Grifantini R, Kumar N, Tzeng YL, Fouts D, et al. 2006. Comparative genomics of *Neisseria meningitidis*: core genome, islands of horizontal transfer and pathogen-specific genes. *Microbiology* 52:3733–49

49. Eisenstein BI, Sox T, Biswas G, Blackman E, Sparling PF. 1977. Conjugal transfer of the gonococcal penicillinase plasmid. *Science* 195:998–1000

50. Evans CM, Pratt CB, Matheson M, Vaughan TE, Findlow J, et al. 2011. Nasopharyngeal colonization by *Neisseria* lactamica and induction of protective immunity against *Neisseria meningitidis*. *Clin. Infect. Dis.* 52:70–77

51. Facinelli B, Varaldo PE. 1987. Plasmid-mediated sulfonamide resistance in *Neisseria meningitidis*. *Antimicrob. Agents Chemother.* 31:1642–43

52. Forest KT, Bernstein SL, Getzoff ED, So M, Tribbick G, et al. 1996. Assembly and antigenicity of the *Neisseria gonorrhoeae* pilus mapped with antibodies. *Infect. Immun.* 64:644–52

53. Francis F, Ramirez-Arcos S, Salimnia H, Victor C, Dillon JR. 2000. Organization and transcription of the division cell wall (*dcw*) cluster in *Neisseria gonorrhoeae*. *Gene* 251:141–51

54. Frye SA, Nilsen M, Tonjum T, Ambur OH. 2013. Dialects of the DNA uptake sequence in *Neisseriaceae*. *PLOS Genet.* 9:e1003458

54a. Gangel H, Hepp C, Muller S, Oldewurtel ER, Aas FE, et al. 2014. Concerted spatio-temporal dynamics of imported DNA and ComE DNA uptake protein during gonococcal transformation. *PLoS Pathog.* 10:e1004043

55. Gold R, Goldschneider I, Lepow ML, Draper TF, Randolph M. 1978. Carriage of *Neisseria meningitidis* and *Neisseria lactamica* in infants and children. *J. Infect. Dis.* 137:112–21

56. Goodman SD, Scocca JJ. 1988. Identification and arrangement of the DNA sequence recognized in specific transformation of *Neisseria gonorrhoeae*. *Proc. Natl. Acad. Sci. USA* 85:6982–86

57. Graves JF, Biswas GD, Sparling PF. 1982. Sequence-specific DNA uptake in transformation of *Neisseria gonorrhoeae*. *J. Bacteriol.* 152:1071–77

58. Grissa I, Vergnaud G, Pourcel C. 2007. The CRISPRdb database and tools to display CRISPRs and to generate dictionaries of spacers and repeats. *BMC Bioinform.* 8:172

59. Haas R, Meyer TF. 1986. The repertoire of silent pilus genes in *Neisseria gonorrhoeae*: evidence for gene conversion. *Cell* 44:107–15

60. Haber JE. 1998. Mating-type gene switching in *Saccharomyces cerevisiae*. *Annu. Rev. Genet.* 32:561–99

61. Hall LM, Henderson-Begg SK. 2006. Hypermutable bacteria isolated from humans: a critical analysis. *Microbiology* 152:2505–14

62. Hamilton HL, Dillard JP. 2006. Natural transformation of *Neisseria gonorrhoeae*: from DNA donation to homologous recombination. *Mol. Microbiol.* 59:376–85

63. Hamilton HL, Dominguez NM, Schwartz KJ, Hackett KT, Dillard JP. 2005. *Neisseria gonorrhoeae* secretes chromosomal DNA via a novel type IV secretion system. *Mol. Microbiol.* 55:1704–21

64. Hamilton HL, Schwartz KJ, Dillard JP. 2001. Insertion-duplication mutagenesis of *Neisseria*: use in characterization of DNA transfer genes in the gonococcal genetic island. *J. Bacteriol.* 183:4718–26

65. Hammerschmidt S, Muller A, Sillmann H, Muhlenhoff M, Borrow R, et al. 1996. Capsule phase variation in *Neisseria meningitidis* serogroup B by slipped-strand mispairing in the polysialyltransferase gene (*siaD*): correlation with bacterial invasion and the outbreak of meningococcal disease. *Mol. Microbiol.* 20:1211–20

66. Hamrick TS, Dempsey JA, Cohen MS, Cannon JG. 2001. Antigenic variation of gonococcal pilin expression in vivo: analysis of the strain FA1090 pilin repertoire and identification of the *pilS* gene copies recombining with *pilE* during experimental human infection. *Microbiology* 147:839–49

67. Harrison LH, Trotter CL, Ramsay ME. 2009. Global epidemiology of meningococcal disease. *Vaccine* 27(Suppl. 2):B51–63

68. Haubold B, Travisano M, Rainey PB, Hudson RR. 1998. Detecting linkage disequilibrium in bacterial populations. *Genetics* 150:1341–48

69. Helm RA, Seifert HS. 2010. Frequency and rate of pilin antigenic variation of *Neisseria meningitidis*. *J. Bacteriol.* 192:3822–23

70. Hill SA, Davies JK. 2009. Pilin gene variation in *Neisseria gonorrhoeae*: reassessing the old paradigms. *FEMS Microbiol. Rev.* 33:521–30

71. Hill SA, Grant CC. 2002. Recombinational error and deletion formation in *Neisseria gonorrhoeae*: a role for RecJ in the production of *pilE* (L) deletions. *Mol. Genet. Genomics* 266:962–72

72. Horrocks P, Pinches R, Christodoulou Z, Kyes SA, Newbold CI. 2004. Variable *var* transition rates underlie antigenic variation in malaria. *Proc. Natl. Acad. Sci. USA* 101:11129–34

73. Hou Z, Zhang Y, Propson NE, Howden SE, Chu LF, et al. 2013. Efficient genome engineering in human pluripotent stem cells using Cas9 from *Neisseria meningitidis*. *Proc. Natl. Acad. Sci. USA* 110:15644–49

74. Howell-Adams B, Seifert HS. 2000. Molecular models accounting for the gene conversion reactions mediating gonococcal pilin antigenic variation. *Mol. Microbiol.* 37:1146–58

75. Howell-Adams B, Wainwright LA, Seifert HS. 1996. The size and position of heterologous insertions in a silent locus differentially affect pilin recombination in *Neisseria gonorrhoeae*. *Mol. Microbiol.* 22:509–22

76. Ison CA, Bellinger CM, Walker J. 1986. Homology of cryptic plasmid of *Neisseria gonorrhoeae* with plasmids from *Neisseria meningitidis* and *Neisseria lactamica*. *J. Clin. Pathol.* 39:1119–23

77. Jain S, Zweig M, Peeters E, Siewering K, Hackett KT, et al. 2012. Characterization of the single stranded DNA binding protein SsbB encoded in the gonoccocal genetic island. *PLOS ONE* 7:e35285

78. Jansen R, Embden JD, Gaastra W, Schouls LM. 2002. Identification of genes that are associated with DNA repeats in prokaryotes. *Mol. Microbiol.* 43:1565–75

79. Jennings MP, Srikhanta YN, Moxon ER, Kramer M, Poolman JT, et al. 1999. The genetic basis of the phase variation repertoire of lipopolysaccharide immunotypes in *Neisseria meningitidis*. *Microbiology* 145(Pt. 11):3013–21

80. Jennings MP, Virji M, Evans D, Foster V, Srikhanta YN, et al. 1998. Identification of a novel gene involved in pilin glycosylation in *Neisseria meningitidis*. *Mol. Microbiol.* 29:975–84

81. Jolley KA, Sun L, Moxon ER, Maiden MC. 2004. Dam inactivation in *Neisseria meningitidis*: prevalence among diverse hyperinvasive lineages. *BMC Microbiol.* 4:34

82. Jonsson AB, Nyberg G, Normark S. 1991. Phase variation of gonococcal pili by frameshift mutation in *pilC*, a novel gene for pilus assembly. *EMBO J.* 10:477–88

83. Jordan P, Snyder LA, Saunders NJ. 2003. Diversity in coding tandem repeats in related *Neisseria* spp. *BMC Microbiol.* 3:23

84. Jordan PW, Snyder LA, Saunders NJ. 2005. Strain-specific differences in *Neisseria gonorrhoeae* associated with the phase variable gene repertoire. *BMC Microbiol.* 5:21

85. Kawai M, Uchiyama I, Kobayashi I. 2006. Genome comparison in silico in *Neisseria* suggests integration of filamentous bacteriophages by their own transposase. *DNA Res.* 12:389–401

86. Kellogg DS Jr, Peacock WL Jr, Deacon WE, Brown L, Pirkle DI. 1963. *Neisseria gonorrhoeae*. I. Virulence genetically linked to clonal variation. *J. Bacteriol.* 85:1274–79

87. Kline KA, Criss AK, Wallace A, Seifert HS. 2007. Transposon mutagenesis identifies sites upstream of the *Neisseria gonorrhoeae pilE* gene that modulate pilin antigenic variation. *J. Bacteriol.* 189:3462–70

88. Kline KA, Sechman EV, Skaar EP, Seifert HS. 2003. Recombination, repair and replication in the pathogenic *Neisseriae*: the 3 R's of molecular genetics of two human-specific bacterial pathogens. *Mol. Microbiol.* 50:3–13

89. Kobayashi I. 2001. Behavior of restriction-modification systems as selfish mobile elements and their impact on genome evolution. *Nucleic Acids Res.* 29:3742–56

90. Koomey M, Gotschlich EC, Robbins K, Bergstrom S, Swanson J. 1987. Effects of *recA* mutations on pilus antigenic variation and phase transitions in *Neisseria gonorrhoeae*. *Genetics* 117:391–98

91. Korch C, Hagblom P, Ohman H, Goransson M, Normark S. 1985. Cryptic plasmid of *Neisseria gonorrhoeae*: complete nucleotide sequence and genetic organization. *J. Bacteriol.* 163:430–38

91a. Kuzminov A. 1999. Recombinational repair of DNA damage in *Escherichia coli* and bacteriophage lambda. *Microbiol. Mol. Biol. Rev.* 63:751–813

92. Kupsch EM, Aubel D, Gibbs CP, Kahrs AF, Rudel T, Meyer TF. 1996. Construction of Hermes shuttle vectors: a versatile system useful for genetic complementation of transformable and non-transformable *Neisseria* mutants. *Mol. Gen. Genet.* 250:558–69

93. Kupsch EM, Knepper B, Kuroki T, Heuer I, Meyer TF. 1993. Variable opacity (Opa) outer membrane proteins account for the cell tropisms displayed by *Neisseria gonorrhoeae* for human leukocytes and epithelial cells. *EMBO J.* 12:641–50

94. Lahra MM, WHO West. Pac. South East Asian Gonococcal Antimicrob. Surveill. Programme. 2012. Surveillance of antibiotic resistance in *Neisseria gonorrhoeae* in the WHO Western Pacific and South East Asian Regions, 2010. *Commun. Dis. Intell. Q. Rep.* 36:95–100

95. LeVan A, Zimmerman LI, Mahle AC, Swanson KV, DeShong P, et al. 2012. Construction and characterization of a derivative of *Neisseria gonorrhoeae* strain MS11 devoid of all *opa* genes. *J. Bacteriol.* 194:6468–78

96. Levinson G, Gutman GA. 1987. High frequencies of short frameshifts in poly-CA/TG tandem repeats borne by bacteriophage M13 in *Escherichia coli* K-12. *Nucleic Acids Res.* 15:5323–38

97. Levinson G, Gutman GA. 1987. Slipped-strand mispairing: a major mechanism for DNA sequence evolution. *Mol. Biol. Evol.* 4:203–21

98. Lin YH, Ryan CS, Davies JK. 2011. Neisserial Correia repeat-enclosed elements do not influence the transcription of *pil* genes in *Neisseria gonorrhoeae* and *Neisseria meningitidis*. *J. Bacteriol.* 193:5728–36

99. Liu SV, Saunders NJ, Jeffries A, Rest RF. 2002. Genome analysis and strain comparison of Correia repeats and Correia repeat-enclosed elements in pathogenic *Neisseria*. *J. Bacteriol.* 184:6163–73

100. Loenen WA, Dryden DT, Raleigh EA, Wilson GG, Murray NE. 2014. Highlights of the DNA cutters: a short history of the restriction enzymes. *Nucleic Acids Res.* 42:3–19

101. Marri PR, Paniscus M, Weyand NJ, Rendon MA, Calton CM, et al. 2010. Genome sequencing reveals widespread virulence gene exchange among human *Neisseria* species. *PLOS ONE* 5:e11835

102. Martin P, Makepeace K, Hill SA, Hood DW, Moxon ER. 2005. Microsatellite instability regulates transcription factor binding and gene expression. *Proc. Natl. Acad. Sci. USA* 102:3800–4

103. Martin P, Sun L, Hood DW, Moxon ER. 2004. Involvement of genes of genome maintenance in the regulation of phase variation frequencies in *Neisseria meningitidis*. *Microbiology* 150:3001–12

104. Masignani V, Giuliani MM, Tettelin H, Comanducci M, Rappuoli R, Scarlato V. 2001. Mu-like prophage in serogroup B *Neisseria meningitidis* coding for surface-exposed antigens. *Infect. Immun.* 69:2580–88

105. Mayer LW. 1982. Rates in vitro changes of gonococcal colony opacity phenotypes. *Infect. Immun.* 37:481–85

106. Mazzone M, De Gregorio E, Lavitola A, Pagliarulo C, Alifano P, Di Nocera PP. 2001. Whole-genome organization and functional properties of miniature DNA insertion sequences conserved in pathogenic *Neisseriae*. *Gene* 278:211–22

107. Mehr IJ, Seifert HS. 1998. Differential roles of homologous recombination pathways in *Neisseria gonorrhoeae* pilin antigenic variation, DNA transformation and DNA repair. *Mol. Microbiol.* 30:697–710

108. Metruccio MM, Pigozzi E, Roncarati D, Berlanda Scorza F, Norais N, et al. 2009. A novel phase variation mechanism in the meningococcus driven by a ligand-responsive repressor and differential spacing of distal promoter elements. *PLOS Pathogens* 5:e1000710

109. Miller F, Phan G, Brissac T, Bouchiat C, Lioux G, et al. 2014. The hypervariable region of meningococcal major pilin PilE controls the host cell response via antigenic variation. *mBio* 5:e01024–13

110. Mirkin SM. 2007. Expandable DNA repeats and human disease. *Nature* 447:932–40

111. Morand PC, Tattevin P, Eugene E, Beretti JL, Nassif X. 2001. The adhesive property of the type IV pilus-associated component PilC1 of pathogenic *Neisseria* is supported by the conformational structure of the N-terminal part of the molecule. *Mol. Microbiol.* 40:846–56

112. Morse SA, Johnson SR, Biddle JW, Roberts MC. 1986. High-level tetracycline resistance in *Neisseria gonorrhoeae* is result of acquisition of streptococcal *tetM* determinant. *Antimicrob. Agents Chemother.* 30:664–70

113. Moxon R, Bayliss C, Hood D. 2006. Bacterial contingency loci: the role of simple sequence DNA repeats in bacterial adaptation. *Annu. Rev. Genet.* 40:307–33

114. Murphy GL, Connell TD, Barritt DS, Koomey M, Cannon JG. 1989. Phase variation of gonococcal protein II: regulation of gene expression by slipped-strand mispairing of a repetitive DNA sequence. *Cell* 56:539–47

115. O'Ryan M, Stoddard J, Toneatto D, Wassil J, Dull PM. 2014. A multi-component meningococcal serogroup B vaccine (4CMenB): the clinical development program. *Drugs* 74:15–30

116. Pachulec E, van der Does C. 2010. Conjugative plasmids of *Neisseria gonorrhoeae*. *PLOS ONE* 5:e9962

117. Pagotto FJ, Salimnia H, Totten PA, Dillon JR. 2000. Stable shuttle vectors for *Neisseria gonorrhoeae*, *Haemophilus* spp. and other bacteria based on a single origin of replication. *Gene* 244:13–19

118. Parker BO, Marinus MG. 1992. Repair of DNA heteroduplexes containing small heterologous sequences in *Escherichia coli*. *Proc. Natl. Acad. Sci. USA* 89:1730–34

119. Parkhill J, Achtman M, James KD, Bentley SD, Churcher C, et al. 2000. Complete DNA sequence of a serogroup A strain of *Neisseria meningitidis* Z2491. *Nature* 404:502–6

120. Phelps LN. 1967. Isolation and characterization of bacteriophages for *Neisseria*. *J. Gen. Virol.* 1:529–36

121. Piekarowicz A, Klyz A, Majchrzak M, Adamczyk-Poplawska M, Maugel TK, Stein DC. 2007. Characterization of the dsDNA prophage sequences in the genome of *Neisseria gonorrhoeae* and visualization of productive bacteriophage. *BMC Microbiol.* 7:66

122. Piekarowicz A, Klyz A, Majchrzak M, Szczesna E, Piechucki M, et al. 2014. *Neisseria gonorrhoeae* filamentous phage Ngoφ6 is capable of infecting a variety of gram-negative bacteria. *J. Virol.* 88:1002–10

123. Piekarowicz A, Majchrzak M, Klyz A, Adamczyk-Poplawska M. 2006. Analysis of the filamentous bacteriophage genomes integrated into *Neisseria gonorrhoeae* FA1090 chromosome. *Pol. J. Microbiol.* 55:251–60

124. Power PM, Roddam LF, Rutter K, Fitzpatrick SZ, Srikhanta YN, Jennings MP. 2003. Genetic characterization of pilin glycosylation and phase variation in *Neisseria meningitidis*. *Mol. Microbiol.* 49:833–47

125. Reynaud CA, Anquez V, Grimal H, Weill JC. 1987. A hyperconversion mechanism generates the chicken light chain preimmune repertoire. *Cell* 48:379–88

126. Richardson AR, Stojiljkovic I. 2001. Mismatch repair and the regulation of phase variation in *Neisseria meningitidis*. *Mol. Microbiol.* 40:645–55

127. Richardson AR, Yu Z, Popovic T, Stojiljkovic I. 2002. Mutator clones of *Neisseria meningitidis* in epidemic serogroup A disease. *Proc. Natl. Acad. Sci. USA* 99:6103–7

128. Roberts M, Piot P, Falkow S. 1979. The ecology of gonococcal plasmids. *J. Gen. Microbiol.* 114:491–94

129. Roberts MC. 1989. Plasmids of *Neisseria gonorrhoeae* and other *Neisseria* species. *Clin. Microbiol. Rev.* 2:S18–23

130. Roberts MC, Knapp JS. 1988. Host range of the conjugative 25.2-megadalton tetracycline resistance plasmid from *Neisseria gonorrhoeae* and related species. *Antimicrob. Agents Chemother.* 32:488–91

131. Roberts RJ, Vincze T, Posfai J, Macelis D. 2010. REBASE—a database for DNA restriction and modification: enzymes, genes and genomes. *Nucleic Acids Res.* 38:D234–36

132. Rohrer MS, Lazio MP, Seifert HS. 2005. A real-time semi-quantitative RT-PCR assay demonstrates that the *pilE* sequence dictates the frequency and characteristics of pilin antigenic variation in *Neisseria gonorrhoeae*. *Nucleic Acids Res.* 33:3363–71

133. Rosenstein NE, Perkins BA, Stephens DS, Popovic T, Hughes JM. 2001. Meningococcal disease. *N. Engl. J. Med.* 344:1378–88

134. Sadarangani M, Hoe JC, Callaghan MJ, Jones C, Chan H, et al. 2012. Construction of Opa-positive and Opa-negative strains of *Neisseria meningitidis* to evaluate a novel meningococcal vaccine. *PLOS ONE* 7:e51045

135. Sadarangani M, Pollard AJ, Gray-Owen SD. 2011. Opa proteins and CEACAMs: pathways of immune engagement for pathogenic *Neisseria*. *FEMS Microbiol. Rev.* 35:498–514

136. Sarkari J, Pandit N, Moxon ER, Achtman M. 1994. Variable expression of the Opc outer membrane protein in *Neisseria meningitidis* is caused by size variation of a promoter containing poly-cytidine. *Mol. Microbiol.* 13:207–17

137. Saunders NJ, Jeffries AC, Peden JF, Hood DW, Tettelin H, et al. 2000. Repeat-associated phase variable genes in the complete genome sequence of *Neisseria meningitidis* strain MC58. *Mol. Microbiol.* 37:207–15

138. Saunders NJ, Snyder LA. 2002. The minimal mobile element. *Microbiology* 148:3756–60

139. Schoen C, Blom J, Claus H, Schramm-Gluck A, Brandt P, et al. 2008. Whole-genome comparison of disease and carriage strains provides insights into virulence evolution in *Neisseria meningitidis*. *Proc. Natl. Acad. Sci. USA* 105:3473–78

140. Schoen C, Joseph B, Claus H, Vogel U, Frosch M. 2007. Living in a changing environment: insights into host adaptation in *Neisseria meningitidis* from comparative genomics. *Int. J. Med. Microbiol.* 297:601–13

141. Schofield MJ, Hsieh P. 2003. DNA mismatch repair: molecular mechanisms and biological function. *Annu. Rev. Microbiol.* 57:579–608

142. Scholten RJ, Kuipers B, Valkenburg HA, Dankert J, Zollinger WD, Poolman JT. 1994. Lipo-oligosaccharide immunotyping of *Neisseria meningitidis* by a whole-cell ELISA with monoclonal antibodies. *J. Med. Microbiol.* 41:236–43

143. Schork S, Schluter A, Blom J, Schneiker-Bekel S, Puhler A, et al. 2012. Genome sequence of a *Neisseria meningitidis* capsule null locus strain from the clonal complex of sequence type 198. *J. Bacteriol.* 194:5144–45

144. Sechman EV, Kline KA, Seifert HS. 2006. Loss of both Holliday junction processing pathways is synthetically lethal in the presence of gonococcal pilin antigenic variation. *Mol. Microbiol.* 61:185–93

145. Sechman EV, Rohrer MS, Seifert HS. 2005. A genetic screen identifies genes and sites involved in pilin antigenic variation in *Neisseria gonorrhoeae*. *Mol. Microbiol.* 57:468–83

146. Seifert HS, Wright CJ, Jerse AE, Cohen MS, Cannon JG. 1994. Multiple gonococcal pilin antigenic variants are produced during experimental human infections. *J. Clin. Investig.* 93:2744–49

147. Serkin CD, Seifert HS. 1998. Frequency of pilin antigenic variation in *Neisseria gonorrhoeae. J. Bacteriol.* 180:1955–58

148. Serkin CD, Seifert HS. 2000. Iron availability regulates DNA recombination in *Neisseria gonorrhoeae. Mol. Microbiol.* 37:1075–86

149. Shafer WM, Rest RF. 1989. Interactions of gonococci with phagocytic cells. *Annu. Rev. Microbiol.* 43:121–45

150. Siddique A, Buisine N, Chalmers R. 2011. The transposon-like Correia elements encode numerous strong promoters and provide a potential new mechanism for phase variation in the meningococcus. *PLOS Genet.* 7:e1001277

151. Skaar EP, Lazio MP, Seifert HS. 2002. Roles of the *recJ* and *recN* genes in homologous recombination and DNA repair pathways of *Neisseria gonorrhoeae. J. Bacteriol.* 184:919–27

152. Skaar EP, Lecuyer B, Lenich AG, Lazio MP, Perkins-Balding D, et al. 2005. Analysis of the Piv recombinase-related gene family of *Neisseria gonorrhoeae. J. Bacteriol.* 187:1276–86

153. Smith HO, Gwinn ML, Salzberg SL. 1999. DNA uptake signal sequences in naturally transformable bacteria. *Res. Microbiol.* 150:603–16

154. Smith JM, Smith NH, O'Rourke M, Spratt BG. 1993. How clonal are bacteria? *Proc. Natl. Acad. Sci. USA* 90:4384–88

155. Snyder LA, Butcher SA, Saunders NJ. 2001. Comparative whole-genome analyses reveal over 100 putative phase-variable genes in the pathogenic *Neisseria* spp. *Microbiology* 147:2321–32

156. Snyder LA, Cole JA, Pallen MJ. 2009. Comparative analysis of two *Neisseria gonorrhoeae* genome sequences reveals evidence of mobilization of Correia repeat enclosed elements and their role in regulation. *BMC Genomics* 10:70

157. Snyder LA, Jarvis SA, Saunders NJ. 2005. Complete and variant forms of the "gonococcal genetic island" in *Neisseria meningitidis. Microbiology* 151:4005–13

158. Snyder LA, McGowan S, Rogers M, Duro E, O'Farrell E, Saunders NJ. 2007. The repertoire of minimal mobile elements in the *Neisseria* species and evidence that these are involved in horizontal gene transfer in other bacteria. *Mol. Biol. Evol.* 24:2802–15

159. Snyder LA, Saunders NJ. 2006. The majority of genes in the pathogenic *Neisseria* species are present in non-pathogenic *Neisseria lactamica*, including those designated as "virulence genes." *BMC Genomics* 7:128

160. Snyder LA, Shafer WM, Saunders NJ. 2003. Divergence and transcriptional analysis of the division cell wall (*dcw*) gene cluster in *Neisseria* spp. *Mol. Microbiol.* 47:431–42

161. Solomon JM, Grossman AD. 1996. Who's competent and when: regulation of natural genetic competence in bacteria. *Trends Genet.* 12:150–55

162. Sparling PF. 1966. Genetic transformation of *Neisseria gonorrhoeae* to streptomycin resistance. *J. Bacteriol.* 92:1364–71

163. Spencer-Smith R, Varkey EM, Fielder MD, Snyder LA. 2012. Sequence features contributing to chromosomal rearrangements in *Neisseria gonorrhoeae. PLOS ONE* 7:e46023

164. Srikhanta YN, Dowideit SJ, Edwards JL, Falsetta ML, Wu HJ, et al. 2009. Phasevarions mediate random switching of gene expression in pathogenic *Neisseria. PLOS Pathog.* 5:e1000400

165. Stein DC. 1991. Transformation of *Neisseria gonorrhoeae*: physical requirements of the transforming DNA. *Can. J. Microbiol.* 37:345–49

166. Stein DC, Gregoire S, Piekarowicz A. 1988. Restriction of plasmid DNA during transformation but not conjugation in *Neisseria gonorrhoeae. Infect. Immun.* 56:112–16

167. Stein DC, Gunn JS, Radlinska M, Piekarowicz A. 1995. Restriction and modification systems of *Neisseria gonorrhoeae. Gene* 157:19–22

168. Stein DC, Silver LE, Clark VL, Young FE. 1983. Construction and characterization of a new shuttle vector, pLES2, capable of functioning in *Escherichia coli* and *Neisseria gonorrhoeae. Gene* 25:241–47

169. Steinberg VI, Hart EJ, Handley J, Goldberg ID. 1976. Isolation and characterization of a bacteriophage specific for *Neisseria perflava. J. Clin. Microbiol.* 4:87–91

170. Stephens DS, Greenwood B, Brandtzaeg P. 2007. Epidemic meningitis, meningococcaemia, and *Neisseria meningitidis. Lancet* 369:2196–210

171. Stern A, Brown M, Nickel P, Meyer TF. 1986. Opacity genes in *Neisseria gonorrhoeae*: control of phase and antigenic variation. *Cell* 47:61–71

172. Stern A, Keren L, Wurtzel O, Amitai G, Sorek R. 2010. Self-targeting by CRISPR: gene regulation or autoimmunity? *Trends Genet.* 26:335–40

173. Stern A, Meyer TF. 1987. Common mechanism controlling phase and antigenic variation in pathogenic *Neisseriae*. *Mol. Microbiol.* 1:5–12

174. Stohl EA, Seifert HS. 2001. The *recX* gene potentiates homologous recombination in *Neisseria gonorrhoeae*. *Mol. Microbiol.* 40:1301–10

175. Stone RL, Culbertson CG, Powell HM. 1956. Studies of a bacteriophage active against a chromogenic *Neisseria*. *J. Bacteriol.* 71:516–20

176. Swanson J. 1973. Studies on gonococcus infection. IV. Pili: their role in attachment of gonococci to tissue culture cells. *J. Exp. Med.* 137:571–89

177. Swanson J. 1978. Studies on gonococcus infection. XIV. Cell wall protein differences among color/opacity colony variants of *Neisseria gonorrhoeae*. *Infect. Immun.* 21:292–302

178. Swanson J, Bergstrom S, Barrera O, Robbins K, Corwin D. 1985. Pilus− gonococcal variants. Evidence for multiple forms of piliation control. *J. Exp. Med.* 162:729–44

179. Swanson J, Bergstrom S, Robbins K, Barrera O, Corwin D, Koomey JM. 1986. Gene conversion involving the pilin structural gene correlates with pilus+ in equilibrium with pilus− changes in *Neisseria gonorrhoeae*. *Cell* 47:267–76

180. Swanson J, Kraus SJ, Gotschlich EC. 1971. Studies on gonococcus infection. I. Pili and zones of adhesion: their relation to gonococcal growth patterns. *J. Exp. Med.* 134:886–906

181. Tabrizi SN, Unemo M, Limnios AE, Hogan TR, Hjelmevoll SO, et al. 2011. Evaluation of six commercial nucleic acid amplification tests for detection of *Neisseria gonorrhoeae* and other *Neisseria* species. *J. Clin. Microbiol.* 49:3610–15

182. Tauseef I, Ali YM, Bayliss CD. 2013. Phase variation of PorA, a major outer membrane protein, mediates escape of bactericidal antibodies by *Neisseria meningitidis*. *Infect. Immun.* 81:1374–80

183. Tettelin H, Saunders NJ, Heidelberg J, Jeffries AC, Nelson KE, et al. 2000. Complete genome sequence of *Neisseria meningitidis* serogroup B strain MC58. *Science* 287:1809–15

184. Thompson SA, Wang LL, West A, Sparling PF. 1993. *Neisseria meningitidis* produces iron-regulated proteins related to the RTX family of exoproteins. *J. Bacteriol.* 175:811–18

185. Tobiason DM, Seifert HS. 2006. The obligate human pathogen, *Neisseria gonorrhoeae*, is polyploid. *PLOS Biol.* 4:e185

186. Tobiason DM, Seifert HS. 2010. Genomic content of *Neisseria* species. *J. Bacteriol.* 192:2160–68

187. Tonjum T, Caugant DA, Dunham SA, Koomey M. 1998. Structure and function of repetitive sequence elements associated with a highly polymorphic domain of the *Neisseria meningitidis* PilQ protein. *Mol. Microbiol.* 29:111–24

188. Trembizki E, Lahra M, Stevens K, Freeman K, Hogan T, et al. 2014. A national quality assurance survey of *Neisseria gonorrhoeae* testing. *J. Med. Microbiol.* 63:45–49

189. Turner CM, Barry JD. 1989. High frequency of antigenic variation in *Trypanosoma brucei rhodesiense* infections. *Parasitology* 99(Pt. 1):67–75

190. van der Ende A, Hopman CT, Dankert J. 2000. Multiple mechanisms of phase variation of PorA in *Neisseria meningitidis*. *Infect. Immun.* 68:6685–90

191. van der Ende A, Hopman CT, Zaat S, Essink BB, Berkhout B, Dankert J. 1995. Variable expression of class 1 outer membrane protein in *Neisseria meningitidis* is caused by variation in the spacing between the −10 and −35 regions of the promoter. *J. Bacteriol.* 177:2475–80

192. van Passel MW, Bart A, Luyf AC, van Kampen AH, van der Ende A. 2006. Identification of acquired DNA in *Neisseria lactamica*. *FEMS Microbiol. Lett.* 262:77–84

193. van Passel MW, van der Ende A, Bart A. 2006. Plasmid diversity in neisseriae. *Infect. Immun.* 74:4892–99

194. Vink C, Rudenko G, Seifert HS. 2012. Microbial antigenic variation mediated by homologous DNA recombination. *FEMS Microbiol. Rev.* 36:917–48

195. Wainwright LA, Frangipane JV, Seifert HS. 1997. Analysis of protein binding to the Sma/Cla DNA repeat in pathogenic *Neisseriae*. *Nucleic Acids Res.* 25:1362–68

196. Wainwright LA, Pritchard KH, Seifert HS. 1994. A conserved DNA sequence is required for efficient gonococcal pilin antigenic variation. *Mol. Microbiol.* 13:75–87

197. WHO. 1998. *Control of Epidemic Meningococcal Disease. WHO Practical Guidelines*. Geneva, Switz.: World Health Organ.

198. WHO. 2012. *Global Incidence and Prevalence of Selected Curable Sexually Transmitted Infections - 2008*. Geneva, Switz: World Health Organ.

199. Wolfgang M, Lauer P, Park HS, Brossay L, Hebert J, Koomey M. 1998. PilT mutations lead to simultaneous defects in competence for natural transformation and twitching motility in piliated *Neisseria gonorrhoeae*. *Mol. Microbiol.* 29:321–30

200. Woodhams KL, Benet ZL, Blonsky SE, Hackett KT, Dillard JP. 2012. Prevalence and detailed mapping of the gonococcal genetic island in *Neisseria meningitidis*. *J. Bacteriol.* 194:2275–85

201. Woods JP, Spinola SM, Strobel SM, Cannon JG. 1989. Conserved lipoprotein H.8 of pathogenic *Neisseria* consists entirely of pentapeptide repeats. *Mol. Microbiol.* 3:43–48

202. Workowski KA, Berman S. 2010. Sexually transmitted diseases treatment guidelines, 2010. *Morb. Mortal. Wkly. Rep.* 59:1–110

203. Workowski KA, Berman SM, Douglas JM Jr. 2008. Emerging antimicrobial resistance in *Neisseria gonorrhoeae*: urgent need to strengthen prevention strategies. *Ann. Intern. Med.* 148:606–13

204. Yang QL, Gotschlich EC. 1996. Variation of gonococcal lipooligosaccharide structure is due to alterations in poly-G tracts in *lgt* genes encoding glycosyl transferases. *J. Exp. Med.* 183:323–27

205. Zhang JR, Norris SJ. 1998. Genetic variation of the *Borrelia burgdorferi* gene *vlsE* involves cassette-specific, segmental gene conversion. *Infect. Immun.* 66:3698–704

206. Zhang Y, Heidrich N, Ampattu BJ, Gunderson CW, Seifert HS, et al. 2013. Processing-independent CRISPR RNAs limit natural transformation in *Neisseria meningitidis*. *Mol. Cell* 50:488–503

207. Zola TA, Strange HR, Dominguez NM, Dillard JP, Cornelissen CN. 2010. Type IV secretion machinery promotes Ton-independent intracellular survival of *Neisseria gonorrhoeae* within cervical epithelial cells. *Infect. Immun.* 78:2429–37

Regulation of Transcription by Long Noncoding RNAs

Roberto Bonasio[1] and Ramin Shiekhattar[2]

[1]Department of Cell and Developmental Biology and Epigenetics Program, Perelman School of Medicine, University of Pennsylvania, Philadelphia, Pennsylvania 19104; email: rbon@mail.med.upenn.edu

[2]Department of Human Genetics, Miller School of Medicine, University of Miami, Miami, Florida 33136; email: rshiekhattar@med.miami.edu

Annu. Rev. Genet. 2014. 48:433–55

First published online as a Review in Advance on September 18, 2014

The *Annual Review of Genetics* is online at genet.annualreviews.org

This article's doi: 10.1146/annurev-genet-120213-092323

Keywords

enhancers, imprinting, transcriptional silencing, chromatin, chromatin-modifying complexes, RNA polymerase II

Abstract

Over the past decade there has been a greater understanding of genomic complexity in eukaryotes ushered in by the immense technological advances in high-throughput sequencing of DNA and its corresponding RNA transcripts. This has resulted in the realization that beyond protein-coding genes, there are a large number of transcripts that do not encode for proteins and, therefore, may perform their function through RNA sequences and/or through secondary and tertiary structural determinants. This review is focused on the latest findings on a class of noncoding RNAs that are relatively large (>200 nucleotides), display nuclear localization, and use different strategies to regulate transcription. These are exciting times for discovering the biological scope and the mechanism of action for these RNA molecules, which have roles in dosage compensation, imprinting, enhancer function, and transcriptional regulation, with a great impact on development and disease.

INTRODUCTION

The RNA world hypothesis, the most widely accepted theory for the origin of life on Earth (20), envisions a world in which RNA constituted the first and only replicative molecule (30, 46). Although not without its opponents, this theory is born from the observation that RNA can replicate via base complementarity, like DNA, and can catalyze chemical reactions, like proteins. A simple hydroxyl group turns the reliable but inert deoxyribose into a less stable but more eclectic ribose, which endows RNA with the ability to fold into complex tertiary structures (176), interact specifically with proteins and other molecules, and catalyze chemical reactions (82). Therefore, it is not surprising that RNA would play central roles in all aspects of molecular biology, weaving an intricate web of regulatory mechanisms in complex multicellular organisms (21). However, for many, the RNA revolution was an unexpected twist in molecular biology. As the race to obtain the complete sequence of the human genome was in full swing, it was widely believed that the functions of RNA were limited to the central dogma. Mainly, messenger RNAs carry genetic information from the nucleus to the cytoplasm, where transfer RNAs read it and ribosomal RNAs coordinate its translation into the more sophisticated language of protein biochemistry.

The past decade has ushered in experimental evidence suggesting a much wider role for RNA in molecular biology. Widespread regulatory functions of noncoding RNAs (ncRNAs) first came to prominence in the 1990s with the discovery of RNA interference (RNAi) and its role in post-transcriptional gene silencing (55). This led to the identification of different types of small ncRNAs that guide protein complexes to mRNA targets and inhibit their translation (161). Having realized that RNA may delineate novel biological pathways, researchers began cataloging other ncRNA species using genome-wide approaches spearheaded by the technological advances in high-throughput sequencing (103). We now know that large expanses of the genome are transcribed into RNA, and only a small portion of it encodes proteins (35, 38, 40).

Given that it is more difficult to infer the existence of a ncRNA based on sequence features alone when compared to protein-coding mRNAs, most established bioinformatic approaches are not well-suited for their analysis (39), leading to greater impetus for development of new computational strategies. We need a clearer understanding of functional roles of ncRNAs before a comprehensive picture of the full scope of the ncRNA repertoire in higher eukaryotes emerges. Many excellent reviews have been written about the various aspects of ncRNA function (18, 50, 67, 89, 99, 119, 126, 136, 165, 171). Here, we focus on the large and complex category of long ncRNAs (lncRNAs) and their role in regulating gene expression by interacting with chromatin and with the protein machinery that controls its structure and function.

Classification of lncRNAs

Long ncRNA have been operationally defined as transcripts that are produced by RNA polymerase II, longer than 200 nucleotides, and devoid of an open reading frame (ORF) that can be translated into a protein (14, 35, 48). It has been difficult to group lncRNAs in different classes because we have scant information regarding their precise scope of function and their overall secondary and tertiary structure. The simplest and most insightful classification of lncRNAs was recently presented by cataloging the spliced and unspliced lncRNAs in human and mouse ES cells according to their loci of origin (150). It was concluded that the large majority of lncRNAs either map to enhancer regions (~20%; termed eRNAs) or correspond to upstream antisense RNAs (uaRNAs) that originate near the transcription start site (TSS) of genic RNAs (60–70%). The remaining lncRNAs derive from antisense transcripts that overlap with annotated gene bodies (~5%) or originate from more distal, unannotated regions (~5%) (**Figure 1**). The latter are commonly referred to as long intervening or

lncRNA class	Names	Examples	References
Promoter-associated 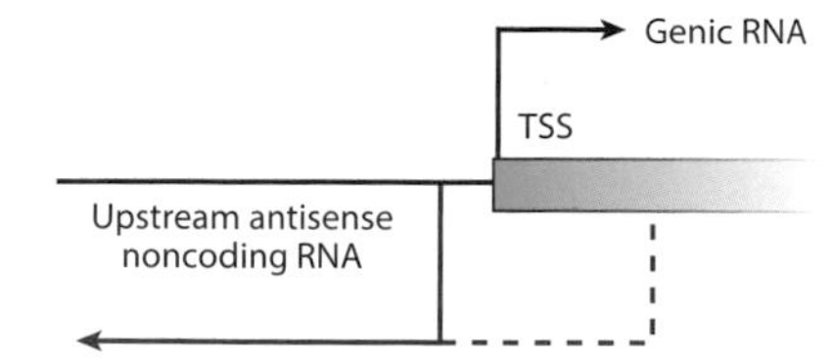	pRNAs (uaRNAs, PROMPTs)	Most genes	150, 167
Gene body-associated (sense) 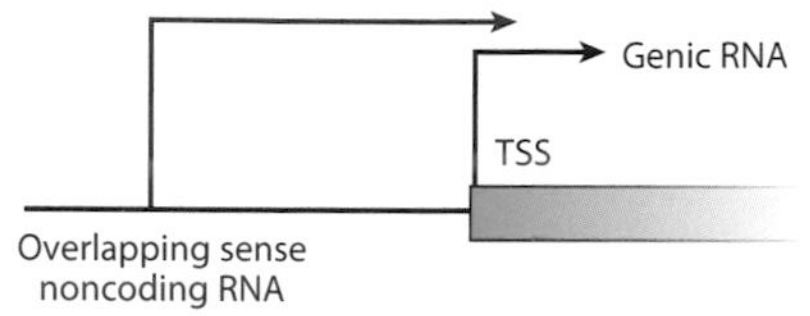	gsRNAs	*DHFR* locus ecCEBPA *fbp1*+ locus	37, 59, 107
Gene body-associated (antisense) 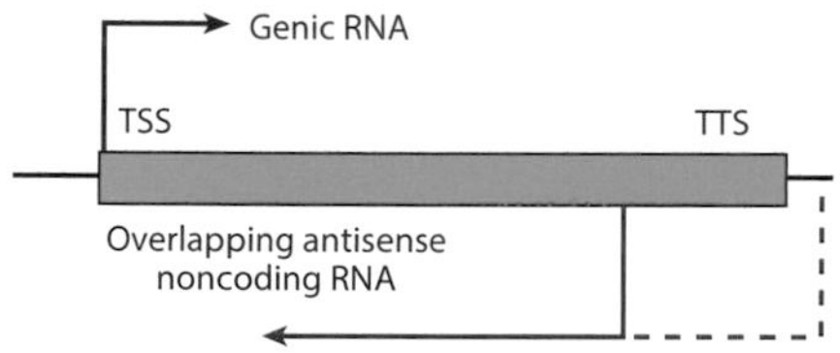	gaRNAs (NATs)	Airn Kcnq1ot1 *BDNF* locus	90a, 112a, 153
Enhancer-associated 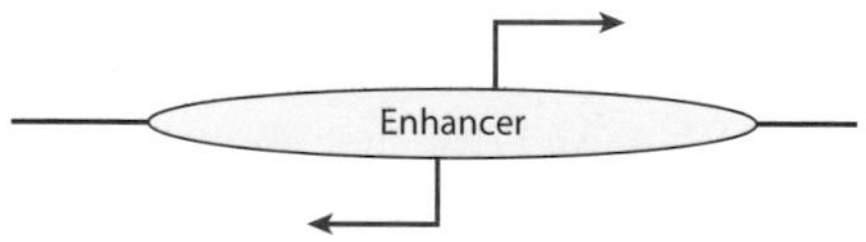	eRNAs (ncRNA-a)	All enhancers	36, 76, 123
Intervening 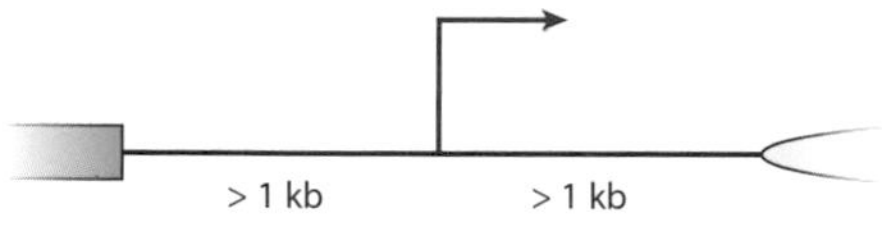	iRNAs (lincRNAs)	HOTAIR MALAT1	14, 48, 137

Figure 1

Classes of lncRNAs (long noncoding RNAs). All known lncRNAs are divided into five classes, on the basis of their relationship with adjacent or overlapping genomic features. The initiation site for the noncoding transcript is shown in red, and the arrow indicates the direction of transcription. Less-common variants within the same class are indicated with dashed arrows. In the second column, existing names for the different classes are listed in gray within parentheses, and their corresponding names from our proposed new nomenclature are in black. Abbreviations: eRNAs, enhancer RNAs; gaRNAs, gene body–associated antisense RNAs; gsRNAs, gene body–associated sense RNAs; iRNAs, intervening long noncoding RNAs; lincRNAs, long intervening noncoding RNAs; NATs, natural antisense transcripts; ncRNAs-a, noncoding RNAs activating; pRNAs, promoter-associated RNAs; PROMPTs, promoter-associated pervasive transcripts; TSS, transcription start site; TTS, transcription termination site; uaRNAs, upstream antisense RNAs.

intergenic ncRNAs (lincRNAs) (48, 166). We note that some intervening and intragenic lncRNAs might be transcribed from as yet annotated enhancer elements (either distal to or within the gene body) and may therefore be eventually recategorized as eRNAs. The finding that a large number of lncRNAs arise from loci with proximity to protein-coding genes is consistent with previous genome-wide analyses of lncRNAs (167). A few cases of lncRNAs that originate upstream of a cognate mRNA gene and are transcribed in the sense direction have also been described (37, 107); in this case, however, the downstream sequence of the mRNA gene is often part of the final RNA, and therefore these overlapping sense RNAs are annotated as alternative isoforms and excluded from lncRNA catalogs. Because of the confusion arising from the phonetically similar lncRNA and lincRNA, for the remainder of this review we refer to the latter as intervening lncRNAs, which we propose to rename iRNAs. Similarly, to unify the nomenclature of lncRNAs, we suggest pRNAs for the promoter-associated uaRNAs, and gsRNAs or gaRNAs for gene body–associated sense and antisense transcripts, respectively (**Figure 1**).

Although many lncRNAs are associated with previously annotated genomic regions, there remains a class of intervening lncRNAs that originate from independent transcriptional units that do not overlap with mRNA genes or enhancers. These loci have their own promoter and are marked by the same chromatin modifications found at protein-coding genes (14, 48). Well-known structural intervening lncRNAs such as MALAT1 and NEAT1 belong to this class. Most intervening lncRNAs are indistinguishable from mRNAs in that they are capped, spliced, and polyadenylated (48), and among the different classes of lncRNAs they are likely the most diverse in sequence features, evolutionary origin, and function. To distinguish intervening lncRNAs from uaRNAs, we favor a more restrictive definition for intervening lncRNAs, which requires their TSS to be located at least 1 kb away from the next closest TSS or enhancer.

It is important to gain an understanding of the properties of lncRNAs in regard to their 5'- and 3'-end processing, which might further aid in their classification. Although all studies agree that the 5' ends of lncRNAs, like those of mRNAs, are capped by methylguanosine, their splicing status and their 3'-end processing have not been fully defined. Initial annotation of lncRNAs favored transcripts that displayed some evidence of splicing and polyadenylation and that did not overlap with protein-coding genes (14, 35, 48). However, recent findings have indicated that a large number of lncRNAs correspond to either eRNAs or uaRNAs and that they are predominantly monoexonic and nonpolyadenylated, particularly in the case of eRNAs (4, 36, 76). Unlike mRNAs, uaRNAs and eRNAs are depleted of splice motifs (2, 4). Therefore, it is likely that splice site recognition occurs with a low frequency at most lncRNA loci and that the predominant form of lncRNAs may be monoexonic and nonpolyadenylated. This contention is consistent with recent findings indicating that uaRNAs and eRNAs display shorter half-lives than do mRNAs and are subject to degradation by the nuclear exosome (4, 122).

Biological Functions of lncRNAs

Although few would doubt that small ncRNAs such as miRNAs (microRNAs) and piRNAs (PIWI-interacting RNAs) possess distinct and fundamental biological functions, understanding of lncRNA biology is at an earlier stage and still elicits controversy (80). At various points in time, it was proposed that pervasive transcription is due to low specificity of RNAPII (157) and that lncRNAs encode cryptic peptides responsible for the observed biological functions (66). Because lncRNAs exhibit low sequence conservation, even between related species (14, 48), and experience frequent gene birth and death (83), their biological significance has been questioned.

Despite early skepticism, the experimental evidence in favor of a direct biological function of lncRNAs has been steadily growing and has convincingly disposed of the criticisms above. As a set, lncRNAs exhibit the imprint of purifying selection (i.e., conservation of sequence above the genomic background), not only in the gene body but also at promoters (132). Phylogenetic comparisons confirmed that thousands of lncRNA families originated more than 90 million years ago and hundreds can be traced back 300 million years to the common ancestor of mammals and birds (120). At a biochemical level, ribosome profiling data are consistent with the notion that the vast majority of lncRNAs are not translated (51), and their localization is predominantly nuclear (35), although the latter conclusion has been challenged (165).

There is also accumulating genetic evidence in favor of the biological importance of lncRNAs. Deleting the entire locus for 5 out of 18 mouse intervening lncRNAs resulted in lethality or developmental defects (144). However, such large genetic deletions might result in the removal of regulatory sequences (e.g., enhancers) of other genes in addition to the removal of the lncRNA gene. In zebrafish, depletion of the intervening lncRNA cyrano caused developmental defects that were rescued by mammalian orthologs, despite the fact that sequence conservation was limited to short patches of nucleotides (166). Importantly, mutations introduced to disrupt the frame of presumed ORFs in the RNA sequence did not affect the rescuing ability of these lncRNAs, providing additional proof that RNA was the functional agent (166).

Taken together, phylogenetic, biochemical, and genetic evidence all support the conclusion that many lncRNAs perform biological functions independent of protein translation. However, most of lncRNAs in yeast are suppressed by a molecular pathway that enforces promoter directionality via Ndr1 and the exosome (146). Similarly, many mammalian lncRNAs that arise from divergent transcription (29, 147, 150) fail to accumulate due to a depletion of splice recognition sites in the antisense direction and removal by the nuclear exosome (2, 122), suggesting that their presence might not confer a fitness advantage (177). In an age when genome editing has become an easily implemented strategy (26), all claims to noncoding functions of RNA molecules should be subjected to rigorous genetic testing. Specifically, to prove that a select biological process is regulated by lncRNAs it should be shown that (*a*) loss-of-function of the lncRNA perturbs the process; (*b*) the defect is rescued by reinstating expression of the lncRNA (in *cis* or *trans*; see below); and (*c*) lncRNAs carrying mutations that would disrupt potential ORFs still rescue the defect (166).

lncRNAs AND TRANSCRIPTION

As lncRNAs can interface with the genome at the sequence level and also fold into tertiary structures capable of specific interactions with proteins, they are particularly well suited to function in regulating gene expression. Indeed, the past decade of investigation has uncovered many specific examples and general classes of lncRNAs that repress or activate transcription.

Transcriptional Interference in Yeast

As in higher eukaryotes, the genome of the simplest eukaryotic model organism, *Saccharomyces cerevisiae*, is extensively transcribed and this includes protein-coding and protein-noncoding regions (33). lncRNAs in yeast originate in large part from bidirectional transcription initiation and most of them are rapidly degraded [cryptic unstable transcripts (CUTs)], although some are stable [stable unannotated transcripts (SUTs)] (178). Early work from Winston and colleagues demonstrated that the regulatory region of *SER3*, an anabolic enzyme involved in serine biosynthesis, is transcribed into a lncRNA (SRG1) in a serine-dependent fashion (106). The presence of SRG1 correlates with repression of the *SER3* ORF; however, the repressive activity was observed even when >90% of the lncRNA sequence was replaced (105), arguing in favor of a transcriptional interference model. According to this model, the act of transcription through the regulatory region is sufficient to mediate repression, and the lncRNA itself does not play a direct role (106). Similar cases have since been reported for other highly regulated loci, including *GAL1–10* (61); master regulators of gametogenesis, *IME4* and *IME1*, which are repressed in *cis* by transcription from the lncRNA loci *RME2* and *IRT1* (60, 169); and *FLO11*, whose expression is regulated by a pair of lncRNAs, ICR1 and PWR1, through a network of mutual transcription interference (13). Whether interference affects gene regulation at the genome-wide level in *S. cerevisiae* remains to be elucidated. Stabilization of Xrn1-sensitive unstable transcripts (XUTs) resulted in global changes in gene expression (168), but stabilization of a different class of lncRNAs [Ndr1-dependent unterminated transcripts (NUTs)], sensitive to the action of the termination factor Nrd1, had relatively minor effects (146).

In addition to causing *cis* interference, select yeast lncRNAs recruit repressive chromatin factors to silence target genes. The *PHO84* locus comprises an antisense lncRNA that participates in the repression of the coding gene by recruiting histone deacetylases specifically to its promoter (17). Interestingly, this lncRNA can also function in *trans* (16), suggesting that transcriptional interference alone cannot explain the repressive action. Similar *trans*-acting lncRNAs have been observed at Ty1 retrotransposons in budding yeast (9) and in fission yeast, where a lncRNA represses the *pho1* gene through RNAi-dependent formation of transient heterochromatin via H3K9me2 deposition (148).

It is tempting to speculate that these recruitment pathways involve specific recognition of lncRNA structure by chromatin modifiers, but an alternative explanation is that some generic feature of overlapping transcription is sufficient to mediate repression (75). For example, a molecular pathway for transcriptome surveillance might have evolved that detects the presence of converging polymerases or of unspliced, unpolyadenylated ncRNAs and guides their removal as well as imposes silencing on the region that transcribes the lncRNAs. Indeed, recent experiments in the fission yeast *Schizosaccharomyces pombe* have delineated a silencing pathway that is coupled to the removal of cryptic introns (91). It is important to point out that the existence of such surveillance pathways does not argue against the fact that lncRNAs might also exert a direct regulatory function. On the contrary, nonfunctional noncoding transcripts might have constituted the raw material that evolutionary forces shaped into regulatory mechanisms.

Annotation and Function of lncRNAs in *Drosophila*

One criticism raised against the notion that lncRNAs fulfill important biological functions argues that forward genetic screens performed in various model organisms, most notably *Drosophila melanogaster*, have failed to identify lncRNAs (109). An answer to this criticism stems from the observation that point mutations induced by ethyl methane sulfonate (155), the most commonly used mutagen in *Drosophila*, are more likely to disrupt protein function by causing missense and nonsense mutations in the protein-coding sequences, whereas similar mutations in ncRNAs may not exact similar functional consequences. Nonetheless, as early as 1919, C.B. Bridges observed a homeotic mutation in *Drosophila* that he named *bithoraxoid* (*bxd*) (95). This mutation maps to a *cis*-regulatory region of the bithorax complex that, we now know, is extensively transcribed into lncRNAs (97, 143).

Beyond the bithorax complex, lncRNAs are a general feature of the *Drosophila* genome. Early annotations identified 14 intervening lncRNAs, some of which were conserved with other *Drosophila* species and led the authors to hypothesize that up to 100 lncRNAs may be present (163). In fact, upward of 1,000 putative lncRNAs loci have since been annotated (140, 184), and many of them show evidence of purifying selection (i.e., sequence conservation) when polymorphisms across different *Drosophila* strains are analyzed (53). What their functions might be remains a topic of intense investigation. So far, little evidence can be summoned in favor of direct, RNA-mediated activities of *Drosophila* lncRNAs. Those transcribed from the *bxd* locus were first shown to activate transcription of the adjacent *Ultrabithorax* gene (142). Later, it was suggested that lncRNAs repress *Ultrabithorax* via transcriptional interference (128). Most recently, genetic inactivation of one such lncRNA resulted in no discernible phenotype (127). Another famous example is *pgc*, which was first proposed to function as lncRNA (108) but later shown to encode a 71-aa protein responsible for gene function (56). Other lncRNAs from the *bxd* and *iab* regions of the bithorax complex appear to function as miRNA precursors or via sequence-independent transcriptional interference (6, 47).

One exception to this litany of results against a direct function of lncRNAs is the process of dosage compensation. In flies, dosage compensation occurs via transcriptional upregulation of the single male X chromosome (27). It requires functional contributions from five proteins (MSL1, MSL2, MSL3, MOF, and MLE) and two lncRNAs (roX1 and roX2) (3) that are transcribed from the X chromosome and coat it in its entirety (111). The two lncRNAs can substitute for each other, but deletion of both is lethal in males (110). Despite this redundancy, very little sequence similarities exist between roX1 and roX2 (125), suggesting that functional similarities might be dictated more by structural homology than by sequence conservation. Chemical and nuclease footprinting, along with biochemical analyses, identified conserved structural motifs required for roX function, supporting a model whereby the lncRNAs participate in the assembly of the dosage compensation complex (65, 100).

Repression by lncRNAs in Plants

Plants need to respond to a changing environment for optimal resource allocation and proper timing of developmental phase transitions, and, because they are sessile, they largely utilize epigenetic mechanisms (1). Consistent with a high degree of epigenetic regulation, the genome of *Arabidopsis thaliana* contains tens of thousands of putative lncRNAs, including more than 30,000 antisense lncRNAs (170) and more than 6,000 intervening lncRNAs (98).

One model locus that has been valuable in dissecting the functions of lncRNA in plants is the flowering control locus *FLC*, which maintains the epigenetic memory of cold exposure that results in vernalization, i.e., the ability to flower in spring (154). The locus gives rise to a number of noncoding transcripts, including various splice isoforms of the downstream antisense lncRNA COOLAIR (160) as well as a nonpolyadenylated intronic lncRNA called COLDAIR (58). Although the mechanism of action of COOLAIR remains unclear, COLDAIR binds CURLY LEAF (CLF) (58), the plant homolog of EZH2 and a subunit of Polycomb repressive complex 2 (PRC2), which mediates epigenetic silencing via trimethylation of histone H3 lysine 27 (H3K27me3) (104). COLDAIR recruits PRC2 to chromatin, thus initiating the epigenetic cascade that leads to vernalization via silencing of *FLC*. This model is similar to those proposed for the mechanism of action of several mammalian lncRNAs, including HOTAIR (137) (see below). CLF also binds to the antisense of COLDAIR in vitro but only to the sense RNA when incubated with nuclear extracts, suggesting that additional factors contribute to specificity in vivo (58).

Plant lncRNAs also contribute to epigenetic silencing via DNA methylation. A subclass of lncRNAs is transcribed by a dedicated RNA polymerase (PolV) and is required for small ncRNAs to recognize their genomic targets in the process of RNA-dependent DNA methylation (174). A current model posits that the lncRNAs transcribed by PolV recruit chromatin remodelers [protein complexes that reposition nucleosomes along the chromatin fiber (15)] via the RNA-binding adapter IDN2. In turn, the repositioned nucleosomes facilitate DNA methylation by

DRM2, thus enforcing epigenetic repression (189). Whether similar mechanisms are at work in metazoans is doubtful, as this particular pathway of DNA methylation, as well as the specialized RNA polymerases that enable it, are specific to plants (85).

lncRNA-Mediated Epigenetic Silencing in Mammals

The initial evidence for direct roles of lncRNAs in epigenetic silencing came from experiments in mammalian systems. The founding member of the lncRNA family is Xist, a ~17-kb transcript that originates from the silent X chromosome in female cells and coats it during early stages of development to establish epigenetic X inactivation (24). The highly conserved repA region of Xist is required for silencing and was reported to interact directly with the PRC2 complex (187). Consistent with this observation, recruitment of PRC2 to the X chromosome tracks that of Xist temporally and spatially, both during initial establishment and during reestablishment following transient Xist depletion (152). However, Xist and PRC2 do not colocalize on chromatin as determined by super-resolution microscopy (22), suggesting that a simple model whereby PRC2 is directly recruited by Xist might not be entirely accurate.

X inactivation is an extreme case of genomic imprinting, the process by which certain genes are expressed from only the maternal or paternal chromosomes (135). To date, parental imprinting has been observed only in plants (78) and mammals (42). There are approximately 150 parentally imprinted loci in the mouse genome (90), most of which harbor lncRNAs that are essential for proper execution of the imprinting program (90, 153). In some notable cases, it has been argued that the imprinted lncRNAs function by recruiting chromatin-modifying factors, such as EHMT2/G9A at the *Air* locus (118) and EHMT2, PRC2, and DNMT1 at *Kcnq1* (113, 124). However, insertion of termination cassettes to truncate the *Airn* lncRNA and shifting of its promoter showed that the RNA itself is dispensable for transcriptional silencing, arguing against the possibility that recruitment of chromatin modifiers is mediated by specific protein-RNA interactions (88). Nonetheless, the presence of lncRNAs at nearly all imprinted loci cannot be coincidental and the fact that one of the most sophisticated types of epigenetic regulation—the allelic discrimination of identical DNA sequences—correlates with the presence of lncRNAs provides strong circumstantial evidence in support of their involvement in epigenetic processes.

Regulation of Polycomb Complexes via lncRNAs

Among the many silencing complexes that have been reported to interact with ncRNAs, the Polycomb group complex PRC2 has been the most intensively studied. Although the biochemical output of the PRC2 complex, methylation of H3K27, is known, the input signals that determine its localization and activity at specific genes in different cells are not fully understood (104, 151). Several histone methyltransferases, including E(Z), the catalytic subunit of PRC2 in *Drosophila*, possess RNA binding activity in vitro (81). Canaani and colleagues ascribed the activity to a fragment of E(Z) that encompasses the SET domain; however, mapping experiments on the mammalian ortholog EZH2 identified an unstructured region comprising residues 342–368 that is required for RNA binding in vitro and whose affinity for RNA is modulated by phosphorylation (70). PRC2 has been proposed to mediate the function of many lncRNAs in mammals: HOTAIR, Kcnq1ot1, Braveheart, and Meg3/Gtl2, to name a few (69, 77, 124, 137, 186). However, the fact that most tested RNAs, including some from unrelated organisms, bind to PRC2 with high affinity and low specificity in vitro and in vivo (34) casts a veil of uncertainty over some of these results (12). Far from being an in vitro artifact, binding to RNA independently of sequence and structure might be a regulatory feature of PRC2, and competing models incorporate this observation into our understanding of PRC2 function. A junk-mail model posits that low-level transcription from escapee Polycomb target genes contributes to recruitment of PRC2 and leads to their resilencing (34), whereas a sensing model proposes that the presence of RNA causes PRC2 to curtail its repressive action on chromatin in a feedback loop that prevents inappropriate silencing of active genes (71). PRC2 also interacts with and is recruited by shorter ncRNAs through its SUZ12 subunit (72). These small ncRNAs originated upstream or near the TSS, and it is unclear to what extent they might correspond to promoter-associated RNAs (73) or to the more recently described TSS miRNAs (185). Thus, the relationship of PRC2 with long and small ncRNAs is complex (**Figure 2**), and more mechanistic studies are required before the full picture comes into focus.

PRC2 is not the only Polycomb complex that has functional links with noncoding RNAs. CBX7 is a mammalian homolog of the Polycomb protein (PC) in *Drosophila*, which is a core component of PRC1 and binds to H3K27me3 via its chromodomain (43). CBX7 is expressed in mouse embryonic stem cells, where it regulates pluripotency, whereas other PC homologs such as CBX2 and CBX4 are expressed during lineage commitment and are required for differentiation (115). The chromodomains of PC, CBX4, and CBX7 bind RNA in vitro, and RNA interactions are required for chromatin localization, at least in the case of CBX7 (8). Importantly, the chromodomain of CBX2 does not bind to RNA, although it retains specificity for H3K27me3, suggesting that the two functions can be uncoupled (8).

In vivo, CBX7 interacts with ANRIL, an antisense lncRNA from the *INK4b-ARF-INK4a* tumor suppressor locus (183). On the basis of the observations that knockdown of ANRIL causes derepression and that RNase A treatment releases PRC1 from this locus, it is possible that ANRIL participates in PRC1 recruitment and Polycomb-mediated repression (183). CBX4 (also known as PC2) binds to at least two lncRNAs in vivo, TUG1 and MALAT1 (180). Selectivity between the two is determined by the methylation status of a lysine residue within the SUMO binding motif of CBX4 and results in differential localization of the protein and its associated chromatin targets within the three-dimensional space of the nucleus (180).

Similar to the case of PRC2, which interacts with RNA not only via core complex subunits [EZH2 (70) and SUZ12 (72)] but also through accessory regulatory components, such as JARID2 (69), PRC1 also appears to make multiple contacts with RNA. We recently discovered an RNA-binding region within SCML2 (11), a homolog of the *Sex comb on midleg* (SCM) protein, which is a substoichiometric subunit of the PRC1 complex in flies (149) and humans (45, 94). SCML2 mutants lacking the RNA-binding region cannot bind to chromatin and result in lower levels of PRC1 recruitment, potentially due to dominant negative effects (11).

Many Repressive Chromatin Components Interact with lncRNAs

Although PRCs have received a large amount of attention, many other chromatin-associated factors and complexes tested have displayed an affinity for RNA in vitro and in vivo (49, 74, 81). One of the methyltransferases that deposit repressive methylation marks on H3K9 (EHMT2/G9A) as well as the reader that recognizes them (HP1) requires an RNA component for proper recruitment to certain target sites (101, 118, 124). Specifically, RNA is required for the accumulation of HP1α at DAPI-rich pericentromeric heterochromatin foci (101), presumably due to the RNA-binding activity of the hinge region located between the chromodomain and the chromo-shadow domain (117). This is consistent with the observation that the HP1 chromodomain alone does not bind RNA in vitro (8). The histone demethylase LSD1 binds to the lncRNA HOTAIR (162) and, judging by RNA immunoprecipitation experiments performed with its complex partner, CoREST, to many more lncRNAs (74), which remain to be thoroughly analyzed. Finally, lncRNAs from the *CCND1* 5′ regulatory sequence stimulate the inhibitory function of the RNA-binding protein

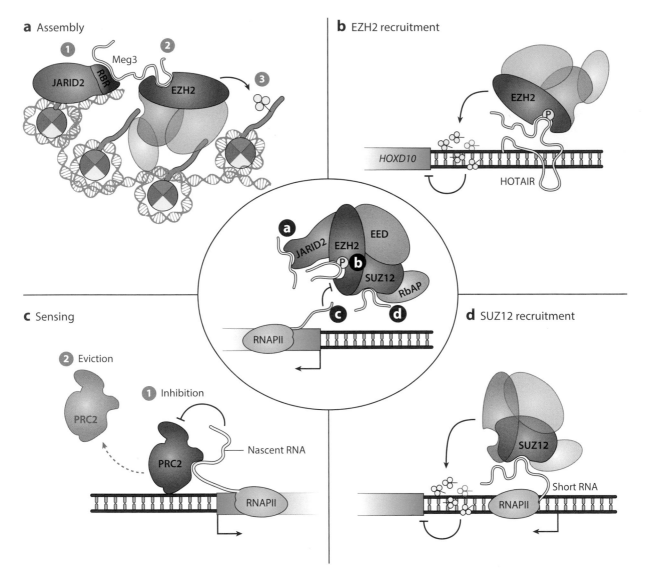

Figure 2

Mechanisms of ncRNA (noncoding RNA)-mediated regulation of PRC2. (*a*) The imprinted lncRNA (long noncoding RNA) Meg3 makes contact with EZH2 and JARID2 and stimulates their interaction on chromatin, facilitating PRC2 assembly, H3K27me3 deposition, and transcriptional repression of a subset of PRC2 targets in human and mouse stem cells (69). (*b*) The lncRNA HOTAIR has been reported to recruit PRC2 in *trans* to the *HOXD* locus in human fibroblasts (137). Later studies revealed that HOTAIR interacts directly with EZH2 and that phosphorylation of EZH2 at T345 stimulates ncRNA binding (70). (*c*) In embryonic stem cells, PRC2 is found at a majority of promoters, including those of active genes. However, interactions with nascent RNAs decrease the amount of deposited H3K27me3 either by inhibiting PRC2 function or by causing its release from chromatin (71). (*d*) SUZ12 was reported to interact with short ncRNAs originating from the region near the transcription start site (TSS) of Polycomb target genes, leading to PRC2 recruitment, H3K27me3 deposition, and silencing (72). SUZ12 also interacts with the lncRNA Braveheart (77). The center image is a schematic of the multiple mechanisms (*a*–*d*) by which ncRNAs have been shown to interact with PRC2.

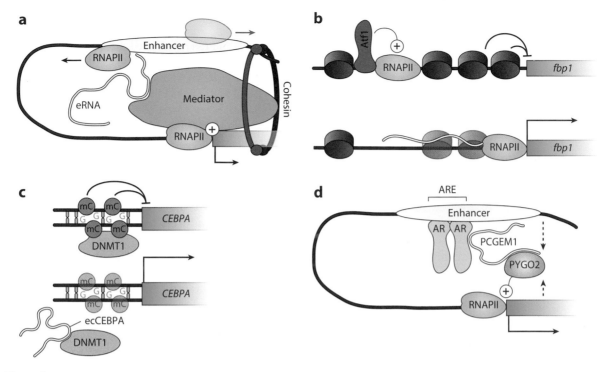

Figure 3

Transcriptional activation by lncRNAs (long noncoding RNAs). (*a*) Several lncRNAs transcribed from a distal enhancer directly interact with subunits of Mediator and facilitate enhancer-promoter communication by promoting loop formation (86), which requires the function of cohesin (96). (*b*) In *Schizosaccharomyces pombe*, activation of *fbp1* requires transcription of lncRNAs originating from an upstream transcription start site (TSS). This is followed by promoter remodeling via nucleosome depletion, which requires the transcribing polymerase (59). (*c*) Transcription of *CEBPA* in human cells is activated by an upstream sense lncRNA (ecCEBPA), which binds DNMT1 and sequesters it from chromatin, causing hypomethylation of the promoter. (*d*) Activation by an androgen receptor (AR) requires a complex cascade of post-translational modifications and lncRNA-protein interactions that culminates with the recruitment of PYGO2 via the lncRNA PCGEM1. In turn, PYGO2 stimulates transcription by strengthening enhancer-promoter looping (*dashed arrows*) (179).

TLS toward the coactivators CBP and p300, thus mediating transcriptional silencing via allosteric regulation of a repressor (173).

lncRNAs and Transcriptional Activation

Although the long list of examples above seems to imply that a majority of lncRNAs regulate transcription via repression, this is most likely the result of a historical tendency toward studying mechanisms involved in epigenetic repression in X inactivation and imprinting. In fact, recent results have indicated an activating function for a large number of lncRNAs (86) (**Figure 3**). Two independent lines of investigation converged to highlight a critical role for lncRNAs in activation of transcription. Functional studies following depletion of a number of lncRNAs revealed a predominant role in activating their neighboring protein-coding genes (123, 172). Many of these lncRNAs were derived from genomic regions that were previously denoted as distal enhancer elements (123, 172). Concomitantly, analysis of stimulus-dependent activation of enhancers in neuronal cells and macrophages revealed the presence of enhancer-derived lncRNAs (eRNAs; see above) (36, 76). Subsequent experiments revealed that active enhancers are transcribed and that eRNAs mediate transcriptional activation of the target promoters (54, 87, 96). In fact, genome-wide analyses of a large set of primary human cells and tissues revealed that the transcription of enhancers is the best diagnostic measure of their activity in any cell type (4). Several lncRNAs have been recovered in association with chromatin-modifying complexes that activate transcription. For example, lncRNAs HOTTIP and Mistral both bind to the MLL complex and activate transcription of neighboring *HOX* genes in *cis* via interactions with WDR5 and MLL, respectively (10, 172, 181). Activating lncRNAs were also shown to associate with the transcriptional coactivator complexes Mediator and Cohesin to promote enhancer-promoter communication through DNA looping (86, 96) (**Figure 3a**).

Noncoding transcripts that traverse promoter chromatin can also participate in the activation of the adjacent coding gene, as seen at the *fbp1*$^{+}$ locus of fission yeast (59) and the *PHO5* gene of *S. cerevisiae* (164) (**Figure 3b**). Similar activating read-through transcripts have been observed at the human *CEBPA* locus, although in this case the molecular mechanism involves sequestration of DNMT1, which results in decreased levels of inhibitory DNA methylation and subsequent activation of *CEBPA* transcription (37) (**Figure 3c**). Finally, lncRNAs can contact transcription factors directly and stimulate their function by recruiting additional coactivators or facilitating enhancer-promoter looping, as seen in the case of the androgen receptor (179) (**Figure 3d**).

MECHANISTIC CONSIDERATIONS

The lack of details on the biogenesis and mechanism of action of most lncRNAs is the major hurdle to progress in the field. We are still at the early stages of understanding the RNA code involved in the activating and repressing functions of lncRNAs. A key to deciphering how RNA nucleotide sequences are utilized to convey functional information is to uncover the precise structural and biochemical requirements for the action of lncRNAs in transcriptional regulation.

Structure and Function

Important steps toward a mechanistic dissection of the function of lncRNAs have been taken by studies of the dosage compensation complex in *Drosophila*. Two reports detailed the network of interactions between roX RNAs and two protein components, MLE and MSL2, of the complex (65, 100). The authors analyzed in detail the secondary structure of these RNAs using traditional footprinting experiments and more modern technologies such as selective 2'-hydroxyl acylation analyzed by primer extension (SHAPE), which determines the RNA structure based on chemical reactivity (112), and individual nucleotide resolution UV cross-linking and immunoprecipitation (iCLIP), which utilizes UV cross-linking to detect protein-RNA contacts (79). Although the two studies reached somewhat different conclusions, they agree that the RNA helicase activity of MLE plays a critical role in the assembly of the dosage compensation ribonucleoprotein complex.

Much work remains to be done to reach a structural understanding of lncRNA function. Several techniques are available to probe RNA structure in vitro and in vivo, each with its own disadvantages. A thorough comparison of SHAPE, dimethyl sulfate (DMS) (28), in-line probing (134), and RNAse footprinting for the analysis of the secondary structure of the lncRNA SRA, revealed that the most accurate information can currently be obtained by combining these techniques (121). Among them, DMS probing might be the most promising, as this chemical easily enters cells and can probe RNA structure in vivo (138), although its reactivity is limited to As and Cs.

Biochemical approaches that investigate the protein side of the interaction are also useful. We and others have argued that the best way to address the function of protein-RNA interactions in the chromatin space is that of defining the protein regions or domains responsible for RNA binding and deleting them (11). Such analysis will be more powerful following the identification of the specific residues required for the RNA contacts and interrogation of point mutations that disrupt such interactions (181, 183). Often these RNA-binding surfaces do not resemble the well-characterized RNA recognition motif (RRM) (25); rather, they are unstructured stretches of charged residues rich in lysines and arginines (11, 69, 70). The lack of predicted secondary structure should not be interpreted as lack of function, as induced fit is a common feature of protein-RNA interactions (175). For example, the fragile X syndrome protein FMR1/FMRP comprises an arginine-rich stretch that is disordered in solution but becomes structured upon binding to RNA (129). An arginine-rich motif is also responsible for binding of the HIV protein REV to its response element (RRE) within the RNA genome of the virus and becomes ordered only upon RNA binding (19). Similar to the case of some RNA-binding regions in chromatin-associated proteins, such as SCML2 (11), the arginine-rich motif of REV binds to a variety of RNA sequences and structures in vitro but is strictly specific for RRE in vivo (31). Structural studies have suggested that this specificity is acquired via its multimerization and precise geometrical orientation in the functional hexamer (32).

Mode of Action: *Trans* versus *Cis*

Another important, and in most cases unanswered, question is whether lncRNAs act in *cis* or *trans* fashion. In the former case, RNA molecules would affect only the structure and function of neighboring chromatin and/or chromatin-associated proteins. This may result from the tethering of the lncRNA to either the transcribing polymerase (cotranscriptional action) or a chromatin component after transcription is complete. Many imprinted lncRNAs, dosage compensation lnc-RNAs, and most lncRNAs with activating functions, such as eRNAs, seem to act in *cis* (96, 123, 172). It has been argued that *cis* function is a specialized task that could be accommodated better by ncRNAs, as they can fold into functional structures during transcription, without requiring translation in the cytoplasm (89). The *cis* mechanism of action requires lncRNAs to function in an allele-specific manner, but to date this critical aspect of lncRNA biology has only been addressed in dosage compensation.

HOTAIR, an intervening lncRNA involved in silencing of *HOX* genes, is thought to act in *trans* (23, 137). *Trans*-mediated function was also reported for linc-p21 (63), linc-Firre (52), and, at least in part, Meg3 (69). The observation that some lncRNAs might be *trans*-acting begs the question of how they reach their intended targets in the genome. It is easy to envision RNA-RNA interactions or protein-RNA interactions, but less so how RNAs could bind to specific sequences within double-stranded genomic DNA. Although formation of a triplex with DNA has been reported (107, 145), strand invasion is also possible (93) and can be guided by ncRNAs, as observed in the CRISPR-Cas9 system (156).

Finally, we note that *cis* and *trans* are not the only two possibilities. lncRNAs might be locally confined by being tethered to their locus of origin and still function at sites that are distant in a two-dimensional representation of DNA sequences but proximal in three-dimensional space, as discussed in the next section.

lncRNAs and Nuclear Organization

A set of recent observations suggests that the function of lncRNAs might be linked to the three-dimensional organization of the mammalian nucleus. The idea that chromatin is not freely diffusing in the nucleoplasmic space and that some structural organization might be required for genome function is not new (7) but has been exceedingly difficult to prove (133). Whether or not a static structural scaffold exists, the three-dimensional organization of chromatin within the nucleus plays a key role in genome function. At the kilobase scale it supports the formation of loops that bring distant regulatory regions, enhancers, and their specific targets into contact (68, 86, 96). At the larger megabase scale, it affects the degree of higher-order compaction of chromosomal domains, varying their accessibility to nuclear factors (114, 131). A proposal has also been put forward that subnuclear positioning of chromatin affects the activation status of embedded genes (188). To date, little is known about how spatial organization of chromatin is achieved, and it is likely that lncRNAs might play a key role in genomic architecture.

Several subnuclear structures contain and/or can be nucleated by well-known lncRNAs, such as NEAT1 (102) and MALAT1 (64). In further support of a connection between lncRNAs and nuclear organization, Xist interacts with HNRNPU (also known as SAFA), a component of the nuclear matrix (57). Moreover, spreading of Xist from its site of transcription during the initial phases of X inactivation is dictated by the three-dimensional organization of the X chromosome rather than by sequence motifs or other molecular features of target sites (41). Similarly, the lncRNA Firre, implicated in mouse adipogenesis (158), participates in nuclear organization by interacting with the nuclear matrix protein HNRNPU and by gathering otherwise distant genes whose products are involved in energy metabolism and adipogenesis, potentially exposing them to a shared local environment that might facilitate coregulation (52).

The Chromatin Organizer Model for lncRNA Function

We propose a "chromatin organizer" model for the function of at least a subset of lncRNAs (**Figure 4**). According to this model, lncRNAs utilize 3D conformation as well as sequence information to spatially organize chromatin and chromatin-associated factors within the nucleus. This function resembles that of cable organizers made of Velcro, which bundle and separate electronic cables by establishing homotypic and heterotypic interactions with themselves and other neighboring organizers.

If true, our model would explain the role of lncRNAs in a variety of contexts. Activating eRNAs might function by looping chromatin locally to bring enhancers in close proximity with promoters and the transcription machinery (**Figure 4a**). This could be accomplished by interactions with protein complexes such as Mediator (86), but also by direct, possibly sequence-specific, interactions with uaRNAs that are produced by divergent transcription at genic TSSs. The prediction that eRNAs might interact with cognate uaRNAs is consistent with the observation that enhancers and the promoters under their control often share sequence motifs (4) and, therefore, patches of complementary sequence.

The chromatin organizer model is also consistent with the observations that Xist and linc-Firre organize chromatin at a large scale to enforce X inactivation or the regulation of functionally related genes, respectively (41, 52). In these cases, the lncRNAs bring together more distant genomic regions with the purpose of subjecting them to a common regulatory environment (**Figure 4b**). A similar scenario might also explain the role of the lncRNA Meg3 in PRC2 regulation. Meg3 binds at least two subunits of PRC2 and helps its assembly on chromatin, but it does so only at a small number of selected sites disseminated in *trans* throughout the genome (69). These Meg3-dependent PRC2 targets, although distant in the primary sequence, might be brought together in the three-dimensional space of the nucleus, possibly to coordinate their regulation. Finally, human CTCF, the insulator protein that functions as master weaver of the genome (130), interacts with RNA directly (141, 159) and indirectly (182) via the DEAD-box helicase p68 (DDX5) (**Figure 4c**). Along the same lines, insulator proteins in *Drosophila* interact

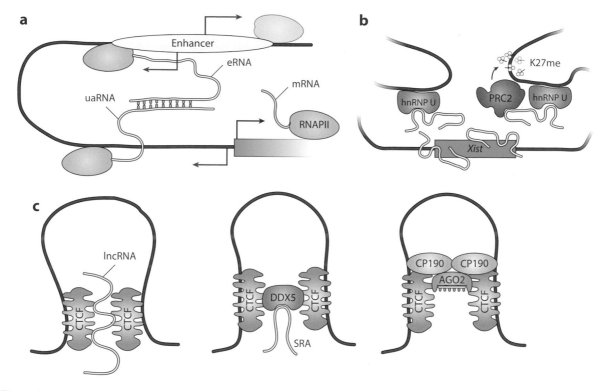

Figure 4

Potential roles of lncRNAs (long noncoding RNAs) in nuclear organization. (*a*) In this speculative model, eRNAs (enhancer RNAs) mediate enhancer-promoter communication directly via base-pair interactions with uaRNAs (upstream antisense RNAs) and possibly facilitate chromatin looping. (*b*) The lncRNA Xist utilizes the three-dimensional organization of chromatin to spread its silencing activity over the whole X chromosome. Interactions with chromatin might be mediated via the matrix protein hnRNP U (41). At the target sites, Xist facilitates H3K27me3 deposition via PRC2 recruitment (187). (*c*) Roles of ncRNAs (noncoding RNAs) in CTCF function. (*Left*) Human CTCF binds RNA directly and RNA interactions mediate multimerization, which might guide chromatin looping and organization (141). (*Middle*) Human CTCF also binds RNA indirectly through its interactions with the RNA helicase DDX5, here bound to the lncRNA SRA (182). (*Right*) *Drosophila* CTCF and CP190 insulator proteins interact physically and genetically with AGO2, a small ncRNA-binding protein and key component of the RNAi machinery (92, 116).

physically and genetically with the RNAi machinery (92, 116) (**Figure 4c**), suggesting a conserved role for ncRNAs in chromatin organization and nuclear structure.

FUTURE DIRECTIONS

The road ahead before we arrive at a full understanding of the molecular mechanisms responsible for lncRNA-mediated regulation of transcription is still very long; the models discussed above will need to be tested and new technologies developed. Current annotations of lncRNAs need to be integrated across different projects (e.g., the human lincRNA catalog of the Broad Institute and the GENCODE annotation) and phylogenetic information should be taken into account whenever possible and should focus on related species and even polymorphism within the same species, in addition to the analysis of conservation across long evolutionary distances (53, 120). Such new annotation should consider the wealth of recent findings, which indicates that a large number of lncRNAs might not be polyadenylated and function predominantly in enhancer-promoter communication. The discovery of this new class of RNAs, mainly eRNAs and uaRNAs, suggests the presence of as yet undiscovered machinery involved in their processing and mode of function. Technologies to detect protein-RNA interactions should also be improved, as they continue to suffer from suboptimal signal-to-noise ratio (12, 44). Possibly the most challenging task ahead is to understand lncRNA function in the context of nuclear organization. New technologies may also help toward this goal. Notable advances have been made in single-molecule RNA-FISH (fluorescence in situ hybridization) (5, 84), proximity labeling (139), and super-resolution microscopy (62). The synergy between these new experimental approaches and tried-and-true biochemical work will ultimately pave the way to a complete understanding of the functional roles of lncRNAs in the nucleus.

DISCLOSURE STATEMENT

The authors are not aware of any affiliations, memberships, funding, or financial holdings that might be perceived as affecting the objectivity of this review.

ACKNOWLEDGMENTS

The authors wish to thank Kristin Ingvarsdottir and Danny Reinberg for their comments on the manuscript. R.S. is supported by R01 GM 078455 and R01 GM 105754.

LITERATURE CITED

1. Allis CD, Jenuwein T, Reinberg D. 2007. *Epigenetics*. Cold Spring Harbor, NY: Cold Spring Harb. Lab. Press. 502 pp.
2. Almada AE, Wu X, Kriz AJ, Burge CB, Sharp PA. 2013. Promoter directionality is controlled by U1 snRNP and polyadenylation signals. *Nature* 499:360–63
3. Amrein H, Axel R. 1997. Genes expressed in neurons of adult male *Drosophila*. *Cell* 88:459–69
4. Andersson R, Gebhard C, Miguel-Escalada I, Hoof I, Bornholdt J, et al. 2014. An atlas of active enhancers across human cell types and tissues. *Nature* 507:455–61
5. Batish M, Raj A, Tyagi S. 2011. Single molecule imaging of RNA in situ. *Methods Mol. Biol.* 714:3–13
6. Bender W. 2008. MicroRNAs in the *Drosophila* bithorax complex. *Genes Dev.* 22:14–19
7. Berezney R, Coffey DS. 1974. Identification of a nuclear protein matrix. *Biochem. Biophys. Res. Commun.* 60:1410–17
8. Bernstein E, Duncan EM, Masui O, Gil J, Heard E, Allis CD. 2006. Mouse polycomb proteins bind differentially to methylated histone H3 and RNA and are enriched in facultative heterochromatin. *Mol. Cell. Biol.* 26:2560–69
9. Berretta J, Pinskaya M, Morillon A. 2008. A cryptic unstable transcript mediates transcriptional *trans*-silencing of the Ty1 retrotransposon in *S. cerevisiae*. *Genes Dev.* 22:615–26
10. Bertani S, Sauer S, Bolotin E, Sauer F. 2011. The noncoding RNA *Mistral* activates *Hoxa6* and *Hoxa7* expression and stem cell differentiation by recruiting MLL1 to chromatin. *Mol. Cell* 43:1040–46
11. Bonasio R, Lecona E, Narendra V, Voigt P, Parisi F, Kluger Y, Reinberg D. 2014. Interactions with RNA direct SCML2 to chromatin where it represses PRC1 target genes. *eLife* 3:e02637
12. Brockdorff N. 2013. Noncoding RNA and polycomb recruitment. *RNA* 19:429–42
13. Bumgarner SL, Dowell RD, Grisafi P, Gifford DK, Fink GR. 2009. Toggle involving *cis*-interfering noncoding RNAs controls variegated gene expression in yeast. *Proc. Natl. Acad. Sci. USA* 106:18321–26
14. Cabili MN, Trapnell C, Goff L, Koziol M, Tazon-Vega B, et al. 2011. Integrative annotation of human large intergenic noncoding RNAs reveals global properties and specific subclasses. *Genes Dev.* 25:1915–27
15. Cairns BR. 2009. The logic of chromatin architecture and remodelling at promoters. *Nature* 461:193–98
16. Camblong J, Beyrouthy N, Guffanti E, Schlaepfer G, Steinmetz LM, Stutz F. 2009. *Trans*-acting antisense RNAs mediate transcriptional gene cosuppression in *S. cerevisiae*. *Genes Dev.* 23:1534–45

17. Camblong J, Iglesias N, Fickentscher C, Dieppois G, Stutz F. 2007. Antisense RNA stabilization induces transcriptional gene silencing via histone deacetylation in *S. cerevisiae*. *Cell* 131:706–17

18. Castel SE, Martienssen RA. 2013. RNA interference in the nucleus: roles for small RNAs in transcription, epigenetics and beyond. *Nat. Rev. Genet.* 14:100–12

19. Casu F, Duggan BM, Hennig M. 2013. The arginine-rich RNA-binding motif of HIV-1 Rev is intrinsically disordered and folds upon RRE binding. *Biophys. J.* 105:1004–17

20. Cech TR. 2012. The RNA worlds in context. *Cold Spring Harb. Perspect. Biol.* 4:a006742

21. Cech TR, Steitz JA. 2014. The noncoding RNA revolution: trashing old rules to forge new ones. *Cell* 157:77–94

22. Cerase A, Smeets D, Tang YA, Gdula M, Kraus F, et al. 2014. Spatial separation of Xist RNA and polycomb proteins revealed by superresolution microscopy. *Proc. Natl. Acad. Sci. USA* 111:2235–40

23. Chu C, Qu K, Zhong FL, Artandi SE, Chang HY. 2011. Genomic maps of long noncoding RNA occupancy reveal principles of RNA-chromatin interactions. *Mol. Cell* 44:667–78

24. Clemson CM, McNeil JA, Willard HF, Lawrence JB. 1996. XIST RNA paints the inactive X chromosome at interphase: evidence for a novel RNA involved in nuclear/chromosome structure. *J. Cell Biol.* 132:259–75

25. Cléry A, Blatter M, Allain FH. 2008. RNA recognition motifs: boring? Not quite. *Curr. Opin. Struct. Biol.* 18:290–98

26. Cong L, Ran FA, Cox D, Lin S, Barretto R, et al. 2013. Multiplex genome engineering using CRISPR/Cas systems. *Science* 339:819–23

27. Conrad T, Akhtar A. 2011. Dosage compensation in *Drosophila melanogaster*: epigenetic fine-tuning of chromosome-wide transcription. *Nat. Rev. Genet.* 13:123–34

28. Cordero P, Kladwang W, VanLang CC, Das R. 2012. Quantitative dimethyl sulfate mapping for automated RNA secondary structure inference. *Biochemistry* 51:7037–39

29. Core LJ, Waterfall JJ, Lis JT. 2008. Nascent RNA sequencing reveals widespread pausing and divergent initiation at human promoters. *Science* 322:1845–48

30. Crick FH. 1968. The origin of the genetic code. *J. Mol. Biol.* 38:367–79

31. Daugherty MD, D'Orso I, Frankel AD. 2008. A solution to limited genomic capacity: using adaptable binding surfaces to assemble the functional HIV Rev oligomer on RNA. *Mol. Cell* 31:824–34

32. Daugherty MD, Liu B, Frankel AD. 2010. Structural basis for cooperative RNA binding and export complex assembly by HIV Rev. *Nat. Struct. Mol. Biol.* 17:1337–42

33. David L, Huber W, Granovskaia M, Toedling J, Palm CJ, et al. 2006. A high-resolution map of transcription in the yeast genome. *Proc. Natl. Acad. Sci. USA* 103:5320–25

34. Davidovich C, Zheng L, Goodrich KJ, Cech TR. 2013. Promiscuous RNA binding by Polycomb repressive complex 2. *Nat. Struct. Mol. Biol.* 20:1250–57

35. Derrien T, Johnson R, Bussotti G, Tanzer A, Djebali S, et al. 2012. The GENCODE v7 catalog of human long noncoding RNAs: analysis of their gene structure, evolution, and expression. *Genome Res.* 22:1775–89

36. De Santa F, Barozzi I, Mietton F, Ghisletti S, Polletti S, et al. 2010. A large fraction of extragenic RNA pol II transcription sites overlap enhancers. *PLOS Biol.* 8:e1000384

37. Di Ruscio A, Ebralidze AK, Benoukraf T, Amabile G, Goff LA, et al. 2013. DNMT1-interacting RNAs block gene-specific DNA methylation. *Nature* 503:371–76

38. Djebali S, Davis CA, Merkel A, Dobin A, Lassmann T, et al. 2012. Landscape of transcription in human cells. *Nature* 489:101–8

39. Eddy SR. 2002. Computational genomics of noncoding RNA genes. *Cell* 109:137–40

40. ENCODE Proj. Consort. 2007. Identification and analysis of functional elements in 1% of the human genome by the ENCODE pilot project. *Nature* 447:799–816

41. Engreitz JM, Pandya-Jones A, McDonel P, Shishkin A, Sirokman K, et al. 2013. The Xist lncRNA exploits three-dimensional genome architecture to spread across the X chromosome. *Science* 341:1237973

42. Ferguson-Smith AC. 2011. Genomic imprinting: the emergence of an epigenetic paradigm. *Nat. Rev. Genet.* 12:565–75

43. Fischle W, Wang Y, Jacobs S, Kim Y, Allis C, Khorasanizadeh S. 2003. Molecular basis for the discrimination of repressive methyl-lysine marks in histone H3 by Polycomb and HP1 chromodomains. *Genes Dev.* 17:1870–81

44. Friedersdorf MB, Keene JD. 2014. Advancing the functional utility of PAR-CLIP by quantifying background binding to mRNAs and lncRNAs. *Genome Biol.* 15:R2

45. Gao Z, Zhang J, Bonasio R, Strino F, Sawai A, et al. 2012. PCGF homologs, CBX proteins, and RYBP define functionally distinct PRC1 family complexes. *Mol. Cell* 45:344–56

46. Gilbert W. 1986. Origin of life: the RNA world. *Nature* 319:618

47. Gummalla M, Maeda RK, Castro Alvarez JJ, Gyurkovics H, Singari S, et al. 2012. abd-A regulation by the iab-8 noncoding RNA. *PLOS Genet.* 8:e1002720

48. Guttman M, Amit I, Garber M, French C, Lin MF, et al. 2009. Chromatin signature reveals over a thousand highly conserved large non-coding RNAs in mammals. *Nature* 458:223–27

49. Guttman M, Donaghey J, Carey BW, Garber M, Grenier JK, et al. 2011. lincRNAs act in the circuitry controlling pluripotency and differentiation. *Nature* 477:295–300

50. Guttman M, Rinn JL. 2012. Modular regulatory principles of large non-coding RNAs. *Nature* 482:339–46

51. Guttman M, Russell P, Ingolia NT, Weissman JS, Lander ES. 2013. Ribosome profiling provides evidence that large noncoding RNAs do not encode proteins. *Cell* 154:240–51

52. Hacisuleyman E, Goff LA, Trapnell C, Williams A, Henao-Mejia J, et al. 2014. Topological organization of multichromosomal regions by the long intergenic noncoding RNA Firre. *Nat. Struct. Mol. Biol.* 21:198–206

53. Haerty W, Ponting CP. 2013. Mutations within lncRNAs are effectively selected against in fruitfly but not in human. *Genome Biol.* 14:R49

54. Hah N, Murakami S, Nagari A, Danko CG, Kraus WL. 2013. Enhancer transcripts mark active estrogen receptor binding sites. *Genome Res.* 23:1210–23

55. Hannon GJ. 2002. RNA interference. *Nature* 418:244–51

56. Hanyu-Nakamura K, Sonobe-Nojima H, Tanigawa A, Lasko P, Nakamura A. 2008. *Drosophila* Pgc protein inhibits P-TEFb recruitment to chromatin in primordial germ cells. *Nature* 451:730–33

57. Hasegawa Y, Brockdorff N, Kawano S, Tsutui K, Tsutui K, Nakagawa S. 2010. The matrix protein hnRNP U is required for chromosomal localization of Xist RNA. *Dev. Cell* 19:469–76

58. Heo JB, Sung S. 2011. Vernalization-mediated epigenetic silencing by a long intronic noncoding RNA. *Science* 331:76–79

59. Hirota K, Miyoshi T, Kugou K, Hoffman CS, Shibata T, Ohta K. 2008. Stepwise chromatin remodelling by a cascade of transcription initiation of non-coding RNAs. *Nature* 456:130–34

60. Hongay CF, Grisafi PL, Galitski T, Fink GR. 2006. Antisense transcription controls cell fate in *Saccharomyces cerevisiae*. *Cell* 127:735–45

61. Houseley J, Rubbi L, Grunstein M, Tollervey D, Vogelauer M. 2008. A ncRNA modulates histone modification and mRNA induction in the yeast GAL gene cluster. *Mol. Cell* 32:685–95

62. Huang B, Bates M, Zhuang X. 2009. Super-resolution fluorescence microscopy. *Annu. Rev. Biochem.* 78:993–1016

63. Huarte M, Guttman M, Feldser D, Garber M, Koziol MJ, et al. 2010. A large intergenic noncoding RNA induced by p53 mediates global gene repression in the p53 response. *Cell* 142:409–19

64. Hutchinson JN, Ensminger AW, Clemson CM, Lynch CR, Lawrence JB, Chess A. 2007. A screen for nuclear transcripts identifies two linked noncoding RNAs associated with SC35 splicing domains. *BMC Genomics* 8:39

65. Ilik IA, Quinn JJ, Georgiev P, Tavares-Cadete F, Maticzka D, et al. 2013. Tandem stem-loops in roX RNAs act together to mediate X chromosome dosage compensation in *Drosophila*. *Mol. Cell* 51:156–73

66. Ingolia NT, Lareau LF, Weissman JS. 2011. Ribosome profiling of mouse embryonic stem cells reveals the complexity and dynamics of mammalian proteomes. *Cell* 147:789–802

67. Jacquier A. 2009. The complex eukaryotic transcriptome: unexpected pervasive transcription and novel small RNAs. *Nat. Rev. Genet.* 10:833–44

68. Kagey MH, Newman JJ, Bilodeau S, Zhan Y, Orlando DA, et al. 2010. Mediator and cohesin connect gene expression and chromatin architecture. *Nature* 467:430–35

69. Kaneko S, Bonasio R, Saldana-Meyer R, Yoshida T, Son J, et al. 2014. Interactions between JARID2 and noncoding RNAs regulate PRC2 recruitment to chromatin. *Mol. Cell* 53:290–300

70. Kaneko S, Li G, Son J, Xu CF, Margueron R, et al. 2010. Phosphorylation of the PRC2 component Ezh2 is cell cycle–regulated and up-regulates its binding to ncRNA. *Genes Dev.* 24:2615–20

71. Kaneko S, Son J, Shen SS, Reinberg D, Bonasio R. 2013. PRC2 binds active promoters and contacts nascent RNAs in embryonic stem cells. *Nat. Struct. Mol. Biol.* 20:1258–64

72. Kanhere A, Viiri K, Araujo CC, Rasaiyaah J, Bouwman RD, et al. 2010. Short RNAs are transcribed from repressed polycomb target genes and interact with polycomb repressive complex-2. *Mol. Cell* 38:675–88

73. Kapranov P, Cheng J, Dike S, Nix DA, Duttagupta R, et al. 2007. RNA maps reveal new RNA classes and a possible function for pervasive transcription. *Science* 316:1484–88

74. Khalil AM, Guttman M, Huarte M, Garber M, Raj A, et al. 2009. Many human large intergenic noncoding RNAs associate with chromatin-modifying complexes and affect gene expression. *Proc. Natl. Acad. Sci. USA* 106:11667–72

75. Kim T, Xu Z, Clauder-Munster S, Steinmetz LM, Buratowski S. 2012. Set3 HDAC mediates effects of overlapping noncoding transcription on gene induction kinetics. *Cell* 150:1158–69

76. Kim TK, Hemberg M, Gray JM, Costa AM, Bear DM, et al. 2010. Widespread transcription at neuronal activity-regulated enhancers. *Nature* 465:182–87

77. Klattenhoff CA, Scheuermann JC, Surface LE, Bradley RK, Fields PA, et al. 2013. Braveheart, a long noncoding RNA required for cardiovascular lineage commitment. *Cell* 152:570–83

78. Kohler C, Wolff P, Spillane C. 2012. Epigenetic mechanisms underlying genomic imprinting in plants. *Annu. Rev. Plant Biol.* 63:331–52

79. Konig J, Zarnack K, Rot G, Curk T, Kayikci M, et al. 2010. iCLIP reveals the function of hnRNP particles in splicing at individual nucleotide resolution. *Nat. Struct. Mol. Biol.* 17:909–15

80. Kowalczyk MS, Higgs DR, Gingeras TR. 2012. Molecular biology: RNA discrimination. *Nature* 482:310–11

81. Krajewski WA, Nakamura T, Mazo A, Canaani E. 2005. A motif within SET-domain proteins binds single-stranded nucleic acids and transcribed and supercoiled DNAs and can interfere with assembly of nucleosomes. *Mol. Cell. Biol.* 25:1891–99

82. Kruger K, Grabowski PJ, Zaug AJ, Sands J, Gottschling DE, Cech TR. 1982. Self-splicing RNA: autoexcision and autocyclization of the ribosomal RNA intervening sequence of Tetrahymena. *Cell* 31:147–57

83. Kutter C, Watt S, Stefflova K, Wilson MD, Goncalves A, et al. 2012. Rapid turnover of long noncoding RNAs and the evolution of gene expression. *PLOS Genet.* 8:e1002841

84. Kwon S. 2013. Single-molecule fluorescence in situ hybridization: quantitative imaging of single RNA molecules. *BMB Rep.* 46:65–72

85. Lahmy S, Bies-Etheve N, Lagrange T. 2010. Plant-specific multisubunit RNA polymerase in gene silencing. *Epigenetics* 5:4–8

86. Lai F, Orom UA, Cesaroni M, Beringer M, Taatjes DJ, et al. 2013. Activating RNAs associate with Mediator to enhance chromatin architecture and transcription. *Nature* 494:497–501

87. Lam MT, Cho H, Lesch HP, Gosselin D, Heinz S, et al. 2013. Rev-Erbs repress macrophage gene expression by inhibiting enhancer-directed transcription. *Nature* 498:511–15

88. Latos PA, Pauler FM, Koerner MV, Senergin HB, Hudson QJ, et al. 2012. *Airn* transcriptional overlap, but not its lncRNA products, induces imprinted *Igf2r* silencing. *Science* 338:1469–72

89. Lee JT. 2012. Epigenetic regulation by long noncoding RNAs. *Science* 338:1435–39

90. Lee JT, Bartolomei MS. 2013. X-inactivation, imprinting, and long noncoding RNAs in health and disease. *Cell* 152:1308–23

90a. Lee MP, DeBaun MR, Mitsuya K, Galonek HL, Brandenburg S, et al. 1999. Loss of imprinting of a paternally expressed transcript, with antisense orientation to KVLQT1, occurs frequently in Beckwith-Wiedemann syndrome and is independent of insulin-like growth factor II imprinting. *Proc. Natl. Acad. Sci. USA* 96:5203–8

91. Lee NN, Chalamcharla VR, Reyes-Turcu F, Mehta S, Zofall M, et al. 2013. Mtr4-like protein co-ordinates nuclear RNA processing for heterochromatin assembly and for telomere maintenance. *Cell* 155:1061–74

92. Lei EP, Corces VG. 2006. RNA interference machinery influences the nuclear organization of a chromatin insulator. *Nat. Genet.* 38:936–41

93. Lesterlin C, Ball G, Schermelleh L, Sherratt DJ. 2014. RecA bundles mediate homology pairing between distant sisters during DNA break repair. *Nature* 506:249–53

94. Levine SS, Weiss A, Erdjument-Bromage H, Shao Z, Tempst P, Kingston RE. 2002. The core of the polycomb repressive complex is compositionally and functionally conserved in flies and humans. *Mol. Cell. Biol.* 22:6070–78

95. Lewis EB. 2003. C.B. Bridges' repeat hypothesis and the nature of the gene. *Genetics* 164:427–31

96. Li W, Notani D, Ma Q, Tanasa B, Nunez E, et al. 2013. Functional roles of enhancer RNAs for oestrogen-dependent transcriptional activation. *Nature* 498:516–20

97. Lipshitz HD, Peattie DA, Hogness DS. 1987. Novel transcripts from the *Ultrabithorax* domain of the bithorax complex. *Genes Dev.* 1:307–22

98. Liu J, Jung C, Xu J, Wang H, Deng S, et al. 2012. Genome-wide analysis uncovers regulation of long intergenic noncoding RNAs in *Arabidopsis. Plant Cell* 24:4333–45

99. Luteijn MJ, Ketting RF. 2013. PIWI-interacting RNAs: from generation to transgenerational epigenetics. *Nat. Rev. Genet.* 14:523–34

100. Maenner S, Muller M, Frohlich J, Langer D, Becker PB. 2013. ATP-dependent roX RNA remodeling by the helicase maleless enables specific association of MSL proteins. *Mol. Cell* 51:174–84

101. Maison C, Bailly D, Peters AHFM, Quivy J-P, Roche D, et al. 2002. Higher-order structure in pericentric heterochromatin involves a distinct pattern of histone modification and an RNA component. *Nat. Genet.* 30:329–34

102. Mao YS, Sunwoo H, Zhang B, Spector DL. 2011. Direct visualization of the co-transcriptional assembly of a nuclear body by noncoding RNAs. *Nat. Cell Biol.* 13:95–101

103. Mardis ER. 2013. Next-generation sequencing platforms. *Annu. Rev. Anal. Chem.* 6:287–303

104. Margueron R, Reinberg D. 2011. The Polycomb complex PRC2 and its mark in life. *Nature* 469:343–49

105. Martens JA, Laprade L, Winston F. 2004. Intergenic transcription is required to repress the *Saccharomyces cerevisiae SER3* gene. *Nature* 429:571–77

106. Martens JA, Wu P-YJ, Winston F. 2005. Regulation of an intergenic transcript controls adjacent gene transcription in *Saccharomyces cerevisiae. Genes Dev.* 19:2695–704

107. Martianov I, Ramadass A, Serra Barros A, Chow N, Akoulitchev A. 2007. Repression of the human dihydrofolate reductase gene by a non-coding interfering transcript. *Nature* 445:666–70

108. Martinho RG, Kunwar PS, Casanova J, Lehmann R. 2004. A noncoding RNA is required for the repression of RNApolII-dependent transcription in primordial germ cells. *Curr. Biol.* 14:159–65

109. Mattick JS. 2009. The genetic signatures of noncoding RNAs. *PLOS Genet.* 5:e1000459

110. Meller VH, Rattner BP. 2002. The roX genes encode redundant male-specific lethal transcripts required for targeting of the MSL complex. *EMBO J.* 21:1084–91

111. Meller VH, Wu KH, Roman G, Kuroda MI, Davis RL. 1997. roX1 RNA paints the X chromosome of male *Drosophila* and is regulated by the dosage compensation system. *Cell* 88:445–57

112. Merino EJ, Wilkinson KA, Coughlan JL, Weeks KM. 2005. RNA structure analysis at single nucleotide resolution by selective 2′-hydroxyl acylation and primer extension (SHAPE). *J. Am. Chem. Soc.* 127:4223–31

112a. Modarresi F, Faghihi MA, Lopez-Toledano MA, Fatemi RP, Magistri M, et al. 2012. Inhibition of natural antisense transcripts in vivo results in gene-specific transcriptional upregulation. *Nat. Biotechnol.* 30:453–59

113. Mohammad F, Mondal T, Guseva N, Pandey GK, Kanduri C. 2010. *Kcnq1ot1* noncoding RNA mediates transcriptional gene silencing by interacting with Dnmt1. *Development* 137:2493–99

114. Morey C, Da Silva NR, Perry P, Bickmore WA. 2007. Nuclear reorganisation and chromatin decondensation are conserved, but distinct, mechanisms linked to *Hox* gene activation. *Development* 134:909–19

115. Morey L, Pascual G, Cozzuto L, Roma G, Wutz A, et al. 2012. Nonoverlapping functions of the Polycomb group Cbx family of proteins in embryonic stem cells. *Cell Stem Cell* 10:47–62

116. Moshkovich N, Nisha P, Boyle PJ, Thompson BA, Dale RK, Lei EP. 2011. RNAi-independent role for Argonaute2 in CTCF/CP190 chromatin insulator function. *Genes Dev.* 25:1686–701

117. Muchardt C, Guilleme M, Seeler J-S, Trouche D, Dejean A, Yaniv M. 2002. Coordinated methyl and RNA binding is required for heterochromatin localization of mammalian HP1α. *EMBO Rep.* 3:975–81

118. Nagano T, Mitchell JA, Sanz LA, Pauler FM, Ferguson-Smith AC, et al. 2008. The Air noncoding RNA epigenetically silences transcription by targeting G9a to chromatin. *Science* 322:1717–20

119. Natoli G, Andrau JC. 2012. Noncoding transcription at enhancers: general principles and functional models. *Annu. Rev. Genet.* 46:1–19

120. Necsulea A, Soumillon M, Warnefors M, Liechti A, Daish T, et al. 2014. The evolution of lncRNA repertoires and expression patterns in tetrapods. *Nature* 505:635–40

121. Novikova IV, Hennelly SP, Sanbonmatsu KY. 2012. Structural architecture of the human long non-coding RNA, steroid receptor RNA activator. *Nucleic Acids Res.* 40:5034–51

122. Ntini E, Jarvelin AI, Bornholdt J, Chen Y, Boyd M, et al. 2013. Polyadenylation site-induced decay of upstream transcripts enforces promoter directionality. *Nat. Struct. Mol. Biol.* 20:923–28

123. Orom UA, Derrien T, Beringer M, Gumireddy K, Gardini A, et al. 2010. Long noncoding RNAs with enhancer-like function in human cells. *Cell* 143:46–58

124. Pandey RR, Mondal T, Mohammad F, Enroth S, Redrup L, et al. 2008. *Kcnq1ot1* antisense noncoding RNA mediates lineage-specific transcriptional silencing through chromatin-level regulation. *Mol. Cell* 32:232–46

125. Park SW, Kang Y, Sypula JG, Choi J, Oh H, Park Y. 2007. An evolutionarily conserved domain of roX2 RNA is sufficient for induction of H4-Lys16 acetylation on the *Drosophila* X chromosome. *Genetics* 177:1429–37

126. Pauli A, Rinn JL, Schier AF. 2011. Non-coding RNAs as regulators of embryogenesis. *Nat. Rev. Genet.* 12:136–49

127. Pease B, Borges AC, Bender W. 2013. Noncoding RNAs of the *Ultrabithorax* domain of the *Drosophila* bithorax complex. *Genetics* 195:1253–64

128. Petruk S, Sedkov Y, Riley KM, Hodgson J, Schweisguth F, et al. 2006. Transcription of *bxd* noncoding RNAs promoted by trithorax represses *Ubx* in *cis* by transcriptional interference. *Cell* 127:1209–21

129. Phan AT, Kuryavyi V, Darnell JC, Serganov A, Majumdar A, et al. 2011. Structure-function studies of FMRP RGG peptide recognition of an RNA duplex-quadruplex junction. *Nat. Struct. Mol. Biol.* 18:796–804

130. Phillips JE, Corces VG. 2009. CTCF: master weaver of the genome. *Cell* 137:1194–211

131. Phillips-Cremins JE, Sauria ME, Sanyal A, Gerasimova TI, Lajoie BR, et al. 2013. Architectural protein subclasses shape 3D organization of genomes during lineage commitment. *Cell* 153:1281–95

132. Ponjavic J, Ponting CP, Lunter G. 2007. Functionality or transcriptional noise? Evidence for selection within long noncoding RNAs. *Genome Res.* 17:556–65

133. Razin SV, Iarovaia OV, Vassetzky YS. 2014. A requiem to the nuclear matrix: from a controversial concept to 3D organization of the nucleus. *Chromosoma* 123:217–24

134. Regulski EE, Breaker RR. 2008. In-line probing analysis of riboswitches. *Methods Mol. Biol.* 419:53–67

135. Reik W, Lewis A. 2005. Co-evolution of X-chromosome inactivation and imprinting in mammals. *Nat. Rev. Genet.* 6:403–10

136. Rinn JL, Chang HY. 2012. Genome regulation by long noncoding RNAs. *Annu. Rev. Biochem.* 81:145–66

137. Rinn JL, Kertesz M, Wang JK, Squazzo SL, Xu X, et al. 2007. Functional demarcation of active and silent chromatin domains in human *HOX* loci by noncoding RNAs. *Cell* 129:1311–23

138. Rouskin S, Zubradt M, Washietl S, Kellis M, Weissman JS. 2014. Genome-wide probing of RNA structure reveals active unfolding of mRNA structures in vivo. *Nature* 505:701–5

139. Roux KJ, Kim DI, Raida M, Burke B. 2012. A promiscuous biotin ligase fusion protein identifies proximal and interacting proteins in mammalian cells. *J. Cell Biol.* 196:801–10

140. Roy S, Ernst J, Kharchenko PV, Kheradpour P, Negre N, et al. 2010. Identification of functional elements and regulatory circuits by *Drosophila* modENCODE. *Science* 330:1787–97

141. Saldana-Meyer R, Gonzalez-Buendia E, Guerrero G, Narendra V, Bonasio R, et al. 2014. CTCF regulates the human *p53* gene through direct interaction with its natural antisense transcript, Wrap53. *Genes Dev.* 28:723–34

142. Sanchez-Elsner T, Gou D, Kremmer E, Sauer F. 2006. Noncoding RNAs of trithorax response elements recruit *Drosophila* Ash1 to *Ultrabithorax*. *Science* 311:1118–23

143. Sánchez-Herrero E, Akam M. 1989. Spatially ordered transcription of regulatory DNA in the bithorax complex of *Drosophila*. *Development* 107:321–29

144. Sauvageau M, Goff LA, Lodato S, Bonev B, Groff AF, et al. 2013. Multiple knockout mouse models reveal lincRNAs are required for life and brain development. *eLife* 2:e01749

145. Schmitz KM, Mayer C, Postepska A, Grummt I. 2010. Interaction of noncoding RNA with the rDNA promoter mediates recruitment of DNMT3b and silencing of rRNA genes. *Genes Dev.* 24:2264–69

146. Schulz D, Schwalb B, Kiesel A, Baejen C, Torkler P, et al. 2013. Transcriptome surveillance by selective termination of noncoding RNA synthesis. *Cell* 155:1075–87

147. Seila AC, Calabrese JM, Levine SS, Yeo GW, Rahl PB, et al. 2008. Divergent transcription from active promoters. *Science* 322:1849–51

148. Shah S, Wittmann S, Kilchert C, Vasiljeva L. 2014. lncRNA recruits RNAi and the exosome to dynamically regulate *pho1* expression in response to phosphate levels in fission yeast. *Genes Dev.* 28:231–44

149. Shao Z, Raible F, Mollaaghababa R, Guyon JR, Wu CT, et al. 1999. Stabilization of chromatin structure by PRC1, a Polycomb complex. *Cell* 98:37–46

150. Sigova AA, Mullen AC, Molinie B, Gupta S, Orlando DA, et al. 2013. Divergent transcription of long noncoding RNA/mRNA gene pairs in embryonic stem cells. *Proc. Natl. Acad. Sci. USA* 110:2876–81

151. Simon JA, Kingston RE. 2009. Mechanisms of polycomb gene silencing: knowns and unknowns. *Nat. Rev. Mol. Cell Biol.* 10:697–708

152. Simon MD, Pinter SF, Fang R, Sarma K, Rutenberg-Schoenberg M, et al. 2013. High-resolution Xist binding maps reveal two-step spreading during X-chromosome inactivation. *Nature* 504:465–69

153. Sleutels F, Zwart R, Barlow DP. 2002. The non-coding Air RNA is required for silencing autosomal imprinted genes. *Nature* 415:810–13

154. Song J, Angel A, Howard M, Dean C. 2012. Vernalization: a cold-induced epigenetic switch. *J. Cell Sci.* 125:3723–31

155. St. Johnston D. 2002. The art and design of genetic screens: *Drosophila melanogaster*. *Nat. Rev. Genet.* 3:176–88

156. Sternberg SH, Redding S, Jinek M, Greene EC, Doudna JA. 2014. DNA interrogation by the CRISPR RNA-guided endonuclease Cas9. *Nature* 507:62–67

157. Struhl K. 2007. Transcriptional noise and the fidelity of initiation by RNA polymerase II. *Nat. Struct. Mol. Biol.* 14:103–5

158. Sun L, Goff LA, Trapnell C, Alexander R, Lo KA, et al. 2013. Long noncoding RNAs regulate adipogenesis. *Proc. Natl. Acad. Sci. USA* 110:3387–92

159. Sun S, Del Rosario BC, Szanto A, Ogawa Y, Jeon Y, Lee JT. 2013. Jpx RNA activates *Xist* by evicting CTCF. *Cell* 153:1537–51

160. Swiezewski S, Liu F, Magusin A, Dean C. 2009. Cold-induced silencing by long antisense transcripts of an *Arabidopsis* Polycomb target. *Nature* 462:799–802

161. Tijsterman M, Ketting RF, Plasterk RH. 2002. The genetics of RNA silencing. *Annu. Rev. Genet.* 36:489–519

162. Tsai MC, Manor O, Wan Y, Mosammaparast N, Wang JK, et al. 2010. Long noncoding RNA as modular scaffold of histone modification complexes. *Science* 329:689–93

163. Tupy JL, Bailey AM, Dailey G, Evans-Holm M, Siebel CW, et al. 2005. Identification of putative noncoding polyadenylated transcripts in *Drosophila melanogaster*. *Proc. Natl. Acad. Sci. USA* 102:5495–500

164. Uhler JP, Hertel C, Svejstrup JQ. 2007. A role for noncoding transcription in activation of the yeast *PHO5* gene. *Proc. Natl. Acad. Sci. USA* 104:8011–16

165. Ulitsky I, Bartel DP. 2013. lincRNAs: genomics, evolution, and mechanisms. *Cell* 154:26–46

166. Ulitsky I, Shkumatava A, Jan CH, Sive H, Bartel DP. 2011. Conserved function of lincRNAs in vertebrate embryonic development despite rapid sequence evolution. *Cell* 147:1537–50

167. van Bakel H, Nislow C, Blencowe BJ, Hughes TR. 2010. Most "dark matter" transcripts are associated with known genes. *PLOS Biol.* 8:e1000371

168. van Dijk EL, Chen CL, d'Aubenton-Carafa Y, Gourvennec S, Kwapisz M, et al. 2011. XUTs are a class of Xrn1-sensitive antisense regulatory non-coding RNA in yeast. *Nature* 475:114–17

169. van Werven FJ, Neuert G, Hendrick N, Lardenois A, Buratowski S, et al. 2012. Transcription of two long noncoding RNAs mediates mating-type control of gametogenesis in budding yeast. *Cell* 150:1170–81

170. Wang H, Chung PJ, Liu J, Jang IC, Kean MJ, et al. 2014. Genome-wide identification of long noncoding natural antisense transcripts and their responses to light in *Arabidopsis. Genome Res.* 24:444–53

171. Wang KC, Chang HY. 2011. Molecular mechanisms of long noncoding RNAs. *Mol. Cell* 43:904–14

172. Wang KC, Yang YW, Liu B, Sanyal A, Corces-Zimmerman R, et al. 2011. A long noncoding RNA maintains active chromatin to coordinate homeotic gene expression. *Nature* 472:120–24

173. Wang X, Arai S, Song X, Reichart D, Du K, et al. 2008. Induced ncRNAs allosterically modify RNA-binding proteins in *cis* to inhibit transcription. *Nature* 454:126–30

174. Wierzbicki AT, Ream TS, Haag JR, Pikaard CS. 2009. RNA polymerase V transcription guides ARGONAUTE4 to chromatin. *Nat. Genet.* 41:630–34

175. Williamson JR. 2000. Induced fit in RNA-protein recognition. *Nat. Struct. Biol.* 7:834–37

176. Wimberly BT, Brodersen DE, Clemons WM Jr., Morgan-Warren RJ, Carter AP, et al. 2000. Structure of the 30S ribosomal subunit. *Nature* 407:327–39

177. Wu X, Sharp PA. 2013. Divergent transcription: a driving force for new gene origination? *Cell* 155:990–96

178. Xu Z, Wei W, Gagneur J, Perocchi F, Clauder-Munster S, et al. 2009. Bidirectional promoters generate pervasive transcription in yeast. *Nature* 457:1033–37

179. Yang L, Lin C, Jin C, Yang JC, Tanasa B, et al. 2013. lncRNA-dependent mechanisms of androgen-receptor-regulated gene activation programs. *Nature* 500:598–602

180. Yang L, Lin C, Liu W, Zhang J, Ohgi KA, et al. 2011. ncRNA- and Pc2 methylation-dependent gene relocation between nuclear structures mediates gene activation programs. *Cell* 147:773–88

181. Yang YW, Flynn RA, Chen Y, Qu K, Wan B, et al. 2014. Essential role of lncRNA binding for WDR5 maintenance of active chromatin and embryonic stem cell pluripotency. *eLife* 3:e02046

182. Yao H, Brick K, Evrard Y, Xiao T, Camerini-Otero RD, Felsenfeld G. 2010. Mediation of CTCF transcriptional insulation by DEAD-box RNA-binding protein p68 and steroid receptor RNA activator SRA. *Genes Dev.* 24:2543–55

183. Yap KL, Li S, Munoz-Cabello AM, Raguz S, Zeng L, et al. 2010. Molecular interplay of the noncoding RNA ANRIL and methylated histone H3 lysine 27 by polycomb CBX7 in transcriptional silencing of INK4a. *Mol. Cell* 38:662–74

184. Young RS, Marques AC, Tibbit C, Haerty W, Bassett AR, et al. 2012. Identification and properties of 1,119 candidate lincRNA loci in the *Drosophila melanogaster* genome. *Genome Biol. Evol.* 4:427–42

185. Zamudio JR, Kelly TJ, Sharp PA. 2014. Argonaute-bound small RNAs from promoter-proximal RNA polymerase II. *Cell* 156:920–34

186. Zhao J, Ohsumi TK, Kung JT, Ogawa Y, Grau DJ, et al. 2010. Genome-wide identification of polycomb-associated RNAs by RIP-seq. *Mol. Cell* 40:939–53

187. Zhao J, Sun BK, Erwin JA, Song J-J, Lee JT. 2008. Polycomb proteins targeted by a short repeat RNA to the mouse X chromosome. *Science* 322:750–56

188. Zhao R, Bodnar MS, Spector DL. 2009. Nuclear neighborhoods and gene expression. *Curr. Opin. Genet. Dev.* 19:172–79

189. Zhu Y, Rowley MJ, Bohmdorfer G, Wierzbicki AT. 2013. A SWI/SNF chromatin-remodeling complex acts in noncoding RNA–mediated transcriptional silencing. *Mol. Cell* 49:298–309

Centromeric Heterochromatin: The Primordial Segregation Machine

Kerry S. Bloom

Department of Biology, University of North Carolina, Chapel Hill, North Carolina 27599-3280; email: kerry_bloom@unc.edu

Annu. Rev. Genet. 2014. 48:457–84

First published online as a Review in Advance on September 18, 2014

The *Annual Review of Genetics* is online at genet.annualreviews.org

This article's doi: 10.1146/annurev-genet-120213-092033

Keywords

centromere, heterochromatin, chromosome segregation, DNA mechanics, molecular springs

Abstract

Centromeres are specialized domains of heterochromatin that provide the foundation for the kinetochore. Centromeric heterochromatin is characterized by specific histone modifications, a centromere-specific histone H3 variant (CENP-A), and the enrichment of cohesin, condensin, and topoisomerase II. Centromere DNA varies orders of magnitude in size from 125 bp (budding yeast) to several megabases (human). In metaphase, sister kinetochores on the surface of replicated chromosomes face away from each other, where they establish microtubule attachment and bi-orientation. Despite the disparity in centromere size, the distance between separated sister kinetochores is remarkably conserved (approximately 1 μm) throughout phylogeny. The centromere functions as a molecular spring that resists microtubule-based extensional forces in mitosis. This review explores the physical properties of DNA in order to understand how the molecular spring is built and how it contributes to the fidelity of chromosome segregation.

INTRODUCTION

The centromere is the primary constriction observed in condensed chromosomes during mitosis and provides the site of assembly for the kinetochore. The primary constriction on condensed chromosomes was first documented by Walther Flemming (44). Almost 100 years later, a description of the specialized disc-shaped kinetochore emerged: a proteinaceous structure found at the periphery of the centromere derived from electron micrographs of fixed specimens (15, 80). The identity of centromere and kinetochore proteins hinged serendipitously on the discovery of centromere-specific autoantibodies in sera of patients with an autoimmune disease scleroderma (14, 46, 103, 104). More than 100 kinetochore proteins have since been identified (10, 159, 174). They are organized minimally into five to six key complexes within the kinetochore and are largely conserved from yeast to mammals. Centromere DNA was first cloned and sequenced in the early 1980s (24, 43). In contrast to the high degree of protein conservation among diverse kinetochores, centromeres range from 125 bp of unique DNA in the budding yeast *Saccharomyces cerevisiae* to several megabases of repetitive DNA in *Homo sapiens* (**Figure 1**) (reviewed in 147). There are numerous features that distinguish centromeric heterochromatin from chromosome arms;

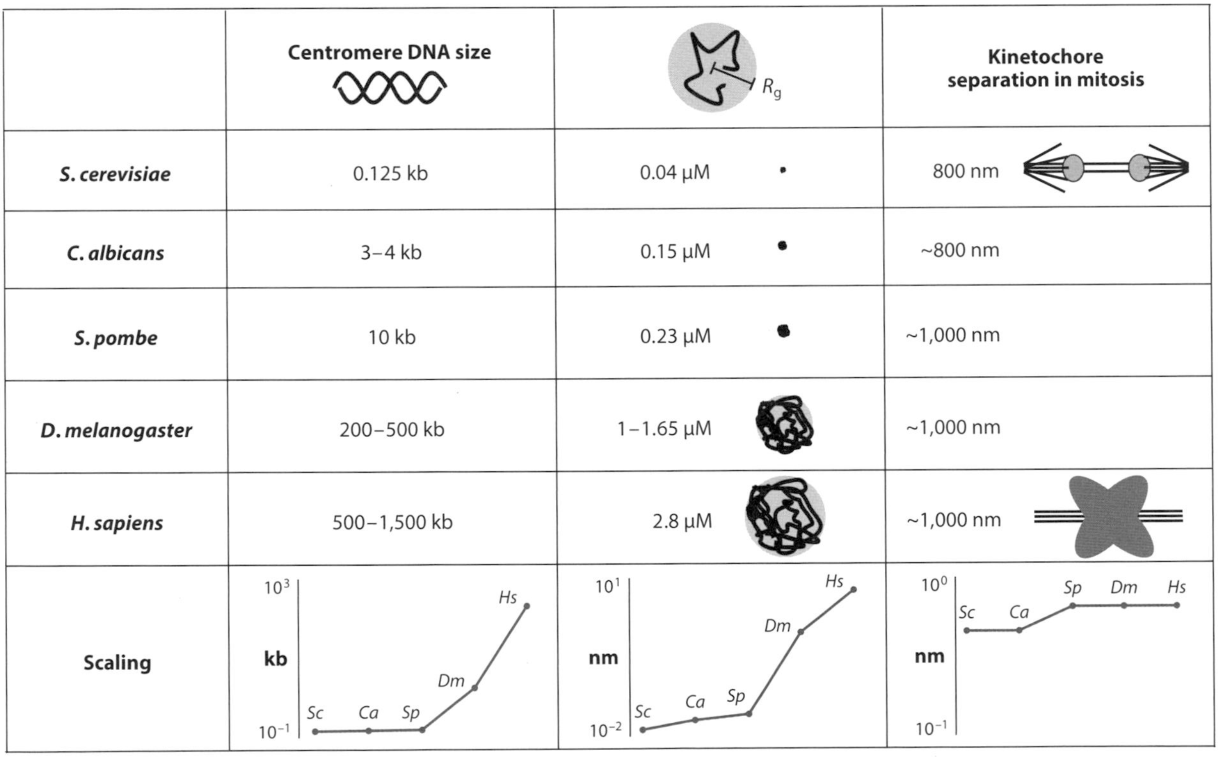

	Centromere DNA size	R_g	Kinetochore separation in mitosis
S. cerevisiae	0.125 kb	0.04 µM	800 nm
C. albicans	3–4 kb	0.15 µM	~800 nm
S. pombe	10 kb	0.23 µM	~1,000 nm
D. melanogaster	200–500 kb	1–1.65 µM	~1,000 nm
H. sapiens	500–1,500 kb	2.8 µM	~1,000 nm
Scaling			

Figure 1

Physical properties of centromere DNA. Centromere DNA size is defined as the region of DNA required for the segregation function in a variety of organisms. R_g is the radius of the random coil (radius of gyration) defined by $R_g{}^2 = n \times (2L_p)^2/6$. L_p is the persistence length of the polymer, and n is the number of segments (total DNA contour length L_c/L_p) (61). Kinetochore separation is the distance between sister kinetochores observed experimentally in a number of organisms [budding yeast, *S. cerevisiae* (118); fission yeast, *S. pombe* (32); worm, *C. albicans* (94); *Drosophila melanogaster* (160); and human, *H. sapiens* (132)]. The bottom graphs highlight the scaling (ordinate) of DNA (four orders of magnitude), radius of gyration (three orders of magnitude), and kinetochore separation (constant) throughout phylogeny (abscissa).

however, the chromatin organization and the divergence of centromeres throughout phylogeny are outstanding questions in the field. In this review, we explore conserved features of centromeric chromatin organization and evidence for the proposal that the centromere is the primordial segregation machine, preceding the evolution of kinetochore and spindle microtubules.

Why Heterochromatin versus Euchromatin at Centromeres

Euchromatin is an open chromatin state that enhances accessibility to transcription complexes. Heterochromatin, far from simply being a compaction state, provides a mechanism for recruiting and spreading components across large distances in DNA space (59, 60). To appreciate this aspect of heterochromatin we need to compare DNA length to nuclear volume. In a typical human diploid cell, two meters of DNA reside in an approximately 10-μm-diameter nucleus. The DNA is also very skinny (2 nm), which works in favor of chromatin-packaging strategies. The volume a random coil of DNA adopts in the absence of cellular material can be estimated from its radius, known as the radius of gyration (R_g) (**Figure 1**). For the mammalian genome, $R_g = (\sqrt{N}/6)2L_p = 130$ nm, equivalent to a volume of $\sim 10^7$ μm^3 ($4/3\pi R_g^3$). In a typical cell, nuclear volume is approximately 500 μm^3 ($4/3\pi$ 5 μm^3). Thus, chromatin is compacted on the order of 2×10^4. The volume of DNA polymer (excluded volume) is approximately 6 μm^3 ($\pi r^2 h = \pi$ *1 nm^2* 6 m) or only 1% of the nuclear volume. These estimates exclude cellular mechanisms that distinguish euchromatin versus heterochromatin. It is not packaging per se that is challenging to understand, it is the organization of functional domains. The subnuclear organization of genes into various domains, or bodies, is indicative of spatial segregation according to function. Active genes are often clustered to the nuclear periphery, whereas tRNA genes are frequently associated with the nucleolus (73, 84). Mechanistically, chromosome domains entrapped by cohesin bring regulatory domains adjacent to transcription start sites (35, 36, 120), whereas condensin binding to polymerase III transcription factors functions to cluster tRNA genes (65). With respect to the centromere, information about microtubule attachment at one kinetochore is transferred across centromeric heterochromatin to its sister kinetochore as a mechanism for force balance and tension sensing that is critical for chromosome segregation fidelity. In this situation, the pair of sister kinetochores and centromeric heterochromatin make up a single unit with structural/mechanical integrity. Remarkably, centromere DNA spans more than four orders of magnitude (from yeast to human), yet the physical separation between kinetochores is highly conserved (**Figure 1**). How the transmission of force is managed over the vast range of DNA size is critical for understanding the basis for faithful chromosome segregation. An evolutionarily conserved pathway [spindle assembly checkpoint (SAC)] (105) monitors the status of the kinetochore microtubule attachment site, including the presence or absence of a microtubule and whether tension is generated between sister kinetochores. In the absence of attachment and/or tension, chemical signals are generated that invoke the SAC. These signals relay the status of occupancy and/or tension at the kinetochore to master regulatory kinases that drive cell cycle progression (105). Additionally, there are mechanisms for correcting erroneous attachments, providing the opportunity for sister kinetochores to establish bi-orientation. The error-correction mechanisms and how the cell promotes stable versus unstable microtubule attachment in response to the state of each sister kinetochore are embedded in the structure of the kinetochore and the centromeric heterochromatin. The folding of megabases of centromeric heterochromatin (in the case of humans) into a highly compact and organized structure is better suited for transmitting mechanical force than the more open, disordered euchromatin. Protein machines acting as compactors, loopers, and topology adjusters are integral to mechanisms that convert the DNA random coil into a mechanical tension sensor suitable for transmission of microtubule-based spindle force.

Heterochromatin Organization: Transcription from Repetitive DNA

Heterochromatin is traditionally thought of as a means for setting boundaries between transcribed regions, as well as for providing an environment conducive to centromere function. One of the hallmarks of centromeric heterochromatin is the enrichment of simple sequences and repetitive DNA (68). In mammals, the centromere is marked by hierarchical repeats of α-satellite DNA. α-Satellites are a tandemly repeated array of a 171-bp monomeric unit. In fission yeast, the centromere region is characterized by large repeat structures (imr and otr) surrounding a unique centromere central core (23, 108). The extent of spreading and degree of compaction of centromeric heterochromatin is highly regulated. Transcription from repeat elements is part of this regulatory mechanism and provides a framework for assembly and maintenance of heterochromatin (59, 90). Increasing the stoichiometry of small noncoding RNAs through increased rates of transcription of mouse satellite leads to cell cycle delay, a decrease in heterochromatin compaction, and, likely, errors in chromosome segregation (75). Centromeric heterochromatin also represses transcription at the microtubule attachment site and is crucial for segregation function (50, 66). In budding yeast, modulation of transcription levels through the 125-bp CEN DNA leads to a marked increase in chromosome missegregation (70, 114, 115). In addition to the regulatory role of small RNAs, the act of transcription and topological consequences thereof may contribute to centromere function. However, as discussed below, phylogenetic studies of evolutionary new centromeres (ENC) reveals that satellite DNA accumulation follows centromere formation (112). It is therefore unlikely that number or physical arrangement of DNA repeats contributes to the mechanical coupling between sister chromatids. The accumulation of repeat sequences in the pericentromeres provides mechanisms through RNA function for creating and maintaining heterochromatin boundaries.

Pericentromeres Are Enriched in Condensin and Cohesin

The structural maintenance of chromosome (SMC) proteins cohesin and condensin are enriched at the pericentric region in a number of organisms (27, 39, 86). SMC proteins assemble into complexes that adopt a ring-like conformation (72, 109). The backbone of the ring is formed by the SMC proteins themselves (MukB in bacteria, Smc2 and Smc4 in *S. cerevisiae* condensin, and Smc1 and Smc3 in *S. cerevisiae* cohesin). In eukaryotes, the SMC monomer is folded in an antiparallel coiled coil. At one end, the two monomers associate to form a hinge, and at the other end is an ATP-binding head domain. Closure of the ring at the head domain is carried out by proteins known as kleisins, including Scc1 (also known as Mcd1) and Brn1. Each dimer is associated with additional proteins [e.g., Ysc4, Ycg1, Scc3 (also known as Irr1), Rad61, and Pds5] at the head domain to form a functional complex in vivo. Condensin, cohesin, and topoisomerase II are approximately threefold enriched in the 30–50 kb of DNA surrounding the budding yeast centromere (6, 27, 39) relative to the bulk of the genome. An attractive hypothesis is that cohesin, by physically entrapping sister chromatids, is responsible for the resistance between bi-oriented sister kinetochores in mitosis. Consistent with this, most cohesin dissociates from mammalian chromatin via the prophase pathway (93, 165), but a small population, enriched at the centromeric region, remains on chromatin until the onset of anaphase. Studies in budding yeast challenged this hypothesis by observing well-separated sister centromere DNA occurring frequently during metaphase (56, 69, 118, 152). How could proteins that specifically bind DNA together be enriched in a region where the DNA is both separated and transiently together? This apparent contradiction can be resolved if sister kinetochore microtubules stretch the centromeres into a cruciform structure in which each sister centromere lies at the apex of an intramolecular DNA loop while sister chromosome arms are paired via intermolecular cohesion. The cruciform structure serves to orient the centromere (and therefore the kinetochore) on the outer-face of the chromosome, facing toward the spindle pole of the segregation apparatus (142, 175). Cohesin is radially displaced from the microtubule axis, indicating a more complex organization that can be accounted for in simple models (**Figure 2**).

To gain insight into how these SMC architectural proteins might be organized to resist spindle forces in mitosis, we turn to their organization in compacting ribosomal DNA in the actively transcribed nucleolus. A major site of repeat sequences in all organisms is the nucleolus. The budding yeast nucleolus contains 100–200 copies of a 9.1-kb rDNA repeat (~1–2 Mb) together with approximately a threefold greater concentration of cohesin and condensin relative to RNA polymerase II transcribed genes (4, 28, 48). The nucleolus is also a repository of several cell cycle regulatory complexes (7, 129). Condensin plays a fundamental role in rDNA segregation (28) and is coupled to one of these cell cycle regulatory complexes (FEAR) to ensure the timely segregation of rDNA prior to cytokinesis. Condensin recruitment to rDNA depends upon monopolin (including Csm1 and Lrs4) Fob1 and is negatively correlated with levels of RNA polymerase I transcription (78, 79). Condensin also binds tRNA gene transcription complexes and promotes clustering of tRNA genes (65). Furthermore, these tRNA genes are clustered to the nucleolus, dependent on *CBF5*, a gene implicated in centromere function (84).

The concentration of cohesin is dynamically equilibrated between the nucleolus and pericentric heterochromatin. Reduction of pericentric cohesin (via *mcm21* deletion) leads to increased nucleolar levels, whereas reduction of nucleolar cohesin (via *sir2* deletion) leads to increased pericentric levels (143). Enrichment to the pericentromere depends upon the kinetochore, but what restricts these proteins to the 30–50-kb region surrounding the centromere is not clear. Because the rDNA repeats are enriched for condensin and cohesin, other repeats might similarly be enriched for these SMC proteins and thereby account for their enrichment in pericentromeric regions. The major repeated DNA sequences in yeast are the subtelomere repeats, LTRs (long terminal repeats of 300–400-bp bracketing retrotransposons, 429 total), and tDNA genes (307 total). LTRs and tDNA genes are enriched 1.8 times in the 50-kb region surrounding the centromere, relative to their concentration in the remainder of the genome. The enrichment of tDNA in the pericentromere and the recruitment of condensin to tRNA transcription factors provide a mechanism for restricting condensin to this functional domain. The enrichment of repeat sequences within the pericentromere may also provide the structural basis for partitioning condensin to the spindle axis and the radial displacement of cohesin (142, 143). Thus, the repeat sequences, although not directly involved in mechanosensing of the kinetochore microtubule attachments, may contribute indirectly to structure via their role in concentrating and partitioning proteins such as cohesin and condensin.

Heterochromatin Function: Making an Elastic Material

Centromeric heterochromatin is distinguished from heterochromatic regions along chromosome arms principally by the degree of compaction and enrichment of cohesin and condensin. Many of the histone modifications are shared with transcriptionally silenced regions (55). The centromere-specific histone H3 variant, CENP-A, resides at the CEN DNA in budding yeast (**Figure 2**) (145) and is interspersed with methylated histone H3 (mono- and dimethylation of lysine 4 and di- and trimethylation of lysine 36) in centromeres of multicellular organisms (8, 9, 13, 53). Centromere heterochromatin provides three functions in chromosome segregation: to serve as the template for kinetochore formation and microtubule attachment to the mitotic spindle; to link separated sister kinetochores; and to translate the state of kinetochore microtubule attachment between sister kinetochores. The kinetochore constituents and specificity of formation have been addressed

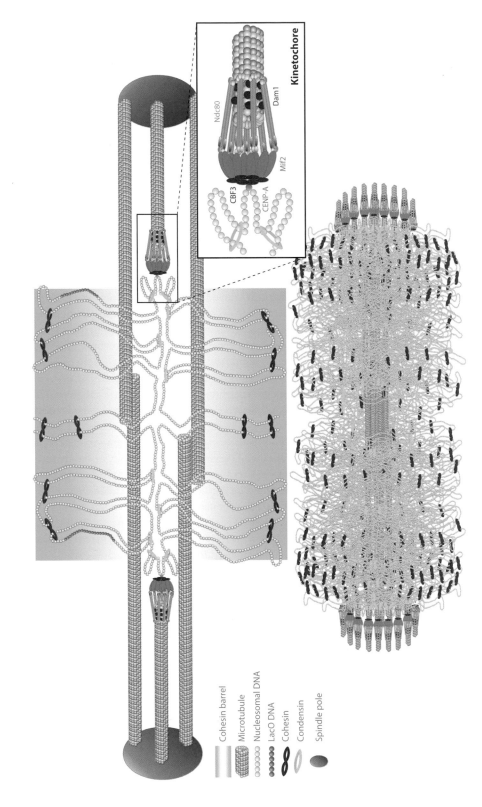

Kinetochore

Ndc80

Dam1

Mif2

CBF3

CENP-A

Cohesin barrel
Microtubule
Nucleosomal DNA
LacO DNA
Cohesin
Condensin
Spindle pole

in a number of recent reviews (10, 19, 172). Recent progress in genomics and mathematical modeling allows us to address the physical properties of centromere chromatin. For the mitotic spindle, kinesin microtubule-based motor proteins act as force production machines that slide the interpolar microtubules apart (**Figure 2**), generating an extensional force on the spindle poles (137, 138). The balance of microtubule-based extensional force and a chromatin-spring contractile force is necessary to produce a steady-state spindle length. Tension at the kinetochore satisfies the spindle checkpoint (11). Although the structure and form of the chromatin spring may vary throughout phylogeny, in all organisms the spring is contractile. To build an understanding of the chromatin spring, we start with the mechanical properties of DNA. DNA exhibits properties of an entropic spring, which may contribute to the contractility of centromeric heterochromatin. We can calculate the spring constant for an entropic spring according to polymer theory (12). The short rigid domains of DNA, linked via flexible joints, adopt a state of greatest disorder (entropy). From Hooke's law, we know that $F = \kappa x$, where κ is the spring constant (newton/meter) and x is the change in distance (meters). For small forces, $F = 3k_B Tx/n(2L_p)^2$. The spring constant of this freely jointed chain is equal to $3k_B T/n(2L_p)^2$, where $k_B T$ is the Boltzmann constant $\times$ T (newton meters), L_p is persistence length (meters), and n is the number of segments. For a DNA length of 10 kb, the spring constant is 0.036 fN/nm, which is small indeed. Note that the larger the DNA [increasing number of segments (n)], the weaker the spring. Thus, the tendency of DNA to adopt a random coil is not sufficient to balance microtubule-based extensional forces in the range of pN (98).

To gain insight into how mechanical force is transmitted between sister centromeres, we must understand the physical organization, starting with the size and shape of the pericentromere and centromere. Upon attachment to kinetochore microtubules, bi-oriented centromeres in everything from budding yeast to flies to mammals exhibit a stereotypic separation ranging from 800 nm to 1,000 nm (**Figure 1**) (118, 133). Although the yeast centromere is defined by a very small region of DNA (125 bp), the 16 centromeres are clustered around spindle microtubules, where they attach to the plus-ends of 16 kinetochore microtubules (kinetochore microtubule plus-ends shown in **Figure 2**). On the basis of the enrichment of cohesin and condensin in the 30–50-kb region of DNA flanking the 125-bp centromere, one can consider this domain as functionally equivalent to centromeric heterochromatin in organisms with multiple-kinetochore microtubule attachment sites, such as flies and mammals. Furthermore, the amount of pericentric chromatin

← _____

Figure 2

Organization of pericentric chromatin and cohesin in metaphase in budding yeast. (*Top*) The yeast segregation apparatus is a composite structure of the kinetochore and interpolar microtubules (*green*), the spindle pole body (*large red sphere*), and pericentric chromatin loops (DNA strands shown as strings of nucleosomes; *gray*). Centromere DNA (CENP-A nucleosome; *pink*) is attached to microtubule plus-ends via the kinetochore (*orange barbells* surrounding the microtubule plus-end). Kinetochore components (*right*) include Ndc80 (*orange barbells*), the Dam1 complex (*small red spheres* interleaved with Ndc80), the Mif2 complex (*blue rods*), and the DNA-binding complex CBF3 (*purple ovals*). Cohesin (*red*) and condensin (*yellow*) are enriched in the pericentromere and surround the central spindle. Cohesin is radially displaced from the spindle axis, whereas condensin is proximal to the spindle axis (142, 143). One pair of sister chromatids is shown for simplicity. Sister chromatids occupy the left and right half-spindle, respectively. Sister chromatids at the mid-spindle position (perpendicular to the spindle axis) are held via cohesin rings (*red*). As sister DNA strands become proximal to the spindle microtubules, they adopt a cruciform-like DNA configuration, with intermolecular sister pairing midway between the spindle poles and intramolecular pairing to the left and right. Condensin rings along the spindle axis contribute to formation of intramolecular loops shown perpendicular to the spindle axis, proximal to the left and right kinetochore. Lac operator DNA (LacO DNA) (nucleosomes; *blue*) is radially displaced from the spindle axis when visualized with LacI-GFP (2), similar in dimension to the cohesin barrel. (*Bottom*) The pericentric chromatin from all 16 chromosomes is shown together with kinetochore microtubules in metaphase. The amount of DNA is drawn to scale and represents the region of DNA from all 16 centromeres that is enriched in the SMC (structural maintenance of chromosome) protein complexes cohesin and condensin. Adapted from figure 4 of Reference 62.

encircling the yeast spindle is ∼500–800 kb (16 chromosomes × 30–50 kb), comparable to the 1–5 megabase lengths of α-satellite DNA in a mammalian kinetochore. The area between the two clusters of 16 microtubule attachment sites in budding yeast ($2\pi rh \approx 0.6\ \mu m^2$) is comparable to estimates of the size of mammalian kinetochores (0.4 μm^2) (20). Thus, the distance between sister centromeres as well as the area of centromeric heterochromatin are conserved features in centromere organization.

The next level of understanding of pericentromere function requires the accurate dissection of spatial features, including the DNA and chromatin proteins within the structure. The advantage of the pericentromere in budding yeast is that, unlike in mammals and most other organisms, the site of microtubule attachment is known with base-pair precision. The point centromere provides a critical reference point for localizing components relative to the site of microtubule-based force transduction. Pericentric DNA can be visualized through lac operators and lacI-GFP, which are introduced at specific sites relative to the centromere (146). Using this system, several investigators have found that sister centromeres are precociously separated during mitosis (56, 68, 118, 152). Because separated sister *lacO* (*Escherichia coli* lac operator) foci are mobile, their position is best captured as a distribution of statistical probabilities. Surprisingly, *lacO* foci at centromere proximal positions are found least often on the spindle axis (**Figure 2**) (2, 142). Pericentric DNA is radially displaced from the position of the kinetochores and kinetochore microtubule plus-ends (**Figure 2**). Thus, forces at one centromere are not linearly transmitted through the DNA to the sister centromere. The localization of cohesin and condensin helps us understand the distribution of pericentric DNA. Data from chromatin immunoprecipitation (ChIP) studies appear to indicate that cohesin and condensin are mostly uniform across the pericentromere (27, 52, 76, 99, 151, 168). In contrast, live-cell imaging studies indicate that cohesin and condensin adopt stereotypic and spatially segregated structures within the pericentric heterochromatin in mitosis (142, 143). Cohesin is radially distributed into an apparent barrel, displaced from the kinetochore and kinetochore microtubules (500-nm diameter versus 250-nm diameter kinetochore microtubules) (**Figure 2**). The barrel is uniform in fluorescence intensity, indicative of the even distribution of cohesin. The degree of displacement of the cohesin barrel is comparable to the displacement of the average distribution of pericentric DNA, indicating that both cohesin and pericentric DNA are, on average, displaced from the spindle axis and the main site of microtubule attachment. In contrast, condensin is localized along the spindle axis. Condensin is heterogeneous and appears as single or multiple foci or linear elements, consistent with its biochemical tendency to aggregate (4, 142). The apparent disparity in localization deduced by ChIP versus live-cell fluorescence microscopy is reconciled by considering DNA fluctuations. If the binding sites for pericentric cohesin and condensin are highly variable, a population method such as ChIP averages out the differences and misses the spatial segregation of these complexes observed in live images of single cells. The spatial segregation of chromatin proteins as well as the DNA within the centromeric heterochromatin reveals how little we understand the chromatin spring that resists outward microtubule-based forces.

Ring-like proteins endow synthetic polymers with several unique properties (57, 116). Slip-links in place of rigid cross-linkers in a synthetic fiber allow polymers to wiggle through the cross-link, resulting in a "sharing of the load" phenomenon (**Figure 3**). In the centromeric heterochromatin, condensin and/or cohesin protein rings may function as staples or slip-links (molecular pulleys) to distribute tension from one microtubule attachment site to the entire network (**Figure 3**). Cohesin and condensin (**Figures 2** and **3**) have the physical attributes to function as slip rings and regulate centromere elasticity. Cohesin acting distally from the spindle axis promotes looping and/or distributes tension throughout the network as a way of averaging fluctuations from the 32 individual kinetochore microtubules. Condensin localized along the spindle axis could generate a spring force through DNA compaction. Microtubule-extensional forces have a tendency to pull DNA,

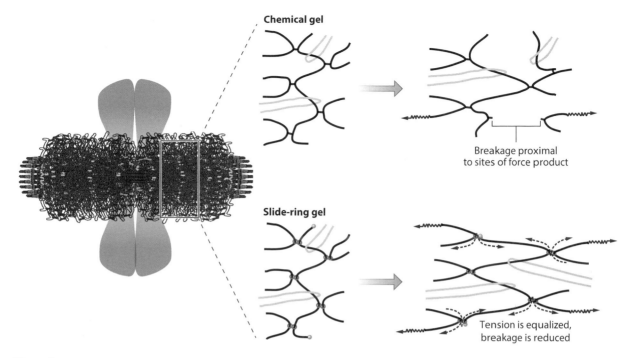

Figure 3

Strategies for cross-linking centromere DNA in the primary constriction of metaphase chromosomes. (*Left*) The organization of centromeric heterochromatin relative to chromosome arms in mitosis. The pericentric chromatin from 16 chromosomes surrounds the spindle axis (see key in **Figure 2**). Sister chromatid arms relative to centromeric heterochromatin are indicated by the blue masses. Pericentric chromatin functions as a contractile element in the spindle, counteracting extensional forces from microtubule-based motor proteins to achieve force balance in metaphase. (*Middle*) Various ways to generate chromatin cross-links. In a chemical gel (*top*), inflexible links (*red*) connect DNA strands (*black*). In a slide-ring gel (*bottom*), molecular pulleys (*green*), through which DNA can slide, connect DNA strands. The light gray strands indicate entanglements in which DNA strands are catenated. (*Right, top*) When force is exerted on DNA strands in a chemical gel, strain is concentrated on inflexible linkages, resulting in rupture. (*Right, bottom*) When force is exerted on DNA strains that can slip through molecular linkages, tension is distributed throughout the network, equalizing strain among all links, reducing rupture (57, 116).

extending its length, whereas condensin decreases DNA length through compaction. In support of this view, loss of either pericentric cohesin or condensin results in increased spindle length in yeast (142) as well as increased kinetochore-kinetochore distance in vertebrate cells (128, 135).

Unique Features of Polymer Springs

The importance of fidelity in chromosome segregation cannot be overstated. It is not enough to build a robust microtubule attachment site. The ability to detect and correct errors in chromosome alignment is paramount in the development of any organism. In multiple microtubule-attached kinetochores, the error detection/correction system must compensate for stochastic growth and shortening of kinetochore microtubules. Unless the system tolerates a large range of tension at the kinetochore microtubule interface, there are no plausible mechanisms for how natural fluctuations in tension are buffered. We turn to the chromatin spring to gain insight into how tension is managed within the centromere. Long chain polymers such as chromosomes are challenging to model because of the wide range of length and timescales in the system. One of the problems for modeling is the vast number and breadth of polymer configurations, even though the large number

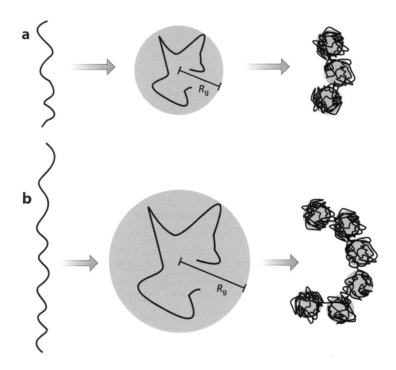

Figure 4

Polymer model of chromatin provides insight into force transduction through the centromere heterochromatin. (*a,b*) Linear DNA (*left*) spontaneously folds into a random coil (*middle*). The size of the random coil is defined by the radius of gyration (R_g). In a confined and crowded environment such as a cell, the chains adopt smaller random coils within which monomer subunits are free to fluctuate, thus maximizing entropy (*right, smaller gray spheres containing red polymer*). These are known as thermal blobs and are defined as regions of the polymer chain in which the monomers are unaffected by their environment. If force is exerted at the ends of the chain, the blobs decrease laterally in size, but within each blob, the chains continue to fluctuate with defined statistical properties (164). Tension blobs create significant consequences for the centromere. When force is exerted on the kinetochore, tension can be transmitted over large distances (800 nm) without reducing the entropy of each and every monomer within the chain. The differing lengths of centromere DNA across phylogeny (small to large, as depicted in panels *a* and *b*) predict that the number of tension blobs varies considerably, but the mechanism of force transduction is highly conserved. The centroids of blobs can be connected to each other through springs, leading to the bead-spring representation of the chromatin polymer.

of configurations contributes significantly to the entropy and therefore the elastic restoring force. Even with robust algorithms, the complexity in the models must necessarily be reduced relative to what is present in vivo. Representation of the restoring force with springs and modeling the polymer as a bead-spring chain have proven valuable to understanding chromosome behavior (156) (**Figure 4**). These polymer models do remarkably well in capturing the large-scale folding and organizational principles derived from population studies.

Chromosome motion varies in predicted ways along the length of the chromosome. To see how the chromosome behaves, hold a slinky with each hand at an end and shake. The tendency for the middle of the spring to move more and the ends of the spring to move less captures experimental observations (162). The centromere behaves as one of the ends in this toy model. The restricted motion of an end reflects the biological feature that the centromere remains attached to the spindle pole body throughout the cell cycle in budding yeast (162). This is called a tether in the polymer world and is an important physical principle that, together with the bead-spring model, is required to generate the observed variation in chromosome mobility (**Figure 4**).

During interphase, the Rabl orientation describes a configuration in which centromeres of each chromosome are associated with the spindle pole and telomeres distal to the pole (37). In mitosis, chromosomes are attached to the mitotic spindle and align along the metaphase plate, whereas in meiotic prophase, telomeres are clustered in a bouquet configuration with the centromeres extended distally (21, 22). Microtubule attachment of the centromere to the spindle poles or telomeres to the nuclear envelope provides a physical linkage that significantly restricts the dynamic range of chromatin motion (162). In budding yeast, detached centromeres exhibit the same degree of freedom as a locus hundreds of kilobase pairs away on the chromosome arm (162) that explores threefold more space than the attached centromere configuration (radius of confinement attached 274 nm versus detached 745 nm). The restricted centromere movement in wild-type cells is observed in all stages of the cell cycle. Likewise, telomere tethering to the nuclear envelope must be invoked to predict their observed motion. In a bead-spring model, sister kinetochores function as tethers restricting chromatin motion on the surface of the chromosome (see **Figure 2**), whereas intervening centromeric heterochromatin fluctuates like the aforementioned slinky. These basic physical principles reveal that in the absence of other features, the chromatin spring constant will be stiffest closest to the kinetochore and softest at the midpoint between sister kinetochores.

A second driving principle to understand how a polymer spring might function involves the excluded volume interaction. Just as we describe the hierarchy of chromosome organization from DNA to nucleosomes to higher-order structures, there are hierarchies in parameter interactions for a polymer. The primary interactions are the covalent bonds. Secondary interactions include the polymer and the solvent or are between monomers that are not nearest neighbors. The volume taken up by one monomer is excluded from that available to any other monomer. For a dynamic polymer, the area the polymer explores is larger than the chain itself, giving rise to an exclusion zone unavailable for other molecules to penetrate. This is known as excluded volume, the magnitude of which depends upon the chemical environment. Excluded volumes can overlap when molecules are immersed in poor solvents or crowded conditions, such as are found inside the cell. Considering the enormous length of the chromosome relative to the nucleus, the excluded volume is a significant fraction of the nuclear space. When two monomers come into proximity (i.e., crowding), their excluded volumes overlap. This results in increased available volume for small molecules (here called depletants). The gain in entropy of the system from the increased available volume to the depletants is known as the depletion (entropic) force. The depletion force favors molecular crowding (163) and, depending on the strength of the attraction force between individual segments of a polymer, can lead to collapse of the polymer into a compact structure known as a globule (61). In this way, the depletion force promotes chromosome compaction from first principles. Students of mitosis may recall that counter to intuition, thermodynamics favors the formation of the microtubule polymer (132). It is the loss of ordered water around tubulin dimers that drives this reaction. The critical issue is that the thermodynamics of any system that involves a polymer in solution must be considered in terms of entropy of the whole system, and this consideration can lead to counterintuitive phenomena, such as microtubule polymerization and chromosome condensation.

Finally, cellular geometry confines the DNA to an area much smaller than the thermodynamically favored R_g (**Figure 4**). For the *E. coli* chromosome, R_g is ~5 μm. Constraints imposed by cellular geometry mean that DNA is no longer hydrodynamically a spherical ball, but rather is more akin to a deformed or anisotropic configuration. Constraints lead to nonintuitive changes in behavior, such as dynamics and/or segregation of multiple polymer chains in a small space. Entropy drives fluctuations of all chains to adopt a random coil. When the chains overlap, the number of conformational states is reduced (resulting in reduced entropy). This entropic penalty can drive segregation of the two chains in a confined space (82, 83). Such a mechanism has been proposed to operate in prokaryotes (82, 83) and might contribute to bulk chromosome segregation in eukaryotes that undergo closed mitosis (such as fungi and some protists).

Spring Dynamics

Chromosomes are dynamic structures that undergo large spatial fluctuations in interphase and mitosis. In mitosis, centromeres exhibit oscillatory motion toward and away from the poles. The behavior of the chromatin fiber underlying this motion is not resolved in a light microscope. Even with super-resolution microscopy modes, one cannot resolve individual chromatin strands (11–30 nm in diameter). To visualize the chromatin fiber, an array of E. coli lacO can be integrated into a chromosome together with lac repressor–GFP fusion proteins (146). In this way, the underlying chromatin dynamics can be directly visualized in the light microscope. Furthermore, in budding yeast, the operator arrays can be introduced at defined positions along a chromosome. To understand the motion of these DNA arrays, we return to the behavior of the polymer chain. Unlike diffusion of a protein or solute that bounces through the nucleus or cytoplasm with the force of thermal energy ($k_B T$), a polymer wiggles and slithers like a snake across sand. The parameters for describing the motion of a polymer are its persistence length (L_p) and contour length (L_c). The persistence length is the length scale over which the polymer, such as naked DNA, is stiff, with the mathematical definition of L_p being based on the correlation of the ends in space. The contour length is the total chain length. The persistence length of a microtubule is 6 mm as compared with only 50 nm for DNA. The consequence of this ∼5 orders of magnitude difference is that microtubules are very stiff and DNA is very floppy. The timescale of motion becomes a variable parameter for a long floppy chain such as DNA. On very short timescales (milliseconds), the motion is dictated by Brownian driven fluctuations of L_p segments. On these timescales, the motion reflects internal conformational changes of the monomers. Even a simple bead-spring model predicts distinct statistical fluctuation of each bead along the chain (162) (**Figure 4**). This leads to a position-dependent effective spring constant as seen by a particular bead relative to a tether point (i.e., centromere). The stiffer the chromatin spring, the less variance in fluctuations, whereas softer chromatin springs exhibit increased variance.

On longer timescales (seconds to minutes) the chain exhibits translational motion in space. Various models, such as the bead spring (**Figure 4**), have been invoked to understand the motion. Rouse models include spring and thermal forces, whereas Zimm models add hydrodynamic interactions to the Rouse model (29, 34, 130, 131, 178). The relaxation time of a polymer segment is defined as $\tau_R = \zeta/k_B T(NR^2)$, where ζ is the drag coefficient, $k_B T$ is the Boltzmann constant × T, R is the end-end distance of the chain, and N is the number of beads. Motion analysis on timescales less than τ_R samples viscoelastic modes, whereas timescales greater than τ_R sample diffusive motion. For purposes of intuition, the diffusion constant of an average protein is 5–15 $\mu m^2/s$. For a chromosome (bacterial or yeast) the apparent diffusion constant is ∼0.0003–0.005 $\mu m^2/s$ (169–171), 3–4 orders of magnitude slower than for a protein or small metabolite. The diffusion time for translation of the chromosome across the E. coli cell or a yeast nucleus is ∼50 s. This is important when we consider that this is substantially less than the timescale of cell division or mitosis [20 min doubling time in bacteria, ∼10–20 min mitosis in yeast (176)]. In other words, polymer diffusion dynamics are sufficient to provide the kinetic motion behind chromosome segregation. From these laws of motion, we infer that the critical function of kinetochore/spindle systems is to direct sisters into daughter cells, ensuring that each cell gets one copy of each chromosome. Thus, consideration of basic physical principles reveals that segregation machines are mainly in the business of accuracy and sorting, and that they just need to act as directional rectifiers to direct kinetic motion toward the spindle pole. As noted by Nicklas (110) many years ago, mitosis is not about speed; rather, it is about fidelity.

Chromatin Proteins That Drive Elasticity

The finding that cohesin is enriched at centromeres (52, 99, 151, 168) raised a paradox, namely how to reconcile sister chromatid cohesion with sister centromere separation in mitosis. The discovery that the pericentromeric region adopts an intramolecular loop structure in vivo provided a solution to this paradox (175). Instead of holding sister chromatids together, cohesin may contribute to the mechanisms that promote centromeric DNA looping and structural features that stiffen the spring and/or promote confinement (**Figures 2** and **3**). Introducing protein linkages into the system could have a profound effect on the stiffness of the chromatin spring. To understand the contribution of proteins to chromatin stiffness we turn to the relationship between force and chain extension. As discussed above, a fluctuating polymer chain adopts a random coil in the absence of an external force (**Figure 4**). When forces are exerted on the polymer, as experienced at the kinetochore microtubule attachment site, the polymer (centromere) is stretched. This reduces the number of states available for fluctuation. The reduction in entropy works against external elongation. A model that captures the relationship between force and extension for flexible polymers (random coils, such as unstructured protein or DNA) is known as the worm-like chain (58). The solution can be approximated according to the formula $F = (k_B T/L_p)\{[(1/4)(1 - x/L_c)^{-2}] - 1/4 + x/L_c\}$, where F is force, $k_B T$ is the Boltzmann constant $\times T$, L_p and L_c are as defined above, and x is the extension. It takes very small forces (on the order of several piconewtons) to reduce the number of available states and extend the chain 60–80% of its contour length. Stretch length (x) approaches the chain length of the polymer (L_c), and extensional forces rise exponentially because of the stretching of covalent bonds. Note that the persistence length for unstructured proteins is less than a nanometer versus 50 nm for DNA (see above). The consequence is that small domains of unstructured proteins in the spring require roughly 50 times the extensional force relative to the same length of DNA. The makeup of centromere heterochromatin is approximately equal mass DNA and histone protein, with protein linkages such as cohesin and condensin distributed in the network. Putting cohesin and/or condensin into the system adds considerable stiffness to the network, analogous to putting pop-rivets along the DNA helix.

Alternatively, the cohesin and condensin rings may contribute in other ways to the elastic properties of the centromere and/or chromosome. When tension is applied to an individual polymer in a cross-linked network, force is exerted on the proximal cross-link that leads to local regions of high stress. Slip rings (or molecular pulleys) provide a mechanism to distribute tension from one location to the entire network (**Figure 3**) (57, 116). In this scenario, the stretched polymer slips through the ring, equalizing the tension cooperatively. Such a topological distribution of stress was originally proposed in the Doi, Edwards, deGennes reptation tube model (29, 34) and was demonstrated in a recent study of the chemistry of polyrotaxanes (57, 116). The ring-like structure of cohesin and condensin (72, 109) is indicative of the physical attributes expected of slip rings to provide the chemistry for distributing microtubule-based tension throughout centromere heterochromatin. SMC protein rings are ~40 nm [cohesin (64)] or 25 nm [condensin (1)] in diameter, sufficient to encircle one or two 11-nm nucleosome fibers. Cohesin links sister chromatids by encircling two chromatin strands (bracelet model) or dimerization of two rings, each of which encircles a single chromatin strand (handcuff model) or both (109, 117). Condensin also links chromatin strands by encircling one strand and binding to another (25, 26). Both cohesin and condensin are mobile in vivo. In budding yeast, cohesin is loaded at the centromere and slides to locations that cover the 50 kb of pericentric heterochromatin (76), whereas RNA polymerase can slide cohesin to the 3′ end of transcribed genes (112). Likewise, condensin sliding and stopping have been shown to be integral to mechanisms of chromosome condensation during anaphase segregation (26). Slip rings provide a mechanism for distributing tension through centromere heterochromatin that is feeling the ensemble of stochastically growing and shortening kinetochore microtubules.

Another mechanism to regulate motion and the distribution of tension throughout the network is through entanglements between the chains. Entanglements arise when a DNA from one chromosome or topological domain encircles another (see entanglements in **Figure 3**). This situation may arise in pericentric heterochromatin, where there is a high packing density chromatin. Increasing the number of entanglements throughout a network increases its elasticity (177). This confers a rubber-like behavior to the chromosome. A rubber band stretches because the polymer chains locally slip past one another. As a rubber band is stretched, the number of configurational states decreases, reducing the entropy. The entropic restoring force, which maximizes the number of configurations the chain can explore, is responsible for recoil when the stretch force is released. Cross-links in the polymer prevent chain slippage, providing a mechanism to tune the strength or stiffness of the network. A counterintuitive property of rubber bands is their contraction when heated. At higher temperature, fluctuations increase and the shorter, random configurations are favored. In the centromere, more energy input in the form of protein machines (e.g., chromatin remodeling complexes and chaperones) increases chain motion, biasing the system toward chain stiffening. Centromere stretch and relaxation are observed in most eukaryotic kinetochores, highlighting this constant interplay between entropic chain fluctuation with its cross-links and entanglements versus the microtubule-based extensional forces that work to stretch the chain and reduce entropy.

Driving Force for Compaction of Pericentric Chromatin

A typical mammalian chromosome is compacted roughly 10,000-fold in mitosis. If we consider the ambient cellular pressure and the energy to pack the chromosome together with molecular crowding, we find that the chromosome is still fairly soft, i.e., loosely packed, even at these seemingly high compaction ratios. Although proteins such as condensin and topoisomerase II play a critical role in compaction, it is important to begin with first principles to determine how much energy is required for compaction. As discussed above, entropic forces favor chromosome compaction because of the increased available volume and therefore gain free energy from the depletants (small molecules) in the cell. Each molecule has an effect of ~1 $k_B T$ (3). Given the ~one million proteins in *E. coli* as an example, the upper limit for the change in free energy available for chromosome compaction is on the order of 10^6 $k_B T$. It has been estimated that ~10^5 $k_B T$ (~100 pN) free energy is stored mechanically in the in vivo bacterial nucleoid (119). The external pressure required to hold the chromosome at this size is therefore 1,000 times smaller than the turgor pressure inside a cell [~1 atmospheric pressure (ATM) for bacteria (31) and 10 ATM for yeast (100) (1 ATM = 1 × 10^5 pN/μm^2)]. The finding that compaction pressure inside the cell is 1/1,000th of the surrounding pressure indicates that the bacterial chromosome is fundamentally very soft. Similarly, measurements of the Young's modulus of the chromosome are on the order of a few hundred pascals (pN/μm^2) (149). This is very different from compaction of DNA in a virus that is subject to 50 ATM (equivalent to 50 × 10^5 pN/μm^2) (102). Thus, although the degree of compaction (10,000-fold) is large at face value, it is not so large relative to the thermodynamics of the living system. In addition to the role structural proteins play in chromatin compaction, they may also provide internal structure in the form of tethers or entanglements that link chromatin domains (121, 122) and promote distant interactions in heritable ways.

The principles derived from a polymer physics view of centromere heterochromatin impinge on issues central to mechanisms of chromosome segregation. Tension at the kinetochore is derived from a balance of microtubule-based extensional forces and chromatin-based inward force. The shear bulk of centromeric heterochromatin (**Figure 1**) precludes a simple understanding of the structure of the spring. Furthermore, the stochastic microtubule growth and shortening at the multiple microtubule attachment sites in many organisms, including ourselves, are difficult to integrate into a tension-sensing mechanism that monitors individual kinetochore-microtubule attachment sites. Tackling the polymeric aspect of chromatin provides testable insights into these basic questions. Starting from thermodynamics, considering the entire system, including polymer and solute, we find that chromosome compaction increases the entropy for the solute. Thus, the force to compact the chromosome may be very small relative to what is intuitively a large problem (10,000 × compaction). Likewise, chromosome diffusion is very slow, but not particularly slow on the timescale of chromosome movement, consistent with suggestions made several decades ago (110, 153) that segregation mechanisms are concerned with accuracy rather than strength or speed. Finally, protein complexes that form rings, such as condensin and cohesin, may impart elastic properties to centromere heterochromatin and therefore, together with the entropic nature of a DNA spring, constitute the centromere spring.

Building a Centromere or "Only If You Build It, Do You Understand It"

The concentration of repeated sequences (satellites and/or tDNA), of cohesins, and of condensin distinguishes the centromere from chromosome arms and euchromatin but does not reveal the initiating event in centromere formation. As voiced by Richard Feynman in the 1950s, "Only if you build it, do you understand it" (41). We turn to different strategies, including synthetic centromeres, de novo centromere formation, and evolutionary centromere lineages, that allow us to peer into critical nucleation events in centromere formation.

Synthetic centromeres. The elucidation of kinetochore structure in yeast, fly, and human guides strategies for synthesis (10, 125, 150). The kinetochore is the protein-DNA machine built at the centromere. There are more than 100 proteins organized into 6–7 complexes, each of which contains between 4 and 12 protein subunits (**Figure 2**). The major force coupler is Ndc80, a tetrameric complex that lies on the microtubule lattice, proximal to its plus-end (158). The Dam1 complex comprises a 12-protein multimer that oligomerizes into a ring encircling the microtubule lattice in vitro (111). The Dam1 complex is essential in budding yeast and is one of the few components that is not evolutionarily conserved. The functional equivalent in mammalian kinetochores may be the Ska1 complex that binds depolymerizing microtubules and enhances the ability of Ndc80 to track microtubule ends (139, 173). The linkage to the chromatin is established via the CCAN complex (45, 150) and the centromere-specific histone H3 variant CENP-A (45, 145). In budding yeast, there is a high-affinity sequence-specific DNA binding complex that recognizes the centromere DNA CBF3 (54, 89). Several recent studies have built microtubule attachment sites for chromosome segregation. A site for a high-affinity DNA-binding protein is one requirement for DNA segregation that has been reconstructed through the use of tandem copies of the *lacO*. The DNA binding protein lac repressor (product of the *LacI* gene) is fused to a protein component of the kinetochore essential for microtubule interaction. In one case, fusion of lac repressor to Ask1, a member of the Dam1 complex, was used; in another case, lac repressor was fused to Dam1, a member of the same complex (85, 88). In these synthetic kinetochores, the fusion protein recruits sufficient kinetochore proteins to promote chromosome segregation. These systems completely bypass the requirement for the authentic centromere and essential DNA binding components of the natural centromere.

Mammalian centromeres, unlike budding yeast, do not have a specific DNA sequence at the site of microtubule attachment. The first synthesis of an artificial human centromere [human artificial chromosome (HAC)] introduced fragments of cloned centromeres to recruit kinetochore components (67, 95). A clever marriage between the human alphoid repeats and site-specific bacterial binding sites allowed Earnshaw and colleagues (107) to build a conditional human centromere. These investigators introduced a tetracycline operator (TetO) every second alphoid monomer. This system provides the capability to direct a range of proteins into the functional kinetochore as tetracycline repressor-fusion proteins. These synthetic kinetochores can be conditionally inactivated, which has significant implications for gene therapy studies, and can be used to study the chromatin requirements for centromere formation and heritability.

More recently, several groups have used these strategies to address the chromatin requirements for human kinetochores. Gascoigne et al. (49) directed two of the DNA binding components of the mammalian kinetochore to an ectopic site and were able to assemble a microtubule attachment site without CENP-A. An exhaustive study of ectopic targeting by the Fukagawa group revealed two modes of assembly (74). One (using CENP-T or CENP -C N terminus) recruits kinetochore proteins and assembles a functional kinetochore lacking CENP-A. A second strategy that recruits the full complement of kinetochore proteins, including CENP-A, utilized full-length HJURP, CENP-C, CENP-I, or the CENP-C C terminus. Thus, the recruitment of CENP-A is separable from the kinetochore function.

The Fukagawa study (74) is a critical advance in understanding centromeres, as it provides the first functional distinction of centromere versus kinetochore. A microtubule attachment site can be built by tethering DNA to a microtubule component within the kinetochore or via recruitment of a DNA-binding protein that recruits microtubule-binding components. The conceptual advance is that we can formally separate CENP-A chromatin, and potentially alphoid repeats, from kinetochore protein recruitment. The ability to assemble a functional kinetochore without underlying centromeric sequences or chromatin suggests, in turn, that the centromere is not just a landing pad or attachment site for the kinetochore and that the centromere has evolved with unique physical properties distinct from microtubule attachment that contribute to optimal kinetochore function.

The role of the centromere-unique histone H3 variant Cse4 (yeast) or CENP-A (mammals) remains enigmatic (30, 40) because these alternative mechanisms for tethering DNA to a microtubule-binding component of the kinetochore suffice for chromosome segregation. Current understanding of the organization of Cse4 itself within chromatin is equivocal (16). In organisms from yeast to mammal, more Cse4 resides in the centromere than lies at the plus-end of the kinetochore microtubule (62, 92). In budding yeast, several molecules of Cse4 reside in the vicinity of the kinetochore but are not bound in a sequence-specific manner (62). Binding of these noncentromere molecules depends upon Pat1, unlike Cse4 binding at CEN DNA (62, 101), indicating that heterogeneity in nucleosome structure is to be expected. Unlike in histone H3 and canonical nucleosomes, attempts to ascribe a single structure to Cse4-containing nucleosomes have been difficult, with both hemisome and nucleosomal states being identified. A parsimonious view is that the Cse4 nucleosome is dynamic (16). Because molecules closest to the plus-end of the microtubule experience more force than canonical nucleosomes, there may be value for the nucleosome to be adept at switching between hemisome and nucleosomal states. Like the unfolding of transcriptionally active nucleosomes (124), the centromere nucleosome may be adapted to rapid changes in pulling and pushing microtubule-based forces.

Features of the centromere and pericentric heterochromatin

Enriched in topology adjusters
  • Cohesin, condensin, topoisomerase II
  • Centromere-specific histone H3 variant Cse4

Epigenetic marker for identity: differential requirements for de novo versus template-directed assembly

Elastic properties that distribute force and resist MT-based extensional forces

Twofold enriched in tRNA genes and LTR sequence DNA

Figure 5

Emergent properties of the centromere. (*Left*) The yeast segregation apparatus, shown in the vertical position, serves as a model for multiple kinetochore-microtubule attachment-site centromeres found in other organisms. The kinetochore microtubules (*green*) can be seen at the top and bottom, emanating from the spindle pole body (*red*). The pericentromere DNA is organized as a network of loops (*yellow*) radially arranged around the spindle axis. Cohesin (*blue*) and condensin (*purple*) are enriched in the pericentromere and surround the central spindle. Abbreviations: LTR, long terminal repeat; MT, microtubule. Adapted from figure 6 in Reference 63.

Neocentromeres. New centromeres have been found to arise in natural populations in a variety of organisms, providing means to ask how nature builds her own centromeres. These neocentromeres, or ENCs, have been reported in a variety of *Equus* species (horse, donkey, and zebra) and in *Drosophila*, *Candida albicans* (yeast), maize, and humans (17, 47, 140). Several key mechanistic steps in de novo centromere function can be deduced. Using in situ hybridization to detect satellite DNA and immunofluorescence to detect kinetochore proteins, Piras et al. (122) have shown that new centromeres (one million years old; in *Equus*) arise in regions free of satellite DNA. Through phylogenetic reconstructions, they show that the acquisition of satellite sequences follows kinetochore formation. These satellite-deficient centromeres can function over millions of years, indicating that segregation fidelity must approximate that of endogenous centromeres. The accumulation of repeated sequences over evolutionary time is indicative of their selective advantage. Lessons from the polymer world indicate that satellite sequences, through promotion of entanglements or recruitment of cohesin and condensin, confer elasticity to the pericentromere (**Figure 5**). Through greater elasticity between sister kinetochores, a higher degree of accuracy in discriminating stochastic microtubule dynamics from bona fide bi-orientation can be achieved.

Neocentromere formation (also known as centromere repositioning) has been studied in systems in which the endogenous centromere can be deleted and replaced with a genetically selectable marker. In *C. albicans* (yeast) and *Gallus gallus* (chicken) DT40 cells, new centromeres were found in several positions, most frequently close to the original centromere (141, 154). It is thought that low levels of CENP-A in regions flanking the centromere may seed these new centromeres. The initiating event in these repositioned centromeres remains unknown. In maize, fluid structures dictated by retrotransposons and local repositioning of CenH3 have been reported as well (47, 155).

Centromere Switching: Epigenetic States

The fluidity of centromeres can be readily appreciated from the perspective of a mechanism required to inactivate a centromere in rare chromosome fusion events. As first described by McClintock (96) in maize in the late 1930s, dicentric chromosomes are genetically unstable and undergo a breakage-fusion-bridge reaction. More recently, using directed methods to create dicentric chromosomes (expression of a dominant negative mutant of telomere protein TRF-2), Stimpson et al. (144) demonstrated centromere inactivation in de novo dicentric chromosomes. Studies in human and budding yeast reveal the critical role of CENP-A histone modification in regulating the functional state of the centromere (5, 134). The switchable nature of a centromere highlights the fact that centromere function arises from more than the sum of the parts. There is an emergent structure that is fluid and more likely to be subjected to constant molding and remolding, dependent on cellular and chromosome physiology.

Evidence for centromere remodeling can be found in both fission and budding yeast. In fission yeast, Sato et al. (136) have demonstrated that epigenetic inactivation is initiated by disassembly of kinetochore components. Mutations that compromise kinetochore protein promote centromere inactivation. This is reminiscent of a finding in budding yeast that mutations in centromere DNA binding proteins lead to the stabilization of dicentric chromosomes (33). However, studies have used a centromere that can be conditionally inactivated via transcription from an adjacent promoter to demonstrate that de novo centromere function depends upon both the Swi2/Snf2 chromatin remodeling factor Fun30 (38) and the kinetochore component Chl4 (106) for its activation. In this way, de novo versus template-directed propagation of a centromere state was distinguished. Thus, even in budding yeast, the centromere can exist in two states, and state switching is likely to involve chromatin remodelers. It should be noted that both Fun30 and a second Swi2/Snf2 homolog, Rdh54/Tid1, have been localized to kinetochores (91, 123).

In species other than budding yeasts, the centromere is a dynamic genomic element, and this lack of sequence-specific requirements makes it difficult to identify the chromatin features that predispose centromere formation. The presence of CENP-A (13, 121, 145), reduced histone H2A.Z (113), and a concentration of tRNA genes (77), retrotransposons, or other mobile genetic elements (126) have all been suggested as contributing to centromere-permissive regions. The absence of satellite sequences in neocentromeres is indicative of their nonessential centromere function. The conclusion from a phylogenetically broad spectrum of plants and animals indicates that there are neocentromere-favorable and -unfavorable sites (154, 161). Unlike the kineto-chore, with its readily identifiable essential proteins, it is difficult (outside of budding yeast) to identify genetic sequences or modifications that definitively mark or exclude centromere formation.

The capability for neocentromere formation must be balanced against the formation of dicentric chromosomes and potential chromosome instability (71, 97). Nonetheless, the cell retains the capacity to build centromeres de novo in rare instances of centromere loss. It is not surprising that de novo centromere formation is nonrandom, but how sites are chosen is less clear. In addition, centromere complexity evolves with time. The late acquisition of satellite sequences into neocentromeres is indicative of advantageous functions that evolve on timescales slower than those for the microtubule attachment mechanism. Enumerating these functions and considering alternative segregation mechanisms might provide insights into the centromere that transcend kinetochore function.

IMPLICATIONS OF NONCANONICAL SEGREGATION MECHANISMS FOR KINETOCHORE-INDEPENDENT CENTROMERE FUNCTION

Velcro Segregation Mechanisms

The two-micron plasmid in budding yeast is a high-copy extrachromosomal element. The plasmid is partitioned with high fidelity in part because of a highly evolved stabilization element (STB) and the DNA binding proteins Rep1 and Rep2. Recent studies have shown that the two-micron plasmid physically associates with the chromosome and recruits cohesin (51), leading to a proposal for a hitchhiking mechanism of segregation (18). The plasmid recruits substoichiometric quantities of the centromere-specific histone (CENP-A), indicating that the potential site of hitchhiking might be the centromere. Centromere chromatin may provide a similar mechanism of association between the individual centromeres in budding yeast, thereby contributing to the fidelity of segregation.

Structural Maintenance of Chromosome-Based Bacterial Segregation

The timing of condensin-mediated compaction of DNA in bacteria is closely linked to the chromosome segregation cycle and cell division (148, 166). In eukaryotes, condensin and cohesin loading is coupled to DNA replication (27, 167), and chromosomes are condensed well before anaphase chromosome segregation. Linking chromosome compaction with segregation in bacteria may reflect a strategy to harness SMC-mediated compaction forces for segregation. Rudner and colleagues (148, 166) argue that the key feature of SMC-mediated compaction is the orderly folding of DNA, thereby facilitating protein-DNA transactions, and that this has a central role in driving segregation. Chromosome individualization (visibly separated sister chromatids) in multicellular organisms likewise reflects the ability of condensation mechanisms to segregate strands. The enrichment of cohesin and condensin in centromeric heterochromatin (27, 86, 99) is indicative of additional functions beyond compaction and cohesion. The analysis of polymer networks makes testable predications for how these proteins contribute to centromere elasticity.

Entropic Recoil in Confinement

Jun and colleagues (82, 83) have shown that the penalty due to polymer repulsion is sufficient to drive chains apart in a confined space, and may drive chromosome segregation in bacterial cells. As soon as the genome becomes fragmented, such as in eukaryotic chromosomes, thermodynamics is completely insufficient, as this mechanism cannot provide a single order or spatial direction. Thus, the polymer repulsion model cannot be the sole mechanism for eukaryotic chromosome segregation but serves as a guide to our intuition. Kleckner and colleagues (81) posit that origins of replication in bacteria are pushed toward the cell poles as a result of intranucleoid confinement. The replicated origins undergo condensation and resolution from each other but remain juxtaposed at specific origin-proximal sites (called snaps). The accumulation of DNA in the confined cellular space generates internal pushing forces. When these forces exceed the strength of the snaps, cohesion is lost, resulting in the abrupt and rapid extrusion of the condensed origin regions toward opposite poles. Because replication probably initiates at the nucleoid periphery, the newly replicated DNA is naturally compartmentalized from the unreplicated chromosome. This helps to prevent entanglements and provides an unimpeded path for segregation of the origins and may explain why yeast centromeres are replicated early. Polymer repulsion may also underlie a mechanism to extrude centromeres and thus prevent them from being buried in the mass of centromeric heterochromatin.

EMERGENT FUNCTION FROM ENRICHMENT AND MOLECULAR ORGANIZATION OF CENTROMERE COMPONENTS

If we build our understanding of the centromere from the biochemical and physical properties of the components, we emerge with a deeper understanding that transcends the kinetochore. The enrichment of topology adjusters, including cohesin, condensin, and topoisomerase II, endows centromeres with elasticity. Entanglement reflects a topological constraint resulting from the fact that DNA strands are unable to cross one another (157) in the absence of topoisomerase II. As a result of entanglements, the motion of one chain influences the other, resulting in an increased elastic modulus of the network (87, 157). In the case of λ DNA at DNA concentrations greater than 0.5 mg/mL, the elastic modulus exceeds the viscous modulus of the network. The network thus exhibits elastic behavior. In the centromere, we can estimate the concentration of DNA on the basis of the region of DNA enriched in SMC proteins (50 kb/chromosome) and the volume of the cylinder between bi-oriented centromeres in mitosis ($\sim$0.15 μm^3). The DNA concentration is approximately 9 mg/mL (about three times greater than bulk DNA in the nucleus). Thus, it is very likely that strands are topologically constrained within the cohesin barrel.

As discussed above, DNA strand sliding within the cohesin ring confers a slip-link property to the network. Tension generated at one strand is distributed among remaining strands because each strand "feels" the other through the equalization of tension via nanopulleys. Gels with slip-links (polyrotaxane gels) have the remarkable property of maintaining their three-dimensional shape over orders of magnitude of size scales (116). The consequence of such a network is that the stochastic growth and shortening of kinetochore microtubules at individual attachment sites can be averaged across the chromatin spring.

Condensin is responsible for chromosome condensation in metaphase as well as recoil of chromosome arms throughout anaphase (127). Condensin is highly enriched in regions of very high transcription, such as the nucleolus, and also at centromeres (4, 27, 28, 86). It has been proposed that condensin in the nucleolus serves to protect the genome from massive rearrangement from recombination between rDNA repeats and potential shuffling thereof. In the centromere, condensin contributes to heterochromatin compaction. However, considering the central role for condensin in prokaryotic chromosome segregation, the enrichment of condensin in eukaryotic chromosomes may be more profound. Compaction of strands naturally (i.e., thermodynamically) promotes strand segregation. Thus, the centromere provides the site of initiation for chromosome strand separation. From this perspective, the spindle apparatus functions as a guidance mechanism to ensure that sisters separate. The driving force for segregation may lie within the centromere proper.

The physical basis for the chromatin spring will ultimately require an understanding of DNA fluctuations, compaction, histone modification and mobility, DNA cross-linking complexes (inter- as well as intramolecular interactions), and topology adjusters. Through DNA strand cross-linking and entanglements, a robust elastic network emerges that maintains mechanical integrity across hundreds to thousands of kilobase pairs of centromere DNA. Polymer physics provide an important framework for integrating protein function with DNA dynamics to comprehend the accuracy of chromosome segregation.

DISCLOSURE STATEMENT

The author is not aware of any affiliations, memberships, funding, or financial holdings that might be perceived as affecting the objectivity of this review.

LITERATURE CITED

1. Anderson DE, Losada A, Erickson HP, Hirano T. 2002. Condensin and cohesin display different arm conformations with characteristic hinge angles. *J. Cell Biol.* 156:419–24

2. Anderson M, Haase J, Yeh E, Bloom K. 2009. Function and assembly of DNA looping, clustering, and microtubule attachment complexes within a eukaryotic kinetochore. *Mol. Biol. Cell* 20:4131–39

3. Asakura S, Oosawa F. 1954. On interaction between two bodies immersed in a solution of macromolecules. *J. Chem. Phys.* 22:1255–56

4. Bachellier-Bassi S, Gadal O, Bourout G, Nehrbass U. 2008. Cell cycle–dependent kinetochore localization of condensin complex in *Saccharomyces cerevisiae*. *J. Struct. Biol.* 162:248–59

5. Bailey AO, Panchenko T, Sathyan KM, Petkowski JJ, Pai PJ, et al. 2013. Posttranslational modification of CENP-A influences the conformation of centromeric chromatin. *Proc. Natl. Acad. Sci. USA* 110:11827–32

6. Baldwin M, Warsi T, Bachant J. 2009. Analyzing Top2 distribution on yeast chromosomes by chromatin immunoprecipitation. *Methods Mol. Biol.* 582:119–30

7. Bardin AJ, Amon A. 2001. Men and sin: What's the difference? *Nat. Rev. Mol. Cell Biol.* 2:815–26

8. Bergmann JH, Jakubsche JN, Martins NM, Kagansky A, Nakano M, et al. 2012. Epigenetic engineering: histone H3K9 acetylation is compatible with kinetochore structure and function. *J. Cell Sci.* 125:411–21

9. Bergmann JH, Rodriguez MG, Martins NM, Kimura H, Kelly DA, et al. 2011. Epigenetic engineering shows H3K4me2 is required for HJURP targeting and CENP-A assembly on a synthetic human kinetochore. *EMBO J.* 30:328–40

10. Biggins S. 2013. The composition, functions, and regulation of the budding yeast kinetochore. *Genetics* 194:817–46

11. Bloom K, Yeh E. 2010. Tension management in the kinetochore. *Curr. Biol.* 20:R1040–48

12. Bloom KS. 2008. Beyond the code: the mechanical properties of DNA as they relate to mitosis. *Chromosoma* 117:103–10

13. Blower MD, Sullivan BA, Karpen GH. 2002. Conserved organization of centromeric chromatin in flies and humans. *Dev. Cell* 2:319–30

14. Brenner S, Pepper D, Berns MW, Tan E, Brinkley BR. 1981. Kinetochore structure, duplication, and distribution in mammalian cells: analysis by human autoantibodies from scleroderma patients. *J. Cell Biol.* 91:95–102

15. Brinkley BR, Stubblefield E. 1966. The fine structure of the kinetochore of a mammalian cell in vitro. *Chromosoma* 19:28–43

16. Bui M, Walkiewicz MP, Dimitriadis EK, Dalal Y. 2013. The CENP-A nucleosome: a battle between Dr Jekyll and Mr Hyde. *Nucleus* 4:37–42

17. Carbone L, Nergadze SG, Magnani E, Misceo D, Francesca Cardone M, et al. 2006. Evolutionary movement of centromeres in horse, donkey, and zebra. *Genomics* 87:777–82

18. Chan KM, Liu YT, Ma CH, Jayaram M, Sau S. 2013. The 2 micron plasmid of *Saccharomyces cerevisiae*: a miniaturized selfish genome with optimized functional competence. *Plasmid* 70:2–17

19. Cheerambathur DK, Desai A. 2014. Linked in: formation and regulation of microtubule attachments during chromosome segregation. *Curr. Opin. Cell Biol.* 26:113–22

20. Cherry LM, Faulkner AJ, Grossberg LA, Balczon R. 1989. Kinetochore size variation in mammalian chromosomes: an image analysis study with evolutionary implications. *J. Cell Sci.* 92(Pt. 2):281–89

21. Chikashige Y, Ding DQ, Funabiki H, Haraguchi T, Mashiko S, et al. 1994. Telomere-led premeiotic chromosome movement in fission yeast. *Science* 264:270–73

22. Chikashige Y, Tsutsumi C, Yamane M, Okamasa K, Haraguchi T, Hiraoka Y. 2006. Meiotic proteins bqt1 and bqt2 tether telomeres to form the bouquet arrangement of chromosomes. *Cell* 125:59–69

23. Clarke L, Baum MP. 1990. Functional analysis of a centromere from fission yeast: a role for centromere-specific repeated DNA sequences. *Mol. Cell. Biol.* 10:1863–72

24. Clarke L, Carbon J. 1980. Isolation of a yeast centromere and construction of functional small circular chromosomes. *Nature* 287:504–9

25. Cuylen S, Metz J, Haering CH. 2011. Condensin structures chromosomal DNA through topological links. *Nat. Struct. Mol. Biol.* 18:894–901

26. Cuylen S, Metz J, Hruby A, Haering CH. 2013. Entrapment of chromosomes by condensin rings prevents their breakage during cytokinesis. *Dev. Cell.* 27:469–78

27. D'Ambrosio C, Schmidt CK, Katou Y, Kelly G, Itoh T, et al. 2008. Identification of *cis*-acting sites for condensin loading onto budding yeast chromosomes. *Genes Dev.* 22:2215–27

28. D'Amours D, Stegmeier F, Amon A. 2004. Cdc14 and condensin control the dissolution of cohesin-independent chromosome linkages at repeated DNA. *Cell* 117:455–69

29. de Gennes PG. 1979. *Scaling Concepts in Polymer Physics*. Ithaca, NY: Cornell Univ. Press

30. De Rop V, Padeganeh A, Maddox PS. 2012. CENP-A: the key player behind centromere identity, propagation, and kinetochore assembly. *Chromosoma* 121:527–38

31. Deng Y, Sun M, Shaevitz JW. 2011. Direct measurement of cell wall stress stiffening and turgor pressure in live bacterial cells. *Phys. Rev. Lett.* 107:158101

32. Ding R, McDonald KL, McIntosh JR. 1993. Three-dimensional reconstruction and analysis of mitotic spindles from the yeast, *Schizosaccharomyces pombe*. *J. Cell Biol.* 120:141–51

33. Doheny KF, Sorger PK, Hyman AA, Tugendreich S, Spencer F, Hieter P. 1993. Identification of essential components of the *S. cerevisiae* kinetochore. *Cell* 73:761–74

34. Doi M, Edwards SF. 1986. *The Theory of Polymer Dynamics*. Oxford, UK: Oxford Univ. Press

35. Dorsett D. 2011. Cohesin: genomic insights into controlling gene transcription and development. *Curr. Opin. Genet. Dev.* 21:199–206

36. Dorsett D, Merkenschlager M. 2013. Cohesin at active genes: a unifying theme for cohesin and gene expression from model organisms to humans. *Curr. Opin. Cell Biol.* 25:327–33

37. Duan Z, Andronescu M, Schutz K, McIlwain S, Kim YJ, et al. 2010. A three-dimensional model of the yeast genome. *Nature* 465:363–67

38. Durand-Dubief M, Will WR, Petrini E, Theodorou D, Harris RR, et al. 2012. SWI/SNF-like chromatin remodeling factor Fun30 supports point centromere function in *S. cerevisiae*. *PLOS Genet.* 8:e1002974

39. Eckert CA, Gravdahl DJ, Megee PC. 2007. The enhancement of pericentromeric cohesin association by conserved kinetochore components promotes high-fidelity chromosome segregation and is sensitive to microtubule-based tension. *Genes Dev.* 21:278–91

40. Falk SJ, Black BE. 2013. Centromeric chromatin and the pathway that drives its propagation. *Biochim. Biophys. Acta* 1819:313–21

41. Feynman RP. 1960. There's plenty of room at the bottom. *Caltech Eng. Sci.* 23:22–36

42. Fisher JK, Bourniquel A, Witz G, Weiner B, Prentiss M, Kleckner N. 2013. Four-dimensional imaging of *E. coli* nucleoid organization and dynamics in living cells. *Cell* 153:882–95

43. Fitzgerald-Hayes M, Clarke L, Carbon J. 1982. Nucleotide sequence comparisons and functional analysis of yeast centromere DNAs. *Cell* 29:235–44

44. Flemming W. 1882. *Zellsubstanz, Kern und Zelltheilung*. Leipzig, Ger.: Vogel. 424 pp.

45. Foltz DR, Jansen LE, Black BE, Bailey AO, Yates JR 3rd, Cleveland DW. 2006. The human CENP-A centromeric nucleosome-associated complex. *Nat. Cell Biol.* 8:458–69

46. Fritzler MJ, Kinsella TD. 1980. The CREST syndrome: a distinct serologic entity with anticentromere antibodies. *Am. J. Med.* 69:520–26

47. Fu S, Lv Z, Gao Z, Wu H, Pang J, et al. 2013. De novo centromere formation on a chromosome fragment in maize. *Proc. Natl. Acad. Sci. USA* 110:6033–36

48. Gartenberg M. 2009. Heterochromatin and the cohesion of sister chromatids. *Chromosome Res.* 17:229–38

49. Gascoigne KE, Takeuchi K, Suzuki A, Hori T, Fukagawa T, Cheeseman IM. 2011. Induced ectopic kinetochore assembly bypasses the requirement for CENP-A nucleosomes. *Cell* 145:410–22

50. Gent JI, Dawe RK. 2012. RNA as a structural and regulatory component of the centromere. *Annu. Rev. Genet.* 46:443–53

51. Ghosh SK, Huang CC, Hajra S, Jayaram M. 2010. Yeast cohesin complex embraces 2 micron plasmid sisters in a tri-linked catenane complex. *Nucleic Acids Res.* 38:570–84

52. Glynn EF, Megee PC, Yu HG, Mistrot C, Unal E, et al. 2004. Genome-wide mapping of the cohesin complex in the yeast *Saccharomyces cerevisiae*. *PLOS Biol.* 2:E259

53. Glynn M, Kaczmarczyk A, Prendergast L, Quinn N, Sullivan KF. 2010. Centromeres: assembling and propagating epigenetic function. *Subcell. Biochem.* 50:223–49

54. Goh PY, Kilmartin JV. 1993. *NDC10*: a gene involved in chromosome segregation in *Saccharomyces cerevisiae*. *J. Cell Biol.* 121:503–12

55. Gopalakrishnan S, Sullivan BA, Trazzi S, Della Valle G, Robertson KD. 2009. DNMT3B interacts with constitutive centromere protein CENP-C to modulate DNA methylation and the histone code at centromeric regions. *Hum. Mol. Genet.* 18:3178–93

56. Goshima G, Yanagida M. 2000. Establishing biorientation occurs with precocious separation of the sister kinetochores, but not the arms, in the early spindle of budding yeast. *Cell* 100:619–33

57. Granick S, Rubinstein M. 2004. Polymers: a multitude of macromolecules. *Nat. Mater.* 3:586–87

58. Greulich KO, Wachtel E, Ausio J, Seger D, Eisenberg H. 1987. Transition of chromatin from the "10 nm" lower order structure, to the "30 nm" higher order structure as followed by small angle X-ray scattering. *J. Mol. Biol.* 193:709–21

59. Grewal SI. 2010. RNAi-dependent formation of heterochromatin and its diverse functions. *Curr. Opin. Genet. Dev.* 20:134–41

60. Grewal SI, Jia S. 2007. Heterochromatin revisited. *Nat. Rev. Genet.* 8:35–46

61. Grosberg AY, Khokhlov AR. 1997. *Giant Molecules Here, There, and Everywhere.* San Diego, CA: Acad. Press. 244 pp.

62. Haase J, Mishra PK, Stephens A, Haggerty R, Quammen C, et al. 2013. A 3D map of the yeast kinetochore reveals the presence of core and accessory centromere-specific histone. *Curr. Biol.* 23:1939–44

63. Haase J, Stephens A, Verdaasdonk J, Yeh E, Bloom K. 2012. Bub1 kinase and Sgo1 modulate pericentric chromatin in response to altered microtubule dynamics. *Curr. Biol.* 22:471–81

64. Haering CH, Lowe J, Hochwagen A, Nasmyth K. 2002. Molecular architecture of SMC proteins and the yeast cohesin complex. *Mol. Cell* 9:773–88

65. Haeusler RA, Pratt-Hyatt M, Good PD, Gipson TA, Engelke DR. 2008. Clustering of yeast tRNA genes is mediated by specific association of condensin with tRNA gene transcription complexes. *Genes Dev.* 22:2204–14

66. Hall LE, Mitchell SE, O'Neill RJ. 2012. Pericentric and centromeric transcription: a perfect balance required. *Chromosome Res.* 20:535–46

67. Harrington JJ, Van Bokkelen G, Mays RW, Gustashaw K, Willard HF. 1997. Formation of de novo centromeres and construction of first-generation human artificial microchromosomes. *Nat. Genet.* 15:345–55

68. Hayden KE, Strome ED, Merrett SL, Lee HR, Rudd MK, Willard HF. 2013. Sequences associated with centromere competency in the human genome. *Mol. Cell. Biol.* 33:763–72

69. He X, Asthana S, Sorger PK. 2000. Transient sister chromatid separation and elastic deformation of chromosomes during mitosis in budding yeast. *Cell* 101:763–75

70. Hill A, Bloom K. 1987. Genetic manipulation of centromere function. *Mol. Cell. Biol.* 7:2397–405

71. Hill A, Bloom K. 1989. Acquisition and processing of a conditional dicentric chromosome in *Saccharomyces cerevisiae*. *Mol. Cell. Biol.* 9:1368–70

72. Hirano T. 2006. At the heart of the chromosome: SMC proteins in action. *Nat. Rev. Mol. Cell Biol.* 7:311–22

73. Hopper AK, Pai DA, Engelke DR. 2010. Cellular dynamics of tRNAs and their genes. *FEBS Lett.* 584:310–17

74. Hori T, Shang WH, Takeuchi K, Fukagawa T. 2013. The CCAN recruits CENP-A to the centromere and forms the structural core for kinetochore assembly. *J. Cell Biol.* 200:45–60

75. Hsieh CL, Lin CL, Liu H, Chang YJ, Shih CJ, et al. 2011. WDHD1 modulates the post-transcriptional step of the centromeric silencing pathway. *Nucleic Acids Res.* 39:4048–62

76. Hu B, Itoh T, Mishra A, Katoh Y, Chan KL, et al. 2011. ATP hydrolysis is required for relocating cohesin from sites occupied by its Scc2/4 loading complex. *Curr. Biol.* 21:12–24

77. Iwasaki O, Tanaka A, Tanizawa H, Grewal SI, Noma K. 2010. Centromeric localization of dispersed Pol III genes in fission yeast. *Mol. Biol. Cell* 21:254–65

78. Johzuka K, Horiuchi T. 2007. RNA polymerase I transcription obstructs condensin association with 35S rRNA coding regions and can cause contraction of long repeat in *Saccharomyces cerevisiae*. *Genes Cells* 12:759–71

79. Johzuka K, Horiuchi T. 2009. The *cis* element and factors required for condensin recruitment to chromosomes. *Mol. Cell* 34:26–35

80. Jokelainen PT. 1967. The ultrastructure and spatial organization of the metaphase kinetochore in mitotic rat cells. *J. Ultrastruct. Res.* 19:19–44

81. Joshi MC, Bourniquel A, Fisher J, Ho BT, Magnan D, et al. 2011. *Escherichia coli* sister chromosome separation includes an abrupt global transition with concomitant release of late-splitting intersister snaps. *Proc. Natl. Acad. Sci. USA* 108:2765–70

82. Jun S, Mulder B. 2006. Entropy-driven spatial organization of highly confined polymers: lessons for the bacterial chromosome. *Proc. Natl. Acad. Sci. USA* 103:12388–93

83. Jun S, Wright A. 2010. Entropy as the driver of chromosome segregation. *Nat. Rev. Microbiol.* 8:600–7

84. Kendall A, Hull MW, Bertrand E, Good PD, Singer RH, Engelke DR. 2000. A CBF5 mutation that disrupts nucleolar localization of early tRNA biosynthesis in yeast also suppresses tRNA gene-mediated transcriptional silencing. *Proc. Natl. Acad. Sci. USA* 97:13108–13

85. Kiermaier E, Woehrer S, Peng Y, Mechtler K, Westermann S. 2009. A Dam1-based artificial kinetochore is sufficient to promote chromosome segregation in budding yeast. *Nat. Cell Biol.* 11:1109–15

86. Kim JH, Zhang T, Wong NC, Davidson N, Maksimovic J, et al. 2013. Condensin I associates with structural and gene regulatory regions in vertebrate chromosomes. *Nat Commun.* 4:2537

87. Kundukad B, van der Maarel JR. 2010. Control of the flow properties of DNA by topoisomerase II and its targeting inhibitor. *Biophys. J.* 99:1906–15

88. Lacefield S, Lau DT, Murray AW. 2009. Recruiting a microtubule-binding complex to DNA directs chromosome segregation in budding yeast. *Nat. Cell Biol.* 11:1116–20

89. Lechner J, Carbon J. 1991. A 240 kd multisubunit protein complex, CBF3, is a major component of the budding yeast centromere. *Cell* 64:717–25

90. Lejeune E, Bayne EH, Allshire RC. 2010. On the connection between RNAi and heterochromatin at centromeres. *Cold Spring Harb. Symp. Quant. Biol.* 75:275–83

91. Lisby M, Barlow JH, Burgess RC, Rothstein R. 2004. Choreography of the DNA damage response: spatiotemporal relationships among checkpoint and repair proteins. *Cell* 118:699–713

92. Liu ST, Rattner JB, Jablonski SA, Yen TJ. 2006. Mapping the assembly pathways that specify formation of the trilaminar kinetochore plates in human cells. *J. Cell Biol.* 175:41–53

93. Losada A, Hirano M, Hirano T. 2002. Cohesin release is required for sister chromatid resolution, but not for condensin-mediated compaction, at the onset of mitosis. *Genes Dev.* 16:3004–16

94. Maddox PS, Portier N, Desai A, Oegema K. 2006. Molecular analysis of mitotic chromosome condensation using a quantitative time-resolved fluorescence microscopy assay. *Proc. Natl. Acad. Sci. USA* 103:15097–102

95. Masumoto H, Ikeno M, Nakano M, Okazaki T, Grimes B, et al. 1998. Assay of centromere function using a human artificial chromosome. *Chromosoma* 107:406–16

96. McClintock B. 1938. The production of homozygous deficient tissues with mutant characteristics by means of the aberrant mitotic behavior of ring-shaped chromosomes. *Genetics* 23:315–76

97. McClintock B. 1953. Induction of instability at selected loci in maize. *Genetics* 38:579–99

98. McIntosh JR, Molodtsov MI, Ataullakhanov FI. 2012. Biophysics of mitosis. *Q. Rev. Biophys.* 45:147–207

99. Megee PC, Mistrot C, Guacci V, Koshland D. 1999. The centromeric sister chromatid cohesion site directs Mcd1p binding to adjacent sequences. *Mol. Cell* 4:445–50

100. Minc N, Boudaoud A, Chang F. 2009. Mechanical forces of fission yeast growth. *Curr. Biol.* 19:1096–101

101. Mishra PK, Ottmann AR, Basrai M. 2013. Structural integrity of centromeric chromatin and faithful chromosome segregation requires Pat1. *Genetics* 195(2):369–79

102. Molineux IJ, Panja D. 2013. Popping the cork: mechanisms of phage genome ejection. *Nat. Rev. Microbiol.* 11:194–204

103. Moroi Y, Hartman AL, Nakane PK, Tan EM. 1981. Distribution of kinetochore (centromere) antigen in mammalian cell nuclei. *J. Cell Biol.* 90:254–59

104. Moroi Y, Peebles C, Fritzler MJ, Steigerwald J, Tan EM. 1980. Autoantibody to centromere (kinetochore) in scleroderma sera. *Proc. Natl. Acad. Sci. USA* 77:1627–31

105. Musacchio A, Ciliberto A. 2012. The spindle-assembly checkpoint and the beauty of self-destruction. *Nat. Struct. Mol. Biol.* 19:1059–61

106. Mythreye K, Bloom KS. 2003. Differential kinetochore protein requirements for establishment versus propagation of centromere activity in *Saccharomyces cerevisiae*. *J. Cell Biol.* 160:833–43

107. Nakano M, Cardinale S, Noskov VN, Gassmann R, Vagnarelli P, et al. 2008. Inactivation of a human kinetochore by specific targeting of chromatin modifiers. *Dev. Cell* 14:507–22

108. Nakaseko Y, Adachi Y, Funahashi S, Niwa O, Yanagida M. 1986. Chromosome walking shows a highly homologous repetitive sequence present in all the centromere regions of fission yeast. *EMBO J.* 5:1011–21

109. Nasmyth K, Haering CH. 2009. Cohesin: its roles and mechanisms. *Annu. Rev. Genet.* 43:525–58

110. Nicklas RB. 1988. The forces that move chromosomes in mitosis. *Annu. Rev. Biophys. Biophys. Chem.* 17:431–49

111. Nogales E, Ramey VH. 2009. Structure-function insights into the yeast Dam1 kinetochore complex. *J. Cell Sci.* 122:3831–36

112. Ocampo-Hafalla MT, Uhlmann F. 2011. Cohesin loading and sliding. *J. Cell Sci.* 124:685–91

113. Ogiyama Y, Ohno Y, Kubota Y, Ishii K. 2013. Epigenetically induced paucity of histone H2A.Z stabilizes fission-yeast ectopic centromeres. *Nat. Struct. Mol. Biol.* 20:1397–406

114. Ohkuni K, Kitagawa K. 2011. Endogenous transcription at the centromere facilitates centromere activity in budding yeast. *Curr. Biol.* 21:1695–703

115. Ohkuni K, Kitagawa K. 2012. Role of transcription at centromeres in budding yeast. *Transcription* 3:193–97

116. Okumura Y, Ito K. 2001. The polyrotaxane gel: a topological gel by figure-of-eight cross-links. *Adv. Mater.* 13:485–87

117. Onn I, Heidinger-Pauli JM, Guacci V, Unal E, Koshland DE. 2008. Sister chromatid cohesion: a simple concept with a complex reality. *Annu. Rev. Cell Dev. Biol.* 24:105–29

118. Pearson CG, Maddox PS, Salmon ED, Bloom K. 2001. Budding yeast chromosome structure and dynamics during mitosis. *J. Cell Biol.* 152:1255–66

119. Pelletier J, Halvorsen K, Ha BY, Paparcone R, Sandler SJ, et al. 2012. Physical manipulation of the *Escherichia coli* chromosome reveals its soft nature. *Proc. Natl. Acad. Sci. USA* 109:E2649–56

120. Phillips-Cremins JE, Corces VG. 2013. Chromatin insulators: linking genome organization to cellular function. *Mol. Cell* 50:461–74

121. Pidoux AL, Allshire RC. 2005. The role of heterochromatin in centromere function. *Philos. Trans. R. Soc. Lond. B* 360:569–79

122. Piras FM, Nergadze SG, Magnani E, Bertoni L, Attolini C, et al. 2010. Uncoupling of satellite DNA and centromeric function in the genus *Equus*. *PLOS Genet.* 6:e1000845

123. Prasad TK, Robertson RB, Visnapuu ML, Chi P, Sung P, Greene EC. 2007. A DNA-translocating Snf2 molecular motor: *Saccharomyces cerevisiae* Rdh54 displays processive translocation and extrudes DNA loops. *J. Mol. Biol.* 369:940–53

124. Prior CP, Cantor CR, Johnson EM, Littau VC, Allfrey VG. 1983. Reversible changes in nucleosome structure and histone H3 accessibility in transcriptionally active and inactive states of rDNA chromatin. *Cell* 34:1033–42

125. Przewloka MR, Glover DM. 2009. The kinetochore and the centromere: a working long distance relationship. *Annu. Rev. Genet.* 43:439–65

126. Qi LL, Wu JJ, Friebe B, Qian C, Gu YQ, et al. 2013. Sequence organization and evolutionary dynamics of *Brachypodium*-specific centromere retrotransposons. *Chromosome Res.* 21:507–21

127. Renshaw MJ, Ward JJ, Kanemaki M, Natsume K, Nedelec FJ, Tanaka TU. 2010. Condensins promote chromosome recoiling during early anaphase to complete sister chromatid separation. *Dev. Cell* 19:232–44

128. Ribeiro SA, Gatlin JC, Dong Y, Joglekar A, Cameron L, et al. 2009. Condensin regulates the stiffness of vertebrate centromeres. *Mol. Biol. Cell* 20:2371–80

129. Rock JM, Amon A. 2009. The FEAR network. *Curr. Biol.* 19:R1063–68

130. Rouse PE. 1953. A theory of the linear viscoelastic properties of dilute solutions of coiling polymer. *J. Chem. Phys.* 21:1272–80

131. Rubinstein M, Colby RH. 2003. *Polymer Physics*. Oxford, UK: Oxford Univ. Press. 441 pp.

132. Salmon ED. 1975. Pressure-induced depolymerization of spindle microtubules. II. Thermodynamics of in vivo spindle assembly. *J. Cell Biol.* 66:114–27

133. Salmon ED, Goode D, Maugel TK, Bonar DB. 1976. Pressure-induced depolymerization of spindle microtubules. III. Differential stability in HeLa cells. *J. Cell Biol.* 69:443–54

134. Samel A, Cuomo A, Bonaldi T, Ehrenhofer-Murray AE. 2012. Methylation of CenH3 arginine 37 regulates kinetochore integrity and chromosome segregation. *Proc. Natl. Acad. Sci. USA* 109:9029–34

135. Samoshkin A, Arnaoutov A, Jansen LE, Ouspenski I, Dye L, et al. 2009. Human condensin function is essential for centromeric chromatin assembly and proper sister kinetochore orientation. *PLOS ONE* 4:e6831

136. Sato H, Masuda F, Takayama Y, Takahashi K, Saitoh S. 2012. Epigenetic inactivation and subsequent heterochromatinization of a centromere stabilize dicentric chromosomes. *Curr. Biol.* 22:658–67

137. Saunders WS, Hoyt MA. 1992. Kinesin-related proteins required for structural integrity of the mitotic spindle. *Cell* 70:451–58

138. Saunders WS, Koshland D, Eshel D, Gibbons IR, Hoyt MA. 1995. *Saccharomyces cerevisiae* kinesin- and dynein-related proteins required for anaphase chromosome segregation. *J. Cell Biol.* 128:617–24

139. Schmidt JC, Arthanari H, Boeszoermenyi A, Dashkevich NM, Wilson-Kubalek EM, et al. 2012. The kinetochore-bound Ska1 complex tracks depolymerizing microtubules and binds to curved protofilaments. *Dev. Cell* 23:968–80

140. Scott KC, Sullivan BA. 2014. Neocentromeres: a place for everything and everything in its place. *Trends Genet.* 30:66–74

141. Shang WH, Hori T, Martins NM, Toyoda A, Misu S, et al. 2013. Chromosome engineering allows the efficient isolation of vertebrate neocentromeres. *Dev. Cell* 24:635–48

142. Stephens AD, Haase J, Vicci L, Taylor RM 2nd, Bloom K. 2011. Cohesin, condensin, and the intramolecular centromere loop together generate the mitotic chromatin spring. *J. Cell Biol.* 193:1167–80

143. Stephens AD, Quammen CW, Chang B, Haase J, Taylor RM 2nd, Bloom K. 2013. The spatial segregation of pericentric cohesin and condensin in the mitotic spindle. *Mol. Biol. Cell* 24:3909–19

144. Stimpson KM, Song IY, Jauch A, Holtgreve-Grez H, Hayden KE, et al. 2010. Telomere disruption results in non-random formation of de novo dicentric chromosomes involving acrocentric human chromosomes. *PLOS Genet.* 6:1–19

145. Stoler S, Keith KC, Curnick KE, Fitzgerald-Hayes M. 1995. A mutation in CSE4, an essential gene encoding a novel chromatin-associated protein in yeast, causes chromosome nondisjunction and cell cycle arrest at mitosis. *Genes Dev.* 9:573–86

146. Straight AF, Belmont AS, Robinett CC, Murray AW. 1996. GFP tagging of budding yeast chromosomes reveals that protein-protein interactions can mediate sister chromatid cohesion. *Curr. Biol.* 6:1599–608

147. Sullivan BA. 2009. *The Centromere*. New York: Springer. 509 pp.

148. Sullivan NL, Marquis KA, Rudner DZ. 2009. Recruitment of SMC by ParB-parS organizes the origin region and promotes efficient chromosome segregation. *Cell* 137:697–707

149. Sun M, Kawamura R, Marko JF. 2011. Micromechanics of human mitotic chromosomes. *Phys. Biol.* 8:015003

150. Takeuchi K, Fukagawa T. 2012. Molecular architecture of vertebrate kinetochores. *Exp. Cell Res.* 318:1367–74

151. Tanaka T, Cosma MP, Wirth K, Nasmyth K. 1999. Identification of cohesin association sites at centromeres and along chromosome arms. *Cell* 98:847–58

152. Tanaka T, Fuchs J, Loidl J, Nasmyth K. 2000. Cohesin ensures bipolar attachment of microtubules to sister centromeres and resists their precocious separation. *Nat. Cell Biol.* 2:492–99

153. Taylor EW. 1965. Brownian and saltatory movements of cytoplasmic granules and the movement of anaphase chromosomes. *Proc. Int. Congr. Rheol. Symp. Biorheol. 4th, Providence, RI*, pp. 175–91. New York: Interscience

154. Thakur J, Sanyal K. 2013. Efficient neocentromere formation is suppressed by gene conversion to maintain centromere function at native physical chromosomal loci in *Candida albicans*. *Genome Res.* 23:638–52

155. Topp CN, Okagaki RJ, Melo JR, Kynast RG, Phillips RL, Dawe RK. 2009. Identification of a maize neocentromere in an oat-maize addition line. *Cytogenet. Genome Res.* 124:228–38

156. Underhill PT, Doyle PS. 2004. On the coarse-graining of polymers into bead-spring chains. *J. Non-Newton. Fluid Mech.* 122:3–31

157. Van der Maarel JR. 2008. *Introduction to Biopolymer Physics*. Hackensack, NJ: World Sci. Publ. 246 pp.

158. Varma D, Salmon ED. 2012. The KMN protein network: chief conductors of the kinetochore orchestra. *J. Cell Sci.* 125:5927–36

159. Varma D, Wan X, Cheerambathur D, Gassmann R, Suzuki A, et al. 2013. Spindle assembly checkpoint proteins are positioned close to core microtubule attachment sites at kinetochores. *J. Cell Biol.* 202:735–46

160. Venkei Z, Przewloka MR, Ladak Y, Albadri S, Sossick A, et al. 2012. Spatiotemporal dynamics of Spc105 regulates the assembly of the *Drosophila* kinetochore. *Open Biol.* 2:110032

161. Ventura M, Weigl S, Carbone L, Cardone MF, Misceo D, et al. 2004. Recurrent sites for new centromere seeding. *Genome Res.* 14:1696–703

162. Verdaasdonk JS, Vasquez PA, Barry RM, Barry T, Goodwin S, et al. 2013. Centromere tethering confines chromosome domains. *Mol. Cell* 52:819–31

163. Verma R, Crocker JC, Lubensky TC, Yodh AG. 1998. Entropic colloidal interactions in concentrated DNA solutions. *Phys. Rev. Lett.* 81:4004–7

164. Waigh TA. 2007. *Applied Biophysics: A Molecular Approach for Physical Scientists*. West Sussex, UK: Wiley. 421 pp.

165. Waizenegger IC, Hauf S, Meinke A, Peters JM. 2000. Two distinct pathways remove mammalian cohesin from chromosome arms in prophase and from centromeres in anaphase. *Cell* 103:399–410

166. Wang X, Tang OW, Riley EP, Rudner DZ. 2014. The SMC condensin complex is required for origin segregation in *Bacillus subtilis*. *Curr. Biol.* 24:287–92

167. Watrin E, Schleiffer A, Tanaka K, Eisenhaber F, Nasmyth K, Peters JM. 2006. Human Scc4 is required for cohesin binding to chromatin, sister-chromatid cohesion, and mitotic progression. *Curr. Biol.* 16:863–74

168. Weber SA, Gerton JL, Polancic JE, DeRisi JL, Koshland D, Megee PC. 2004. The kinetochore is an enhancer of pericentric cohesin binding. *PLOS Biol.* 2:E260

169. Weber SC, Spakowitz AJ, Theriot JA. 2010. Bacterial chromosomal loci move subdiffusively through a viscoelastic cytoplasm. *Phys. Rev. Lett.* 104:238102

170. Weber SC, Spakowitz AJ, Theriot JA. 2012. Nonthermal ATP-dependent fluctuations contribute to the in vivo motion of chromosomal loci. *Proc. Natl. Acad. Sci. USA* 109:7338–43

171. Weber SC, Theriot JA, Spakowitz AJ. 2010. Subdiffusive motion of a polymer composed of subdiffusive monomers. *Phys. Rev. E* 82:011913

172. Welburn JP, Cheeseman IM. 2008. Toward a molecular structure of the eukaryotic kinetochore. *Dev. Cell* 15:645–55

173. Welburn JP, Grishchuk EL, Backer CB, Wilson-Kubalek EM, Yates JR 3rd, Cheeseman IM. 2009. The human kinetochore Ska1 complex facilitates microtubule depolymerization-coupled motility. *Dev. Cell* 16:374–85

174. Westhorpe FG, Straight AF. 2013. Functions of the centromere and kinetochore in chromosome segregation. *Curr. Opin. Cell Biol.* 25:334–40

175. Yeh E, Haase J, Paliulis LV, Joglekar A, Bond L, et al. 2008. Pericentric chromatin is organized into an intramolecular loop in mitosis. *Curr. Biol.* 18:81–90

176. Yeh E, Skibbens RV, Cheng JW, Salmon ED, Bloom K. 1995. Spindle dynamics and cell cycle regulation of dynein in the budding yeast, *Saccharomyces cerevisiae*. *J. Cell Biol.* 130:687–700

177. Zhu X, Kundukad B, van der Maarel JR. 2008. Viscoelasticity of entangled λ-phage DNA solutions. *J. Chem. Phys.* 129:185103

178. Zimm BH. 1956. Dynamics of polymer molecules in dilute solution: viscoelasticity, flow birefringence and dielectric loss. *J. Chem. Phys.* 24:269–78

Nonadditive Gene Expression in Polyploids

Mi-Jeong Yoo,[1] Xiaoxian Liu,[1] J. Chris Pires,[2]
Pamela S. Soltis,[3,4] and Douglas E. Soltis[1,3,4,]*

[1]Department of Biology, University of Florida, Gainesville, Florida 32611-8525;
email: ymj@ufl.edu, xiaoxianliu@ufl.edu, dsoltis@ufl.edu

[2]Division of Biological Sciences, University of Missouri, Columbia, Missouri 65211;
email: piresjc@missouri.edu

[3]Florida Museum of Natural History, University of Florida, Gainesville, Florida 32611-7800;
email: psoltis@flmnh.ufl.edu

[4]Genetics Institute, University of Florida, Gainesville, Florida 32610

Annu. Rev. Genet. 2014. 48:485–517

The *Annual Review of Genetics* is online at
genet.annualreviews.org

This article's doi:
10.1146/annurev-genet-120213-092159

*Corresponding author

Keywords

genome duplication, additivity, expression-level dominance, transgressive
expression, homeolog expression bias, alternative splicing

Abstract

Allopolyploidy involves hybridization and duplication of divergent parental
genomes and provides new avenues for gene expression. The expression levels of duplicated genes in polyploids can show deviation from parental additivity (the arithmetic average of the parental expression levels). Nonadditive
expression has been widely observed in diverse polyploids and comprises at
least three possible scenarios: (*a*) The total gene expression level in a polyploid is similar to that of one of its parents (expression-level dominance);
(*b*) total gene expression is lower or higher than in both parents (transgressive expression); and (*c*) the relative contribution of the parental copies
(homeologs) to the total gene expression is unequal (homeolog expression
bias). Several factors may result in expression nonadditivity in polyploids, including maternal-paternal influence, gene dosage balance, *cis-* and/or *trans-*
regulatory networks, and epigenetic regulation. As our understanding of
nonadditive gene expression in polyploids remains limited, a new generation of investigators should explore additional phenomena (i.e., alternative
splicing) and use other high-throughput "omics" technologies to measure the
impact of nonadditive expression on phenotype, proteome, and metabolome.

INTRODUCTION

Homeologs: each pair of chromosomes or genes that are derived from two different parental species in an allopolyploid species

Subfunctionalization: partitioning of ancestral functions or expression patterns of duplicated genes so that both genes are retained following gene or genome duplication

Polyploidy, or whole-genome duplication (WGD), generally results in instant speciation, increasing biodiversity and providing new genetic material for evolution (e.g., 102, 103). New tetraploids are expected to have duplicate gene copies that share expression patterns and functions. However, polyploids experience the combined challenge and potential of having two or more genomes together in the same nucleus, and for survival, a polyploid individual must balance the potential benefits of extra heterozygosity and biochemical diversity with the cost of carrying and expressing multiple genomes. Through time, duplicate gene copies (homeologs) have several possible fates. Both genes may retain their original function, one copy may be silenced, or they may diversify in function or expression (e.g., 113, 131). One copy may be lost by deletion, or the copy number may be altered through homeologous exchanges. Duplicated genes may also interact via interlocus recombination, gene conversion, or concerted evolution. Furthermore, analyses of synthetic allopolyploids, crops, and *Arabidopsis* show that polyploidy can evoke rapid responses in both genome structure and gene expression, including concerted evolution of rDNA (48, 83, 94, 185), loss and restructuring of low-copy DNA (51, 57, 85, 100, 134, 135, 161, 167), activation of genes and retrotransposons (85, 86), gene silencing (35, 36, 41, 57, 99), epigenetic changes (104, 110, 112, 114, 132, 153), chromosomal rearrangement (88, 109, 138, 140, 141, 176), and organ-specific subfunctionalization of gene expression (1–3, 5). In this review, we focus on patterns of gene expression associated with polyploidy in plants, the factors that govern shifts in gene expression in polyploids, and the implications of altered gene expression for genome and organismal evolution.

Types of Polyploids

Two general types of polyploidy have long been recognized: those involving the multiplication of one chromosome set, autopolyploidy (auto = same), and those resulting from the merger of structurally different chromosome sets, allopolyploidy (allo = different) (89). This convention has long been employed (39, 46, 124) and is thought to represent a fundamental distinction between polyploid types (63, 168). However, nature has produced a continuum of polyploid types, many of which defy clear placement in either of these two groups (105, 168). Hence, there has been debate for more than 70 years as to the types of polyploids that should be recognized in nature and the proper definitions of autopolyploidy and allopolyploidy. For example, Garsmeur et al. (60) use a genomic approach and suggest that allopolyploids have genome dominance and biased fractionation, whereas autopolyploids do not have these features. However, it is easy to envision scenarios in which an allopolyploid would lack genome dominance.

Allopolyploids are derived from hybridization and chromosome doubling between two different species (usually congeners) and hence have two divergent parental genomes (*AABB* in an allotetraploid) (**Figure 1**). An allotetraploid therefore contains two of each pair of the counterpart chromosomes derived from two different species: These are called homeologous chromosomes, or homeologs (**Figure 1**).

Autopolyploids are formed by genome duplication within a species: This could involve doubling of one individual genotype or, more likely, hybridization between different plants/populations within a species and genome doubling (*AAAA*) (**Figure 1**). Autopolyploids exhibit little morphological or cytogenetic divergence from the diploid progenitor and harbor more than two copies of each chromosome (i.e., four in an autotetraploid); multivalents may therefore form during meiosis. Another consequence of autopolyploidy is polysomic inheritance, in which multiple alleles (i.e., more than two) of a given gene segregate randomly at meiosis and can be detected in the offspring.

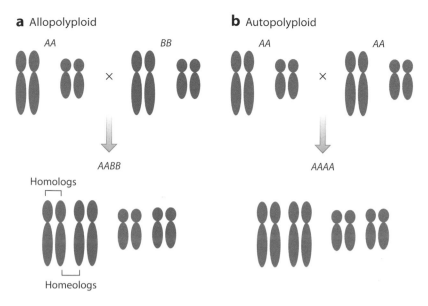

Figure 1

Two types of polyploids. (*a*) Allopolyploids are derived from hybridization between two related species and chromosome doubling. (*b*) Autopolyploids are formed by genome duplication involving a single species. Thus, an allotetraploid contains homologous and homeologous chromosomes, with the latter indicated by two of each pair of the corresponding chromosomes derived from the two different parental species.

Segmental allopolyploidy represents an intermediate condition between auto- and allopolyploidy (63, 168). The genotype of a segmental allotetraploid may be represented as $A_sA_sA_tA_t$ because the parental genomes are somewhat diverged.

On the basis of this background, discussion of nonadditive gene expression in polyploids requires the ability to differentiate the parental genomes. This ability does not apply to autopolyploids, and nonadditive gene expression is therefore a phenomenon applicable to the study of allopolyploid genomes in which the parental genomes are clearly distinct. One may also attempt to address nonadditive gene expression in segmental allopolyploids, but this may be problematic depending on how well-differentiated the parental genomes are in the study system.

Defining Nonadditive Expression Patterns in Polyploids

When two parental genomes are present in an allopolyploid nucleus, the expression levels between duplicated genes can show deviations from parental additivity (i.e., typically considered to be the arithmetic average of the expression levels of parental genes). Additivity can be referred to as expression conservation of parental genes, parental legacy, or the vertical transmission of preexisting expression patterns of parents (22). However, not all duplicated gene pairs exhibit additivity; such nonadditive expression pattern has been the focus of studies of polyploids with respect to relative expression of the respective parental copies (homeologs) or the total expression level of homeologs (**Figure 2**). The former can be considered as the preferential expression of one homeolog relative to the other and is often called transcriptome dominance (33), bias (54), nucleolar dominance (which refers to rRNA expression only) (36), genome dominance (60, 158), or gene dominance (188). However, recently the term homeolog expression bias was proposed to synthesize

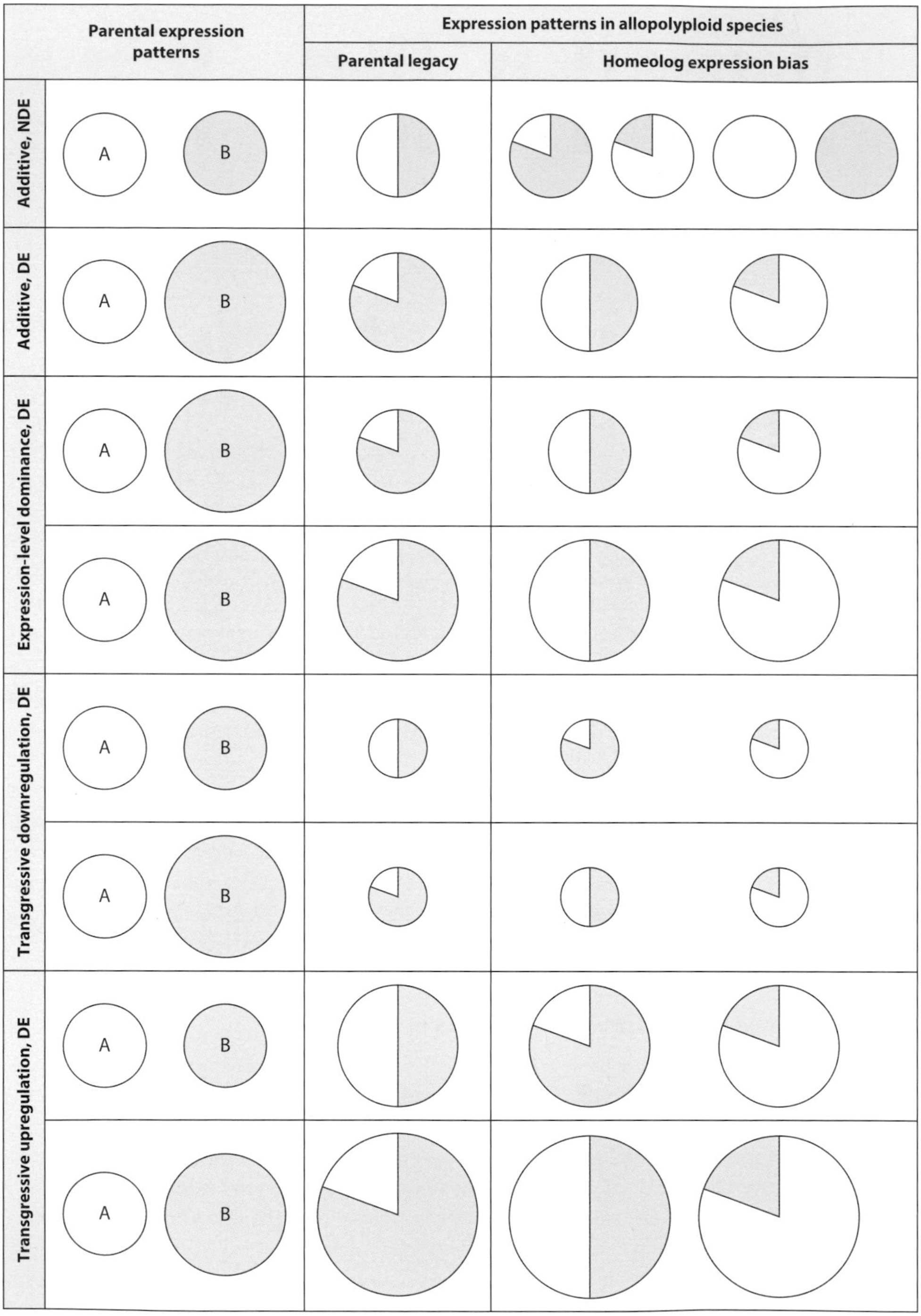
Parental expression patterns
Expression patterns in allopolyploid species
Parental legacy
Homeolog expression bias
Additive, NDE
Additive, DE
Expression-level dominance, DE
Transgressive downregulation, DE
Transgressive upregulation, DE
A
B

these unequal expression patterns of homeologs in allopolyploids (64). Loss of expression of one homeolog could perhaps be considered the extreme case of this homeolog expression bias.

Considering the total expression of a duplicated gene pair in an allopolyploid compared with its parents, two nonadditive scenarios are possible. First, the overall expression level in the allopolyploid may be similar to that of one of its parents. This case was originally described in cotton as genomic expression dominance (145) and also genomic dominance (54), and as parental dominance in *Spartina anglica* (31). However, this phenomenon is derived from data at the transcriptomic level, so a more appropriate term is expression-level dominance (64). Alternatively, the overall expression level may be lower or higher than those of both parents (transgressive expression).

Nonadditive expression patterns may therefore derive from expression-level dominance, transgressive expression across all duplicated gene pairs, and homeolog expression bias at the level of individual homeologs (**Figure 2**). Although the latter reflects the preferential expression of one homeolog over the other, we must consider whether observed homeolog expression bias originates from differential expression between parental genes. For example, if one parental gene has a higher expression level than the other parental gene, and there is no regulatory change in the allopolyploid species, we expect differential expression between the two homeologs in the allopolyploid [**Figure 2**; additive, differentially expressed (DE)]. This does not mirror homeolog expression bias, but rather vertical transmission of preexisting patterns of the parents (i.e., parental legacy; see 22). Therefore, when considering homeolog expression bias, we must consider the initial parental gene expression patterns.

Investigating nonadditive expression on a genome-wide basis is problematic for most organisms. Because of the lack of a complete genome sequence for most species and the low sequence divergence between the parental genomes of many allopolyploids, only small-scale experiments have typically been conducted to assess homeolog expression bias. For example, only 13 genes were examined in coffee [*Coffea arabica* (42)] and 10 to 144 genes in *Tragopogon miscellus* and *Tragopogon mirus* (23, 92). Large-scale gene surveys of allopolyploids have been based on expressed sequence tags (ESTs) (1,318 duplicated gene pairs in cotton) (54), RNA-sequencing (RNA-Seq) technology and Sequenom [<2,000 transcripts in *Tragopogon* (20)], and RNA-Seq only [~25,000 in cotton (194); ~10,000 genes in coffee (43)]. The paucity of studies limits our understanding of how homeologs are regulated in allopolyploid species. Because of the advantages of RNA-Seq technology (see below), studies of homeolog expression bias can now be conducted without prior information on the complete (or nearly so) parental genomes. Technological advances have thus promoted easier investigation of homeolog-specific expression patterns in nonmodel allopolyploids.

←

Figure 2

Nonadditive expression patterns in allopolyploids. Considering the total expression level of homeologs, expression could be additive for either nondifferentially expressed (NDE) or differentially expressed (DE) genes between an allopolyploid and its diploid parents (*first* and *second rows*). Nonadditive expression can be classified into expression-level dominance, where the total gene expression of homeologs resembles one of the parents regardless of downregulation or upregulation (*third* and *fourth rows*), and transgressive expression, where genes are downregulated or upregulated relative to the diploid parents (*fifth*, *sixth*, *seventh*, and *eighth rows*). As for the relative expression of homeologs, two homeologs in an allopolyploid may inherit preexisting parental expression differences (parental legacy), or one homeolog may be preferentially expressed compared to the other one, which is known as homeolog expression bias. Homeolog loss has the effect of resulting in extreme homeolog expression bias, shown in the first row only, although the mechanisms of reduced expression are clearly different.

Several studies show expression-level dominance in allopolyploids formed in the past ~1 million years or less, including cotton (54, 145, 194), *Spartina anglica* (31), *Triticum aestivum* (26, 143), and coffee (12, 43). Expression-level dominance was originally described from global gene expression profiles in cotton as genomic expression dominance (145). All of these studies investigated expression-level dominance in a genome-wide manner, but all but two (43, 194) employed microarrays, an approach with limitations (see below). Using RNA-Seq (see below), three studies have attempted to elucidate how nonadditive expression patterns are related, especially the relationship between expression-level dominance and homeolog expression bias. Two studies showed no correlation between expression-level dominance and homeolog expression bias (43, 144), whereas the third study revealed that expression-level dominance might be caused by the upregulation or downregulation of the nondominant parental copy (194). Given that more studies using RNA-Seq will be conducted on additional allopolyploids (e.g., in *Tragopogon*, *Glycine*, *Senecio*, and *Spartina*), we may soon have a better view of how homeolog expression bias has affected global gene expression patterns.

Examining Expression in Polyploids

Most studies of expression nonadditivity in polyploids have used either microarrays or Sequenom, and more recently, transcriptome sequencing using RNA-Seq. Microarrays have been widely leveraged in global gene expression comparisons of allopolyploids and their diploid parents [e.g., in *Arabidopsis* (179), *Gossypium* (cotton) (54, 145), *Spartina* (31), *Senecio* (69, 70), *Triticum* (wheat) (26, 143), and *Coffea* (coffee) (12)]. These allopolyploids were globally differentiated from their progenitors with respect to overall gene expression levels. However, microarrays require prior sequence information for probe design, which limits global transcriptome profiling in nonmodel allopolyploids. Also, microarrays rely on hybridization between probes and target sequences, which is problematic in that microarrays may not be sufficiently sensitive to distinguish between homeologs.

Sequenom MassARRAY is a powerful tool for studying nonadditive expression in allopolyploids. It requires prior information on genome-specific single nucleotide polymorphisms (SNPs); genome-specific SNP variants are then amplified via PCR, and their relative homeolog transcript abundance quantitatively measured through mass spectrometry. Sequenom is very efficient and sensitive in detecting homeolog expression bias but is limited in the number of genes that can be examined because of the requirement of homeolog-specific SNPs. Sequenom has been applied to synthetic allotetraploid cotton (29) and *Gossypium hirsutum* (53) and *Tragopogon miscellus* (20, 21). These studies targeted only 11 to 70 duplicated gene pairs but included 18 to 24 tissues in cotton and 59 individuals from five natural populations in *T. miscellus*. Importantly, these studies showed how homeologs are temporally and spatially regulated relative to each other.

RNA-Seq has several advantages for gene expression profiling, including no need for a priori information on genome sequences, a wide range of quantification, and higher accuracy and reproducibility in measuring expression levels of homeolog-specific transcripts (181). Thus, RNA-Seq enables investigators to examine all nonadditive expression patterns at once and dissect the relationship between homeolog expression bias and gene-level nonadditive expression. Several transcriptome profiling studies have now been conducted, including analyses of *Glycine* (soybean) (79, 150), *Arabidopsis* (7), cotton (144, 194), coffee (43), and *Tragopogon* (M.J. Yoo, J. Koh, A.M. Morse, J.L. Boatwright, L.M. McIntyre, W.B. Barbazuk, D.E. Soltis, P.S. Soltis, S. Chen, unpublished results). Studies on cotton and coffee focused not only on global gene expression patterns in the allopolyploids relative to the parental species in a genome-wide manner but also on how homeolog expression bias is linked to expression-level dominance (43, 144, 194).

NONADDITIVE EXPRESSION IN ALLOPOLYPLOIDS: CASE STUDIES

Several allopolyploid systems have now been examined in detail for patterns of gene expression across much of the genome. Below we summarize expression data for some of the best-studied allopolyploid systems.

Arabidopsis suecica

Arabidopsis suecica ($2n = 26$) (Brassicaceae) originated 12,000 to 300,000 years ago, with *Arabidopsis thaliana* ($2n = 10$) and *Arabidopsis arenosa* ($2n = 16$) as the maternal and paternal parents, respectively. Morphologically, *A. suecica* is more similar to *A. arenosa* than to *A. thaliana*. Wang et al. (179), using microarrays, found that 5.2% (1,362) and 5.6% (1,469) of the genes displayed nonadditive expression in two synthetic allotetraploids; more than 65% of these nonadditively expressed genes are repressed (**Figure 3a**). Most repressed genes (>94%) in the allotetraploids matched those genes that are expressed at higher levels in *A. thaliana* than *A. arenosa*. That is, allotetraploid plants show homeolog expression bias toward *A. arenosa*. *Trans*-effects might drive these homeolog expression patterns (162).

Chang et al. (27) used microarrays and RNA-Seq to investigate homeolog-specific retention and expression in *A. suecica*. Again, global expression favored *A. arenosa*: 3,458 genes preferentially expressed *A. thaliana* homeologs, whereas 4,150 expressed *A. arenosa* homeologs. Several hypotheses might explain this *A. arenosa* bias: (*a*) The *A. thaliana* subgenome is less capable of purging mildly deleterious mutations than the *A. arenosa* subgenome; (*b*) *A. arenosa* homeologs are a better fit for the environment in which *A. suecica* occurs; and (*c*) the *A. arenosa* transcription machinery is preferentially expressed in F_1 plants of *A. suecica*, and this initial pattern was then enhanced by homeolog-specific methylation (27).

Other research on *A. suecica* has focused on small RNAs, including microRNAs (miRNAs) (67, 90, 127, 175). In *A. suecica*, many small interfering RNAs (siRNAs) were nonadditively expressed; this nonadditive expression was related to tissue, growth stage, and developmental changes (67). Ng et al. (127) showed that the expression of *A. suecica* miR163 is biased toward *A. arenosa* rather than *A. thaliana*. These expression changes may reflect divergence between promoters in the two parents and the absence of putative *trans*-acting repressors in *A. thaliana*. In contrast, the miR172 loci from *A. arenosa* were repressed in the allotetraploid, whereas expression of *A. thaliana* loci was maintained after allopolyploidization (175). This nonadditive expression might be associated with the nonadditive expression of miR172 targets.

Brassica napus

Allotetraploid *Brassica napus* ($2n = 38$) (Brassicaceae) formed 5,000 to 10,000 years ago from *Brassica rapa* ($2n = 20$) and *Brassica oleracea* ($2n = 18$) (126); both parents are themselves ancient polyploids (137). Genetic changes caused by homeologous chromosome rearrangement are common in newly resynthesized *B. napus* (57, 167). Gaeta et al. (57) analyzed genetic, epigenetic, gene expression, and phenotypic changes in 50 independently resynthesized lines of *B. napus*. Nonadditive expression, measured by amplified fragment length polymorphism (AFLP)-cDNA and single-strand conformation polymorphism (SSCP)-cDNA, correlated with chromosomal rearrangements (57). Analysis of the newly synthesized polyploids (S_0 generation) revealed that genetic changes were rare and methylation changes frequent. Most of the S_0 methylation changes remained fixed in later-generation (S_5) progeny. Genetic changes were much more frequent in the S_5 generation, occurring in every line. Exchanges among homeologous chromosomes are a major mechanism for generating novel allele combinations and phenotypic variation (190). Fluorescence in situ hybridization (FISH) and genomic in situ hybridization (GISH) revealed

microRNA (miRNA): a small noncoding RNA molecule (approximately 22 nucleotides) that is involved in transcriptional and post-transcriptional regulation of gene expression

Small interfering RNA (siRNA): a double-stranded short RNA molecule (20~25 nucleotides) that interferes with gene expression through RNA interference (RNAi); also known as short interfering RNA or silencing RNA

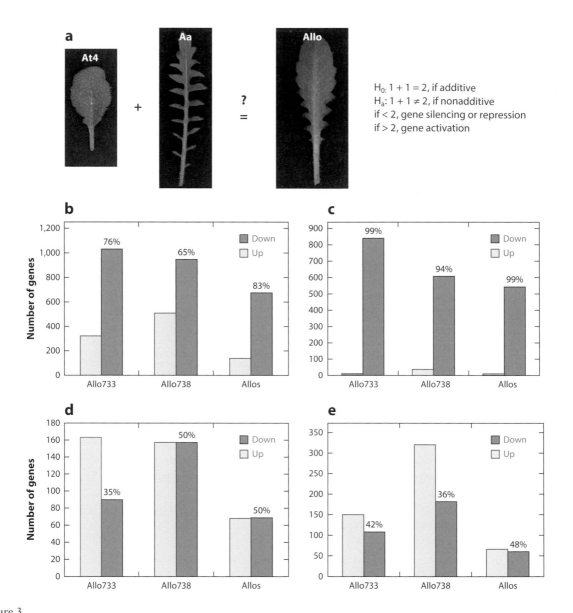

Figure 3

(*a*) Hypotheses for testing additive and nonadditive gene regulation in *Arabidopsis* allotetraploids. The null hypothesis (H$_0$) is that gene expression levels in an allotetraploid (Allo) are equal to the sum of the two progenitors, *Arabidopsis thaliana* autotetraploid (At4) and *Arabidopsis arenosa* (Aa). Typical seedling leaves in At4, Aa, and Allo are shown. Abbreviation: H$_a$, alternative hypothesis. Modified from Reference 33. (*b–e*) Downregulation of *A. thaliana* genes in the synthetic allotetraploids. (*b*) Distribution of nonadditively expressed genes detected in each allotetraploid (Allo733 or Allo738) or both allotetraploids (Allos). (*c*) The nonadditively expressed genes in each allotetraploid matched the genes that were highly expressed in the *A. thaliana* autotetraploid. (*d*) The nonadditively expressed genes in each allotetraploid matched the genes that were highly expressed in *A. arenosa*. (*e*) The nonadditively expressed genes matched the genes that were equally expressed in both parents. The percentages of downregulated genes are indicated above the columns in each histogram. Modified from Reference 179.

extensive chromosome rearrangements, aneuploidy, and homeologous chromosome compensation (190), similar to what has been reported in *T. miscellus* (38). Recent studies have been conducted in *B. napus* on alternative splicing (AS) (198) and the proteome (115), and current investigations on both resynthesized and natural *B. napus* are using a wide array of high-throughput "omics" approaches.

Coffea arabica

Coffea arabica (coffee; Rubiaceae) is an allotetraploid ($2n = 4x = 44$; ~50,000 years old) formed from diploids *Coffea eugenioides* and *Coffea canephora* (25, 97). The two diploid parents exhibit different ecological adaptations. As a result, most studies on coffee have involved investigation of environmental conditions. For example, how does growing temperature affect subgenome patterns of expression in allopolyploid coffee (12, 42, 43)? Genome-wide expression patterns of *C. arabica* relative to its parental species were investigated for ~15,000 duplicated gene pairs (12); evidence was obtained for the modulation of expression-level dominance in the allopolyploid based on growing conditions. Bardil et al. (12) showed that 67% of the genes examined showed nonadditive expression patterns in the allopolyploid, including expression-level dominance and transgressive expression (**Table 1**). However, the authors used microarrays and therefore could not examine homeolog expression bias. Combes et al. (42) surveyed the relative expression of homeologs of *C. arabica* using 13 genes; homeolog expression patterns were highly variable across plant organs and conditions.

Based on RNA-Seq data, 56,000 transcripts were assembled and quantitatively assessed; of these, ~10,000 duplicated gene pairs were investigated for homeolog expression bias and their influence on expression-level dominance on a genome-wide scale (43). This study revealed that the relative contributions of each subgenome to the transcriptome seemed to be altered by growing conditions, suggesting that the polyploid may tolerate a broader range of environmental conditions than do the diploid parents, although this appeared to be unrelated to differential use of homeologs.

Glycine dolichocarpa

Coate et al. (40) used polysome profiling and RNA-Seq to quantify translational regulation of gene expression in the ~100,000-year-old allotetraploid *Glycine dolichocarpa* (Fabaceae), whose diploid parents are *Glycine tomentella* and *Glycine syndetika*. There was a slight homeolog bias toward *G. syndetika* and close agreement between the allopolyploid transcriptome and what Coate et al. term the translatome, but ~25% of the transcriptome is translationally regulated. Homeolog expression bias observed at the transcriptional level was largely preserved in the translatome. Coate et al. (40) found that translational regulation preferentially targets genes involved in transcription, translation, and photosynthesis, causing regional and possibly whole-chromosome shifts in expression bias between homeologs, and reduces transcriptional differences between the polyploid and its diploid progenitors. Translational regulation correlates positively with long-term retention of homeologs from a paleopolyploidy event, suggesting that it plays a significant role in polyploid evolution.

Gossypium hirsutum

Allopolyploid cotton, *Gossypium hirsutum* ($2n = 4x = 52$) (Malvaceae), formed ~1–2 Mya via hybridization between A-genome (similar to *Gossypium arboreum* and *Gossypium herbaceum*) and D-genome (much like *Gossypium raimondii*) progenitors. This polyploid event ultimately yielded five polyploid species (183). Studies on nonadditive expression have focused on homeolog

Table 1 Comparison of previous studies of genome-wide nonadditive expression patterns in allopolyploids[a]

Expression pattern in allopolyploids		Cotton leaf								Cotton petal				Coffee (22–26C)		Coffee (26–30C)		Tragopogon leaf			
		TX2094	%	Maxxa	%	TX2094	%	Maxxa	%	Maxxa	%	Maxxa	%	Java	%	Java	%	Tragopogon mirus	%	Tragopogon miscellus	%
Nondifferential expression		11,401	46.1	11,032	44.6	13,657	69.1	13,209	65.0	18,323	43.2	4,869	35.4	2,686	23.7	5,112	54.1	22,178	65.8	22,858	64.8
Additivity		3,480	14.1	3,418	13.8	1,390	7.0	1,242	6.1	3,328	7.8	484	3.5	1,799	15.9	289	3.1	4,383	13.0	4,718	13.4
Maternal ELD		3,525	14.2	3,946	15.9	1,519	7.7	1,885	9.3	7,940	18.7	1,365	9.9	1,234	10.9	1,859	19.7	3,348	9.9	3,435	9.7
Paternal ELD		2,919	11.8	2,470	10.0	1,051	5.3	806	4.0	7,256	17.1	1,340	9.8	2,110	18.6	316	3.3	1,873	5.6	2,342	6.6
Transgressive down		1,896	7.66	2,145	8.7	1,023	5.2	1,487	7.3	2,973	7.0	86	0.6	1,916	16.9	1,530	16.2	895	2.7	538	1.5
Transgressive up		1,529	6.18	1,731	7.0	1,135	5.7	1,686	8.3	2,639	6.2	5,592	40.7	1,578	13.9	348	3.7	1,029	3.1	1,368	3.9
Total number of genes investigated		24,750		24,742		19,775		20,315		42,459		13,736		11,323		9,454		33,706		35,259	
Total nonadditive expression			39.9		41.6		23.9		28.9		49.0		61.0		60.4		42.9		21.2		21.8
Technology used (reference)		RNA-Seq (cotton 46A)				RNA-Seq (cotton D)				Microarray (cotton 46A)		RNA-Seq (cotton D)		Microarray		RNA-Seq		RNA-Seq			
Literature		194				Reanalysis of Reference 194				54		144		12				M.J. Yoo, J. Koh, A.M. Morse, J.L. Boatwright, L.M. McIntyre, W.B. Barbazuk, D.E. Soltis, P.S. Soltis, S. Chen, unpublished results			

[a] Red text shows biased expression-level dominance (ELD) toward one of the diploid parental species. In the line graphs, each dot represents the maternal parent, allopolyploid, and paternal parent, in order.

expression bias, including homeolog silencing, biased expression, and organ-specific expression differences, variously using a focused approach and a few (1, 2, 4, 53) to many (~1,400 genes) (29, 52, 54, 74) genes. All studies revealed differential homeolog expression in different organs (1, 2, 4, 53), developmental time points (29, 74), and evolutionary stages (29, 52).

Global transcriptome profiling was investigated with respect to expression-level dominance employing microarray [genome dominance (145) and genomic dominance (54)] and RNA-Seq techniques (144, 194). Rapp et al. (145) first introduced the concept of expression-level dominance in allopolyploids and showed that there was a bias in expression-level dominance toward one of the diploid parents. Flagel & Wendel (54) supported the presence of unbalanced expression-level dominance in petals and reported unbalanced homeolog expression (**Table 1**). Transgressive expression was greater in the five allopolyploid species than in a synthetic diploid F_1 hybrid, suggesting that long-term evolutionary processes might play a role in establishing transgressive expression (54). However, these two pioneering studies could not show how expression-level dominance and homeolog expression bias are related because of the limitations of microarrays.

Global expression-level dominance and homeolog expression bias were subsequently surveyed using RNA-Seq technology in leaf (194) and petal tissues (144). Yoo et al. (194) provided support for unbalanced expression-level dominance and homeolog expression bias; 40% of the genes investigated exhibited nonadditive expression in leaf tissue (**Table 1**). Furthermore, the degree of nonadditive expression, including transgressive and novel gene expression, increased over time. Expression-level dominance may be caused by upregulation or downregulation of the homeolog from the nondominant parent (194). However, another study using RNA-Seq (144) did not find unbalanced expression-level dominance or homeolog expression bias, indicating that nonadditive expression might be tissue-specific. They also reported more nonadditive expression (144) than did Flagel & Wendel (54), who investigated the same tissue using microarrays (e.g., 49% versus 61% of the genes examined in microarrays versus RNA-Seq, respectively, showed nonadditive expression; **Table 1**). The data suggest that use of different technologies, as well as different databases, might affect the degree or direction of nonadditive expression reported in an allopolyploid.

Senecio cambrensis

Senecio cambrensis (an allohexaploid; $2n = 60$) (Compositae) originated in the United Kingdom within the past 100 years following hybridization between diploid *Senecio squalidus* ($2n = 20$) and tetraploid *Senecio vulgaris* ($2n = 40$). Genome-wide expression patterns were examined using microarrays (68). Studies of resynthesized *S. cambrensis* revealed that hybridization initially results in genome-wide, nonadditive alterations to parental patterns of gene expression and DNA methylation; genome duplication then results in a secondary burst of transcriptional and epigenetic modification. In natural populations, different origins of the polyploid show genome-wide nonadditive patterns comparable to those seen in synthetics, so it appears that polyploid expression changes may be repeated across multiple origins. In synthetic *S. cambrensis*, phenotypic changes become apparent from the second to fifth generations, and, again, different origins of the polyploid and synthetics have similar patterns of change; evolutionary patterns are repeated.

Spartina anglica

Spartina anglica (Poaceae) is an allo-dodecaploid ($2n = 120–124$) that formed in the past 200 years from the hexaploid parents *Spartina alterniflora* ($2n = 60$) and *Spartina maritima* ($2n = 62$). The parents hybridized both in the United Kingdom and in France, yielding two hybrids (*Spartina* × *townsendii* in the United Kingdom and *Spartina* × *neyrautii* in France), but only the event in the United Kingdom yielded the allopolyploid *Spartina anglica*.

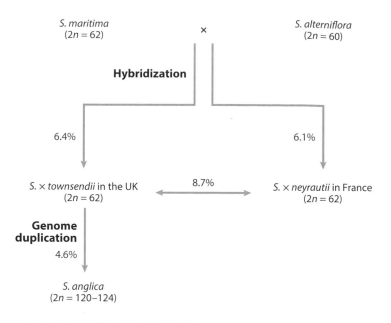

Figure 4

Nonadditive expression in allopolyploid *Spartina anglica*. Red text indicates transcriptomic changes, both parental expression-level dominance and transgressive expression, inferred using microarrays (30, 31). Modified from Reference 6.

Genome-wide patterns of expression change were examined in *Spartina* via microarrays (30, 31). *Spartina* is clonal, and populations of both hybrids are still living. The parental species *S. maritima* and *S. alterniflora* exhibited 1,247 differentially expressed genes (31), most of which were upregulated in *S. alterniflora*. Similar levels of nonadditive parental patterns of gene expression were observed in both of the natural hybrids *S.* × *townsendii* and *S.* × *neyrautii* (6.1% and 6.4% of the analyzed genes were nonadditive, respectively; **Figure 4**). Maternal (*S. alterniflora*) expression-level dominance and transgressively expressed genes were observed in both F$_1$ hybrids (31). However, maternal expression-level dominance was more pronounced in *S.* × *townsendii* than in *S.* × *neyrautii*, with ∼8.7% of the genes differentially expressed between these two F$_1$ hybrids (**Figure 4**) (6). Hence, the two independent hybridizations yielded different consequences in terms of gene expression. There are also phenotypic differences between the two hybrids.

Genome duplication in *S. anglica* resulted in additional transcriptome changes (**Figure 4**) (6), with an attenuation of the maternal expression-level dominance observed in the F$_1$ hybrid as well as an increase in the number of transgressively overexpressed genes (31). Hence, both hybridization and genome duplication have been important, but with different effects on the transcriptome: Hybridization has had a greater impact on gene silencing than genome doubling per se.

Tragopogon

Tragopogon (Compositae) provides textbook examples of recent polyploidy (133, 163–165), with two recently (∼80 years old; 40 generations in these biennials) and repeatedly formed natural allotetraploids (*T. mirus* and *T. miscellus*) and their diploid parents (*Tragopogon dubius*, *Tragopogon pratensis*, and *Tragopogon porrifolius*). In both allotetraploids, homeolog-specific patterns were examined using Sequenom (23), and expression-level dominance and transgressive expression patterns were investigated using RNA-Seq (**Table 1**) (M.J. Yoo, J. Koh, A.M. Morse, J.L.

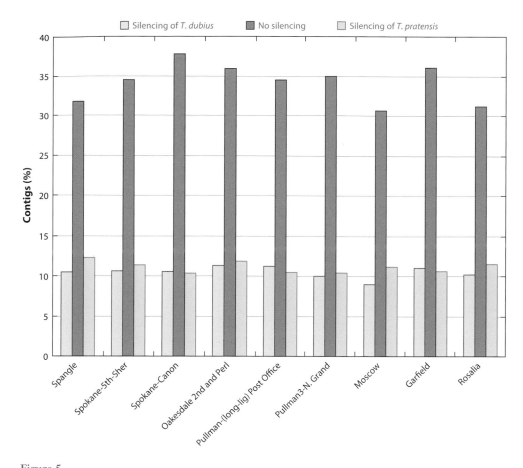

Figure 5

Evolution of genome-wide expression is repeated across natural populations of *Tragopogon miscellus*. Approximately 40–45% of all loci (contigs) show evidence of homeolog loss in *T. miscellus* (not shown); of the remaining 55–60% of all loci, most show equal expression of both parental homeologs, with nearly equal silencing of the two parental homeologs. Overall, approximately 35% of all loci express both parental homeologs, with about 10% of all loci showing silencing of the *Tragopogon dubius* homeolog and about 10% showing silencing of the *Tragopgon pratensis* homeolog. Likewise, populations of *Tragopogon mirus* show similar patterns of gene expression (not shown); however, in *T. mirus*, the *Tragopogon porrifolius* homeologs are silenced to a greater extent (∼5–9% of all loci expressed) than the *T. dubius* homeologs (∼10–12% of all loci expressed) (I.E. Jordon-Thaden, R.J.A. Buggs, L.F. Viccini, J. Tate, J. Combs, M. Chester, A.V.C. Silva, R. Sanford, S. Chamala, R. Davenport, B. Jordon-Thaden, W. Wu, C.T. Yeh, A. Hu, P.S. Schnable, W.B. Barbazuk, D.E. Soltis, P.S. Soltis, unpublished results).

Boatwright, L.M. McIntyre, W.B. Barbazuk, D.E. Soltis, P.S. Soltis, S. Chen, unpublished results). In *T. miscellus* the homeolog expression of one parent dominates, and changes occur immediately with hybridization, with a smaller impact of polyploidy per se (22). Significantly, evolution is repeated across natural populations and synthetic lines (**Figure 5**), with similar results for *T. mirus* (I.E. Jordon-Thaden, R.J.A. Buggs, L.F. Viccini, J. Tate, J. Combs, M. Chester, A.V.C. Silva, R. Sanford, S. Chamala, R. Davenport, B. Jordon-Thaden, W. Wu, C.T. Yeh, A. Hu, P.S. Schnable, W.B. Barbazuk, D.E. Soltis, P.S. Soltis, unpublished results).

Genome-wide nonadditive expression patterns were investigated using de novo assembled transcriptomes via RNA-Seq (M.J. Yoo, J. Koh, A.M. Morse, J.L. Boatwright, L.M. McIntyre,

W.B. Barbazuk, D.E. Soltis, P.S. Soltis, S. Chen, unpublished results). As in cotton and coffee, *Tragopogon* allopolyploids exhibited unbalanced expression-level dominance (toward the maternal parent): Approximately 21–22% of transcripts investigated were nonadditively expressed in leaf tissue of the two recent allopolyploids (**Table 1**) (M.J. Yoo, J. Koh, A.M. Morse, J.L. Boatwright, L.M. McIntyre, W.B. Barbazuk, D.E. Soltis, P.S. Soltis, S. Chen, unpublished results), mirroring results for homeolog-specific analyses of *T. miscellus*. However, the degree of nonadditive expression was lower in these recent allopolyploids compared to cotton and coffee, both of which are much older, suggesting nonadditive expression may increase over time, via selection and modulation of regulatory networks (54).

Triticum aestivum

T. aestivum (wheat; Poaceae) is an allohexaploid ($2n = 42$) that arose ~8,000 years ago from cultivated allotetraploid *Triticum turgidum* ($2n = 28$, AABB genome) and diploid ($2n = 14$) goatgrass, *Aegilops tauschii* (D genome). On the basis of ESTs and SNPs, 11 of 90 genes examined initially exhibit homeolog silencing (122); six are from the A genome, two from the B genome, and three from the D genome. In a study of 236 single-copy genes using SSCPs, homeolog silencing was found in 27% of the genes in leaf tissue and 26% of the genes in roots (19). An SSCP-cDNA analysis showed that ~13% of 30 homeologs were differentially expressed in synthetic hexaploid *T. aestivum* relative to the parents of this synthetic line (*A. tauschii* and *T. turgidum*); microarray analysis showed that ~16% of the genes displayed nonadditive expression in synthetic hexaploid wheat (142), and 2.9% of nonadditively expressed genes exhibited transgressive expression, indicating that allopolyploidization per se results in rapid changes in homeolog and gene expression.

In an Affymetrix-based study, 34,000 parent-specific features were detected in wheat, 19% of which showed evidence of nonadditive expression (8). Among those nonadditive expression genes, a bias toward *T. turgidum* parental expression was detected in hexaploid wheat. This bias might be caused by divergent mutations in *cis*- and *trans*-acting regulatory elements. Two other studies also used Affymetrix wheat genome arrays (26, 143), but the synthesized hexaploid wheat lines differed from those used in Reference 8 (see above). Using different lines of the diploid parent, Chague et al. (26) found no global gene expression bias or dominance toward either of the progenitor genomes (S_0 and S_7 generations). However, Qi et al. (143) used different lines of the tetraploid parent and found that the transcriptomes of the synthetic allohexaploid lines (S_4 and S_5 generations) exhibited maternal expression-level dominance. The different results of these studies (26, 143) may reflect the use of different parental lines in producing the synthetic allohexaploid lines. The results for wheat also contrast with *Tragopogon* polyploids, where changes were consistent across different polyploid lines.

Based on RNA-Seq data (98), 650 of 2,356 genes on chromosomes 1 and 5 exhibited differential expression among the three parental homeologs; 55% of those genes showed predominant expression of one homeolog, whereas 45% of the genes were co-upregulated by the other two homeologs (98). However, no global bias in homeolog expression toward one parent was detected, although B-genome homeologs tended to contribute more to gene expression (98). Consistent with a previous study (19), Leach et al. (98) suggested that *T. aestivum* has undergone extensive functional diploidization through homeolog loss and silencing.

Zea mays

Zea mays ($2n = 20$) (maize; Poaceae) is functionally diploid, but a polyploid origin was proposed more than a century ago (96) and is supported by diverse data (149). Although its parental species are unclear, the presence within maize of two subgenomes indicates ancient allotetraploidy (154, 182). The parental genomes diverged ~12 Mya, but the age of maize is much younger (~5 Mya) (18, 157, 170, 191).

The maize genome has undergone extensive fractionation, or loss of duplicate gene copies (156, 158, 188). However, rather than random losses of one parental homeolog or the other across the polyploid genome, authors argue for biased loss, resulting from differential expression of parental homeologs (see 157 in particular). The pattern of genome dominance (homeolog expression bias), i.e., the biased retention of one subgenome over the other in an allotetraploid, is hypothesized to arise through differential selection against loss of homeologs exhibiting differential expression. For example, loss of a homeolog that contributes more than its share to the RNA pool has a more severe effect on the overall phenotype than loss of a homeolog with lower expression. Consequently, selection will favor loss of lower-expressing homeologs. If the low-expressing homeologs all originate from one parental genome, there is biased gene expression (homeolog expression bias), with the hypothesized outcome being genome dominance following loss of the low-expressing subgenome. Despite evidence of high- and low-expressing homeologs, it is not completely clear that all high-expressing (or low-expressing) homeologs trace to the same parental genome because they are unknown, thus complicating the interpretation.

Repeated patterns of gene expression change. Recent as well as ongoing studies of several polyploid systems beg the question: Are there repeated patterns of gene expression change that are observed? That is, are aspects of evolution repeated following polyploidization? Based on the data now available (reviewed above), the answer appears to be yes and no, depending on the system investigated and how gene expression is measured. In recent natural *Tragopogon* allopolyploids as well as synthetic lines, patterns of gene expression as well as homeolog loss are repeated (163). A smaller data set using older approaches (microarrays) suggests repeated patterns of expression change in independently formed *Senecio cambrensis* lines (68). In natural and synthetic lines of *Brassica napus*, some patterns of expression dominance are correlated with patterns of chromosomal exchanges, whereas other changes in expression may be due to epigenetic or other phenomena (57). Significantly, in synthetic lines of wheat, patterns of expression-level dominance may vary based on the parental lines employed.

Hybridization versus genome doubling. Recent studies of several polyploid systems also permit us to evaluate the critical question: What is more important, hybridization or genome doubling? Based on the studies to date, it appears that both hybridization and genome doubling per se have important consequences in terms of subsequent gene expression. However, only a few studies have been conducted in such a way to permit these two processes to be disentangled, in part because F_1 hybrids are often sterile and short lived. The use of synthetics is crucial for examining hybridization effects relative to immediate genome-doubling effects. Some investigations indicate a particularly important role for hybridization as a major determinant of changes in expression, as in *Arabidopsis* (162), *B. napus* (58) cotton (29, 52), *Senecio* (68), *Spartina* (6), and *Tragopogon* (23, 163). However, in cotton (54, 194) the degree of nonadditive expression was higher in natural polyploid species compared to synthetic diploid and polyploid accessions, suggesting that nonadditive expression might be responsive to long-term evolutionary alteration. Therefore, not only the immediate impact of hybridization and genome doubling must be considered, but also subsequent evolutionary history.

UNDERLYING CAUSAL FACTORS FOR NONADDITIVE GENE EXPRESSION

Nonadditive gene expression in allopolyploids may be due to many controlling factors acting separately or in concert. Among these is parental legacy, or the extent to which differences in gene expression between duplicate copies in an allopolyploid are a legacy of expression differences inherited from the progenitor diploid species (22). This concept, developed decades ago (62), has implications for interpretation of gene expression data today; i.e., differential homeolog expression may not necessarily reflect a departure from parental patterns and should not be attributed to post-polyploidization shifts in expression without careful analysis of parental expression. Thus, observed expression patterns in polyploids may derive from the combined effects of parental legacy and other influences, such as those described below.

Maternal-Paternal Influences

Nonadditive expression may be expected to reflect maternal influences, as cytonuclear incompatibilities have long been known (28, 101, 186, 187). Because organellar genomes are typically maternally inherited in plants (with paternal inheritance of plastids in conifers a notable exception), greater compatibility is observed between the nuclear and maternal, rather than paternal, genomes, at least in angiosperms. Certainly, genes that encode proteins assembled from nuclear and plastid genome components must undergo coordinated expression, and this cytonuclear balance is observed for *rbcS* and *rbcL*, the nuclear and plastid genes responsible for the small and large subunits, respectively, of RUBISCO (61, 159). However, reports attributing large-scale expression differences to maternal effects are rare, most likely because so few data sets have been explored.

Although the degree and direction of nonadditive expression are variable depending on the parentage, tissues, and technology employed, most studies revealed unbalanced expression-level dominance toward the maternal parent, as in cotton (145, 194), coffee grown under hot conditions (12), *S. anglica* (31), *T. aestivum* (143), and *Tragopogon* (21, 23; I.E. Jordon-Thaden, R.J.A. Buggs, L.F. Viccini, J. Tate, J. Combs, M. Chester, A.V.C. Silva, R. Sanford, S. Chamala, R. Davenport, B. Jordon-Thaden, W. Wu, C.T. Yeh, A. Hu, P.S. Schnable, W.B. Barbazuk, D.E. Soltis, P.S. Soltis, unpublished results; M.J. Yoo, J. Koh, A.M. Morse, J.L. Boatwright, L.M. McIntyre, W.B. Barbazuk, D.E. Soltis, P.S. Soltis, S. Chen, unpublished results). Interestingly, these studies examined gene expression in leaf tissue, which contains many chloroplasts, suggesting that nonadditive expression could be affected by maternal-paternal influence. However, there are exceptions: One synthetic line of cotton (145) and coffee grown under cold conditions (12) both showed unbalanced expression-level dominance toward the paternal parent. Further studies are required, involving more polyploid systems and considering temporal and spatial factors.

The Role of Gene Function

Just as genes encoding proteins with nuclear and organellar components may exhibit biased gene expression, so, too, may genes of different functional classes. Given the cytonuclear patterns described above, it may be that photosynthetic genes, which are composed of nuclear and plastid-encoded subunits, exhibit nonadditive patterns, although these would be due more to their origin than their function per se. However, other classes of genes may consistently exhibit nonrandom gene expression in polyploids. Again, studies that have specifically addressed this question are few, but data may be available to pursue the issue in more detail.

In *A. thaliana*, certain classes of genes, based on Gene Ontology (GO) categories, have been either consistently retained as duplicates or consistently returned to singleton status (reviewed in 56). If expression shifts represent a prerequisite for duplicate gene loss (see below), then we may posit that nonadditive expression characterizes early duplicates of genes encoding structural genes, such as ribosomal protein genes, proteasomal protein genes, and transcription factors, at least in *Arabidopsis* and perhaps its relatives (17, 160). Similar patterns of duplicate gene loss and retention are reported for rice, puffer fish, and yeast (139); species of Brassicaceae (123); and species of *Paramecium* (119), suggesting possible generalities for factors controlling which genes are retained and which are lost from polyploid genomes. However, alternative patterns of gene retention and loss are observed for ancient duplicates in Compositae (13, 21), in which structural genes are retained in duplicate and transcription factors are typically singletons. Preliminary evidence from *Tragopogon* suggests that these alternative patterns in Compositae may begin early in polyploidization. However, these results are based on a small number of genes in *Tragopogon*, a very young polyploid in which loss and expression changes are ongoing. The possible differences in patterns of retention in Compositae compared to other systems certainly require more thorough analysis.

Gene Dosage Balance Hypothesis

Nonadditive expression patterns observed in allopolyploids can be explained by the gene-balance hypothesis (**Figure 6**), which was formulated from classic observations and more recent studies of gene expression modulation in aneuploid and polyploid species (reviewed in 16). This hypothesis states that the stoichiometric differences among members of macromolecular complexes affect the function of the whole complex, eventually leading to phenotypic effects (16). In particular, it has been reported that organisms that experience genome fractionation (diploidization) after WGD (polyploidization) exhibit nonrandom distribution of genes that are lost during diploidization. That is, genes belonging to specific functional classes, such as ribosomal protein genes and transcription factors, were more often retained in duplicate (16, 56, 119, 123, 139).

This concept could be applied to nonadditive expression patterns observed in allopolyploids, specifically homeolog expression bias. Preferential expression of one homeolog can be explained by biased fractionation of subgenomes (**Figure 6b**). This has been shown in several species, including *A. suecica* (27, 179), *Z. mays* (158), and *B. rapa* (37, 180, 188); the less-expressed parental genome or subgenome experienced more sequence deletion (60), resulting in biased homeolog expression in the allopolyploid (27, 179). However, only one example (*A. suecica*) involves a relatively recent allopolyploid, whereas the others are diploidized, ancient polyploids that experienced extensive fractionation. Thus, more studies are needed to link biased fractionation and homeolog expression bias via detailed analysis of more allopolyploid species, including paleo-, neo-, and synthetic polyploid species (188).

Transgressive expression can be viewed in terms of gene balance. If the expression level of one homeolog is higher than that of the other homeolog, that gene may not function properly because of the imbalance between two members of the gene pair if genes are under functional constraint. Somehow, two homeologs are regulated to be similarly expressed by upregulating or downregulating the counterpart homeolog (**Figure 6c,d**), which can affect gene expression modulation in allopolyploids. However, investigation of the functional categories of these genes is needed to determine whether they belong to specific functions, such as transcription, and structures, such as the signal transduction pathway, the ribosome, and the proteasome, that are known to be under functional constraint.

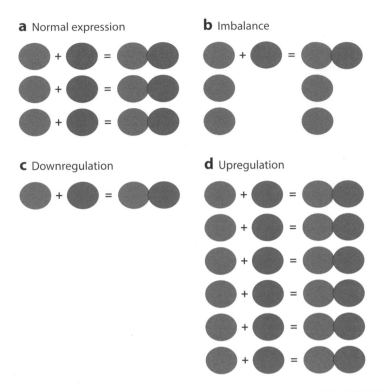

a Normal expression   **b** Imbalance

c Downregulation   **d** Upregulation

Figure 6

Gene balance hypothesis. Stoichiometry among members of macromolecular complexes is critical for the proper function of the whole complex; for example, (*a*) three dimers of Blue-Red are required for normal expression under the gene balance hypothesis. If these molecules are not constrained under gene dosage, they might be (*b*) relatively easily lost or (*c*) downregulated (transgressive-down), which leads to a decrease in the amount of the dimer. This dimer cannot function properly or find a new function. (*d*) The upregulation of both members (transgressive-up) can cause an increase in the dimer, possibly leading to a new function.

cis- and *trans*-Regulation

Among nonadditive gene expression patterns found in allopolyploids, homeolog expression bias can result from modulation of *cis*- and/or *trans*-regulatory elements. Genome merger combines two divergent regulatory systems and can therefore lead to different patterns of homeolog activation and repression. Two parental genomes that exhibit low sequence divergence, such as those found in coffee [1.35% in 60,000 coding nucleotides (25)], cotton (65, 66), and *Tragopogon* [1.45% in 50,766 coding nucleotides (M.J. Yoo, J. Koh, A.M. Morse, J.L. Boatwright, L.M. McIntyre, W.B. Barbazuk, D.E. Soltis, P.S. Soltis, S. Chen, unpublished results)], may enable cross talk between parental copies of *trans*-elements in resulting allopolyploids. Thus, it is possible that homeologs might be regulated by intertwined mechanisms of *cis*- and *trans*-elements that originated from both parents.

    Only a few polyploids have been investigated for *cis*- and/or *trans*-regulatory factors and homeolog expression bias [e.g., cotton (29, 194), *A. suecica* (162), and coffee (43)]. Although *cis*-regulatory changes explain most homeolog expression differentiation in *Arabidopsis* [2,775 of 14,713 gene pairs (19%) (162)] and cotton [15 out of 30 genes (50%) (29)], a combined regulation of *cis*- and *trans*-elements explains the differential expression between homeologs in coffee [774 of 1,434 gene pairs (54%) (43)]. Interestingly, these two different polyploid species exhibit a similar extent of nondifferential expression between homeologs, 68% and 69% in *A. suecica* and cotton, respectively (162, 194). However, these studies did not examine how *cis*- and/or *trans*-regulatory mechanisms affect overall gene expression levels, such as expression-level dominance and transgressive expression patterns. In addition, studies have so far investigated a relatively small number of genes, ranging from 30 (29) to ~3,000 genes (43). Therefore, a genome-wide approach on the effects of *cis*- and/or *trans*-regulatory elements needs to be conducted in more allopolyploid species.

Regulation by Transposable Elements

McClintock (118) noted that interspecific hybridization can bring about genome shock, which results in activation of transposable elements (TEs) and genomic instability. The merged TEs affect gene expression modulation: specifically, nonadditive expression patterns. So far, there is little evidence for gene expression changes by TE transposition or deletion in polyploids. Kashkush et al. (86) showed that gene expression was altered by transcriptional activation of retrotransposons, indicating that epigenetic changes in the expression of TEs might play an important role in gene expression modulation in allopolyploids (136).

Recent studies of TEs in *A. thaliana* have shown that methylated TE insertions can be a cause of downregulation of nearby genes, implying TE methylation not only reduces TE activity but also leads to perturbed gene expression (72, 73). siRNA-targeted TEs are related to reduced gene expression in *A. thaliana* and *Arabidopsis lyrata*, the latter with two or three times more TEs than the former, as well as to gene expression differences between orthologs (73). Study of *B. rapa*, a diploidized paleopolyploid, also suggested that small RNA-mediated silencing of TEs can affect the regulation of nearby genes (188). These authors hypothesized that genes in allopolyploids can have novel expression balance between homeologs based on the coverage of parental transposons. Epigenetic regulation of parental TEs merged in one polyploid nucleus may therefore affect gene expression modulation in allopolyploids.

Frequent methylation around points of TE insertion was reported in allopolyploid *Spartina* (136), and massive methylation changes were detected in specific TEs in synthetic allopolyploid wheat (95), but these studies did not examine how methylated TEs affect gene expression around them. Further study is needed to assess how methylated TEs are associated with gene expression modulation in allopolyploids in a genome-wide manner.

INTEGRATING MODELS OF GENE RETENTION

How Does Nonadditive Gene Expression Relate to Gene Retention? A Case for Pluralism

As described above, nonadditive gene expression can be caused by a wide array of phenomena. One intriguing corollary to the gene balance hypothesis is that these phenomena can be genome wide and inherited through time, with nonadditive expression starting out as a bias in epigenetic silencing and leading to a bias in gene retention patterns (60, 125, 188). To address this scenario, we need to consider what forces might preserve duplicate genes, including neofunctionalization, subfunctionalization, and absolute dosage (selection to increase copy number of gene products) (45). Recently, another means of duplicate retention [relative dosage (selection on relative copy number of gene products)] has been recognized (15, 16, 56). Under relative dosage, genes that are in stoichiometric balance are maintained by selection to avoid the deleterious consequences of having dosage-sensitive genes out of balance. Relative dosage effects of aneuploidy and polyploidy

Neofunctionalization: one member of a duplicate gene pair takes on a new function

Absolute dosage: more protein concentration (more gene product) is beneficial, so duplicated gene pairs are retained, opposing the loss of a duplicated copy

Relative dosage: stoichiometric balance is important for genes involved in networks or multisubunit protein complexes; thus, duplicated gene pairs are maintained

on phenotype have been known for almost a century, but only recently has it been hypothesized that the selection for stoichiometry is more widespread and may be linked by a unifying hypothesis: the kinetics of multisubunit protein complex formation (16, 177).

Many researchers currently attribute nonadditive gene expression to one of these models (neofunctionalization, subfunctionalization, absolute dosage, or relative dosage). But over time, duplicate genes may have had several of these phenomena acting to preserve them in duplicate, defying strict classification systems. An even more fundamental classification problem is that these models are often difficult to distinguish. For example, although the duplication, degeneration, and complementation (DDC) version of subfunctionalization is nonadaptive (55), the escape from adaptive conflict (EAC) version of subfunctionalization is adaptive (47). This is problematic, as adaptive sequence evolution is also used to classify models of neofunctionalization. In addition, sequence divergence is also compatible with gene balance hypotheses because new effective balances or dosages may evolve. Thus, there need not be only a single force acting on a duplicate gene pair. Bekaert et al. (15) suggested that several mechanisms of preserving duplicate genes could be at play, speculating that relative dosage may be important immediately after polyploidization, and other mechanisms such as subfunctionalization, neofunctionalization, and absolute dosage could be operating later in evolutionary time. Although some researchers continue to explain nonadditive gene expression primarily in terms of neofunctionalization and subfunctionalization (148, 150), we think it important to avoid this simple dichotomy to explain the mechanisms that retain duplicate genes. A more pluralistic framework not only considers additional phenomena (i.e., absolute and relative dosage) but also allows for more complex relationships and interactions of multiple mechanisms that may overlap or change over time (14–16, 44, 77, 117, 177).

Elucidating Mechanisms for Nonadditive Gene Expression Requires More Than Transcriptomes and Genomes: Emergence of Systems Biology

As reviewed above, transcriptomic (RNA-Seq) and genomic data are providing an abundance of evidence concerning nonadditive gene expression and gene loss in polyploids. One common result of these studies is that duplicate genes are often found to have one copy exhibiting higher expression than the other copy at the same tissue or time point or that one copy is expressed and the other copy is not expressed at all in the same tissue or time point, and this evidence is taken to implicate subfunctionalization as the cause (148, 150, 194). In our view, if there are no expression data on related species for inferring the ancestral state, extreme caution should be taken before concluding that neofunctionalization, subfunctionalization, absolute dosage, relative dosage, or some other mechanism is the sole cause of nonadditive gene expression. Although nonadditive gene expression changes from transcriptome data are often used as defining characteristics of subfunctionalization, there are alternative reasons as to why there may be tissue-specific or time-point-specific expression. For example, in diploids, variation in expression may be due to neutral divergence or variation among individuals within a population or across ecotypes. A parallel situation would be similar in paralogs in polyploids and therefore could be due to neutral divergence and not necessarily subfunctionalization, as has been recently claimed (e.g., 148, 150). Again, the divergence among paralogs for subfunctionalization is not inconsistent with dosage balance.

In the context of allopolyploidy, nonadditive gene expression has recently been hypothesized to be a genome-wide phenomenon in which one of the parental diploids is epigenetically dominant over the other diploid, and this sets the stage for longer-term patterns of gene loss, biased fractionation, and genome evolution (60, 125, 188). The hypothesized scenario links how biased epigenetic silencing in recent allopolyploids induces biased changes in gene expression to patterns of biased gene retention. Additional evidence for this view may perhaps be emerging with possible associations with the evolution of conserved noncoding sequences.

As we now have multiple genomes all possessing the same polyploid event, it is possible to phylogenetically date both nonadditive gene expression and gene losses. Plant biologists can then begin using systems biology tools already being deployed in polyploid yeast and *Paramecium*, where dosage balance has also been found to play a role in duplicate gene retention.

In sum, neofunctionalization and subfunctionalization are often considered as the only two alternatives to consider for duplicate gene retention and nonadditive gene expression. However, the mechanisms for duplicate gene retention include an array of mechanisms that span various types of selection and neutral evolution. In addition to various mechanisms being difficult to distinguish, they may also overlap over time or even be complementary. For example, retention of duplicate genes right after a polyploid event may be due to relative dosage balance, which would allow for a longer period of retention for other forces such as absolute dosage, subfunctionalization, or neofunctionalization to occur (15, 120). This pluralistic framework is timely, given that data from genomes and transcriptomes alone are now being seen as insufficient for explaining nonadditive gene expression, and new systems biology methods and frameworks will allow for new types of investigations and novel explanations of nonadditive gene expression following polyploidy (44).

ALTERNATIVE SPLICING AND NONADDITIVE EXPRESSION

RNA AS occurs after a precursor mRNA (pre-mRNA) transcript forms from template DNA (11, 87, 146, 171). In this process, introns in the pre-mRNA are removed, and exons are reconnected in multiple ways (11, 146) (**Figure 7**). The frequency of AS varies greatly. In humans, >95% of genes are alternatively spliced, whereas in *A. thaliana*, 61% of intron-containing genes show AS. AS can influence gene expression on several levels: (*a*) AS creates multiple forms of mRNA from a single gene, which then create multiple types of protein isoforms; (*b*) studies in *Arabidopsis* (50, 84, 130) indicate that AS could influence mRNA stability through the nonsense-mediated decay pathway; and (*c*) studies in *Arabidopsis* and rice suggest that AS can modulate mRNA stability and translation through miRNA regulation (71, 121, 172, 192, 193).

Studies of AS in plants have progressed rapidly with the development of high-throughput approaches, such as RNA-Seq and large-scale microarrays. In *Arabidopsis*, AS frequency was first estimated using EST data as 1.2% of total intron-containing genes (199), whereas in a recent genome-wide analysis, this frequency increased to 61% (116). Plant species for which AS analyses are under way include *A. thaliana* (116), *Oryza sativa* (195), *Z. mays* (107), *Solanum tuberosum* (111), *Physcomitrella patens* (189), *T. aestivum* (174), *Brachypodium distachyon* (178), *B. napus* (198), *G. raimondii* (108), and *Glycine max* (152).

Although the important role of AS is now appreciated in eukaryotes (e.g., 146, 147, 171), few studies have analyzed the impact of either gene duplication or polyploidy on AS. Fractionation, neofunctionalization (131), and subfunctionalization (55, 78) are important processes that occur following polyploidy, but the impact of these on alternative transcript processing after WGD is unknown. Early studies suggest a negative correlation between AS and genome duplication and that alternatively spliced isoforms between gene duplicates may differ dramatically (80, 169, 173). More recent investigations suggest a more complex correlation between AS and gene duplication (32, 82, 151). AS frequency decreases significantly with the increase of family size, whereas among singletons and small families, AS frequency may increase. Consistent with the notion that duplication may affect AS potential, Zhang et al. (197) found that exonic splicing enhancers and silencers rapidly diverge after gene duplication, whereas Santos et al. (155) present evidence of isoform loss

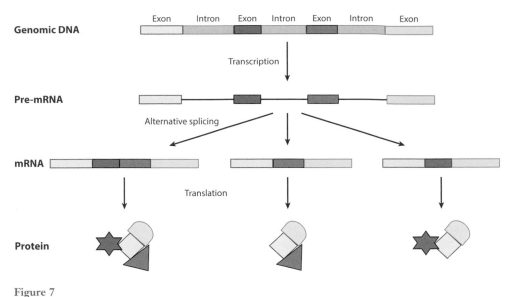

Figure 7

RNA alternative splicing process, from single gene to multiple types of protein, using exon skipping as an example. Colored and gray bars indicate exon and intron, respectively.

and neofunctionalization after duplication. Zhang et al. (196) provide evidence of divergence of AS patterns following gene and genome duplication in *Arabidopsis*; some of the differences occur in an organ- or stress-specific manner.

Allopolyploidization could inhibit the efficient splicing of the *DREB2* homolog *WDREB2* in hexaploid wheat (174). In *B. napus*, 16 of 82 AS events (20%) showed AS changes associated with polyploidy, indicating that AS patterns can change rapidly after genome doubling (198). Also, two independently synthesized tetraploid lines showed parallel loss of AS events after polyploidy, which indicated that some changes may be repeated after polyploidization (198).

More investigations of AS following polyploidy are clearly needed. A new set of major questions can now be posed: Given that AS increases proteomic flexibility, what occurs following

Figure 8

(*a*) No alternative splicing (AS) events are present in the diploids at a given gene, but a new AS isoform is present after polyploidization: gain of AS in homeologs from at least one parent. T1–T3: gain of AS in one or both homeologs without homeolog loss/silencing. T4–T5: AS pattern changes associated with homeolog loss; only one homeolog could be found in the genome, constitutive splicing (CS) mRNA products, and AS mRNA products. T6–T7: AS pattern changes associated with homeolog silencing; homeologs from both parents could be found in the genome, but only one homeolog could be found in CS mRNA products and AS mRNA products. (*b*) AS events were observed in both diploids at a given gene, and the AS pattern changes after polyploidization: gains or losses of AS in the homeologs from the different parents. T1: no change in homeolog AS patterns; homeologs from both parents could be found in CS mRNA products and in AS mRNA products. T2: gain of novel AS isoform; homeologs from both parents could be found in CS mRNA products and in AS mRNA products. T3–T5: loss of AS in one or both homeologs without homeolog loss/silencing; homeologs from both parents could be found in the genome and CS mRNA products, but only one homeolog could be found in AS mRNA products. T6–T9: AS pattern changes due to homeolog loss; only one homeolog could be found in the genome, CS mRNA products, and AS mRNA products. T10–T13: AS pattern changes due to homeolog silencing; homeologs from both parents could be found in the genome, but only one homeolog could be found in CS mRNA products and AS mRNA products.

polyploidy? Are both parental AS profiles maintained? Does one parent dominate? How much novel AS occurs? Considering the differing homeologs contributed by the diploid parents to an allotetraploid, several possible AS patterns might occur in an allopolyploid (**Figure 8**). If no AS event is detected in the diploid parents, new AS events could occur in one or both homeologs after polyploidization (**Figure 8a**, T1–T7). If the diploid parents have AS events, the polyploid may retain, lose, or gain novel AS events in one or both homeologs (**Figure 8b**, T1–T13). Homeolog loss/silencing after polyploidization could also change AS patterns, and the splicing transcripts caused by homeolog loss and homeolog silencing could be the same (e.g., T4 versus T6 in **Figure 8a**; T6 versus T10 in **Figure 8b**; transcripts are showed in dashed boxes in **Figure 8**).

a No AS in diploids, but AS after polyploidization

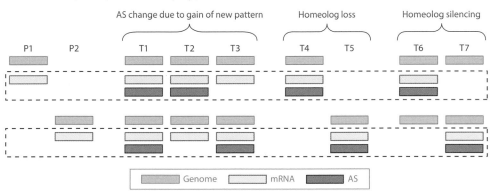

b AS changes after polyploidization

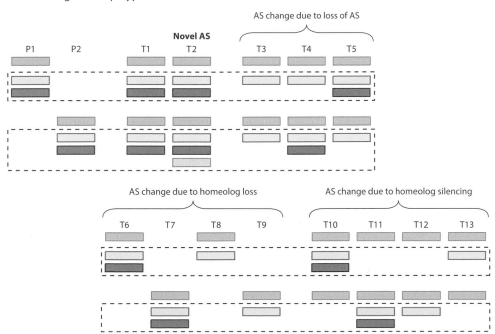

FUTURE STUDIES

More Homeolog-Specific Data and More Taxa

Although numerous studies have provided many new insights into the genetic and genomic consequences of polyploidy, we still know very little about patterns of nonadditive gene expression. In large part, this is because appropriate data that distinguish the patterns of individual homeologs are available only for a few polyploid species. Furthermore, the data available represent only a few angiosperms: We need more data for angiosperms, and we have no data for nonangiosperm lineages. Given the importance of polyploidy in ferns and lycophytes, these lineages also require study. With new technologies, it should be possible to obtain homeolog-specific expression data for more polyploids, and not just for a few well-studied models, enabling the search for broader patterns.

Do Differences in Expression Translate to Phenotype (i.e., the Proteome)?

Many studies reveal that polyploid genomes undergo major chromosomal, genomic, and genetic changes (e.g., 6, 20, 21, 23, 34, 38, 49, 57, 68, 81, 184, 190). Despite great progress in clarifying the genomic and transcriptomic changes that accompany polyploidization, few studies have explicitly correlated these to changes in phenotype (but see 57). New "omics" technologies can now be used in a high-throughput manner to investigate phenotypic changes, including the proteome, metabolome, and important agronomic traits that can be measured in phenotyping facilities.

The impact of WGD on the proteome has been one of the first applications of these technologies (9, 10, 24, 59, 75, 76, 91, 93, 128). Given that the functional states of proteins directly affect molecular and biochemical events that determine phenotype, investigating how changes in genomes, gene expression profiles, and AS events relate to protein-level changes is essential for understanding more fully the molecular and evolutionary consequences of polyploidy, including molecular, biochemical, and physiological mechanisms that ultimately result in evolutionary change. However, the proteome is not necessarily a complete reflection of what is transcribed. A comparison of transcriptomes and proteomes revealed that ~62% of the transcript changes did not reflect differential protein abundance in resynthesized *B. napus* relative to its diploid progenitors (115). This discordance between transcriptome and proteome was also reported from other polyploids, including cotton (76) and synthetic *A. suecica* (128). However, recent studies suggest that an underestimation of mRNA and/or protein abundance may have contributed to the poor correlation between transcriptome and proteome (106, 129). Indeed, the cited studies examined the steady-state levels of RNA and protein. It is possible that the rates of synthesis of proteins might be the critical aspect for assembly of enzymatic and structural features of the cell that ultimately impact how WGD fractionates over time. Improved technology will soon help to better quantify the transcriptome versus proteome, providing new insights into the correlation between gene and protein expression.

What Is the Role of Alternative Splicing?

Studies of diverse eukaryotes reveal the enormous importance of AS in increasing protein flexibility. However, few data are available for polyploids: This is a crucial area of future research. Compared with singletons, a duplicated gene could result in a further increase in AS (32, 82), thus yielding additional genetic and protein flexibility. Extensive gene duplication via polyploidy could provide even greater opportunity for AS and protein diversity. Data on AS in polyploids are limited; a single study has found slightly decreased AS in synthetic *B. napus* (198), but more studies are needed to determine the generality of this result.

During polyploidization, genomes from different species are merged to form one new genome. During this process, immediate changes in the abundance, composition, and activity of splicing factors could occur (11, 146, 171). Such changes and further changes to splicing sites could affect the presence and absence of AS in polyploids. Predicting the presence and absence of AS isoforms after polyploidization becomes complicated. First, AS could create new transcript isoforms and hence change the protein's amino acid sequence and domain arrangement. As a result, subcellular localization, stability, and function of the resulting protein may differ (11, 147). Second, AS can regulate transcript levels through the nonsense-mediated decay (NMD) pathway (146, 171). After AS, new premature termination codons could be introduced into transcript isoforms; these then trigger NMD pathway degradation. Third, AS could also modulate miRNA-mediated regulation of gene expression via retention or loss of miRNA target sites in some AS isoforms, and via the regulation of splicing of pri-miRNAs (primary transcripts of miRNAs) (146).

All of these consequences may affect the adaptation and evolution of a polyploid species in response to different environmental conditions (171). Although more work is needed to determine the frequency of AS events after polyploidization, current studies suggest that AS change is a non-random response to polyploidization (198). There may be some parental homeolog preferences, as well as tissue-specific and stress-specific AS patterns, in a polyploid plant. To investigate the role of AS after polyploidy, more quantitative and functional data are necessary. Currently, such data for a given AS isoform are not available for polyploid plants. One obvious reason for these limitations is the lack of well-annotated genome references for most polyploid plants, especially nonmodel species. By applying high-throughput sequencing and computational approaches, and perhaps a plant in vitro splicing system, AS may be more readily addressed in the near future. We have only scratched the surface in terms of the role of AS in gene expression following polyploidy.

Taking Complexity into Account

As reviewed, multiple factors are involved in nonadditive gene expression; these are not easily distinguishable because they may overlap, interact, shift, and vary temporally and spatially in regulating gene expression. This complexity challenges researchers to elucidate an explicit mechanism for nonadditive gene expression in allopolyploids—this represents one of the major avenues of future research needed. Therefore, a pluralistic view is very important for exploring nonadditive gene expression in which duplicate genes might have been preserved by interaction of multiple overlapping and shifting mechanisms over time.

DISCLOSURE STATEMENT

The authors are not aware of any affiliations, memberships, funding, or financial holdings that might be perceived as affecting the objectivity of this review.

LITERATURE CITED

1. Adams KL, Cronn R, Percifield R, Wendel JF. 2003. Genes duplicated by polyploidy show unequal contributions to the transcriptome and organ-specific reciprocal silencing. *Proc. Natl. Acad. Sci. USA* 100:4649–54
2. Adams KL, Percifield R, Wendel JF. 2004. Organ-specific silencing of duplicated genes in a newly synthesized cotton allotetraploid. *Genetics* 168:2217–26

3. Adams KL, Wendel JF. 2004. Exploring the genomic mysteries of polyploidy in cotton. *Biol. J. Linn. Soc.* 82:573–81

4. Adams KL, Wendel JF. 2005. Allele-specific, bidirectional silencing of an alcohol dehydrogenase gene in different organs of interspecific diploid cotton hybrids. *Genetics* 171:2139–42

5. Adams KL, Wendel JF. 2005. Novel patterns of gene expression in polyploid plants. *Trends Genet.* 21:539–43

6. Ainouche M, Chelaifa H, Ferreira J, Bellot S, Ainouche A, Salmon A. 2012. Polyploid evolution in *Spartina*: dealing with highly redundant hybrid genomes. See Ref. 166, pp. 225–44

7. Akama S, Shimizu-Inatsugi R, Shimizu KK, Sese J. 2014. Genome-wide quantification of homeolog expression ratio revealed nonstochastic gene regulation in synthetic allopolyploid *Arabidopsis*. *Nucleic Acids Res.* 42:e46

8. Akhunova AR, Matniyazov RT, Liang H, Akhunov ED. 2010. Homoeolog-specific transcriptional bias in allopolyploid wheat. *BMC Genomics* 11:505

9. Albertin W, Alix K, Balliau T, Brabant P, Davanture M, et al. 2007. Differential regulation of gene products in newly synthesized *Brassica napus* allotetraploids is not related to protein function nor subcellular localization. *BMC Genomics* 8:56

10. Albertin W, Balliau T, Brabant P, Chevre AM, Eber F, et al. 2006. Numerous and rapid nonstochastic modifications of gene products in newly synthesized *Brassica napus* allotetraploids. *Genetics* 173:1101–13

11. Barbazuk WB, Fu Y, McGinnis KM. 2008. Genome-wide analyses of alternative splicing in plants: opportunities and challenges. *Genome Res.* 18:1381–92

12. Bardil A, de Almeida JD, Combes MC, Lashermes P, Bertrand B. 2011. Genomic expression dominance in the natural allopolyploid *Coffea arabica* is massively affected by growth temperature. *New Phytol.* 192:760–74

13. Barker MS, Kane NC, Matvienko M, Kozik A, Michelmore RW, et al. 2008. Multiple paleopolyploidizations during the evolution of the Compositae reveal parallel patterns of duplicate gene retention after millions of years. *Mol. Biol. Evol.* 25:2445–55

14. Bekaert M, Edger PP, Hudson CM, Pires JC, Conant GC. 2012. Metabolic and evolutionary costs of herbivory defense: systems biology of glucosinolate synthesis. *New Phytol.* 196:596–605

15. Bekaert M, Edger PP, Pires JC, Conant GC. 2011. Two-phase resolution of polyploidy in the *Arabidopsis* metabolic network gives rise to relative and absolute dosage constraints. *Plant Cell* 23:1719–28

16. Birchler JA, Veitia RA. 2012. Gene balance hypothesis: connecting issues of dosage sensitivity across biological disciplines. *Proc. Natl. Acad. Sci. USA* 109:14746–53

17. Blanc G, Wolfe KH. 2004. Functional divergence of duplicated genes formed by polyploidy during *Arabidopsis* evolution. *Plant Cell* 16:1679–91

18. Blanc G, Wolfe KH. 2004. Widespread paleopolyploidy in model plant species inferred from age distributions of duplicate genes. *Plant Cell* 16:1667–78

19. Bottley A, Xia GM, Koebner RM. 2006. Homoeologous gene silencing in hexaploid wheat. *Plant J.* 47:897–906

20. Buggs RJ, Chamala S, Wu W, Gao L, May GD, et al. 2010. Characterization of duplicate gene evolution in the recent natural allopolyploid *Tragopogon miscellus* by next-generation sequencing and Sequenom iPLEX MassARRAY genotyping. *Mol. Ecol.* 19(Suppl. 1):132–46

21. Buggs RJ, Chamala S, Wu W, Tate JA, Schnable PS, et al. 2012. Rapid, repeated, and clustered loss of duplicate genes in allopolyploid plant populations of independent origin. *Curr. Biol.* 22:248–52

22. Buggs RJ, Wendel JF, Doyle JJ, Soltis DE, Soltis PS, Coate JE. 2014. The legacy of diploid progenitors in allopolyploid gene expression patterns. *Philos. Trans. R. Soc. B* 369:pi:20130354

23. Buggs RJ, Zhang L, Miles N, Tate JA, Gao L, et al. 2011. Transcriptomic shock generates evolutionary novelty in a newly formed, natural allopolyploid plant. *Curr. Biol.* 21:551–56

24. Carpentier SC, Panis B, Renaut J, Samyn B, Vertommen A, et al. 2011. The use of 2D-electrophoresis and de novo sequencing to characterize inter- and intra-cultivar protein polymorphisms in an allopolyploid crop. *Phytochemistry* 72:1243–50

25. Cenci A, Combes MC, Lashermes P. 2012. Genome evolution in diploid and tetraploid *Coffea* species as revealed by comparative analysis of orthologous genome segments. *Plant Mol. Biol.* 78:135–45

26. Chague V, Just J, Mestiri I, Balzergue S, Tanguy AM, et al. 2010. Genome-wide gene expression changes in genetically stable synthetic and natural wheat allohexaploids. *New Phytol.* 187:1181–94

27. Chang PL, Dilkes BP, McMahon M, Comai L, Nuzhdin SV. 2010. Homoeolog-specific retention and use in allotetraploid *Arabidopsis suecica* depends on parent of origin and network partners. *Genome Biol.* 11:R125

28. Chase CD. 2007. Cytoplasmic male sterility: a window to the world of plant mitochondrial-nuclear interactions. *Trends Genet.* 23:81–90

29. Chaudhary B, Flagel L, Stupar RM, Udall JA, Verma N, et al. 2009. Reciprocal silencing, transcriptional bias and functional divergence of homeologs in polyploid cotton (*Gossypium*). *Genetics* 182:503–17

30. Chelaifa H, Mahe F, Ainouche M. 2010. Transcriptome divergence between the hexaploid salt-marsh sister species *Spartina maritima* and *Spartina alterniflora* (Poaceae). *Mol. Ecol.* 19:2050–63

31. Chelaifa H, Monnier A, Ainouche M. 2010. Transcriptomic changes following recent natural hybridization and allopolyploidy in the salt marsh species *Spartina × townsendii* and *Spartina anglica* (Poaceae). *New Phytol.* 186:161–74

32. Chen TW, Wu TH, Ng WV, Lin WC. 2011. Interrogation of alternative splicing events in duplicated genes during evolution. *BMC Genomics* 12(Suppl. 3):S16

33. Chen ZJ. 2007. Genetic and epigenetic mechanisms for gene expression and phenotypic variation in plant polyploids. *Annu. Rev. Plant Biol.* 58:377–406

34. Chen ZJ. 2010. Molecular mechanisms of polyploidy and hybrid vigor. *Trends Plant Sci.* 15:57–71

35. Chen ZJ, Pikaard CS. 1997. Epigenetic silencing of RNA polymerase I transcription: a role for DNA methylation and histone modification in nucleolar dominance. *Genes Dev.* 11:2124–36

36. Chen ZJ, Pikaard CS. 1997. Transcriptional analysis of nucleolar dominance in polyploid plants: Biased expression/silencing of progenitor rRNA genes is developmentally regulated in *Brassica*. *Proc. Natl. Acad. Sci. USA* 94:3442–47

37. Cheng F, Wu J, Fang L, Sun S, Liu B, et al. 2012. Biased gene fractionation and dominant gene expression among the subgenomes of *Brassica rapa*. *PLOS ONE* 7:e36442

38. Chester M, Gallagher JP, Symonds VV, Cruz da Silva AV, Mavrodiev EV, et al. 2012. Extensive chromosomal variation in a recently formed natural allopolyploid species, *Tragopogon miscellus* (Asteraceae). *Proc. Natl. Acad. Sci. USA* 109:1176–81

39. Clausen J, Keck DD, Hiesey WM. 1945. *Experimental Studies on the Nature of Species. II. Plant Evolution Through Amphiploidy and Autopolyploidy, with Examples from the Madiinae.* Washington, DC: Carnegie Inst. Wash.

40. Coate JE, Bar H, Doyle JJ. 2014. Extensive translational regulation of gene expression in an allopolyploid (*Glycine dolichocarpa*). *Plant Cell* 26:136–50

41. Comai L, Tyagi AP, Winter K, Holmes-Davis R, Reynolds SH, et al. 2000. Phenotypic instability and rapid gene silencing in newly formed *Arabidopsis* allotetraploids. *Plant Cell* 12:1551–67

42. Combes MC, Cenci A, Baraille H, Bertrand B, Lashermes P. 2012. Homeologous gene expression in response to growing temperature in a recent allopolyploid (*Coffea arabica* L.). *J. Hered.* 103:36–46

43. Combes MC, Dereeper A, Severac D, Bertrand B, Lashermes P. 2013. Contribution of subgenomes to the transcriptome and their intertwined regulation in the allopolyploid *Coffea arabica* grown at contrasted temperatures. *New Phytol.* 200:251–60

44. Conant GC, Birchler JA, Pires JC. 2014. Dosage, duplication, and diploidization: clarifying the interplay of multiple models for duplicate gene evolution over time. *Curr. Opin. Plant Biol.* 19:91–98

45. Conant GC, Wolfe KH. 2008. Turning a hobby into a job: how duplicated genes find new functions. *Nat. Rev. Genet.* 9:938–50

46. Darlington CD. 1937. *Recent Advances in Cytology.* Philadelphia, PA: P. Blakiston's Son Co.

47. Des Marais DL, Rausher MD. 2008. Escape from adaptive conflict after duplication in an anthocyanin pathway gene. *Nature* 454:762–65

48. Doyle JJ, Doyle JL, Rauscher JT, Brown AHD. 2004. Evolution of the perennial soybean polyploid complex (*Glycine* subgenus *Glycine*): a study of contrasts. *Biol. J. Linn. Soc.* 82:583–97

49. Doyle JJ, Flagel LE, Paterson AH, Rapp RA, Soltis DE, et al. 2008. Evolutionary genetics of genome merger and doubling in plants. *Annu. Rev. Genet.* 42:443–61

50. Drechsel G, Kahles A, Kesarwani AK, Stauffer E, Behr J, et al. 2013. Nonsense-mediated decay of alternative precursor mRNA splicing variants is a major determinant of the *Arabidopsis* steady state transcriptome. *Plant Cell* 25:3726–42

51. Feldman M, Liu B, Segal G, Abbo S, Levy AA, Vega JM. 1997. Rapid elimination of low-copy DNA sequences in polyploid wheat: a possible mechanism for differentiation of homoeologous chromosomes. *Genetics* 147:1381–87

52. Flagel L, Udall J, Nettleton D, Wendel J. 2008. Duplicate gene expression in allopolyploid *Gossypium* reveals two temporally distinct phases of expression evolution. *BMC Biol.* 6:16

53. Flagel LE, Chen L, Chaudhary B, Wendel JF. 2009. Coordinated and fine-scale control of homoeologous gene expression in allotetraploid cotton. *J. Hered.* 100:487–90

54. Flagel LE, Wendel JF. 2010. Evolutionary rate variation, genomic dominance and duplicate gene expression evolution during allotetraploid cotton speciation. *New Phytol.* 186:184–93

55. Force A, Lynch M, Pickett FB, Amores A, Yan YL, Postlethwait J. 1999. Preservation of duplicate genes by complementary, degenerate mutations. *Genetics* 151:1531–45

56. Freeling M. 2009. Bias in plant gene content following different sorts of duplication: tandem, whole-genome, segmental, or by transposition. *Annu. Rev. Plant Biol.* 60:433–53

57. Gaeta RT, Pires JC, Iniguez-Luy F, Leon E, Osborn TC. 2007. Genomic changes in resynthesized *Brassica napus* and their effect on gene expression and phenotype. *Plant Cell* 19:3403–17

58. Gaeta RT, Yoo SY, Pires JC, Doerge RW, Chen ZJ, Osborn TC. 2009. Analysis of gene expression in resynthesized *Brassica napus* allopolyploids using *Arabidopsis* 70mer oligo microarrays. *PLOS ONE* 4:e4760

59. Gancel AL, Grimplet J, Sauvage FX, Ollitrault P, Brillouet JM. 2006. Predominant expression of diploid mandarin leaf proteome in two citrus mandarin-derived somatic allotetraploid hybrids. *J. Agric. Food Chem.* 54:6212–18

60. Garsmeur O, Schnable JC, Almeida A, Jourda C, D'Hont A, Freeling M. 2014. Two evolutionarily distinct classes of paleopolyploidy. *Mol. Biol. Evol.* 31:448–54

61. Gong L, Salmon A, Yoo MJ, Grupp KK, Wang Z, et al. 2012. The cytonuclear dimension of allopolyploid evolution: an example from cotton using rubisco. *Mol. Biol. Evol.* 29:3023–36

62. Gottlieb LD. 2003. Plant polyploidy: gene expression and genetic redundancy. *Heredity* 91:91–92

63. Grant V. 1981. *Plant Speciation*. New York: Columbia Univ. Press

64. Grover CE, Gallagher JP, Szadkowski EP, Yoo MJ, Flagel LE, Wendel JF. 2012. Homoeolog expression bias and expression level dominance in allopolyploids. *New Phytol.* 196:966–71

65. Grover CE, Kim H, Wing RA, Paterson AH, Wendel JF. 2004. Incongruent patterns of local and global genome size evolution in cotton. *Genome Res.* 14:1474–82

66. Grover CE, Kim H, Wing RA, Paterson AH, Wendel JF. 2007. Microcolinearity and genome evolution in the AdhA region of diploid and polyploid cotton (*Gossypium*). *Plant J.* 50:995–1006

67. Ha M, Kim ED, Chen ZJ. 2009. Duplicate genes increase expression diversity in closely related species and allopolyploids. *Proc. Natl. Acad. Sci. USA* 106:2295–300

68. Hegarty MJ, Abbott RJ, Hiscock SJ. 2012. Allopolyploid speciation in action: the origins and evolution of *Senecio cambrensis*. See Ref. 166, pp. 245–70

69. Hegarty MJ, Barker GL, Brennan AC, Edwards KJ, Abbott RJ, Hiscock SJ. 2008. Changes to gene expression associated with hybrid speciation in plants: further insights from transcriptomic studies in *Senecio*. *Philos. Trans. R. Soc. Lond. B.* 363:3055–69

70. Hegarty MJ, Barker GL, Wilson ID, Abbott RJ, Edwards KJ, Hiscock SJ. 2006. Transcriptome shock after interspecific hybridization in *Senecio* is ameliorated by genome duplication. *Curr. Biol.* 16:1652–59

71. Hirsch J, Lefort V, Vankersschaver M, Boualem A, Lucas A, et al. 2006. Characterization of 43 non-protein-coding mRNA genes in *Arabidopsis*, including the MIR162a-derived transcripts. *Plant Physiol.* 140:1192–204

72. Hollister JD, Gaut BS. 2009. Epigenetic silencing of transposable elements: a trade-off between reduced transposition and deleterious effects on neighboring gene expression. *Genome Res.* 19:1419–28

73. Hollister JD, Smith LM, Guo YL, Ott F, Weigel D, Gaut BS. 2011. Transposable elements and small RNAs contribute to gene expression divergence between *Arabidopsis thaliana* and *Arabidopsis lyrata*. *Proc. Natl. Acad. Sci. USA* 108:2322–27

74. Hovav R, Udall JA, Chaudhary B, Rapp R, Flagel L, Wendel JF. 2008. Partitioned expression of duplicated genes during development and evolution of a single cell in a polyploid plant. *Proc. Natl. Acad. Sci. USA* 105:6191–95

75. Hu G, Houston NL, Pathak D, Schmidt L, Thelen JJ, Wendel JF. 2011. Genomically biased accumulation of seed storage proteins in allopolyploid cotton. *Genetics* 189:1103–15

76. Hu G, Koh J, Yoo MJ, Grupp K, Chen S, Wendel JF. 2013. Proteomic profiling of developing cotton fibers from wild and domesticated *Gossypium barbadense*. *New Phytol.* 200:570–82

77. Hudson CM, Puckett EE, Bekaert M, Pires JC, Conant GC. 2011. Selection for higher gene copy number after different types of plant gene duplications. *Genome Biol. Evol.* 3:1369–80

78. Hughes AL. 1994. The evolution of functionally novel proteins after gene duplication. *Proc. Biol. Sci.* 256:119–24

79. Ilut DC, Coate JE, Luciano AK, Owens TG, May GD, et al. 2012. A comparative transcriptomic study of an allotetraploid and its diploid progenitors illustrates the unique advantages and challenges of RNA-Seq in plant species. *Am. J. Bot.* 99:383–96

80. Irimia M, Rukov JL, Penny D, Garcia-Fernandez J, Vinther J, Roy SW. 2008. Widespread evolutionary conservation of alternatively spliced exons in *Caenorhabditis*. *Mol. Biol. Evol.* 25:375–82

81. Jackson S, Chen ZJ. 2010. Genomic and expression plasticity of polyploidy. *Curr. Opin. Plant Biol.* 13:153–59

82. Jin L, Kryukov K, Clemente JC, Komiyama T, Suzuki Y, et al. 2008. The evolutionary relationship between gene duplication and alternative splicing. *Gene* 427:19–31

83. Joly S, Rauscher JT, Sherman-Broyles SL, Brown AHD, Doyle JJ. 2004. Evolutionary dynamics and preferential expression of homeologous 18S-5.8S-26S nuclear ribosomal genes in natural and artificial glycine allopolyploids. *Mol. Biol. Evol.* 21:1409–21

84. Kalyna M, Simpson CG, Syed NH, Lewandowska D, Marquez Y, et al. 2012. Alternative splicing and nonsense-mediated decay modulate expression of important regulatory genes in *Arabidopsis*. *Nucleic Acids Res.* 40:2454–69

85. Kashkush K, Feldman M, Levy AA. 2002. Gene loss, silencing and activation in a newly synthesized wheat allotetraploid. *Genetics* 160:1651–59

86. Kashkush K, Feldman M, Levy AA. 2003. Transcriptional activation of retrotransposons alters the expression of adjacent genes in wheat. *Nat. Genet.* 33:102–6

87. Kazan K. 2003. Alternative splicing and proteome diversity in plants: the tip of the iceberg has just emerged. *Trends Plant Sci.* 8:468–71

88. Kenton A, Parokonny AS, Gleba YY, Bennett MD. 1993. Characterization of the *Nicotiana tabacum* L. genome by molecular cytogenetics. *Mol. Gen. Genet.* 240:159–69

89. Kihara H, Ono T. 1926. Chromosomenzahlen und Systematische Gruppierung der Rumex-Arten. *Zeitschr. Zellf. Mikrosk. Anat.* 4:475–81

90. Kim ED, Chen ZJ. 2011. Unstable transcripts in *Arabidopsis* allotetraploids are associated with nonadditive gene expression in response to abiotic and biotic stresses. *PLOS ONE* 6:e24251

91. Koh J, Chen S, Zhu N, Yu F, Soltis PS, Soltis DE. 2012. Comparative proteomics of the recently and recurrently formed natural allopolyploid *Tragopogon mirus* (Asteraceae) and its parents. *New Phytol.* 196:292–305

92. Koh J, Soltis PS, Soltis DE. 2010. Homeolog loss and expression changes in natural populations of the recently and repeatedly formed allotetraploid *Tragopogon mirus* (Asteraceae). *BMC Genomics* 11:97

93. Kong F, Mao S, Jiang J, Wang J, Fang X, Wang Y. 2011. Proteomic changes in newly synthesized *Brassica napus* allotetraploids and their early generations. *Plant Mol. Biol. Rep.* 29:927–35

94. Kovarik A, Matyasek R, Lim KY, Skalicka K, Koukalova B, et al. 2004. Concerted evolution of 18–5.8–26S rDNA repeats in *Nicotiana* allotetraploids. *Biol. J. Linn. Soc.* 82:615–25

95. Kraitshtein Z, Yaakov B, Khasdan V, Kashkush K. 2010. Genetic and epigenetic dynamics of a retrotransposon after allopolyploidization of wheat. *Genetics* 186:801–12

96. Kuwada Y. 1911. Meiosis in the pollen mother cells of *Zea mays* L. *Bot. Mag. Tokyo* 25:163–81

97. Lashermes P, Paczek V, Trouslot P, Combes MC, Couturon E, Charrier A. 2000. Single-locus inheritance in the allotetraploid *Coffea arabica* L. and interspecific hybrid *C. arabica* × *C. canephora*. *J. Hered.* 91:81–85

98. Leach LJ, Belfield EJ, Jiang C, Brown C, Mithani A, Harberd NP. 2014. Patterns of homoeologous gene expression shown by RNA sequencing in hexaploid bread wheat. *BMC Genomics* 15:276

99. Lee HS, Chen ZJ. 2001. Protein-coding genes are epigenetically regulated in *Arabidopsis* polyploids. *Proc. Natl. Acad. Sci. USA* 98:6753–58

100. Leitch IJ, Bennett MD. 2004. Genome downsizing in polyploid plants. *Biol. J. Linn. Soc.* 82:651–63

101. Leon P, Arroyo A, Mackenzie S. 1998. Nuclear control of plastid and mitochondrial development in higher plants. *Annu. Rev. Plant Physiol. Plant Mol. Biol.* 49:453–80

102. Levin DA. 1983. Polyploidy and novelty in flowering plants. *Am. Nat.* 122:1–25

103. Levin DA. 2002. *The Role of Chromosomal Change in Plant Evolution*. Oxford: Oxford Univ. Press

104. Levy AA, Feldman M. 2004. Genetic and epigenetic reprogramming of the wheat genome upon allopolyploidization. *Biol. J. Linn. Soc.* 82:607–13

105. Lewis WH. 1980. Polyploidy in angiosperms: dicotyledons. In *Polyploidy: Biological Relevance*, ed. WH Lewis, pp. 241–68. New York: Plenum Press

106. Li JJ, Bickel PJ, Biggin MD. 2014. System wide analyses have underestimated protein abundances and the importance of transcription in mammals. *PeerJ.* 2:e270

107. Li P, Ponnala L, Gandotra N, Wang L, Si Y, et al. 2010. The developmental dynamics of the maize leaf transcriptome. *Nat. Genet.* 42:1060–67

108. Li Q, Xiao G, Zhu YX. 2014. Single-nucleotide resolution mapping of the *Gossypium raimondii* transcriptome reveals a new mechanism for alternative splicing of introns. *Mol. Plant.* 7:829–40

109. Lim KY, Matyasek R, Kovarik A, Leitch AR. 2004. Genome evolution in allotetraploid *Nicotiana*. *Biol. J. Linn. Soc.* 82:599–606

110. Liu B, Wendel JF. 2003. Epigenetic phenomena and the evolution of plant allopolyploids. *Mol. Phylogenet. Evol.* 29:365–79

111. Lozano R, Ponce O, Ramirez M, Mostajo N, Orjeda G. 2012. Genome-wide identification and mapping of NBS-encoding resistance genes in *Solanum tuberosum* group Phureja. *PLOS ONE* 7:e34775

112. Lukens LN, Pires JC, Leon E, Vogelzang R, Oslach L, Osborn T. 2006. Patterns of sequence loss and cytosine methylation within a population of newly resynthesized *Brassica napus* allopolyploids. *Plant Physiol.* 140:336–48

113. Lynch M, Force AG. 2000. The origin of interspecific genomic incompatibility via gene duplication. *Am. Nat.* 156:590–605

114. Madlung A, Tyagi AP, Watson B, Jiang HM, Kagochi T, et al. 2005. Genomic changes in synthetic *Arabidopsis* polyploids. *Plant J.* 41:221–30

115. Marmagne A, Brabant P, Thiellement H, Alix K. 2010. Analysis of gene expression in resynthesized *Brassica napus* allotetraploids: transcriptional changes do not explain differential protein regulation. *New Phytol.* 186:216–27

116. Marquez Y, Brown JW, Simpson C, Barta A, Kalyna M. 2012. Transcriptome survey reveals increased complexity of the alternative splicing landscape in *Arabidopsis*. *Genome Res.* 22:1184–95

117. Mayfield-Jones D, Washburn JD, Arias T, Edger PP, Pires JC, Conant GC. 2013. Watching the grin fade: tracing the effects of polyploidy on different evolutionary time scales. *Semin. Cell Dev. Biol.* 24:320–31

118. McClintock B. 1984. The significance of responses of the genome to challenge. *Science* 226:792–801

119. McGrath CL, Gout JF, Johri P, Doak TG, Lynch M. 2014. Differential retention and divergent resolution of duplicate genes following whole-genome duplication. *Genome Res.* doi:10.1101/gr.173740.114

120. McGrath C, Lynch M. 2012. Evolutionary significance of whole-genome duplication. See Ref. 166, pp. 1–20

121. Meng Y, Shao C, Ma X, Wang H. 2013. Introns targeted by plant microRNAs: a possible novel mechanism of gene regulation. *Rice* 6:8

122. Mochida K, Yamazaki Y, Ogihara Y. 2003. Discrimination of homoeologous gene expression in hexaploid wheat by SNP analysis of contigs grouped from a large number of expressed sequence tags. *Mol. Genet. Genomics* 270:371–77

123. Moghe GD, Hufnagel DE, Tang H, Xiao Y, Dworkin I, et al. 2014. Consequences of whole-genome triplication as revealed by comparative genomic analyses of the wild radish *Raphanus raphanistrum* and three other Brassicaceae species. *Plant Cell* 26(5):1925–37

124. Müntzing A. 1936. The evolutionary significance of autopolyploidy. *Hereditas* 21:263–378

125. Murat F, Zhang R, Guizard S, Flores R, Armero A, et al. 2014. Shared subgenome dominance following polyploidization explains grass genome evolutionary plasticity from a seven protochromosome ancestor with 16K protogenes. *Genome Biol. Evol.* 6:12–33

126. Nagaharu U. 1935. Genome analysis in *Brassica* with special reference to the experimental formation of *B. napus* and peculiar mode of fertilization. *Jpn. J. Bot.* 7:389–452

127. Ng DW, Lu J, Chen ZJ. 2012. Big roles for small RNAs in polyploidy, hybrid vigor, and hybrid incompatibility. *Curr. Opin. Plant Biol.* 15:154–61

128. Ng DW, Zhang C, Miller M, Shen Z, Briggs SP, Chen ZJ. 2012. Proteomic divergence in *Arabidopsis* autopolyploids and allopolyploids and their progenitors. *Heredity* 108:419–30

129. Ning K, Fermin D, Nesvizhskii AI. 2012. Comparative analysis of different label-free mass spectrometry based protein abundance estimates and their correlation with RNA-Seq gene expression data. *J. Proteome Res.* 11:2261–71

130. Nyiko T, Sonkoly B, Merai Z, Benkovics AH, Silhavy D. 2009. Plant upstream ORFs can trigger nonsense-mediated mRNA decay in a size-dependent manner. *Plant Mol. Biol.* 71:367–78

131. Ohno S. 1970. *Evolution by Gene Duplication*. New York: Springer-Verlag

132. Osborn TC, Pires JC, Birchler JA, Auger DL, Chen ZJ, et al. 2003. Understanding mechanisms of novel gene expression in polyploids. *Trends Genet.* 19:141–47

133. Ownbey M. 1950. Natural hybridization and amphiploidy in the genus *Tragopogon. Am. J. Bot.* 37:487–99

134. Ozkan H, Levy AA, Feldman M. 2001. Allopolyploidy-induced rapid genome evolution in the wheat (*Aegilops-Triticum*) group. *Plant Cell* 13:1735–47

135. Ozkan H, Levy AA, Feldman M. 2002. Rapid differentiation of homeologous chromosomes in newly-formed allopolyploid wheat. *Isr. J. Plant Sci.* 50:S65–76

136. Parisod C, Salmon A, Zerjal T, Tenaillon M, Grandbastien MA, Ainouche M. 2009. Rapid structural and epigenetic reorganization near transposable elements in hybrid and allopolyploid genomes in *Spartina. New Phytol.* 184:1003–15

137. Parkin IA, Gulden SM, Sharpe AG, Lukens L, Trick M, et al. 2005. Segmental structure of the *Brassica napus* genome based on comparative analysis with *Arabidopsis thaliana. Genetics* 171:765–81

138. Parkin IAP, Sharpe AG, Keith DJ, Lydiate DJ. 1995. Identification of the A and C genomes of amphidiploid *Brassica napus* (oilseed rape). *Genome* 38:1122–31

139. Paterson AH, Chapman BA, Kissinger JC, Bowers JE, Feltus FA, Estill JC. 2006. Many gene and domain families have convergent fates following independent whole-genome duplication events in *Arabidopsis, Oryza, Saccharomyces* and *Tetraodon. Trends Genet.* 22:597–602

140. Pires JC, Zhao JW, Schranz ME, Leon EJ, Quijada PA, et al. 2004. Flowering time divergence and genomic rearrangements in resynthesized *Brassica* polyploids (Brassicaceae). *Biol. J. Linn. Soc.* 82:675–88

141. Pontes O, Neves N, Silva M, Lewis MS, Madlung A, et al. 2004. Chromosomal locus rearrangements are a rapid response to formation of the allotetraploid *Arabidopsis suecica* genome. *Proc. Natl. Acad. Sci. USA* 101:18240–45

142. Pumphrey M, Bai J, Laudencia-Chingcuanco D, Anderson O, Gill BS. 2009. Nonadditive expression of homoeologous genes is established upon polyploidization in hexaploid wheat. *Genetics* 181:1147–57

143. Qi B, Huang W, Zhu B, Zhong X, Guo J, et al. 2012. Global transgenerational gene expression dynamics in two newly synthesized allohexaploid wheat (*Triticum aestivum*) lines. *BMC Biol.* 10:3

144. Rambani A, Page JT, Udall JA. 2014. Polyploidy and the petal transcriptome of *Gossypium. BMC Plant Biol.* 14:3

145. Rapp RA, Udall JA, Wendel JF. 2009. Genomic expression dominance in allopolyploids. *BMC Biol.* 7:18

146. Reddy AS, Marquez Y, Kalyna M, Barta A. 2013. Complexity of the alternative splicing landscape in plants. *Plant Cell* 25:3657–83

147. Reddy AS, Rogers MF, Richardson DN, Hamilton M, Ben-Hur A. 2012. Deciphering the plant splicing code: experimental and computational approaches for predicting alternative splicing and splicing regulatory elements. *Front. Plant Sci.* 3:18

148. Renny-Byfield S, Gallagher JP, Grover CE, Szadkowski E, Page JT, et al. 2014. Ancient gene duplicates in *Gossypium* (cotton) exhibit near-complete expression divergence. *Genome Biol. Evol.* 6:559–71

149. Rhoades MM. 1951. Duplicate genes in maize. *Am. Nat.* 85:105–10

150. Roulin A, Auer PL, Libault M, Schlueter J, Farmer A, et al. 2012. The fate of duplicated genes in a polyploid plant genome. *Plant J.* 73:143–53

151. Roux J, Robinson-Rechavi M. 2011. Age-dependent gain of alternative splice forms and biased duplication explain the relation between splicing and duplication. *Genome Res.* 21:357–63

152. Sagasti S, Bernal M, Sancho D, del Castillo MB, Picorel R. 2014. Regulation of the chloroplastic copper chaperone (CCS) and cuprozinc superoxide dismutase (CSD2) by alternative splicing and copper excess in *Glycine max*. *Funct. Plant Biol.* 41:144–55

153. Salmon A, Ainouche ML, Wendel JF. 2005. Genetic and epigenetic consequences of recent hybridization and polyploidy in *Spartina* (Poaceae). *Mol. Ecol.* 14:1163–75

154. Salse J, Bolot S, Throude M, Jouffe V, Piegu B, et al. 2008. Identification and characterization of shared duplications between rice and wheat provide new insight into grass genome evolution. *Plant Cell* 20:11–24

155. Santos ME, Athanasiadis A, Leitao AB, DuPasquier L, Sucena E. 2011. Alternative splicing and gene duplication in the evolution of the *FoxP* gene subfamily. *Mol. Biol. Evol.* 28:237–47

156. Schnable JC, Freeling M. 2011. Genes identified by visible mutant phenotypes show increased bias toward one of two subgenomes of maize. *PLOS ONE* 6:e17855

157. Schnable JC, Freeling M. 2012. Maize (*Zea mays*) as a model for studying the impact of gene and regulatory sequence loss following whole-genome duplication. See Ref. 166, pp. 137–45

158. Schnable JC, Springer NM, Freeling M. 2011. Differentiation of the maize subgenomes by genome dominance and both ancient and ongoing gene loss. *Proc. Natl. Acad. Sci. USA* 108:4069–74

159. Sehrish T, Symonds VV, Soltis DE, Soltis PS, Tate JA. 2014. Gene silencing via DNA methylation in naturally occurring *Tragopogon miscellus* (Asteraceae) allopolyploids. *BMC Genomics* 15:701

160. Seoighe C, Gehring C. 2004. Genome duplication led to highly selective expansion of the *Arabidopsis thaliana* proteome. *Trends Genet.* 20:461–64

161. Shaked H, Kashkush K, Ozkan H, Feldman M, Levy AA. 2001. Sequence elimination and cytosine methylation are rapid and reproducible responses of the genome to wide hybridization and allopolyploidy in wheat. *Plant Cell* 13:1749–59

162. Shi X, Ng DW, Zhang C, Comai L, Ye W, Chen ZJ. 2012. *Cis*- and *trans*-regulatory divergence between progenitor species determines gene-expression novelty in *Arabidopsis* allopolyploids. *Nat. Commun.* 3:950

163. Soltis DE, Buggs RJA, Barbazuk WB, Chamala S, Chester M, et al. 2012. Rapid and repeated evolution in the early stages of polyploidy: genomic and cytogenetic studies of recent polyploidy in *Tragopogon*. See Ref. 166, pp. 271–92

164. Soltis DE, Buggs RJA, Barbazuk WB, Schnable PS, Soltis PS. 2009. On the origins of species: Does evolution repeat itself in polyploid populations of independent origin? *Cold Spring Harb. Symp. Quant. Biol.* 74:215–23

165. Soltis DE, Soltis PS, Pires JC, Kovarik A, Tate JA. 2004. Recent and recurrent polyploidy in *Tragopogon* (Asteraceae): genetic, genomic, and cytogenetic comparisons. *Biol. J. Linn. Soc.* 82:485–501

166. Soltis PS, Soltis DE, eds. 2012. *Polyploidy and Genome Evolution*. Heidelberg, Ger.: Springer

167. Song K, Lu P, Tang K, Osborn TC. 1995. Rapid genome change in synthetic polyploids of *Brassica* and its implications for polyploid evolution. *Proc. Natl. Acad. Sci. USA* 92:7719–23

168. Stebbins GL. 1950. *Variation and Evolution in Plants*. New York: Columbia Univ. Press

169. Su Z, Wang J, Yu J, Huang X, Gu X. 2006. Evolution of alternative splicing after gene duplication. *Genome Res.* 16:182–89

170. Swigonova Z, Lai J, Ma J, Ramakrishna W, Llaca V, et al. 2004. Close split of sorghum and maize genome progenitors. *Genome Res.* 14:1916–23

171. Syed NH, Kalyna M, Marquez Y, Barta A, Brown JW. 2012. Alternative splicing in plants: coming of age. *Trends Plant Sci.* 17:616–23

172. Szarzynska B, Sobkowiak L, Pant BD, Balazadeh S, Scheible WR, et al. 2009. Gene structures and processing of *Arabidopsis thaliana* HYL1-dependent pri-miRNAs. *Nucleic Acids Res.* 37:3083–93

173. Talavera D, Vogel C, Orozco M, Teichmann SA, de la Cruz X. 2007. The (in)dependence of alternative splicing and gene duplication. *PLOS Comput. Biol.* 3:e33

174. Terashima A, Takumi S. 2009. Allopolyploidization reduces alternative splicing efficiency for transcripts of the wheat DREB2 homolog, WDREB2. *Genome* 52:100–5

175. Tian L, Li X, Ha M, Zhang C, Chen ZJ. 2014. Genetic and epigenetic changes in a genomic region containing MIR172 in *Arabidopsis* allopolyploids and their progenitors. *Heredity* 112:207–14

176. Udall JA, Quijada PA, Osborn TC. 2005. Detection of chromosomal rearrangements derived from homeologous recombination in four mapping populations of *Brassica napus* L. *Genetics* 169:967–79

177. Veitia RA, Bottani S, Birchler JA. 2013. Gene dosage effects: nonlinearities, genetic interactions, and dosage compensation. *Trends Genet.* 29:385–93

178. Walters B, Lum G, Sablok G, Min XJ. 2013. Genome-wide landscape of alternative splicing events in *Brachypodium distachyon*. *DNA Res.* 20:163–71

179. Wang J, Tian L, Lee HS, Wei NE, Jiang H, et al. 2006. Genomewide nonadditive gene regulation in *Arabidopsis* allotetraploids. *Genetics* 172:507–17

180. Wang X, Wang H, Wang J, Sun R, Wu J, et al. 2011. The genome of the mesopolyploid crop species *Brassica rapa*. *Nat. Genet.* 43:1035–39

181. Wang Z, Gerstein M, Snyder M. 2009. RNA-Seq: a revolutionary tool for transcriptomics. *Nat. Rev. Genet.* 10:57–63

182. Wei F, Coe E, Nelson W, Bharti AK, Engler F, et al. 2007. Physical and genetic structure of the maize genome reflects its complex evolutionary history. *PLOS Genet.* 3:e123

183. Wendel JF, Cronn RC. 2003. Polyploidy and the evolutionary history of cotton. *Adv. Agron.* 78:139–86

184. Wendel JF, Flagel LE, Adams KL. 2012. Jeans, genes, and genomes: cotton as a model for studying polyploidy. See Ref. 166, pp. 181–207

185. Wendel JF, Schnabel A, Seelanan T. 1995. Bidirectional interlocus concerted evolution following allopolyploid speciation in cotton (*Gossypium*). *Proc. Natl. Acad. Sci. USA* 92:280–84

186. Wolf JB. 2009. Cytonuclear interactions can favor the evolution of genomic imprinting. *Evolution* 63:1364–71

187. Wolf JB, Hager R. 2006. A maternal-offspring coadaptation theory for the evolution of genomic imprinting. *PLOS Biol.* 4:e380

188. Woodhouse MR, Cheng F, Pires JC, Lisch D, Freeling M, Wang X. 2014. Origin, inheritance, and gene regulatory consequences of genome dominance in polyploids. *Proc. Natl. Acad. Sci. USA* 111:5283–88

189. Wu HP, Su YS, Chen HC, Chen YR, Wu CC, et al. 2014. Genome-wide analysis of light-regulated alternative splicing mediated by photoreceptors in *Physcomitrella patens*. *Genome Biol.* 15:R10

190. Xiong Z, Gaeta RT, Pires JC. 2011. Homoeologous shuffling and chromosome compensation maintain genome balance in resynthesized allopolyploid *Brassica napus*. *Proc. Natl. Acad. Sci. USA* 108:7908–13

191. Xu JH, Messing J. 2008. Organization of the prolamin gene family provides insight into the evolution of the maize genome and gene duplications in grass species. *Proc. Natl. Acad. Sci. USA* 105:14330–35

192. Yan K, Liu P, Wu CA, Yang GD, Xu R, et al. 2012. Stress-induced alternative splicing provides a mechanism for the regulation of microRNA processing in *Arabidopsis thaliana*. *Mol. Cell* 48:521–31

193. Yang X, Zhang H, Li L. 2012. Alternative mRNA processing increases the complexity of microRNA-based gene regulation in *Arabidopsis*. *Plant J.* 70:421–31

194. Yoo MJ, Szadkowski E, Wendel JF. 2013. Homoeolog expression bias and expression level dominance in allopolyploid cotton. *Heredity* 110:171–80

195. Zhang G, Guo G, Hu X, Zhang Y, Li Q, et al. 2010. Deep RNA sequencing at single base-pair resolution reveals high complexity of the rice transcriptome. *Genome Res.* 20:646–54

196. Zhang PG, Huang SZ, Pin AL, Adams KL. 2010. Extensive divergence in alternative splicing patterns after gene and genome duplication during the evolutionary history of *Arabidopsis*. *Mol. Biol. Evol.* 27:1686–97

197. Zhang Z, Zhou L, Wang P, Liu Y, Chen X, et al. 2009. Divergence of exonic splicing elements after gene duplication and the impact on gene structures. *Genome Biol.* 10:R120

198. Zhou RC, Moshgabadi N, Adams KL. 2011. Extensive changes to alternative splicing patterns following allopolyploidy in natural and resynthesized polyploids. *Proc. Natl. Acad. Sci. USA* 108:16122–27

199. Zhu W, Schlueter SD, Brendel V. 2003. Refined annotation of the *Arabidopsis* genome by complete expressed sequence tag mapping. *Plant Physiol.* 132:469–84

Lineage Sorting in Apes

Thomas Mailund,[1] Kasper Munch,[1] and Mikkel Heide Schierup[1,2]

[1]Bioinformatics Research Centre, Aarhus University, DK-8000 Aarhus C, Denmark;
email: mailund@birc.au.dk, kaspermunch@birc.au.dk, mheide@birc.au.dk

[2]Department of Bioscience, Aarhus University, DK-8000 Aarhus C, Denmark

Annu. Rev. Genet. 2014. 48:519–35

First published online as a Review in Advance on
September 19, 2014

The *Annual Review of Genetics* is online at
genet.annualreviews.org

This article's doi:
10.1146/annurev-genet-120213-092532

Keywords

population genomics, great apes evolution, speciation, incomplete lineage
sorting, phylogenetics

Abstract

Recombination allows different parts of the genome to have different genealogical histories. When a species splits in two, allelic lineages sort into the two descendant species, and this lineage sorting varies along the genome. If speciation events are close in time, the lineage sorting process may be incomplete at the second speciation event and lead to gene genealogies that do not match the species phylogeny. We review different recent approaches to model lineage sorting along the genome and show how it is possible to learn about population sizes, natural selection, and recombination rates in ancestral species from application of these models to genome alignments of great ape species.

SPECIES DIVERGENCE AND GENOME DIVERGENCE

DNA sequencing has had a dramatic effect on the field of molecular phylogenetics. The rate of DNA substitutions often adheres quite closely to a molecular clock, i.e., differences accumulate linearly since the point at which a pair of sequences had a common ancestor. It is therefore tempting to sample pieces or entire genomes from a set of species of interest, build phylogenies, enforce the molecular clock, and assign the dates of speciation events from the resulting tree. The literature is abundant with examples of such practice. For examples of application to primates, see References 16, 17, 45, and 65. Divergence times, however, do not represent speciation times. Assuming for now that speciation occurs as a simple split of one species into two, with no subsequent gene flow, then the most recent common ancestor of sequences from two species is always further back in time than the speciation event. Tracing the ancestry of a human and a chimpanzee sequence back in time to the exact time of speciation, their ancestral sequences will be found in separate individuals in the ancestral population. We would thus need to look further back in time to find their common ancestor.

The difference between species divergence and genome divergence depends on the population size and generation time in the ancestor, as detailed below. In primates, the relative difference between genome divergence time and speciation time is appreciable, as shown in **Table 1**. Generally, more ancient speciation times result in a smaller relative difference between the species split time and the time to most recent common ancestry inferred from genome sequences, as the speciation time takes up more of the total time. However, the divergence time of human and orangutan genome sequences may still exceed the speciation time by 50% because of a large ancestral population size. Using fossil calibration points to date speciation times and then equate speciation and divergence times can thus lead to grossly inaccurate dates.

The divergence time of two sequences from separate species varies along the genome because each sequence consists of many segments with different ancestry separated by recombination. **Table 1** shows that human and chimpanzee genome divergence is on average further back in time than the species divergence of human and gorilla, implying that the common ancestor of many such segments of human and chimpanzees genomes are found in the ancestor to human, chimpanzee, and gorilla. In these cases, a human segment may find a most recent common ancestor

Table 1   The difference between speciation time (population divergence time) and average genomic divergence time for pairs of great ape species[a]

Species 1	Species 2	Generation time	Ancestral population size	Speciation time	Divergence time	Reference
Human	Neanderthal	29	15,000	0.55	1.42	(49)
Human	Chimpanzee	25	86,000	5.6	9.9	(46) ILS model
Human	Gorilla	25	67,000	8.6	12.0	(46) ILS model
Human	Orangutan	25	167,000	18.5	26.9	(46) CTMC model
Human	Macaque	25	196,667	44.0	53.8	(8)
Chimpanzee	Bonobo	25	32,000	1.2	2.8	(46) ILS model
Bornean orangutan	Sumatran orangutan	23	63,000	1	3.9	(46) CTMC model
West gorilla	East gorilla	20	64,000	0.4	2.9	(46) CTMC model

[a]All values assume a mutation rate of 0.6×10^{-9} per base pair per year. The ancestral population size estimate for the human-Neanderthal ancestor is visually estimated from Reference 49.
Abbreviations: CTMC, continuous-time Markov chain; ILS, incomplete lineage sorting.

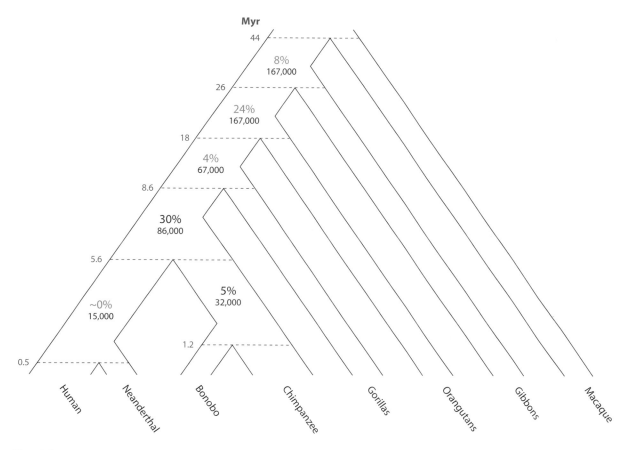

Figure 1

Species tree with speciation times (*blue numbers*) and ancestral population sizes (*red numbers*) as in **Table 1**. Human-gibbon speciation time is derived from an estimated population size shared with the human-orangutan ancestor and the human-gibbon divergence (35a). Black numbers are empirical estimates of incomplete lineage sorting from CoalHMM (coalescent hidden Markov model) analyses of species alignments, and gray numbers represent theoretical expectations from the estimated species divergence times and ancestral population sizes. These numbers are associated with uncertainty. All values assume a substitution rate of 0.6×10^{-9} per base pair per year.

with a gorilla sequence and thus display a genealogy different from the species phylogeny. This phenomenon is termed incomplete lineage sorting (ILS). In the ancestral lineage to human, ILS is expected to affect all the internal branches to various extents, as shown in **Figure 1**. The extent of ILS warrants caution in relying on small DNA fragments, such as single genes, for phylogenetic analysis. However, ILS also provides valuable information about speciation processes and ancestral population processes, including natural selection, population size changes, and recombination patterns from an alignment of genomes from different apes. Before we show how this can be done we introduce the relevant population genetics theory.

GENE TREES WITHIN SPECIES TREES: THE MULTISPECIES COALESCENT

Coalescence theory (18, 70) describes the ancestry of a sample of genes and assigns probabilities to the timing and order in which sampled sequences find common ancestors. This framework has been successful because it keeps track of the ancestry of only a sample and not the whole population backward in time. In the following, we consider time in this backward fashion, starting at present and reaching into the past. In a sample of n genes in a randomly mating population of constant size N, we let T_n denote the number of generations into the past when the first pair of genes shares a common ancestor. This time is well approximated by the exponential distribution with density

$$f(T_n = t) = \frac{\binom{n}{2}}{2N} e^{-t\binom{n}{2}/2N}$$

Any of the $\binom{n}{2}$ possible pairs in the sample are equally likely to be the first (most recent) to coalesce into one lineage. After the first pair coalesces, the waiting time until the next pair coalesces is given by the same formula for a sample of $n-1$ genes. In many populations, the assumptions of a random mating is violated by population structure or mating patterns. Nevertheless, the framework usually provides a very good description of the genealogical process if the census population size is substituted by the effective population size (N_e), which represents the size of a random mating population giving rise to the pattern of ancestry observed in the population. For humans this effective population size is usually estimated to be approximately 10,000, despite a census population size in the billions. This is because the human population only very recently rose to large numbers and the effective size represents the human population size over the complete genealogical history ($\sim$0.5 million years), which includes periods with a small number of breeding individuals (population bottlenecks).

Coalescence theory allows us to compute how many genes in the sample of size n have coalesced at some time τ or, equivalently, how many lineages represent the original sample at that time. **Figure 2** shows the probability that k lineages remain at time τ when the original sample was of size n. The rate at which coalescences occur when k lineages are left is of the order k^2. This means that while k is large, coalescence happens very quickly, and therefore the number of lineages is rapidly reduced to a few lineages. In fact, for any sample size, the amount of time during which only two lineages remain is expected to take up more than half of the time to the most recent common ancestor of the entire sample. The probability that only a single lineage is left as a function of time and initial sample size is shown in **Figure 3**, showing that initial sample size primarily affects this probability for recent times, whereas the probability that all lineages have coalesced is largely independent of n as time increases.

By analyzing sequences sampled from multiple species, we find that lineage sorting is determined by the number of lineages that have not coalesced at the time of a particular speciation event. If a species is represented by only a single lineage at the time of a speciation event then lineage sorting is complete. However, when more than one lineage remains at the time of speciation, ILS may result, producing genetic relationships that do not reflect the species relationship. Consider samples of n_1 and n_2 genes from two different species that split apart at time τ in the past and let these be coalesced into k_1 and k_2 lineages at the time of speciation. If $k_1 = 2$ and $k_2 = 1$ this corresponds to a sample of three lineages from the ancestral species. The ancestry of these lineages is determined by the coalescence process in the ancestral population and may result in three different gene-tree topologies, only one of which corresponds to the species relationship. In the other two cases, sequences from different species are most closely related.

At the modern human-Neanderthal split [$\sim$0.5 Myr ago (49)] or the bonobo-chimpanzee split [1–1.5 Myr ago (20, 34, 48, 73, 76)], we do not expect to observe complete lineage sorting but rather that some polymorphism is shared between the different species (**Figure 3**). At the human-chimpanzee split, however, where even the most recent estimates put the event more than four million years in the past (22), we would not expect any shared polymorphism.

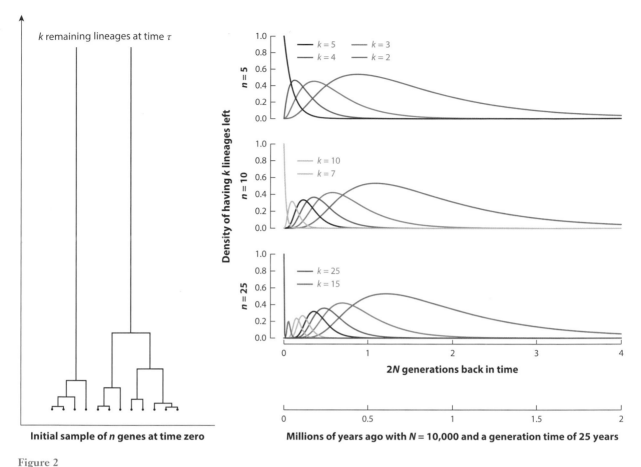

k remaining lineages at time τ

Density of having k lineages left

$n = 5$

k = 5	k = 3
k = 4	k = 2

$n = 10$

| k = 10 |
| k = 7 |

$n = 25$

| k = 25 |
| k = 15 |

2N generations back in time

Initial sample of n genes at time zero

Millions of years ago with N = 10,000 and a generation time of 25 years

Figure 2

The probability of having exactly k lineages remaining at time τ for a coalescence process that starts with a sample of size n at time 0. Time is shown in two units, so-called coalescence units corresponding to $2N_e$ generations and the corresponding number of years if the effective population size is 10,000 and the generation time is 25 years, with the values roughly corresponding to modern humans.

If a speciation event follows so shortly after another that lineage sorting does not have time to complete, ILS between the three species becomes fixed for some segments of the genome. Gorilla split from the ancestor of human and chimpanzee approximately two to three millions years before human and chimpanzee separated. When the human and chimpanzee ancestors split into distinct non-interbreeding species, some genes in the emerging populations of one species were more closely related to sequences from the other species. Most of these have been lost by genetic drift, but some are fixed in humans, making them more closely related to gorilla than to chimpanzee. So even when ancestral polymorphism is no longer shared, ILS is a permanent part of our genome.

A sample of one genome from each of three species allows us to compute the probability that a gene tree is incongruent with the species tree. The probability that a lineage from species A and one from species B have not coalesced at time t_2 is $e^{-(t_2-t_1)/2N}$, where t_1 is the number of generations since speciation of species A and B, and t_2 is the number of generations between C and the ancestral species AB. In this event, lineages from all three species coexist in one species and coalesce with equal probability, which results in three different gene-tree topologies, two of

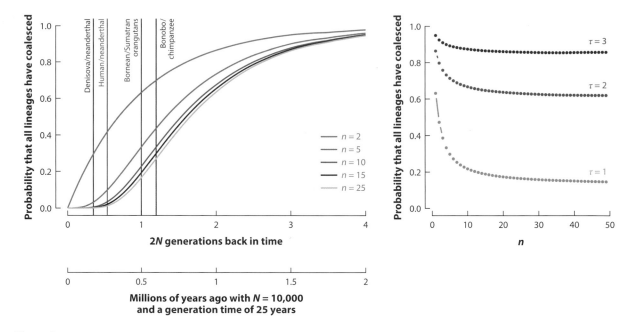

Figure 3

The probability of having completed lineage sorting for five different sample sizes assuming an effective population size and generation time close to modern humans, i.e., $N_e = 10,000$ and a generation time of 25 years. The figure shows the probability of having exactly one lineage back at time τ when starting with a sample of size n. On the left the x axis is time and the different lines are different initial sample sizes, whereas on the right the x axis depicts a range of initial sample sizes and the different lines show different time points. On the left are also shown recent speciation times taken from **Table 1**.

which represent ILS. The proportion of genomic sequences that are incongruent with the species tree is simply the above formula multiplied by two-thirds (**Figure 4**).

INCOMPLETE LINEAGE SORTING AND PHYLOGENETIC INFERENCE

ILS renders phylogenetic inference more complex because local gene trees in a genome are likely to produce conflicting results. A phylogenetic tree based on a single or a few genes may therefore not reflect the true species tree. This may be true for all internal nodes in a phylogeny (**Figure 1**) and should also be of concern when constructing phylogenies for distantly related species. Although this complication has long been acknowledged (32, 41, 77), it has been fully appreciated only since the recent availability of whole-genome sequences. Sampling more taxa paradoxically exacerbates this problem, as each new taxon splits a branch in two, further reducing distances between speciation events in the taxon group.

Fitting a single tree to data with ILS results in excess homoplasies because informative sites supporting discordant gene trees mimic the effects of multiple mutations on individual branches. Even rare unique genomic events such as LINE (long interspersed element) insertions and specific gene duplications cannot be used as proof for a specific phylogenetic relationship because of ILS. False homoplasies bias the number of nonsynonymous and synonymous events, possibly to different degrees, leading to biases in estimates of dN/dS, the rate of nonsynonymous to synonymous substitutions, and to biased estimates of positive selection through branch-specific models (84). Likewise, ancestral genome sequence reconstructions based on a single tree are biased (7).

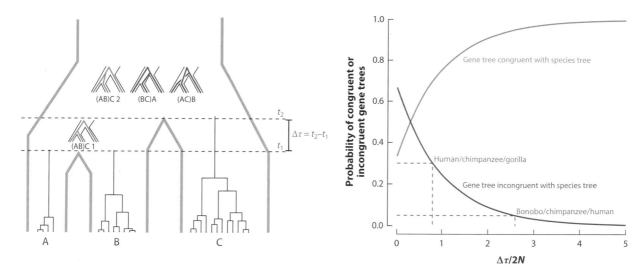

Figure 4

The probability that a gene tree is congruent or incongruent with the species tree (see also Reference 15a). Consider three species, A, B, and C, with A and B being more closely related by a time period $\Delta\tau$ and with all speciation events far enough back in time that we can ignore ancestral polymorphism. Regardless of which gene we pick in the present day species, at the time of the speciation events we have one from each species. The possible gene trees can be split into four categories, two of which are congruent with the species tree and two of which are incongruent. Either the genes from A and B coalesce in the time between the two speciation events, a case we call (AB)C1, or the two lineages reach the common ancestor of all three species, in which case either A and B coalesce first [(AB)C2], B and C coalesce first [(BC)A], or A and C coalesce first [(AC)B]. (AB)C1 and (AB)C2 are congruent with the species tree, whereas (BC)A and (AC)B are not. The probability of seeing (AB)C1 depends on $\Delta\tau$, which is determined by the actual time between the two speciation events divided by the effective population size in the A and B ancestor. The other three cases all have the same probability. The probability of seeing a congruent or incongruent gene tree is shown on the right as a function of this $\Delta\tau$. The human-chimpanzee-gorilla triplet, in which approximately one-third of the gene trees are incongruent (48), and the bonobo-chimpanzee-human triplet, in which approximately 5% of the gene trees are incongruent (58), are shown.

A common approach is to assume that the most frequently occurring gene tree supports the true phylogeny, although this is not always true (12, 13). Widely used approaches include reconciliation by concatenation of data (54; but see also 55), voting procedures (9), and modeling frameworks (51, 79). A recent suggestion is to choose data where ILS is expected to occur less frequently because of a smaller expected effective population size in the ancestral species (44). This can be the X chromosome, with an expected population size that is three-quarters that of autosomes, or mitochondria or the Y chromosome, with an expected population size one-quarter that of autosomes. However, mitochondria and Y chromosomes do not recombine and are therefore single manifestations of an evolutionary relationship. In recent human evolution, mitochondria appear to not be very reliable for phylogenetic inference. Mitochondrial sequences group human and Neanderthals, with Denisovans as an outgroup, but the nuclear genome puts Neanderthals and Denisovans as sister taxa (28, 38, 53). A recently published, more divergent hominoid mitochondrial sequence (37) should be interpreted cautiously until nuclear sequences are reported from the same sample.

LINEAGE SORTING AND ANCESTRAL POPULATION GENOMICS

Although ILS can be considered a nuisance that must be treated carefully in phylogenetic analysis, it is a blessing for our ability to infer evolutionary processes in ancestral species. Incorporating multiple loci in phylogenetic inference improves the inference but requires explicit modeling of ILS to be generally applicable. More importantly, such models describe population genetics processes of ancestral populations and allow inference of the parameters of these processes. In an early example of explicit modeling of ILS to resolve a trichotomy, Wu (77) derived the probabilities of congruent versus incongruent gene trees for a three-species model (**Figure 4**). He developed a maximum likelihood test to resolve the species tree from the number of gene trees, exposing each of the three topologies, and provided a formal method for testing whether the fraction of different gene trees significantly prefers one species tree. Chen & Li (9) later used this method to resolve the human-chimpanzee-gorilla phylogeny using sequence data from 53 autosomal segments.

Takahata et al. (67) took this beyond merely resolving phylogenies in the presence of incongruent gene trees. They developed a maximum likelihood method that could not only resolve the phylogeny but also separate species divergence times from gene divergence times and infer the effective population size of ancestral species. The method considers the four cases of gene trees from **Figure 4**, (AB)C1, (AB)C2, (BC)A, and (AC)B, and computes the probabilities of observing them, conditional on the branch lengths of the trees and the effective population sizes in the two ancestral species, and then computes the likelihood of the data given these gene trees. The method was used to infer the speciation time between humans, chimpanzees, and gorillas, the effective size of the human-chimpanzee ancestor, and the ancestor of all three species.

Building upon Takahata et al. (67), Yang and colleagues developed both maximum likelihood and Bayesian methods for inferring speciation times and ancestral effective population sizes from three species (8, 50, 81, 82). These methods also consider the four cases from Takahata et al. (67) and integrate over the probabilities of each gene tree and over the branch lengths to compute the parameters of interest, speciation times, and population sizes. Wall (71) took a similar approach, using Monte Carlo simulations of the coalescence process in the ancestral species to compute the likelihood of speciation times and ancestral population parameters.

These methods assume allopatric speciation, in which a panmictic population splits instantaneously into two populations with no subsequent gene flow. Recent evidence, however, suggests that in great apes (34, 46, 58, 68) and archaic hominins (38, 49, 53, 57), this scenario is the exception rather than the rule.

Innan & Watanabe (26) constructed a test for comparing a clean population split to a split followed by gene flow by modeling the coalescence density as a continuous-time Markov model and integrating over this density. This method is also applied in References 21 and 72 and later generalized to larger sample sizes (2, 74). Innan & Watanabe (26) applied their method to the human-chimpanzee speciation but were unable to reject the null hypothesis of no gene flow after an initial split.

Patterson et al. (42) analyzed the mutation patterns and the spatial patterns of segregating sites supporting different gene topologies in alignments of human, chimpanzee, gorilla, orangutan, and macaque. They concluded that the variation in coalescence time of the human-chimpanzee ancestor was extremely large, spanning up to four million years. For the X chromosome they observed a much smaller variation in coalescence time as well as near absence of gene-tree incongruence with the species tree, and suggested that this would be compatible with a complex speciation process in which proto-humans and chimpanzees initially split apart but after one or two million years hybridized before splitting into the two species we have today. This scenario was met with critiques arguing that the variation in coalescence times would also be consistent with an allopatric speciation provided that the ancestral effective population size was sufficiently large (5, 69). Others suggested that the reduced variation in the X chromosome could be caused by male-biased mutation and sperm competition (47) and that the observations could be caused by selection on X (58). Yamamichi et al. (80) built and tested an explicit model of the hybridization scenario in a maximum likelihood framework and did not reject the null model of instantaneous speciation.

Wu & Ting (78) argued that speciation genes in the presence of gene flow could leave a signal in the divergence patterns along the genome. Genes adapting in different directions in two ancestral populations would cause reproductive isolation by slowly decreasing heterozygotic fitness as they evolve, and this should be visible as a deeper divergence between the two species. Osada & Wu (40) compared coding and noncoding SNPs and rejected the hypothesis that they had the same (speciation) divergence time, and Yang also rejected a genome-wide split time using an extension of his previous model to allow different speciation times along the genome (83). Zhu & Yang (85) extended this further with an explicit model of gene flow and again found a preference for the gene-flow model. All of this suggests that the human and chimpanzee speciation could have occurred in the presence of gene flow.

Incorporating Recombination in Full-Genome Analyses

The methods reviewed in the previous section model coalescence times and changing gene trees by assuming that the loci considered are sufficiently far apart to be practically independent and that they are sufficiently small for intralocus recombination to be ignored. These assumptions are problematic because the rates of recombination and mutation are on the same order of magnitude. Loci without recombination are either too short to have enough mutations to inform us of the divergence or are biased toward more recent ancestry.

Becquet & Przeworski (6) introduced intralocus recombination in a Markov chain Monte Carlo framework to model speciation in the presence of gene flow. The model estimates the likelihood at any parameter point by simulating genealogies from the coalescent process with recombination and computing the probability of the genealogies producing the same summary statistics as observed in the real data. Simulating data makes models with recombination more computationally tractable and for short loci the method is able to include intralocus recombination while still assuming free recombination between loci. Simulating data with recombination for very long sequences, however, is not a feasible approach, as the simulation time is expected to grow exponentially as a function of recombination rate. For long sequences, some approximation seems necessary.

Focusing on independent loci imparts a great reduction in the amount of data available to explore the speciation events and ancestral population genetics. Effective analysis of complete genomes cannot ignore dependencies between loci, requiring that models take recombination fully into account. The sequential Markov coalescent has recently received a lot of attention. The coalescence process with recombination was originally described as a process running backward in time (24), but Wiuf & Hein (75) showed that it could also be modeled as a process running along a sequence alignment. This process, however, is computationally intractable for long sequences because it requires keeping track of all local gene trees along the sequence.

The sequential process can be approximated by one that is Markovian along the sequence. McVean & Cardin (36) constructed such a model, the sequential Markov coalescence (SMC), by restricting the model to allow only coalescence events between lineages that share ancestral material. This implies that two lineages arising from a recombination event cannot coalesce back to remove any effect of the recombination. A large fraction of the recombination and coalescence events deep in the genealogy are, however, of this type, and Chen et al. (10) and Marjoram & Wall (35) removed the restriction creating the alternative sequential Markov coalescence SMC′. The theoretical differences between these two models were recently explored by Hobolth & Jensen (23).

Hobolth et al. (22) built a model to analyze four species alignments that combine the coalescence with hidden Markov models (HMMs) to infer the speciation times of human and chimpanzee and of human and gorilla. The HMM used the four genealogies of Takahata et al. (**Figure 4**) as the hidden states in the HMM and inferred coalescence parameters indirectly from fitted transition probabilities of the HMM. Dutheil et al. (15) later combined the SMC model with the HMM from Reference 22 with a model deriving transition probabilities between gene trees from coalescence theory. This coalescent HMM has been applied to genome alignments of the great apes to disentangle speciation time from divergence time and obtain many of the estimates in **Table 1** and **Figure 1**. Using an alignment of human, chimpanzee, and gorilla (and orangutan as an outgroup), the human-chimpanzee and human-gorilla speciations were dated to 5.2 and 8.6 Myr, respectively, using full-genome data (58). This analysis showed that the ancestral population sizes of humans and chimpanzees were much larger than those of living great apes (86,000), producing 30% ILS between these three species. Applying the same methodology, the bonobo-chimpanzee split (48) was dated to 1.2 Myr, but these species show only 3% ILS owing to the much longer time between the bonobo-chimpanzee and human-chimpanzee speciation events and to a smaller population size of the bonobo-chimpanzee ancestor. A more recent study (46), however, suggests that the proportion of ILS between these three species may be closer to 5%. Only a small amount of ILS in a full-genome alignment is required to obtain confident estimates of split times and ancestral population sizes, evidenced by an analysis of human, chimpanzee, and orangutan showing only ~1% ILS (31).

The sequential Markov coalescent approximation allows a much wider range of different population genetics applications recently developed in parallel by many groups. Li & Durbin (30) presented the now widely used pairwise sequential Markov coalescent (PSMC) approach, which models changing coalescence times for a diploid genome to infer changing population sizes over time. This approach can also indirectly infer speciation times when comparing the trajectories of population sizes for individuals from different species or by analyzing an artificial diploid individual consisting of haploid genomes from individuals of two different species (46). Song and others (43, 61, 63, 64) combined the sequential Markov coalescent with conditional sampling approaches to extend the model beyond what is computationally feasible if all possible gene trees are modeled. Rasmussen et al. (52) combined the sequential Markov coalescence with a Markov Chain Monte Carlo approach to sample a distribution of gene trees. For inference of speciation events, Mailund et al. (33, 34) modeled changing coalescence times. They used a continuous-time Markov chain similar to Simonsen & Churchill's (62) to construct HMM models of simple divergence either with no gene flow or that allowed for a period of restricted gene flow, thus modeling allopatric and sympatric speciation, respectively. Applying these models to pairs of great ape species, speciation with a period of gene flow provides the best fit to the divergence of the eastern and western gorilla and to the Bornean and Sumatran orangutan, but not to the split between bonobo and chimpanzee, which appears to be instantaneous (34). The same study also finds evidence for a slow divergence process of human and chimpanzee in line with References 40, 83, and 85, and discussed above.

Patterns of ILS inferred along a genomic alignment can identify the genomic segments where pairs of species are more closely related and the genomic breakpoints where the shift in gene genealogy approximately occurs. An HMM allows posterior decoding of the probability for each possible gene genealogy for each base pair in the alignment. **Figure 5** shows posterior decoding for a typical example for the human, chimpanzee, and gorilla analysis based on the coalescent HMM approach (15). The definition of regions sharing the same genealogy is subsequently possible by defining an appropriate threshold of, e.g., 50% support. In general, for the case of human, chimpanzee, and gorilla, the median lengths of segments where gorillas are more closely related to either humans or chimpanzees are approximately 200 bp, whereas segments supporting the species phylogeny are longer (median: approximately 2 kb). The mixed ancestry induced by ILS may also be reflected in functional differences between species. A comparison of transcriptome

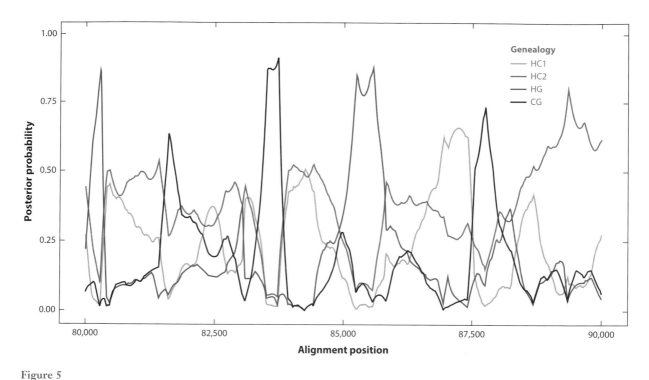

Figure 5

Posterior probabilities of the four genealogies representing states in the CoalHMM (coalescent hidden Markov model) shown for a segment of aligned chromosomes 2 from human, chimpanzee, gorilla, and orangutan. HC, human-chimpanzee; HG, human gorilla; and CG, chimpanzee-gorilla gene genealogy. HC1 and HC2 denote the two different cases for the human-chimp genealogy, see **Figure 4**.

variation among 1:1:1 orthologous genes among human, chimpanzee, and gorilla (58) showed that patterns in relative divergence of expression and splicing significantly overlapped the patterns of inferred ILS.

DETECTING SELECTION IN ANCESTRAL POPULATIONS

Speciation times generally apply equally to the entire genome, although exceptions include regions involved in speciation or subsequent admixture. In contrast, the effective size of an ancestral population varies along the genome as a result of population genetics processes such as selection. Coalescent HMMs can estimate N_e but require at least one megabase of genomic alignment to fit all model parameters. However, the pattern of ILS offers information about N_e on a finer scale. By identifying individual segments of the genomic alignment showing ILS, the proportion of ILS can be identified for any genomic region. Given two speciation times $\Delta\tau$ generations apart, the effective population size expected to produce an observed proportion of ILS, p, can be computed as $N_e = \Delta\tau/2\ln(2/3p)$.

Both positive and negative selection in the ancestral population is expected to reduce the effective population size in regions linked to selected sites. Indeed, the proportion of ILS is reduced near coding genes (22, 48, 58). Consecutive regions devoid of ILS may be evidence of strong positive selection given that a selective sweep can force all lineages to coalesce, leaving none available for ILS. One such region was found on chromosome 3 in the bonobo-chimpanzee ancestor spanning 6.1 Mb, twice the length of the second longest such region found on autosomes

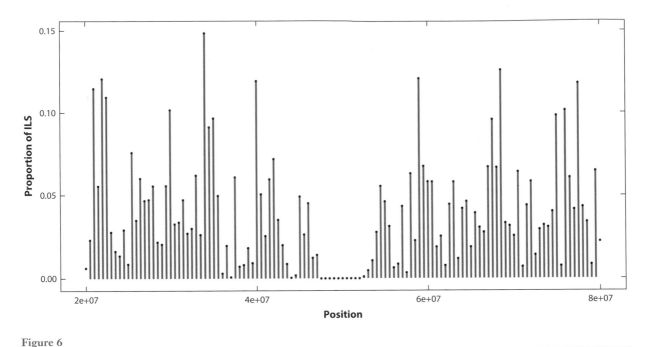

Figure 6

Proportion of incomplete lineage sorting (ILS) between bonobo, chimpanzee, and human in windows of 500 kb around a region of chromosome 3 that is completely devoid of ILS.

(48; see **Figure 6)**. The region contains the *CCR5* gene involved in HIV resistance (56) as well as a cluster of immunity-related genes shown to be under positive selection in humans (4). It is also devoid of polymorphism in central chimpanzees (25) and shows a strong signature of introgression from Neanderthals to modern humans (14). However, the efficacy by which selection reduces the local effective population size, and with it the probability of ILS, is increased in regions where the recombination rate is low, and in the chromosome 3 region the human recombination rate is an order of magnitude lower than the genome average (27). The dependence on recombination rate applies to both positive and negative selection, and a positive correlation between recombination rate and proportion of ILS has been reported in great apes (22, 48, 58), suggesting that selection is a prominent force in shaping patterns of ILS. This example illustrates that patterns of ILS can aid interpretations of patterns of genetic diversity along the genome. It has long been discussed whether the positive correlation between recombination rate and genetic diversity in the human genome (19) can be attributed to an association between recombination and mutation. However, because ILS patterns are not influenced by mutation, the results support the conclusion from the 1,000 Genomes Project that the correlation between recombination and diversity is not caused by mutational processes alone (1).

Balancing selection has the opposite effect on the local effective population size, making lineages coalesce further back in time than under neutrality. Regions targeted by balancing selection should thus be more variable and the variation should be of more intermediate frequency (3, 11). As a result of longer coalescence times, the proportion of ILS should also increase under balancing selection. If sufficiently strong, balancing selection should lead to shared lineages even between human and chimpanzees, and mutations occurring in the common ancestor would potentially manifest themselves as shared polymorphism. Leffler et al. (29) used this idea to scan human and chimpanzee genome-wide polymorphism data and found 125 cases in which sets of linked variants were polymorphic in both species. These regions tended to show association with proteins presumed to be involved with pathogen interaction. One was even shared with gorilla, showing how extensive ILS can be under balancing selection. These new regions add to well known cases of balancing selection at the MHC (major histocompatibility complex) (59, 66) and the ABO blood group system (60).

ANCESTRAL RECOMBINATION

The spatial pattern of genealogies along an alignment also holds information about past recombination. Segments of genomic alignment representing different genealogies are separated by points of crossover where a historical recombination event has allowed two new lineages to follow individual histories. The rate of change of genealogies along an alignment of many individuals from the same species is summarized by linkage disequilibrium and has been used to estimate genetic maps. In the same way, change in genealogies along an alignment of different species is evidence of past recombination events (39). For species showing ILS, a subset of gene trees represents topologies different from the species tree. Recombination events that mark the transition to such an ILS topology are of particular interest because these events must occur so far back in time that the resulting lineages have the opportunity to coalesce with lineages other than the one representing the sister species and thus form alternative topologies. Such recombination events can be identified by observing transitions between segments of the alignment supporting alternative topologies (see **Figure 5**). An ancestral recombination map can then be derived from the rate of such transitions along an alignment.

Although the detectable number of recombination events causing topology change depends on the amount of ILS, and thus the relative size of $\Delta \tau$ and ancestral N_e, their distribution in time and among the branches of the species tree are determined by the relative size of the three extant populations and the two ancestral populations. The effective size of the human-chimpanzee population is large compared with those of the human, chimpanzee, gorilla, and the human-gorilla ancestor. This serves to concentrate more than 60% of recombination events in the human-chimpanzee ancestor with the remaining events occurring in the proximal part of the three adjoining branches. Ancestral recombination maps can be created from any three species showing ILS. The approach allows differences in recombination rate between species to be resolved into the change that occurred in each species in the course of their divergence, allowing us to study whether the rate of evolution is constant across primate evolution and to directly link evolution of genomic sequence to evolution of recombination rate to better understand how it is controlled.

DISCLOSURE STATEMENT

The authors are not aware of any affiliations, memberships, funding, or financial holdings that might be perceived as affecting the objectivity of this review.

ACKNOWLEDGMENTS

The authors thank Freddy B. Christiansen for comments on a previous version of the manuscript. This research was supported by grants from the Danish Council for Independent Research–Natural Sciences to Thomas Mailund and Mikkel H. Schierup.

LITERATURE CITED

1. Abecasis GR, Altshuler D, Auton A, Brooks LD, Durbin RM, et al. 2010. A map of human genome variation from population-scale sequencing. *Nature* 467(7319):1061–73
2. Andersen LN, Mailund T, Hobolth A. 2014. Efficient computation in the IM model. *J. Math. Biol.* 68(6):1423–51
3. Andrés AM, Hubisz MJ, Indap A, Torgerson DG, Degenhardt JD, et al. 2009. Targets of balancing selection in the human genome. *Mol. Biol. Evol.* 26(12):2755–64
4. Barreiro L, Quintana-Murci L. 2010. From evolutionary genetics to human immunology: how selection shapes host defence genes. *Nat. Rev. Genet.* 11(1):17–30
5. Barton NH. 2006. Evolutionary biology: How did the human species form? *Curr. Biol.* 16(16):R647–50
6. Becquet C, Przeworski M. 2007. A new approach to estimate parameters of speciation models with application to apes. *Genome Res.* 17(10):1505–19
7. Blanchette M, Diallo AB, Green ED, Miller W, Haussler D. 2008. Computational reconstruction of ancestral DNA sequences. *Methods Mol. Biol.* 422:171–84
8. Burgess R, Yang Z. 2008. Estimation of hominoid ancestral population sizes under Bayesian coalescent models incorporating mutation rate variation and sequencing errors. *Mol. Biol. Evol.* 25(9):1979–94
9. Chen FC, Li WH. 2001. Genomic divergences between humans and other hominoids and the effective population size of the common ancestor of humans and chimpanzees. *Am. J. Hum. Genet.* 68(2):444–56
10. Chen GK, Marjoram P, Wall JD. 2009. Fast and flexible simulation of DNA sequence data. *Genome Res.* 19(1):136–42
11. DeGiorgio M, Lohmueller KE, Nielsen R. 2014. A model-based approach for identifying signatures of ancient balancing selection in genetic data. *PLOS Genet.* 10(8):e1004561
12. Degnan JH, Rosenberg NA. 2006. Discordance of species trees with their most likely gene trees. *PLOS Genet.* 2(5):e68
13. Degnan JH, Rosenberg NA. 2009. Gene tree discordance, phylogenetic inference and the multispecies coalescent. *Trends Ecol. Evol.* 24(6):332–40
14. Ding Q, Hu Y, Xu S, Wang J, Jin L. 2014. Neanderthal introgression at chromosome 3p21.31 was under positive natural selection in East Asians. *Mol. Biol. Evol.* 31(3):683–95
15. Dutheil JY, Ganapathy G, Hobolth A, Mailund T, Uyenoyama MK, Schierup MH. 2009. Ancestral population genomics: the coalescent hidden Markov model approach. *Genetics* 183(1):259–74
15a. Dutheil JY, Hobolth A.. 2012. Ancestral population genomics. *Methods Mol. Biol.* 856:293–13
16. Fabre P-H, Rodrigues A, Douzery EJP. 2009. Patterns of macroevolution among Primates inferred from a supermatrix of mitochondrial and nuclear DNA. *Mol. Phylogenet. Evol.* 53(3):808–25
17. Glazko GV, Nei M. 2003. Estimation of divergence times for major lineages of primate species. *Mol. Biol. Evol.* 20(3):424–34
18. Hein J, Schierup M, Wiuf C. 2005. *Gene Genealogies, Variation and Evolution: A Primer in Coalescent Theory.* Oxford: Oxford Univ. Press
19. Hellmann I, Ebersberger I, Ptak SE, Pääbo S, Przeworski M. 2003. A neutral explanation for the correlation of diversity with recombination rates in humans. *Am. J. Hum. Genet.* 72(6):1527–35
20. Hey J. 2010. The divergence of chimpanzee species and subspecies as revealed in multipopulation isolation-with-migration analyses. *Mol. Biol. Evol.* 27(4):921–33
21. Hobolth A, Andersen LN, Mailund T. 2011. On computing the coalescence time density in an isolation-with-migration model with few samples. *Genetics* 187(4):1241–43
22. Hobolth A, Christensen OF, Mailund T, Schierup MH. 2007. Genomic relationships and speciation times of human, chimpanzee, and gorilla inferred from a coalescent hidden Markov model. *PLOS Genet.* 3(2):e7
23. Hobolth A, Jensen JL. 2014. Markovian approximation to the finite loci coalescent with recombination along multiple sequences. *Theor. Popul. Biol.* In press
24. Hudson RR. 1983. Properties of a neutral allele model with intragenic recombination. *Theor. Popul. Biol.* 23(2):183–201
25. Hvilsom C, Qian Y, Bataillon T, Li Y, Mailund T, et al. 2012. Extensive X-linked adaptive evolution in central chimpanzees. *Proc. Natl. Acad. Sci. USA* 109(6):2054–59

26. Innan H, Watanabe H. 2006. The effect of gene flow on the coalescent time in the human-chimpanzee ancestral population. *Mol. Biol. Evol.* 23(5):1040–47

27. Kong A, Thorleifsson G, Gudbjartsson DF, Masson G, Sigurdsson A, et al. 2010. Fine-scale recombination rate differences between sexes, populations and individuals. *Nature* 467(7319):1099–103

28. Krause J, Fu Q, Good JM, Viola B, Shunkov MV, et al. 2010. The complete mitochondrial DNA genome of an unknown hominin from southern Siberia. *Nature* 464(7290):894–97

29. Leffler EM, Gao Z, Pfeifer S, Ségurel L, Auton A, et al. 2013. Multiple instances of ancient balancing selection shared between humans and chimpanzees. *Science* 339(6127):1578–82

30. Li H, Durbin R. 2011. Inference of human population history from individual whole-genome sequences. *Nature* 475(7357):493–96

31. Locke DP, Hillier LW, Warren WC, Worley KC, Nazareth LV, et al. 2011. Comparative and demographic analysis of orangutan genomes. *Nature* 469(7331):529–33

32. Maddison WP. 1997. Gene trees in species trees. *Syst. Biol.* 46(3):523–36

33. Mailund T, Dutheil JY, Hobolth A, Lunter G, Schierup MH. 2011. Estimating divergence time and ancestral effective population size of Bornean and Sumatran orangutan subspecies using a coalescent hidden Markov model. *PLOS Genet.* 7(3):e1001319

34. Mailund T, Halager AE, Westergaard M, Dutheil JY, Munch K, et al. 2012. A new isolation with migration model along complete genomes infers very different divergence processes among closely related great ape species. *PLOS Genet.* 8(12):e1003125

35. Marjoram P, Wall JD. 2006. Fast "coalescent" simulation. *BMC Genet.* 7:16

35a. Matsudaira K, Ishida T. 2010. Phylogenetic relationships and divergence dates of the whole mitochondrial genome sequences among three gibbon genera. *Mol. Phylogenet. Evol.* 55(2):454–59

36. McVean GA, Cardin NJ. 2005. Approximating the coalescent with recombination. *Philos. Trans. R. Soc. Lond. Ser. B* 360(1459):1387–93

37. Meyer M, Fu Q, Aximu-Petri A, Glocke I, Nickel B, et al. 2014. A mitochondrial genome sequence of a hominin from Sima de los Huesos. *Nature* 505(7483):403–6

38. Meyer M, Kircher M, Gansauge M-T, Li H, Racimo F, et al. 2012. A high-coverage genome sequence from an archaic Denisovan individual. *Science* 338(6104):222–26

39. Munch K, Mailund T, Dutheil JY, Schierup MH. 2014. A fine-scale recombination map of the human-chimpanzee ancestor reveals faster change in humans than in chimpanzees and a strong impact of GC-biased gene conversion. *Genome Res.* 24(3):467–74

40. Osada N, Wu C-I. 2005. Inferring the mode of speciation from genomic data: a study of the great apes. *Genetics* 169(1):259–64

41. Pamilo P, Nei M. 1988. Relationships between gene trees and species trees. *Mol. Biol. Evol.* 5(5):568–83

42. Patterson N, Richter DJ, Gnerre S, Lander ES, Reich D. 2006. Genetic evidence for complex speciation of humans and chimpanzees. *Nature* 441(7097):1103–8

43. Paul JS, Steinrücken M, Song YS. 2011. An accurate sequentially Markov conditional sampling distribution for the coalescent with recombination. *Genetics* 187(4):1115–28

44. Pease JB, Hahn MW. 2013. More accurate phylogenies inferred from low-recombination regions in the presence of incomplete lineage sorting. *Evol. Int. J. Org. Evol.* 67(8):2376–84

45. Perelman P, Johnson WE, Roos C, Seuánez HN, Horvath JE, et al. 2011. A molecular phylogeny of living primates. *PLOS Genet.* 7(3):e1001342

46. Prado-Martinez J, Sudmant PH, Kidd JM, Li H, Kelley JL, et al. 2013. Great ape genetic diversity and population history. *Nature* 499(7459):471–75

47. Presgraves DC, Yi SV. 2009. Doubts about complex speciation between humans and chimpanzees. *Trends Ecol. Evol.* 24(10):533–40

48. Prüfer K, Munch K, Hellmann I, Akagi K, Miller JR, et al. 2012. The bonobo genome compared with the chimpanzee and human genomes. *Nature* 486(7404):527–31

49. Prüfer K, Racimo F, Patterson N, Jay F, Sankararaman S, et al. 2014. The complete genome sequence of a Neanderthal from the Altai Mountains. *Nature* 505(7481):43–49

50. Rannala B, Yang Z. 2003. Bayes estimation of species divergence times and ancestral population sizes using DNA sequences from multiple loci. *Genetics* 164(4):1645–56

51. Rasmussen MD, Kellis M. 2012. Unified modeling of gene duplication, loss, and coalescence using a locus tree. *Genome Res.* 22(4):755–65

52. Rasmussen MD, Hubisz MJ, Gronau I, Siepel A. 2014. Genome-wide inference of ancestral recombination graphs. *PLOS Genet.* 10(5):e1004342

53. Reich D, Green RE, Kircher M, Krause J, Patterson N, et al. 2010. Genetic history of an archaic hominin group from Denisova Cave in Siberia. *Nature* 468(7327):1053–60

54. Rokas A, Williams BL, King N, Carroll SB. 2003. Genome-scale approaches to resolving incongruence in molecular phylogenies. *Nature* 425(6960):798–804

55. Salichos L, Rokas A. 2013. Inferring ancient divergences requires genes with strong phylogenetic signals. *Nature* 497(7449):327–31

56. Samson M, Libert F, Doranz BJ, Rucker J, Liesnard C, et al. 1996. Resistance to HIV-1 infection in Caucasian individuals bearing mutant alleles of the CCR-5 chemokine receptor gene. *Nature* 382(6593):722–25

57. Sankararaman S, Mallick S, Dannemann M, Prüfer K, Kelso J, et al. 2014. The genomic landscape of Neanderthal ancestry in present-day humans. *Nature* 507(7492):354–57

58. Scally A, Dutheil JY, Hillier LW, Jordan GE, Goodhead I, et al. 2012. Insights into hominid evolution from the gorilla genome sequence. *Nature* 483(7388):169–75

59. Schierup MH, Mikkelsen AM, Hein J. 2001. Recombination, balancing selection and phylogenies in MHC and self-incompatibility genes. *Genetics* 159(4):1833–44

60. Ségurel L, Thompson EE, Flutre T, Lovstad J, Venkat A, et al. 2012. The ABO blood group is a trans-species polymorphism in primates. *Proc. Natl. Acad. Sci. USA* 109(45):18493–98

61. Sheehan S, Harris K, Song YS. 2013. Estimating variable effective population sizes from multiple genomes: a sequentially Markov conditional sampling distribution approach. *Genetics* 194(3):647–62

62. Simonsen KL, Churchill GA. 1997. A Markov chain model of coalescence with recombination. *Theor. Popul. Biol.* 52(1):43–59

63. Steinrücken M, Paul JS, Song YS. 2013. A sequentially Markov conditional sampling distribution for structured populations with migration and recombination. *Theor. Popul. Biol.* 87:51–61

64. Steinrücken M, Wang YX, Song YS. 2013. An explicit transition density expansion for a multi-allelic Wright-Fisher diffusion with general diploid selection. *Theor. Popul. Biol.* 83:1–14

65. Steiper ME, Young NM. 2006. Primate molecular divergence dates. *Mol. Phylogenet. Evol.* 41(2):384–94

66. Takahata N, Satta Y, Klein J. 1992. Polymorphism and balancing selection at major histocompatibility complex loci. *Genetics* 130(4):925–38

67. Takahata N, Satta Y, Klein J. 1995. Divergence time and population size in the lineage leading to modern humans. *Theor. Popul. Biol.* 48(2):198–221

68. Thalmann O, Fischer A, Lankester F, Pääbo S, Vigilant L. 2007. The complex evolutionary history of gorillas: insights from genomic data. *Mol. Biol. Evol.* 24(1):146–58

69. Wakeley J. 2008. Complex speciation of humans and chimpanzees. *Nature* 452(7184):E3–4; discussion E4

70. Wakeley J. 2009. *Coalescent Theory: An Introduction*. Greenwood Village, CO: Roberts & Company Publ.

71. Wall JD. 2003. Estimating ancestral population sizes and divergence times. *Genetics* 163(1):395–404

72. Wang Y, Hey J. 2010. Estimating divergence parameters with small samples from a large number of loci. *Genetics* 184(2):363–79

73. Wegmann D, Excoffier L. 2010. Bayesian inference of the demographic history of chimpanzees. *Mol. Biol. Evol.* 27(6):1425–35

74. Wilkinson-Herbots HM. 2012. The distribution of the coalescence time and the number of pairwise nucleotide differences in a model of population divergence or speciation with an initial period of gene flow. *Theor. Popul. Biol.* 82(2):92–108

75. Wiuf C, Hein J. 1999. Recombination as a point process along sequences. *Theor. Popul. Biol.* 55(3):248–59

76. Won YJ, Hey J. 2005. Divergence population genetics of chimpanzees. *Mol. Biol. Evol.* 22(2):297–307

77. Wu CI. 1991. Inferences of species phylogeny in relation to segregation of ancient polymorphisms. *Genetics* 127(2):429–35

78. Wu C-I, Ting C-T. 2004. Genes and speciation. *Nat. Rev. Genet.* 5(2):114–22

79. Wu Y-C, Rasmussen MD, Bansal MS, Kellis M. 2013. Most parsimonious reconciliation in the presence of gene duplication, loss, and deep coalescence using labeled coalescent trees. *Genome Res.* 24(3):475–86

80. Yamamichi M, Gojobori J, Innan H. 2012. An autosomal analysis gives no genetic evidence for complex speciation of humans and chimpanzees. *Mol. Biol. Evol.* 29(1):145–56

81. Yang Z. 1997. On the estimation of ancestral population sizes of modern humans. *Genet. Res.* 69(2):111–16

82. Yang Z. 2002. Likelihood and Bayes estimation of ancestral population sizes in hominoids using data from multiple loci. *Genetics* 162(4):1811–23

83. Yang Z. 2010. A likelihood ratio test of speciation with gene flow using genomic sequence data. *Genome Biol. Evol.* 2:200–11

84. Yang Z, Nielsen R. 2000. Estimating synonymous and nonsynonymous substitution rates under realistic evolutionary models. *Mol. Biol. Evol.* 17(1):32–43

85. Zhu T, Yang Z. 2012. Maximum likelihood implementation of an isolation-with-migration model with three species for testing speciation with gene flow. *Mol. Biol. Evol.* 29(10):3131–42

Messenger RNA Degradation in Bacterial Cells

Monica P. Hui,* Patricia L. Foley,* and Joel G. Belasco

Kimmel Center for Biology and Medicine at the Skirball Institute and Department of Microbiology, New York University School of Medicine, New York, NY 10016; email: joel.belasco@med.nyu.edu

Annu. Rev. Genet. 2014. 48:537–59

First published online as a Review in Advance on October 1, 2014

The *Annual Review of Genetics* is online at genet.annualreviews.org

This article's doi: 10.1146/annurev-genet-120213-092340

*These authors contributed equally to this manuscript.

Keywords

mRNA stability, ribonuclease, gene regulation, translation, sRNA

Abstract

mRNA degradation is an important mechanism for controlling gene expression in bacterial cells. This process involves the orderly action of a battery of cellular endonucleases and exonucleases, some universal and others present only in certain species. These ribonucleases function with the assistance of ancillary enzymes that covalently modify the 5′ or 3′ end of RNA or unwind base-paired regions. Triggered by initiating events at either the 5′ terminus or an internal site, mRNA decay occurs at diverse rates that are transcript specific and governed by RNA sequence and structure, translating ribosomes, and bound sRNAs or proteins. In response to environmental cues, bacteria are able to orchestrate widespread changes in mRNA lifetimes by modulating the concentration or specific activity of cellular ribonucleases or by unmasking the mRNA-degrading activity of cellular toxins.

INTRODUCTION

Critical to survival for all living organisms is the ability to precisely regulate the expression of genetic information in order to produce the proteins needed to navigate the assorted challenges posed by an ever-changing environment. This principle holds true for multicellular organisms and for bacteria. In bacterial cells, protein synthesis can be controlled at any of three stages: transcription, translation, or mRNA degradation.

The capacity of cells to degrade mRNA is an evolutionary imperative. The energetic costs of translation and the benefit of recycling ribonucleotides demand a mechanism for rapidly destroying transcripts that are no longer useful. Equally important, rapid mRNA turnover confers a distinct advantage by allowing cells to quickly adapt protein synthesis to sudden environmental challenges, as response times would be much slower if dependent solely on modulating transcription.

Because mRNA degradation is not indiscriminate, it makes an important contribution to differential gene expression. The longevity of individual transcripts can differ significantly, with half-lives ranging from seconds to approximately an hour in bacterial cells, causing proportionate effects on protein synthesis. Translational units within the same polycistronic transcript can also differ in stability, allowing cotranscribed genes to be expressed at distinct levels. The ability to alter rates of mRNA degradation is often crucial for the response of cells to environmental cues. This review focuses on mRNA turnover in bacterial cells, including the ribonucleases and RNA elements that govern mRNA decay, the various pathways by which messages are degraded, and the mechanisms for controlling the lifetimes of transcripts individually and collectively.

HISTORICAL PERSPECTIVE

Initial models for mRNA degradation in bacteria were based on a limited number of observations in *Escherichia coli*. For example, no 5′ exoribonuclease activity was ever detected in *E. coli* (40). Furthermore, the action of 3′ exoribonucleases appeared to be blocked by the nearly ubiquitous presence of a stem-loop at the 3′ end of mRNAs (118), such that differences in rates of mRNA decay did not seem to be governed by characteristics of the 3′-terminal stem-loop (12). Therefore, logic dictated that mRNA degradation must begin endonucleolytically (3, 12). With the discovery that the endonuclease RNase E controls the decay of most transcripts in *E. coli* (7, 112, 119, 126, 151), a model for endonucleolytic initiation coalesced with this enzyme as the centerpiece. For a majority of mRNAs, degradation was envisioned to begin with internal cleavage by RNase E to yield two decay intermediates. Freed of its protective 3′-terminal stem-loop, the 5′ fragment would be rapidly degraded by 3′ exonucleases, and the 3′ fragment would be degraded through further rounds of RNase E cleavage and 3′ exonuclease degradation.

Although this model accounted for many observations, a number of phenomena remained unexplained. How are stem-loops and other base-paired regions degraded? Why are the 3′ fragments generated by endonucleolytic cleavage typically less stable than their full-length precursors (155)? And if decay begins internally, why was degradation observed to be impeded by base pairing at the 5′ end of transcripts (15, 48)? Equally curious was the discovery that the genomes of a significant number of bacterial species do not encode an RNase E homolog. The realization that there is no universally conserved set of ribonucleolytic enzymes that all bacteria rely upon for mRNA turnover meant that *E. coli* could not be treated as a paradigm for understanding mRNA degradation in all species. Explaining these phenomena requires a fuller knowledge of the enzymes responsible for mRNA degradation.

BACTERIAL RIBONUCLEASES

Bacteria utilize a large arsenal of ribonucleolytic enzymes to carry out mRNA degradation, many of which are present only in certain bacterial clades.

Endoribonucleases

RNase E, RNase Y, and RNase III are the principal endonucleases that have so far been implicated in bacterial mRNA turnover.

RNase E and its homolog RNase G. Among bacterial ribonucleases, RNase E is one of the most important for governing rates of mRNA decay. Initially discovered for its role in ribosomal RNA maturation in *E. coli* (4), this endonuclease was later implicated in mRNA degradation when it was observed that bulk mRNA stability and the half-lives of many individual transcripts increase significantly when a temperature-sensitive RNase E mutant is shifted to nonpermissive temperatures (7, 112, 119, 126, 151).

Each subunit of an *E. coli* RNase E homotetramer consists of a well-conserved amino-terminal domain that houses the catalytic site and a poorly conserved carboxy-terminal domain that includes a membrane-binding helix, two arginine-rich RNA-binding domains, and a region that serves as a scaffold for the assembly of a ribonucleolytic complex called the RNA degradosome (**Figure 1**) (78, 108, 153). RNase E cuts RNA internally within single-stranded regions that are AU rich, but with little sequence specificity (110). Despite being an endonuclease that can cleave RNA far from the 5' terminus, RNase E displays a marked preference for RNAs whose 5' end is monophosphorylated and unpaired (99). Comparison of monophosphorylated RNAs with their triphosphorylated counterparts has shown their difference in reactivity in vitro to typically be greater than an order of magnitude (76). This phenomenon is explained by the presence of a discrete 5'-end binding pocket in the catalytic domain, which serves as a phosphorylation sensor able to accommodate a 5' monophosphate but not a 5' triphosphate (20).

The essential nature of RNase E makes it difficult to determine the full extent of its role in mRNA turnover, but it appears that the vast majority of *E. coli* mRNAs decay by an RNase E–dependent mechanism. Interestingly, in addition to RNase E, *E. coli* also contains a nonessential paralog, RNase G. RNase G closely resembles the amino-terminal catalytic domain of RNase E, sharing almost 50% similarity in amino acid sequence as well as a comparable 5'-monophosphate dependence and cleavage site preference (76, 109, 152). Nevertheless, overexpression of RNase G cannot fully compensate for the absence of RNase E (34, 84). The effect on the *E. coli* transcriptome of deleting the RNase G gene is rather modest, likely because of the relatively low cellular concentration of this enzyme [only 1% as abundant as RNase E (84)].

RNase Y. In species that lack an RNase E homolog, RNase Y can fulfill the role of an endonuclease that mediates mRNA degradation. This enzyme consists of a transmembrane domain, a disordered coiled-coil domain, an RNA-binding KH domain, and a catalytic HD domain (**Figure 1**) (86). Although RNase Y is structurally distinct from RNase E, the two ribonucleases share certain characteristics. For example, both are membrane associated and cleave RNA internally and with little sequence specificity within single-stranded regions that are AU rich (141). However, unlike RNase E, the membrane-binding domain is essential for RNase Y function (86).

Multiple studies have implicated RNase Y as a major regulator of RNA metabolism. In *Bacillus subtilis*, which lacks RNase E, a large percentage of the transcriptome is affected by RNase Y depletion (44, 82, 88). Furthermore, in *Streptococcus pyogenes* and *Staphylococcus aureus*, RNase Y has been shown to be important for controlling the expression of virulence genes (27, 101).

Endonucleases

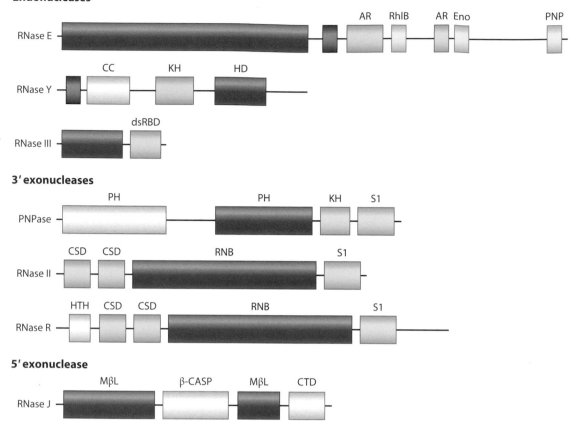

Figure 1

Constituent domains of mRNA-degrading ribonucleases. The domains of ribonucleases from representative species (*Escherichia coli* and *Bacillus subtilis*) are shown. Structural domains are depicted as colored rectangles: red, catalytic domain; blue, RNA-binding domain; yellow, protein-binding domain; purple, membrane-binding domain; and gray, miscellaneous domain. The sites at which RhlB, enolase (Eno), and PNPase (PNP) bind to RNase E are marked, as are the two arginine-rich RNA-binding domains (AR) of RNase E. Of the two PH domains in PNPase, only the second is catalytically active. The single metallo-β-lactamase domain (MβL) of RNase J comprises two noncontiguous segments of the polypeptide. The catalytic domains of RNase E and RNase III have not been named. Abbreviations: β-CASP, metallo-β-lactamase-associated CPSF Artemis SNM1/PSO2 domain; CC, coiled-coil domain; CSD, cold-shock domain; CTD, carboxy-terminal domain; dsRBD, double-stranded RNA-binding domain; HD, hydrolytic domain; HTH, helix-turn-helix domain; KH, hnRNP K homology domain; PH, RNase PH-like domain; RNB, RNase II-like catalytic domain; S1, ribosomal protein S1-like domain.

RNase III. Unlike RNase E/G and RNase Y, RNase III cuts RNA within double-stranded regions (138). By this means, RNase III plays a general role in the maturation of ribosomal RNA and a more selective role in the processing and degradation of mRNAs, sRNAs, and CRISPR RNAs (38, 106).

RNase III is a dimer of identical subunits, each comprising an endonucleolytic domain and a double-stranded RNA-binding domain (**Figure 1**) (14). The two centrally located catalytic sites function independently of one another to cleave each strand of the RNA duplex, yielding products that have a characteristic 2-bp overhang at the 3′ end (56, 113). Although cleavage at a reduced rate has been observed in vitro for substrates as short as 11 bp (83, 129), biological substrates typically span a minimum of two turns of an RNA helix or ~20 bp (137). Consequently, most natural stem-loop structures are too short to be targeted by RNase III in vivo. No consensus sequence has been identified for RNase III cleavage sites, but certain sequence features in and around that site appear to influence the ease with which an RNA duplex is cut (129, 162). RNase III is also able to target certain double-stranded RNAs that contain an internal loop, sometimes cleaving only one of the two strands (19).

RNase III has a more limited role in gene regulation than RNase E and RNase Y. Tiling array studies in *E. coli* and *B. subtilis* show that a small but significant portion of the transcriptome is affected, either directly or indirectly, by the absence of RNase III (44, 146). Consistent with its limited regulatory influence, RNase III is not generally essential for viability, except in *B. subtilis*, where it serves as part of a defense mechanism against chromosomally encoded toxins (45).

Minor endonucleases. Other endoribonucleases that function primarily in tRNA biogenesis have also been implicated in the decay of certain mRNAs. For example, RNase P, a ribonucleoprotein complex critical for the maturation of tRNA 5′ ends, targets noncoding regions within some messages (93). RNase Z (RNase BN), which removes aberrant tRNA 3′ ends in *E. coli* and appears to have both endonuclease and 3′ exonuclease activity, has also been implicated in the decay of a few mRNAs (47, 130).

Exoribonucleases

To complement the activity of cellular endonucleases, bacteria rely on a panel of exoribonucleases to rapidly degrade decay intermediates that lack protection at one or the other terminus. For the most part, these exonucleases act processively with little or no sequence specificity.

Phosphorolytic 3′ exonucleases. Bacterial 3′ exoribonucleases function by one of two mechanisms, either hydrolytically and irreversibly to yield nucleoside monophosphate products or phosphorolytically (i.e., using orthophosphate as a nucleophile) to produce nucleoside diphosphates in a reversible reaction.

To date, all known phosphorolytic 3′ exonucleases are members of the PDX family of enzymes (163). Prototypical representatives of this family are polynucleotide phosphorylase (PNPase) and RNase PH. The former is heavily involved in the turnover of mRNA, whereas the latter has principally been studied in the context of tRNA maturation and appears to have only a minor role in mRNA decay (41, 73).

True to the nature of the reversible phosphorolytic reaction it catalyzes, PNPase has both degradative and synthetic capabilities. In vitro, it can degrade RNA from 3′ to 5′ as well as add a heteropolymeric tail to the 3′ end (61). In vivo, both of these activities contribute to mRNA degradation. As an exonuclease, PNPase preferentially degrades RNAs with a single-stranded 3′ end (26, 156). As a polymerase, PNPase is capable of adding single-stranded adenine-rich tails that can facilitate the 3′-exonucleolytic degradation of structured regions of RNA (156) (see section mRNA Degradation Pathways below).

Our understanding of how PNPase degrades RNA exonucleolytically is shaped by a combination of biochemical, structural, and genetic studies. The enzyme is a trimer of identical subunits, each of which consists of two PH domains, a KH domain, and an S1 domain (**Figure 1**). The trimer forms a ring-shaped structure with the KH and S1 domains, which are critical for substrate binding, surrounding one end of the central channel (148, 150). The PH domains, although homologous to one another, are not identical, and in each subunit only one such domain (the second) is catalytically active (150). Because the active sites are located inside the channel, the 3′ end of RNA must thread partway through the channel to reach them. PNPase degrades RNA processively from the 3′ end until it encounters a base-paired structure of significant thermodynamic stability (26), whereupon it dissociates several nucleotides downstream of the stem-loop, likely due to the inability of the stem-loop to enter the narrow channel (145, 150). In *E. coli*, PNPase functions in association with the ATP-dependent RNA helicase RhlB, which can assist PNPase by unwinding internal stem-loops that are encountered (132). When unimpeded, PNPase degrades RNA almost completely, releasing a 5′-terminal dinucleotide as its final product (29).

Hydrolytic 3′ exonucleases. The principal hydrolytic 3′ exoribonucleases in bacterial cells are members of the RNR superfamily. As catalysts of an irreversible reaction, they function exclusively as degradative enzymes. Like most other γ-proteobacteria, *E. coli* contains two such exonucleases, RNase II and RNase R. It tolerates the absence of either of these enzymes or of PNPase individually, but paired mutations that eliminate PNPase in combination with either RNase II or RNase R are synthetically lethal (30, 42).

RNase II resembles PNPase in terms of its intrinsic substrate selectivity. A single-stranded 3′ end is required for RNase II to engage and degrade its target (145). The enzyme stalls upon encountering a stable stem-loop (145). However, whereas PNPase is able to slowly navigate through such structural impediments with the aid of its associated helicase (95, 132), RNase II cannot do so and dissociates a few nucleotides downstream of the stem-loop (145).

RNase II is a monomeric enzyme comprising one catalytic RNB domain flanked on both sides by RNA-binding domains (two cold-shock domains and one S1 domain) (**Figure 1**) (54). To reach the catalytic center, the 3′ end of RNA substrates threads through a narrow channel, where five 3′-terminal nucleotides make intimate contact with the enzyme (54), thereby explaining why unimpeded digestion by RNase II requires an unpaired 3′ end and generates a 5′-terminal oligonucleotide as the final reaction product (28). Additional nucleotides further upstream associate with the three RNA-binding domains, which function as an anchoring region where sustained contact with the RNA ensures degradative processivity with substrates ≥10 nucleotides long (2, 54).

The other RNR family member, RNase R, shares many structural and catalytic properties with RNase II (28). However, a key distinguishing characteristic of RNase R is its intrinsic ability to unwind double-stranded RNA, which enables it to degrade highly structured RNAs nearly to completion without the aid of a helicase or an external source of energy such as ATP, provided that a single-stranded 3′ end is initially available for binding (6, 29). This property of RNase R has been attributed to unique features of its catalytic domain, S1 domain, and carboxy-terminal tail (105, 154).

5′ exonucleases. The longstanding belief that 5′ exoribonucleases do not exist in bacteria was overturned by the discovery that RNase J is able to remove nucleotides sequentially from the 5′ end of RNA, with a strong preference for 5′ monophosphorylated substrates (103, 134). Absent from *E. coli* and initially identified in *B. subtilis* as an endonuclease (50), this enzyme is a dimer of dimers in which every subunit contains a bipartite metallo-β-lactamase domain, a β-CASP domain, and a carboxy-terminal domain (**Figure 1**). At each dimer interface, an RNA-binding channel leads deep inside the protein to a catalytic active site, where a monophosphorylated, but not a triphosphorylated, 5′ end can bind so as to position the 5′-terminal nucleotide for hydrolytic removal (43, 91). The channel continues past the catalytic center and emerges on the other side of the enzyme, thus explaining the ability of RNase J to act not only as a 5′ exonuclease but also as an endonuclease.

The impact of RNase J on global mRNA decay has been best studied in *B. subtilis*, which encodes two paralogs (J1 and J2) that assemble to form a heterotetramer in vivo (104). Of the two, only RNase J1 has significant 5′ exonuclease activity, and its absence markedly slows *B. subtilis* cell growth (52, 104). Severely depleting RNase J1 affects a large portion the *B. subtilis* transcriptome, suggesting that this enzyme plays a major role in *B. subtilis* mRNA degradation (44). The presence of two RNase J paralogs is common in Firmicutes, but in many other species only a single RNase J ortholog is present (18).

Oligoribonucleases

A hydrolytic 3′ exoribonuclease, oligoribonuclease differs from other bacterial exonucleases in one fundamental aspect: This enzyme displays a marked preference for RNA substrates no more than five nucleotides long (33). It plays a vital role in RNA degradation. Because the structures and mechanisms of PNPase, RNase II, and RNase R prevent them from completely degrading their substrates, they generate 5′-terminal oligonucleotides ranging from two to five nucleotides in length as reaction products (28, 29). Oligoribonuclease converts these remnants into mononucleotides, thus replenishing the cellular pool of RNA precursors (58), while also preventing the misincorporation of these oligonucleotides at the 5′ end of new transcripts (59).

Oligoribonuclease is essential in *E. coli* (58), where it is the only ribonuclease that can efficiently degrade oligonucleotides, but a sequence homolog of the *E. coli* enzyme (Orn) is not present in all bacterial species. Some species that lack this enzyme have been shown to contain a distinct ribonuclease (NrnA/B or NrnC) with similar properties (51, 96, 111). Other species may contain as yet unidentified ribonucleases that can perform this function.

RNA Degradosomes

Presumably to enhance their degradative efficiency, enzymes important for mRNA decay often assemble to form a multimeric complex called an RNA degradosome. These degradosomes commonly contain one or more ribonuclease(s) and an RNA helicase.

The degradosome studied most extensively is that of *E. coli*, where PNPase, RhlB, and the glycolytic enzyme enolase bind to discrete sites in the noncatalytic carboxy-terminal half of RNase E (23, 132, 153). The association of PNPase with RNase E may facilitate the exonucleolytic degradation of decay intermediates produced by endonucleolytic cleavage. Likewise, the ability of the RNA helicase RhlB to disrupt RNA base pairing can both expose internal sites to RNase E cleavage and aid PNPase when significant 3′-terminal structure is encountered (79, 132). Less clear is the role of enolase in the RNA degradosome, where it may play a role in sensing the metabolic state of cells (116). Two-hybrid and co-immunoprecipitation studies suggest that similar degradosome complexes may be present in a number of other Proteobacteria (1, 49, 67, 72). Although the formation of degradosomes is not essential, an *E. coli* strain harboring a truncated form of RNase E that cannot nucleate degradosome assembly grows more slowly and degrades many mRNAs less swiftly than its wild-type counterpart (13, 89). Thus, the ability of components of the ribonucleolytic machinery to associate with one another is of no small consequence.

In bacterial species lacking RNase E, ribonucleolytic counterparts may associate with each other in a similar fashion. Notably, two-hybrid studies in *B. subtilis* and *S. aureus* have detected the interaction of RNase J, PNPase, the RNA helicase CshA, and other proteins (32, 87, 139). In *B. subtilis*, one of those other proteins is RNase Y, which may serve as the scaffold for assembly of a complex (86). However, unlike the RNA degradosome of *E. coli*, a heteromultimer containing stoichiometric amounts of each of these proteins has yet to be verified by purification from cells.

Phylogenetic Distribution of Ribonucleases

As noted above, no universal set of mRNA-degrading enzymes is present in all bacteria. However, some unifying principles are evident upon examining the phylogenetic distribution of ribonucleases (**Table 1**). Two ribonucleases, RNase III and PNPase, are encoded by almost all bacterial genomes annotated to date. Other ribonucleases, such as RNase E/G, RNase Y, RNase J, and RNase II/R, are conserved in many species but notably absent in a number of others. All told, nearly all bacteria (>90%) contain a low-specificity endonuclease that cuts single-stranded RNA (RNase E/G and/or RNase Y), an endonuclease specific for double-stranded RNA (RNase III), one or more 3′ exonucleases (PNPase, RNase II, and/or RNase R), and an oligoribonuclease (Orn, NrnA/B, and/or NrnC), and more than half also contain a 5′ exonuclease (RNase J). Most species (>75%) contain both PNPase and one or more hydrolytic 3′ exonucleases, and a significant number (~20%) contain both RNase E/G and RNase Y. The fact that very few species other than Spirochaetales lack both RNase E/G and RNase J, two 5′-monophosphate-stimulated ribonucleases, suggests that a 5′-end-dependent degradation pathway may be nearly universal in bacteria.

mRNA DEGRADATION PATHWAYS

Despite the diverse sets of ribonucleases found in bacteria, the basic pathways of mRNA degradation are remarkably similar across species. There appear to be two mechanisms for initiating mRNA decay. In one (direct access), degradation begins with ribonuclease attack, whereas in the other (5′-end-dependent access), the 5′-terminal triphosphate is first converted to a monophosphate.

Direct-Access Pathway

The first degradative event in the direct-access pathway is internal cleavage by an endonuclease (**Figure 2**). In *E. coli* and related species, this step is usually catalyzed by RNase E (7, 112, 119, 126, 151), but for some mRNAs it has been shown that other endonucleases initiate decay (75, 93, 106, 146). By contrast, in species that lack RNase E, such as *B. subtilis*, degradation often begins instead with internal cleavage by RNase Y (44, 82, 88, 159). Regardless of the endonuclease, this initial cleavage produces 5′- and 3′-terminal mRNA fragments, each of which is typically shorter lived than the full-length transcript.

In most cases, the 5′ fragment produced by endonucleolytic cleavage no longer has a protective stem-loop at its 3′ end and is therefore susceptible to rapid 3′-exonucleolytic degradation (**Figure 2**). Such degradation often proceeds to completion despite various obstacles that the 3′ exonucleases may encounter. Although thermodynamically robust base pairing typically impedes exonucleolytic degradation, such barriers can eventually be overcome with the aid of an enzyme that appends a single-stranded tail downstream of the impediment (**Figure 3**). In *E. coli*, tailing is achieved primarily by the action of poly(A) polymerase (PAP), which can polyadenylate the 3′ end of decay intermediates from which a 3′ exonuclease has disengaged (21, 62, 157). In bacterial species that lack a dedicated poly(A) polymerase, A-rich tails can be added by the template-independent polymerase activity of a phosphorolytic exonuclease such as PNPase (114). Successive rounds of poly(A) addition and removal downstream of a base-paired structure provide repeated opportunities for penetration of the barrier by PNPase (with assistance from RhlB) or RNase R, thereby allowing exonucleolytic degradation to proceed past the structured region. In contrast, because of its strict specificity for single-stranded 3′ ends, RNase II can impede the exonucleolytic destruction of stem-loop structures by unproductively removing the poly(A) tail on which PNPase and RNase R rely without ever damaging the stem-loop itself (64). Consequently, 3′-exonucleolytic

Table 1 Phylogenetic distribution of major bacterial mRNA-degrading ribonucleases[a]

Taxon	Endos			5' Exo	3' Exos		Oligoribonucleases		
	RNase E/G	RNase Y	RNase III	RNase J	PNPase	RNase II/R	Orn	NrnA/B	NrmC
Proteobacteria									
Alpha	91/91	0/91	91/91	91/91	91/91	64/91	0/91	0/91	91/91
Beta	53/53	0/53	53/53	0/53	53/53	53/53	52/53	0/53	0/53
Gamma	122/122	0/122	122/122	0/122	122/122	118/122	122/122	0/122	0/122
Delta	29/29	24/29	29/29	25/29	29/29	28/29	5/29	20/29	0/29
Epsilon	0/16	15/16	16/16	16/16	16/16	16/16	0/16	16/16	0/16
Firmicutes									
Bacilli									
Bacillales	14/32	32/32	32/32	32/32	32/32	32/32	0/32	32/32	0/32
Lactobacillales	0/30	29/30	30/30	30/30	13/30	30/30	0/30	30/30	0/30
Clostridia	34/37	37/37	36/37	37/37	37/37	31/37	0/37	37/37	1/37
Actinobacteria	65/70	18/70	70/70	69/70	70/70	32/70	65/70	35/70	0/70
Cyanobacteria	25/25	0/25	24/25	25/25	25/25	25/25	0/25	0/25	25/25
Bacteroidetes and Chlorobi	31/31	29/31	31/31	0/31	31/31	31/31	0/31	31/31	0/31
Tenericutes	0/22	14/22	22/22	22/22	5/22	18/22	0/22	22/22	0/22
Chlamydiae and Verrucomicrobia	10/10	2/10	10/10	0/10	10/10	9/10	0/10	3/10	0/10
Spirochaetales	0/13	10/13	13/13	1/13	13/13	6/13	0/13	10/13	3/13
Chloroflexi	0/12	12/12	12/12	12/12	12/12	0/12	0/12	8/12	0/12
Thermotogae	11/11	11/11	11/11	0/11	11/11	11/11	0/11	11/11	0/11
Deinococcus–Thermus	0/4	4/4	0/4	4/4	4/4	4/4	0/4	4/4	0/4
Aquificae	5/5	5/5	5/5	0/5	5/5	5/5	0/5	5/5	0/5
Fusobacteria	1/4	4/4	4/4	4/4	4/4	4/4	0/4	4/4	0/4
Fibrobacteres	4/4	0/4	4/4	3/4	4/4	4/4	1/4	3/4	0/4

Legend: 0–10% | 10–50% | 50–90% | 90–100%

[a]Numbers (colors) correspond to the fraction (approximate percentage) of analyzed species in each taxon that contain a particular ribonuclease. Homologs of major bacterial ribonucleases were identified in the sequenced and annotated genomes of 621 distinct bacterial species by using the program STRING v9.1 (53). Initial hits were defined as those having a minimum bit score of 60. Individual sequence alignments with bit scores above 120 were accepted as homologs, and those with bit scores between 60 and 120 were re-analyzed with BLAST (Basic Local-Alignment Search Tool) to identify true homologs with the proper protein length, a good sequence alignment, and an E value below 1×10^{-8}. The distribution of ribonucleases was then mapped by using iTOL (Interactive Tree of Life) (90) and tallied as a fraction of the number of analyzed species in each taxonomic clade. Phyla are listed in boldface type, classes in regular type, and orders in italics. Abbreviations: Endo, endonuclease; Exo, exonuclease.

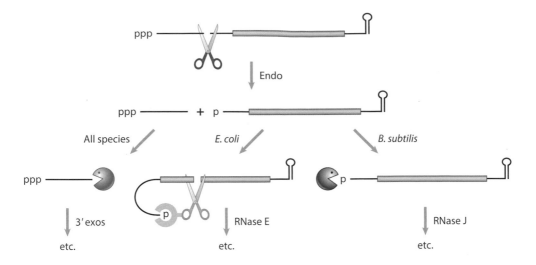

Figure 2

Direct-access pathway for mRNA degradation. An endonuclease (Endo; *black-handled scissors*), usually but not always RNase E or RNase Y, cleaves the primary transcript internally to generate two fragments. Unprotected at its 3′ end, the 5′ fragment is quickly attacked by 3′ exonucleases (3′ exos; *blue Pac-Man*) in all bacterial species. The fate of the monophosphorylated 3′ fragment depends on the ribonucleases present in the cell. In some species, such as *Escherichia coli*, this fragment undergoes further endonucleolytic cleavage by RNase E (*gray-handled scissors*), which rapidly degrades such intermediates by selectively binding the monophosphorylated 5′ terminus in a discrete pocket on the surface of the catalytic domain and cutting downstream. In others, such as *Bacillus subtilis*, the fragment undergoes rapid 5′-exonucleolytic digestion by RNase J (*red Pac-Man*), whose exonuclease activity aggressively degrades RNAs bearing a single phosphate at the 5′ end. Abbreviations: etc., additional degradative steps (not shown); p, phosphate.

penetration of such structures may often be slower than endonucleolytic cleavage upstream, especially when they are thermodynamically robust and located in an untranslated region.

As they degrade 5′-terminal mRNA fragments, 3′ exonucleases may also encounter translating ribosomes that are moving in the opposite direction. To rescue ribosomes stalled at the 3′ end of degradation intermediates that lack a termination codon, a specialized bacterial RNA (tmRNA) that has features of both tRNA and mRNA is recruited together with its protein escort (SmpB) (77). SmpB facilitates ribosome template switching from the truncated mRNA to the tmRNA, which contains a termination codon that allows the ribosome to be released. RNase R subsequently degrades the mRNA fragment from its now exposed 3′ end (136).

Although the 3′ fragment generated by the initial endonucleolytic cleavage ends with a stem-loop that protects it from 3′-exonucleolytic degradation, it too is typically quite labile because of

Figure 3

3′-exonucleolytic degradation of decay intermediates. Cleavage of mRNA by an endonuclease (Endo; *scissors*) generates two intermediates: a 5′-terminal fragment whose single-stranded 3′ end is trimmed by 3′ exonucleases (3′ exos; *blue Pac-Man*) until a structural barrier is encountered, and a 3′-terminal fragment whose 3′ end is protected from exonucleolytic digestion by a terminator stem-loop. Subsequent polyadenylation by poly(A) polymerase or PNPase provides the 3′ exonucleases PNPase (operating with help from the RNA helicase RhlB) and RNase R (operating alone) an opportunity to overcome the barriers by creating a single-stranded binding site from which they can launch an attack. The process of poly(A) addition and removal is repeated until the structural barriers are breached. By contrast, RNase II can degrade poly(A) and other unpaired 3′ ends but not structured 3′ ends. Degradation by these 3′ exonucleases eventually generates fragments that are too small for them to shorten further and are instead degraded by an oligoribonuclease (*yellow Pac-Man*). Abbreviation: p, phosphate.

its monophosphorylated 5′ terminus (**Figure 2**). In bacterial species that contain RNase J, the presence of only one phosphate at that end exposes such intermediates to swift 5′-exonucleolytic degradation (36, 160). In species that lack RNase J, these decay intermediates are rapidly destroyed by RNase E, whose ribonucleolytic potency is greatly enhanced when the 5′ end of a substrate is monophosphorylated (99). Repeated cleavage by this endonuclease yields mRNA fragments susceptible to exonucleolytic degradation from an unprotected 3′ end or, in the case of the 3′-terminal fragment bearing the terminator stem-loop of the original transcript, to degradation by a mechanism involving polyadenylation followed by 3′-exonucleolytic attack (**Figure 3**) (64, 156, 157).

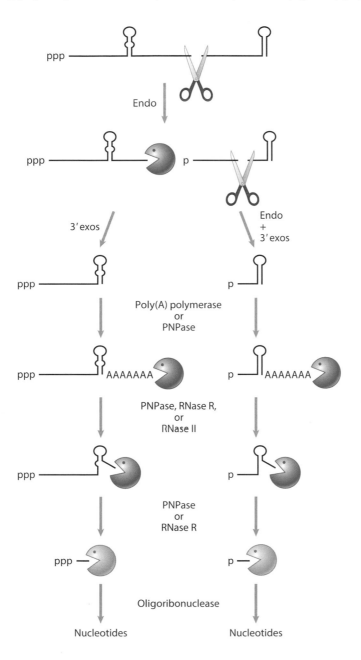

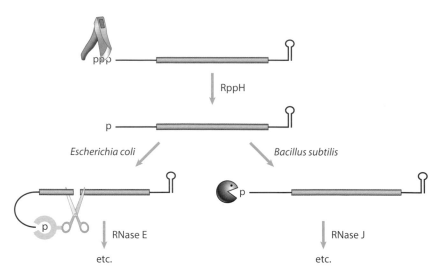

Figure 4

5′-end-dependent pathway for initiating mRNA degradation. The RNA pyrophosphohydrolase RppH (*hatchet*) converts the 5′-terminal triphosphate of the primary transcript to a monophosphate. The resulting full-length decay intermediate is then rapidly degraded by either RNase E (*scissors*) or RNase J (*Pac-Man*), depending on which of these enzymes is present in the host species. Abbreviations: etc., additional degradative steps (not shown); p, phosphate.

5′-End-Dependent Pathway

Although pertinent to the decay of a large percentage of primary transcripts, the direct-access pathway for endonucleolytic initiation does not explain the ability of a 5′-terminal stem-loop to stabilize many transcripts (9, 15, 48, 65, 143). This observation led to the discovery and characterization of a distinct, 5′-end-dependent pathway for mRNA degradation in which endonucleolytic cleavage is not the initial event. Instead, decay by this pathway is triggered by a prior non-nucleolytic event that marks transcripts for rapid turnover: the conversion of the 5′ terminus from a triphosphate to a monophosphate (**Figure 4**). Catalyzed by the RNA pyrophosphohydrolase RppH, this modification greatly increases the susceptibility of mRNA to degradation by RNase E or RNase J (25, 35, 134), both of which aggressively attack monophosphorylated RNA substrates. In *E. coli*, the steady-state concentration of hundreds of mRNAs increases significantly when the *rppH* gene is deleted, indicating that a significant portion of the transcriptome is degraded via the 5′-end-dependent pathway (98).

The discovery of the mechanism of 5′-end-dependent degradation explained the protective effect of 5′-terminal stem-loops, as RppH, RNase E, and RNase J can interact only with 5′ ends that are single-stranded. Indeed, biochemical studies of RppH from *B. subtilis* and *E. coli* indicate that it requires at least two and preferably three or more unpaired nucleotides at the 5′ end of its substrates (70) (P.-K. Hsieh & J.G. Belasco, unpublished results). In addition, *B. subtilis* RppH, but not *E. coli* RppH, has a strict requirement for guanylate as the second nucleotide. However, 5′-end-dependent mRNA degradation in *B. subtilis* does not rely entirely on the identity of the second nucleotide or even on RppH, apparently because of the presence of another, as yet unidentified, RNA pyrophosphohydrolase in that species (70, 134). By contrast, there is no evidence for an alternative pyrophosphate-removing enzyme in *E. coli*.

3′-Exonucleolytic Initiation of Decay

mRNA decay in *E. coli* is retarded but not abolished upon inactivation of RNase E, indicating that alternative, RNase E–independent degradation pathways exist. Indeed, several transcripts whose degradation is impeded by RNase E inactivation are further stabilized when cells lack PAP or PNPase in addition to RNase E (62, 64, 125). Taken together, these findings suggest that poly(A)-dependent 3′-exonucleolytic degradation can sometimes initiate mRNA decay. However, the fact that the influence of PAP and PNPase is generally meager when RNase E is present indicates that 3′-exonucleolytic initiation of decay is ordinarily much slower than other degradation mechanisms.

mRNA FEATURES THAT GOVERN STABILITY

Because of the low sequence specificity of RNase E and RNase Y, a typical protein-encoding transcript is likely to possess many potential cleavage sites, none of which alone is critical for degradation. Therefore, the diversity of bacterial mRNA lifetimes suggests that the susceptibility of individual transcripts to degradation depends instead on the ease with which RNase E or RNase Y gains access to those sites, as governed by the sequence and/or structure of each transcript and the cellular factors with which the mRNA interacts.

Ribosome Binding and Translation

Among the most important non-nucleolytic *trans*-acting factors that influence mRNA stability are ribosomes. In *E. coli*, the lifetime of a monocistronic message can usually be prolonged or abbreviated by increasing or decreasing, respectively, the ribosome-binding affinity of the Shine-Dalgarno element (5, 16, 161). Such effects are observed irrespective of whether the transcript is degraded by a direct-access or 5′-end-dependent mechanism (5, 135). Efficient ribosome binding and translation are thought to stabilize mRNA by sterically masking RNase E cleavage sites within the message. However, several lines of evidence suggest that the mechanism by which ribosomes protect mRNA is more complex, including the relatively modest effect of reducing the frequency of translation initiation by replacing an AUG initiation codon with a less efficient GUG or CUG codon (5) and the variable effect of premature translation termination, which is both transcript and position dependent (66, 122). Furthermore, although influenced by ribosome binding, mRNA decay rates appear to be less sensitive to premature translation termination in *B. subtilis* (142), which lacks RNase E but contains another low-specificity endonuclease, RNase Y, and the 5′ exonuclease RNase J.

Rates of mRNA degradation can also be affected by ribosomes that stall during translation elongation or termination because of the sequence of the nascent polypeptide or the scarcity of a required aminoacyl-tRNA. In *E. coli*, such events can trigger cleavage of the mRNA in or adjacent to the ribosomal A-site (68, 92) or upstream of the stalled ribosome (97) by mechanisms that have not yet been fully delineated. Conversely, in *B. subtilis* a stalled ribosome can act as a barrier that protects mRNA downstream of the stall site from 5′-exonucleolytic degradation by RNase J (11, 103, 140).

Intramolecular Base Pairing

Another major influence on bacterial mRNA degradation is RNA structure, which can impact rates of mRNA decay either directly by determining the accessibility of an entire transcript or a segment thereof to ribonuclease attack or indirectly by governing the binding of ribosomes or other non-nucleolytic factors that affect degradation. Some of these structural influences are ubiquitous, such as the stem-loops at the 3′ ends of nearly all full-length bacterial transcripts. Present as a component of an intrinsic transcription terminator or as a result of exonucleolytic trimming from an unpaired 3′ end, these 3′-terminal structures protect mRNA from 3′-exonuclease attack and thereby force degradation to begin elsewhere (12, 118). Less common is a stem-loop at the 5′ end of mRNA, where it can prevent 5′-end-dependent degradation by inhibiting conversion of the 5′-terminal triphosphate to a monophosphate (35, 134).

Of course, intramolecular base pairing in bacterial mRNAs is not confined to the 5′ or 3′ end. In a number of cases, an internal stem-loop structure has been shown to play a pivotal role in the differential expression of genes within a polycistronic transcript. Whether such a stem-loop confers greater stability on the upstream or downstream RNA segment depends on the location of the stem-loop relative to the initial site of endonucleolytic cleavage. For example, a large intercistronic stem-loop between the *malE* and *malF* segments of the *E. coli malEFG* transcript protects the upstream *malE* segment against 3′-exonucleolytic propagation of decay from a downstream site of initial endonucleolytic cleavage. As a consequence, a comparatively stable 5′-terminal decay intermediate encompassing only *malE* accumulates, resulting in substantially greater production of maltose-binding protein (MalE) than of the membrane-bound subunits of the maltose transporter (MalF and MalG) (120). The large number of *E. coli* operons that contain palindromic sequences in intercistronic regions suggests that stem-loop structures of this kind may have a widespread role in differential gene expression (121, 147). Conversely, the presence of a stem-loop immediately downstream of a site of endonucleolytic cleavage can protect the 3′ fragment from 5′-monophosphate-stimulated RNase E cleavage, as observed for the dicistronic *papBA* transcript, which encodes a low-abundance transcription factor (PapB) and a major pilus protein (PapA) in uropathogenic strains of *E. coli*. RNase E cleavage at a site two nucleotides upstream of an intercistronic stem-loop structure contributes to swift 3′-exonucleolytic degradation of the *papB* segment of that transcript without exposing the 5′-monophosphorylated *papA* intermediate to rapid degradation by RNase E (8, 17). An interesting combination of both phenomena is illustrated by the degradation of the *pufQBALMX* photosynthesis transcript of *Rhodobacter capsulatus*, where intercistronic stem-loop structures flanking the internal *pufBA* segment enable this fragment to accumulate as a long-lived decay intermediate that survives the rapid degradation of the surrounding portions of the transcript (69).

Alternatively, intramolecular base pairing can instead act, indirectly or directly, to destabilize a transcript. In *E. coli*, indirect destabilization by such base pairing is usually a consequence of an RNA conformation that prevents ribosome binding by sequestering the site of translation initiation (5, 16, 161), whereas direct destabilization by intramolecular base pairing is often attributable to formation of a cleavage site for RNase III, an endonuclease specific for lengthy double-stranded regions of RNA (106, 144). Less frequently, the destabilizing structured element is a metabolite-binding riboswitch that can accelerate mRNA degradation upon undergoing a conformational change in response to an increase or decrease in the concentration of its ligand. It may do so by occluding the ribosome-binding site (123), by unmasking nearby RNase E cleavage sites (22), or, in the case of a catalytic riboswitch, by activating or repressing an intrinsic self-cleavage activity (31).

sRNA Binding

Small noncoding RNAs (sRNAs) are among the most common means by which bacteria regulate mRNA abundance post-transcriptionally in response to environmental cues. Expressed from regions of the genome that are usually distinct from the genes they regulate, sRNAs bind specific transcripts within segments to which they are partially or fully complementary and in doing so can influence the translation and/or decay rate of those mRNAs (39, 149). Typically, a single sRNA targets multiple transcripts so as to coordinately modulate the production of several proteins.

sRNAs can destabilize or stabilize a target transcript, depending on the nature of their interaction. Sometimes the mRNA-sRNA duplex itself is cleaved by RNase III (24). More frequently, sRNA binding stimulates mRNA degradation indirectly by interfering with ribosome binding and translation initiation (39). As a result, the repressive effect of the sRNA becomes irreversible. Interestingly, sRNA binding can also stimulate RNase E cleavage by mechanisms apparently unrelated to translation. For example, in *Salmonella enterica*, binding of the sRNA MicC to a site deep inside the coding region of *ompD* mRNA induces RNase E cleavage four to five nucleotides downstream of the sRNA-mRNA complex without affecting translation initiation (131). A mechanism has been proposed wherein *ompD* cleavage by RNase E is stimulated in *trans* by a monophosphate at the 5′ end of MicC (10). Although demonstrable with purified components in vitro, it is unclear whether this mechanism explains the destabilizing effect of MicC in *Salmonella*, where <1% of MicC is monophosphorylated (P.L. Foley & J.G. Belasco, unpublished results).

In a number of other cases, sRNAs have been shown to upregulate the expression of the mRNAs they target. Often they do so by disrupting an inhibitory stem-loop that would otherwise sequester the Shine-Dalgarno element (107). By exposing the ribosome binding site, the sRNA facilitates translation initiation and, as a consequence, prolongs the lifetime of the transcript. In addition, sRNAs sometimes act directly to protect mRNA from degradation by masking RNase E cleavage sites without help from ribosomes (55, 128) or by sequestering the 5′ terminus so as to prevent mRNA degradation via a 5′-end-dependent pathway (133).

In many species such as *E. coli*, sRNAs usually act in concert with the RNA chaperone protein Hfq. Hfq has a multifaceted role in sRNA-mediated regulation. It not only protects sRNAs from degradation by cellular ribonucleases (102) but also facilitates sRNA-mRNA base pairing (115). Hfq also has been shown to associate directly with RNase E, and this binding may play a role in mRNA degradation by facilitating RNase E recruitment to sRNA-associated transcripts (117). Finally, Hfq can stimulate the activity of poly(A) polymerase, an enzyme important for 3′-exonucleolytic degradation (63).

CONTROL OF mRNA-DEGRADING ENZYMES

Needing at times to alter the abundance of a great many transcripts simultaneously, bacteria have several ways to coordinate changes in mRNA stability. These include altering the concentration or specific activity of cellular ribonucleases or activating bacterial toxins. In addition, bacteriophages have evolved mechanisms to protect their transcripts from rapid degradation by host enzymes.

Regulation of Ribonuclease Concentration and Activity

Bacteria maintain precise control over the cellular activity of many of the ribonucleases most important for mRNA decay by regulating either their concentration or their specific activity. For instance, to achieve homeostasis, RNase E, RNase III, and PNPase autoregulate their synthesis in *E. coli* by modulating the decay rates of their respective mRNAs as a function of the cellular activity of the corresponding enzymes (74, 75, 106). The concentration of other ribonucleases is growth-phase dependent. During stationary phase or upon cold shock, RNase R is three- to tenfold more abundant in *E. coli* than during unimpeded exponential growth due to its diminished susceptibility to proteolysis (94). *B. subtilis* RNase Y also exhibits growth-phase-dependent changes in abundance by an undetermined mechanism (88).

In addition to concentration changes, the cellular activity of RNase E, RNase III, and PNPase in *E. coli* can also be modulated in response to environmental signals by altering the specific activity of these enzymes. These changes in catalytic potency result from binding either a cellular metabolite or a protein. For example, PNPase activity is inhibited by ATP and citrate, suggesting that RNA degradation may be sensitive to cellular energy levels and to central metabolism (37, 124). RNase III activity is regulated by the protein YmdB, which is expressed upon cold shock or entry into stationary phase and acts by preventing RNase III dimerization (80). Similarly, RNase E activity can be inhibited by the proteins RraA and RraB, which bind to its carboxy-terminal domain and are thought to stabilize distinct sets of mRNAs under certain stress conditions (57, 60, 85). RraA can also interact directly with the RNA degradosome helicase RhlB and impair its function (60).

Toxin-Antitoxin Systems in Which the Toxin Degrades mRNA

Bacterial genomes encode several toxin-antitoxin (TA) systems, some of which have an impact on mRNA degradation. A TA system consists of a toxin-antitoxin pair in which the deleterious effect of the toxin protein is neutralized by the presence of its cognate antitoxin. The toxin of many type II or type III TA systems is a ribonuclease that normally is inhibited by the tight binding of a protein or RNA antitoxin (158). When triggered by stress, such as amino acid starvation, DNA damage, or heat shock, the unstable antitoxin is degraded, freeing the more stable toxin to attack cellular RNAs. The endonuclease toxins of these TA systems are of two kinds: those that cleave RNA at specific sequences (MazF- and VapC-like toxins) and those that cleave ribosome-associated RNAs within the coding region (RelE-like toxins). Because the specificity of MazF-like toxins is defined by a rather short sequence motif (typically 3–5 nt), they degrade mRNAs fairly indiscriminately (158), as do RelE-like toxins (71). The consequent reduction in protein synthesis is thought to help cells become dormant for the duration of the stress.

Effects of Phage Infection

Infecting bacteriophages utilize a variety of mechanisms to manipulate mRNA degradation in host cells to their advantage. For example, the protein product of phage T7 gene 0.7 phosphorylates RNase E and RhlB, among other *E. coli* proteins, thereby selectively inhibiting endonucleolytic cleavage of nascent T7 transcripts that are transiently ribosome-deficient due to the ability of T7 RNA polymerase to outpace ribosomes (100). Another *E. coli* endonuclease implicated in mRNA degradation in phage-infected cells is RNase LS (RnlA), the toxin component of a TA system (81). Owing to its short lifetime, the cognate antitoxin RnlB is quickly degraded upon global inhibition of host gene expression by phage T4. As a result, RNase LS becomes activated. To prevent RNase LS from degrading T4 transcripts, the bacteriophage encodes its own antitoxin, Dmd, which neutralizes RNase LS (127). In addition, the ability of T4 polynucleotide kinase to monophosphorylate the 5′-hydroxyl termini of decay intermediates generated by the T4-encoded endonuclease RegB accelerates their degradation by RNase E (46).

CONCLUDING REMARKS

The interaction of bacterial ribonucleases with their mRNA targets and the mechanisms that bacteria use to govern rates of mRNA degradation have profound implications for gene regulation, environmental adaptation, cell growth and survival, and pathogenesis. Although studies to date have provided an informative glimpse of the many strategies utilized by bacteria to control mRNA turnover, our understanding is far from complete. Consequently, the field of bacterial mRNA decay remains an area ripe for further study.

DISCLOSURE STATEMENT

The authors are not aware of any affiliations, memberships, funding, or financial holdings that might be perceived as affecting the objectivity of this review.

ACKNOWLEDGMENTS

We are grateful to Ciarán Condon for his helpful comments on the manuscript. The writing of this review was supported by a fellowship to M.P.H. (F32GM101962) and a research grant to J.G.B. (R01GM035769) from the National Institutes of Health.

LITERATURE CITED

1. Aït-Bara S, Carpousis AJ. 2010. Characterization of the RNA degradosome of *Pseudoalteromonas haloplanktis*: conservation of the RNase E-RhlB interaction in the gammaproteobacteria. *J. Bacteriol.* 192:5413–23

2. Amblar M, Barbas A, Fialho AM, Arraiano CM. 2006. Characterization of the functional domains of *Escherichia coli* RNase II. *J. Mol. Biol.* 360:921–33

3. Apirion D. 1973. Degradation of RNA in *Escherichia coli*. A hypothesis. *Mol. Gen. Genet.* 122:313–22

4. Apirion D. 1978. Isolation, genetic mapping and some characterization of a mutation in *Escherichia coli* that affects the processing of ribonucleic acid. *Genetics* 90:659–71

5. Arnold TE, Yu J, Belasco JG. 1998. mRNA stabilization by the *ompA* 5′ untranslated region: two protective elements hinder distinct pathways for mRNA degradation. *RNA* 4:319–30

6. Awano N, Rajagopal V, Arbing M, Patel S, Hunt J, et al. 2010. *Escherichia coli* RNase R has dual activities, helicase and RNase. *J. Bacteriol.* 192:1344–52

7. Babitzke P, Kushner SR. 1991. The Ams (altered mRNA stability) protein and ribonuclease E are encoded by the same structural gene of *Escherichia coli*. *Proc. Natl. Acad. Sci. USA* 88:1–5

8. Båga M, Göransson M, Normark S, Uhlin BE. 1988. Processed mRNA with differential stability in the regulation of *E. coli* pilin gene expression. *Cell* 52:197–206

9. Baker KE, Mackie GA. 2003. Ectopic RNase E sites promote bypass of 5′-end-dependent mRNA decay in *Escherichia coli*. *Mol. Microbiol.* 47:75–88

10. Bandyra KJ, Said N, Pfeiffer V, Gorna MW, Vogel J, Luisi BF. 2012. The seed region of a small RNA drives the controlled destruction of the target mRNA by the endoribonuclease RNase E. *Mol. Cell* 47:943–53

11. Bechhofer DH, Zen KH. 1989. Mechanism of erythromycin-induced *ermC* mRNA stability in *Bacillus subtilis*. *J. Bacteriol.* 171:5803–11

12. Belasco JG, Nilsson G, von Gabain A, Cohen SN. 1986. The stability of *E. coli* gene transcripts is dependent on determinants localized to specific mRNA segments. *Cell* 46:245–51

13. Bernstein JA, Lin PH, Cohen SN, Lin-Chao S. 2004. Global analysis of *Escherichia coli* RNA degradosome function using DNA microarrays. *Proc. Natl. Acad. Sci. USA* 101:2758–63

14. Blaszczyk J, Gan J, Tropea JE, Court DL, Waugh DS, Ji X. 2004. Noncatalytic assembly of ribonuclease III with double-stranded RNA. *Structure* 12:457–66

15. Bouvet P, Belasco JG. 1992. Control of RNase E-mediated RNA degradation by 5′-terminal base pairing in *E. coli*. *Nature* 360:488–91

16. Braun F, Le Derout J, Regnier P. 1998. Ribosomes inhibit an RNase E cleavage which induces the decay of the *rpsO* mRNA of *Escherichia coli*. *EMBO J.* 17:4790–97

17. Bricker AL, Belasco JG. 1999. Importance of a 5′ stem-loop for longevity of *papA* mRNA in *Escherichia coli*. *J. Bacteriol.* 181:3587–90

18. Britton RA, Wen T, Schaefer L, Pellegrini O, Uicker WC, et al. 2007. Maturation of the 5′ end of *Bacillus subtilis* 16S rRNA by the essential ribonuclease YkqC/RNase J1. *Mol. Microbiol.* 63:127–38

19. Calin-Jageman I, Nicholson AW. 2003. Mutational analysis of an RNA internal loop as a reactivity epitope for *Escherichia coli* ribonuclease III substrates. *Biochemistry* 42:5025–34

20. Callaghan AJ, Marcaida MJ, Stead JA, McDowall KJ, Scott WG, Luisi BF. 2005. Structure of *Escherichia coli* RNase E catalytic domain and implications for RNA turnover. *Nature* 437:1187–91

21. Cao GJ, Sarkar N. 1992. Identification of the gene for an *Escherichia coli* poly(A) polymerase. *Proc. Natl. Acad. Sci. USA* 89:10380–84

22. Caron MP, Bastet L, Lussier A, Simoneau-Roy M, Masse E, Lafontaine DA. 2012. Dual-acting riboswitch control of translation initiation and mRNA decay. *Proc. Natl. Acad. Sci. USA* 109:E3444–53

23. Carpousis AJ, Van Houwe G, Ehretsmann C, Krisch HM. 1994. Copurification of *E. coli* RNAase E and PNPase: evidence for a specific association between two enzymes important in RNA processing and degradation. *Cell* 76:889–900

24. Case CC, Simons EL, Simons RW. 1990. The *IS10* transposase mRNA is destabilized during antisense RNA control. *EMBO J.* 9:1259–66

25. Celesnik H, Deana A, Belasco JG. 2007. Initiation of RNA decay in *Escherichia coli* by 5′ pyrophosphate removal. *Mol. Cell* 27:79–90

26. Chen LH, Emory SA, Bricker AL, Bouvet P, Belasco JG. 1991. Structure and function of a bacterial mRNA stabilizer: analysis of the 5′ untranslated region of *ompA* mRNA. *J. Bacteriol.* 173:4578–86

27. Chen Z, Itzek A, Malke H, Ferretti JJ, Kreth J. 2013. Multiple roles of RNase Y in *Streptococcus pyogenes* mRNA processing and degradation. *J. Bacteriol.* 195:2585–94

28. Cheng ZF, Deutscher MP. 2002. Purification and characterization of the *Escherichia coli* exoribonuclease RNase R. Comparison with RNase II. *J. Biol. Chem.* 277:21624–29

29. Cheng ZF, Deutscher MP. 2005. An important role for RNase R in mRNA decay. *Mol. Cell* 17:313–18

30. Cheng ZF, Zuo Y, Li Z, Rudd KE, Deutscher MP. 1998. The *vacB* gene required for virulence in *Shigella flexneri* and *Escherichia coli* encodes the exoribonuclease RNase R. *J. Biol. Chem.* 273:14077–80

31. Collins JA, Irnov I, Baker S, Winkler WC. 2007. Mechanism of mRNA destabilization by the *glmS* ribozyme. *Genes Dev.* 21:3356–68

32. Commichau FM, Rothe FM, Herzberg C, Wagner E, Hellwig D, et al. 2009. Novel activities of glycolytic enzymes in *Bacillus subtilis*: interactions with essential proteins involved in mRNA processing. *Mol. Cell Proteomics* 8:1350–60

33. Datta AK, Niyogi K. 1975. A novel oligoribonuclease of *Escherichia coli*. II. Mechanism of action. *J. Biol. Chem.* 250:7313–19

34. Deana A, Belasco JG. 2004. The function of RNase G in *Escherichia coli* is constrained by its amino and carboxyl termini. *Mol. Microbiol.* 51:1205–17

35. Deana A, Celesnik H, Belasco JG. 2008. The bacterial enzyme RppH triggers messenger RNA degradation by 5′ pyrophosphate removal. *Nature* 451:355–58

36. Deikus G, Condon C, Bechhofer DH. 2008. Role of *Bacillus subtilis* RNase J1 endonuclease and 5′-exonuclease activities in *trp* leader RNA turnover. *J. Biol. Chem.* 283:17158–67

37. Del Favero M, Mazzantini E, Briani F, Zangrossi S, Tortora P, Dehò G. 2008. Regulation of *Escherichia coli* polynucleotide phosphorylase by ATP. *J. Biol. Chem.* 283:27355–59

38. Deltcheva E, Chylinski K, Sharma CM, Gonzales K, Chao Y, et al. 2011. CRISPR RNA maturation by *trans*-encoded small RNA and host factor RNase III. *Nature* 471:602–7

39. Desnoyers G, Bouchard MP, Masse E. 2013. New insights into small RNA-dependent translational regulation in prokaryotes. *Trends Genet.* 29:92–98

40. Deutscher MP. 1985. *E. coli* RNases: making sense of alphabet soup. *Cell* 40:731–32

41. Deutscher MP, Marshall GT, Cudny H. 1988. RNase PH: an *Escherichia coli* phosphate-dependent nuclease distinct from polynucleotide phosphorylase. *Proc. Natl. Acad. Sci. USA* 85:4710–14

42. Donovan WP, Kushner SR. 1986. Polynucleotide phosphorylase and ribonuclease II are required for cell viability and mRNA turnover in *Escherichia coli* K-12. *Proc. Natl. Acad. Sci. USA* 83:120–24

43. Dorleans A, Li de la Sierra-Gallay I, Piton J, Zig L, Gilet L, et al. 2011. Molecular basis for the recognition and cleavage of RNA by the bifunctional 5′-3′ exo/endoribonuclease RNase J. *Structure* 19:1252–61

44. Durand S, Gilet L, Bessières P, Nicolas P, Condon C. 2012. Three essential ribonucleases—RNase Y, J1, and III—control the abundance of a majority of *Bacillus subtilis* mRNAs. *PLOS Genet.* 8:e1002520

45. Durand S, Gilet L, Condon C. 2012. The essential function of *B. subtilis* RNase III is to silence foreign toxin genes. *PLOS Genet.* 8:e1003181

46. Durand S, Richard G, Bontems F, Uzan M. 2012. Bacteriophage T4 polynucleotide kinase triggers degradation of mRNAs. *Proc. Natl. Acad. Sci. USA* 109:7073–78

47. Dutta T, Deutscher MP. 2009. Catalytic properties of RNase BN/RNase Z from *Escherichia coli*: RNase BN is both an exo- and endoribonuclease. *J. Biol. Chem.* 284:15425–31

48. Emory SA, Bouvet P, Belasco JG. 1992. A 5′-terminal stem-loop structure can stabilize mRNA in *Escherichia coli*. *Genes Dev.* 6:135–48

49. Erce MA, Low JK, March PE, Wilkins MR, Takayama KM. 2009. Identification and functional analysis of RNase E of *Vibrio angustum* S14 and two-hybrid analysis of its interaction partners. *Biochim. Biophys. Acta* 1794:1107–14

50. Even S, Pellegrini O, Zig L, Labas V, Vinh J, et al. 2005. Ribonucleases J1 and J2: two novel endoribonucleases in *B. subtilis* with functional homology to *E. coli* RNase E. *Nucleic Acids Res.* 33:2141–52

51. Fang M, Zeisberg WM, Condon C, Ogryzko V, Danchin A, Mechold U. 2009. Degradation of nanoRNA is performed by multiple redundant RNases in *Bacillus subtilis*. *Nucleic Acids Res.* 37:5114–25

52. Figaro S, Durand S, Gilet L, Cayet N, Sachse M, Condon C. 2013. *Bacillus subtilis* mutants with knockouts of the genes encoding ribonucleases RNase Y and RNase J1 are viable, with major defects in cell morphology, sporulation, and competence. *J. Bacteriol.* 195:2340–48

53. Franceschini A, Szklarczyk D, Frankild S, Kuhn M, Simonovic M, et al. 2013. STRING v9.1: protein-protein interaction networks, with increased coverage and integration. *Nucleic Acids Res.* 41:D808–15

54. Frazao C, McVey CE, Amblar M, Barbas A, Vonrhein C, et al. 2006. Unravelling the dynamics of RNA degradation by ribonuclease II and its RNA-bound complex. *Nature* 443:110–14

55. Fröhlich KS, Papenfort K, Fekete A, Vogel J. 2013. A small RNA activates CFA synthase by isoform-specific mRNA stabilization. *EMBO J.* 32:2963–79

56. Gan J, Tropea JE, Austin BP, Court DL, Waugh DS, Ji X. 2006. Structural insight into the mechanism of double-stranded RNA processing by ribonuclease III. *Cell* 124:355–66

57. Gao J, Lee K, Zhao M, Qiu J, Zhan X, et al. 2006. Differential modulation of *E. coli* mRNA abundance by inhibitory proteins that alter the composition of the degradosome. *Mol. Microbiol.* 61:394–406

58. Ghosh S, Deutscher MP. 1999. Oligoribonuclease is an essential component of the mRNA decay pathway. *Proc. Natl. Acad. Sci. USA* 96:4372–77

59. Goldman SR, Sharp JS, Vvedenskaya IO, Livny J, Dove SL, Nickels BE. 2011. NanoRNAs prime transcription initiation in vivo. *Mol. Cell* 42:817–25

60. Górna MW, Pietras Z, Tsai YC, Callaghan AJ, Hernández H, et al. 2010. The regulatory protein RraA modulates RNA-binding and helicase activities of the *E. coli* RNA degradosome. *RNA* 16:553–62

61. Grunberg-Manago M. 1963. Enzymatic synthesis of nucleic acids. *Prog. Biophys. Mol. Biol.* 13:175–239

62. Hajnsdorf E, Braun F, Haugel-Nielsen J, Regnier P. 1995. Polyadenylylation destabilizes the *rpsO* mRNA of *Escherichia coli*. *Proc. Natl. Acad. Sci. USA* 92:3973–77

63. Hajnsdorf E, Régnier P. 2000. Host factor Hfq of *Escherichia coli* stimulates elongation of poly(A) tails by poly(A) polymerase I. *Proc. Natl. Acad. Sci. USA* 97:1501–5

64. Hajnsdorf E, Steier O, Coscoy L, Teysset L, Regnier P. 1994. Roles of RNase E, RNase II and PNPase in the degradation of the *rpsO* transcripts of *Escherichia coli*: stabilizing function of RNase II and evidence for efficient degradation in an *ams pnp rnb* mutant. *EMBO J.* 13:3368–77

65. Hambraeus G, Karhumaa K, Rutberg B. 2002. A 5′ stem-loop and ribosome binding but not translation are important for the stability of *Bacillus subtilis aprE* leader mRNA. *Microbiology* 148:1795–803

66. Hansen MJ, Chen LH, Fejzo ML, Belasco JG. 1994. The *ompA* 5′ untranslated region impedes a major pathway for mRNA degradation in *Escherichia coli*. *Mol. Microbiol.* 12:707–16

67. Hardwick SW, Chan VS, Broadhurst RW, Luisi BF. 2011. An RNA degradosome assembly in *Caulobacter crescentus*. *Nucleic Acids Res.* 39:1449–59

68. Hayes CS, Sauer RT. 2003. Cleavage of the A site mRNA codon during ribosome pausing provides a mechanism for translational quality control. *Mol. Cell* 12:903–11

69. Heck C, Balzer A, Fuhrmann O, Klug G. 2000. Initial events in the degradation of the polycistronic *puf* mRNA in *Rhodobacter capsulatus* and consequences for further processing steps. *Mol. Microbiol.* 35:90–100

70. Hsieh PK, Richards J, Liu Q, Belasco JG. 2013. Specificity of RppH-dependent RNA degradation in *Bacillus subtilis*. *Proc. Natl. Acad. Sci. USA* 110:8864–69

71. Hurley JM, Cruz JW, Ouyang M, Woychik NA. 2011. Bacterial toxin RelE mediates frequent codon-independent mRNA cleavage from the 5′ end of coding regions in vivo. *J. Biol. Chem.* 286:14770–78

72. Jäger S, Fuhrmann O, Heck C, Hebermehl M, Schiltz E, et al. 2001. An mRNA degrading complex in *Rhodobacter capsulatus*. *Nucleic Acids Res.* 29:4581–88

73. Jain C. 2012. Novel role for RNase PH in the degradation of structured RNA. *J. Bacteriol.* 194:3883–90

74. Jain C, Belasco JG. 1995. RNase E autoregulates its synthesis by controlling the degradation rate of its own mRNA in *Escherichia coli*: unusual sensitivity of the *rne* transcript to RNase E activity. *Genes Dev.* 9:84–96

75. Jarrige AC, Mathy N, Portier C. 2001. PNPase autocontrols its expression by degrading a double-stranded structure in the *pnp* mRNA leader. *EMBO J.* 20:6845–55

76. Jiang X, Diwa A, Belasco JG. 2000. Regions of RNase E important for 5′-end-dependent RNA cleavage and autoregulated synthesis. *J. Bacteriol.* 182:2468–75

77. Karzai AW, Roche ED, Sauer RT. 2000. The SsrA-SmpB system for protein tagging, directed degradation and ribosome rescue. *Nat. Struct. Biol.* 7:449–55

78. Khemici V, Poljak L, Luisi BF, Carpousis AJ. 2008. The RNase E of *Escherichia coli* is a membrane-binding protein. *Mol. Microbiol.* 70:799–813

79. Khemici V, Poljak L, Toesca I, Carpousis AJ. 2005. Evidence in vivo that the DEAD-box RNA helicase RhlB facilitates the degradation of ribosome-free mRNA by RNase E. *Proc. Natl. Acad. Sci. USA* 102:6913–18

80. Kim KS, Manasherob R, Cohen SN. 2008. YmdB: a stress-responsive ribonuclease-binding regulator of *E. coli* RNase III activity. *Genes Dev.* 22:3497–508

81. Koga M, Otsuka Y, Lemire S, Yonesaki T. 2011. *Escherichia coli rnlA* and *rnlB* compose a novel toxin-antitoxin system. *Genetics* 187:123–30

82. Laalami S, Bessières P, Rocca A, Zig L, Nicolas P, Putzer H. 2013. *Bacillus subtilis* RNase Y activity in vivo analysed by tiling microarrays. *PLOS ONE* 8:e54062

83. Lamontagne B, Elela SA. 2004. Evaluation of the RNA determinants for bacterial and yeast RNase III binding and cleavage. *J. Biol. Chem.* 279:2231–41

84. Lee K, Bernstein JA, Cohen SN. 2002. RNase G complementation of *rne* null mutation identifies functional interrelationships with RNase E in *Escherichia coli*. *Mol. Microbiol.* 43:1445–56

85. Lee K, Zhan X, Gao J, Qiu J, Feng Y, et al. 2003. RraA: a protein inhibitor of RNase E activity that globally modulates RNA abundance in *E. coli*. *Cell.* 114:623–34

86. Lehnik-Habrink M, Newman J, Rothe FM, Solovyova AS, Rodrigues C, et al. 2011. RNase Y in *Bacillus subtilis*: a natively disordered protein that is the functional equivalent of RNase E from *Escherichia coli*. *J. Bacteriol.* 193:5431–41

87. Lehnik-Habrink M, Pförtner H, Rempeters L, Pietack N, Herzberg C, Stülke J. 2010. The RNA degradosome in *Bacillus subtilis*: identification of CshA as the major RNA helicase in the multiprotein complex. *Mol. Microbiol.* 77:958–71

88. Lehnik-Habrink M, Schaffer M, Mader U, Diethmaier C, Herzberg C, Stulke J. 2011. RNA processing in *Bacillus subtilis*: identification of targets of the essential RNase Y. *Mol. Microbiol.* 81:1459–73

89. Leroy A, Vanzo NF, Sousa S, Dreyfus M, Carpousis AJ. 2002. Function in *Escherichia coli* of the non-catalytic part of RNase E: role in the degradation of ribosome-free mRNA. *Mol. Microbiol.* 45:1231–43

90. Letunic I, Bork P. 2011. Interactive Tree Of Life v2: online annotation and display of phylogenetic trees made easy. *Nucleic Acids Res.* 39:W475–78

91. Li de la Sierra-Gallay I, Zig L, Jamalli A, Putzer H. 2008. Structural insights into the dual activity of RNase J. *Nat. Struct. Mol. Biol.* 15:206–12

92. Li X, Hirano R, Tagami H, Aiba H. 2006. Protein tagging at rare codons is caused by tmRNA action at the 3′ end of nonstop mRNA generated in response to ribosome stalling. *RNA* 12:248–55

93. Li Y, Altman S. 2003. A specific endoribonuclease, RNase P, affects gene expression of polycistronic operon mRNAs. *Proc. Natl. Acad. Sci. USA* 100:13213–18

94. Liang W, Malhotra A, Deutscher MP. 2011. Acetylation regulates the stability of a bacterial protein: growth stage-dependent modification of RNase R. *Mol. Cell* 44:160–66

95. Liou GG, Chang HY, Lin CS, Lin-Chao S. 2002. DEAD box RhlB RNA helicase physically associates with exoribonuclease PNPase to degrade double-stranded RNA independent of the degradosome-assembling region of RNase E. *J. Biol. Chem.* 277:41157–62

96. Liu MF, Cescau S, Mechold U, Wang J, Cohen D, et al. 2012. Identification of a novel nanoRNase in *Bartonella*. *Microbiology* 158:886–95

97. Loomis WP, Koo JT, Cheung TP, Moseley SL. 2001. A tripeptide sequence within the nascent DaaP protein is required for mRNA processing of a fimbrial operon in *Escherichia coli*. *Mol. Microbiol.* 39:693–707

98. Luciano DJ, Hui MP, Deana A, Foley PL, Belasco KJ, Belasco JG. 2012. Differential control of the rate of 5′-end-dependent mRNA degradation in *Escherichia coli*. *J. Bacteriol.* 194:6233–39

99. Mackie GA. 1998. Ribonuclease E is a 5′-end-dependent endonuclease. *Nature* 395:720–23

100. Marchand I, Nicholson AW, Dreyfus M. 2001. Bacteriophage T7 protein kinase phosphorylates RNase E and stabilizes mRNAs synthesized by T7 RNA polymerase. *Mol. Microbiol.* 42:767–76

101. Marincola G, Schafer T, Behler J, Bernhardt J, Ohlsen K, et al. 2012. RNase Y of *Staphylococcus aureus* and its role in the activation of virulence genes. *Mol. Microbiol.* 85:817–32

102. Massé E, Escorcia FE, Gottesman S. 2003. Coupled degradation of a small regulatory RNA and its mRNA targets in *Escherichia coli*. *Genes Dev.* 17:2374–83

103. Mathy N, Benard L, Pellegrini O, Daou R, Wen T, Condon C. 2007. 5′-to-3′ exoribonuclease activity in bacteria: role of RNase J1 in rRNA maturation and 5′ stability of mRNA. *Cell* 129:681–92

104. Mathy N, Hebert A, Mervelet P, Benard L, Dorleans A, et al. 2010. *Bacillus subtilis* ribonucleases J1 and J2 form a complex with altered enzyme behaviour. *Mol. Microbiol.* 75:489–98

105. Matos RG, Barbas A, Gomez-Puertas P, Arraiano CM. 2011. Swapping the domains of exoribonucleases RNase II and RNase R: conferring upon RNase II the ability to degrade dsRNA. *Proteins* 79:1853–67

106. Matsunaga J, Simons EL, Simons RW. 1996. RNase III autoregulation: structure and function of *rncO*, the posttranscriptional "operator." *RNA* 2:1228–40

107. McCullen CA, Benhammou JN, Majdalani N, Gottesman S. 2010. Mechanism of positive regulation by DsrA and RprA small noncoding RNAs: pairing increases translation and protects *rpoS* mRNA from degradation. *J. Bacteriol.* 192:5559–71

108. McDowall KJ, Cohen SN. 1996. The N-terminal domain of the *rne* gene product has RNase E activity and is non-overlapping with the arginine-rich RNA-binding site. *J. Mol. Biol.* 255:349–55

109. McDowall KJ, Hernandez RG, Lin-Chao S, Cohen SN. 1993. The *ams-1* and *rne-3071* temperature-sensitive mutations in the *ams* gene are in close proximity to each other and cause substitutions within a domain that resembles a product of the *Escherichia coli mre* locus. *J. Bacteriol.* 175:4245–49

110. McDowall KJ, Lin-Chao S, Cohen SN. 1994. A+U content rather than a particular nucleotide order determines the specificity of RNase E cleavage. *J. Biol. Chem.* 269:10790–96

111. Mechold U, Fang G, Ngo S, Ogryzko V, Danchin A. 2007. YtqI from *Bacillus subtilis* has both oligoribonuclease and pAp-phosphatase activity. *Nucleic Acids Res.* 35:4552–61

112. Melefors O, von Gabain A. 1991. Genetic studies of cleavage-initiated mRNA decay and processing of ribosomal 9S RNA show that the *Escherichia coli ams* and *rne* loci are the same. *Mol. Microbiol.* 5:857–64

113. Meng W, Nicholson AW. 2008. Heterodimer-based analysis of subunit and domain contributions to double-stranded RNA processing by *Escherichia coli* RNase III in vitro. *Biochem. J.* 410:39–48

114. Mohanty BK, Kushner SR. 2000. Polynucleotide phosphorylase functions both as a 3′→5′ exonuclease and a poly(A) polymerase in *Escherichia coli*. *Proc. Natl. Acad. Sci. USA* 97:11966–71

115. Moller T, Franch T, Hojrup P, Keene DR, Bachinger HP, et al. 2002. Hfq: a bacterial Sm-like protein that mediates RNA-RNA interaction. *Mol. Cell* 9:23–30

116. Morita T, Kawamoto H, Mizota T, Inada T, Aiba H. 2004. Enolase in the RNA degradosome plays a crucial role in the rapid decay of glucose transporter mRNA in the response to phosphosugar stress in *Escherichia coli*. *Mol. Microbiol.* 54:1063–75

117. Morita T, Maki K, Aiba H. 2005. RNase E-based ribonucleoprotein complexes: mechanical basis of mRNA destabilization mediated by bacterial noncoding RNAs. *Genes Dev.* 19:2176–86

118. Mott JE, Galloway JL, Platt T. 1985. Maturation of *Escherichia coli* tryptophan operon mRNA: evidence for 3′ exonucleolytic processing after rho-dependent termination. *EMBO J.* 4:1887–91

119. Mudd EA, Krisch HM, Higgins CF. 1990. RNase E, an endoribonuclease, has a general role in the chemical decay of *Escherichia coli* mRNA: evidence that *rne* and *ams* are the same genetic locus. *Mol. Microbiol.* 4:2127–35

120. Newbury SF, Smith NH, Higgins CF. 1987. Differential mRNA stability controls relative gene expression within a polycistronic operon. *Cell* 51:1131–43

121. Newbury SF, Smith NH, Robinson EC, Hiles ID, Higgins CF. 1987. Stabilization of translationally active mRNA by prokaryotic REP sequences. *Cell* 48:297–310

122. Nilsson G, Belasco JG, Cohen SN, von Gabain A. 1987. Effect of premature termination of translation on mRNA stability depends on the site of ribosome release. *Proc. Natl. Acad. Sci. USA* 84:4890–94

123. Nou X, Kadner RJ. 1998. Coupled changes in translation and transcription during cobalamin-dependent regulation of *btuB* expression in *Escherichia coli*. *J. Bacteriol.* 180:6719–28

124. Nurmohamed S, Vincent HA, Titman CM, Chandran V, Pears MR, et al. 2011. Polynucleotide phosphorylase activity may be modulated by metabolites in *Escherichia coli*. *J. Biol. Chem.* 286:14315–23

125. O'Hara EB, Chekanova JA, Ingle CA, Kushner ZR, Peters E, Kushner SR. 1995. Polyadenylylation helps regulate mRNA decay in *Escherichia coli*. *Proc. Natl. Acad. Sci. USA* 92:1807–11

126. Ono M, Kuwano M. 1979. A conditional lethal mutation in an *Escherichia coli* strain with a longer chemical lifetime of messenger RNA. *J. Mol. Biol.* 129:343–57

127. Otsuka Y, Yonesaki T. 2012. Dmd of bacteriophage T4 functions as an antitoxin against *Escherichia coli* LsoA and RnlA toxins. *Mol. Microbiol.* 83:669–81

128. Papenfort K, Sun Y, Miyakoshi M, Vanderpool CK, Vogel J. 2013. Small RNA–mediated activation of sugar phosphatase mRNA regulates glucose homeostasis. *Cell* 153:426–37

129. Pertzev AV, Nicholson AW. 2006. Characterization of RNA sequence determinants and antideterminants of processing reactivity for a minimal substrate of *Escherichia coli* ribonuclease III. *Nucleic Acids Res.* 34:3708–21

130. Perwez T, Kushner SR. 2006. RNase Z in *Escherichia coli* plays a significant role in mRNA decay. *Mol. Microbiol.* 60:723–37

131. Pfeiffer V, Papenfort K, Lucchini S, Hinton JC, Vogel J. 2009. Coding sequence targeting by MicC RNA reveals bacterial mRNA silencing downstream of translational initiation. *Nat. Struct. Mol. Biol.* 16:840–46

132. Py B, Higgins CF, Krisch HM, Carpousis AJ. 1996. A DEAD-box RNA helicase in the *Escherichia coli* RNA degradosome. *Nature* 381:169–72

133. Ramirez-Peña E, Treviño J, Liu Z, Perez N, Sumby P. 2010. The group A *Streptococcus* small regulatory RNA FasX enhances streptokinase activity by increasing the stability of the *ska* mRNA transcript. *Mol. Microbiol.* 78:1332–47

134. Richards J, Liu Q, Pellegrini O, Celesnik H, Yao S, et al. 2011. An RNA pyrophosphohydrolase triggers 5′-exonucleolytic degradation of mRNA in *Bacillus subtilis*. *Mol. Cell* 43:940–49

135. Richards J, Luciano DJ, Belasco JG. 2012. Influence of translation on RppH-dependent mRNA degradation in *Escherichia coli*. *Mol. Microbiol.* 86:1063–72

136. Richards J, Mehta P, Karzai AW. 2006. RNase R degrades non-stop mRNAs selectively in an SmpB-tmRNA-dependent manner. *Mol. Microbiol.* 62:1700–12

137. Robertson HD. 1982. *Escherichia coli* ribonuclease III cleavage sites. *Cell* 30:669–72

138. Robertson HD, Webster RE, Zinder ND. 1968. Purification and properties of ribonuclease III from *Escherichia coli*. *J. Biol. Chem.* 243:82–91

139. Roux CM, DeMuth JP, Dunman PM. 2011. Characterization of components of the *Staphylococcus aureus* mRNA degradosome holoenzyme-like complex. *J. Bacteriol.* 193:5520–26

140. Sandler P, Weisblum B. 1989. Erythromycin-induced ribosome stall in the *ermA* leader: a barricade to 5′-to-3′ nucleolytic cleavage of the *ermA* transcript. *J. Bacteriol.* 171:6680–88

141. Shahbabian K, Jamalli A, Zig L, Putzer H. 2009. RNase Y, a novel endoribonuclease, initiates riboswitch turnover in *Bacillus subtilis*. *EMBO J.* 28:3523–33

142. Sharp JS, Bechhofer DH. 2003. Effect of translational signals on mRNA decay in *Bacillus subtilis*. *J. Bacteriol.* 185:5372–79

143. Sharp JS, Bechhofer DH. 2005. Effect of 5′-proximal elements on decay of a model mRNA in *Bacillus subtilis*. *Mol. Microbiol.* 57:484–95

144. Sim SH, Yeom JH, Shin C, Song WS, Shin E, et al. 2010. *Escherichia coli* ribonuclease III activity is downregulated by osmotic stress: consequences for the degradation of *bdm* mRNA in biofilm formation. *Mol. Microbiol.* 75:413–25

145. Spickler C, Mackie GA. 2000. Action of RNase II and polynucleotide phosphorylase against RNAs containing stem-loops of defined structure. *J. Bacteriol.* 182:2422–27

146. Stead MB, Marshburn S, Mohanty BK, Mitra J, Pena Castillo L, et al. 2011. Analysis of *Escherichia coli* RNase E and RNase III activity in vivo using tiling microarrays. *Nucleic Acids Res.* 39:3188–203

147. Stern MJ, Ames GF, Smith NH, Robinson EC, Higgins CF. 1984. Repetitive extragenic palindromic sequences: a major component of the bacterial genome. *Cell* 37:1015–26

148. Stickney LM, Hankins JS, Miao X, Mackie GA. 2005. Function of the conserved S1 and KH domains in polynucleotide phosphorylase. *J. Bacteriol.* 187:7214–21

149. Storz G, Vogel J, Wassarman KM. 2011. Regulation by small RNAs in bacteria: expanding frontiers. *Mol. Cell* 43:880–91

150. Symmons MF, Jones GH, Luisi BF. 2000. A duplicated fold is the structural basis for polynucleotide phosphorylase catalytic activity, processivity, and regulation. *Structure* 8:1215–26

151. Taraseviciene L, Miczak A, Apirion D. 1991. The gene specifying RNase E (*rne*) and a gene affecting mRNA stability (*ams*) are the same gene. *Mol. Microbiol.* 5:851–55

152. Tock MR, Walsh AP, Carroll G, McDowall KJ. 2000. The CafA protein required for the 5′-maturation of 16 S rRNA is a 5′-end-dependent ribonuclease that has context-dependent broad sequence specificity. *J. Biol. Chem.* 275:8726–32

153. Vanzo NF, Li YS, Py B, Blum E, Higgins CF, et al. 1998. Ribonuclease E organizes the protein interactions in the *Escherichia coli* RNA degradosome. *Genes Dev.* 12:2770–81

154. Vincent HA, Deutscher MP. 2009. The roles of individual domains of RNase R in substrate binding and exoribonuclease activity. The nuclease domain is sufficient for digestion of structured RNA. *J. Biol. Chem.* 284:486–94

155. von Gabain A, Belasco JG, Schottel JL, Chang AC, Cohen SN. 1983. Decay of mRNA in *Escherichia coli*: investigation of the fate of specific segments of transcripts. *Proc. Natl. Acad. Sci. USA* 80:653–57

156. Xu F, Cohen SN. 1995. RNA degradation in *Escherichia coli* regulated by 3′ adenylation and 5′ phosphorylation. *Nature* 374:180–83

157. Xu F, Lin-Chao S, Cohen SN. 1993. The *Escherichia coli pcnB* gene promotes adenylylation of antisense RNAI of ColE1-type plasmids in vivo and degradation of RNAI decay intermediates. *Proc. Natl. Acad. Sci. USA* 90:6756–60

158. Yamaguchi Y, Park JH, Inouye M. 2011. Toxin-antitoxin systems in bacteria and archaea. *Annu. Rev. Genet.* 45:61–79

159. Yao S, Bechhofer DH. 2010. Initiation of decay of *Bacillus subtilis rpsO* mRNA by endoribonuclease RNase Y. *J. Bacteriol.* 192:3279–86

160. Yao S, Sharp JS, Bechhofer DH. 2009. *Bacillus subtilis* RNase J1 endonuclease and 5′ exonuclease activities in the turnover of Λ*ermC* mRNA. *RNA* 15:2331–39

161. Yarchuk O, Jacques N, Guillerez J, Dreyfus M. 1992. Interdependence of translation, transcription and mRNA degradation in the *lacZ* gene. *J. Mol. Biol.* 226:581–96

162. Zhang K, Nicholson AW. 1997. Regulation of ribonuclease III processing by double-helical sequence antideterminants. *Proc. Natl. Acad. Sci. USA* 94:13437–41

163. Zuo Y, Deutscher MP. 2001. Exoribonuclease superfamilies: structural analysis and phylogenetic distribution. *Nucleic Acids Res.* 29:1017–26

Population Genomics of Transposable Elements in *Drosophila*

Maite G. Barrón,[1] Anna-Sophie Fiston-Lavier,[2] Dmitri A. Petrov,[3,*] and Josefa González[1,*]

[1] Institute of Evolutionary Biology (CSIC-Universitat Pompeu Fabra), Barcelona, Spain 08003; email: maite.barron@ibe.upf-csic.es, josefa.gonzalez@ibe.upf-csic.es

[2] Institut des Sciences de l'Evolution (ISEM), UMR5554 CNRS-Université Montpellier 2, France 34090; email: asfiston@univ-montp2.fr

[3] Department of Biology, Stanford University, Stanford, California 94305; email: dpetrov@stanford.edu

Annu. Rev. Genet. 2014. 48:561–81

First published online as a Review in Advance on October 1, 2014

The *Annual Review of Genetics* is online at genet.annualreviews.org

This article's doi: 10.1146/annurev-genet-120213-092359

*Contributed equally to this review

Keywords

evolutionary models, next-generation sequencing, adaptation

Abstract

Studies of the population dynamics of transposable elements (TEs) in *Drosophila melanogaster* indicate that consistent forces are affecting TEs independently of their modes of transposition and regulation. New sequencing technologies enable biologists to sample genomes at an unprecedented scale in order to quantify genome-wide polymorphism for annotated and novel TE insertions. In this review, we first present new insights gleaned from high-throughput data for population genomics studies of *D. melanogaster*. We then consider the latest population genomics models for TE evolution and present examples of functional evidence revealed by genome-wide studies of TE population dynamics in *D. melanogaster*. Although most of the TE insertions are deleterious or neutral, some TE insertions increase the fitness of the individual that carries them and play a role in genome adaptation.

TRANSPOSABLE ELEMENTS ARE AN ABUNDANT, DIVERSE, AND ACTIVE COMPONENT OF GENOMES

Transposable elements (TEs) are mobile DNA sequences that encode an ability to copy themselves to other sites in the genome and increase their copy number in the process. Certain TEs, known as nonautonomous, rely on the enzymatic machinery of autonomous copies to move around the genome. Owing to mobility and self-replication ability, TEs can be abundant, diverse, and active components of genomes.

TEs are present in virtually all eukaryotic organisms studied to date and in 80% of the sequenced prokaryotes (112). In all of these organisms, TEs represent a sizable portion of the genome that can vary from ~1% (e.g., in the filamentous fungus *Fusarium graminearum*) to ~85% in *Zea mays* and *Zea luxurians* (32, 111).

TEs are diverse components of genomes. They are classified into two different classes on the basis of whether their transposition mechanism is DNA-based or RNA-based; into different orders on the basis of structural relationships; and into families on the basis of sequence similarities (62, 116). Within each TE family, order, and class, the age and the number of copies can also drastically vary.

TEs are also active components of genomes that generate mutations both when they transpose from one genomic location to another and when they induce structural rearrangements, most commonly via ectopic recombination between TE copies. TE-induced mutations vary greatly in size, ranging from small-scale nucleotide changes, e.g., when a few nucleotides are left behind after a TE excises, to large chromosomal rearrangements. TE-induced mutations are also diverse in terms of their molecular effect: TEs can inactivate or duplicate genes, add or remove regulatory regions, induce new patterns of alternative splicing, and cause epigenetic changes, affecting the expression and/or structure of nearby genes (2).

TE abundance, diversity, and activity are highly variable from one species to another. For example, mammalian genomes tend to harbor a small number of high copy-number families comprising few currently active TEs and thus primarily old TEs, whereas organisms such as plants and insects tend to harbor a much larger number of smaller copy-number families composed of very young TEs (114). TE content also varies among individuals from the same species: Whereas the total copy number might be similar between different individuals, the location of particular TE insertions can vary substantially. Understanding the dynamics of TEs in populations, i.e., which factors explain the number of TEs that belong to a particular TE family, order, or class, the diversity of TEs present in a given genome, and the frequency distribution of individual TE insertions, is crucial if we want to understand the complex organization, function, and evolution of genomes.

DROSOPHILA MELANOGASTER AS A MODEL ORGANISM TO STUDY TRANSPOSABLE ELEMENT DYNAMICS

Drosophila melanogaster has been used as a model organism for the study of TE population dynamics for more than 25 years. The possibility of physically mapping, by in situ hybridization into polytene chromosomes, the location of individual TE copies in the genome provided the first insights into TE dynamics in populations (24, 90). The small genome size of *D. melanogaster* and its relatively small TE content (approximately 20% of the genome) made it an obvious choice for obtaining the first genome sequence of a complex animal (1). *D. melanogaster* is still to date one of the best sequenced, assembled and annotated eukaryotic genomes (87). Additionally, the wealth of functional information available for *D. melanogaster* and the multiple genetic manipulation techniques that are available make this organism ideal for the study of the evolutionary forces shaping genetic variation in natural populations, including TE-induced variation. The availability of the reference genome sequence and of new sequencing techniques has accelerated our understanding of TE population dynamics.

NEW INSIGHTS OFFERED BY NEXT-GENERATION SEQUENCING TECHNOLOGIES

Next-generation sequencing (NGS) technologies represent a quantum leap in our ability to study TE population dynamics. It is now possible to sequence several individuals or pooled fly samples collected in different geographical locations and/or at different time points and analyze TE dynamics across space and time. Several tools have been designed to discover and annotate TEs in an assembled genome or in raw NGS data (84, 94). More recently, new tools were designed specifically to call (i.e. define whether a TE is present and/or absent in individual strains or pooled NGS data) TEs and to estimate their population frequencies when NGS data sets are available for multiple strains and/or pooled samples (43, 68). The analysis of numerous genomes from the same population also offers the possibility to improve the annotation of TEs in a species and the accuracy of the frequency estimates of TEs in the population (42, 43) (**Figure 1a**).

One of the challenges of analyzing the TE content of whole-genome data sets is the identification of novel TE insertions, i.e., TE copies not present in the reference genome. Such TE copies can belong to known or novel TE families. During the past two years, numerous tools were designed to discover and annotate TE insertions [ngs_te_mapper (77), RelocaTE (103), RetroSeq (63), PoPoolationTE (68), TEA (transposable element analyzer) (73), TE-locate (100), and T-lex2 (42)]. All of these approaches analyze the mapping of the NGS data on reference sequences in order to search for evidence of TE presence and/or absence. In order to specifically identify nonreference (i.e., novel) insertions, the approach consists of searching for discordant mapping of pair-ends (specifically, pairs in which only one read of the pair is mapped) and single reads that are partially mapped (i.e., soft-clipped reads) (**Figure 1b**). To distinguish among the insertions corresponding to TE insertions, the approach relies on the availability of a library of representative known TE sequences. All the evidence of the presence of an insertion can be then aligned against the TE library. However, because of that, only the variation in the number of individual TE copies from known families can be reported, not the variation in the number of TE families or TE orders. New approaches that allow thorough annotation of all the TE insertions present in a given genome, from known and unknown TE families and/or orders, are being generated by several groups (44, 86).

One of the limitations of the currently available methodologies is the short length of the reads, which limits our ability to reconstruct individual TE sequences (64). TE length is one of the parameters that influence TE population dynamics (see below), and as such it is crucial to get accurate estimates of individual TE sizes and to identify TEs of all lengths with similar accuracy. Being able to reconstruct individual TE sequences allows us to determine the number of full-length active copies, relic copies evolving as pseudogenes, and nonautonomous copies. The relative proportion of these three functional categories is likely to affect the dynamics of TE families, as has been shown in simulation studies (13, 72). New technologies that allow one to obtain much longer reads, such as Illumina TruSeq synthetic long-read (87, 115) and Pacific Biosciences' single molecule real-time (SMRT) technologies (58), hold great promise for the improved understanding of TE dynamics.

However, even with the limitations of the currently available analytical and technological approaches, we have moved from having small data sets focused on a limited number of families and a limited number of population samples to genome-wide data for all TEs in a genome and frequency estimates based on hundreds of genomes. As expected, these new data sets have confirmed how diverse TE content is and have prompted new studies of TE dynamics.

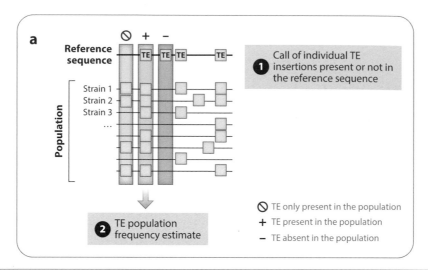

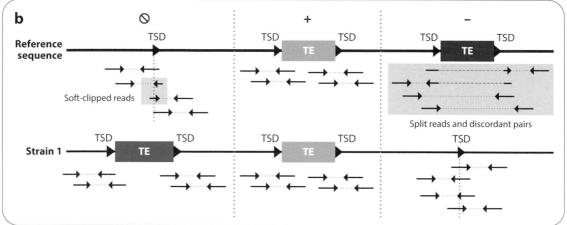

Figure 1

Short-read sequencing technologies offer new insights for transposable element (TE) population genomics study. (*a*) Nonreference TE insertions (i.e., absent from a reference sequence), reference TE insertions (i.e., annotated in a reference sequence) present in the population, and reference TE insertions absent in the population. Short reads in individual strains or those pooled can be used in profiling the TE polymorphism. Nonreference and reference TE insertions could be discovered and then found in other strains or population data. By combining the presence and/or absence detection of TEs from individual strains from the same population (or a pooled sample), the TE population frequency can be estimated. (*b*) Short-read data are represented by pairs of reads pointing in opposite directions and a distance defined during the library preparation (also called insert size). All the pairs represented here come from strain 1 (or a pooled sample). Mapping the reads as single ends or pairs on a reference sequence allows the detection of the TE presence and/or absence. The presence of a TE is supported by the presence of TSDs (target site duplications), short and direct repeats flanking the TE that are usually created after the TE insertion. If a TE is not present in the reference sequence, the reads spanning the TE junction are partially mapped (soft-clipped reads). If a reference TE insertion is also present in the strain, all the reads/pairs from the strain map correctly along the TE sequence, whereas no read maps onto the TE sequence if the reference TE insertion is absent from the strain. The absence is represented by discordant pairs, which refer to a read and its mate, with the insert size greater than the expected insert size distribution of the data set, and by split-reads.

DISTRIBUTION OF TRANSPOSABLE ELEMENTS IN THE *DROSOPHILA MELANOGASTER* GENOME

Several genome population projects provided sequencing data of *D. melanogaster* populations: a European population (68), a North American population [*Drosophila* Genetic Reference Panel (DGRP)] (31, 81), and a laboratory population [*Drosophila* Synthetic Population Resource (DSRP)] (67). The comparison between the TE annotations of these three NGS data sets and the most recent release of the reference genome (Flybase v5.49) yielded a number of insights. The number of TEs identified in each one of these data sets is different. On top of the 5,434 TE insertions annotated in the reference genome, they discovered 10,208 TE insertions in the European population, 17,639 TEs in the North American populations (DGRP data set) and 7,104 TEs in the laboratory population (DSRP data set). The differences in the total number of TEs among the different populations are most likely explained by the different number of strains used for each population: 113 and 131 for European and North American populations, respectively, whereas the analyzed laboratory population contained only 15 strains. However, in this last population (DSRP) the number of TE insertions per strain is larger compared with the other populations, most probably because of the higher sequence coverage and/or the increased amount of time that these strains have remained in laboratory conditions (31, 93). The latter possibility would suggest that TEs continue to transpose in laboratory conditions but are less subject to selection against new copies.

In the reference genome, the TE density per chromosome is similar for all chromosomes (~4%–~10%) except for the fourth chromosome that shows a much higher density (~66%) (**Figure 2**). When NGS data sets are considered, a homogeneous TE density between all the chromosomes is detected but not with the fourth chromosome. The TE density of the fourth chromosome is lower compared with the other chromosomes in the DGRP and DSRP populations, suggesting

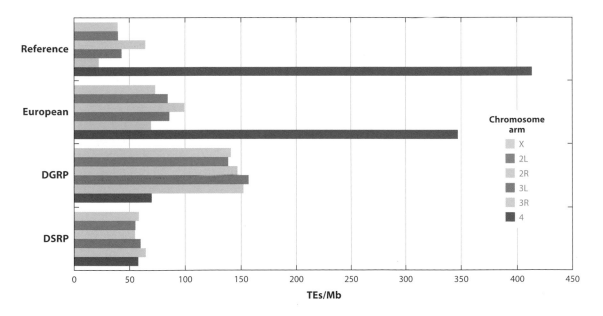

Figure 2

Number of transposable elements (TEs) per megabase (Mb) for each chromosome arm (X, 2L, 2R, 3L, 3R, 4) for the reference genome (Flybase, version 5.49), for a European population (68), for a North American population [*Drosophila* Genetic Resource Panel (DGRP)] (31), and for a laboratory population [*Drosophila* Synthetic Population Resource (DSRP)] (31).

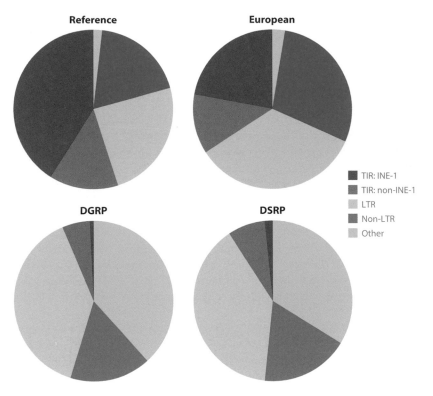

Figure 3

Distribution of transposable element (TE) groups for reference genome and three genome-wide studies that included nonreference insertions: a European population (68), a North American population [*Drosophila* Genetic Resource Panel (DGRP)] (31), and a laboratory population [*Drosophila* Synthetic Population Resource (DSRP)] (31). The five TE groups represent the three main TE orders [i.e., long terminal repeat (LTR), non-LTR, and terminal inverted repeat (TIR)]. The TIR elements are represented in two groups that distinguish the highest copy-number repeats in *D. melanogaster*, which are called INE-1, from the other elements from this order. The last TE group, called "Other," includes TEs that are not part of the main TE orders, except for the North American population, which also includes nonclassified TEs.

that new approaches can underestimate the detection of TEs, specifically the old, fixed, partial, and/or nested TE insertions that are mainly observed in low-recombining regions of the genome such as the fourth chromosome. In contrast to initial findings and theoretical predictions (5, 90), there is no evidence for a reduction in TE density on the X chromosome relative to autosomes (**Figure 2**) (60, 97), and this pattern does not change when nonreference TEs are included (31, 68).

Similarly, most of the TEs in the reference genome belong to the DNA transposon class (**Figure 3**). The INE-1 family, known to have been inactive for approximately the last three million years, contains 2,235 of the 2,986 DNA transposons and is thus composed of old fixed TEs (61, 108). The proportion of DNA transposons in the NGS data sets is smaller. This is probably explained at least in part by the nondetection of old and fixed TEs by the new approaches and the considerable proportion of TEs that cannot be classified (**Figure 3**).

Overall, a substantially increased number of individual TE insertions detected with the approaches that detect nonreference TE insertions highlights the relevance of these techniques to obtain a global view of the TE content in the populations and in the genome. These data sets also show that current methodologies detect TEs with a skew toward younger, non-nested, euchromatic TEs. Hence, future studies that allow annotation of all individual TE copies are needed to get a complete view of the distribution of TEs along the genome in different populations.

Burst: movement of large numbers of TE sequences through the genome in a short evolutionary time

NEW INSIGHTS ON TRANSPOSABLE ELEMENT DYNAMICS IN *DROSOPHILA MELANOGASTER* POPULATIONS

The general model to address the study of TE dynamics assumes that each TE is transposed at a given rate, and it is subsequently removed by a combined effect of a given excision rate and purifying selection (21). An example of this logic is the well-known transposition-selection balance model, in which the maintenance of TEs in the population is explained by equilibrium between the increase in copy number by a constant transposition rate and elimination from the population by natural selection acting against the deleterious effects of TEs (22). This model allows us to make predictions about the changes in the copy number per genome per generation as well as about TE frequency distribution in the populations under different evolutionary hypotheses (see below). However, the assumption of constant transposition rate has been questioned, given that TE transpositions are known to occur in bursts (34, 35, 66, 69, 80, 104). Hence, the burst transposition model, which relaxes the assumption of transposition-selection balance, has also been proposed to explain TE population dynamics (7, 11, 18).

Analysis of TE dynamics under the assumption of transposition-selection balance starts with the estimate of TE population frequencies (10, 11, 48, 68, 90, 96, 117). If TE insertions are neutral, then their frequency in the population should be indicative of their time of insertion in the genome (i.e., age), with rare TEs expected to be young and frequent and fixed TEs expected to be old. When selection is acting, this logic remains true with some caveats. Deleterious TEs should primarily be rare and should still be young; however, the adaptive TEs might reach high frequencies or even fix quickly and thus might be either young (if we analyzed them soon after their increase in frequency) or old (if we analyzed them much later). Thus, the distribution of TE frequencies should inform us about how selection is acting on the TE families, with families composed primarily of rare TEs likely under purifying selection and/or having just undergone a transposition burst, whereas unusually frequent TEs and TEs that are too young for their high frequency might be suspected of having an adaptive effect.

Other factors, such as horizontal TE transfer and host regulation and/or self-regulation of transposition, do affect TE dynamics as well. However, horizontal transfer of TEs is not likely to be a very common event in *Drosophila* and as such is expected to play a limited role in TE dynamics on a short timescale (79, 89). Host control mechanisms, such as the piRNA pathway (51, 65) and self-regulation of transposition by TEs (23, 54, 59), cannot explain the observed patterns of TE frequencies, i.e., the majority of TEs are present at low population frequencies. In the following section, we summarize the most recent insights on TE dynamics that have been obtained from genome-wide analyses of TEs in *D. melanogaster*.

Transposition-Selection Balance Model

Attempts to accurately estimate TE population frequencies have been carried out for more than 25 years. Early experimental analysis performed in a limited number of TE families showed that most TEs are present at low population frequencies (9, 10, 24, 25, 90, 109). Recent genome-wide TE frequency estimates have confirmed these initial findings: A large proportion of TEs, ranging from 47.9% to 76% in the different studies, are present at low frequencies, suggesting that purifying selection plays a major role in TE population dynamics (**Table 1**) (11, 31, 48, 68).

Table 1 Genome-wide studies of transposable element (TE) population dynamics

Study	Approach[a]	Population	Data set	Number of strains	Number of TEs	Number of filtered TEs[b]	Percentage of TEs detected at low frequency
Petrov et al. 2011 (97)	Pooled-PCR	5 NA, 1 Af	Approximately 50% of all euchromatic reference, non-nested, non-INE-1	75	755	755	75%
Kofler et al. 2012 (68)	Pair-end sequencing[c]	1 EU	Reference and nonreference, non-nested, in NGS pooled DNA	113	10,208	7,843	47.9%
Cridland et al. 2013 (31)	Pair-end sequencing[c]	1 NA, 1 lab strain[d]	Reference and nonreference in NGS individual strains	146	23,087	–	>83.3%
Blumenstiel et al. 2014 (11)	Pooled-PCR	1 NA, 1 Af	Pseudogene-like evolving TEs	24	190	190	70%, 76%

[a]Approach: Approach to estimate TE population frequencies.

[b]Number of TE insertions for which population frequency has been estimated.

[c]Bioinformatical approach based on pair-end sequencing data in which a TE is identified if one read of the pair-end fragment maps to the unique region of a reference genome and the other maps to a TE.

[d]DSPR (*Drosophila* Synthetic Population Resource) strains.

Abbreviations: Af, African; EU, European Union; NA, North American; NGS, next-generation sequencing; PCR, polymerase chain reaction.

Three nonmutually exclusive hypotheses have been described to explain the nature of purifying selection acting against TE insertions (reviewed in 92): (*a*) the gene-disruption hypothesis (41, 88); (*b*) the deleterious TE-product expression hypothesis (92); and (*c*) the ectopic recombination hypothesis (90).

Gene-disruption hypothesis. The gene-disruption hypothesis is a widely accepted model in which purifying selection is assumed to be strongly against TE insertions when they are inside a gene or regulatory region (29, 41, 88, 101, 106). In *D. melanogaster*, the analyses of laboratory-induced TE mutations show that TEs do not exclusively transpose outside of coding regions (4, 6). However, in the first in-depth analysis of the euchromatic reference genome (release 3), among the 1,572 TEs identified, none was annotated in coding regions (60). However, if we take into account that 18.3% of the genome corresponds to exons, under the null hypothesis of homogenous insertions genome-wide, we would expect 283 TE insertions to be found in exons. Hence, strong deleterious selection can be invoked to explain this pattern. Follow-up analysis of the reference genome identified only one of these TEs inserted in a protein-coding region (3, 82, 96). Lipatov and colleagues (78) specifically looked for transcripts containing TE and host-gene sequences and found only four that were part of protein-coding regions. Finally, in a genome-wide study in which TEs not present in the reference genome were additionally analyzed, Kofler and colleagues (68) found 249 TEs in coding regions (~2.5% of all TEs), but only 16 of them were fixed (<0.2%). Overall, there are fewer TE insertions in exons and untranslated regions than expected based on the proportion of these sequences in the genome (31, 36, 60, 68, 78, 97), suggesting that selection acting against the deleterious effects of TE insertions inside genes is strong.

Deleterious transposable element–product expression hypothesis. The deleterious TE-product expression model is based on the assumption that the replication of active and inactive copies and the translation of TE-encoded proteins can have a metabolic cost for the cell (16, 92). Additionally TE-encoded proteins could be deleterious because they can disrupt cellular processes (92). Petrov and colleagues (97) made an attempt to test this hypothesis genome-wide by comparing the population frequency of full-length TEs versus truncated TEs (>90% size of the canonical element) in transcriptionally active TE families [according to Deloger and colleagues (36)]. Full-length TEs should be transcribed at higher levels or at least more often than incomplete copies, and, consequently, deleterious selection against them should result in decreased population frequency. However, they were not able to observe this effect and hence, at present there is no direct evidence of selection against the expression of TE-encoded proteins at a genome-wide scale.

Ectopic recombination hypothesis. The previous models alone cannot explain why TEs are also observed at low frequency in nonfunctional regions of the genome. An alternative, nonexclusive hypothesis is the ectopic recombination model. Ectopic recombination between TE copies that belong to the same family, and thus share sequence identity, and are located in different genomic regions can generate chromosomal rearrangements that often lead to inviable gametes (45, 70, 91). Under this model, we expect that (*a*) meiotic recombination rate, (*b*) size of the TE insertion, and (*c*) family copy number should affect the probability of ectopic recombination and therefore the intensity of selection acting against TE insertions. Under the ectopic recombination model, we expect a negative correlation between TE frequency and rate of meiotic recombination. It is assumed that meiotic recombination is correlated with ectopic recombination, and, consequently, meiotic recombination rate should be a good estimator of the ectopic recombination intensity (45, 46, 70, 91). As expected according to this model, low-recombining regions, such as pericentromeric regions and the fourth chromosome, are highly enriched in TEs and harbor most fixed TE insertions (31, 68, 74). After removing heterochromatic regions, a negative correlation is still observed between recombination rate and TE population frequency (5, 68, 97), even among polymorphic TEs (68, 97), suggesting that ectopic recombination is an important factor affecting TE dynamics.

However, there are at least two alternative hypotheses to the ectopic recombination model that could also explain the observed negative correlation between recombination rate and TE population frequency: Hill-Robertson interference (see sidebar, Hill-Robertson Interference)

HILL-ROBERTSON INTERFERENCE

The Hill-Robertson interference is the reduction in the efficiency of selection operating on a locus as a consequence of simultaneous selection operating on linked loci. Two distinct scenarios can be described: (*a*) Two or more adaptive mutations appear in two different haplotypes in the population. If the recombination rate is low, both haplotypes compete against each other until one of them is fixed and the other disappears from the population. With recombination, the two haplotypes could exchange alleles and generate a new haplotype that carries both adaptive alleles. Hence, having a low rate of recombination reduces the rate of adaptive fixation. (*b*) Slightly deleterious and adaptive mutations are found in a haplotype. Some slightly deleterious mutations become fixed owing to the selective sweep of adjacent adaptive mutations. In addition, the elimination of deleterious alleles by selection also eliminates adjacent, weakly adaptive mutations from the population. Overall, the lack of recombination reduces the efficiency of selection, increasing the fixation of slightly deleterious mutations and decreasing the rate of adaptive substitutions.

and gene density. Recombination rate correlates with the efficiency of selection due to Hill-Robertson interference (56). Indeed, genomic regions with reduced recombination rate show an excess of deleterious mutations and a dearth of adaptive substitutions among different *Drosophila* species consistent with reduced efficiency of selection in these regions (8, 17, 27, 53, 57, 107, 118). However, according to computer simulations performed by Dolgin & Charlesworth (38), the observed pattern of TE fixation in low-recombining regions could be explained only by Hill-Robertson interference under a highly unlikely combination of conditions: When recombination rate is extremely low, excision rate is effectively absent and synergism among TEs is weak.

However, low-recombination regions also have low gene density (1). This could decrease negative selection pressures in low-recombination regions, allowing higher TE density or higher TE population frequencies. Under this hypothesis, the observed correlation between recombination rate and TE population frequency beyond these specific regions will need additional evidence, such as a positive correlation between recombination rate and gene density. Therefore, although it is possible that the negative correlation between recombination rate and TE population frequency could be explained by the Hill-Robertson effect and/or gene density, these two alternative hypotheses are both unlikely to explain the global observed correlation.

TE insertion size is also expected to affect the probability of undergoing ectopic recombination: Longer TEs should recombine more often, as they represent longer targets for homologous pairing (39, 96). Among TEs annotated in the reference genome, a negative correlation has been described between TE length and TE population frequency (68, 97), implying that longer TEs are more deleterious and tend to be removed more efficiently from the genome. Although reliable length estimates for nonreference detected TEs cannot be obtained, indirect estimates based on the canonical sequence length also suggest that TE length and population frequency are negatively correlated (68).

Another prediction of the ectopic recombination model is that population frequency should negatively correlate with family copy number. Indeed, ectopic recombination is more likely to happen when TEs are heterozygous and hence, the probability of undergoing ectopic recombination should increase with the copy number of polymorphic TEs (90, 91). As expected, a negative correlation has been described between TE population frequencies and the copy number of polymorphic TEs (68, 97).

Note that a significant statistical interaction between TE length and family TE copy number indicates that families with longer TEs tend to have a larger copy number in *Drosophila* (97). If long TEs exerted a more deleterious effect on its nearby genes, correlation between length and copy number of TEs within a family and population frequency of these TEs could be explained without additional need to invoke the ectopic recombination model. Under this alternative hypothesis, TEs of similar length or similar recombination backgrounds should suffer similar intensity of negative selection independently of the family to which they belong. Petrov and colleagues (97) tested this alternative explanation and found that it does not hold: Two TEs within the same family are more likely to have similar frequencies over and above the frequency predicted by their length and local recombination rate compared with TEs belonging to different families.

Finally, another testable prediction of the ectopic recombination model is that we should observe lower TE density, and TEs at lower population frequencies in the X chromosome compared with autosomes, because of the higher rate of recombination on the X chromosome (28). Higher ectopic recombination rates increase the strength of negative selection in the X chromosome, leading to a faster elimination of TEs and preventing their increase in frequency. This idea is reinforced by data showing higher efficiency of selection in the X chromosome (17, 71, 81), probably due to the faster X hypothesis (see sidebar, Faster X Hypothesis) (20).

FASTER X HYPOTHESIS

Genes on the X chromosomes evolve more rapidly than genes on autosomes. In males, new recessive X-linked mutations are hemizygous and are directly exposed to selection, whereas new recessive autosomal mutations are masked from expression in heterozygotes. This results in increased efficiency of selection for novel X-linked mutations. The X chromosome also has a higher recombination rate (28), which reduces Hill-Robertson interference and increases the efficiency of selection. Several lines of analysis of SNP data are consistent with increased efficiency of selection in the X chromosome (17, 81). Faster X evolution also implies that genes for reproductive isolation might have a higher probability of being X-linked, and in fact this is generally true.

However, TE density is higher in the X chromosome compared with autosomes (31, 68). When differences in the amount of low-recombining regions among chromosomes are taken into account, no differences between X and autosomes are observed (68, 97). Additionally, there is a clear family effect: Although some families show a higher number of insertions on the X compared with autosomes, other families show the opposite pattern (31). These observations are in contrast with the ectopic recombination model. To reconcile them, assumptions can be made: Some families may have an insertion bias toward the X chromosome, and/or meiotic recombination is not a good predictor of ectopic recombination. Vázquez and colleagues (113) found that indeed the transposition rate in the X chromosome was higher, but they analyzed only one specific TE family, *roo*. To date, there are no genome-wide transposition rate estimates that would allow us to shed light on this issue. Additionally, as mentioned above, the relationship between meiotic and ectopic recombination is not yet fully understood.

Overall, in our opinion, the negative correlation observed between TE population frequency and recombination rate, TE length, and family copy number, together with the observed main family effect, tips the scales in favor of ectopic recombination playing a relevant role in explaining TE dynamics in different *Drosophila* populations.

Transposition Burst Model

As mentioned above, one of the assumptions of the transposition-selection balance hypothesis is that transposition rate is constant over time. Some authors (7, 11) claim that this is unlikely to occur on the basis of evidence that some families can undergo periods of transposition bursts (34, 35, 66, 69, 80, 92, 104). Hence, a relaxation in transposition-selection equilibrium has been proposed to explain TE dynamics: the transposition burst model.

Most of the observed features of TE dynamics explained under the ectopic recombination hypothesis can also hold under the transposition burst model. First, the main effect of the family could be explained by bursts of transposition activity. Insertions from the same family tend to happen together in time (7, 14, 75, 92), and thus share a population frequency. This additionally generates a positive correlation between TE frequency and TE age that was not expected under the ectopic recombination hypothesis by itself, i.e., recently active families should be at low population frequencies and long-time inactive families should be fixed. A positive correlation between TE age, based on sequence diversity, and TE frequency is indeed observed for some families (68), although this is expected under any model of TE population dynamics. Second, the relationship between TE frequency and TE copy number or TE length is the same as under the ectopic recombination hypothesis. We expect that recent TEs will have larger family copy number and that older TEs will have lower copy number owing to the fact that some old TEs would have been removed. Moreover, newly inserted TEs tend to be longer than old TEs because of the deletion bias observed in *Drosophila* (12, 98, 99). Hence, under this burst transposition activity model we also expect to observe a negative correlation between TE population frequency and TE length and TE family copy number, exactly as predicted by the ectopic recombination model. However, short but young TEs within a family are not expected to be more frequent, especially for families in which transposition itself commonly generates short TE copies (such as non-LTR elements), and this pattern is observed in the empirical data (96, 97). Furthermore, the observed negative correlation with recombination rate cannot be explained by this model.

Under the burst transposition model, recently inserted TE families have not had enough time to reach equilibrium and hence they are expected to be at low frequency even under a strictly neutral model. Older insertions can reach this equilibrium and be affected by negative selection, as explained under selection-transposition balance. Blumenstiel and colleagues (11) developed a method to test, based on insertion time, whether a strictly neutral model can explain the observed TE frequency pattern in the genome. TEs undergoing purifying selection are outliers of this model if they have lower frequency than expected under a neutral model, as are putatively adaptive TEs if they have higher frequency than expected under a neutral model. Blumenstiel et al. (11) analyzed in North American and African populations 190 LTR and non-LTR insertions previously shown to have a pseudogene-like, or unconstrained, sequence. Therefore, these analyzed TEs should evolve neutrally. However, in both populations a model that includes negative selection better fits the observed TE frequencies than a strictly neutral model. The authors argued that the bottleneck suffered by North American populations could be biasing their method, and when they corrected for demographic effect, their results suggested that young TEs (67 insertions) can be exclusively explained by a strictly neutral model, whereas middle-aged TEs (87 insertions) and old TE insertions (36 insertions) fit better to a model that includes purifying selection.

For these 190 pseudogene-like putatively neutral TEs, adding their age as a variable seems to be an important factor explaining ∼72% of the observed TE frequency variation (11). Hence, at least for some families, TE age seems to be an important factor in explaining TE dynamics. TE age has also been considered as a variable to explain TE population dynamics of a random sample of 671 TEs (68). Together with other explanatory variables, such as recombination and TE length or distance to nearest genes, age can explain ∼13.6% of the variance of TE frequency. However, these authors argue that the regression coefficient is not reliable owing to the fact that some variables are not independent among them. Moreover, interpretation of TE age can have a confounding effect because it is based on sequence identity. Having few mutations makes a TE look young, and its expected frequency will appear to be low. However, having fewer mutations also increases the probability of suffering ectopic recombination. By consequence such young TEs may tend to be at a lower frequency because of the deleterious effects of ectopic recombination. Hence, under both hypotheses (strict neutrality and negative selection via ectopic recombination model) the same observations are expected.

Cridland and colleagues (31) analyzed 6,613 relatively young TE insertions (>75% of the canonical length) in euchromatic regions of a North American population. The distribution of TE insertions showed an excess of rare variants compared with SNPs of small introns (<86 bp), which are considered the best candidates for neutral sites in the genome (52). Hence, these results suggest that negative selection is acting among TEs despite the effect of the bottleneck.

Moreover, the model of strict neutrality implicitly assumes that families with TE insertions at low frequency should have recently suffered a burst of activity in *Drosophila* populations. Moreover, the observation that most LTR families are at a lower frequency than non-LTR families (7, 11) implies that families in this order have coordinately invaded the *Drosophila* genome, and the other orders are older invaders. There is no known evidence suggesting such an extreme scenario.

Hence, overall, although the frequency distribution of some TE families can be explained without taking into account purifying selection, for the majority of TEs purifying selection is essential to fully explain their frequency distribution. Finally, although the transposition-selection equilibrium is a simplification, it is remarkable how well it can explain the observed patterns of TE frequencies in genomes. Although there are some families in which a recent burst of transposition activity could certainly explain their low frequency, evidence of purifying selection acting along genome-wide TE insertions is overwhelming. TE family characteristics, such as copy number and TE length and age, seem to be important factors in explaining the dynamics of TEs. Although the ectopic recombination model seems to explain the TE frequency distribution observed outside coding regions, it may not provide a complete picture. Further analysis and exploration are required to fully understand the family copy number, frequency distribution, and diversity, i.e., the dynamics of TEs. This will allow us to discard alternative models and to clarify observations that still remain to be explained, such as the higher, or similar, TE density observed in X versus autosomes.

FUNCTIONAL ROLES OF TRANSPOSABLE ELEMENT INSERTIONS

Although most of the TE insertions are deleterious or neutral, some TE insertions are expected to increase the fitness of the individual that carries them. Genome-wide studies of TE dynamics have made an attempt to list these putatively adaptive TEs using several distinct criteria. González and colleagues (47, 48) selected candidate TEs that are located in high-recombining regions, not located inside inversions, and present at low frequency in ancestral populations (Africa) and at high frequency in derived populations (North America and Australia). In addition, they applied a maximum likelihood approach that allowed them to identify TE families likely to evolve under strong purifying selection and families likely to evolve neutrally (48, 49).

Kofler and colleagues (68) selected insertions fixed in high-recombining regions and in the regions showing five percent lowest quantile genome-wide Tajima's D values (110) [One of the possible signatures of a sweep by positive selection is the generation of an excess of rare mutations (15) that can be detected by negative values of Tajima's D]. Finally, Blumenstiel and colleagues (11) selected insertions at high-recombining regions that have higher population frequency than expected according to their age. Using the previously characterized *Doc1420* (*FBti0019430*) in *CHKov1* gene insertion (3, 96), Blumenstiel and colleagues establish a cut-off value to detect new putatively adaptive TE insertions. The concordance of these findings is low, suggesting that each method detects different putatively adaptive TE insertions (**Table 2**).

To date, only a limited number of TE insertions in *Drosophila* have been unequivocally connected to their relevant fitness effects: an *Accord* LTR retrotransposon (26, 33, 105), a *Doc* non-LTR retrotransposon (3, 83), a *Bari1* transposon (48–50), and a *pogo* transposon (85).

Daborn and colleagues (33) identified an *Accord* element inserted in the 5′ regulatory region of *Cyp6g1*, a gene involved in the detoxification of multiple insecticides. The presence of this TE insertion was associated with increased *Cyp6g1* expression and increased resistance to insecticides such as DDT. Further analyses demonstrated that this *Accord* element carries regulatory sequences that specifically increase *Cyp6g1* expression in tissues important for detoxification (26). When several *D. melanogaster* strains were analyzed, it was discovered that successive mutations, including gene duplications and additional TE insertions, have occurred in the *Cyp6g1* locus (105). Interestingly, there is an increase in resistance level for at least three of the five mutant alleles described, suggesting that this allelic succession could have been driven by selection removing fitness costs associated with the preceding resistance allele (105).

Similarly, an allelic series affecting the gene region in which the adaptive *Doc1420* is inserted has also been reported (83). Aminetzach and colleagues (3) first described the putatively adaptive

Table 2 Putatively adaptive transposable element (TE) insertions identified in different genome-wide studies

Flybase ID	TE family/superfamily	Reference
FBti0018880	*Bari1*/TIR	47
FBti0019056	*pogo*/TIR	47
FBti0019065	*pogo*/TIR	47
FBti0019144	*Rt1b*/non-LTR	47
FBti0019164	*X*-element/non-LTR	47
FBti0019170	*F*-element/non-LTR	47
FBti0019372	*S*-element/TIR	47
FBti0019386	*Invader4*/LTR	47
FBti0019430	*Doc*/non-LTR	47, 68, 11
FBti0019443	*Rt1b*/non-LTR	47
FBti0019624	*Hopper*/TIR	47
FBti0019627	*pogo*/TIR	47
FBti0019679	1731/LTR	47
FBti0019747	*F*-element/non-LTR	47
FBti0020042	*Jockey*/non-LTR	47
FBti0020046	*Doc*/non-LTR	47, 11
FBti0020091	*Rt1a*/non-LTR	47
FBti0020119	*S*-element/TIR	47
FBti0019564	*Mdg1*/LTR	68
FBti0060479	*HMS-Beagle*/LTR	68
FBti0062283	*Ninja-Dsim*/LTR	68
FBti0019082	*Rt1b*/non-LTR	68
FBti0019655	*3S18*/LTR	68
FBti0060388	*S*-element/TIR	68
FBti0061742	*rooA*/LTR *Accord*/LTR *roo*/LTR	68
FBti0059793	*hobo*/TIR	68
FBti0063191	*gypsy12*/LTR	68
FBti0020329	*G5*/non-LTR	68
FBti0019200	*Doc*/non-LTR	11
FBti0020082	*412*/LTR	11
FBti0020086	*17.6*/LTR	11
FBti0020149	*BS*/non-LTR	11
FBti0019354	*17.6*/LTR	11
FBti0019199	*Doc*/non-LTR	11
FBti0020125	*BS*/non-LTR	11

Abbreviations: LTR, long terminal repeat; TIR, terminal inverted repeat.

insertion of this *Doc* element into the coding region of *CHKov1*. This insertion generates two sets of altered transcripts, and it is associated with increased resistance to an organophosphate insecticide (AZM). However, the authors estimated that the allele containing the *Doc1420* insertion was 90,000 years old. Because insecticides were first used only a few decades ago, the original reasons for the fast evolution and persistence in natural populations of the *Doc1420*-containing allele must be related to some other phenotypic effect. As anticipated by Aminetzach and colleagues (3), polymorphisms in the *CHKov1* gene region were later found to be associated with a different phenotype: resistance to viral infection (83). Magwire and colleagues (83) showed that although the truncation of *CHKov1* coding region by insertion of the *Doc1420* element confers resistance to the sigma virus, an allele containing two duplications resulting in three copies of the truncated allele of both *CHKov1* and *CHKov2* (one of which is also truncated) caused increased resistance.

Recently, a third TE insertion has been connected to its ecologically relevant fitness effect (50). A full-length *Bari1* insertion was identified as being putatively adaptive based on its population frequency and on the detection of a selective sweep in its flanking regions (48). This insertion, named *Bari-Jheh*, is located in the intergenic region of juvenile hormone epoxy hydrolase (*Jheh*) genes, and it was found to affect the level of expression of its two nearby genes (49). Phenotypic effects consistent with the reduced level of expression of *Jheh3* and *Jheh2* genes were also found. However, the phenotypic effects identified, reduced viability and increased developmental time, are likely to represent the cost of selection of this insertion, whose adaptive effect remains unknown. A detailed analysis of the *Bari-Jheh* sequence revealed that this TE adds extra antioxidant response elements (AREs) to the upstream regions of *Jheh2* and *Jheh1* genes. AREs are highly conserved sequences, found in organisms from flies to humans, which mediate response to stress by upregulating the expression of downstream genes. As expected, we found that flies with *Bari-Jheh* showed increased levels of expression of *Jheh2* and *Jheh1* under oxidative stress conditions and increased resistance to this stress (50). Furthermore, we also found that TEs other than *Bari-Jheh* add extra AREs to the upstream region of several genes, suggesting a more general role of *Bari* elements in response to oxidative stress (50).

Finally, a *pogo* transposon has been shown to mediate resistance to xenobiotics (85). This TE affects the polyadenylation signal choice of its nearby gene CG11699. As a result, only one of the two CG11699 transcripts is produced and the expression of CG11699 increases. Mateo et al. (85) further showed that increased CG11699 expression leads to increased aldehyde dehydrogenase III enzymatic activity that results in increased resistance to xenobiotic stress.

These four examples show the variety of molecular mechanisms underlying TE-driven adaptation: from adding tissue-specific or response-to-stress regulatory regions, to the generation of new transcripts, to inactivation of genes. However, the number of adaptive TEs whose adaptive phenotypic effects are known is still too small, and many more TEs need to be characterized to get a more general picture of the adaptive process.

Other than the adaptive effects of particular TE-induced mutations, evidence for the functional impact of TEs in a diversity of cellular processes is starting to accumulate in a range of organisms (19, 30, 37). Recently in *Drosophila*, substantiation has been provided for a role of TEs in (*a*) the establishment of dosage compensation (40), (*b*) heterochromatin assembly (106), and (*c*) brain genomic heterogeneity (95).

1. Dosage compensation in *Drosophila* is achieved by upregulating X-linked genes by approximately twofold in males. In *D. melanogaster*, the male-specific lethal (MSL) complex binds to MSL recognition elements (MREs) located on the male X chromosome, inducing a local change in the chromatin state that promotes the increase in gene expression. Ellison & Bachtrog (40) recently discovered that two related families of *Helitron* elements have

been independently domesticated to provide MRE sites in *Drosophila miranda* at two different evolutionary time points. In both cases, the acquisition of the TE was followed by fine-tuning mutations that refine the ability of the TEs to recruit the MSL complex and by the amplification of this modified TE on the X chromosome. Further secondary fine-tuning mutations and erosion of the TE sequences that are not required for binding MSL continued after the expansion. This secondary refinement and erosion may eventually lead to degradation of the signatures of the TE origins, suggesting that rewiring of regulatory networks by domesticated TEs may eventually be undetectable (40).

2. Heterochromatin assembly. It is known that TEs from the *1360* transposon family induce silencing of nearby genes by promoting the accumulation of Heterochromatin Protein 1a (HP1a) (55). In a recent work, Sentmanat & Elgin (106) narrowed down the specific region of the *1360* element responsible for heterochromatin assembly, leading to repression of transcription. It turns out that silencing does not require the terminal inverted repeats nor the internal transcription start sites, but the sequences that are bound by PIWI-interacting RNAs (piRNAs). Furthermore, the authors extended this observation to the *Invader4* LTR-retrotransposon family, suggesting that this silencing mechanism is not restricted to TEs from a single family and is likely a broadly applicable mechanism (106).

3. Brain genomic heterogeneity. Transposon expression has been found to be more abundant in $\alpha\beta$ neurons of the mushroom body, a brain structure critical for olfactory memory, than in neighboring neurons (95). Perrat and colleagues further determined that the piRNA proteins Aubergine and Argonaute 3 were less abundant in $\alpha\beta$ neurons and that increased expression of TEs in these neurons resulted in nonreference transposon insertions, some of which are located in memory-relevant loci. Perrat and colleagues (95) suggested that it is possible that mushroom body neurons differentially use piRNA to control the expression of memory-relevant loci and that the observed transposon mobilization is an associated cost. Although disruptive insertions may lead to neural decline and cognitive dysfunction (76), it is also possible that allowing transposition produces genetic variability that may contribute to normal brain function (95).

Overall, it is now clear that TEs are a major source of genetic variation. Although most TEs are present at low population frequencies, strongly suggesting that they are deleterious, a significant fraction of TEs have been recruited to perform cellular functions.

DISCLOSURE STATEMENT

The authors are not aware of any affiliations, memberships, funding, or financial holdings that might be perceived as affecting the objectivity of this review.

ACKNOWLEDGMENTS

J.G. is a Ramon y Cajal fellow (RYC-2010-07306). This work was supported by grants BFU-2011-24397 and PCIG-2011-293860 awarded to J.G. and grant R01GM089926 awarded to D.A.P.

LITERATURE CITED

1. Adams MD, Celniker SE, Holt RA, Evans CA, Gocayne JD, et al. 2000. The genome sequence of *Drosophila melanogaster*. *Science* 287:2185–95
2. Akagi K, Li J, Symer DE. 2013. How do mammalian transposons induce genetic variation? A conceptual framework: The age, structure, allele frequency, and genome context of transposable elements may define their wide-ranging biological impacts. *BioEssays* 35:397–407

3. Aminetzach YT, Macpherson JM, Petrov DA. 2005. Pesticide resistance via transposition-mediated adaptive gene truncation in *Drosophila*. *Science* 309:764–67

4. Ashburner M, Bergman CM. 2005. *Drosophila melanogaster*: a case study of a model genomic sequence and its consequences. *Genome Res.* 15:1661–67

5. Bartolome C, Maside X, Charlesworth B. 2002. On the abundance and distribution of transposable elements in the genome of *Drosophila melanogaster*. *Mol. Biol. Evol.* 19:926–37

6. Bellen HJ, Levis RW, He Y, Carlson JW, Evans-Holm M, et al. 2011. The *Drosophila* gene disruption project: progress using transposons with distinctive site specificities. *Genetics* 188:731–43

7. Bergman CM, Bensasson D. 2007. Recent LTR retrotransposon insertion contrasts with waves of non-LTR insertion since speciation in *Drosophila melanogaster*. *Proc. Natl. Acad. Sci. USA* 104:11340–45

8. Betancourt AJ, Presgraves DC. 2002. Linkage limits the power of natural selection in *Drosophila*. *Proc. Natl. Acad. Sci. USA* 99:13616–20

9. Biemont C. 1992. Population genetics of transposable DNA elements. A *Drosophila* point of view. *Genetica* 86:67–84

10. Biemont C, Lemeunier F, Garcia Guerreiro MP, Brookfield JF, Gautier C, et al. 1994. Population dynamics of the *copia*, *mdg1*, *mdg3*, *gypsy*, and *P* transposable elements in a natural population of *Drosophila melanogaster*. *Genet. Res.* 63:197–212

11. Blumenstiel JP, Chen X, He M, Bergman CM. 2014. An age-of-allele test of neutrality for transposable element insertions. *Genetics* 196:523–38

12. Blumenstiel JP, Noll AC, Griffiths JA, Perera AG, Walton KN, et al. 2009. Identification of EMS-induced mutations in *Drosophila melanogaster* by whole-genome sequencing. *Genetics* 182:25–32

13. Boutin TS, Le Rouzic A, Capy P. 2012. How does selfing affect the dynamics of selfish transposable elements? *Mob. DNA* 3:5

14. Bowen NJ, McDonald JF. 2001. *Drosophila* euchromatic LTR retrotransposons are much younger than the host species in which they reside. *Genome Res.* 11:1527–40

15. Braverman JM, Lazzaro BP, Aguade M, Langley CH. 2005. DNA sequence polymorphism and divergence at the erect wing and suppressor of sable loci of *Drosophila melanogaster* and *D. simulans*. *Genetics* 170:1153–65

16. Burt A, Trivers R. 2006. *Genes in Conflict*. Cambridge, MA: Belknap Press

17. Campos JL, Halligan DL, Haddrill PR, Charlesworth B. 2014. The relation between recombination rate and patterns of molecular evolution and variation in *Drosophila melanogaster*. *Mol. Biol. Evol.* 31:1010–28

18. Carr M, Bensasson D, Bergman CM. 2012. Evolutionary genomics of transposable elements in *Saccharomyces cerevisiae*. *PLOS ONE* 7:e50978

19. Casacuberta E, Gonzalez J. 2013. The impact of transposable elements in environmental adaptation. *Mol. Ecol.* 22:1503–17

20. Charlesworth B, Coyne JA, Barton NH. 1987. The relative rates of evolution of sex chromosomes and autosomes. *Am. Nat.* 130:113–46

21. Charlesworth B, Charlesworth D. 1983. The population dynamics of transposable elements. *Genet. Res.* 42:1–27

22. Charlesworth B, Jarne P, Assimacopoulos S. 1994. The distribution of transposable elements within and between chromosomes in a population of *Drosophila melanogaster*. III. Element abundances in heterochromatin. *Genet. Res.* 64:183–97

23. Charlesworth B, Langley CH. 1986. The evolution of self-regulated transposition of transposable elements. *Genetics* 112:359–83

24. Charlesworth B, Langley CH. 1989. The population genetics of *Drosophila* transposable elements. *Annu. Rev. Genet.* 23:251–87

25. Charlesworth B, Lapid A, Canada D. 1992. The distribution of transposable elements within and between chromosomes in a population of *Drosophila melanogaster*. I. Element frequencies and distribution. *Genet. Res.* 60:103–14

26. Chung H, Bogwitz MR, McCart C, Andrianopoulos A, Ffrench-Constant RH, et al. 2007. *Cis*-regulatory elements in the *Accord* retrotransposon result in tissue-specific expression of the *Drosophila melanogaster* insecticide resistance gene *Cyp6g1*. *Genetics* 175:1071–77

27. Comeron JM, Kreitman M. 2000. The correlation between intron length and recombination in *Drosophila*. Dynamic equilibrium between mutational and selective forces. *Genetics* 156:1175–90

28. Comeron JM, Ratnappan R, Bailin S. 2012. The many landscapes of recombination in *Drosophila melanogaster*. *PLOS Genet.* 8:e1002905

29. Cooley L, Kelley R, Spradling A. 1988. Insertional mutagenesis of the *Drosophila* genome with single P elements. *Science* 239:1121–28

30. Cowley M, Oakey RJ. 2013. Transposable elements re-wire and fine-tune the transcriptome. *PLOS Genet.* 9:e1003234

31. Cridland JM, Macdonald SJ, Long AD, Thornton KR. 2013. Abundance and distribution of transposable elements in two *Drosophila* QTL mapping resources. *Mol. Biol. Evol.* 30:2311–27

32. Cuomo CA, Guldener U, Xu JR, Trail F, Turgeon BG, et al. 2007. The *Fusarium graminearum* genome reveals a link between localized polymorphism and pathogen specialization. *Science* 317:1400–2

33. Daborn PJ, Yen JL, Bogwitz MR, Le Goff G, Feil E, et al. 2002. A single p450 allele associated with insecticide resistance in *Drosophila*. *Science* 297:2253–56

34. Daniels SB, Chovnick A, Boussy IA. 1990. Distribution of *hobo* transposable elements in the genus *Drosophila*. *Mol. Biol. Evol.* 7:589–606

35. de la Chaux N, Wagner A. 2011. BEL/Pao retrotransposons in metazoan genomes. *BMC Evol. Biol.* 11:154

36. Deloger M, Cavalli FM, Lerat E, Biemont C, Sagot MF, Vieira C. 2009. Identification of expressed transposable element insertions in the sequenced genome of *Drosophila melanogaster*. *Gene* 439:55–62

37. de Souza FS, Franchini LF, Rubinstein M. 2013. Exaptation of transposable elements into novel *cis*-regulatory elements: Is the evidence always strong? *Mol. Biol. Evol.* 30:1239–51

38. Dolgin ES, Charlesworth B. 2008. The effects of recombination rate on the distribution and abundance of transposable elements. *Genetics* 178:2169–77

39. Dray T, Gloor GB. 1997. Homology requirements for targeting heterologous sequences during P-induced gap repair in *Drosophila melanogaster*. *Genetics* 147:689–99

40. Ellison CE, Bachtrog D. 2013. Dosage compensation via transposable element mediated rewiring of a regulatory network. *Science* 342:846–50

41. Finnegan DJ. 1992. Transposable elements. *Curr. Opin. Genet. Dev.* 2:861–67

42. Fiston-Lavier AS, Barrón M, Petrov DA, González J. 2014. T-lex2: genotyping, frequency estimation and re-annotation of transposable elements using single or pooled next-generation sequencing data. *BioRxiv* doi: **http://dx.doi.org/10.1101/002964**

43. Fiston-Lavier AS, Carrigan M, Petrov DA, Gonzalez J. 2011. T-lex: a program for fast and accurate assessment of transposable element presence using next-generation sequencing data. *Nucleic Acids Res.* 39:e36

44. Flutre T, Duprat E, Feuillet C, Quesneville H. 2011. Considering transposable element diversification in de novo annotation approaches. *PLOS ONE* 6:e16526

45. Goldman AS, Lichten M. 1996. The efficiency of meiotic recombination between dispersed sequences in *Saccharomyces cerevisiae* depends upon their chromosomal location. *Genetics* 144:43–55

46. Goldman AS, Lichten M. 2000. Restriction of ectopic recombination by interhomolog interactions during *Saccharomyces cerevisiae* meiosis. *Proc. Natl. Acad. Sci. USA* 97:9537–42

47. González J, Karasov TL, Messer PW, Petrov DA. 2010. Genome-wide patterns of adaptation to temperate environments associated with transposable elements in *Drosophila*. *PLOS Genet.* 6:e1000905

48. González J, Lenkov K, Lipatov M, Macpherson JM, Petrov DA. 2008. High rate of recent transposable element–induced adaptation in *Drosophila melanogaster*. *PLOS Biol.* 6:e251

49. González J, Macpherson JM, Petrov DA. 2009. A recent adaptive transposable element insertion near highly conserved developmental loci in *Drosophila melanogaster*. *Mol. Biol. Evol.* 26:1949–61

50. Guio L, Barron MG, Gonzalez J. 2014. The transposable element *Bari-Jheh* mediates oxidative stress response in *Drosophila*. *Mol. Ecol.* 23:2020–30

51. Guzzardo PM, Muerdter F, Hannon GJ. 2013. The piRNA pathway in flies: highlights and future directions. *Curr. Opin. Genet. Dev.* 23:44–52

52. Haddrill PR, Charlesworth B, Halligan DL, Andolfatto P. 2005. Patterns of intron sequence evolution in *Drosophila* are dependent upon length and GC content. *Genome Biol.* 6:R67

53. Haddrill PR, Halligan DL, Tomaras D, Charlesworth B. 2007. Reduced efficacy of selection in regions of the *Drosophila* genome that lack crossing over. *Genome Biol.* 8:R18

54. Hartl DL, Lohe AR, Lozovskaya ER. 1997. Regulation of the transposable element mariner. *Genetica* 100:177–84

55. Haynes KA, Caudy AA, Collins L, Elgin SC. 2006. Element 1360 and RNAi components contribute to HP1-dependent silencing of a pericentric reporter. *Curr. Biol.* 16:2222–27

56. Hill WG, Robertson A. 1966. The effect of linkage on limits to artificial selection. *Genet. Res.* 8:269–94

57. Hilton HR, Kliman M, Hey J. 1994. Using hitchhiking genes to study adaptation and divergence during speciation within the *Drosophila melanogaster* species complex. *Evolution* 48:1900–13

58. Huddleston J, Ranade S, Malig M, Antonacci F, Chaisson M, et al. 2014. Reconstructing complex regions of genomes using long-read sequencing technology. *Genome Res.* 24:688–96

59. Jaillet J, Genty M, Cambefort J, Rouault JD, Auge-Gouillou C. 2012. Regulation of mariner transposition: the peculiar case of Mos1. *PLOS ONE* 7:e43365

60. Kaminker JS, Bergman CM, Kronmiller B, Carlson J, Svirskas R, et al. 2002. The transposable elements of the *Drosophila melanogaster* euchromatin: a genomics perspective. *Genome Biol.* 3:research0084.1–0084.20

61. Kapitonov VV, Jurka J. 2003. Molecular paleontology of transposable elements in the *Drosophila melanogaster* genome. *Proc. Natl. Acad. Sci. USA* 100:6569–74

62. Kapitonov VV, Jurka J. 2008. A universal classification of eukaryotic transposable elements implemented in Repbase. *Nat. Rev. Genet.* 9:411–12; author reply 4

63. Keane TM, Wong K, Adams DJ. 2013. RetroSeq: transposable element discovery from next-generation sequencing data. *Bioinformatics* 29:389–90

64. Kelley JL, Peyton JT, Fiston-Lavier A-S, Teets NM, Yee MC. 2014 Compact genome of the Antarctic midge is likely an adaptation to an extreme environment. *Nat. Commun.* 5:4611

65. Khurana JS, Theurkauf W. 2010. piRNAs, transposon silencing, and *Drosophila* germline development. *J. Cell Biol.* 191:905–13

66. Kidwell MG. 1983. Hybrid dysgenesis in *Drosophila melanogaster*: factors affecting chromosomal contamination in the P-M system. *Genetics* 104:317–41

67. King EG, Macdonald SJ, Long AD. 2012. Properties and power of the *Drosophila* synthetic population resource for the routine dissection of complex traits. *Genetics* 191:935–49

68. Kofler R, Betancourt AJ, Schlotterer C. 2012. Sequencing of pooled DNA samples (Pool-Seq) uncovers complex dynamics of transposable element insertions in *Drosophila melanogaster*. *PLOS Genet.* 8:e1002487

69. Lander ES, Linton LM, Birren B, Nusbaum C, Zody MC, et al. 2001. Initial sequencing and analysis of the human genome. *Nature* 409:860–921

70. Langley CH, Montgomery E, Hudson R, Kaplan N, Charlesworth B. 1988. On the role of unequal exchange in the containment of transposable element copy number. *Genet. Res.* 52:223–35

71. Langley CH, Stevens K, Cardeno C, Lee YC, Schrider DR, et al. 2012. Genomic variation in natural populations of *Drosophila melanogaster*. *Genetics* 192:533–98

72. Le Rouzic A, Boutin TS, Capy P. 2007. Long-term evolution of transposable elements. *Proc. Natl. Acad. Sci. USA* 104:19375–80

73. Lee E, Iskow R, Yang L, Gokcumen O, Haseley P, et al. 2012. Landscape of somatic retrotransposition in human cancers. *Science* 337:967–71

74. Lee YC, Langley CH. 2010. Transposable elements in natural populations of *Drosophila melanogaster*. *Philos. Trans. R. Soc. Lond. Ser. B* 365:1219–28

75. Lerat E. 2010. Identifying repeats and transposable elements in sequenced genomes: how to find your way through the dense forest of programs. *Heredity* 104:520–33

76. Li W, Prazak L, Chatterjee N, Gruninger S, Krug L, et al. 2013. Activation of transposable elements during aging and neuronal decline in *Drosophila*. *Nat. Neurosci.* 16:529–31

77. Linheiro RS, Bergman CM. 2012. Whole genome resequencing reveals natural target site preferences of transposable elements in *Drosophila melanogaster*. *PLOS ONE* 7:e30008

78. Lipatov M, Lenkov K, Petrov DA, Bergman CM. 2005. Paucity of chimeric gene-transposable element transcripts in the *Drosophila melanogaster* genome. *BMC Biol.* 3:24

79. Loreto EL, Carareto CM, Capy P. 2008. Revisiting horizontal transfer of transposable elements in *Drosophila*. *Heredity* 100:545–54

80. Lu C, Chen J, Zhang Y, Hu Q, Su W, Kuang H. 2012. Miniature inverted-repeat transposable elements (MITEs) have been accumulated through amplification bursts and play important roles in gene expression and species diversity in *Oryza sativa*. *Mol. Biol. Evol.* 29:1005–17

81. Mackay TF, Richards S, Stone EA, Barbadilla A, Ayroles JF, et al. 2012. The *Drosophila melanogaster* Genetic Reference Panel. *Nature* 482:173–78

82. Magwire ML. 2011. Addressing barriers to insulin therapy: the role of insulin pens. *Am. J. Ther.* 18:392–402

83. Magwire MM, Bayer F, Webster CL, Cao C, Jiggins FM. 2011. Successive increases in the resistance of *Drosophila* to viral infection through a transposon insertion followed by a duplication. *PLOS Genet.* 7:e1002337

84. Makalowski W, Pande A, Gotea V, Makalowska I. 2012. Transposable elements and their identification. *Methods Mol. Biol.* 855:337–59

85. Mateo L, Ullastres A, González JA. 2014. A transposable element insertion confers xenobiotic resistance in *Drosophila*. *PLOS Genet.* 10(8):e1004560

86. Maumus F, Quesneville H. 2014. Deep investigation of *Arabidopsis thaliana* junk DNA reveals a continuum between repetitive elements and genomic dark matter. *PLOS ONE* 9:e94101

87. McCoy RC, Taylor RW, Blauwkamp TA, Kelley JL, Kertesz M, et al. 2014. Illumina TruSeq synthetic long-reads empower de novo assembly and resolve complex, highly repetitive transposable elements. *PLOS ONE* 9(9):e106689

88. McDonald JF, Matyunina LV, Wilson S, Jordan IK, Bowen NJ, Miller WJ. 1997. LTR retrotransposons and the evolution of eukaryotic enhancers. *Genetica* 100:3–13

89. Modolo L, Picard F, Lerat E. 2014. A new genome-wide method to track horizontally transferred sequences: application to *Drosophila*. *Genome Biol. Evol.* 6:416–32

90. Montgomery E, Charlesworth B, Langley CH. 1987. A test for the role of natural selection in the stabilization of transposable element copy number in a population of *Drosophila melanogaster*. *Genet. Res.* 49:31–41

91. Montgomery EA, Huang SM, Langley CH, Judd BH. 1991. Chromosome rearrangement by ectopic recombination in *Drosophila melanogaster*: genome structure and evolution. *Genetics* 129:1085–98

92. Nuzhdin SV. 1999. Sure facts, speculations, and open questions about the evolution of transposable element copy number. *Genetica* 107:129–37

93. Nuzhdin SV, Pasyukova EG, Mackay TF. 1997. Accumulation of transposable elements in laboratory lines of *Drosophila melanogaster*. *Genetica* 100:167–75

94. Permal E, Flutre T, Quesneville H. 2012. Roadmap for annotating transposable elements in eukaryote genomes. *Methods Mol. Biol.* 859:53–68

95. Perrat PN, DasGupta S, Wang J, Theurkauf W, Weng Z, et al. 2013. Transposition-driven genomic heterogeneity in the *Drosophila* brain. *Science* 340:91–95

96. Petrov DA, Aminetzach YT, Davis JC, Bensasson D, Hirsh AE. 2003. Size matters: non-LTR retrotransposable elements and ectopic recombination in *Drosophila*. *Mol. Biol. Evol.* 20:880–92

97. Petrov DA, Fiston-Lavier AS, Lipatov M, Lenkov K, Gonzalez J. 2011. Population genomics of transposable elements in *Drosophila melanogaster*. *Mol. Biol. Evol.* 28:1633–44

98. Petrov DA, Hartl DL. 1998. High rate of DNA loss in the *Drosophila melanogaster* and *Drosophila virilis* species groups. *Mol. Biol. Evol.* 15:293–302

99. Petrov DA, Lozovskaya ER, Hartl DL. 1996. High intrinsic rate of DNA loss in *Drosophila*. *Nature* 384:346–49

100. Platzer A, Nizhynska V, Long Q. 2012. TE-locate: a tool to locate and group transposable element occurrences using paired-end next-generation sequencing data. *Biology* 1(2):395–410

101. Puig M, Caceres M, Ruiz A. 2004. Silencing of a gene adjacent to the breakpoint of a widespread *Drosophila* inversion by a transposon-induced antisense RNA. *Proc. Natl. Acad. Sci. USA* 101:9013–18

102. Quesneville H, Bergman CM, Andrieu O, Autard D, Nouaud D, et al. 2005. Combined evidence annotation of transposable elements in genome sequences. *PLOS Comput. Biol.* 1:166–75

103. Robb SM, Lu L, Valencia E, Burnette JM 3rd, Okumoto Y, et al. 2013. The use of RelocaTE and unassembled short reads to produce high-resolution snapshots of transposable element generated diversity in rice. *G3* 3:949–57

104. SanMiguel P, Gaut BS, Tikhonov A, Nakajima Y, Bennetzen JL. 1998. The paleontology of intergene retrotransposons of maize. *Nat. Genet.* 20:43–45

105. Schmidt JM, Good RT, Appleton B, Sherrard J, Raymant GC, et al. 2010. Copy number variation and transposable elements feature in recent, ongoing adaptation at the *Cyp6g1* locus. *PLOS Genet.* 6:e1000998

106. Sentmanat MF, Elgin SC. 2012. Ectopic assembly of heterochromatin in *Drosophila melanogaster* triggered by transposable elements. *Proc. Natl. Acad. Sci. USA* 109:14104–9

107. Shapiro JA, Huang W, Zhang C, Hubisz MJ, Lu J, et al. 2007. Adaptive genic evolution in the *Drosophila* genomes. *Proc. Natl. Acad. Sci. USA* 104:2271–76

108. Singh ND, Petrov DA. 2004. Rapid sequence turnover at an intergenic locus in *Drosophila*. *Mol. Biol. Evol.* 21:670–80

109. Sniegowski PD, Charlesworth B. 1994. Transposable element numbers in cosmopolitan inversions from a natural population of *Drosophila melanogaster*. *Genetics* 137:815–27

110. Tajima F. 1989. Statistical method for testing the neutral mutation hypothesis by DNA polymorphism. *Genetics* 123:585–95

111. Tenaillon MI, Hufford MB, Gaut BS, Ross-Ibarra J. 2011. Genome size and transposable element content as determined by high-throughput sequencing in maize and *Zea luxurians*. *Genome Biol. Evol.* 3:219–29

112. Touchon M, Rocha EP. 2007. Causes of insertion sequences abundance in prokaryotic genomes. *Mol. Biol. Evol.* 24:969–81

113. Vázquez JF, Albornoz J, Dominguez A. 2007. Direct determination of the effects of genotype and extreme temperature on the transposition of *roo* in long-term mutation accumulation lines of *Drosophila melanogaster*. *Mol. Genet. Genomics* 278:653–64

114. Venner S, Feschotte C, Biemont C. 2009. Dynamics of transposable elements: towards a community ecology of the genome. *Trends Genet.* 25:317–23

115. Voskoboynik A, Neff NF, Sahoo D, Newman AM, Pushkarev D, et al. 2013. The genome sequence of the colonial chordate, *Botryllus schlosseri*. *Elife* 2:e00569

116. Wicker T, Sabot F, Hua-Van A, Bennetzen JL, Capy P, et al. 2007. A unified classification system for eukaryotic transposable elements. *Nat. Rev. Genet.* 8:973–82

117. Yang HP, Nuzhdin SV. 2003. Fitness costs of Doc expression are insufficient to stabilize its copy number in *Drosophila melanogaster*. *Mol. Biol. Evol.* 20:800–4

118. Zhang Z, Parsch J. 2005. Positive correlation between evolutionary rate and recombination rate in *Drosophila* genes with male-biased expression. *Mol. Biol. Evol.* 22:1945–47

Genetic, Epigenetic, and Environmental Contributions to Neural Tube Closure

Jonathan J. Wilde,[1,2,*] Juliette R. Petersen,[1,3,*] and Lee Niswander[1,*]

[1]Department of Pediatrics, University of Colorado Anschutz Medical Campus, Children's Hospital Colorado, Aurora, Colorado 80045; email: lee.niswander@ucdenver.edu

[2]Cell Biology, Stem Cells and Development Graduate Program, University of Colorado Anschutz Medical Campus, Aurora, Colorado 80045; email: jonathan.wilde@ucdenver.edu

[3]Molecular Biology Graduate Program, University of Colorado Anschutz Medical Campus, Aurora, Colorado 80045; email: juliette.petersen@ucdenver.edu

Annu. Rev. Genet. 2014. 48:583–611

First published online as a Review in Advance on October 6, 2014

The *Annual Review of Genetics* is online at genet.annualreviews.org

This article's doi:
10.1146/annurev-genet-120213-092208

*Co-first authors and equal contribution

Keywords

embryonic brain, embryonic spinal cord, gene-environment interactions, neural tube defects, epigenetics

Abstract

The formation of the embryonic brain and spinal cord begins as the neural plate bends to form the neural folds, which meet and adhere to close the neural tube. The neural ectoderm and surrounding tissues also coordinate proliferation, differentiation, and patterning. This highly orchestrated process is susceptible to disruption, leading to neural tube defects (NTDs), a common birth defect. Here, we highlight genetic and epigenetic contributions to neural tube closure. We describe an online database we created as a resource for researchers, geneticists, and clinicians. Neural tube closure is sensitive to environmental influences, and we discuss disruptive causes, preventative measures, and possible mechanisms. New technologies will move beyond candidate genes in small cohort studies toward unbiased discoveries in sporadic NTD cases. This will uncover the genetic complexity of NTDs and critical gene-gene interactions. Animal models can reveal the causative nature of genetic variants, the genetic interrelationships, and the mechanisms underlying environmental influences.

OVERVIEW OF EMBRYONIC NEURAL TUBE CLOSURE

Neural Tube Closure: The Embryonic Beginning of the Brain and Spinal Cord

The embryonic brain and spinal cord start with the formation of a simple tube. The neural tube (NT) forms when the flat sheet of neuroepithelial cells, bordered by cells that become the epidermis, rolls up and seals together to form the closed neural tube covered by a layer of surface ectoderm (**Figure 1a–d**). Closure occurs in a progressive manner, starting at the caudal

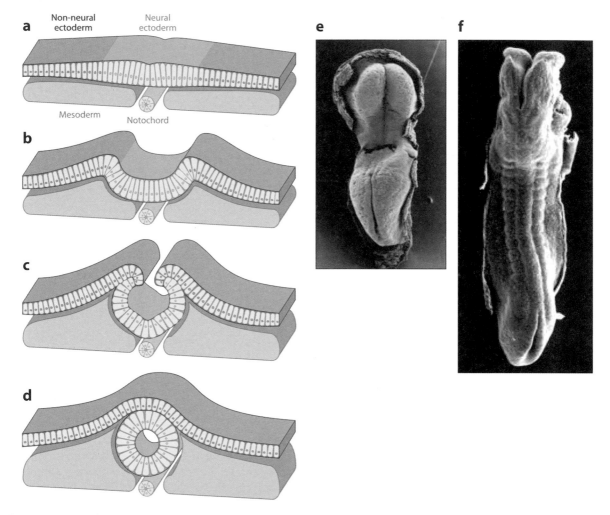

Figure 1

Neural tube closure. (*a–d*) Schematic transverse sections that illustrate the (*a*) flat neural plate stage, (*b*) hinge-point formation and neural-fold elevation, (*c*) apposition of the neural folds with the neural ectoderm covered by the non-neural ectoderm (NNE), and (*d*) remodeling of the neural ectoderm and NNE and cohesion to form a closed neural tube covered by a single layer of NNE. (*e–f*) Carnegie stages 9–11 human embryos (∼20–24 days of gestation) just prior to the (*e*) initiation of neural tube closure and (*f*) closure of most of the spinal region. Panels *a–d* adapted from Reference 133; panels *e* and *f* taken from scanning electron micrographs of early human embryos by Dr. K. Sulik, of embryos collected by Dr. Vekemans and T. Attie-Bitacha, and presented on **http://embryology.med.unsw.edu.au/embryology/index.php?title=Embryonic_Development**.

hindbrain/rostral spinal cord and then proceeding through the head and down along the spinal cord between week 3 and 4 of human gestation (**Figure 1e–g**) (embryonic day E8.75–10 in mouse).

The deceptively simple tube formation requires highly orchestrated activities, including dramatic tissue movements, especially in the brain region with its large neural folds, tight coordination between numerous cellular and molecular processes, and extensive interactions between the neural ectoderm, mesenchyme, and surface ectoderm across time and space. A few examples of individual activities are presented here. Cell proliferation must be tightly regulated and coordinated with cell differentiation: Failure of NT closure is associated with gene mutations that cause either too little or too much cell division or with alterations in the timing of differentiation [for example, *Neurofibromin*, *Pax3*, *Phactr4*, *Jumonji*, *Notch* pathway (36, 69, 70, 171)]. Cell movements and changes in cell shape critically drive the process of NT closure [see planar cell polarity (PCP) pathway below and cytoskeletal proteins such as SHROOM3, VINCULIN, COFILIN, and MENA (36, 69)]. Patterning of the neural tissue occurs during NT closure, and disruption of key patterning pathways is associated with failure of NT closure [e.g., Sonic Hedgehog signaling, bone morphogenetic protein (BMP) signaling, and retinoid signaling (36, 69)]. Movement of the neural folds toward one another requires physical forces generated by the neural, mesenchymal, and surface ectoderm tissues (3, 111, 121). The complexity of these tissue interactions is perhaps best highlighted by the final step in NT closure. As the neural folds rise up and approach each other across the physical gap, the neural ectoderm and surface ectoderm must release their contact with one another and then rapidly reestablish contact with their cognate tissue layer on the other neural fold to seal the neural tube and cover the NT with a single sheet of ectoderm (133). Concomitantly, the neural crest cells, which arise at the border between the neural and surface ectoderm, undergo an epithelial to mesenchymal transition and migrate away, whereas the neighboring neural ectoderm and surface ectoderm must maintain their epithelial character. Overall, the actions of many genes that regulate multiple cell biological and molecular events need to be tightly coordinated in time and space for proper NT closure.

Neural Tube Defects

As might be expected for such an exquisitely coordinated and dynamic morphogenetic process, NT closure is highly sensitive to perturbations, and these can result in neural tube defects (NTDs), a common and devastating birth defect (**Figure 2**). Failure to close the NT in the cranial region (exencephaly or called anencephaly after degradation of the exposed neural tissue) leads to death before or at birth. Infants born with caudal NTDs (e.g., myelomeningocele or spina bifida) have increased risk of mortality, and those that survive often face life-long disabilities and neurologic, cognitive, urologic, and gastrointestinal complications. NTDs occur in ~1 in 1,000 live births in the United States and resulted in 71,000 deaths globally in 2010 (28, 105). Therefore, understanding neural tube development and the causes of NTDs are among the most important health-related studies today.

The embryo grows rapidly during the period of NT closure. If the neural folds fail to rise up and come together within the correct time frame, it is likely that continued growth of the embryo will result in neural folds that are too far apart to close. It is possible that even minor changes in the timing of gene function could affect the movement and meeting of the neural folds and result in NTDs. Moreover, the highly orchestrated activities during NT closure indicate the involvement of numerous genes in this critical embryonic process. Indeed, the great majority of NTDs in humans are thought to have a multifactorial and complex etiology in which disturbances in more than one gene affect closure. Moreover, environmental factors can also alter the risk of NTDs,

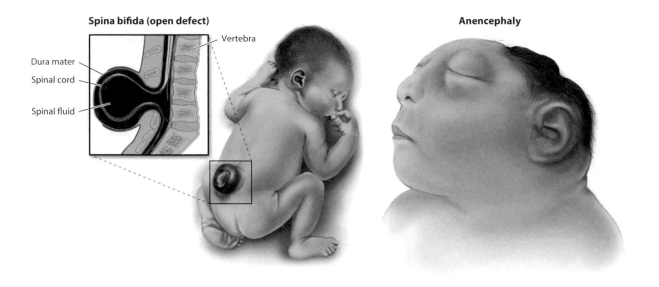

Figure 2

Two types of open neural tube defects (NTDs). Spina bifida occurs in the spinal region, whereas failure of cranial neural tube closure is initially called exencephaly but after exposure and degradation of the brain tissue is called anencephaly. Both caudal and cranial NTDs can vary in extent of the opening and the rostral-caudal level. Adapted from Centers for Disease Control and Prevention, National Center on Birth Defects and Developmental Disabilities: spina bifida, **http://www.cdc.gov/ncbddd/spinabifida/facts.html**; anencephaly, **http://www.cdc.gov/ncbddd/birthdefects/anencephaly.html**.

and this is discussed later in the review. Next, we discuss the current state of understanding of the genetic causes of NTDs, which has largely been driven by studies in animal models.

GENETIC REGULATION OF NEURAL TUBE CLOSURE

A few genetic syndromes are associated with NTDs in humans, including trisomy for chromosome 3, 18, and the X chromosome, although the responsible gene or genes that are dosage sensitive on these chromosomes are unknown (57). In the general population, the genetic risk for having a second child with an NTD is only 2–5% (40, 81). This suggests that differences in genetic makeup can affect the penetrance of a phenotype. Indeed, in mouse models, genetic background can strongly affect NTD risk, and the search for modifier loci has been an ongoing effort (69, 70, 86, 93, 95).

This genetic complexity is acknowledged, but as a starting point to understand the genetic contributions to NT closure, the efforts to date have largely focused on single gene mutations. NT closure in the mouse is similar to human neurulation at the cellular, tissue, and genetic levels, and is the favored animal model for extrapolation to humans. Research in the mouse, from classical mouse NTD mutants to targeted mutations and random mutagenesis screens, has provided considerable insights into the genes that are critically required for NT closure. Indeed, several hundred genes have been identified whose functions are required for NT closure in the mouse embryo (61, 69, 70), and this number is expected to increase as research in this field expands. These data have informed human studies, starting from the search for candidate genes in small cohorts of NTD patients to recent whole-genome or exome sequencing of NTD-affected individuals. Some prominent examples are the genes involved in the formation and signaling of cilia and the PCP pathway (see below). We also discuss the folic acid (FA) metabolism pathway, which, despite years of study and because of the action of FA in preventing NTDs, has provided very few genes that have been conclusively associated with human or mouse NTD etiology. As outlined below, it is critical to move between human and animal cell studies to functionally interpret the human data and to move toward an understanding of complex gene interactions.

Cilia and the Hedgehog Signal Transduction Pathway

Cilia are hair-like extensions from the cell surface that regulate signal transduction by mechanical and molecular cues. The evidence for a role of genes involved in cilia formation and function in NT closure converged from both human and mouse studies. In humans, NTDs have been associated with several syndromes, including Meckel-Gruber Syndrome and Joubert Syndrome (94, 103). Mapping and cloning of the genes causative for these syndromes, and the creation of mouse genetic models that recapitulate many of the clinical features, has highlighted the importance of cilia in key embryonic and adult processes, including NT closure. Functional studies in human cells and animal systems also showed the requirement for cilia in key developmental signaling pathways, most prominently the Sonic Hedgehog pathway during NT formation (2, 91, 103, 177). In the mouse, approximately 5% of the known NTD genes are related to cilia formation and function, providing a rich source of gene candidates for analysis in human NTD samples that is only beginning to be explored (34).

Planar Cell Polarity Pathway

Over the past few years, a candidate gene approach has driven the identification of human NTD genes. A prominent focus has been on the PCP or noncanonical WNT pathway. The PCP pathway provides a prime example to illustrate insights gained, as well as the challenges and open questions discovered, from studies of human NTDs and animal models. PCP core proteins, such as VANGL1/2, DISHEVELLED, CELSR1, and SCRIBBLE, and PCP effector proteins, such as FUZZY and INTURNED, have been widely implicated in driving critical cell movements during NT closure. Research in frog, mouse, and chick embryos has shown that the PCP pathway controls convergent extension cell movements, which help to lengthen the embryo along the rostral-caudal axis and narrow the neural plate (170, 185). Disruptions in the PCP pathway result in a wide neural plate, which prevents the neural folds from contacting each other. This results in the most severe NTD, which is called craniorachischisis, wherein the neural tube remains open from the brain to the spinal cord. On the basis of this severe phenotype in animal models, studies have sought to determine whether craniorachischisis in humans is also associated with mutations in the PCP pathway. Interestingly, mutations in PCP genes have been identified in human fetuses with craniorachischisis (82, 137), highlighting the importance of animal models in informing human NTD studies.

Craniorachischisis may be the rare case in which a specific pathway can be implicated in the etiology of an NTD. However, it is becoming apparent that mutations in PCP genes are also associated with a broader range of NTDs, including sporadic cases of cranial, spinal, and open and closed NTDs. Sequencing of PCP genes in a larger cohort of patients with NTDs has identified rare variants in PCP genes (82, 84). This highlights the challenges in predicting genotype/phenotype relationships. In general, if failed NT closure is the ultimate observational consequence and the initiating event is unknown, it is unlikely that strong predictions can be made as to the causative genetic defect. Similarly, NTDs that occur at a specific rostral-caudal level may not mean that the gene has a particular function in that region. There is evidence to argue for this, as well as evidence that argues for a more complex situation, including cases in which the same genetic mutation in the same genetic background may result in cranial NTD in one animal and caudal NTD in another (69, 70). Even the evidence in favor of region-specific function may be overturned as additional NTD samples are analyzed, new alleles are found, and gene-gene interactions are discovered. In reality, it may remain difficult to predict precise genotype/phenotype relationships. Nonetheless, the gene candidates revealed by animal studies provide a solid foundation for human genetic studies, although the penetrance, expressivity, and regional specificity may depend on other genetic and environmental factors.

Unbiased Determination of the Genetics Underlying Neural Tube Defects in Humans

Considering the large spectrum of genes identified in animal models and the multigenic nature of human NTDs, significant advances in understanding the genetics of NTDs will require moving beyond single candidate gene studies to discovery-based approaches, including high-throughput sequencing of large cohorts of patients. With the decreasing costs of exome and whole-genome sequencing, investigators are beginning to explore the genomes of NTD-affected individuals and families. It is not yet known how copy number variants (CNVs) may contribute to NTDs, beyond a few genetic syndromes associated with NTDs (trisomy for chromosome 3 or 18, or the X chromosome) and the recent association between CNVs in cilia genes in NTD samples (34). Protein coding mutations and potential regulatory mutations will likely form the first wave of functional analyses. However, as can be expected from studies of other human diseases, numerous gene variants will be identified and the challenge will be to evaluate the data and define possible causative mutations. Notably, the set of genes identified in animal NTD models will provide an invaluable framework to begin to decipher these large genomic data sets. Moreover, it will be important to return to animal models or simpler systems, such as cultured cells, to functionally validate gene variants.

The greater promise of the comprehensive view of the genome is the ability to reveal potential gene-gene interactions and to open avenues to interrogate the combinatorial effects of polymorphisms in NTD pathway genes across the genome. However, the challenges increase almost exponentially because of the potential for combinatorial interactions. Importantly, new technologies such as CRISPR will allow rapid genome editing, such that multiple genetic variants can be created in cells or animals to begin to decipher genetic relationships. Overall, in sporadic cases of NTDs, it is unlikely that the NTD results from a single gene mutation but rather may be attributable to the combinatorial effect of multiple genes, each with small contributions, including changes in gene and/or protein levels in time and space, in conjunction with epigenetic and environmental effects, which are discussed below.

Creation of a Publicly Available Repository of Genes Whose Function Is Required for Neural Tube Closure

One goal of this review is to establish an online, publicly available repository to provide a current and continually updated list of genes implicated in NT closure based on mutations that cause NTDs. The ultimate goal is for the NTD scientific community to contribute their information and insights to the online database to enhance the flow of knowledge between researchers, geneticists, clinicians, and epidemiologists, and to inform ongoing studies of human NTD patients. We gratefully acknowledge Dr. Muriel Harris and Dr. Diana Juriloff for their exceptional reviews (69, 70), which have provided the initial foundation of the online database, and Dr. Claudia Kappen and Dr. Michael Salbaum for providing an extraction of the MGI database related to NTDs (141). The link for the NTD Wiki database is **http://ntdwiki.wikispaces.com**.

With more than 300 genes already identified as critically required for NT closure, it may seem like a daunting prospect to attempt to understand this complex embryonic process. However, through the use of well-defined molecular markers, histology, and even newly developed live-imaging systems to visualize the dynamic cell and tissue movements of NT closure (111), it is possible to define the function of even novel genes in the process of NT closure. Thus, a framework is established to categorize the functions of new genes, and we predict that the major pathways and processes involved in NT closure have been identified. The field is coming to the point at which it should be possible to take a systems-level approach to help decipher the genomic information that is forthcoming from genome sequencing efforts.

EPIGENETIC REGULATION OF NEURAL TUBE CLOSURE

Genome-Wide Contributions to Human Neural Tube Defects

A prominent mechanism in which gene expression can be modified globally is through the action of epigenetic regulators. The mammalian epigenome consists of all modifications to the genome that confer heritable changes in gene function independently of nucleotide sequence. This includes, but is not limited to, DNA methylation, chromatin modification, and nucleosome repositioning. Mutations in a growing number of epigenetic regulators in mice have been shown to result in NTDs (**Table 1**), suggesting that coordinate regulation of transcriptional networks is required for proper NT closure. These observations highlight the potential of animal models to uncover pathways that will inform a global view of human NTD etiology. This, in combination with advancing technologies in whole-genome sequencing, gene expression analysis, and epigenetic profiling, provides a means for better understanding of the genetic causes of NTDs in humans. Moreover, epigenetic-related hypotheses extend to the mechanisms of FA-induced NTD prevention (discussed below) and the NTD risk conferred by disease treatments with the histone deacetylase (HDAC)-inhibitor valproic acid (136). Thus, the role of the epigenome in NT closure is of particular interest. Here, we focus on the mechanisms of epigenetic modification as contributing factors to proper closure of the mammalian neural tube.

Ground-Level Changes: DNA Methylation

DNA methylation is a direct chemical modification of DNA that alters transcription of target genes via proposed mechanisms involving changes in stabilization of transcription factors and alteration of chromatin structure (115). Methylation occurs at CpG dinucleotides, is mediated by a class of enzymes known as DNA methyltransferases, and is dynamically regulated during mammalian development. Prior to implantation of the embryo, the majority of the embryonic genome, with the exception of imprinted regions, undergoes complete demethylation and is subsequently remethylated by the de novo DNA methylases DNMT3A and DNMT3B, and is maintained by DNMT1 (19). These events result in stable methylation of the majority (>90%) of CpG dinucleotides across the genome in all cell types (161). However, regions with high concentrations of CpG dinucleotides, called CpG islands, remain relatively unmethylated and are the main targets of DNA methylation-mediated transcriptional regulation (115).

Although there is a wealth of data supporting the role of CpG island methylation in transcriptional repression, recent data have painted a more complex picture. In addition to acting as a strong epigenetic suppressor of transcription through destabilization of transcription factors with their

Table 1 Epigenetic regulators required for neural tube closure

Gene	Protein	Function	Observed Neural Tube Defect	Reference
Ppm1g	PPM1G	Chromatin remodeling	Exencephaly	(50)
Dnmt3a	DNMT3A	DNA methylation	Exencephaly	(123)
Dnmt3b	DNMT3B	DNA methylation	Exencephaly	(123)
Dnmt3L	DNMT3L	DNA methylation	Exencephaly	(71)
CBP	CBP	Histone acetylation	Exencephaly	(128)
Kat2a	GCN5	Histone acetylation	Exencephaly	(22)
Ep300	p300	Histone acetylation	Exencephaly	(128)
Hdac4	HDAC4	Histone deacetylation	Exencephaly	(166)
Sirt1	SIRT1	Histone deacetylation	Exencephaly	(35)
Kdm2b	FBXL10	Histone demethylation	Exencephaly	(52)
Kdm6a	UTX	Histone demethylation	Exencephaly	(152, 179)
Uty	UTY	Histone demethylation	Exencephaly	(152)
Alkbh1	ALKBH1	Histone methylation	Exencephaly	(126)
Jmj	JARID2/Jumonji	Histone methylation	Exencephaly	(162)
Cited2	CITED2	Co-regulator of CBP/p300	Exencephaly	(187)
Smarcc1	BAF155	Nucleosome remodeling	Exencephaly	(67, 85)
Smarca4	BRG1	Nucleosome remodeling	Exencephaly	(23)
Cecr2	CECR2	Nucleosome remodeling	Exencephaly	(10)
mIR-124a	N/A	Nucleosome remodeling	Spina bifida	(178)
mIR-9*	N/A	Nucleosome remodeling	Spina bifida	(178)
Nap1L2	NAP1L2	Nucleosome assembly	Exencephaly	(138)

target promoters, CpG methylation of distal regulatory elements and distant promoters has been linked to binding of sequence-specific transcription factors (17). These data together support an integral role of CpG methylation in genome-wide transcriptional regulation. Moreover, due to its reversible nature, DNA methylation allows for dynamic tuning of transcriptional programs. Indeed, numerous human diseases have been linked to defects in DNA methylation and its dynamics, including Prader-Willi syndrome [loss of genomic imprinting, which confers monoallelic, parent-of-origin specific expression of genes (26)] and Rett syndrome [loss of Mecp2, which reads the methylation state of DNA (4)]. To date, DNA methylation defects have been identified in developmental disorders, cancer, diabetes, and heart failure, suggesting that they may underlie many common human disorders (16, 63).

With respect to NTDs, mouse models lacking the de novo DNA methyltransferases DNMT3A and DNMT3B exhibit cranial NTDs, demonstrating that proper remethylation of the genome after implantation is essential for NT closure (123). The closely related DNMT3L is also required for NT closure; however, its function in DNA methylation remains poorly understood beyond its involvement in maternal imprinting (71). Other null mouse models for methyltransferases such as DNMT1, however, have been less informative about the role of DNA methylation in NT closure, largely because of early embryonic death (98). Several genes involved in one-carbon metabolism (OCM), which provides methyl groups for DNA methylation, have been studied in human NTDs; however, we only touch on these below, as they have been extensively reviewed by

Greene et al. (62). Together, these data, in conjunction with the knowledge that the majority of CpG dinucleotides in the genome outside of CpG islands are constitutively methylated, suggest that methylation-related NTDs may be due to disruption of methylation at specific CpG islands. In future studies of mouse models and human NTD patients, it will be important to utilize next-generation sequencing technologies in conjunction with methylation-specific analysis to uncover the epigenetic code that helps to drive NT closure.

Altering Genomic Topology: Histone Modifications

Within the nucleus, DNA is packaged via its association with histone octamers called nucleosomes. Each nucleosome is wrapped by 147 base pairs of DNA and consists of four homodimers of the core histone proteins H2A, H2B, H3, and H4 that extend N-terminal tails from the core of the nucleosome into the nucleus (9). Histone tails can be modified with eight different covalent post-translational modifications, and more than 60 different modification sites on histone tails have been identified, but this underestimates the repertoire of histone modifications, as many residues undergo di- and trivalent modification, adding to the total possible number of alterations (88). The most commonly studied histone modifications, acetylation and methylation, are generally involved in transcriptional activation and repression, respectively. However, the actual effect of any given modification depends upon the modified residue, as well as the state of neighboring residues. The predominant theory as to how these modifications affect downstream transcription is that they alter the way that nucleosomes interact with each other (88). More specifically, they either promote or antagonize interactions between nucleosomes, therefore altering the packaging of DNA and the accessibility of transcriptional machinery to its target genomic regions (9). It is therefore reasonable to surmise that defects in the enzymatic pathways responsible for adding and removing these chemical modifications can lead to severe cellular and developmental phenotypes.

Mouse NTD models have implicated several chromatin-modifying enzymes in NT closure. Most notably, the histone acetyltransferases (HATs) GCN5 and CBP/p300, which play critical roles in transcriptional activation, are each required for NT closure (22, 128). Conversely, inhibition of HDACs with the drugs valproic acid and trichostatin A (TSA) disrupts NT closure in animal models and humans (118, 136). The H3K27me3 histone demethylase UTX is required for NT closure (179), as is the histone demethylase FBXL10 (52), but there is no direct evidence that histone methyltransferases are needed for NT closure (70). However, the polycomb repressive complex 2 (PRC2) component Jumonji, which plays a role in loading PRC2 onto target loci, is required for NT closure (162), and PRC2 has known roles in transcriptional repression through histone methylation (148). Nonetheless, the bias toward mutation of genes involved in histone acetylation and demethylation in NTDs is interesting, as histone acetylation is thought to be far more dynamic than histone methylation. The requirement for histone acetylation and histone demethylation, which often precede transcriptional activation, suggests that regulation of dynamic changes in chromatin structure and transcriptional activation is more important for proper NT closure than stable repression of gene expression through histone methylation. This is consistent with the highly dynamic yet coordinated cellular and molecular programs that drive changes in morphology, patterning, proliferation, and differentiation during NT closure.

It has been difficult to identify the underlying molecular mechanisms by which mutations in the aforementioned enzymes lead to NTDs. In the case of CBP/p300, it is possible that its role in neurulation is, in part, related to regulation of metabolic genes, as the interacting protein CITED2 is also required for NT closure and is thought to act through mechanisms that regulate cellular responses to hypoxia and energy requirements (187). Interestingly, CITED2 can regulate glucose homeostasis through PGC-1α (140), a direct target of acetylation by GCN5 (96). Additionally, p300 has been linked to regulation of hepatic glycogen storage (72). Although the underlying cause of NTDs in mice that lack GCN5 acetyltransferase activity (*Gcn5^{hat}*) has yet to be uncovered, GCN5 plays important roles in glucose homeostasis and stress response (27, 96, 119), and the convergence of these data with those related to CBP/p300 and CITED2 and NTDs suggests that these complexes may play essential roles in regulating specific metabolic pathways that are required for NT closure. This is especially intriguing considering the NTD risk associated with maternal diabetes, which is discussed below.

Histone acetylation also appears to control the expression of patterning molecules that are essential for NT closure. By treating mouse embryos with the HDAC inhibitor TSA, it was discovered that class I and II HDAC activity is required to inhibit BMP2/4 signaling in the developing forebrain (149). Embryos treated with TSA display a distinct shift in the fate of neural progenitor cells, and defects in neurogenesis can cause NTDs (33). Our unpublished microarray data from *Gcn5$^{hat/hat}$* mice also suggest defects in programs of neural differentiation, indicating that HATs and HDACs may play multiple roles within the developing neural tube and that combinatorial fine-tuning of these transcriptional programs is required to coordinate the complex morphological events that are the basis of primary neurulation. The developing NT seems particularly susceptible to changes in chromatin architecture and its dynamic remodeling, as HAT mutants are exencephalic but otherwise largely anatomically normal at the time of NT closure (22, 128).

Moving Roadblocks: Nucleosome Positioning

In addition to regulation of nucleosome aggregation via histone modification, transcriptional output is also modified via deposition and repositioning of nucleosomes along chromosomes. Nucleosome repositioning is carried out by ATP-dependent chromatin-remodeling complexes. These can shift nucleosomes to different chromosomal positions to regulate transcription by protecting the nucleosome-associated DNA from binding with transcriptional machinery (157). Genome-wide studies of nucleosome positioning have revealed that positioning can vary greatly, even within a single cell type (157). Nonetheless, it is clear that nucleosome positioning plays an important role in transcriptional regulation in conjunction with DNA methylation and histone modifications. It is therefore of no surprise that enzymes responsible for nucleosome localization are required for embryonic development and, more specifically, NT closure.

The SWI/SNF-related nucleosome remodeling BAF complex appears to play a specific role in NT closure, as mutations in multiple subunits result in NTDs in mouse. Both the core catalytic component BRG1 and the complex subunit BAF155 are required for NT closure, suggesting that proper nucleosome positioning by this complex is required for neurulation (23, 67). Complete loss of *Brg1* causes early embryonic death, highlighting the importance of nucleosome positioning in general development, but loss of a single allele results in, among other defects, exencephaly (23). An allele of BAF155 that forms an assembled but malfunctional BAF chromatin-remodeling complex results in exencephaly and decreased neural cell survival and proliferation, yet a surprisingly small number of genes were dysregulated (67). The function of BAF155 is still not well understood. Although it is thought to be a core component of the BAF complex in all cells, it is unclear whether this BAF155 allele causes defects in nucleosome repositioning. Other studies of BAF155 function in ES cells, however, have demonstrated a direct role in regulating repression of self-renewal genes, which may help explain its roles in regulating neural development, proliferation, and cell survival (146).

Additional components of nucleosome remodeling complexes, as well as miRNAs that regulate such complexes, have also been implicated in NTDs. CECR2, which forms a specific chromatin-remodeling complex with SNF2L, causes exencephaly when mutated (10). Additionally, NAP1L2, a neuron-specific protein closely related to nucleosome assembly factors, promotes proliferation of neuroepithelial cells and is required for cranial NT closure (138). As is the case with DNA methylation and histone modification, dynamic changes in nucleosome positioning are also important for cellular function and these changes can be mediated by switching of subunits in chromatin-remodeling complexes (66). As such, changes in the activity of miR-9* and miR-124a cause defects in subunit switching of ATP-dependent chromatin-remodeling complexes, resulting in spina bifida in rats (178).

As may be expected, epigenetic regulators do not function independently of one another. NAP1L2, for example, can promote acetylation of H3K9 and H3K14, and loss of NAP1L2 causes overproliferation of neural stem cells, resulting in inadequate neural differentiation that has been attributed to NAP1L2 promotion of histone acetylation at the *Cdkn1c* promoter (8). Although the exact mechanistic attributes of NAP1L2 have not been discovered, it does associate with nucleosome assembly proteins such as NAP1L1 and NAP1L4 (7), suggesting that nucleosome assembly and positioning are tightly correlated with histone modification at specific loci. As another example, local demethylation of DNA is directed by the acetylation state of specific residues on locally positioned nucleosomes (31). Altogether, these concepts strongly support highly regulated mechanisms for altering genomic structure and function through multiple interconnections between epigenetic regulators.

Distal Regulatory Elements and Their Functions

In addition to gene regulatory elements located in close proximity to gene bodies, transcription is modified by distal regulatory elements. These elements include enhancers, locus control regions, insulators, and silencing elements (73). Recent data from the ENCODE project suggest that these elements are highly promiscuous, with some distal elements interacting with up to 10 different transcriptional start sites (144). Promoters also interact with numerous distal regulatory elements, even within a single cell type (144). It is therefore reasonable to hypothesize that coordinate regulation of genomic architecture is required within each cell type to bring the proper regulatory elements in close proximity to their target promoters. A theme that emerges when examining chromatin-remodeling complexes is the diversity of subunits seen within a given complex. Recent data suggest that these variable subunits may play important roles in recognizing these regulatory elements. For example, ChIP-seq experiments demonstrate that SAGA and ATAC, the two complexes in which the HAT GCN5 is the major enzymatic subunit, have significant overlap in their preferences for promoters, but significant divergence with respect to their regulatory element localization (89). Thus, SAGA and ATAC may function similarly to promote transcription via acetylation of histone residues near active promoters, but the two complexes may have tissue-specific gene regulatory functions mediated by specific preferences for distal regulatory elements. As another example, the HAT PCAF can compensate for GCN5, although these two HATs have differing affinities for their cognate regulatory complexes, and this may partially explain why $Gcn5^{hat/hat}$ mutants show such specific defects in NT closure (22).

The importance of distal regulatory elements in development may also be an underlying reason why it has been difficult to identify causes for human NTDs. The discovery of genomic elements such as super enhancers that control complex traits (78) suggests that sequence variability within distal regulatory elements that control multiple aspects of neurulation could contribute to NTD risk. Unfortunately, the promiscuity of these elements, as well as data showing that distal elements interact with the nearest promoter only 7% of the time (144), makes it extremely difficult to infer the downstream effects of sequence variability within a human regulatory element. Moreover, distal regulatory elements coincide with CpG dinucleotides whose methylation states are crucial for regulation of broad gene regulatory networks and maintenance of cellular health (5). This could help explain how distal element-specific localization of histone modification complexes interfaces with DNA methylation to regulate large transcriptional networks that are required for NT closure.

Transcriptional Variability in Development and Disease

Although the above-mentioned mechanisms are robust and redundant, transcriptional output is still subject to significant noise. This transcriptional noise might be expected to have profound negative effects upon development, but it has been postulated that transcriptional stochasticity actually plays an important role in proper development (130). During the progression from embryonic stem cells to differentiated cells, the mammalian genome undergoes massive rearrangements that require active repression of stemness genes, temporary maintenance of poised epigenetic states for differentiative genes, and activation of genes that give differentiated cells their individual traits. This process has been well-characterized in the developing nervous system, where neural stem cells undergo multiple differentiative events to give rise to more committed progenitors that eventually contribute to the diverse neuronal and glial cell populations (76). It therefore seems necessary for mechanisms that regulate stochastic gene expression within stem and progenitor cells both to allow for differentiation into multiple cell types and to repress unwanted cellular characteristics. It is hypothesized that chromatin modifiers are the main regulatory elements used to maintain order within the required transcriptional noise of development (130).

Variability of gene expression during development poses a significant problem, however, if elements that keep stochasticity in check are disrupted, which can lead to phenotypic changes. This concept was beautifully demonstrated in a paper by Raj et al. (131) that identified increased variability in gene expression as the underlying cause of incomplete phenotypic penetrance. Interestingly, gene expression data sets from the *Baf155* (67) and *Gcn5^bat* (J. Wilde & L. Niswander, unpublished results) mouse models of NTDs indicate that individual mutant embryos display significantly different gene expression patterns from one another, suggesting that GCN5 and the BAF complexes play important roles in controlling variability of gene expression during NT closure. As reviewed by Pujadas & Feinberg (130), the epigenome is particularly susceptible to variability, leading to the hypothesis that tight regulation of chromatin structure and dynamics is required during neurulation. This may be especially true in the developing cranial neural tube, as functional loss of almost all epigenetic modifiers identified in animal models of NTDs causes exencephaly (**Table 1**). This could be due to the fact that the closing cranial NT simultaneously undergoes some of the most complex and dynamic morphogenetic, patterning, and differentiative events during embryogenesis. These processes may necessitate that specific regions of the genome remain extremely plastic, through modulation of chromatin structure and inherent transcriptional variability. Future studies, therefore, may warrant a deeper look into the relationships between transcriptional variability and the regulation of chromatin structure in mouse models of NTDs.

ENVIRONMENTAL CONTRIBUTIONS TO NEURAL TUBE CLOSURE

Environmental Alteration of the Genomic Landscape

Growing evidence now suggests that environmental factors have the ability to alter the epigenetic landscape and, therefore, transcriptional activity. Several lines of study using the agouti viable yellow (A^{vy}) mouse have shown that exposure to endocrine disruptors, dietary changes, and toxic compounds can cause phenotypic changes in offspring owing to disrupted DNA methylation (47).

Interestingly, gestation appears to be particularly sensitive to environmental changes, as maternal folate intake during gestation induces differential DNA methylation and phenotypic changes in A^{vy} offspring but not in their mothers (176). Even in normal animals, long-term methyl-enriched diets can lead to profound changes within the epigenome, specifically, increased transcriptional variability (97). These studies have added to a growing amount of literature suggesting that maternal diet is tightly linked to the epigenome of the offspring. This link is not just important for development, however, as rats born from mothers with suboptimal nutrition have a significantly increased risk for developing type II diabetes (143), and mice born from mothers on a low-protein diet show disrupted *Ppara* expression (102); both phenotypes were shown to be associated with alterations in the epigenome. Together, these data strongly support a significant epigenetic role for maternal nutrition in the proper development of offspring and suggest that epigenetic fidelity is imperative for proper NT closure.

Folic Acid Fortification: Reducing the Incidence of Neural Tube Defects

Although there is clearly a genetic component to NTDs, numerous environmental factors have also been strongly implicated in NTD etiology. One of the earliest records correlating environmental risks with NTDs comes from an eighteenth-century midwife, Catherina Schrader. Her exceptional records showed two clusters of increased NTD incidence: the first following unusually poor crop yields and the second primarily in lower-class, urban families, implicating socioeconomic status, nutrition, and possibly folate deficiency in the incidence of NTDs in eighteenth-century Holland (113). Since then, gene-environment interactions have been implicated in multiple diseases, including diabetes, metabolic syndrome, neurological diseases, and numerous cancers (156). As mentioned above, maternal diet can affect the developing embryo, and one of the best-studied dietary factors relative to NT closure is maternal folate.

Folate deficiency was suggested as a risk factor for early fetal death in the 1950s (163, 164), when the folate antagonist, aminopterin, was used to induce therapeutic abortions. Women who failed to abort with aminopterin treatment gave birth to babies with numerous defects, including NTDs (175). Work by Hibbard & Smithells (75) implicated maternal folate deficiency as a risk for NTD-affected pregnancies in the 1960s, and several landmark human trials in the late 1980s and early 1990s supported these findings (38, 117). The Medical Research Council Vitamin Study, published in 1991, was the first large, randomized trial to show a strong protective effect of FA supplementation. For women who had already had one NTD-affected pregnancy, supplementation with 4 mg/day of FA decreased the risk of a second NTD-affected pregnancy by 72% (6 of 593 NTDs in the FA-supplemented group versus 21 of 602 in the unsupplemented group) (117). A second large-scale randomized trial in Hungary provided evidence that FA supplementation could reduce the risk of a first NTD-affected pregnancy. Supplementation with 0.8 mg/day of FA for one month prior to conception and for the first 8 weeks of pregnancy significantly decreased the risk of having a first NTD-affected pregnancy (0 of 2,104 NTD cases in the vitamin-supplement group, compared to 6 of 2,052 NTD cases in the control group) (38).

As a result of these studies, the US Centers for Disease Control and Prevention (CDC) made the official recommendation that "[a]ll women of childbearing age in the United States who are capable of becoming pregnant should consume 0.4 mg of folic acid per day for the purpose of reducing their risk of having a pregnancy affected with spina bifida or other NTDs" (29). This supplementation campaign proved unsuccessful, however, possibly in part because physicians and patients were not sufficiently educated on the importance of supplementation (65), and less than a third of women were reported to be taking a FA supplement in 1997, despite increased public awareness (30). Given the limited effectiveness of the supplementation campaign and the fact

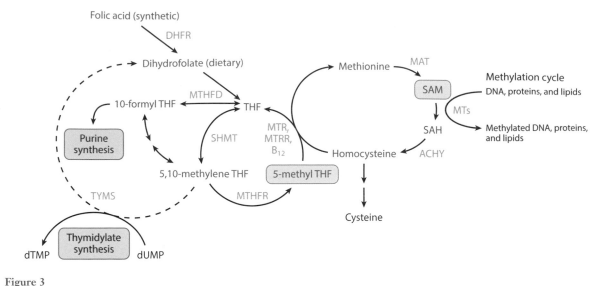

Figure 3

Schematic of folate in one-carbon metabolism. Synthetic and dietary folates are reduced to tetrahydrofolates (THFs), which are converted to the biologically active 5-methyltetrahydrofolate (5-methyl-THF) by SHMT and MTHFR. 5-Methyl-THF acts a methyl donor for purine and thymidylate synthesis and generates the major cellular methylation donor S-adenosylmethionine (SAM). Abbreviations: AHCY, S-adenosylhomocysteine hydrolase; DHFR, dihydrofolate reductase; MAT, methionine adenosyltransferase; MTHFD, methylenetetrahydrofolate dehydrogenase; MTHFR, methylenetetrahydrofolate reductase; MTR, methionine synthase; MTRR, methionine synthase reductase; MTs, methyltransferases; SAH, S-adenosylhomocysteine; SAM, S-adenosylmethionine; SHMT, serine-hydroxymethyltransferase.

that an estimated 50% of pregnancies in the United States are unplanned (87), the United States implemented a fortification campaign in 1998 to ensure that women of childbearing age received the daily recommended dose of 400 μg/day (46). Since fortification, the NTD incidence in the US has decreased by almost 20% overall (31% reduction in spina bifida and 19% reduction in anencephaly) (122). Similar fortification campaigns in other countries have yielded comparable results (41, 42, 104). Overall, more than 75 countries have implemented mandatory FA fortification campaigns, primarily in wheat flour, and it is estimated that FA fortification has prevented 15–25% of FA-responsive NTDs worldwide (188). Interestingly, the incidence of spina bifida did not continue to decline after the initial decrease but rates of anencephaly did (21).

Folate is a water-soluble vitamin that is found naturally in many foods, especially green, leafy vegetables (18). Dietary folates, in the form of tetrahydrofolates (THFs), play an important role in OCM (**Figure 3**). FA is a synthetic form of folate that is more stable than the naturally occurring form. Both FA and folate must be reduced by dihydrofolate reductase (DHFR) and then converted to a biologically active form, 5-methyl-tetrahydrofolate (5-meTHF), by serine-hydroxymethyl-transferase (SHMT) and 5,10-methylenetetrahydrofolate reductase (MTHFR) (60). 5-meTHF is then used in the biosynthesis of purines and thymidylate, the synthesis of methionine from homocysteine, and the biosynthesis of S-adenosylmethionine (SAM), the universal methyl donor for cellular methylation reactions, including DNA and protein methylation (14).

Drugs that act as folate antagonists have continued to be implicated as risk factors for NTDs. Women who took folate antagonists in their first trimester were more than sixfold more likely to have an NTD-affected pregnancy. These included DHFR inhibitors, such as methotrexate, which inhibit the conversion of folate to its active form, and the antiepileptic drug valproic acid (an HDAC inhibitor) (6, 112).

After the Brownsville Cluster, in which the incidence of NTDs was four- to fivefold higher along the Texas-Mexico border than in the rest of the United States, attention turned to other potential environmental risk factors. The Mexican-American population in the region was known to consume high levels of corn, and cornmeal samples harvested around the time of the Cluster contained high levels of fumonisin contamination. Fumonisin is a mycotoxin that inhibits the biosynthesis of sphingolipids, ultimately interfering with the cellular uptake of 5-meTHF (159). In mouse studies, FA supplementation partially rescued fumonisin-induced NTDs (55, 139). Furthermore, much of the population affected in the Brownsville cluster was also folate deficient, suggesting that fumonisin is more likely to compound folate deficiency in a human population and less likely to be a significant risk factor in a folate-sufficient population.

Given the strong epidemiological evidence for the requirement of adequate FA during development, and especially at the time of NT closure, attention has focused on genes in the folate OCM pathway. More than 40 genes involved in FA metabolism have been studied in NTD patient samples (reviewed in 61). Numerous studies have focused on a common polymorphism in the *MTHFR* gene, 677C→T. *MTHFR* is required to reduce 5,10-methylenetetrahydrofolate (5,10-methylene-THF) to 5-methyl THF, which is the main circulating form of folate and is required to generate SAM (18). The 677C→T mutation generates a more thermolabile enzyme with significantly reduced activity that results in low plasma folate levels and higher levels of homocysteine (51, 165). Several meta-analyses of the epidemiological data indicate that the 677T→C mutation is an overall risk factor, with either the CT or TT allelic combinations conferring a 1.2–2-fold increased risk of an NTD-affected pregnancy (181, 191). Furthermore, the risk associated with the *MTHFR* 677C→T mutation can be alleviated by increased maternal folate (92), again highlighting the importance of gene-environment interactions. Despite the strong association with the *MTHFR* polymorphism and the intensive focus on OCM gene variants in cohorts of NTD patients, there is little additional evidence for a major causative association between genetic variants in this pathway and NTDs (61).

Consistent with the human NTD data, very few genes involved in OCM have been implicated in mouse NTDs (68), and folate deficiency in the absence of a genetic insult fails to cause spontaneous murine NTDs. A handful of mouse NTD models have been used for FA supplementation studies, with very mixed results (**Table 2**). For some mouse models, FA has a preventative effect (12, 13, 68, 116), whereas others are nonresponsive (68, 120), much like in the human population. Surprisingly, several models have shown a detrimental response to FA, including increased NTDs and increased fetal loss (59, 110). These studies raise an important question: In some cases, is the decreased incidence of NTDs due to an increase in early fetal loss rather than a rescue of NTDs, a possibility first raised nearly two decades ago (79)? It will be important to understand which genetic variants are favorably influenced by FA fortification, and it is expected that animal models can provide genetic, experimental, and mechanistic insights into FA-responsive and nonresponsive NTDs. Some experts have raised the idea that the dose of FA should be increased to try to further decrease the incidence of NTDs in the population (168). Indeed, FA fortification in conjunction with an increase in vitamin usage has resulted in a trend toward higher FA levels throughout the US population (80, 129). However, it is important to note that there will likely be certain genotypes that are resistant to FA, and the potential for long-term epigenetic changes remains an important consideration.

The mechanism by which FA affects NT closure is unknown, but as discussed above, changes in DNA methylation likely play a role, as do alterations in synthesis of nucleotide precursors. Methylation of the insulin-like growth factor gene 2 differentially methylated region (*IGF2* DMR) was significantly higher in children whose mothers took FA supplements (400 μg/day) than in children of unsupplemented mothers (154). In mice, genome-wide CpG methylation patterns

Table 2 Mouse NTD models studied for responsiveness to folic acid

Gene Name	Folic acid response			Process	Reference
	Beneficial	Detrimental	None		
Amt null			x	Involved in glycine cleavage system, part of the mitochondrial folate metabolism; implicated in human and mouse neural tube defects; responsive to methionine	(120)
Axd mutant			x	Unknown function; axial defects; responsive to methionine	(68)
Cart1 null	x			*Paired*-class homeobox domain-containing gene, aka *Alx1*	(68)
Cited2 null	x			Required for heart morphogenesis and left-right axis determination in mouse development	(68)
Curly tail (*Grhl3* mutant)			x	Transcription factor; *Ct* mutation is responsive to inositol and retinoic acid	(68)
Fkbp8 null			x	Member of immunophilin protein family; involved in dorsal-ventral patterning of neural tube; inositol-resistant	(68)
Folr1 null	x			Folate receptor; involved in transport of 5-meTHF into the cell	(68)
Frem2^{m1Nisw} allele			x	Extracellular matrix protein	(110)
Grhl2^{m1Nisw} allele	x (het)		x (null)	Transcription factor	(110)
Grhl3 null			x	Transcription factor	(68)
Kat2a hypomorph (GCN5)	x			Histone acetyltransferase	(68)
L3P		x		Ciliogenesis	(110)
Lrp6 null		x		Receptor or coreceptor (with *Frizzled*) of WNT; canonical WNT/β-catenin signaling	(59)
Lrp6^{Cd} allele	x				(68)
Map3k4 null			x	MAPK pathway	(68)
Mthfd1l^{F} allele				Untested with folic acid; partial rescue observed upon sodium formate supplementation	(116)
Nog null			x	Bone morphogenetic protein antagonist; resistant to inositol and pifithrin-α (cell-death inhibitor)	(68)
SELH/*Bc* strain			x	Inbred mouse strain with a high frequency of nonsyndromic neural tube defects	(68)
Shmt null	x			Thymidylate biosynthesis pathway; choline-resistant	(12, 13)
Shroom3^{m1Nisw} allele	x (short-term)		x (long-term)	Involved in actin remodeling, especially apical constriction and apicobasal elongation	(110)
Slc19a1 null (RFC1)	x			Reduced folate carrier	(68)
Sp and *Sp^{2H}* mutant at *Pax3* gene	x			Transcription factor involved in ear, eye, and facial development as well as dorsal patterning of the neural tube	(68, 110)
Zic2^{m1Nisw} allele	x			Transcriptional activator and repressor involved in organogenesis of the central nervous system	(110)

were found to be significantly different in the cerebral hemispheres in offspring of wild-type dams exposed to an FA-fortified diet compared with nonfortified controls (11). Furthermore, differential methylation of important developmental genes was correlated with altered levels of gene expression. Future studies are needed to determine what the link may be between FA status, changes in DNA methylation, and NTD incidence. Moreover, the OCM cycle also produces nucleotides, which are needed for proliferation, and the rate of proliferation is extremely important in NT closure. Hence, adequate folate levels likely play an important role in maintaining sufficient proliferation levels during NT closure.

Maternal Diabetes and Obesity

Maternal diabetes has long been identified as a risk factor for congenital birth defects, including NTDs (190). SNPs in maternal genes related to glucose metabolism, such as *FTO*, *LEP*, *TCF7L2*, *LEPR*, *GLUT1* and *HK1*, have been linked to an increased risk for NTDs (39, 107), and potentially informative maternal-fetal SNP interactions have also been correlated to altered NTD risk. For example, SNPs in maternal *ENPP1* and fetal *SLC2A2* correlate with increased risk of NTDs (2–3.5-fold), whereas the same fetal *SLC2A2* SNPs interact with maternal SNPs in *LEP* to provide a protective effect (0.5-fold risk) (108). Fetal SNPs in *SLC2A2* alone do not appear to confer altered risks of NTDs (107). It is thought that altered maternal glucose metabolism leads to an altered intrauterine environment, which is not tolerated by the developing embryo, as pancreatic function in human embryos begins at approximately week 7 of pregnancy. Diabetic embryopathies are known to particularly affect the neural tube and heart, both of which form prior to week 7 (107).

Animal models have proven invaluable in elucidating the molecular pathophysiology of diabetic embryopathies. In mice, high glucose levels cause decreased proliferation and increased apoptosis in neural progenitor cells in the developing spinal cord (53), suggesting that a reduction in cell number contributes to diabetic embryopathies. Hyperglycemia can activate several related pathways, including altered lipid metabolism, increased generation of reactive oxygen species (ROS), and activation of apoptotic pathways (reviewed in 135). Briefly, hyperglycemia alters arachidonic acid metabolism, leading to the generation of altered levels of prostaglandins, which have been linked to adverse pregnancy outcomes. Hyperglycemia also leads to increased glucose transport into the cell, which activates the polyol pathway, causing increased accumulation of sorbitol and ultimately increased ROS levels. Increased sorbitol and the hyperglycemia-induced stress response in the endoplasmic reticulum (ER) have both been shown to activate the JNK1/2 pathway, which has been implicated in diabetes-induced NTDs (101, 183). Oxidative and ER stress have been shown to activate the PKC pathway (77), which can also activate the JNK1/2 pathway (24, 25) and induce apoptosis via Caspase-8 (192). Alleviating oxidative stress by overexpression of the antioxidant enzyme superoxide dismutase 1 (SOD1) can rescue diabetes-induced NTDs coincident with decreased activation of the PKC and JNK1/2 pathways (99, 100, 172). Moreover, ROS in mice activates apoptosis signal-regulating kinase 1 (ASK1), which activates the transcription factor FoxO3a and Caspase-8, as well as increases levels of the apoptosis-promoting adaptor protein, TRADD, which is a direct target of FoxO3a. Interestingly, increased ASK1 phosphorylation, decreased phosphorylation of FoxO3a, increased TRADD expression, and increased cleaved Caspase-8 were also observed in neural tissue from human NTD-affected fetuses (182).

Intriguingly, a model of transcriptional variability as the underlying cause of partial penetrance has been proposed, supported by recent data from diabetic mouse models. In both a chemically induced diabetes model and a nonobese diabetic (NOD) strain, NTD-affected embryos had significantly higher variability in gene expression compared with their non-NTD littermates.

Furthermore, the pathways enriched in the class of highly variable genes were different than those enriched in the differentially expressed but less-variable gene sets, suggesting a role for environmentally induced variable gene expression as a mechanism of increased NTD risk (83). Mechanistically, this may fit with recent data showing that hyperglycemia alters the epigenetic landscape. Shyamasundar et al. (153) found that neural stem cells isolated from chemically induced diabetic pregnancies had increased global levels of H3K9 trimethylation, increased DNA methylation, and decreased H3K9 acetylation compared with neural stem cells derived from normal pregnancies. As discussed above, disruption of the epigenetic landscape can increase transcriptional variability and may be an underlying factor of NTDs.

Given what is known about the molecular mechanisms involved in diabetes-induced NTDs, several measures have been proposed to protect developing embryos. There is evidence that nutritional supplements can alleviate the risks. Diabetic mice supplemented with trehalose had a much lower incidence of NTD-affected embryos than unsupplemented mice (10% versus 28%) (180). Diabetic rats supplemented with FA showed a decreased incidence of NTDs (supplemented, 6%; unsupplemented, 28%) (54). However, diabetic mice also showed premature neuronal and astrocyte differentiation when given high doses of FA compared with a moderate dose, suggesting that too much FA may cause unintended consequences (189). Human epidemiological data also suggest that FA supplementation may reduce the risk of NTDs in a diabetic pregnancy (37, 127), with a larger protective effect observed against anencephaly than spina bifida (37).

Maternal obesity, defined as a body mass index (BMI) ≥ 30, has also been linked to an increased risk for NTDs. Meta-analyses showed an overall 1.7–1.8-fold increase in NTD-affected pregnancies for women who are obese and 3.1-fold increase for women who are severely obese (132, 150, 155). Interestingly, a prefortification study in the United States showed that the risk associated with increased BMI was independent of maternal folate status (74, 150), a finding confirmed in a Canadian study spanning pre- and postfortification, which showed that FA fortification did not alleviate the obesity-related risk of NTDs (134, 169).

Maternal Hyperthermia

Maternal hyperthermia can arise from either febrile illnesses or external exposure to heat, such as prolonged periods in a hot tub or sauna. Epidemiological studies have strongly implicated maternal fever in early pregnancy as a risk factor for NTD-affected pregnancies. A recent meta-analysis evaluated 46 studies ranging from 1990–2013 and found that the NTD risk was increased almost threefold in cases of maternal fever during the first trimester The risk for oral clefts and congenital heart defects also increased (43). Interestingly, studies of Mexican-American women along the Texas-Mexico border (160), a more general US population (48), and Hungarian women (1) indicate that use of antifever medication can significantly alleviate this risk. Furthermore, the use of either a multivitamin or FA supplementation also appears to abrogate the risk associated with maternal febrile illness (1, 20). Surprisingly, animal models of maternal hyperthermia show a mixed response to FA supplementation. In the golden hamster, FA supplementation does not appear to rescue heat-induced NTDs (58), whereas studies indicate that FA can rescue heat-induced NTDs in mice (151).

Early studies linking maternal hyperthermia as a result of prolonged external exposure, such as in a hot tub or sauna, to increased risk for NTDs are mixed. An increased risk was found in several studies (114, 142), but another study found no correlation with sauna use and NTD risk in a Finnish cohort (145). Data collected from the large, multisite National Birth Defects Prevention Study from 1997 to 2005 show the risk for anencephaly increased by 1.7-fold for women who reported using hot tubs during early pregnancy. However, only hot tub sessions lasting more than 30 minutes had a significant effect (44). Overall, in animal and human models, the key factor appears to be raising the core temperature more than 2°C above baseline (32).

The mechanism by which heat may affect the developing embryo is unknown, but there appears to be a genetic component. In one study, five different mouse strains were exposed to hyperthermic treatment on embryonic day 8.5, just as the neural tube is closing. One strain had a 44% incidence of heat-induced NTDs in their offspring, whereas the other four strains showed no more than a 14% penetrance of NTDs (49). A follow-up study confirmed these findings and suggested the importance of maternal genotype in susceptibility to heat-induced NTDs (106).

Although it has been proposed that hyperthermia-induced NTDs may arise from changes in metabolism, research in animal models indicates that it is the heat that adversely affects development, likely as a result of increased cell death, decreased proliferation, disruption of gene expression, and damage to the embryonic vasculature (reviewed in 45). Recent work on mouse neural stem cells showed a narrow temperature range of responsiveness to heat shock (124), leading to induction of apoptosis, inhibition of proliferation, and delayed differentiation. These data highlight the importance of maintaining thermal homeostasis during the critical period of NT closure.

Maternal and Embryonic Exposure to Alcohol and Drugs

Although there is no question that exposure to alcohol in utero is detrimental to the developing fetus, the epidemiology linking maternal alcohol consumption to NTDs is less clear. A small Italian study found that mothers of children with spina bifida were 3 times more likely to have consumed 0.5 liters/day or more of alcohol in the 3 months leading up to and immediately following conception (41). Two American studies, however, showed no correlation between periconceptional maternal alcohol consumption and NTDs (109, 158). However, both American studies were retrospective, and the authors suggest that the social stigma against alcohol during pregnancy may have led to under-reporting. Animal models have shown adverse response to ethanol exposure, evidenced by increased NTDs in mice injected with ethanol around the time of NT closure (64). The authors proposed that ethanol interferes with polyamine synthesis in the developing embryo. As polyamines are known to regulate cell growth and proliferation, this may be one mechanism by which alcohol disrupts embryonic development. Alcohol can also interfere with folate in OCM, although the mechanisms are not well understood (reviewed in 90). Ethanol treatment of pigs results in decreased expression of methionine synthase (MS), a key enzyme in OCM (167), and it has been suggested that ethanol-induced inactivation of MS could lead to a pseudofolate deficiency.

Maternal smoking and maternal exposure to environmental tobacco smoke (ETS) have also been studied epidemiologically with respect to risk for an NTD-affected pregnancy, with conflicting conclusions. Some find that smoking shows no correlation and some show an increased risk of NTD-affected pregnancies. Recent meta-analyses indicate that maternal smoking confers a small increased risk for spina bifida, and a moderately increased NTD risk in cases of maternal ETS exposure (173, 174). Interestingly, the data suggest regional differences in susceptibility, with maternal smoking being a higher risk factor in Europe than in the United States (173). Both smoking and ETS are associated with decreased serum folate levels (125), so it is possible that the increased risk in Europe compared with the United States may be related to differences in FA fortification policies. It has been proposed that the risk associated with maternal smoking may be underestimated, as there is evidence that both maternal smoking and ETS are associated with a significantly increased risk of spontaneous abortion (56), and early fetal loss is rarely included in retrospective studies.

Given the prevalence of caffeine use, several studies have looked at whether periconceptional caffeine use is a risk factor for NTDs. One US study found no correlation between maternal caffeine

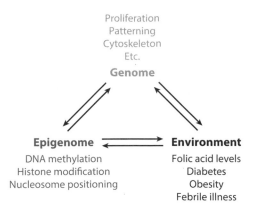

Proliferation
Patterning
Cytoskeleton
Etc.
Genome

Epigenome ⇄ **Environment**

DNA methylation
Histone modification
Nucleosome positioning

Folic acid levels
Diabetes
Obesity
Febrile illness

Figure 4

Interrelationship between the genome, epigenome, and environment, with respect to neural tube closure.

intake and NTD-affected pregnancy (15), whereas another small study suggested that there might be a slightly elevated risk (147). Moreover, the susceptibility to caffeine as a risk factor varied by maternal race/ethnicity, with non-Hispanic whites showing no correlation and Hispanic and all other ethnicities showing a 1.5–2-fold increased risk (147). An Italian study, however, showed a strong association between maternal coffee intake (≥3 cups/day) and spina bifida–associated pregnancy (41), but it was a relatively small study, and the overall conclusions from the study focused on folate deficiency as the most important risk factor. Tea has also been implicated in NTDs, as tea contains catechins that can inhibit DHFR activity. In northern China, daily tea consumption was shown to increase the risk of NTD-affected pregnancy by more than threefold (186), although most of these women were also FA deficient, confounding the interpretation. However, in the United States, no correlation was seen between tea consumption and NTDs (184).

CONCLUSIONS AND FUTURE OUTLOOK

As highlighted in this review, NTDs are a complex disease impacted by genetic susceptibility, epigenetic influences, and environmental insults (**Figure 4**). The tools are now available to identify the genetic contributions in humans using unbiased methods to evaluate the genome and the epigenome. This, combined with expanded NTD cohorts that include familial and sporadic cases, should lead to a much broader understanding of this complex genetic trait. It may be discovered that most sporadic cases of NTD in humans are not due to strong single gene contributions but instead result from a combination of small genetic perturbations in a number of genes. Animal models will be necessary to provide experimental validation of coding and regulatory polymorphisms, the functional intersections between multiple genetic components, and mechanistic insights into the normal action of these genes and why NT closure fails upon genetic disruption. Finally, there is little insight into the mechanisms that underlie environmental contributions, whether disruptive or preventative. Research has lagged behind in using animal models to decipher how environmental factors impact NT closure and in determining genetic susceptibility to environmental influences. Moreover, despite the clear beneficial impact of FA fortification on NTD risk in the general population, there are still a large number of NTD-affected pregnancies. By understanding the cellular and molecular mechanisms of neural tube closure and metabolic factors, it may be possible to identify other measures to help prevent this devastating birth defect.

DISCLOSURE STATEMENT

The authors are not aware of any affiliations, memberships, funding, or financial holdings that might be perceived as affecting the objectivity of this review.

ACKNOWLEDGMENTS

We thank A. Copp, R. Finnell, N. Greene, M. Harris, D. Juriloff, C. Kappen, L. Mitchell, M. Salbaum, and the NTD community for their contributions to the NTD Wiki site and for discussions at the International Conferences on Neural Tube Defects, sponsored by the National Institute of Child Health and Human Development. This work was supported by NIH R01NS058979, and L.N. was an investigator at the Howard Hughes Medical Institute.

LITERATURE CITED

1. Acs N, Bánhidy F, Puhó E, Czeizel AE. 2005. Maternal influenza during pregnancy and risk of congenital abnormalities in offspring. *Birth Defects Res. A Clin. Mol. Teratol.* 73(12):989–96
2. Aguilar A, Meunier A, Strehl L, Martinovic J, Bonniere M, et al. 2012. Analysis of human samples reveals impaired SHH-dependent cerebellar development in Joubert syndrome/Meckel syndrome. *Proc. Natl. Acad. Sci. USA.* 109:16951–56
3. Alvarez IS, Schoenwolf GC. 1992. Expansion of surface epithelium provides the major extrinsic force for bending of the neural plate. *J. Exp. Zool.* 261:340–48
4. Amir RE, Van den Veyver IB, Wan M, Tran CQ, Francke U, Zoghbi HY. 1999. Rett syndrome is caused by mutations in X-linked *MECP2*, encoding methyl-CpG-binding protein 2. *Nat. Genet.* 23(2):185–88
5. Aran D, Sabato S, Hellman A. 2013. DNA methylation of distal regulatory sites characterizes dysregulation of cancer genes. *Genome Biol.* 14(3):R21
6. Artama M, Auvinen A, Raudaskoski T, Isojärvi I, Isojärvi J. 2005. Antiepileptic drug use of women with epilepsy and congenital malformations in offspring. *Neurology* 64:1874–78
7. Attia M, Murko C, Förster A, Lagger S, Rachez C, et al. 2011. Interaction between nucleosome assembly protein 1-like family members. *J. Mol. Biol.* 407(5):647–60
8. Attia M, Rachez C, De Pauw A, Avner P, Rogner UC. 2007. Nap1l2 promotes histone acetylation activity during neuronal differentiation. *Mol. Cell. Biol.* 27(17):6093–102
9. Bannister AJ, Kouzarides T. 2011. Regulation of chromatin by histone modifications. *Cell Res.* 21(3):381–95
10. Banting GS. 2004. CECR2, a protein involved in neurulation, forms a novel chromatin remodeling complex with SNF2L. *Hum. Mol. Genet.* 14(4):513–24
11. Barua S, Kuizon S, Chadman KK, Flory MJ, Brown WT, Junaid MA. 2014. Single-base resolution of mouse offspring brain methylome reveals epigenome modifications caused by gestational folic acid. *Epigenet. Chromatin* 7(1):1–15
12. Beaudin AE, Abarinov EV, Malysheva O, Perry CA, Caudill M, Stover PJ. 2011. Dietary folate, but not choline, modifies neural tube defect risk in *Shmt1* knockout mice. *Am. J. Clin. Nutr.* 95:109–14
13. Beaudin AE, Abarinov EV, Noden DM, Perry CA, Chu S, et al. 2011. *Shmt1* and de novo thymidylate biosynthesis underlie folate-responsive neural tube defects in mice. *Am. J. Clin. Nutr.* 93(4):789–98
14. Beaudin AE, Stover PJ. 2007. Folate-mediated one-carbon metabolism and neural tube defects: balancing genome synthesis and gene expression. *Birth Defects Res. C Embryo Today* 81(3):183–203
15. Benedum C, Yazdy M, Mitchell A, Werler M. 2013. Risk of spina bifida and maternal cigarette, alcohol, and coffee use during the first month of pregnancy. *Int. J. Environ. Res. Public Health* 10(8):3263–81
16. Bergman Y, Cedar H. 2013. DNA methylation dynamics in health and disease. *Nat. Struct. Mol. Biol.* 20(3):274–81
17. Blattler A, Ronan JL, Farnham PJ, Wu W, Crabtree GR. 2013. Cross-talk between site-specific transcription factors and DNA methylation states. *J. Biol. Chem.* 288(48):34287–94

18. Blom HJ, Shaw GM, den Heijer M, Finnell RH. 2006. Neural tube defects and folate: case far from closed. *Nat. Rev. Neurosci.* 7(9):724–31

19. Borgel J, Guibert S, Li Y, Chiba H, Schübeler D, et al. 2010. Targets and dynamics of promoter DNA methylation during early mouse development. *Nat. Genet.* 42(12):1093–100

20. Botto LD, Erickson JD, Mulinare J, Lynberg MC, Liu Y. 2002. Maternal fever, multivitamin use, and selected birth defects: evidence of interaction? *Epidemiology* 13(4):485–88

21. Boulet SL, Yang Q, Mai C, Kirby RS, Collins JS, et al. 2008. Trends in the postfortification prevalence of spina bifida and anencephaly in the United States. *Birth Defects Res. A Clin. Mol. Teratol.* 82(7):527–32

22. Bu P, Evrard YA, Lozano G, Dent SYR. 2007. Loss of Gcn5 acetyltransferase activity leads to neural tube closure defects and exencephaly in mouse embryos. *Mol. Cell. Biol.* 27(9):3405–16

23. Bultman SJ, Herschkowitz JI, Godfrey V, Gebuhr TC, Yaniv M, et al. 2007. Characterization of mammary tumors from *Brg1* heterozygous mice. *Oncogene* 27(4):460–68

24. Cao Y, Zhao Z, Eckert RL, Reece EA. 2011. Protein kinase Cβ2 inhibition reduces hyperglycemia-induced neural tube defects through suppression of a caspase 8–triggered apoptotic pathway. *Am. J. Obstet.* 204:226; e1–5

25. Cao Y, Zhao Z, Eckert RL, Reece EA. 2012. The essential role of protein kinase Cδ in diabetes-induced neural tube defects. *J. Matern. Fetal Neonatal Med.* 25(10):2020–24

26. Cassidy SB, Schwartz S. 1998. Prader-Willi and Angelman syndromes: disorders of genomic imprinting. *Medicine* 77(2):140–51

27. Caton PW, Nayuni NK, Kieswich J, Khan NQ, Yaqoob MM, Corder R. 2010. Metformin suppresses hepatic gluconeogenesis through induction of SIRT1 and GCN5. *J. Endocrinol.* 205(1):97–106

28. Cent. Dis. Control Prev. 2011. *CDC Birth Defects Data/Statistics Registry*. Atlanta, GA: Cent. Dis. Control Prev.

29. Cent. Dis. Control Prev. 1992. Recommendations for the use of folic acid to reduce the number of cases of spina bifida and other neural tube defects. *MMWR Recomm. Rec.* 41:1–77

30. Cent. Dis. Control Prev. 1998. Use of folic acid–containing supplements among women of childbearing age: United States, 1997. *MMWR Morb. Mortal. Wkly. Rep.* 47(7):131–34

31. Cervoni N. 2001. Demethylase activity is directed by histone acetylation. *J. Biol. Chem.* 276(44):40778–87

32. Chambers CD. 2006. Risks of hyperthermia associated with hot tub or spa use by pregnant women. *Birth Defects Res. A Clin. Mol. Teratol.* 76(8):569–73

33. Chen J, Lai F, Niswander L. 2012. The ubiquitin ligase mLin41 temporally promotes neural progenitor cell maintenance through FGF signaling. *Genes Dev.* 26(8):803–15

34. Chen X, Shen Y, Gao Y, Zhao H, Sheng X, et al. 2013. Detection of copy number variants reveals association of cilia genes with neural tube defects. *PLoS ONE* 8:e54492

35. Cheng H-L, Mostoslavsky R, Saito S, Manis JP, Gu Y, et al. 2003. Developmental defects and p53 hyperacetylation in Sir2 homolog (SIRT1)-deficient mice. *Proc. Natl. Acad. Sci. USA.* 100(19):10794–99

36. Copp AJ, Greene NDE. 2010. Genetics and development of neural tube defects. *J. Pathol.* 220(2):217–30

37. Correa A, Gilboa SM, Botto LD, Moore CA, Hobbs CA, et al. 2012. Lack of periconceptional vitamins or supplements that contain folic acid and diabetes mellitus–associated birth defects. *Am. J. Obstet. Gynecol.* 206(3):218; e1–13

38. Czeizel AE, Dudás I. 1992. Prevention of the first occurrence of neural-tube defects by periconceptional vitamin supplementation. *N. Engl. J. Med.* 327(26):1832–35

39. Davidson CM, Northrup H, King TM, Fletcher JM, Townsend I, et al. 2008. Genes in glucose metabolism and association with spina bifida. *Reprod. Sci.* 15(1):51–58

40. Deak KL, Siegel DG, George TM, Gregory S, Ashley-Koch A, et al. 2008. Further evidence for a maternal genetic effect and a sex-influenced effect contributing to risk for human neural tube defects. *Birth Defects Res. A Clin. Mol. Teratol.* 82(10):662–69

41. De Marco P, Merello E, Calevo MG, Mascelli S, Pastorino D, et al. 2011. Maternal periconceptional factors affect the risk of spina bifida–affected pregnancies: an Italian case-control study. *Child's Nerv. Syst.* 27(7):1073–81

42. De Wals P, Tairou F, Van Allen MI, Uh S-H, Lowry RB, et al. 2007. Reduction in neural-tube defects after folic acid fortification in Canada. *N. Engl. J. Med.* 357(2):135–42

43. Dreier JW, Andersen A-MN, Berg-Beckhoff G. 2014. Systematic review and meta-analyses: fever in pregnancy and health impacts in the offspring. *Pediatrics* 133(3):e674–88

44. Duong HT, Shahrukh Hashmi S, Ramadhani T, Canfield MA, Scheuerle A, et al. 2011. Maternal use of hot tub and major structural birth defects. *Birth Defects Res. A Clin. Mol. Teratol.* 91(9):836–41

45. Edwards MJ, Saunders RD, Shiota K. 2003. Effects of heat on embryos and foetuses. *Int. J. Hyperthermia* 19(3):295–324

46. FDA. 1996. *Food Standards: Amendment of Standards of Identity for Enriched Grain Products to Require Addition of Folic Acid*, Vol. 61. Washington, DC: FDA

47. Feil R, Waterland RA, Fraga MF, Jirtle RL. 2012. Epigenetics and the environment: emerging patterns and implications. *Nat. Rev. Genet.* 13(2):97–109

48. Feldkamp ML, Meyer RE, Krikov S, Botto LD. 2010. Acetaminophen use in pregnancy and risk of birth defects. *Obstet. Gynecol.* 115(1):109–15

49. Finnell RH, Moon SP, Abbott LC, Golden JA, Chernoff GF. 1986. Strain differences in heat-induced neural tube defects in mice. *Teratology* 33(2):247–52

50. Foster WH, Langenbacher A, Gao C, Chen J, Wang Y. 2013. Nuclear phosphatase PPM1G in cellular survival and neural development. *Dev. Dyn.* 242(9):1101–9

51. Frosst P, Blom HJ, Milos R, Goyette P, Sheppard CA, et al. 1995. A candidate genetic risk factor for vascular disease: a common mutation in methylenetetrahydrofolate reductase. *Nat. Genet.* 10(1):111–13

52. Fukuda T, Li E, Tokunaga A, Bestor TH, Sakamoto R, et al. 2011. Fbxl10/Kdm2b deficiency accelerates neural progenitor cell death and leads to exencephaly. *Mol. Cell. Neurosci.* 46(3):614–24

53. Gao Q, Gao Y-M. 2007. Hyperglycemic condition disturbs the proliferation and cell death of neural progenitors in mouse embryonic spinal cord. *Int. J. Dev. Neurosci.* 25(6):349–57

54. Gäreskog M, Eriksson UJ, Wentzel P. 2006. Combined supplementation of folic acid and vitamin E diminishes diabetes-induced embryotoxicity in rats. *Birth Defects Res. A Clin. Mol. Teratol.* 76(6):483–90

55. Gelineau-van Waes J, Starr L, Maddox J, Aleman F, Voss KA, et al. 2005. Maternal fumonisin exposure and risk for neural tube defects: mechanisms in an in vivo mouse model. *Birth Defects Res. A Clin. Mol. Teratol.* 73(7):487–97

56. George L, Granath F, Johansson ALV, Annerén G, Cnattingius S. 2006. Environmental tobacco smoke and risk of spontaneous abortion. *Epidemiology* 17(5):500–5

57. Goetzinger KR, Stamilio DM, Dicke JM, Macones GA, Odibo AO. 2008. Evaluating the incidence and likelihood ratios for chromosomal abnormalities in fetuses with common central nervous system malformations. *Am. J. Obstet. Gynecol.* 199(3):285; e1–6

58. Graham JM, Ferm VH. 1985. Heat- and alcohol-induced neural tube defects: interactions with folate in a golden hamster model. *Pediatr. Res.* 19(2):247–51

59. Gray JD, Nakouzi G, Slowinska-Castaldo B, Dazard J-E, Rao JS, et al. 2010. Functional interactions between the LRP6 WNT co-receptor and folate supplementation. *Hum. Mol. Genet.* 19(23):4560–72

60. Greenberg JA, Bell SJ, Guan Y, Yu Y-H. 2011. Folic acid supplementation and pregnancy: more than just neural tube defect prevention. *Rev. Obstet. Gynecol.* 4(2):52–59

61. Greene NDE, Stanier P, Copp AJ. 2009. Genetics of human neural tube defects. *Hum. Mol. Genet.* 18(R2):R113–29

62. Greene NDE, Stanier P, Moore GE. 2011. The emerging role of epigenetic mechanisms in the etiology of neural tube defects. *Epigenetics* 6(7):875–83

63. Haas J, Frese KS, Park YJ, Keller A, Vogel B, et al. 2013. Alterations in cardiac DNA methylation in human dilated cardiomyopathy. *EMBO Mol. Med.* 5(3):413–29

64. Haghighi Poodeh S, Alhonen L, Salonurmi T, Savolainen MJ. 2014. Ethanol-induced impairment of polyamine homeostasis: a potential cause of neural tube defect and intrauterine growth restriction in fetal alcohol syndrome. *Biochem. Biophys. Res. Commun.* 446:173–78

65. Hall J, Solehdin F. 1998. Folic acid for the prevention of congenital anomalies. *Eur. J. Pediatr.* 157(6):445–50

66. Hargreaves DC, Crabtree GR. 2011. ATP-dependent chromatin remodeling: genetics, genomics and mechanisms. *Cell Res.* 21:396–420

67. Harmacek L, Watkins-Chow DE, Chen J, Jones KL, Pavan WJ, et al. 2014. A unique missense allele of BAF155, a core BAF chromatin remodeling complex protein, causes neural tube closure defects in mice. *Dev. Neurobiol.* 74:483–97

68. Harris MJ. 2009. Insights into prevention of human neural tube defects by folic acid arising from consideration of mouse mutants. *Birth Defects Res. A Clin. Mol. Teratol.* 85(4):331–39

69. Harris MJ, Juriloff DM. 2007. Mouse mutants with neural tube closure defects and their role in understanding human neural tube defects. *Birth Defects Res. A Clin. Mol. Teratol.* 79(3):187–210

70. Harris MJ, Juriloff DM. 2010. An update to the list of mouse mutants with neural tube closure defects and advances toward a complete genetic perspective of neural tube closure. *Birth Defects Res. A Clin. Mol. Teratol.* 88(8):653–69

71. Hata K, Okano M, Lei H, Li E. 2002. Dnmt3L cooperates with the Dnmt3 family of de novo DNA methyltransferases to establish maternal imprints in mice. *Development* 129(8):1983–93

72. He L, Cao J, Meng S, Ma A, Radovick S, Wondisford FE. 2013. Activation of basal gluconeogenesis by coactivator p300 maintains hepatic glycogen storage. *Mol. Endocrinol.* 27(8):1322–32

73. Heintzman ND, Ren B. 2009. Finding distal regulatory elements in the human genome. *Curr. Opin. Genet. Dev.* 19(6):541–49

74. Hendricks KA, Nuno OM, Suarez L, Larsen R. 2001. Effects of hyperinsulinemia and obesity on risk of neural tube defects among Mexican Americans. *Epidemiology* 12(6):630–35

75. Hibbard ED, Smithells RW. 1965. Folic acid metabolism and human embryopathy. *Lancet* 285(7398):1254

76. Hirabayashi Y, Schaniel C, Pujadas E, Banting GS, Gotoh Y, et al. 2010. Epigenetic control of neural precursor cell fate during development. *Nat. Rev. Neurosci.* 11:377–88

77. Hiramatsu Y, Sekiguchi N, Hayashi M, Isshiki K, Yokota T, et al. 2002. Diacylglycerol production and protein kinase C activity are increased in a mouse model of diabetic embryopathy. *Diabetes* 51(9):2804–10

78. Hnisz D, Abraham BJ, Lee TI, Lau A, Saint-André V, et al. 2013. Super-enhancers in the control of cell identity and disease. *Cell* 155(4):934–47

79. Hook EB, Czeizel AE. 1997. Can terathanasia explain the protective effect of folic-acid supplementation on birth defects? *Lancet* 350(9076):513–15

80. Hoyo C, Murtha AP, Schildkraut JM, Forman MR, Calingaert B, et al. 2011. Folic acid supplementation before and during pregnancy in the Newborn Epigenetics STudy (NEST). *BMC Public Health* 11(1):46

81. Joó JG, Beke A, Papp C, Tóth-Pál E, Csaba A, et al. 2007. Neural tube defects in the sample of genetic counselling. *Prenat. Diagn.* 27(10):912–21

82. Juriloff DM, Harris MJ. 2012. A consideration of the evidence that genetic defects in planar cell polarity contribute to the etiology of human neural tube defects. *Birth Defects Res. A Clin. Mol. Teratol.* 94(10):824–40

83. Kappen C, Salbaum JM. 2014. Gene expression in teratogenic exposures: a new approach to understanding individual risk. *Reprod. Toxicol.* 45:94–104

84. Kibar Z, Torban E, McDearmid JR, Reynolds A, Berghout J, et al. 2007. Mutations in *VANGL1* associated with neural-tube defects. *N. Engl. J. Med.* 356(14):1432–37

85. Kim JK, Huh SO, Choi H, Lee KS, Shin D, et al. 2001. Srg3, a mouse homolog of yeast SWI3, is essential for early embryogenesis and involved in brain development. *Mol. Cell Biol.* 21(22):7787–95

86. Korstanje R, Desai J, Lazar G, King B, Rollins J, et al. 2008. Quantitative trait loci affecting phenotypic variation in the vacuolated lens mouse mutant, a multigenic mouse model of neural tube defects. *Physiol. Genomics* 35(3):296–304

87. Kost K. 2013. *Unintended Pregnancy Rates at the State Level: Estimates For 2002, 2004, 2006 and 2008.* New York: Guttmacher Instit.

88. Kouzarides T. 2007. Chromatin modifications and their function. *Cell* 128(4):693–705

89. Krebs AR, Karmodiya K, Lindahl-Allen M, Struhl K, Tora L. 2011. SAGA and ATAC histone acetyl transferase complexes regulate distinct sets of genes and ATAC defines a class of p300-independent enhancers. *Mol. Cell* 44(3):410–23

90. Kruman II, Fowler A-K. 2014. Impaired one carbon metabolism and DNA methylation in alcohol toxicity. *J. Neurochem.* 192:770–80

91. Kyttala M, Tallila J, Salonen R, Kopra O, Kohlschmidt N, et al. 2006. *MKS1*, encoding a component of the flagellar apparatus basal body proteome, is mutated in Meckel syndrome. *Nat. Genet.* 38:155–57

92. Lacasaña M, Blanco-Muñoz J, Borja-Aburto VH, Aguilar-Garduño C, Rodríguez-Barranco M, et al. 2012. Effect on risk of anencephaly of gene-nutrient interactions between methylenetetrahydrofolate reductase C677T polymorphism and maternal folate, vitamin B12 and homocysteine profile. *Public Health Nutr.* 15(8):1419–28

93. Lakkis MM, Golden JA, O'Shea KS, Epstein JA. 1999. Neurofibromin deficiency in mice causes exencephaly and is a modifier for *Splotch* neural tube defects. *Dev. Biol.* 212(1):80–92

94. Lee JE, Gleeson JG. 2011. Cilia in the nervous system: linking cilia function and neurodevelopmental disorders. *Curr. Opin. Neurol.* 24:98–105

95. Lemos MC, Harding B, Reed AAC, Jeyabalan J, Walls GV, et al. 2009. Genetic background influences embryonic lethality and the occurrence of neural tube defects in *Men1* null mice: relevance to genetic modifiers. *J. Endocrinol.* 203(1):133–42

96. Lerin C, Rodgers JT, Kalume DE, Kim S-H, Pandey A, Puigserver P. 2006. GCN5 acetyltransferase complex controls glucose metabolism through transcriptional repression of PGC-1α. *Cell Metab.* 3(6):429–38

97. Li CCY, Cropley JE, Cowley MJ, Preiss T, Martin DIK, Suter CM. 2011. A sustained dietary change increases epigenetic variation in isogenic mice. *PLOS Genet.* 7(4):e1001380

98. Li E, Bestor TH, Jaenisch R. 1992. Targeted mutation of the DNA methyltransferase gene results in embryonic lethality. *Cell* 69(6):915–26

99. Li X, Weng H, Reece EA, Yang P. 2011. SOD1 overexpression in vivo blocks hyperglycemia-induced specific PKC isoforms: substrate activation and consequent lipid peroxidation in diabetic embryopathy. *Am. J. Obstet. Gynecol.* 205(1):84; e1–6

100. Li X, Weng H, Xu C, Reece EA, Yang P. 2012. Oxidative stress-induced JNK1/2 activation triggers proapoptotic signaling and apoptosis that leads to diabetic embryopathy. *Diabetes* 61(8):2084–92

101. Li X, Xu C, Yang P. 2013. c-Jun NH$_2$-terminal kinase 1/2 and endoplasmic reticulum stress as interdependent and reciprocal causation in diabetic embryopathy. *Diabetes* 62(2):599–608

102. Lillycrop KA, Phillips ES, Jackson AA, Hanson MA, Burdge GC. 2005. Dietary protein restriction of pregnant rats induces and folic acid supplementation prevents epigenetic modification of hepatic gene expression in the offspring. *J. Nutr.* 135(6):1382–86

103. Logan CV, Abdel-Hamed Z, Johnson CA. 2011. Molecular genetics and pathogenic mechanisms for the severe ciliopathies: insights into neurodevelopment and pathogenesis of neural tube defects. *Mol. Neurobiol.* 43:12–26

104. López-Camelo JS, Orioli IM, da Graça Dutra M, Nazer-Herrera J, Rivera N, et al. 2005. Reduction of birth prevalence rates of neural tube defects after folic acid fortification in Chile. *Am. J. Med. Genet. A* 135(2):120–25

105. Lozano R, Naghavi M, Foreman K, Lim S, Shibuya K, et al. 2012. Global and regional mortality from 235 causes of death for 20 age groups in 1990 and 2010: a systematic analysis for the Global Burden of Disease Study 2010. *Lancet* 380(9859):2095–128

106. Lundberg YW, Wing MJ, Xiong W, Zhao J, Finnell RH. 2003. Genetic dissection of hyperthermia-induced neural tube defects in mice. *Birth Defects Res. A Clin. Mol. Teratol.* 67(6):409–13

107. Lupo PJ, Canfield MA, Chapa C, Lu W, Agopian AJ, et al. 2012. Diabetes and obesity-related genes and the risk of neural tube defects in the national birth defects prevention study. *Am. J. Epidemiol.* 176(12):1101–9

108. Lupo PJ, Mitchell LE, Canfield MA, Shaw GM, Olshan AF, et al. 2014. Maternal-fetal metabolic gene-gene interactions and risk of neural tube defects. *Mol. Genet. Metab.* 111(1):46–51

109. Makelarski JA, Romitti PA, Sun L, Burns TL, Druschel CM, et al. 2013. Periconceptional maternal alcohol consumption and neural tube defects. *Birth Defects Res. A Clin. Mol. Teratol.* 97(3):152–60

110. Marean A, Graf A, Zhang Y, Niswander L. 2011. Folic acid supplementation can adversely affect murine neural tube closure and embryonic survival. *Hum. Mol. Genet.* 20(18):3678–83

111. Massarwa R, Niswander L. 2013. In toto live imaging of mouse morphogenesis and new insights into neural tube closure. *Development* 140(1):226–36

112. Matok I, Gorodischer R, Koren G, Landau D, Wiznitzer A, Levy A. 2009. Exposure to folic acid antagonists during the first trimester of pregnancy and the risk of major malformations. *Br. J. Clin. Pharmacol.* 68(6):956–62

113. Michie CA. 1991. Neural tube defects in 18th century. *Lancet* 337:504

114. Milunsky A, Ulcickas M, Rothman KJ, Willett W, Jick SS, Jick H. 1992. Maternal heat exposure and neural tube defects. *J. Am. Med. Assoc.* 268(7):882–85

115. Miranda TB, Bassuk AG, Jones PA, Kibar Z. 2007. DNA methylation: the nuts and bolts of repression. *J. Cell. Physiol.* 213(2):384–90

116. Momb J, Lewandowski JP, Bryant JD, Fitch R, Surman DR, et al. 2013. Deletion of *Mthfd1l* causes embryonic lethality and neural tube and craniofacial defects in mice. *Proc. Natl. Acad. Sci. USA* 110(2):549–54

117. MRC Vitam. Study Res. Group. 1991. Prevention of neural tube defects: results of the Medical Research Council Vitamin Study. *Lancet* 338(8760):131–37

118. Murko C, Lagger S, Steiner M, Seiser C, Schoefer C, Pusch O. 2013. Histone deacetylase inhibitor trichostatin A induces neural tube defects and promotes neural crest specification in the chicken neural tube. *Differentiation* 85(1–2):55–66

119. Nagy Z, Riss A, Romier C, le Guezennec X, Dongre AR, et al. 2009. The human SPT20-containing SAGA complex plays a direct role in the regulation of endoplasmic reticulum stress-induced genes. *Mol. Cell. Biol.* 29(6):1649–60

120. Narisawa A, Komatsuzaki S, Kikuchi A, Niihori T, Aoki Y, et al. 2012. Mutations in genes encoding the glycine cleavage system predispose to neural tube defects in mice and humans. *Hum. Mol. Genet.* 21(7):1496–503

121. Nishimura T, Takeichi M. 2008. Shroom3-mediated recruitment of Rho kinases to the apical cell junctions regulates epithelial and neuroepithelial planar remodeling. *Development* 135:1493–502

122. Obican SG, Finnell RH, Mills JL, Shaw GM, Scialli AR. 2010. Folic acid in early pregnancy: a public health success story. *FASEB J.* 24(11):4167–74

123. Okano M, Bell DW, Haber DA, Li E. 1999. DNA methyltransferases Dnmt3a and Dnmt3b are essential for de novo methylation and mammalian development. *Cell* 99(3):247–57

124. Omori H, Otsu M, Suzuki A, Nakayama T, Akama K, et al. 2014. Effects of heat shock on survival, proliferation and differentiation of mouse neural stem cells. *Neurosci. Res.* 79:13–21

125. Ortega RM, Requejo AM, López-Sobaler AM, Navia B, Mena MC, et al. 2004. Smoking and passive smoking as conditioners of folate status in young women. *J. Am. Coll. Nutr.* 23(4):365–71

126. Ougland R, Lando D, Jonson I, Dahl JA, Moen MN, et al. 2012. ALKBH1 is a histone H2A dioxygenase involved in neural differentiation. *Stem Cells* 30(12):2672–82

127. Parker SE, Yazdy MM, Tinker SC, Mitchell AA, Werler MM. 2013. The impact of folic acid intake on the association among diabetes mellitus, obesity, and spina bifida. *Am. J. Obstet. Gynecol.* 209(3):239; e1–8

128. Partanen M, Motoyama J, Hui CC. 1999. Developmentally regulated expression of the transcriptional cofactors/histone acetyltransferases CBP and p300 during mouse embryogenesis. *Int. J. Dev. Biol.* 43:487–94

129. Pfeiffer CM, Hughes JP, Lacher DA, Bailey RL, Berry RJ, et al. 2012. Estimation of trends in serum and RBC folate in the U.S. population from pre- to postfortification using assay-adjusted data from the NHANES 1988–2010. *J. Nutr.* 142(5):886–93

130. Pujadas E, Feinberg AP. 2012. Regulated noise in the epigenetic landscape of development and disease. *Cell* 148(6):1123–31

131. Raj A, Rifkin SA, Andersen E, van Oudenaarden A. 2011. Variability in gene expression underlies incomplete penetrance. *Nature* 463(7283):913–18

132. Rasmussen SA, Chu SY, Kim SY, Schmid CH, Lau J. 2008. Maternal obesity and risk of neural tube defects: a metaanalysis. *Am. J. Obstet. Gynecol.* 198(6):611–19

133. Ray HJ, Niswander L. 2012. Mechanisms of tissue fusion during development. *Development* 139(10):1701–11

134. Ray JG, Wyatt PR, Vermeulen MJ, Meier C, Cole DEC. 2005. Greater maternal weight and the ongoing risk of neural tube defects after folic acid flour fortification. *Obstet. Gynecol.* 105(2):261–65

135. Reece EA. 2012. Diabetes-induced birth defects: What do we know? What can we do? *Curr. Diab. Rep.* 12(1):24–32

136. Robert E, Smith-Roe SL, Guibaud P, Bultman SJ. 1982. Maternal valproic acid and congenital neural tube defects. *Lancet* 320(8304):937

137. Robinson A, Escuin S, Doudney K, Vekemans M, Stevenson RE, et al. 2012. Mutations in the planar cell polarity genes *CELSR1* and *SCRIB* are associated with the severe neural tube defect craniorachischisis. *Hum. Mutat.* 33(2):440–47

138. Rogner UC, Spyropoulos DD, Le Novère N, Changeux JP, Avner P. 2000. Control of neurulation by the nucleosome assembly protein-1-like 2. *Nat. Genet.* 25(4):431–35

139. Sadler TW, Merrill AH, Stevens VL, Sullards MC, Wang E, Wang P. 2002. Prevention of fumonisin B1-induced neural tube defects by folic acid. *Teratology* 66(4):169–76

140. Sakai M, Tujimura T, Yongheng C, Noguchi T, Inagaki K, et al. 2012. CITED2 links hormonal signaling to PGC-1α; acetylation in the regulation of gluconeogenesis. *Nat. Med.* 18:612–17

141. Salbaum JM, Kappen C. 2010. Neural tube defect genes and maternal diabetes during pregnancy. *Birth Defects Res. A Clin. Mol. Teratol.* 88(8):601–11

142. Sandford MK, Kissling GE, Joubert PE. 1992. Neural tube defect etiology: new evidence concerning maternal hyperthermia, health and diet. *Dev. Med. Child Neurol.* 34(8):661–75

143. Sandovici I, Harris MJ, Smith NH, Juriloff DM, Nitert MD, et al. 2011. Maternal diet and aging alter the epigenetic control of a promoter-enhancer interaction at the *Hnf4a* gene in rat pancreatic islets. *Proc. Natl. Acad. Sci. USA* 108(13):5449–54

144. Sanyal A, Lajoie BR, Jain G, Dekker J. 2012. The long-range interaction landscape of gene promoters. *Nature* 489(7414):109–13

145. Saxén L, Holmberg PC, Nurminen M, Kuosma E. 1982. Sauna and congenital defects. *Teratology* 25(3):309–13

146. Schaniel C, Ang Y-S, Ratnakumar K, Cormier C, James T, et al. 2009. Smarcc1/Baf155 couples self-renewal gene repression with changes in chromatin structure in mouse embryonic stem cells. *Stem Cells* 27:2979–91

147. Schmidt RJ, Romitti PA, Burns TL, Browne ML, Druschel CM, et al. 2009. Maternal caffeine consumption and risk of neural tube defects. *Birth Defects Res. A Clin. Mol. Teratol.* 85(11):879–89

148. Schwartz YB, Pirrotta V. 2007. Polycomb silencing mechanisms and the management of genomic programmes. *Nat. Rev. Genet.* 8(1):9–22

149. Shakèd M, Weissmüller K, Svoboda H, Hortschansky P, Nishino N, et al. 2008. Histone deacetylases control neurogenesis in embryonic brain by inhibition of BMP2/4 signaling. *PLOS ONE* 3(7):e2668

150. Shaw GM, Todoroff K, Finnell RH, Lammer EJ. 2000. Spina bifida phenotypes in infants or fetuses of obese mothers. *Teratology* 61(5):376–81

151. Shin J-H, Shiota K. 1999. Folic acid supplementation of pregnant mice suppresses heat-induced neural tube defects in the offspring. *J. Nutr.* 129:2070–73

152. Shpargel KB, Sengoku T, Yokoyama S, Magnuson T. 2012. UTX and UTY demonstrate histone demethylase-independent function in mouse embryonic development. *PLoS Genet.* 8(9):e1002964

153. Shyamasundar S, Jadhav SP, Bay BH, Tay SSW, Kumar SD, et al. 2013. Analysis of epigenetic factors in mouse embryonic neural stem cells exposed to hyperglycemia. *PLOS ONE* 8(6):e65945

154. Steegers-Theunissen RP, Obermann-Borst SA, Kremer D, Lindemans J, Siebel C, et al. 2009. Periconceptional maternal folic acid use of 400 μg per day is related to increased methylation of the *IGF2* gene in the very young child. *PLOS ONE* 4(11):e7845

155. Stothard KJ, Tennant PWG, Bell R, Rankin J. 2009. Maternal overweight and obesity and the risk of congenital anomalies: a systematic review and meta-analysis. *J. Am. Med. Assoc.* 301(6):636–50

156. Stover PJ, Caudill MA. 2008. Genetic and epigenetic contributions to human nutrition and health: managing genome-diet interactions. *J. Am. Diet. Assoc.* 108(9):1480–87

157. Struhl K, Rogner UC, Segal E, Spyropoulos DD, Le Novère N, et al. 2013. Determinants of nucleosome positioning. *Nat. Struct. Mol. Biol.* 20(3):267–73

158. Suarez L, Felkner M, Brender JD, Canfield M, Hendricks K. 2007. Maternal exposures to cigarette smoke, alcohol, and street drugs and neural tube defect occurrence in offspring. *Matern. Child Health J.* 12(3):394–401

159. Suarez L, Felkner M, Brender JD, Canfield M, Zhu H, Hendricks KA. 2012. Neural tube defects on the Texas-Mexico border: what we've learned in the 20 years since the Brownsville cluster. *Birth Defects Res. A Clin. Mol. Teratol.* 94(11):882–92

160. Suarez L, Felkner M, Hendricks K. 2004. The effect of fever, febrile illnesses, and heat exposures on the risk of neural tube defects in a Texas-Mexico border population. *Birth Defects Res. A Clin. Mol. Teratol.* 70(10):815–19

161. Suzuki MM, Zhang Y, Salbaum JM, Bamforth SD, Bird A, et al. 2008. DNA methylation landscapes: provocative insights from epigenomics. *Nat. Rev. Genet.* 9(6):465–76

162. Takeuchi T, Yamazaki Y, Katoh-Fukui Y, Tsuchiya R, Kondo S, et al. 1995. Gene trap capture of a novel mouse gene, *jumonji*, required for neural tube formation. *Genes Dev.* 9(10):1211–22

163. Thiersch JB. 1952. Therapeutic abortions with a folic acid antagonist, 4-aminopteroylglutamic acid (4-amino P.G.A) administered by the oral route. *Am. J. Obstet. Gynecol.* 63(6):1298–304

164. Thiersch JB, Philips FS. 1950. Effect of 4-amino-pteroylglutamic acid (aminopterin) on early pregnancy. *Proc. Soc. Exp. Biol. Med.* 74(1):204–8

165. van der Put NM, Steegers-Theunissen RP, Frosst P, Trijbels FJ, Eskes TK, et al. 1995. Mutated methylenetetrahydrofolate reductase as a risk factor for spina bifida. *Lancet* 346(8982):1070–71

166. Vega RB, Matsuda K, Oh J, Barbosa AC, Yang X, et al. 2004. Histone deacetylase 4 controls chondrocyte hypertrophy during skeletogenesis. *Cell* 119(4):555–66

167. Villanueva JA, Halsted CH. 2004. Hepatic transmethylation reactions in micropigs with alcoholic liver disease. *Hepatology* 39(5):1303–10

168. Wald NJ, Law MR, Morris JK, Wald DS. 2001. Quantifying the effect of folic acid. *Lancet* 358(9298):2069–73

169. Waller DK, Shaw GM, Rasmussen SA, Hobbs CA, Canfield MA, et al. 2007. Prepregnancy obesity as a risk factor for structural birth defects. *Arch. Pediatr. Adolesc. Med.* 161(8):745–50

170. Wallingford JB. 2012. Planar cell polarity and the developmental control of cell behavior in vertebrate embryos. *Annu. Rev. Cell Dev. Biol.* 28:627–53

171. Wallingford JB, Niswander LA, Shaw GM, Finnell RH. 2013. The continuing challenge of understanding, preventing, and treating neural tube defects. *Science* 339(6123):1222002

172. Wang F, Reece EA, Yang P. 2013. Superoxide dismutase 1 overexpression in mice abolishes maternal diabetes-induced endoplasmic reticulum stress in diabetic embryopathy. *Am. J. Obstet. Gynecol.* 209(4):345; e1–7

173. Wang M, Wang Z-P, Gong R, Zhao Z-T. 2014. Maternal smoking during pregnancy and neural tube defects in offspring: a meta-analysis. *Child's Nerv. Syst.* 30(1):83–89

174. Wang M, Wang Z-P, Zhang M, Zhao Z-T. 2014. Maternal passive smoking during pregnancy and neural tube defects in offspring: a meta-analysis. *Arch. Gynecol. Obstet.* 289(3):513–21

175. Warkany J, Beaudry PH, Hornstein S. 1959. Attempted abortion with aminopterin (4-amino-pteroylglutamic acid); malformations of the child. *Am. Med. Assoc. J. Dis. Child* 97(3):274–81

176. Waterland RA, Jirtle RL. 2003. Transposable elements: targets for early nutritional effects on epigenetic gene regulation. *Mol. Cell. Biol.* 23(15):5293–300

177. Weatherbee SD, Niswander LA, Anderson KV. 2009. A mouse model for Meckel syndrome reveals Mks1 is required for ciliogenesis and hedgehog signaling. *Hum. Mol. Genet.* 18:4565–75

178. Wei X, Li H, Miao J, Liu B, Zhan Y, et al. 2013. miR-9*- and miR-124a-mediated switching of chromatin remodelling complexes is altered in rat spina bifida aperta. *Neurochem. Res.* 38(8):1605–15

179. Welstead GG, Creyghton MP, Bilodeau S, Cheng AW, Markoulaki S, et al. 2012. X-linked H3K27me3 demethylase Utx is required for embryonic development in a sex-specific manner. *Proc. Natl. Acad. Sci. USA* 109(32):13004–9

180. Xu C, Li X, Wang F, Weng H, Yang P. 2013. Trehalose prevents neural tube defects by correcting maternal diabetes-suppressed autophagy and neurogenesis. *Am. J. Physiol. Endocrinol. Metab.* 305(5):E667–78

181. Yan L, Zhao L, Long Y, Zou P, Ji G, et al. 2012. Association of the maternal MTHFR C677T polymorphism with susceptibility to neural tube defects in offsprings: evidence from 25 case-control studies. *PLOS ONE* 7(10):e41689

182. Yang P, Li X, Xu C, Eckert RL, Reece EA, et al. 2013. Maternal hyperglycemia activates an ASK1-FoxO3a-caspase 8 pathway that leads to embryonic neural tube defects. *Sci. Signal.* 6(290):ra74

183. Yang P, Zhao Z, Reece EA. 2007. Involvement of c-Jun N-terminal kinases activation in diabetic embryopathy. *Biochem. Biophys. Res. Commun.* 357(3):749–54
184. Yazdy MM, Tinker SC, Mitchell AA, Demmer LA, Werler MM. 2012. Maternal tea consumption during early pregnancy and the risk of spina bifida. *Birth Defects Res. A Clin. Mol. Teratol.* 94(10):756–61
185. Ybot-Gonzalez P, Savery D, Gerrelli D, Signore M, Mitchell CE, et al. 2007. Convergent extension, planar-cell-polarity signalling and initiation of mouse neural tube closure. *Development* 134(4):789–99
186. Ye R, Ren A, Zhang L, Li Z, Liu J, et al. 2011. Tea drinking as a risk factor for neural tube defects in northern China. *Epidemiology* 22(4):491–96
187. Yin Z, Haynie J, Yang X, Han B, Kiatchoosakun S, et al. 2002. The essential role of *Cited2*, a negative regulator for HIF-1α, in heart development and neurulation. *Proc. Natl. Acad. Sci. USA* 99(16):10488–93
188. Youngblood ME, Williamson R, Bell KN, Johnson Q, Kancherla V, Oakley GP Jr. 2013. 2012 Update on global prevention of folic acid–preventable spina bifida and anencephaly. *Birth Defects Res. A Clin. Mol. Teratol.* 97(10):658–63
189. Yuan Q, Zhao S, Liu S, Zhang Y, Fu J, et al. 2013. Folic acid supplementation changes the fate of neural progenitors in mouse embryos of hyperglycemic and diabetic pregnancy. *J. Nutr. Biochem.* 24(7):1202–12
190. Zabihi S, Loeken MR. 2010. Understanding diabetic teratogenesis: Where are we now and where are we going? *Birth Defects Res. A Clin. Mol. Teratol.* 88(10):779–90
191. Zhang T, Lou J, Zhong R, Wu J, Zou L, et al. 2013. Genetic variants in the folate pathway and the risk of neural tube defects: a meta-analysis of the published literature. *PLOS ONE* 8(4):e59570
192. Zhao Z, Yang P, Eckert RL, Reece EA. 2009. Caspase-8: a key role in the pathogenesis of diabetic embryopathy. *Birth Defects Res. B Dev. Reprod. Toxicol.* 86(1):72–77